AF412958

FUNDAMENTALS and APPLICATIONS

Volume I

Fundamentals and Applications

Edited by

SUBHAS K. SIKDAR
Cincinnati, Ohio

ROBERT L. IRVINE
Department of Civil Engineering and Geological Sciences
University of Notre Dame
Notre Dame, Indiana

TECHNOMIC
PUBLISHING CO., INC.
LANCASTER · BASEL

Fundamentals and Applications, Volume I

aTECHNOMIC ®publication

Published in the Western Hemisphere by
Technomic Publishing Company, Inc.
851 New Holland Avenue, Box 3535
Lancaster, Pennsylvania 17604 U.S.A.

Distributed in the Rest of the World by
Technomic Publishing AG
Missionsstrasse 44
CH-4055 Basel, Switzerland

Main entry under title:
   Bioremediation: Principles and Practice—Fundamentals and
    Applications, Volume I

A Technomic Publishing Company book
Bibliography: p.
Includes index p. 751

Library of Congress Catalog Card No. 97-60884
ISBN No. 1-56676-308-8

To the memory of Arthur Winston Busch, 1926–1994: Arthur W. Busch
served as Professor Irvine's mentor from the moment that they met. His
thoughts and ideas created an atmosphere which fostered creative research and development. The Sequencing Batch Reactor (SBR) is a
technology that emerged from this environment and a technology which
demanded fundamental understandings from science and creative applications from engineers. Art Busch's memory serves as a beacon lighting
a path on which challenge will bring us closer to truth.

To Professor Alan D. Randolph, who despite severe physical difficulties
imposed by multiple sclerosis still keeps his mind engaged in the pursuit
of advancing the science and engineering analysis of particulate processes, in which field he is the most significant pioneer. Subhas K. Sikdar
fondly remembers the graduate research experience under Professor
Randolph, who still remains an inspiring figure in his (SKS) quest for
innovative solutions to engineering problems.

Contents

Preface

This three-volume series, *Bioremediation: Principles and Practice,* will provide a state-of-the-art description of advances in pollution treatment and reduction using biological means; identify and address, at a fundamental level, broad scientific and technological areas that are unique to the subject or theme and that must be understood if advances are to be made; and provide a comprehensive overview of new developments at the regulatory, desk-top, bench-scale, pilot-scale, and full-scale levels. The series will cover all media—air, water, and soil/sediment—and will blend the talents, knowledge, and know-how of academic, industrial, governmental, and international contributors.

Developing this series has been a lengthy process because of the size of the project and because of the standard of excellence that has been set. We wish to thank Technomic Publishing Company for its support throughout the process.

Glancing through the tables of contents, you will see that our objective for this series was to provide the theoretician and practitioner with an overview of bioremediation that will allow new research programs to be formulated and bioremediation technologies to be improved. This series is intended for the student and consulting engineer and the scientist and industrialist alike. Accordingly, we have addressed the removal of both hazardous and nonhazardous contaminants from the liquid, solid, and gas phase using biological processes. This includes the biological treatment of wastes of municipal and industrial origin; bioremediation of leachates, soils, and sediments; and biofiltration for contaminated gases.

As you read through many of the excellent chapters, you may wonder why some chapters were selected. In fact, one chapter that did not make the publisher's desk is one that dealt with incineration. We wanted a chapter on incineration so that we could better demonstrate, by contrast, how and when bioremediation is the appropriate choice. But as was the case with several other chapters, we simply could not convince the best authors to prepare chapters for what they perceived to be unrelated topics. Because of that, we would like to add a few words here about biological processes.

As we all know, microorganisms will grow if provided the proper conditions. In fact, it is often difficult to stop microorganisms from growing. Consider your soured milk, moldy week-old bread, or slippery shower room floors as examples. In fact, if you put a match to a substance, place your hands over the resulting flame, and keep your hands warm, that substance is very likely biodegradable, even if someone has labeled the substance as hazardous! The best that we can determine, unless the substance is drinking alcohol or rubbing alcohol, is that the compound is likely to be found on one hazardous substance list or another or that someone simply forgot to put it there.

Microorganisms need a carbon and energy source. As just discussed, both hazardous and nonhazardous contaminants satisfy either or both of these requirements. The design engineer must ensure that all of the necessary ingredients are supplied in sufficient quantities during any bioremediation project. Just like the grass on our public golf courses, microorganisms need nutrients such as nitrogen, phosphorus, and trace metals. Aerobic organisms need a source of oxygen. The temperature and pH must be controlled as needed, and sometimes, substances that are toxic to the organism (e.g., heavy metals) must be removed. In general, it is not difficult for the designer to engineer a system that will meet all of these requirements for the biological treatment of contaminated liquids, soil or sediments, and gases.

A basic message is that most hazardous and nonhazardous organics can be treated biologically, provided that the proper organism distribution can be established. This is how our authors fit in. Some will tell us how a substance that is hazardous for one group of organisms is a valuable food source for another group. It is only by viewing biological treatment systems as a dynamic population of microbes whose composition is subject to various selection pressures that proper systems can be designed, maintained, and corrected when problems arise. As a result, these chapters emphasize how fundamental principles can be used to develop commonsense approaches to establishing the desired microbial consortium, either in biological reactors designed for the treatment of heavily contaminated leachates or, in soils, using in situ and on-site decontamination techniques or, in gases, using biofilters or biotrickling filters.

Using 20 chapters, Volume I of the series, *Fundamentals and Applications,* begins with the law, the "fundamental force" that drives remediation; continues with some basic principles of mathematics that help us understand the present and predict the future; dwells on soils and the complexity that they add to the overall remediation process; fixes on the microbes and biochemical pathways that give bioremediation its uniqueness; and ends with an academic understanding of practice.

Volume II, *Biodegradation Technology Developments,* uses 20 chapters to lead us through a plethora of subjects ranging from the control of filaments, removal of phosphorus, and the treatment of chemical warfare agents in activated sludge processes to exotic biological systems, including those that use the white rot fungus, halophiles, and higher forms of plants, and fixed-film systems that treat chlorinated organics.

Volume III, *Bioremediation Technologies,* is divided into two sections: "International Bioremediation Agenda" and "Pilot-, Full-Scale, and Commercial Demonstration." In 22 chapters, this volume provides insight into how bioremediation is viewed throughout the world and gives full-scale examples of how it has been applied for the treatment of contaminated liquids, soils, and gases.

All of the chapters in this series have been peer-reviewed. To these reviewers, we owe our deepest gratitude. We would also like to thank the graduate students at the University of Notre Dame, who spent many hours helping select authors (that they would like to see in the book), editing, and organizing.

SUBHAS K. SIKDAR
ROBERT L. IRVINE

Editors

PROJECT EDITORS

Subhas K. Sikdar, Ph.D.
11043 Huntwicke Place
Cincinnati, OH 45241

Robert L. Irvine, Ph.D.
Department of Civil Engineering and Geological Sciences
University of Notre Dame
Notre Dame, IN 46556-0767

EDITORIAL BOARD

Motoyuki Suzuki
Institute of Industrial Science
University of Tokyo
7-22-1, Roppongi, Minato-Ku
Tokyo 106, Japan

Dennis D. Focht
Soils and Environmental Sciences Department
University of California, Riverside
Riverside, CA 92521

Jost O. L. Wendt
Department of Chemical Engineering
University of Arizona
Tucson, AZ 85721

Peter A. Wilderer
Technical University of Munich
Water Quality Management
D-85748 Garching, Germany

Howard Klee, Jr.
Amoco Corporation
200 E. Randolph Drive
Chicago, IL 60601

List of Reviewers

Carolyn Acheson, *USEPA*
Michael D. Aitken, *University of North Carolina*
Robert Arnold, *University of Arizona*
J. K. Bhattacharjee, *Miami University*
Dolloff F. Bishop, *USEPA*
Alec Breen, *University of Cincinnati*
Richard Brenner, *USEPA*
Daniel P. Cassidy, *University of Notre Dame*
Harlal Chaudhuri, *USEPA*
Steve Chiesa, *Santa Clara University*
Robert Chozick, *Foster-Wheeler Environmental Corporation*
Al Cunningham, *Montana State University*
Chamindra Y. Dassanayake, *University of Notre Dame*
Kate Devine, *Devoe Enterprises*
Larry E. Erickson, *Kansas State University*
Dennis Focht, *University of California, Riverside*
Sybren Gerbens, *University of Notre Dame*
John Glaser, *USEPA*
Rakesh Govind, *University of Cincinnati*
Janet Grimsley, *Texas A&M University*
James Heidman, *USEPA*

Michael A. Heitkamp, *Monsanto Company*
Ahmad Hilaly, *Archer Daniel Midlands*
Patrick Hirl, *University of Notre Dame*
Sa V. Ho, *Monsanto Company*
Robert L. Irvine, *University of Notre Dame*
C. Y. Jeng, *Amoco Corporation*
Douglas E. Jerger, *OHM Remediation*
David L. Kaplan, *U.S. Army Natick R&D Center*
Dhinakar Kompala, *University of Colorado*
Radha Krishnan, *IT Corporation*
Suzanne Lesage, *Environment Canada*
Ronald Lewis, *USEPA*
K. Y. Li, *Lamar University*
Bill Liu, *IGT*
Thomas Lupo, *Seyfarth, Shaw, Fairweather & Geraldson*
Ping-Ching Maness, *NREL*
Casey Miler, *University of Notre Dame*
William M. Moe, *University of Notre Dame*
Frank Mondello, *GE R&D Center*
Carlo Montemagno, *Cornell University*
Brij Moudgil, *University of Florida*
Debdas Mukerji, *USEPA*
Kimberly Ogden, *University of Arizona*
John Pellegrino, *NIST*
Jairaj Pothuluri, *FDA*
Vipin Rastogi, *Texas A&M University*
David Reisman, *USEPA*
Diane Saber, *IGT*
Steve Safferman, *University of Dayton*
Michael Shannon, *Envirogen*
Kanij Siddique, *City of Cincinnati*
Subhas Sikdar, *USEPA*
Rajib Sinha, *IT Corporation*
Makram Suidan, *University of Cincinnati*
David Szlag, *USEPA*
Henry Tabak, *USEPA*

Arthur Umble, *University of Notre Dame*
Ronald Unterman, *Envirogen*
Albert Venosa, *USEPA*
Walter Weber, *University of Michigan*
Daniel M. White, *University of Alaska, Fairbanks*

How the Law Drives Site Cleanups

THOMAS D. LUPO, ESQ. AND LISA L. HARRIS, ESQ.
Seyfarth, Shaw, Fairweather & Geraldson
55 E. Monroe Street, Suite, 4200
Chicago, IL 60603, USA

INTRODUCTION

Most people agree that individuals or entities rarely undertake site cleanups because of an individual's or an entity's moral or responsible nature. Rather, site cleanups are prompted by the compulsion of a federal or state law, concern for one's reputation for responsible corporate citizenship, the desire to restore a property's value or marketability, or another party's use of the legal system to compel a cleanup or the redress of cleanup costs. Once a site cleanup has begun, the law continues to compel and guide the process until, and often even after, completion of the site's remediation. In short, the law drives the site cleanup process.

Federal and state statutes, rules, regulations, and policies create legal schemes that empower regulatory agencies to compel and/or oversee site cleanups, define an individual's or entity's rights vis-à-vis the government and relative to one another, and define the incidents and activities that need to be reported and when. Whether an attorney advises a client on cleanup obligations or its rights against another potentially responsible party, commonly called a "PRP," or a consultant or environmental engineer advises someone on technical standards or monitoring requirements, the law establishes the scheme that will control and drive the process to its end result. The remediation process, whether the contemplated technology is bioremediation or otherwise, achieves the end result, but it is brought about and prompted by the law, as shown in the "True Scenario" set forth later.

The number of cleanups undertaken has steadily risen over the years. As stated throughout this book, bioremediation occurs naturally and can be prompted and facilitated. As a controlled alternative, it has evolved into an accepted option. Its use has and will continue to increase as its

1

specific roles and capabilities are defined. For petroleum releases alone, the prospects are almost immeasurable. With the controlled and natural bioremediation of recent well-known oil spills such as the Exxon Valdez release further establishing the credibility of the process and increasing scientific education, acceptance and application will grow. For example, the United States Environmental Protection Agency (U.S. EPA) reports that, as of October 31, 1994, the existence of more than 270,000 sites of petroleum contamination from leaking or overfilled underground storage tanks has been reported in the United States alone, each a potential bioremediation site.[1]

This chapter is not directed solely at bioremediation activities, but at the context in which treatment or remediation activities arise in site cleanups. This chapter is intended to provide a framework and to be a continuing reference tool for attorneys, consultants, and parties involved in site cleanups and faced with determining where they are in the cleanup process—from the initial recognition that contamination exists to the final closure of a project. Frequently, consultants, engineers, business-people and attorneys understand distinct aspects of the numerous technical or legal issues related to site cleanups but may knowingly or unknowingly be hindered by an ignorance of other steps or aspects unfamiliar to them. For example, one might immediately know that bioremediation is the appropriate technology for a project but may not know the requirements of what and when something must be reported to a regulatory agency, whether permits are necessary for particular activities, the relevant agencies' ability to control the process and set deadlines, and finally, who a client should look to as solely or jointly responsible for cleanup costs.

While each aspect of the increasingly complex statutory schemes are not addressed in this chapter, general roadmaps or examples are provided. In summary, this chapter sets forth the general framework for the parties considering potential site cleanup scenarios to determine where they are in the process and the legal requirements that may affect their specific cleanup scenarios. One such scenario follows.

A TRUE SCENARIO

In the 1960s and 1970s, two enterprising brothers living in a rural, semi-residential area decided that they could make a living by collecting and reconditioning pails and drums. They bought and collected pails and drums from businesses in neighboring locales. The contents of the ink, paint, solvent, and asphalt containers were dumped on their twenty-acre parcel of land ("the Site"). The remaining contents were flushed with the collected solvent and burned out. They conducted this business on their property for twenty years.

In 1980, the Site was discovered in connection with a state agency's investigation of another drum reconditioning operation. The state environmental agency determined that the brothers' container reclaiming operation had resulted in a 6-inch-thick layer of waste covering approximately 10 acres of the Site. Numerous trenches, approximately 20 feet in width and depth, were found containing crushed pails and drums. Later that year, the state's attorney general's office obtained a court order requiring the brothers to close their business and clean up the Site. The brothers failed to comply with the court order, citing a lack of finances. The attorney general's office amended its complaint to include a number of large corporations with whom the brothers had allegedly conducted business. One of the corporations successfully prosecuted a motion to dismiss the amended complaint.

In 1983, the Site was listed on the Comprehensive Environmental Response, Compensation and Liability Act's (CERCLA)[2] National Priorities List (NPL), "the EPA's list of the most serious uncontrolled or abandoned hazardous waste sites identified for possible long-term remedial action under Superfund."[3]

The state environmental agency, in cooperation with the U.S. EPA, conducted a remedial investigation/feasibility study (RI/FS) of the Site. A remedial investigation is an "in-depth study designed to gather data needed to determine the nature and extent of contamination at a Superfund site, establish site cleanup criteria, identify preliminary alternatives for remedial action, and support technical and cost analyses of alternatives."[4] A feasibility study is an analysis of the practicability of cleanup alternatives.[5] The state agency alone conducted the removal of surficial and buried waste materials and conducted a hydrogeological study/feasibility study. A number of volatile organic compounds, chlorinated solvents, and metals were identified in the Site's soils and groundwater.

The state's attorney general's office filed a lawsuit against the brothers and the previously referenced corporations seeking cost recovery and a ruling that the defendants were jointly and severally liable for response costs incurred by the state in relation to the Site. The U.S. Department of Justice (U.S. DOJ), on behalf of the U.S. EPA, filed a cost recovery action to recover response costs incurred by the United States relative to the Site. The defendants asserted cross-claims against one another for contribution towards cleanup costs. Working with the surviving brother, the defendants identified numerous additional potentially responsible parties, whom they added to the cost recovery actions as third-party defendants.

In September 1989, the U.S. EPA issued a Record of Decision (ROD), a document describing which cleanup methods will be used at National Priorities List sites where Trust Fund money will pay for the cleanup regarding the Site, and asked the corporations to perform remedial design/remedial action activities at the Site. The corporations declined. The U.S. EPA then issued a unilateral administrative order pursuant to CERCLA Section 106(a) to the Site owner and the previously identified corporations (Respondents).[6] U.S. EPA directed the unilateral administrative order respondents to undertake remedial design/remedial action activities to abate an imminent and substantial endangerment to the public health or welfare or the environment as identified in the ROD, subject to potential $25,000 penalties per day of violation and the tripling of response costs incurred by U.S. EPA and its agents.

In response to the unilateral administrative order, the respondents began spending millions of dollars on remedial activities. At one point, compliance with the unilateral administrative order ceased. U.S. DOJ threatened an enforcement action. Cleanup work resumed. At the same time, the defendants and third-party defendants litigated and negotiated with one another and with U.S. EPA and the state agency to resolve cost issues.

Each party ultimately financed the Site cleanup. However, if not for the compulsion of law, the cleanup might never have commenced.

THE ROLE OF FEDERAL AND STATE AGENCIES AND THEIR ENFORCEMENT TOOLS

Federal and state agencies empowered to address environmental issues are known by a number of names, most commonly Environmental Protection Agency or as departments of "environmental management" or "environmental quality." Within a particular jurisdiction, the discrete responsibilities may be divided or overlapping. For example, on the federal level, the U.S. EPA possesses broad authority over a wide range of hazardous, solid, and industrial waste issues. At the same time, the United States Department of Transportation regulates how wastes are transported, and the Department of the Interior addresses natural resources issues.

Responsibility and authority also overlap between the federal and state levels. Federal initiatives frequently establish regulatory schemes. Some programs initiated at the federal level may be delegated to qualified state or regional programs. A state or region also may create its own additional requirements.

Without federal and state agencies, federal and state environmental statutes would be almost meaningless. Federal and state agencies give the statutes practical meaning by promulgating rules that specify how the broad goals of the statutes will be implemented. Interestingly, federal agencies do not always promulgate the rules that implement the federal statutes. Sometimes, federal statutes give states the power to promulgate their own rules, as long as they meet the guidelines specified by the statute. As a general rule, state programs will receive a federal delegation of authority if the state program is as stringent or more stringent than a federal program or its guidelines.

In general, the rules promulgated by the agencies specify when contaminated property must be cleaned up; to what level it must be cleaned up; cleanup standards; what must be reported, including requirements for written plans; deadlines for deliverables, the commencement, and completion of a cleanup; and who shall be responsible for the cleanup. In order to determine when, what, and who, the agencies have the power to collect information about certain businesses or properties. The agencies often are empowered to collect information through mechanisms such as inspections, permit applications, and specific information requests.

Agencies have many tools to prompt remediation activities and to facilitate voluntary projects. On the civil and administrative enforcement side, state and federal governments have a number of effective enforcement tools available. First, under certain statutes, U.S. EPA and other empowered governmental entities may assess administrative penalties for a party's failure to comply with regulatory requirements, including information requests, reporting requirements, and performance deadlines. Some statutes permit assessments of up to $25,000 per day of noncompliance.

Second, U.S. EPA may issue administrative orders directing immediate compliance, assessment of civil penalties, or both. In some instances, the administrative order provisions afford the recipient a public hearing, including the opportunity to introduce evidence and examine witnesses, with some discovery. Others allow the recipient to submit written comments prior to an administrative order-taking effect. If the recipient of the order fails to act, a civil lawsuit to enforce the order or to recover penalties may be filed.

Third, an agency may file a civil lawsuit requesting injunctive relief. If a request for an injunction is granted, a judge will order a party to undertake or cease certain activities to comply with the applicable law.

Fourth, the government may file a lawsuit to recover costs incurred for response activities, including investigation and cleanup, from liable parties on a strict, joint, and several liability basis. Governmental entities pursuing enforcement or cost recovery actions generally can recover attorneys' fees and other litigation costs incurred in their efforts. Private parties may also file cost recovery actions against other liable parties seeking contribution toward costs incurred for response activities, albeit without the recovery of attorneys' fees and litigation costs.

Fifth, almost "anyone" may file a "Citizen's Suit" to enforce certain environmental provisions, even against the federal government. Citizen's suits permit enforcement where a lack of technical ability, political will, or resources hinders a more diligently prosecuted state or federal action.

Finally, the threat of criminal enforcement serves to prevent pollution and to punish polluters.

Agencies are frequently the ultimate decision maker on issues concerning site cleanups. Remedy selection and site cleanup subjected to agency input and approval will generally be somewhat slower, considering the formal submittal and decision criteria. For example, the CERCLA remedy selection process described below considers at least nine criteria.

When not subjected to formal agency review, the decision-making process frequently considers the same criteria, although in a more informal manner. This type of remedy selection process simply requires less formal reporting and written consideration of the remedial alternatives. The ultimate goal remains the same: a successful cleanup project.

U.S. EPA has publicly worked toward a streamlined decision-making process—applying preferred remedies for particular types of sites as the applicability and the effectiveness of various remedies become more predictable. Because bioremediation is shown to be a controllable, cost-effective, and timely method for achieving cleanup goals on certain types of sites, it too may enter this protocol. Its cost-effectiveness for in situ activity, in contrast to the previously commonplace "dig and remove" for off-site disposal or destruction approach, has become increasingly accepted for petroleum, wood treatment, and volatile organic compound residue sites as further described in this book.

THE LAWS THAT DRIVE SITE CLEANUPS

The following federal laws affect site cleanups. Many analogous state laws exist throughout the country.

CERCLA

In 1980, Congress enacted the Comprehensive Environmental, Response, Compensation and Liability Act, frequently called CERCLA, or the Superfund (42 U.S.C. §§ 9601, et seq.). Six years later, Congress expanded CERCLA through the Superfund Amendments and Reauthorization Act of 1986. CERCLA presently serves the dual purposes of remediation and prevention: (1) CERCLA provides the federal government with funding and enforcement authority for the cleanup of hazardous waste dumps and hazardous substance spills or other threatened releases, and (2) provides private parties with incentives for preventing the release of hazardous substances. CERCLA establishes a priority list for the nation's worst sites, known as the National Priorities List, or NPL. A contaminated site need not be on the NPL for CERCLA's provisions to apply.

CERCLA authorizes the federal government to respond to any release or "threat of" release of a "hazardous substance" or a "pollutant or contaminant." The breadth of CERCLA's applicability is even more impressive when one considers the subparts of this description. CERCLA broadly defines "release" to include "any spilling, leaking, pumping, pouring, emitting, emptying, discharging, injecting, escaping, leaching, dumping, or disposing into the environment . . ." [42 U.S.C. § 9601(22)].

In addition, CERCLA's definition of *hazardous substance* encompasses the regulated substance definitions of the various federal environmental statutes, including "hazardous waste" under Subtitle C of the Resource Conservation and Recovery Act (RCRA); *toxic water pollutant* regulated under Section 307 of the Clean Water Act; *imminently hazardous chemical* regulated under Section 7 of the Toxic Substances Control Act, substances subject to Section 311 of the Clean Water Act; *hazardous air pollutant* listed under Section 112 of the Clean Air Act; and additional substances designated by the U.S. EPA [42 U.S.C. § 9601(14)]. 40 C.F.R. § 302 identifies the hundreds of hazardous substances specifically fitting CERCLA's definition, as well as setting forth the reportable quantities for actionable releases of each. Additional materials may be CERCLA hazardous substances if subjected to the RCRA's toxicity, ignitability, corrosivity, and reactivity standards. Finally, CERCLA's definition of *hazardous substances* excludes petroleum.

CERCLA Section 104 authorizes the U.S. EPA to respond to the release of a "hazardous substance" at a site through two types of cleanup actions: a removal action or a remedial action. Removal actions provide a short-term solution to the immediate health and environmental threats posed by a site, generally the removal of the source or "hot spot." For example, a removal action might address leaking drums or tanks on a site or a lagoon or other surface contamination. A removal action may be performed after a preliminary assessment of a site.

Remedial actions are to provide long-term solutions to any health and environmental threats posed by a site. Remedial actions may be performed only after an extensive remedial investigation and feasibility study, or RI/FS, assesses the long-term threat posed by the hazardous substances and the feasibility of alternative cleanup measures. After the RI/FS is completed, the U.S. EPA issues a Record of Decision (ROD) specifying the cleanup measures that have been chosen for the site. The U.S. EPA or performing parties then develop a remedial design (RD), which outlines the specific work or remedial activity to be performed at the site. This remedial action (RA) proceeds until the work is completed.

On occasion, U.S. EPA will direct or approve interim remedial activities in furtherance of site objectives. For example, an impermeable clay cover may be applied to a heavily contaminated area to prevent further leaching to groundwater as a groundwater treatment remedy is designed and implemented. Either a state or federal agency may provide the technical lead on a site's activities.

CERCLA gives the U.S. EPA broad discretion in choosing the cleanup method and determining site cleanup standards. Section 121 requires the U.S. EPA to select an "appropriate" remedy that, to the extent practicable, complies with the National Contingency Plan and that is cost-effective. In addition, Section 121 establishes a preference for on-site treatment, rather than the off-site disposal of waste materials.

The National Contingency Plan (NCP) sets forth the procedures for responding to releases of hazardous substances, including remedy selection and performance standards (40 C.F.R. Part 300). Under the NCP, alternative methods for hazardous waste cleanup are evaluated according to nine criteria for evaluation: (1) overall protectiveness of human health and the environment; (2) compliance with the applicable or relevant and appropriate requirements (ARARs) under federal and state environmental laws; (3) long-term effectiveness and permanence; (4) reduction of toxicity, mobility, or volume through treatment; (5) short-term effectiveness; (6) implementability; (7) cost; (8) state acceptance; and (9) community acceptance [40 C.F.R. Part 300.430(e)(f)].

Initially, the NCP directs the U.S. EPA to consider whether the cleanup method provides overall protection of health and the environment and whether it is consistent with the ARARs. Then, the U.S. EPA balances the next five factors, with primary emphasis on the permanence and reduction of toxicity, mobility, or volume. Finally, the U.S. EPA determines whether the selection of the cleanup method should be influenced by state and community acceptance.

Since the Superfund Amendments and Reauthorization Act of 1986, the U.S. EPA has encouraged the use of innovative technologies in the cleanup of Superfund sites. In pertinent part, CERCLA Section 311(b), "Alternative or Innovative Technology Research and Demonstration Program," authorizes and directs the U.S. EPA:

to carry out a program of research, evaluation, testing, development, and demonstration of alternative or innovative treatment technologies (hereinafter in this subsection referred to as the "program") which may be utilized in response actions to achieve more permanent protection of human health and welfare and the environment. [42 U.S.C. 9660(b)(1)]

The U.S. EPA implemented Section 311(b) through a program called the Superfund Innovative Technology and Evaluation Program (SITE), which evaluates remedies for contamination at Superfund sites. In 1991, statistical studies of U.S. EPA sites showed a trend away from the traditional technologies. Innovative technologies made up 37 percent of the technologies selected at sites on the National Priorities List. At that time, 8.3 percent of the NPL sites were formally using the same aspect of bioremediation.

As an initial premise, when the federal government or a private party responds to a release or a threat of a release, a broad class of parties becomes liable for the response costs. Generally, waste generators, past and present owners and operators of waste disposal facilities, and transporters of hazardous wastes may be held liable for U.S. EPA's enforcement and cleanup costs. CERCLA provides a number of scenarios affecting site cleanups. First, a property owner, or otherwise responsible party, may voluntarily commence a site cleanup, rather than wait for U.S. EPA to initiate notice or enforcement activities. The acting party will have to comply with NCP requirements if it will seek to recover money from the Superfund Trust, set up under CERCLA to help pay for the cleanup of hazardous waste sites and to fund legal action against those responsible for hazardous waste sites, or from other responsible parties.

Second, U.S. EPA may itself undertake the cleanup process, either on its own or by prompting potentially responsible parties (PRPs) to undertake cleanup activities. If U.S. EPA undertakes the cleanup process, it likely does so where time is of the essence or where PRPs have refused to perform. In the meantime, U.S. EPA will seek to determine potentially responsible parties from whom to seek reimbursement.

Third, U.S. EPA may prompt PRPs to fund or perform cleanup activities utilizing a number of tools, some of which were addressed briefly above. U.S. EPA may send a PRP either a general notice of liability or a special notice of liability, alerting the PRP to its potential connection to a site's contamination. Such notices will briefly describe the perceived problem and an intended course of action and request that the PRP fund or perform cleanup-related activities. Frequently, the notices are accompanied by information requests relative to a PRP's possible relationship with a site.

If a PRP or PRPs elect to fund or perform the requested activity, they generally will negotiate an Administrative Order on Consent, detailing the rights, obligations, and repercussions of all involved. In such instances, U.S. EPA begins negotiations with its "Model AOC." If more than one PRP is involved, the funding or performing PRPs allocate costs and responsibilities among themselves. If the requested activity is the RD/RA phase of a site cleanup, a judicially approved consent decree will be required.

If the PRPs elect not to perform voluntarily, U.S. EPA may perform a remedial investigation/feasibility study with Superfund monies and later pursue cost recovery litigation against the PRPs. At the removal or the RD/RA phases, CERCLA Section 106 provides U.S. EPA with the authority to unilaterally issue an administrative order directing the PRPs to perform certain activities (commonly referred to as "106 Orders"), subject to penalties of up to $25,000 per day for nonperformance and the trebling of any costs that U.S. EPA incurs to perform in the PRPs' recalcitrance.

Finally, private parties that fund or perform cleanup activities, either voluntarily or by compulsion, are entitled to contribution toward response costs that are necessary and that are incurred consistent with the NCP. Contribution or cost recovery actions are filed in federal court against nonperforming PRPs to force the allocation of cleanup costs among the performing and recalcitrant PRPs. PRPs funding or performing activities pursuant to an AOC generally receive "contribution protection" from U.S. EPA, which denies the recalcitrant PRPs the ability to, in turn, seek contribution from them.

RCRA

The Resource Conservation and Recovery Act of 1976, as amended by the Hazardous and Solid Waste Amendments of 1984, is generally known as "RCRA." RCRA sets up a "cradle to grave" hazardous waste management system that ensures that hazardous waste is handled, stored, and disposed of to minimize the present and future threat that hazardous waste poses to human health and the environment (42 U.S.C. 6901, et seq.). The primary policy goal of RCRA is to decrease or eliminate the generation of hazardous waste. With respect to the hazardous waste that is produced, RCRA specifically states that the least preferable method for managing hazardous waste is land disposal.

The basic statutory structure of RCRA includes a system for identifying and listing hazardous waste; a permit program for facilities that treat, store, or dispose of hazardous waste; standards for facilities that generate and transport hazardous waste; and a mechanism for delegating to states the administration of the permit program. In contrast to CERCLA, RCRA mainly addresses the prevention of hazardous waste contamination, rather than the cleanup of hazardous waste contamination. However, CERCLA and RCRA overlap to a certain extent. Under Section 7003 of RCRA, the federal government can seek an injunction against any activity causing "imminent and substantial endangerment" and, thus, compel a cleanup. Further, under RCRA's corrective action requirements, facilities may have to perform a cleanup as a condition of maintaining a current operating permit.

When corrective action or cleanup activities are necessary, the range of RCRA-regulated materials is broad. Section 1004(27) broadly defines solid waste as:

> any garbage, refuse, sludge, from a waste treatment plant, water supply treatment plant or air pollution control facility and other discarded material, including solid, liquid, semi-solid, or containing gaseous materials resulting from industrial, commercial, mining and agriculture activities and from community activities but does not include solid or dissolved materials in domestic sewage, or solid or dissolved materials in irrigation return flows or industrial discharges which are point sources subject to permits under section 402 of the Federal Water Pollution Control Act, as amended (86 Stat. 880), or source, special nuclear, or byproduct material as defined by the Atomic Energy Act of 1954, as amended. (68 Stat. 923)

Thus, RCRA potentially applies to any waste, regardless of its physical form.

RCRA defines *hazardous waste* as a subset of "solid waste." Under RCRA Section 1004(5), a solid waste is hazardous if, because of its quantity, concentration, or physical, chemical, or infectious characteristics, it may:

> (A) cause, or significantly contribute to an increase in mortality or an increase in serious irreversible, or incapacitating reversible illness; or

> (B) pose a substantial present or potential hazard to human health or the environment when improperly treated, stored, transported, or disposed of, or otherwise managed.

The U.S. EPA has specifically listed covered hazardous wastes, which fall into three categories: (1) hazardous wastes from nonspecific sources; (2) hazardous wastes from specific sources; and (3) discarded commercial chemical products and all off-specification species, containers, and spill residues thereof (40 C.F.R. 261.31; 40 C.F.R. 261.32; 40 C.F.R. 261.33). In addition, the U.S. EPA has developed a list of solid wastes that are exempted from RCRA's requirements. The following solid wastes are excluded from the definition of hazardous wastes:

(1) household wastes;
(2) agricultural wastes used as fertilizer;
(3) mining overburden returned to the mine Site;
(4) utility wastes from coal combustion;
(5) oil and natural gas exploration drilling waste;
(6) wastes from extraction, beneficiation, and processing of ores and minerals, including coal;
(7) cement kiln dust wastes;
(8) arsenical-treated wood wastes generated by end users of such wood;
(9) certain chromium-bearing wastes [40 C.F.R. 261.4(b)]

If the U.S. EPA has not listed a waste as hazardous, it may still fall within the definition of hazardous wastes if it exhibits one of the four characteristics of hazardous wastes: ignitability, corrosivity, reactivity, or toxicity. If a listed hazardous waste is mixed with a solid waste, U.S. EPA still considers it to be a hazardous waste under the "mixture rule," unless it is exempted from coverage.[7] Additionally, if a solid waste is generated from the treatment, storage, or disposal of a listed hazardous waste, it is considered a hazardous waste under the "derived from" rule.[8]

In a nutshell, the test for regulation as a hazardous waste under RCRA is whether the 40 C.F.R. §261.11 exclusions apply, whether the waste is listed under Subpart 40 C.F.R. Part 261, or if not listed, whether the waste exhibits a hazardous characteristic under Subpart C of Part 261.

Facilities must have a corrective action plan to address any releases from solid waste management units located on site. Corrective action is based on three subsections in RCRA. Section 3004(u) requires that every hazardous waste permit issued after November 8, 1984, impose "corrective action for all releases of hazardous waste or constituents from any solid waste management unit at a treatment, storage, or disposal facility." Section 3004(v) requires an owner or operator of a facility to undertake corrective action beyond the facility boundary, provided the adjacent landowner gives permission to enter.

The term *release* is not defined in RCRA. To carry out the broad purposes of the corrective action program, the U.S. EPA has said the term should be "at least as broad as the definition of release" under CERCLA [50 Fed. Reg. 28,702, 28,713 (1985)].

The term *solid waste management unit* also is not defined in RCRA's regulations. Relying upon legislative history, the U.S. EPA has defined the term to include any unit at the facility "from which hazardous constituents might migrate, irrespective of whether the units were intended for the management of solid and/or hazardous wastes" [50 Fed. Reg. 28,712, quoting H.R. Rep. No. 198, 98th Cong., 1st Sess., Part 1, 60 (1983)].

All off-site facilities also must have a closure plan and, if the facility includes a land disposal unit, a postclosure plan as well. A closure plan describes the actions the facility's owner will take to close the facility once waste acceptance activities cease. Finally, the facility's owner or operator must furnish financial assurance to demonstrate its ability to undertake the activities set forth in the closure plan and, if applicable, the postclosure plan. The purpose of a postclosure plan is to minimize the potential for releases of hazardous wastes from disposal units after closure and postclosure. Requirements exist for each type of waste management unit [40 C.F.R. § 264.197 (tank systems); § 264.228 (surface impoundments); § 264.258 (waste piles); § 264.280 (land treatment); § 264.310 (landfills); and § 264.351 (incinerators)].

Closure activities for nonland disposal units generally involve decontaminating all equipment and structures and removing hazardous wastes. For disposal units such as landfills, however, hazardous waste will necessarily remain on site (§ 264.310). To guard against potential releases, the U.S. EPA has set a time period of 30 years for these landfills to conduct postclosure care (§ 264.117). Where releases occur despite monitoring or corrective action plans, cleanup obligations may be undertaken voluntarily or may be prompted by U.S. EPA, a state agency, an owner or subsequent owner, an affected party, or citizen's suit.

The Underground Storage Tank Program

Pursuant to subtitle I of RCRA, U.S. EPA has developed a regulatory program for underground storage tanks. RCRA defines an underground storage tank (UST) as:

> Any one or combination of tanks (includes underground pipes connected thereto) which is used to contain an accumulation of regulated substances, and the volume of which (including the volume of the underground pipes connected thereto) is 10 percent or more beneath the surface of the ground. [42 U.S.C. § 6991(1)]

The final regulations apply to USTs that contain a "regulated substance" (40 C.F.R. § 280.12). RCRA defines a regulated substance as any hazardous substance that falls within the requirements of CERCLA, except for hazardous substances regulated under RCRA Subtitle C and petroleum, including crude oil or any fraction thereof, that is liquid at standard conditions of temperature and pressure (40 C.F.R. § 280.12). Literally millions of USTs are believed to have existed over the past century.

RCRA Section 9002 requires an owner of a UST system to notify the relevant state of the existence of the system. Pursuant to Section 9003, the U.S. EPA must subject owners and operators of UST systems to requirements that must include the following at a minimum:

1. New tank standards dealing with design, construction, and installation
2. Leak detection system standards and associated record keeping requirements
3. Release reporting standards
4. Corrective action requirements for confirmed releases
5. Closure standards for existing tanks

If owners or operators fail to comply with the promulgated regulations, they may be subject to civil penalties of up to $10,000 per tank for each day of the violation.

Most important to our purposes here, RCRA authorizes the U.S. EPA to clean up petroleum releases or compel owners and operators to clean up petroleum releases through administrative orders or civil actions. If owners or operators fail to comply with an administrative order, they may be subject to fines of up to $25,000 per day. Citizen suits or threats thereof, addressed below, are commonly relied upon by private parties to prompt UST or other petroleum spill cleanups.

Citizen Suit Provisions

While each of the major federal environmental statutes contain Citizen's Suits, RCRA Citizen's Suit provision, 42 U.S.C. § 6972, is the Citizen's Suit provision most often used to affect a site cleanup. Citizen's suit provisions typically authorize suits against any person or corporate entity that is violating a particular statute by failing to comply with the required permits and regulations. A typical citizen's suit provision authorizes any person to file a lawsuit (1) to enforce the requirements of the statute against any person alleged to be in violation and (2) to require the government to perform a mandatory duty specified in the statute. Courts are authorized to award injunctive relief, directing performance of a particular requirement. Some statutes permit the recovery of civil penalties as well.

Citizen suits also may be brought against U.S. EPA to force it to perform "nondiscretionary duties" such as rule makings and record keeping. Any person with a "sufficient interest or personal stake to furnish the necessary concrete case or controversy" may display the necessary distinct injury traceable to the alleged misconduct of the potential defendant. The plaintiff must demonstrate injury that is likely to be redressed by a favorable decision.

Prior to filing a citizen's suit, the prospective plaintiff must give a minimum of 60 days' notice, and longer periods for certain allegations, to federal and state agencies, as well as to the alleged violator. The notice period provides governmental agencies with the opportunity to take responsibility for enforcing environmental regulations and gives the alleged violator the opportunity to correct the violation, thereby avoiding the need for a citizen's suit filing. Citizen's suits are barred where a federal or state agency has commenced an enforcement action qualifying as a "court" proceeding, and the action is diligently prosecuted under the appropriate statute.

Unique to the citizen's suit provisions is the availability to a prevailing party of reasonable attorney fees. This provides a greater incentive for nongovernmental entities to pursue enforcement actions. Indeed, citizen's suit actions are a favored approach of citizen's groups and environmental activists.

Criminal Provisions

Each of the federal and many state environmental statutes contain criminal provisions applicable to most compliance provisions, potentially resulting in prison sentences, criminal forfeiture, and fines. Perhaps most relevant to site cleanups are the RCRA criminal provisions used to pursue individuals and entities illegally transporting or disposing of wastes. Generally, this results from the transport or disposal of wastes without a permit or in violation of a permit condition.

Punishments can be severe. For example, a person who, without a permit, knowingly transfers, treats, stores, or disposes of any hazardous waste will be subject to up to $50,000 per day of violation and up to 5 years of imprisonment. A "knowing" violation merely requires that an individual knows the facts that comprise the forbidden act. Knowledge that the act violates the applicable law is not required or relevant where dangerous or hazardous substances are involved. After a first conviction, the maximum punishment doubles. If the knowing violation places an individual in imminent danger of death or serious bodily injury, the maximum punishments rise to $250,000 and 15 years in prison.

The use of criminal enforcement tools, such as imprisonment, forfeitures, and fines, is now being applied to cases beyond the classic "midnight dumper." Individuals and companies, ranging from Fortune 25 corporations to the management of small, closely held companies, are being prosecuted for their failure to comply with applicable environmental laws and regulations. In addition to environmental provisions, the government frequently resorts to general criminal code provisions to prosecute individuals for false statements and for conspiracy to violate environmental laws.

State and federal government agencies are increasing the amount of resources dedicated to supporting environmental crimes investigations and prosecutions. Pursuant to the Pollution Prosecution Act of 1990, the number of investigators dedicated to environmental crimes investigations has more than tripled from below 60 to over 200. Additionally, the Act established the National Enforcement Training Institute and increased funding for these activities. At all levels of government, state and federal agencies and other interested parties are forming task forces focused in geographical areas or on environmental media to further police environmental regulatory provisions.

This effort has resulted in prosecutors increasing their enforcement efforts in the environmental criminal realm. In 1985, the federal government returned 40 indictments, resulting in 37 guilty pleas or convictions, $585,850 in fines, and 5 years and 5 months in total prison sentences related to environmental crimes. In fiscal 1994, new records in federal criminal enforcement were set: 220 matters were referred to the U.S. Department of Justice, and 250 individuals and corporations were indicted. Convictions resulted in 99 years in prison sentences and $36.8 million in criminal fines.[9]

The almost threefold increase in indictments has been facilitated by the standards of proof that Congress and the state legislatures have applied to the environmental statutes. For example, under the Clean Water Act, an individual may be found guilty of a misdemeanor (resulting in a sentence of up to one year) when a prosecutor proves beyond a reasonable doubt that the defendant failed to exercise due care and reasonably should have known of or foreseen the violation. Actual knowledge of the violation is unnecessary.

While generally not a specific threat to affect a site cleanup, criminal provisions make the various regulatory schemes more effective by preventing unnecessary cleanup situations, punishing corporations and individuals who cause the cleanup scenarios, and forcing parties to toe the mark of the applicable statutes and regulations as a cleanup progresses.

CONCLUSION

Clearly, significant legal tools and authority are available to governmental agencies, individuals, businesses, and property owners to prompt a site cleanup. Often, the threat of legal action alone will start and perpetuate the cleanup process. Whether a lawsuit is filed, an administrative or judicial order is issued, or remediation efforts have commenced, the law will drive the relevant parties to the end result.

ENDNOTES

1. 42 U.S.C. Section 9601, et seq.
2. United States Environmental Protection Agency, *Information Resources Directory*, Spring 1994, at 26. The NPL is authorized at 42 U.S.C. Section 9605.
3. United States Environmental Protection Agency, *Information Resources Directory*, Spring 1994, at 34.
4. United States Environmental Protection Agency, *Information Resources Directory*, Spring 1994, at 16.
5. CERCLA Section 106(a) permits U.S. EPA to unilaterally direct parties that it deems to be potentially responsible to take action to abate an imminent and substantial endangerment to the public health or welfare or the environment because of an actual or threatened release of a hazardous substance from a facility, subject to $25,000 per day in penalties for failure to comply without sufficient cause [42 U.S.C. § 9606(a) (1995)].
6. 40 C.F.R. Section 261.3 (a) (2) (iii)–(iv).
7. 40 C.F.R. Section 261.3 (c) (2).
8. U.S. EPA, Office of Enforcement and Compliance Assurance, Enforcement and Compliance Assurance Accomplishments Report, Fiscal Year 1994, May 1995.
9. U.S. EPA Office of Solid Waste and Emergency Response, Office of Underground Storage Tanks, Directive No. 9610.17.

Basic Equations of Flow and Transport in Porous Media

S. Majid Hassanizadeh*
Department of Water Management, Environmental and Sanitary Engineering
Faculty of Civil Engineering
Delft University of Technology
P.O. Box 5048
2600GA Delft, the Netherlands

William G. Gray
Department of Civil Engineering and Geological Sciences
University of Notre Dame
Notre Dame, IN 46556-0767, USA

INTRODUCTION

Bioremediation **strategies are** applied to decrease or eliminate contaminants that are distributed in the subsurface. For a remediation exercise to be successful, it must build on the response of the subsurface fluids to changes in pressure that might cause immobile fluids to be mobilized or fluids moving along one flow path to be diverted to another. If a cleanup effort is to be conducted, in situ procedures must be employed that will allow the bacteria to reach the contaminant with appropriate nutrients for the degradation reaction to occur. The flow fields must be manipulated to allow juxtaposition of bacteria and the target contaminants. If a cleanup activity is to be on site following a pump and treat strategy, some knowledge of the expected response of the fluid reservoir to pumping stresses must be employed. This is important, not only to ensure that the pumping will bring the contaminants to the treatment facility, but also to assess how much contaminant will remain trapped in the soil over the short term, but providing a long-term pollution threat.

*Also with: RIVM, P.O. Box 1, 3720BA Bilthoven, the Netherlands.

Therefore, a comprehensive approach to biodegradation and bioremediation must include, not only information about kinetics, reaction pathways, enzyme reactions, and remediation technologies, but also principles of mass conservation, momentum conservation, dispersion mechanisms, and the mathematical equations that govern flow in the subsurface environment. It is the purpose of the present chapter to provide the principles that govern porous media flow and to express those principles as differential equations that may be solved to determine the movement of fluid phases in the subsurface. Attention is focused on accounting for the appropriate mechanisms of flow, rather than on solving the governing equation. The equations of flow and transport contain parameters that provide a formidable challenge for experimental determination. Furthermore, the equations are nonlinear and lend themselves, at best, to computer-based numerical solutions. Nevertheless, it is essential that the mechanisms that govern fluid movement be accounted for and understood as an essential ingredient of any comprehensive effort to decontaminate a subsurface location. The equations provided in this chapter and the simulation approaches considered in the following chapter by Celia provide some insights into the mathematics and physics of flow and transport phenomena.

SUBSURFACE REGION, SCALE, AND CONTINUUM DESCRIPTION

System Components and Contaminant Types

An unconsolidated solid is composed of solid grains. These grains are generally considered to be incompressible themselves, although their relative positions may change, and the solid as a whole may be compressed. In the environment, the simplest porous medium system is composed of a single fluid phase interacting with a solid. This situation is found naturally in water-bearing geologic formations at sufficient depths below the surface such that the water saturates the interconnected space within the solid matrix. This region is referred to as the saturated or groundwater zone. A saturated zone that extends upward to the surface of the earth interacts directly with and may underlie rivers, lakes, and estuaries. In other cases, the water-saturated zone is bounded above by a flow zone in which an air phase occupies a portion of the pore space and forms a separate vapor phase. When organic chemicals are introduced into the subsurface at sufficient concentration such that they do not dissolve completely in the water or vaporize completely into the air phase, a third fluid phase may be formed. This phase is commonly referred to as a nonaqueous phase liquid (NAPL) and is further classified as dense (DNAPL) or light (LNAPL), depending on whether its density is greater or less than that of water. The three-phase system consisting of air, water, and solid, when studied as a unit, forms the general naturally occurring porous medium system. In addition to these basic components, contaminants consisting of chemical species that dissolve in the fluid phases, adhere to the solid phase surface, or form a separate phase are of importance.

The solid, water, and vapor phases may each contain dissolved constituents that are background components. These components are typically present at low or trace concentration levels that are nondetrimental to the intended use of the subsurface water. Dissolved constituents present in the water phase are referred to as contaminants when they are present in the groundwater system, due either to natural processes or as a result of human activities, at concentrations unacceptable for the intended uses of groundwater. Thus, groundwater contamination is related to the degradation of the water quality. A myriad of compounds may be considered as water contaminants, including mineral salts, heavy metals, other inorganic materials, hydrocarbons, other organic materials, microorganisms, dissolved organic and inorganic vapors, and dissolved natural gases such as radon (Domenico and Schwartz, 1990).

In the natural solid-water-air system, contamination can be a problem in each of the phases, not just the water phase. Typically, the presence of an additional organic phase is also an undesirable contamination of the system. In general, four distinct categories of contaminants may be recognized:

- airborne (gaseous components or particles)
- waterborne (soluble components or colloidal particles)
- separate fluid phase (immiscible with air and water)
- surface-resident components (adsorbing and absorbing matter, colloidal particles, and surfactant-type molecules or organisms)

Often, contaminants are complex enough that they may exist in more than one form or category. In the study of the spreading of subsurface contaminants, two main types of displacement are distinguished: miscible and immiscible. Miscible displacement refers to spreading of airborne and/or waterborne components within the system, whereas immiscible displacement refers to the movement of a separate fluid phase relative to the water, air, and solid phases. The spreading processes and governing equations for these two types of displacements are fundamentally different, as will be discussed later in this chapter.

Sources and Classification of Contamination

The sources of contamination are as varied as the many forms they may take (Domenico and Schwartz, 1990). Various types of classification are possible. One system of classification distinguishes between contamination that is distributed throughout space, such as agricultural chemicals that have infiltrated into the subsurface, and contamination at a single point, such as gasoline entering the subsurface due to the rupture of a storage tank. Distributed sources may be considered to be volumetrically distributed, distributed over an area, or distributed along a line or curve. Whether a given case of contamination is classified as emanating from a distributed or point source actually depends on the scale of observation and/or modeling employed to describe the event.

Another system of classification of contaminants distinguishes between those that have a natural origin and those whose origin is a result of human activities. Examples of natural contamination are salinity of water in coastal aquifers and minerals present in groundwater due to dissolution of the solid matrix. Examples of contamination sources related to human activities include landfills, agricultural chemicals, and accidental industrial spills. In some cases, natural contamination can be exaggerated by human activity, such as groundwater extraction in coastal aquifers that causes an increase in saltwater intrusion.

Spreading Processes

The spatial distribution and temporal evolution of subsurface contamination is governed by a number of complex and interactive processes. Difficulty in accurately describing the spread of contamination is due to two major factors: heterogeneities and interaction among the various chemical elements. Heterogeneities of properties exist in the fluids and solids over many different length scales (Cushman, 1990). Such heterogeneities include fluid composition, chemical and ionic character of the solid surfaces, the size and spacing of grains, variation in solid matrix types, and the presence of fractures. The interactions among the various chemicals and the phases may be mechanical, biological, and chemical. These can occur within a phase or between phases. The various processes that affect the spreading of contaminants include:

1. Mechanical processes
   * advection (fluid flow due to a gradient in fluid potential)
   * diffusion (spreading as a result of a concentration gradient)
   * hydrodynamic dispersion (spreading as a result of a

concentration gradient coupled with small-scale variations in the
fluid flow velocity)
- filtration (straining of colloids and microorganisms)
- solid phase movement (subsidence of the solid matrix, swelling of
  clays)

2. Thermochemical processes
   - homogeneous chemical reactions (among constituents within the
     same phase)
   - heterogeneous chemical reactions (among constituents in
     different phases occurring at the phase boundaries)
   - dissolution and volatilization of fluids (transfer of constituents
     between fluid phases)
   - adsorption and desorption (transfer of constituents between fluid
     and solid phases)

3. Biological processes
   - transformation of chemicals (aerobic and anaerobic
     transformation of constituents by microbial action)
   - growth of microbial communities (development of microbes on
     the solid surfaces, clogging of void space)

All these processes are influenced by and give rise to heterogeneities in
the system. Proper accounting for heterogeneities is an important area
for research that includes statistical approaches as well as the deterministic considerations being presented here (Dagan, 1989; De Marsily, 1986,
1993). Accounting for the interaction between the identification of heterogeneities and the scale at which the system is observed and physically
sampled is very important for accurate modeling.

Scale

Study of the various processes listed above may be carried out at a
variety of scales. The choice of scale depends on the scope and purpose of
the study. Also, the scale of description of processes is closely related to
the scale of observation of those processes. The qualitative question,
"What is the make-up of a porous medium?" can have many different
answers. A hydrologist would talk of groundwater and the rock as if they
are present everywhere as overlapping continua. Variations in material
properties may be considered negligible over "short" distances on the
order of one meter. Hydrologic observations are carried out over scales of
tens to hundreds of meters. The observation instruments have characteristic dimensions of at least several centimeters. A geochemist, on the
other hand, often considers a porous medium as being made of pore water

and solid grains that are interacting along definite boundaries. Moreover, a geochemist considers large variations in material properties to take place over distances on the order of millimeters. Observations carried out over distances on the order of a few centimeters are considered to cover a "large" length scale. Other serviceable descriptions are possible. For example, a porous medium may be considered to be composed of solid and fluid molecules that are moving around in regions where there are no definite "boundaries," only gradients in molecular types. From this perspective, zones of transition exist where molecular properties change considerably over "large" distances of the order of a few microns.

The three scales described above are commonly called macro- (or field) scale, micro- (or pore) scale, and molecular scale, respectively. Quantities appropriate for one scale are actually average values of corresponding quantities at a lower scale. Average quantities are defined and/or measured over averaging volumes, commonly referred to as representative elementary volumes, or REVs. Thus, for example, properties of fluid phases present within pores (that is, properties defined at the microscale) are the averages of molecular scale properties obtained for an REV that contains a large number of molecules. The microscale REV must be of sufficiently large size to contain enough molecules such that the average quantities may be considered to be representative of the phase. At the same time, it must be small enough to capture the variations in fluid properties within the pore spaces. Similarly, properties of fluid phases at the macroscale are averages of pore-scale properties obtained for an REV that contains a large number of pores. The averaging volume must be of sufficiently large size and contain enough pores to provide an average behavior that may be considered representative of the phase at the macroscale as discussed in Hassanizadeh and Gray (1979a). This REV must also be small enough to capture the variations in macroscale fluid properties occurring within an aquifer or over the depth of the solid phase profile. The use of the REV implies that, indeed, such a volume exists for sampling and field experimentation. In most circumstances, this is a reasonable hypothesis, and a characteristic length of the REV common to all variables being considered may be found. In actuality, the REV provides a reasonable tool for developing a theoretical framework for the study of porous media flow. However, situations exist where the many scales of heterogeneity make parameterization of the REV-scale equations subject to significant errors.

To illustrate the concept of the REV, consider the following thought experiment. Suppose the average density of water occupying the pores of a saturated porous medium is to be determined using a spherical averaging volume with diameter d. A mass density may be obtained as the total mass of water in the volume at a particular instant in time divided by the size of the volume. To make these measurements, a series of averaging spheres will be used, each with a different diameter. Measurements are to be made by centering an averaging volume at a particular point in space and determining the effect of the actual size of the volume on the value of density obtained.

For consideration of the measurements of density, spheres will be used with diameters ranging in size from the microscale continuum length scale up to the characteristic length of the system. Figure 1 provides a typical schematic plot of the mass density versus the diameter of the averaging volume. Three distinct regions can be identified. In the first region, the diameter of the averaging volume is smaller than or equal to the size of a pore. In this case three possibilities exist when measuring the mass in the volume. First, the volume may completely fall within the pore space so that the mass density is essentially equal to the mass of water divided by the volume that water occupies. This is the maximum density that could be achieved and is given the normalized value of 1.0 in Figure 1. Second, the volume may lie completely within a solid grain so that no water mass is in the volume and the normalized water density is equal to 0.0. Third, the volume may partly occupy a pore with the remainder lying in a solid grain. For this case, the normalized measured mass density will take on a value between 0.0 and 1.0. As the size of the averaging volume is increased, fluctuations will appear in the measured density. The fluctuations are caused by the fact that added volume increments may happen to contain a preponderance of additional pores, causing an increase in the water density; or an increment might include

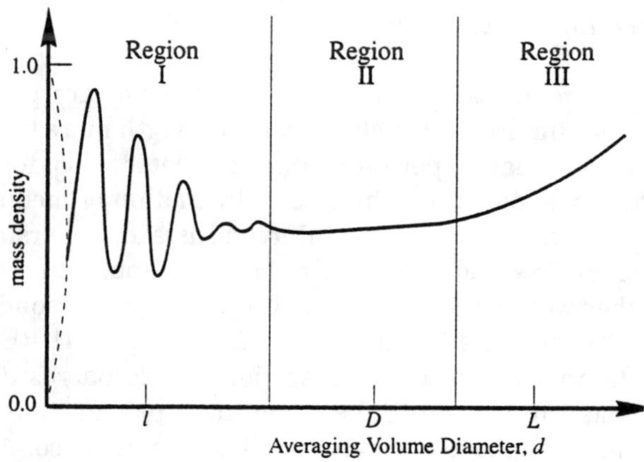

Figure 1. Normalized mass density as a function of the averaging volume size.

primarily solid grains, thus decreasing the water mass density. These fluctuations, however, tend to diminish as the size of the averaging volume increases, and eventually within some interval, the average water density and, in general, all average quantities will be insensitive to the increase in REV size. As a result, a plateau in the graph of mass density versus d will be obtained as depicted in region II in Figure 1. A representative value of d within region II is denoted as D. If, and only if, the plateau in average values is obtained for each point occupied by the material, the material is said to possess a representative elementary volume, an REV, with a characteristic length, D.

In the continuation of the thought experiment, let the averaging volume diameter d, be increased to sizes larger than D with the averaging volume still wholly within the medium under consideration. Thus, calculation of density will be based on measurements made in region III of Figure 1. Although deviation of density from the value measured in region II will not occur if the system is homogeneous, a deviation is depicted in the figure to indicate the possible effects of large-scale inhomogeneities in material properties. Gross inhomogeneities of the medium may cause the values of average quantities to become sensitive to the size of the averaging volume when the characteristic length of the REV is much greater than D. Therefore, to have meaningful average quantities, the size of the averaging volume must be much smaller than the characteristic length dimension of the medium, indicated by L. It follows from the foregoing discussion that the restriction must hold that (Hassanizadeh and Gray, 1979a)

$$l \ll D \ll L \tag{1}$$

The Macroscale Continuum

Typical measurements of groundwater properties are performed on porous medium samples with a characteristic length more than 20 times larger than a solid matrix pore (or grain) diameter. Also, for almost all practical hydrologic purposes, the detailed variation of medium properties within a pore is not of primary interest. Instead, understanding and quantification of flow and transport processes through the medium are sought. In other words, the study and observation of the solid phase and groundwater are commonly carried out at the region II REV scale, or macroscale, described in the previous section. At this macroscale, a porous medium is modeled as a multiphase, multicomponent system with all components and all phases present at all points of space. Accordingly, properties of phases and components are defined continuously at all points and all times. Thus, when viewed at the macroscale, phases and components form overlapping continua with no precisely identifiable or locatable boundaries among them. At the microscale, immiscible phases are separated by distinct interfacial boundaries, and certain components are extant only to specific phases or to interfaces. These conditions will be reflected in the macroscale governing equations that describe processes of interest. These equations are formulated in terms of macroscale, REV-averaged quantities. Various mechanical, chemical, and biological interactions occurring at the phase boundaries evidence themselves as source or sink terms in the macroscale governing equations.

Porosity, Volume Fractions, and Saturation

Averaged quantities for phases and components may be defined as the average of corresponding pore-scale quantities in a manner described previously. In addition to the traditional continuum properties, new quantities accounting for the separateness of phases will appear. For example, one may define the phase average mass density, $\langle \rho \rangle_\alpha$ [ML^{-3}], as the mass of phase α per unit volume of REV. In general, however, the α phase occupies only a portion of the space. Thus, alternatively, an intrinsic average mass density, ρ^α, may be defined as the mass of phase α per total volume of phase α present within REV. The ratio of the two mass density functions is equal to the fraction of the volume of the REV occupied by phase α. This ratio is called the volume fraction and is denoted by ε^α. Thus, the following relationship applies:

$$\langle \rho \rangle_\alpha = \varepsilon^\alpha \rho^\alpha \tag{2}$$

Since each phase occupies a fraction of the REV, the sum of all the volume fractions at any position must be unity:

$$\sum_\alpha \varepsilon^\alpha = 1 \tag{3}$$

where Σ_α denotes a summation over all phases. The fraction of the REV that is not occupied by the solid matrix is referred to as the porosity, n, such that

$$\varepsilon^s = 1 - n \tag{4}$$

where ε^s is the solid phase volume fraction. For saturated conditions such that water occupies all the void space, the water phase volume fraction,

ε^w, is equal to the porosity, n. For unsaturated conditions, where air or other immiscible fluid phases are present in addition to the water, porosity is equal to the sum of volume fractions of all fluid phases (but not the solid phase):

$$n = \sum_{\alpha \neq s} \varepsilon^\alpha \tag{5}$$

Then, for each fluid phase, a saturation function, s^α, is defined to be equal to the fraction of the void space within an REV occupied by phase α such that

$$s^\alpha = \varepsilon^\alpha / n \tag{6a}$$

and

$$\sum_\alpha s^\alpha = 1 \tag{6b}$$

where the summation is over all the fluid phases.

Macroscale Velocities

When considered from a microscale perspective, the velocity of a fluid in the pore is the velocity at which a tracer would be carried through the system. At the macroscale, one velocity of importance is the average pore velocity, denoted for fluid phase α as $\mathbf{v}^\alpha$, which again provides the average velocity at which this fluid actually moves. This velocity may be thought of as the volumetric flux of the α phase obtained as the volumetric flow divided by the cross-sectional area of α phase. An alternative volumetric flux is the volumetric flow divided by the total cross-sectional area of the system. For example, if water is flowing through a homogeneous saturated porous medium in a pipe with cross-sectional area A and porosity n at a rate of Q cubic units per unit time, the intrinsic phase average velocity, or the average pore velocity, along the tube axis may be obtained as $v^w = Q/(nA)$. The phase average velocity would be given as $nv^w = Q/A$. Analogous to the average fluid pore velocity is the average solid grain velocity designated as $\mathbf{v}^s$.

In the description of porous media flow, of particular interest is the velocity of the fluid phases relative to the velocity of the solid phase. For example, for the water phase, this would be designated as $\mathbf{v}^w - \mathbf{v}^s$. This relative velocity is important in that it is the flow velocity at which a contaminant would be carried, relative to the solid phase, through a porous medium. Of historical, as well as practical, significance is the Darcy velocity, or the phase average relative velocity. This velocity is usually designated as $\mathbf{q}$ with an appropriate superscript to designate the fluid phase. For saturated flow of water in a porous medium, the Darcy velocity is given as

$$\mathbf{q}^w = n(\mathbf{v}^w - \mathbf{v}^s) \tag{7a}$$

If the flow is not saturated, the Darcy velocity of any phase is given as

$$\mathbf{q}^\alpha = \varepsilon^\alpha(\mathbf{v}^\alpha - \mathbf{v}^s) \tag{7b}$$

or

$$\mathbf{q}^\alpha = s^\alpha n(\mathbf{v}^\alpha - \mathbf{v}^s) \tag{7c}$$

The Darcy velocity is also referred to as the superficial velocity. In many porous media problems, the medium is essentially at rest, and the rate of deformation of the solid skeleton is negligibly small such that Equation (7c) may be approximated as

$$\mathbf{q}^\alpha = s^\alpha n \mathbf{v}^\alpha \tag{8}$$

EQUATIONS OF MASS BALANCE

To be able to model the distribution of fluids and their chemical components in a porous medium, the equation of mass conservation must be formulated and solved. The general form of the macroscale equation of mass balance for a phase α is (Hassanizadeh and Gray, 1979b)

$$\frac{\partial(\varepsilon^\alpha \rho^\alpha)}{\partial t} + \nabla \bullet (\varepsilon^\alpha \rho^\alpha \mathbf{v}^\alpha) = \hat{r}^\alpha + \hat{M}^\alpha \tag{9}$$

where $\hat{r}^\alpha$ [ML^{-3}T^{-1}] is the apparent rate of production of mass in phase α per unit volume of the porous medium, and $\hat{M}^\alpha$ [ML^{-3}T^{-1}] accounts for pumping.

The production term $\hat{r}^\alpha$, in fact, provides the measure of the mass transferred between the α phase and the neighboring phases. Note that chemical reactions involving species within a phase do not contribute to $\hat{r}^\alpha$ because they do not produce mass, but merely transform it. Thus, $\hat{r}^\alpha$ is negligible if the amount of mass transferred between phases is small. In the case of saturated porous media, unless the solid grains are highly soluble or a significant part of the fluid phase precipitates or is adsorbed onto the solid, the mass exchange between grains and water does not affect the movement of water through the porous medium. In these cases, the term $\hat{r}^\alpha$ may be neglected. Also, in the case of unsaturated flow, the production term is negligible when evaporation of water or volatilization of solutes does not alter the densities or saturation of the fluid phases. One situation where $\hat{r}^\alpha$ is significant is when the uptake of water by plant roots is simulated as a spatially distributed process. In cases where a third fluid phase is present, $\hat{r}^\alpha$ may become significant for those fluids that exist at low saturations and are volatile and/or (partially) soluble in water. For instance, steam injection to remove volatile organics is designed to provide cleanup of a contaminated site precisely by volatilizing the organic components of the liquid water phase. For such a process, accurate accounting for the change in phase is important.

Finally, the quantity $\hat{M}^\alpha$ is used to account for point sources/sinks of mass for the α phase due to well pumping. Pumping may also be treated as a boundary condition when solving the differential equation. When pumping at point $\mathbf{x}_p$ is to be accounted for by the source term on the right side of Equation (9):

$$\hat{M}^\alpha = M_p^\alpha(t)\delta(\mathbf{x} - \mathbf{x}_p) \tag{10}$$

where $M_p^\alpha(t)$ [MT^{-1}], is the time-dependent pumping rate, positive for injection, and $\delta(\mathbf{x} - \mathbf{x}_p)$ is the Dirac delta function acting at position $\mathbf{x}_p$. If more than one well is in the system, $\hat{M}^\alpha$ will be a summation of terms with form similar to the right side of Equation (10) written for each well location.

The general form of the macroscale equation of mass balance for chemical constituents of a phase is (Nguyen et al., 1982; Hassanizadeh, 1986)

$$\frac{\partial(\varepsilon^\alpha \rho^\alpha \omega^{i\alpha})}{\partial t} + \nabla \bullet (\varepsilon^\alpha \rho^\alpha \omega^{i\alpha} \mathbf{v}^{i\alpha}) = \varepsilon^\alpha \hat{r}^{i\alpha} + \hat{M}^{i\alpha} \tag{11}$$

where $\omega^{i\alpha}$ [-] is the mass fraction of the ith component of phase α (i.e., the mass of component i per unit mass of α phase), $\mathbf{v}^{i\alpha}$ [LT^{-1}] is the average velocity of component i, $\hat{r}^{i\alpha}$ [ML^{-3}] is the rate of production of mass of component i per unit volume of phase α, and $\hat{M}^{i\alpha}$ [ML^{-3}] is the addition of species i to the system due to pumping. The production term $\hat{r}^{i\alpha}$ consists of both chemical and biological transformations within the phase and transfer of the species into the phase from other phases. Note that for pumping or injection at point $\mathbf{x}_p$,

$$\hat{M}^{i\alpha} = \omega^{ip} M_p^{\alpha}(t)\delta(\mathbf{x} - \mathbf{x}_p) \tag{12}$$

When M_p^{α} is negative such that pumping is occurring, ω^{ip} is the mass fraction of species i in the porous medium at position $\mathbf{x}_p$, $\omega^{ip} = \omega^{i\alpha}(\mathbf{x}_p,t)$. When the pumping is positive with injection occurring, ω^{ip} is the mass fraction of species i in the α phase fluid that is being injected, not necessarily the mass fraction in the α phase that is in the porous medium at the location of the well.

The term $\varepsilon^{\alpha}\rho^{\alpha}\omega^{i\alpha}\mathbf{v}^{i\alpha}$ is the total mass flux of the component. This mass flux may be broken into two parts: one part due to the advection of solute with the average phase velocity, $\varepsilon^{\alpha}\rho^{\alpha}\omega^{i\alpha}\mathbf{v}^{\alpha}$, and an additional part due to advection relative to this mean velocity, the dispersion of the component, denoted by $\mathbf{J}^{i\alpha}$, so that

$$\varepsilon^{\alpha}\rho^{\alpha}\omega^{i\alpha}\mathbf{v}^{i\alpha} = \varepsilon^{\alpha}\rho^{\alpha}\omega^{i\alpha}\mathbf{v}^{\alpha} + \mathbf{J}^{i\alpha} \tag{13}$$

Therefore, the mass balance equation for a solute may be written as

$$\frac{\partial(\varepsilon^{\alpha}\rho^{\alpha}\omega^{i\alpha})}{\partial t} + \nabla \bullet (\varepsilon^{\alpha}\rho^{\alpha}\omega^{i\alpha}\mathbf{v}^{\alpha}) + \nabla \bullet \mathbf{J}^{i\alpha} = \varepsilon^{\alpha}\hat{r}^{i\alpha} + \hat{M}^{i\alpha} \tag{14}$$

Note that because Equation (14) may be written for each chemical species, i in the α phase, the summation of this equation over all species must lead to Equation (9). This condition confirms that

$$\sum_i \omega^{i\alpha} = 1 \tag{15a}$$

$$\sum_i \mathbf{J}^{i\alpha} = 0 \tag{15b}$$

$$\sum_i \varepsilon^{\alpha}\hat{r}^{i\alpha} = \hat{r}^{\alpha} \tag{15c}$$

and

$$\sum_i \hat{M}^{i\alpha} = \hat{M}^{\alpha} \tag{15d}$$

Although the species mass production terms, $\varepsilon^{\alpha}\hat{r}^{i\alpha}$, sum to the overall phase production $\hat{r}^{\alpha}$, a term that has been identified as often being negligible, the individual species generation terms may be important in the species balance equation when chemical reactions are occurring that transform species within the α phase.

In principle, the equations of mass balance may be seen as equations that can be solved to give the distribution of mass density of phases or mass fraction of components. However, examination of these equations reveals that they contain a number of additional unknowns for which appropriate relationships must be provided. These are the velocity vectors of phases, dispersive mass flux vectors, and production terms of components. The velocity vector for a phase is typically obtained from a simplified momentum balance known as Darcy's law. The dispersive mass flux vector of a component is given by an equation similar to Fick's law of diffusion. In the remainder of this chapter, appropriate relationships for the unknown quantities will be presented, and the equations that govern the movement of phases and components in a porous medium will be obtained.

SATURATED GROUNDWATER FLOW

Darcy's Law

Processes in soil and groundwater are influenced by the movement of water (and other fluids when present). Fluids are driven from regions of higher energy potential to areas of lower potential. The porous medium offers resistance to such movement. The rate at which the fluid is displaced depends on the resistance to flow provided by medium, the difference in energy between the two regions, the distance separating the regions, and the available cross section area of flow. The following mathematical formulation of the above statement for a vertical column of porous medium saturated with a single fluid (see Figure 2) was given by Darcy (1856):

$$Q = K^f \frac{\phi_1^f - \phi_2^f}{L_1 - L_2} A = K^f \frac{\Delta\phi^f}{\Delta L} A \tag{16}$$

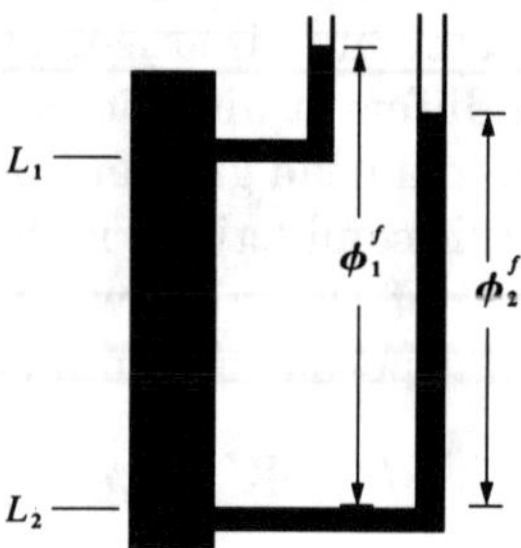

Figure 2. Schematic diagram of Darcy's experimental column.

where Q [L^{3}T^{-1}] is volumetric flow rate, K^f [LT^{-1}] is hydraulic conductivity (the inverse of resistivity) specific to the fluid saturating the particular medium, A [L^2] is the cross-sectional area, ϕ^f [L] is hydraulic head (energy per unit weight) of the fluid, and ΔL is the distance over which the head drop $\Delta\phi^f$ takes place.

In writing this equation, the implicit assumptions are made that Q, K^f, and A are constant over the whole length of the column and that ϕ^f varies linearly. These assumptions are probably acceptable for a short column packed with a homogeneous medium, but not over larger distances. Thus, a more general form is needed. A generalization of Equation (16) for flow in an arbitrary direction l, within an aquifer, is given in the form of a differential equation:

$$q_l^f = -K^f \frac{\partial \phi^f}{\partial l} \tag{17}$$

where q_l^f [MT^{-1}] is the superficial flow velocity in direction l per unit area ($q_l^f = Q/A$), and $\partial\phi^f/\partial l$ [-] is the rate of drop of hydraulic head in the direction of flow, also called hydraulic gradient. In three-dimensional space, Equation (17) can be written in each direction independently of the other directions. Therefore, due to the principle of superposition for linear processes, a generalization of Darcy's law in three dimensions becomes

$$\mathbf{q}^f = -K^f \nabla \phi^f \tag{18}$$

where ∇ is the three-dimensional gradient operator. For a heterogeneous medium, K^f will not be a constant but may be a function of position; $K^f = K^f(\mathbf{x})$. Also, in writing Equation (18), the conductivity of the medium is considered to be the same in all directions. This is true only in the case of isotropic media. When a medium is anisotropic, the hydraulic conductivity will be different in different directions. Moreover, flow may take place in one direction due to a head gradient in an orthogonal direction. The corresponding hydraulic conductivity is thus a tensor, denoted as $\mathbf{K}^f$. Then the appropriate form of Darcy's law for describing the flow of a single fluid in an anisotropic porous medium is given by

$$\mathbf{q}^f = -\mathbf{K}^f \cdot \nabla \phi^f \tag{19}$$

The hydraulic conductivity tensor is commonly assumed to be symmetric.

Another assumption implicit in these formulas is that a hydraulic head for the fluid can be defined. This is possible, however, only if the fluid density is constant or if it is a function only of the pressure in the fluid. In the case of constant-density fluids, the hydraulic head is defined by

$$\phi^f = z + \frac{p^f - p^0}{\rho^f g} \tag{20}$$

where p^f [ML^{-1}T^{-2}] is the fluid pressure, p^0 is a reference pressure, g [LT^{-2}] is the magnitude of the gravitational acceleration, and z [L] is the vertical coordinate assumed to be positive upward. For compressible fluids where ρ^f is a function of pressure, a generalization of the definition (20) is given by Hubbert (1940):

$$\varphi^f = z + \frac{1}{g} \int_{p^0}^{p^f} \frac{1}{\rho^f(p)} \, dp \tag{21}$$

In general, however, when the fluid density is affected by temperature gradients in the system or by the concentration of dissolved constituents, a total hydraulic head may not be defined. For these cases, the appropriate generalization of Darcy's law for a saturated porous medium is given in terms of pressure instead of the hydraulic head:

$$\mathbf{q}^f = -\frac{\mathbf{k}}{\mu^f} \cdot (\nabla p^f - \rho^f \mathbf{g}) \tag{22}$$

where μ^f [ML^{-1}T^{-1}] is the fluid dynamic viscosity, $\mathbf{g}$ [LT^{-2}] denotes the gravity vector, and $\mathbf{k}$ [L^2] is the intrinsic permeability tensor. The intrinsic permeability is a property of the solid, independent of the fluid properties. For the case of constant-density fluids, the hydraulic conductivity tensor can be defined in terms of the intrinsic permeability tensor through the following relationship:

$$\mathbf{K}^f = \frac{\rho^f g \mathbf{k}}{\mu^f} \tag{23}$$

Equations of State and Equation of Stress Equilibrium for the Porous Medium

In principle, an equation of motion is also needed for the solid skeleton. However, any movement of the porous medium due to normal fluid flow (i.e., excluding earthquakes or sinkhole formation) is so slow that the skeleton may be considered to be always at a state of equilibrium such that all inertial terms are negligible. The equation of stress equilibrium for the whole medium, composed of the solid and the single fluid phase, reads

$$\nabla \bullet \mathbf{T}^s + \nabla p^f - \rho^b \mathbf{g} = 0 \tag{24}$$

where $\mathbf{T}^s$ [ML^{-1}T^{-2}], is the effective stress tensor of the solid skeleton, assumed to be positive in compression, and ρ^b is the bulk density of the porous medium (i.e., total mass per unit volume). Because gravity is a constant, and with the assumptions that the solid stress tensor varies only vertically and the bulk mass density is constant, Equation (24) may be integrated to obtain the result that the total stress tensor is constant in a horizontal plane, and its vertical component acting on a horizontal surface is equal to the weight of the column of solid material:

$$T_{zz}^s + p^f = -\rho^b g(z - z_0) \tag{25}$$

where T_{zz}^s is the vertical component of the stress tensor acting on surface whose normal is parallel to the gravitational vector. Next T_{zz}^s is assumed to be given by Hooke's law of elasticity such that:

$$T_{zz}^s = -\left(K + \frac{4}{3}G\right)e \tag{26}$$

where K [ML^{-1}T^{-2}] and G [ML^{-1}T^{-2}] are the solid phase, elastic moduli (bulk and shear moduli, respectively) and e is the solid-solid dilatation (change in volume per unit volume of soil). From kinematic considerations, one can show that (Eringen, 1980):

$$\frac{\partial e}{\partial t} = \nabla \bullet \mathbf{v}^s \tag{27}$$

Substitute Equation (26) into Equation (25) and take the time derivative of the result while considering the variation of the right side of Equation (25) to be negligible. Then use Equation (27) to obtain

$$\alpha \frac{\partial p^f}{\partial t} = \nabla \bullet \mathbf{v}^s \tag{28}$$

where $\alpha = (K + 4G/3)^{-1}$ is the solid skeleton compressibility. This equation is used as the equation of motion of the solid phase.

In addition to Equations (22) and (28), equations of state for the solid and fluid phases are needed. These equations relate the state of stress of a material to its state of energy. For example, for a fluid, the pressure is always equal to the change in free energy of the fluid as a result of change in specific volume of the fluid. Commonly, an inverse relationship is used that relates the specific volume or, equivalently, the mass density to the fluid pressure. Such a relationship takes the form:

$$\frac{\partial \rho^f}{\partial p^f} = \rho^f \beta^f \tag{29}$$

where β^f [LT^{2}M^{-1}] is the compressibility of the fluid phase. Similarly, the compressibility of solid grains must be defined in terms of the state of stress of the solid phase. However, given Equation (25), a relationship may be written in terms of fluid pressure:

$$\frac{\partial \rho^s}{\partial p^f} = \rho^s \beta^s \tag{30}$$

where β^s [LT^{2}M^{-1}] is the compressibility of the solid grains.

Saturated Flow Equation

The equations of mass balance for the fluid and solid phases of a saturated porous medium are, respectively,

$$\frac{\partial(n\rho^f)}{\partial t} + \nabla \bullet (n\rho^f \mathbf{v}^f) = \hat{M}^f \tag{31}$$

$$\frac{\partial[(1-n)\rho^s]}{\partial t} + \nabla \bullet [(1-n)\rho^s \mathbf{v}^s] = 0 \tag{32}$$

where it is assumed that mass exchanges between the phases is negligible but pumping of the fluid phase is included. One could now substitute from either Equation (19) or (22) for the fluid phase velocity $\mathbf{q}^f = n(\mathbf{v}^f - \mathbf{v}^s)$. However, either of these equations introduces a new variable: the hydraulic head ϕ or the pressure p^f. Also, the porosity n is unknown. Therefore, before proceeding, the solid phase mass balance equation will be manipulated to obtain a relation for the porosity.

Application of the product rule for differentiation to Equation (32) yields

$$\rho^s \frac{\partial n}{\partial t} = (1-n)\frac{\partial \rho^s}{\partial t} + \rho^s \nabla \bullet [(1-n)\mathbf{v}^s] + (1-n)\mathbf{v}^s \bullet \nabla \rho^s \tag{33}$$

From this equation, it is clear that the change in porosity could be due to two processes: the compressibility of solid grains and the movement of solid grains with respect to each other. Both of these effects are determined by the state of stress within the solid skeleton and the stress-strain relationship for the solid skeleton. Substitution of Equation (33) into (31) to remove the time derivative of n, application of the chain rule, use of Equations (28) to (30), and collection of terms yield

$$[n\beta^f + (1-n)\beta^s + \alpha]\frac{\partial p^f}{\partial t} + \nabla \bullet \mathbf{q}^f + n\beta^f \mathbf{v}^f \bullet \nabla p^f$$

$$+ (1-n)\beta^s \mathbf{v}^s \bullet \nabla p^f = \hat{Q}^f \tag{34}$$

where $\hat{Q}^f = \hat{M}^f/\rho^f$ [T^{-1}]. Finally, note that both velocities $\mathbf{v}^f$ and $\mathbf{v}^s$, as well as the compressibilities, are very small. Thus, the last two terms on the left side of this equation represent higher order effects and may be neglected. Then, after substitution of the pressure-form of Darcy's law, Equation (22), into this equation, the fluid flow equation is obtained:

$$S_0 \frac{\partial p^f}{\partial t} = \nabla \bullet \left[\frac{\mathbf{k}}{\mu^f} \bullet (\nabla p^f - \rho^f \mathbf{g})\right] + \hat{Q}^f \tag{35}$$

where S_0 [M^{-1}LT^2] is the porous medium compressibility coefficient defined by

$$S_0 = n\beta^f + (1-n)\beta^s + \alpha \tag{36}$$

It should be noted that the solid grain compressibility, β^s, is very small and is often neglected in calculating the medium compressibility coefficient. Equation (35) is the pressure form of the saturated flow equation. The flow equation may also be formulated in terms of hydraulic head when density depends only on pressure by means of definition (21). The spatial and temporal derivatives of that equation are, respectively,

$$\rho^f g \nabla \varphi^f = \nabla p^f - \rho^f \mathbf{g} \tag{37a}$$

$$\rho^f g \frac{\partial \varphi^f}{\partial t} = \frac{\partial p^f}{\partial t} \tag{37b}$$

Substitution of these two relations into Equation (35) yields

$$S_s \frac{\partial \varphi^f}{\partial t} = \nabla \bullet (\mathbf{K}^f \bullet \nabla \varphi^f) + \hat{Q}^f \tag{38}$$

where $\mathbf{K}^f$ is the hydraulic conductivity tensor defined in Equation (23) and S_s [L^{-1}] is called the specific storativity, or specific storage, defined by (Jacob, 1940; Domenico and Schwartz, 1990)

$$S_s = \rho^f g S_0 = \rho^f g [n\beta^f + (1-n)\beta^s + \alpha] \tag{39}$$

Note that in Equations (35) and (38), if the medium is isotropic, the tensors become scalars.

A special form of Equation (38) may be derived for the case of lateral flow where variation of hydraulic head in the vertical direction is considered to be unimportant. For this case with the thickness of the flow region designated as b, which may vary as a function of position and time, the flow equation becomes

$$S \frac{\partial \varphi^f}{\partial t} = \nabla_{xy} \bullet (\mathbf{T}^f \bullet \nabla_{xy} \varphi^f) + \hat{Q}^f_{xy} \tag{40}$$

where $S = bS_s$ [-] is the aquifer storativity, ∇_{xy} is the two-dimensional divergence operator in the lateral directions, $\mathbf{T}^f = b\mathbf{K}^f$ [L^{2}T^{-1}] is the aquifer transmissivity tensor, and $\hat{Q}_{xy}^f$ accounts for well pumping and leakage from the top or bottom of the aquifer. For this two-dimensional equation to apply, the pumping must not induce a significant vertical head gradient. An example of such a situation would be when the well is screened through the thickness of a groundwater aquifer being exploited. Note that, if the x and y directions are the principal axes of the transmissivity (such as in a stratified medium), Equation (40) simplifies to

$$S\frac{\partial \varphi^f}{\partial t} = \frac{\partial}{\partial x}\left(T_{xx}^f \frac{\partial \varphi^f}{\partial x}\right) + \frac{\partial}{\partial y}\left(T_{yy}^f \frac{\partial \varphi^f}{\partial y}\right) + \hat{Q}_{xy}^f \tag{41}$$

MULTIPHASE FLUID FLOW

There are two major differences between single-phase (saturated) porous media flow and multiphase flow. First, the water phase has to share the pore space with one or more fluids. This results in new mechanisms for exchange of mass, momentum, and energy. Second, fluid-fluid interfaces are created. These interfaces possess their own dynamics and properties (such as surface forces and surface energy). The interfacial properties strongly influence the flow of the fluid phases and the distribution of various components. In general, the flow velocity of an individual fluid phase is determined by the forces acting within the phase (pressure and gravity), as well as the forces acting on the pore-scale boundaries of the phase (e.g., surface tension). In other words, a unique relationship between the hydraulic head gradient and the flow velocity, as suggested by Darcy's equation for a saturated porous medium, may not hold any more. An appropriate relationship is bound to be affected by the ratio of the two kinds of forces just described and by the distribution of the fluid within the pore space at any saturation. Nevertheless, as a first approximation, the form of Darcy's equation for saturated flow has been used to describe the movement of the individual phases in a multifluid porous medium flow. The equation is generalized by letting the permeability and hydraulic conductivity depend on fluid properties such that the actual form employed for each phase reads

$$\mathbf{q}^\alpha = -\frac{\mathbf{k}^\alpha}{\mu^\alpha} \bullet (\nabla p^\alpha - \rho^\alpha \mathbf{g}) \tag{42}$$

where the superscript α stands for any of the fluid phases and the Darcy velocity is defined as in Equation (7c).

One important consequence of the simultaneous presence of two or more fluid phases is that the permeability of the porous medium to a given fluid phase will be lower than the saturated permeability of that phase. The reduced permeability effect is commonly modeled for isotropic media through scaling down the components of the saturated permeability k by a coefficient called relative permeability, $k^{r\alpha}$, such that

$$k^\alpha = k^{r\alpha} k \qquad (43)$$

where the relative permeability is a function of the phase saturation. Bear et al. (1987) have indicated that this scaling approach is not appropriate for anisotropic media. Although other properties such as the ratio of phase surface area to the phase volume may also affect the permeability of the phase, such additional effects are not yet accounted for in the mathematical description of multiphase flow.

Introduction of the relative permeability coefficient is a major assumption in the empirical extension of Darcy's equation to multiphase flow. Commonly, relative permeability is found to be a highly nonlinear function of saturation. Typical relative permeability curves for a two-fluid system where one fluid, w, preferentially wets the solid in comparison to the other, n, are given in Figure 3. Note that two curves are given for each phase. This is because the relative permeability has been found to have hysteretic character such that its value depends on whether the measurement is made for the case of drying or imbibition. Also, the curves do not span the complete range of saturations from 0 to 1. As the saturation of a fluid is decreased, a point is reached where the relative permeability

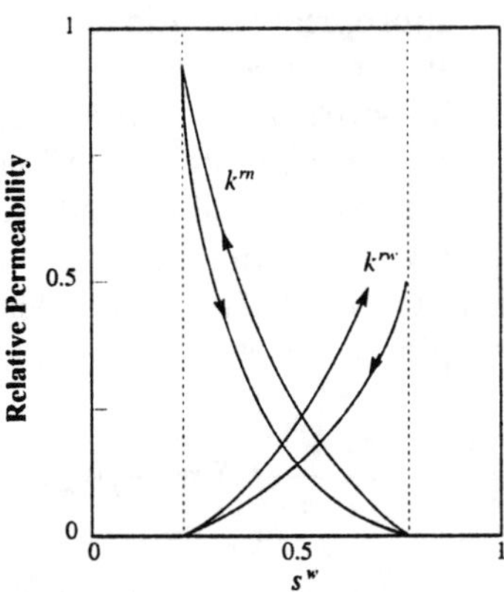

Figure 3. Typical plot of relative permeability vs. saturation exhibiting hysteresis.

for that phase goes to zero and the fluid becomes immobile. This point is referred to as the irreducible saturation for the w phase and the residual saturation for the n and is indicated in the figure.

The effect of surface forces on the motion of a phase is modeled by assuming that the difference in the macroscopic pressures of two neighboring fluid phases is related to the interfacial forces, in the form

$$p^\alpha - p^\beta = \frac{2\sigma^{\alpha\beta}}{R^{\alpha\beta}} \tag{44}$$

where $\sigma^{\alpha\beta}$ is the interfacial tension between the α and β phases and $R^{\alpha\beta}$ is the average radius of curvature of the interface. This equation is sometimes alleged to be a simple direct extension of the Young-Laplace equation, which looks the same as this expression but holds at points on fluid-fluid interfaces at equilibrium when the pressures in Equation (44) are microscale properties (Nitao and Bear, 1996). However, this interpretation is incorrect. As a macroscale approximation, Equation (44) mixes scales in a manner that is rigorously unacceptable. Furthermore, its use requires that dynamic effects be accounted for by letting $R^{\alpha\beta}$ be an effective radius of curvature. In practice, the right side of Equation (44) is replaced with an empirical function that has to be determined experimentally as a function of the saturation of the phases (Gray and Hassanizadeh, 1991). This function is called the capillary pressure between the α and β phases and is defined as

$$p^\alpha - p^\beta = p^{c\alpha\beta}(s^\alpha, s^\beta) \tag{45}$$

When only two fluid phases are present, their saturations are related by $s^\alpha = 1 - s^\beta$ so that dependence on either saturation, rather than both, is required in Equation (45). This function is defined in such a way that it is positive. Thus, the phase designated as the β phase should preferentially wet the solid in comparison to the α phase. In the remainder of this section, three special cases will be discussed. These are unsaturated soil, distribution of oil products in the saturated zone, and distribution of oil products in unsaturated soil. The first two of these cases involve two fluid phases, while the third is a three-phase flow system.

Unsaturated Flow

The name *unsaturated flow* is used to designate flow of water through soil when air occupies some of the pore space. Commonly, the air phase is considered to be infinitely mobile, such that its distribution is determined solely and instantaneously in response to the water phase distribution. The air phase is assumed to be continuous and always in contact with the atmosphere so that the air phase pressure is constant at the ambient value, p^a. Even if this assumption is not precise, the pressure gradient needed for the movement of air is so small that air pressure fluctuations may be neglected in studying the distribution of water in the unsaturated zone (Celia and Binning, 1992). Therefore, to model unsaturated flow, the mass balance and Darcy flow equation need to be considered only for the water phase. The conservation of mass equation is obtained from Equation (9) with the Darcy velocity as given in Equation (7c):

$$\frac{\partial(ns^w\rho^w)}{\partial t} + \nabla \bullet (\rho^w\mathbf{q}^w) + \nabla \bullet (\rho^w s^w n \mathbf{v}^s) = \hat{r}^w + \hat{M}^w \tag{46}$$

where $\hat{r}^w$ accounts for distributed sources/sinks of water such as plant uptake and evaporation and $\hat{M}^w$ accounts for point sources and sinks. Darcy's law is obtained by combining Equations (42) and (43) to yield

$$\mathbf{q}^w = -\frac{k^{rw}k}{\mu^w}(\nabla p^w - \rho^w\mathbf{g}) \tag{47}$$

Now, by use of definition (36) and following the development leading to Equation (35), Equation (46) can be recast into

$$S\frac{\partial p^w}{\partial t} + n\frac{\partial s^w}{\partial t} + \nabla \bullet \mathbf{q}^w = \hat{Q}^w \tag{48}$$

where $\hat{Q}^w = \hat{M}^w/\rho^w$, and $\hat{r}^w$ has been assumed to be unimportant. For unsaturated flow, the wetting phase pressure may be expressed as a function of saturation since

$$p^w = p^a - p^{caw}(s^w) \tag{49}$$

Therefore, since the air phase pressure is constant,

$$\frac{\partial p^w}{\partial t} = -\frac{dp^{caw}}{ds^w}\frac{\partial s^w}{\partial t} \tag{50}$$

and Equation (48) becomes

$$\left(n - S\frac{dp^{caw}}{ds^w}\right)\frac{\partial s^w}{\partial t} + \nabla \bullet \mathbf{q}^w = \hat{Q}^w \tag{51}$$

The porosity is much greater than Sdp^{caw}/ds^w except at very small saturation (Bear and Verruijt, 1987). Therefore, the continuity equation for unsaturated flow is commonly written as

$$n\frac{\partial s^w}{\partial t} + \nabla \bullet \mathbf{q}^w = \hat{Q}^w \tag{52}$$

Now substitution of Equation (47) into Equation (52) provides the flow equation for unsaturated flow:

$$n\frac{\partial s^w}{\partial t} - \nabla \bullet \left[\frac{k^{rw}k}{\mu^w}(\nabla p^w - \rho^w \mathbf{g})\right] = \hat{Q}^w \tag{53}$$

For a rigid porous medium, this equation can also be written in terms of water content, $\theta = ns^w$ as:

$$\frac{\partial \theta}{\partial t} - \nabla \bullet \left[\frac{k^{rw}k}{\mu^w}(\nabla p^w - \rho^w \mathbf{g})\right] = \hat{Q}^w \tag{54}$$

Equations (53) and (54) are referred to as mixed formulations, in that they contain a "mix" of dependent variables: s^w and p^w in one and θ and p^w in the other. In addition to Equations (53) or (54), coupled with Equation (49), two expressions are needed to relate capillary pressure and relative permeability to saturation. These relations may be obtained as experimental data. Many efforts have also been made to provide empirical correlations that depend on properties of the medium. For example, one popular set of equations has been provided by Brooks and Corey (1964) who suggested relations for capillary pressure and relative permeability, respectively, as follows:

$$p^{caw}(s^w) = p^a - p^w = p_b s_{\text{eff}}^{-1/\lambda} \tag{55}$$

$$k^{rw}(s^w) = k_0^w s_{\text{eff}}^{3+2/\lambda} \tag{56}$$

with

$$s_{\text{eff}} = \frac{s^w - s_{\text{irr}}^w}{1 - s_{\text{irr}}^w} \tag{57}$$

where s_{irr}^w is the irreducible saturation of the water. In Equation (55), p_b is the air entry pressure (the amount the air pressure must exceed the water pressure for air to be able to enter a saturated medium), and λ is a dimensionless coefficient that is related to the pore size distribution. In Equation (56), k_0^w is the value of the water phase relative permeability when the nonwetting phase is at its irreducible saturation (i.e., when $s^w = 1 - s_{\text{irr}}^n$). Both p_b and λ depend on the properties of the soil.

An alternative set of correlations that is widely employed is given by van Genuchten (1980). For capillary pressure this expression is

$$s_{\text{eff}} = \left[1 + \left(\alpha \left| \frac{p^{caw}}{\rho^w g} \right| \right)^b \right]^{-m} \tag{58}$$

where $b = 1/(1 - m)$ and a and m are parameters chosen to fit measured data for a particular system. The relative permeability correlation is given as

$$k^{rw} = s_{\text{eff}}^{1/2} [1 - (1 - s_{\text{eff}}^{1/m})^m]^2 \tag{59}$$

Other correlations may also be found throughout the literature.

One feature of the infiltration problem that makes it particularly difficult to solve is the fact that $p^{caw}(s^w)$ and $k^{rw}(s^w)$ are not unique functions of s^w, as represented in Figure 4. At a particular saturation, the capillary pressure measured following some drainage of water is higher than that measured following imbibition of water. This phenomenon of multivaluedness is called hysteresis and is evident in the relative permeability as well (Figure 3), although it is more important in the $p^{caw}(s^w)$ relation. If only drying or only imbibition is considered, the hysteresis does not provide a significant obstacle to modeling. However, when both wetting and drying cycles are simulated, hysteresis makes accurate modeling very difficult as the appropriate value of p^{caw} for a particular value of s^w could be the imbibition curve value, the drainage curve value, or some value in between, depending on the history of the flow event. Correlations for p^{caw} and k^{rw} as functions of s^w, such as those given above, are not able to take hysteresis into account.

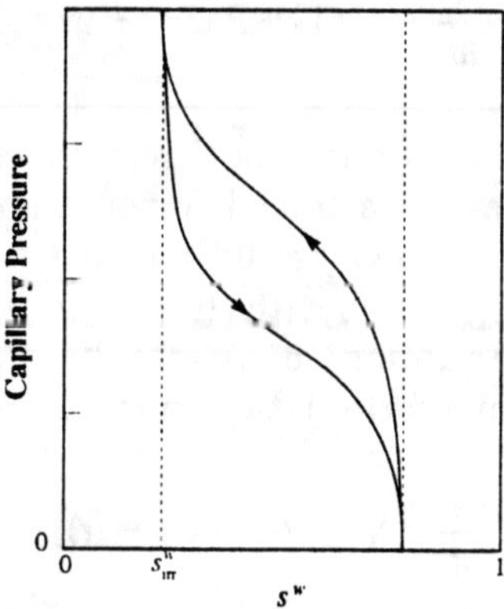

Figure 4. Schematic plot of capillary pressure vs. saturation illustrating hysteresis.

FORMULATION IN TERMS OF SUCTION HEAD

Often, the flow problem is formulated in terms of suction head instead of pressure. The suction head, also called matric potential, is negative in magnitude and is defined as

$$\psi = \frac{p^w - p^a}{p^w g} = \psi(s^w) \tag{60a}$$

or

$$\psi = \psi(\theta) \tag{60b}$$

Because these relationships are assumed to be monotonic, they may be inverted to obtain, for Equation (60a),

$$s^w = s^w(\psi) \tag{61a}$$

or for Equation (60b)

$$\theta = \theta(\psi) \tag{61b}$$

Combination of Equations (53) and (60a) leads to an alternative formulation of the unsaturated flow equation:

$$n\frac{\partial s^w}{\partial t} - \nabla \bullet [k^{rw}K(\nabla\psi - \hat{\mathbf{g}}] = \hat{Q}^w \tag{62}$$

where $\hat{\mathbf{g}}$ is the unit vector in the direction of gravity (i.e., positive downwards). This is called a mixed formulation of the flow equation because it contains both s^w and ψ as unknown dependent variables. Thus, it must be solved in conjunction with the constraint provided by Equation (61a). Alternatively, Equation (61a) may be directly substituted into Equation (62) to obtain a formulation in terms of the matric head ψ:

$$C\frac{\partial\psi}{\partial t} = \nabla \bullet [k^{rw}K(\nabla\psi - \hat{\mathbf{g}})] + \hat{Q}^w \tag{63a}$$

where C [L^{-1}] is the specific moisture capacity and is given by

$$C = n\frac{ds^w}{d\psi} = C(\psi) \tag{63b}$$

Instead, when the porosity is independent of time, Equation (62) may be written in terms of the moisture content θ by making use of Equation (61b) to obtain

$$\frac{\partial\theta}{\partial t} = \nabla \bullet (D\nabla\theta - k^{rw}K\hat{\mathbf{g}}) + \hat{Q}^w \tag{64a}$$

where D [L^{2}T^{-1}] is the hydraulic diffusivity defined as

$$D = k^{rw}K\frac{d\psi}{d\theta} = D(\theta) \tag{64b}$$

Because of the nonlinear character of C, D, and k^{rw}, analytic solution of Equation (63a) or (64a) for the water distribution in the unsaturated system is generally not possible. Although numerical solution schemes have been developed for these equations that are somewhat successful, proper accounting for hysteresis, heterogeneities in the soil, anisotropy, and scale effects make this an area in need of additional research.

Two-Liquid Flow

Many liquids that enter the subsurface environment are organics that are not soluble in water, such that, when they reach the saturated groundwater zone, they form a separate fluid phase from the water. These nonaqueous phase liquids are often identified collectively as NAPLs. In contrast to the unsaturated flow case considered in the last section where the dynamics of the nonaqueous phase could be neglected, the dynamics of both liquid phases must be accounted for to model the spreading and fate of contaminants that form the nonaqueous phase and that are transferred to the water. Therefore, equations of mass conservation and motion must be written for each liquid phase separately. If the compressibilities of the medium and the two liquids exert a negligible influence on the fluid distributions in comparison to changes in saturation caused by flow, the combination of continuity equations and Darcy's law yields equations for each phase that are analogous to Equation (53):

$$n\frac{\partial s^w}{\partial t} - \nabla \bullet \left[\frac{k^{rw}k}{\mu^w}(\nabla p^w - \rho^w \mathbf{g})\right] = \hat{Q}^w + \frac{\hat{r}^{wn}}{\rho^w} + \frac{\hat{r}^{ws}}{\rho^w} \tag{65}$$

$$-n\frac{\partial s^w}{\partial t} - \nabla \bullet \left[\frac{k^{rn}k}{\mu^n}(\nabla p^n - \rho^n \mathbf{g})\right] = \hat{Q}^n - \frac{\hat{r}^{wn}}{\rho^n} + \frac{\hat{r}^{ns}}{\rho^n} \tag{66}$$

where use has been made of the fact that the sum of the saturations in the two liquid phases must be 1 (i.e., $s^w + s^n = 1$ and no vapor phase is present). Additionally, the terms on the right side account for point sources/sinks and the volumetric rate of transfer of mass between phases, as denoted by the paired superscripts, per unit volume. Note that all mass transferred between the w and the n phases must balance. Therefore, the mass-based transfer term between these phases, r^{wn}, appears in both Equations (65) and (66) but with a different sign. For simplicity, these interphase exchange terms are typically neglected in the flow simulations. This is a reasonable approximation if the mass exchange is small in comparison to the total mass of fluids present. In cases where the exchange terms must be retained, they must be parameterized in terms of concentration and other system variables. Additionally, species transport equations will have to be solved in conjunction with the flow equations for the complete phases.

When the phase exchange terms on the right side of Equations (65) and (66) are zero and the point source/sink terms have been specified, these two equations still contain three unknowns: the two fluid-phase pressures and the water-phase saturation. Thus, one additional equation is needed for a complete description of the system. This is provided by the capillary pressure-saturation relation:

$$p^n - p^w = p^{cnw}(s^w) \tag{67}$$

The capillary pressure is measured as a hysteretic function of saturation for the particular pair of fluids and the medium under consideration. In reservoir engineering, the so-called Leverett (1941) J-function is often used to correlate the relations between capillary pressure and saturation such that

$$p^n - p^w = \frac{\sigma^{nw}}{\sqrt{k/n}} J(s^w) \tag{68}$$

where σ^{nw} [MT^{-2}] is the surface tension of the interface between the two fluids, k is saturated permeability, n is porosity, and J is considered to be a unique function of saturation for any given medium independent of the fluid present. This relation is unable to account for hysteresis effects. However, this limitation is not serious in reservoir engineering where the goal is imbibition of water to replace the oil in a formation; and reversal of the process to drain water does not occur.

Computer solution of Equations (65) through (67) is hindered by the nonlinearities and hysteresis in the capillary pressure versus saturation relation, by nonlinearities and uncertainty involved with both the relative permeabilities, and by medium heterogeneities. These equations do not apply for an anisotropic situation. Furthermore, even if the physics were well described by the equations and parameters, complexities inherent in the numerical simulation of sharp fronts or large changes in saturation over short distances make computer solution of the equations a challenging problem.

Immiscible Flow of Two Fluids and a Vapor

The previous section provides equations for flow of an aqueous phase and NAPL in a porous system when no vapor phase is present. These equations are useful for modeling water flooding in oil reservoir extraction processes and for modeling the movement of a nonaqueous phase liquid once it has entered the groundwater zone following a pollution event. One additional system that is of importance is the movement of an NAPL spill within the unsaturated zone. If the NAPL spill is large enough, it will remain a separate phase and move through the unsaturated zone toward the groundwater or the geologic formation that underlies the surface soil. In the unsaturated zone, three phases will be present: water, organic, and air. Some exchange of mass among the phases will occur. For example, if the spill is composed of highly volatile organics, transfer to the air phase will be important. Indeed, pumping of air into such a formation through a well system to strip the organics is an important developing cleanup technology. Modeling of such a system is extremely complex. In extracting oil from petroleum reservoirs, as the pressure in the reservoir is reduced due to oil removal, natural gas may move out of the liquid phase and form a separate gas phase that is under a pressure higher than atmospheric pressure. Because of heterogeneities in the reservoir and the large array of organic chemicals that form the oil, the movement of a gas phase of this type is difficult to model, especially if a third aqueous phase is present. In these instances, all phases must be modeled using the equations of mass conservation and Darcy's law for each phase.

Another slightly simpler situation that is of interest is when the air phase is considered to be connected and at atmospheric pressure, such that it interacts with the two fluid phases in the manner described for unsaturated flow. This scenario occurs when an accidental or illegal spill on the land surface causes organics to seep into the ground for a large period of time. Leaching of chemicals from drums buried in landfills also provides a source of NAPL contamination of this sort. In these cases, biological activity within and between the phases may occur to transform the chemicals. To model the dynamic distribution of the phases within the three-phase region, equations of mass conservation and Darcy's law analogues must be formulated for the liquids while the air phase is treated as being at atmospheric pressure and infinitely mobile. If dynamic pressure changes due to liquid and solid phase compressibilities are negligible in comparison to those caused by changes in saturation, the flow equations and capillary pressure relations obtained are analogous to Equations (65) through (67):

$$n\frac{\partial s^w}{\partial t} - \nabla \cdot \left[\frac{k^{rw}k}{\mu^w}(\nabla p^w - \rho^w \mathbf{g})\right] = \hat{Q}^w + \frac{\hat{r}^{wn}}{\rho^w} + \frac{\hat{r}^{ws}}{\rho^w} + \frac{\hat{r}^{wa}}{\rho^w} \qquad (69)$$

$$n\frac{\partial s^n}{\partial t} - \nabla \cdot \left[\frac{k^{rn}k}{\mu^n}(\nabla p^n - \rho^n \mathbf{g})\right] = \hat{Q}^n - \frac{\hat{r}^{wn}}{\rho^n} + \frac{\hat{r}^{ns}}{\rho^n} + \frac{\hat{r}^{na}}{\rho^n} \qquad (70)$$

$$p^{cnw}(s^w, s^n) = p^n - p^w \qquad (71)$$

$$p^{caw}(s^w, s^n) = p^a - p^w \qquad (72)$$

Note that, although p^{caw} depends on s^w and s^a according to Equation

(45), this dependence may be expressed as in Equation (72) on s^w and s^n because $s^a = 1 - s^w - s^n$. In addition to the expressions for capillary pressure, correlations are also needed for the relative permeabilities in terms of the two fluid phase saturations. Information must also be supplied concerning the right side terms accounting for point sources and exchanges between phases.

Although the set of Equations (69) through (72) is not much more imposing than the set provided for flow of two liquid phases in the absence of a vapor phase, the requirement of three-phase relative permeabilities and capillary pressures is a serious complicating factor. Reliable data of this sort does not exist and is extremely difficult to obtain. Although effort has been made to scale two-phase data to the three-phase case, the validity of this approach is controversial. Solution of the nonlinear equation set is both a numerical challenge and a practical challenge in that sophisticated numerical algorithms must be used, and, even then, utility of the solution as a predictive or assessment tool for a field situation is limited.

SOLUTE TRANSPORT IN POROUS MEDIA

The spreading of solutes in water or any other fluid in a saturated or unsaturated porous medium is governed by the mass balance Equation (14). First, consider the case of dispersion of solutes in saturated groundwater flow. In general, the solutes may be present in both water and solid phases. Thus, two mass balance equations may be written for each species i:

$$\frac{\partial(n\rho^w\omega^{iw})}{\partial t} + \nabla \bullet (n\rho^w\omega^{iw}\mathbf{v}^w) + \nabla \bullet \mathbf{J}^{iw} = n\hat{r}^{iw} + \hat{M}^{iw} \tag{73}$$

$$\frac{\partial[(1-n)\rho^s\omega^{is}]}{\partial t} + \nabla \bullet [(1-n)\rho^s\omega^{is}\mathbf{v}^s] + \nabla \bullet \mathbf{J}^{is} = (1-n)\hat{r}^{is} + \hat{M}^{is}$$

$$\tag{74}$$

When N different chemical species are present in the system, these equations are written separately for each species. Thus, a set of $2N$ transport equations would exist. Recall that the sum over all species of the transport equations for a phase leads to the mass conservation equation for the total phase, Equation (9). Thus it is common to select $N - 1$ of the species balance equations and the phase mass conservation equation as the N independent mass conservation equations for a phase. If the species do not interact with each other through transformation reactions, the equations can be solved independently of each other. Otherwise, they have to be solved simultaneously.

Commonly, in order to make Equations (73) and (74) suitable to the study of solute dispersion, a number of simplifying assumptions based on practical considerations are introduced. The most important assumptions involve mass flux dispersion vectors $\mathbf{J}^{iw}$ and $\mathbf{J}^{is}$. The dispersive flux accounts for the collective effects of spreading processes responsible for movement of a chemical species relative to the average macroscale velocity. These are molecular diffusions of solutes within the pores and velocity variations among the pores due to microscale heterogeneities of the medium. For the dissolved solutes, the form of Fick's law of molecular diffusion is assumed to hold in approximating the dispersion flux if the molecular diffusion coefficient is replaced with a dispersion tensor such that

$$\mathbf{J}^{iw} = -\rho^w n \mathbf{D}^{iw} \cdot \nabla \omega^{iw} \tag{75}$$

where the dispersion tensor, $\mathbf{D}^{iw}$ [L^{2}T^{-1}] is a function of the Darcy flow velocity (Scheidegger, 1974):

$$n\mathbf{D}^{iw} = n\tau \mathcal{D}^{iw}\mathbf{I} + \alpha_T |\mathbf{q}^w|\mathbf{I} + (\alpha_L - \alpha_T)\frac{\mathbf{q}^w \mathbf{q}^w}{|\mathbf{q}^w|} \tag{76}$$

where α_T [L] and α_L [L] are transverse and longitudinal dispersivities, assumed to be medium properties and independent of the solute type, $|\mathbf{q}^w|$ denotes the magnitude of the Darcy velocity in phase w, $\mathcal{D}^{iw}$ [L^{2}T^{-1}] is the molecular diffusivity, and τ is the tortuosity coefficient, with $0 < \tau < 1$ that accounts for the apparent reduction in diffusion due to the tortuous nature of the pore space.

Next, it is assumed that the solutes are immobile in the solid phase. That is to say, the movement of grains (and thus the movement of adsorbed solutes) and the diffusion of solutes across grains or along grain surfaces makes a negligible contribution to the mass flux of the solute within the system. Therefore, the last two terms on the left side of Equation (74) that account for advection and dispersion of a solid phase solute can be neglected. The time rate of change of the porosity is also considered to contribute negligibly to species transport when the solid phase velocity is small. Additionally, because the solid phase velocity is small, Equation (8) may be used to replace the fluid velocity in Equation (73) with the Darcy velocity.

In problems involving high concentrations of contaminant, the transport equations for the dissolved species must be solved simultaneously with the phase flow equations because the advection and dispersion processes are coupled. However, a simpler situation commonly encountered is that the mass fraction of a solute is small enough that its variation through the system of interest has no significant effect on the phase density. Under this condition, the flow of the phase is not affected by density changes due to solute movement. This allows a decoupling of the flow problem from the transport problem to occur such that, in numerical simulation over a time step, the flow equations may be solved first to determine the flow velocity fields and then the solute transport equations may be solved. Under these conditions, for the dissolved species, the mass fraction is commonly replaced with solute mass concentration $c^{i\alpha}$ [ML^{-3}] defined by

$$c^{i\alpha} = \rho^{\alpha}\omega^{i\alpha} \tag{77}$$

Consistent with the convention widely adopted in the hydrologic community, $c^{i\alpha}$ is in terms of mass per volume, not moles per volume as would be the convention in chemical engineering.

With these assumptions, Equations (73) and (74) are recast into the following forms:

$$n\frac{\partial c^{iw}}{\partial t} + \nabla \bullet (c^{iw}\mathbf{q}^{w}) - \nabla \bullet (n\mathbf{D}^{iw} \bullet \nabla c^{iw}) = n\hat{r}^{iw} + \hat{M}^{iw} \tag{78}$$

$$\frac{\partial[(1-n)\rho^{s}\omega^{is}]}{\partial t} = (1-n)\hat{r}^{is} \tag{79}$$

where the point source term for the solid has been set to zero. Note that mass fraction is still used for the adsorbed species because of the traditional measurement methods.

Next, consider the reaction terms $\hat{r}^{iw}$ and $\hat{r}^{is}$. These terms account for homogeneous reactions among the components within a phase, heterogeneous reactions involving components of different phases, and radioactive or biological decay. Also, when the medium has a macropore-micropore structure, these terms account for the exchange of dissolved species between micropores and macropores. An important mechanism of mass exchange is adsorption of solutes. In the remainder of this section, the most common ways of incorporating adsorption into transport equations are discussed.

For convenience, the appropriate formulation of the right side reaction terms of Equations (78) and (79) for a particular solute will be considered one mechanism at a time. The full term would be a sum of the mechanisms that are appropriate for the species of interest.

The decay process is commonly modeled by a first- or second-order reaction formula. In the case of first-order decay of species i, the reaction takes the form in the fluid phase:

$$\hat{r}^{iw}_{\text{dec}} = -\lambda^{iw} c^{iw} \tag{80a}$$

and in the solid phase:

$$\hat{r}^{is}_{\text{dec}} = -\lambda^{is} \rho^s \omega^{is} \tag{80b}$$

where $\lambda^{i\alpha}$ [T^{-1}] is the decay rate constant for species i in the α-phase. In the special case of radioactive decay, the decay rate constant for a species is the same, regardless of the phase.

The adsorption process is commonly modeled either as a kinetic process or as an equilibrium process. In the case of kinetic adsorption, the rate of exchange of solute mass between the fluid and solid is assumed to be a function of the concentrations of the solute in the two phases. A general kinetic relationship for adsorption may be written as

$$n\hat{r}^{iw}_{\text{ads}} = -(1-n)\hat{r}^{is}_{\text{ads}} = n\hat{r}^{iw}_{\text{ads}}(\omega^{is}, c^{iw}) \tag{81a}$$

where the actual form of $\hat{r}^{iw}_{\text{ads}}$ needs to be determined experimentally. A simplified form of this equation is a kinetic relationship linear in ω^{is}:

$$n\hat{r}^{iw}_{\text{ads}} = -(1-n)\hat{r}^{is}_{\text{ads}} = \kappa^i \rho^s [\omega^{is} - f(c^{iw})] \tag{81b}$$

where κ^i [T^{-1}] is the kinetic rate coefficient that may depend on medium properties. Substitution of Equations (80a), (80b), and (81b) in Equations (78) and (79) with adsorption and decay being the only reaction mechanisms, yields

$$n\frac{\partial c^{iw}}{\partial t} + \nabla \bullet (c^{iw} \mathbf{q}^w) - \nabla \bullet (n\mathbf{D} \bullet \nabla c^{iw})$$

$$= \kappa^i \rho^s [\omega^{is} - f(c^{iw})] - n\lambda^{iw} c^{iw} + \hat{M}^{iw} \tag{82}$$

$$(1-n)\rho^s \frac{\partial \omega^{is}}{\partial t} = -\kappa^i \rho^s [\omega^{is} - f(c^{iw})] - (1-n)\lambda^{is} \rho^s \omega^{is} \tag{83}$$

To determine the distribution of species i in the system, these two equations would have to be solved simultaneously for c^{iw} and ω^{is}.

In many practical applications, because of the slow rate of flow of groundwater, the adsorption process is assumed to occur fast enough that an equilibrium between fluid and solid phase concentrations of a chemical species is established at all times. This approach is attractive because it leads to a simplification of the equations and eliminates the difficult experimental task of determining the kinetic rate constants. However, the need to consider kinetic rates in some cases remains. Under the assumption of sorption equilibrium, the kinetic constant in Equation (81b) is considered to be so large that, for finite values of mass exchange, the term $\omega^{is} - f(c^{iw})$ approaches zero such that

$$\omega^{is} = f(c^{iw}) \tag{84}$$

The function $f(c^{iw})$ is called an adsorption isotherm because it is commonly measured in the laboratory under constant temperature conditions. Now, summation of Equations (82) and (83) and substitution from Equation (84) yields a single solute transport equation:

$$nR^{iw}\frac{\partial c^{iw}}{\partial t} + \nabla \bullet (c^{iw}\mathbf{q}^{w}) - \nabla \bullet (n\mathbf{D} \bullet \nabla c^{iw}) = -n\lambda^{iw}R^{i}_{\text{dec}}c^{iw} + \hat{M}^{iw} \tag{85}$$

where R^{iw} is the retardation factor defined by

$$R^{iw} = 1 + \frac{(1-n)\rho^{s}}{n}\frac{df}{dc^{iw}} \tag{86}$$

and R^{i}_{dec} is the retardation factor for the decay process defined by

$$R^{i}_{\text{dec}} = 1 + \frac{(1-n)\rho^{s}}{n}\frac{\lambda^{is}}{\lambda^{iw}}\frac{f(c^{iw})}{c^{iw}} \tag{87}$$

In general, adsorption isotherms are nonlinear functions of c^{iw}. Perhaps the best known examples are the Freundlich or Langmuir isotherms. Each of these isotherms requires two constants to be determined. A simpler and widely used isotherm is a linear relationship involving the so-called distribution coefficient K^{di} [L^{3}M^{-1}] so that

$$\omega^{is} = f(c^{iw}) = K^{di}c^{iw} \tag{88}$$

When the adsorption isotherm is linear, both the retardation factors are constant. Examination of Equations (86) and (87) reveals that, if the decay rate constants for the fluid and solid phases λ^{iw} and λ^{is} are equal, then for the case of a linear isotherm, $R^i_{\text{dec}} = R^{iw}$. For this case, division of Equation (85) by the retardation factor yields

$$n\frac{\partial c^{iw}}{\partial t} + \nabla \bullet \left[\frac{1}{R^{iw}}(c^{iw}\mathbf{q}^w - n\mathbf{D}\bullet\nabla c^{iw}) \right] = -n\lambda^{iw}c^{iw} + \frac{\hat{M}^{iw}}{R^{iw}} \qquad (89)$$

Comparison of Equations (78) and (89) reveals that the effect of adsorption is to cause a reduction in the rate at which a chemical species advects and diffuses through a porous medium. If a phase contains a number of adsorbing species, they will travel at different apparent velocities dependent on their retardation factors. Furthermore, if a chemical species is to be used as a tracer for determination of flow velocity, it is important that this species be inert and not interact with neighboring phases.

SUMMARY

In this chapter, basic equations describing the flow of fluids and spreading of contaminants in the unsaturated region and in groundwater have been presented. The central equations in all cases are the equations of mass balance for fluids and contaminants. In the case of saturated flow of groundwater, the mass balance equation must be supplemented with Darcy's law for the flow velocity and an equation of state for the fluid density. The resulting equation is a parabolic type equation for pressure or hydraulic head, which has to be solved subject to appropriate initial and boundary conditions.

In the case of flow of two or more immiscible fluids, the Darcy equation is applied to each fluid phase by introducing a relative permeability coefficient that scales down the saturated permeability tensor. This coefficient is assumed to be a function of the fluid saturation. In addition to this extension of Darcy's law, the presence of interfaces between phases and the discontinuity in pressure that occurs at these interfaces must be accounted for. Thus, a macroscale capillary pressure is introduced as the difference in pressure between any two fluids and is specified to be a function of the saturations of the two adjacent fluids. In the case of two-phase flow, only one saturation function is independent.

A special case of multiphase flow is water movement in unsaturated soil such that a vapor phase occupies a portion of the void space. In modeling this system, the air phase is assumed to be infinitely mobile so that its dynamics need not be studied. Then, only a flow equation for water needs to be specified. This equation is obtained as a combination of the mass conservation equation and Darcy's law. The unsaturated flow equation is commonly solved to determine the distribution of either water saturation or water suction head. When more than one liquid phase is present, flow equations must be solved for each phase. In this chapter, a gas or vapor phase was considered only to be an infinitely mobile, passive phase that is in contact with the atmosphere. However, when the vapor phase movement is important to transport as in a volatilization study or in gas sparging, flow equations must also be obtained for the vapor.

Equations for the spreading of soluble pollutants within a phase are also presented. The distribution of a solute is affected by advection, dispersion, and mass exchange processes. Advection transports the solutes with the average flow velocity of the phase. In the case of low-concentration solutes, the average velocity can be determined by solving the flow equations described above. The dispersion transport is modeled by an extension of Fick's law in which the scalar diffusion coefficient that accounts for the random motion of molecules at the microscale is replaced by a dispersion tensor that accounts for the variations in the flow velocity as well, due to the tortuous flow path and the interaction of the fluid with the medium. The two most important processes accounting for mass exchange between phases are biodegradation and adsorption. Both adsorption and biodegradation are often modeled using first-order kinetics, but more complex expressions are also employed. Adsorption is also often modeled as if the adsorbing species is at an equilibrium distribution between the solid and fluid phases. This is a simplification over the kinetic case that requires the use of an equilibrium isotherm. In fact the adsorption and degradation processes are so complex in real systems consisting of multiple species that the sorption and reaction expressions required for insertion in the transport equations are not well-parameterized. The main thrust of this chapter has been the description of the physics of flow and transport. However, accurate modeling of the spread and biodegradation of contaminants requires mathematical description of the chemical and biological kinetic and equilibrium behavior of those species.

REFERENCES

Bear, J., 1972. *Dynamics of Fluids in Porous Media*. New York, NY: American Elsevier.

Bear, J., C. Braester, and P. C. Meiner, 1987. "Effective and relative permeabilities in anisotropic porous media." *Transport in Porous Media*, 2:301–316.

Bear, J. and A. Verruijt, 1987. *Modeling Groundwater Flow and Pollution*. Dordrecht, the Netherlands: D. Reidel Publishing Co.

Brooks, R. H. and A. T. Corey, 1964. *Hydraulic Properties of Porous Media*. Fort Collins, CO: Colorado State University Hydrology Papers #3.

Celia, M. A. and P. Binning, 1992. "A mass conservative numerical solution for two-phase flow in porous media with application to unsaturated flow." *Water Resources Research*, 28(10):2819–2828.

Cushman, J. H., 1990. "An introduction to hierarchical porous media." In: *Dynamics of Fluids in Hierarchical Porous Media*. London: Academic Press, pp. 1–6.

Dagan, G., 1989. *Flow and Transport in Porous Formations*. Berlin: Springer-Verlag.

Darcy, H., 1856. *Les Fontaines Publiques de la Ville de Dijon*. Paris: Dalmont.

De Marsily, G., 1986. *Quantitative Hydrogeology: Groundwater Hydrology for Engineers*. Orlando, FL: Academic Press.

De Marsily, G., 1993. *Flow and Contaminant Transport in Fractured Rock*. San Diego, CA: Academic Press.

Domenico, P. A. and F. W. Schwartz, 1990. *Physical and Chemical Hydrogeology*. New York, NY: John Wiley & Sons.

Eringen, A. C., 1980. *Mechanics of Continua*, second ed., Huntington, New York: Krieger.

Gray, W. G. and S. M. Hassanizadeh, 1991. "Unsaturated flow theory including interfacial phenomena." *Water Resources Research*, 27(8): 1855–1863.

Hassanizadeh, S. M., 1986. "Derivation of basic equations of mass transport in porous media, Part 1. Macroscopic balance laws." *Adv. Water Resour.*, 9:196–206.

Hassanizadeh, S. M., and W. G. Gray, 1979a. "General conservation equations for multi-phase systems: 1. Averaging procedure." *Adv. Water Resour.*, 2(3): 131–144.

Hassanizadeh, S. M. and W. G. Gray, 1979b. "General conservation equations for multi-phase systems: 2. Mass, momenta, energy, and entropy equations." *Adv. Water Resour.*, 2(4):191–203.

Hubbert, M. K., 1940. "The theory of ground water motion," *J. Geol.*, 48:785–944.

Jacob, C. E., 1940. "On the flow of water in an elastic artesian aquifer." *Trans. American Geophysical Union*, 22:574–586.

Leverett, M. C., 1941. "Capillary behavior in porous media." *Trans. A.I.M.E.*, 142:341–358.

Nguyen, V. V., W. G. Gray, G. F. Pinder, J. F. Botha, and D. A. Crerar, 1982. "A theoretical investigation on the transport of chemicals in reactive porous media." *Water Resources Research*, 18(4):1149–1156.

Nitao, J. J. and J. Bear, 1996. "Potentials and their role in transport in porous media." *Water Resources Research*, 32(2):225–250.

Scheidegger, A. E., 1974. *The Physics of Flow through Porous Media*, Third Edition, University of Toronto Press, Toronto.

van Genuchten, M. Th., 1980. "A closed-form equation for predicting the hydraulic conductivity of unsaturated soils." *Soil Science Society of America Journal*, 44:892–898.

Fundamental Concepts for Numerical Simulation of Contaminant Transport and Biodegradation

Michael A. Celia
Department of Civil Engineering and Operations Research
Water Resources Program
Princeton University
Princeton, NJ 08544, USA

INTRODUCTION

The partial differential equations that govern subsurface contaminant transport in the presence of biodegradation are usually too complex to solve analytically. Therefore, numerical solution methods are employed to provide approximate solutions to the equations. These numerical methods are based on replacement of the differential equations by an associated set of algebraic equations. These algebraic equations are then solved on a computer by standard matrix methods.

From a practical point of view, it is important to understand the basic concepts of numerical approximations, especially the common sources of errors and associated procedures to control and alleviate these errors. This chapter presents fundamental concepts that underlie common numerical approximations, with an emphasis on those methods that apply most directly to the problem of reactive contaminant transport. From a foundation based on simple model problems, numerical approximations are described for sets of nonlinear, coupled partial differential equations that describe contaminant transport and biodegradation.

The general equation form to be considered is

$$\frac{\partial(R\theta c)}{\partial t} + \nabla \cdot [\mathbf{q}c - \theta \mathbf{D} \cdot \nabla c] + \lambda \theta c = \theta \Gamma \tag{1}$$

59

where R is a retardation coefficient [L^0], θ is moisture content [L^3/L^3], c is solute concentration [M/L^3], $\mathbf{q}$ is the volumetric flux vector [L^3/L^2T], $\mathbf{D}$ is the diffusion/dispersion tensor [L^2/T], λ is the first-order reaction coefficient [$1/T$] (λ may be a nonlinear function of c), and Γ is a zero-order reaction term or a source/sink term [M/L^3T]. Equation (1) serves as a fairly general model for contaminant transport and biodegradation, with concentration c being the unknown to be determined. If numerical methods are developed to solve Equation (1) effectively, then those methods might reasonably be expected to perform well when solving more complex systems of transport equations.

FUNDAMENTALS OF NUMERICAL METHODS

Numerical approximations for differential equations may be developed in a variety of ways. In standard finite difference approximations, mass balances over finite volumes provide a basis for transforming partial derivatives into simple algebraic expressions. In other methods, more involved mathematics is required, sometimes coupled with theoretical analysis of the governing equation itself. Because this chapter is meant to be introductory, finite difference methods will be stressed. However, some concepts related to finite elements will also be presented. In addition, concepts related to certain special characteristics of partial differential equations will be presented to provide adequate context to understand an important class of modern solution methods, referred to as characteristic methods. These methods include so-called Eulerian-Lagrangian methods and Operator Splitting techniques. Rudimentary error analysis will be used to explain some of the important sources of approximation error.

A differential equation such as Equation (1) requires auxiliary information in order to complete the mathematical statement of the problem. This auxiliary information includes specification of the domain within which the equation applies (in both space and time), as well as certain boundary and initial conditions that must be specified along the boundary of the domain. Most numerical methods break the domain into a set of nonoverlapping subdomains (usually referred to as blocks or elements) whose union covers the entire domain, with possible exceptions along irregularly shaped boundaries. Finite difference methods almost always use rectangular blocks, while finite element methods allow other shapes, such as triangles, to be used. The unknown in the governing equation (concentration c in this case) is approximated by simple functions within each block or element, for example, a constant or perhaps a linear variation across the block. For the case of a block-wise constant approximation for c, the number of unknowns within the domain is equal to the number of blocks; for other representations, there might be several unknowns per block. One might expect intuitively that, as more and more blocks are used in the discretization of the domain, the approximate solution will become more accurate because spatial variability of the solution will be captured more effectively. At the same time, more blocks imply more unknowns, which implies greater computational effort. This trade-off between accuracy and computational cost occurs with all numerical methods and is an important consideration in many practical applications.

A Finite Difference Approximation in Three Dimensions

To demonstrate the nature of numerical approximations, consider a domain as shown in Figure 1a. Let the discretization use rectangular blocks, as shown in Figure 1b. While the figure shows a two-dimensional domain for ease of presentation, most domains of interest are three-dimensional. In three-dimensional space, the blocks are usually numbered along each coordinate axis, such that the block in the lower left front corner is block $(1,1,1)$, that in the lower right front corner is $(N_x,1,1)$, that in the upper right back corner is (N_x,N_y,N_z), and so on. This implies N_x blocks in the x-direction, N_y blocks in the y-direction, and N_z blocks in the z-direction. A simple finite difference approximation may be derived by application of a mass balance to each block in the domain, coupled with certain approximations to derivatives. For a typical block (i,j,k), the mass balance statement for the dissolved component of interest may be written mathematically as follows:

$$\left[(c\theta)\big|_{x_i,y_j,z_k}^{t+\Delta t} - (c\theta)\big|_{x_i,y_j,z_k}^{t}\right]\delta x_i \delta y_j \delta z_k$$

$$= \left(m_x\big|_{x_i-(\delta x_i/2)} - m_x\big|_{x_i+(\delta x_i/2)}\right)\Big|^{t+\varepsilon\Delta t} \delta y_j \delta z_k \Delta t$$

$$+ \left(m_y\big|_{y_j-(\delta y_j/2)} - m_y\big|_{y_j+(\delta y_j/2)}\right)\Big|^{t+\varepsilon\Delta t} \delta x_i \delta z_k \Delta t$$

$$+ \left(m_z\big|_{z_k-(\delta z_k/2)} - m_z\big|_{z_k+(\delta z_k/2)}\right)\Big|^{t+\varepsilon\Delta t} \delta x_i \delta y_j \Delta t$$

$$+ \left[(rxn)_{i,j,k} + (src)_{i,j,k}\right]\Big|^{t+\varepsilon\Delta t} \delta x_i \delta y_j \delta z_k \Delta t \tag{2}$$

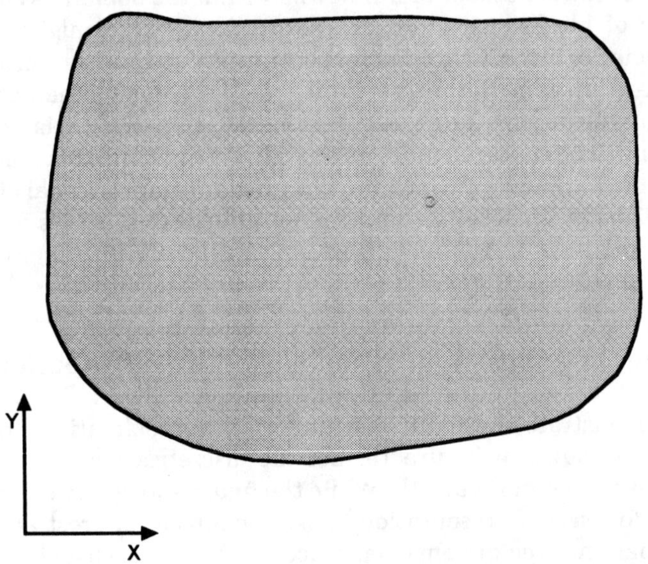

Figure 1a. Typical two-dimensional domain.

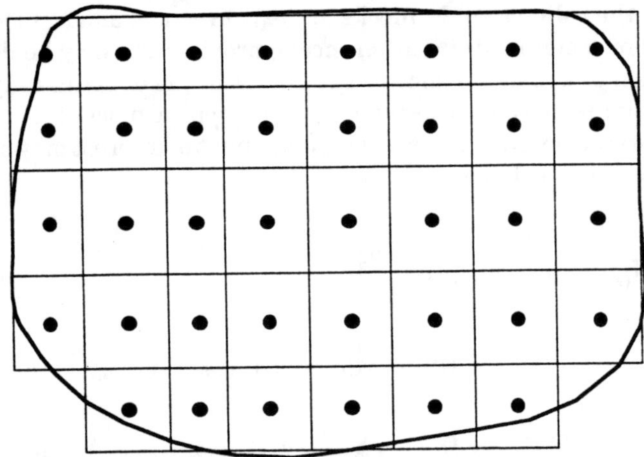

Figure 1b. Finite difference discretization of the domain.

This equation states that the change of mass in block (i,j,k) in time period $[t,t + \Delta t]$ is balanced by the sum of the net mass flux of component i through the boundary of the block plus the net gain of mass of the component due to both reactions (negative for losses) and other sources (negative for sinks). In the equation, m_x, m_y, and m_z are the x-, y-, and z-components, respectively, of the mass flux vector $\mathbf{m}$ (units of mass per area per time); location (x_i,y_j,z_k) corresponds to the center of block (i,j,k); times t and $t + \Delta t$ are discrete values of time, with Δt referred to as the time-step size; time level $t + \varepsilon\Delta t$ is the time level at which the fluxes, reaction, and source/sink terms are evaluated, with $0 \leq \varepsilon \leq 1$; $(rxn)_{i,j,k}$ denotes reaction within block (i,j,k) (units of mass per volume per time); and $(src)_{i,j,k}$ denotes a source/sink function within (i,j,k) (units of mass per volume per time). The mass flux vector is taken as the sum of the advective flux and the diffusive/dispersive flux:

$$\mathbf{m} = \mathbf{q}c - \theta\mathbf{D} \cdot \nabla c$$

with the x-component of the vector having the form

$$m_x = q_x c - \theta D_{xx}\frac{\partial c}{\partial x} - \theta D_{xy}\frac{\partial c}{\partial y} - \theta D_{xz}\frac{\partial c}{\partial z} \tag{3}$$

and the y- and z-components having analogous forms.

For simplicity in this initial derivation, assume that $\mathbf{D}$ is a diagonal tensor, so that $D_{xy} = D_{yx} = D_{xz} = D_{zx} = D_{yz} = D_{zy} = 0$. Let c within each block be approximated as locally constant, let terms evaluated at time $t + \varepsilon\Delta t$ be approximated as weighted averages of terms at t and $t + \Delta t$ [see Equation (6)], let concentration at block boundaries be calculated by linear interpolation between adjacent nodal values [see Equation (4)], and let derivatives at block boundaries be approximated by simple differences [see Equation (5)],

$$c\big|_{x_i+(\delta x_i/2),y_j,z_k} \equiv c_{i+(1/2),j,k} \cong \frac{(\delta x_{i+1})c_{i,j,k} + (\delta x_i)c_{i+1,j,k}}{\delta x_i + \delta x_{i+1}} \tag{4}$$

$$\frac{\partial c}{\partial x}\bigg|_{i+(1/2),j,k} \cong \frac{c_{i+1,j,k} - c_{i,j,k}}{x_{i+1} - x_i} = \frac{c_{i+1,j,k} - c_{i,j,k}}{\Delta x_i} \tag{5}$$

$$c_{i,j,k}^{n+\varepsilon} \cong \varepsilon c_{i,j,k}^{n+1} + (1 - \varepsilon)c_{i,j,k}^{n} \tag{6}$$

In Equations (4)–(6), integer subscripts and superscripts are used for simplicity, with subscripts (i,j,k) referring to discrete spatial locations and superscript n referring to discrete time levels. The operator Δx_i (Δ is a "forward difference" operator) is defined as the distance between x_{i+1} and x_i, $\Delta x_i \equiv x_{i+1} - x_i$. Substitution of Equations (4) and (5) into (3), with the assumption of diagonal $\mathbf{D}$, leads to the following expression for mass flux m_x in terms of discrete values of c:

$$m_x|_{i+(1/2),j,k} = \left(qc - \theta D_{xx} \frac{\partial c}{\partial x} \right)\Bigg|_{i+(1/2),j,k}$$

$$\cong q_{i+(1/2),j,k}\left(\frac{(\delta x_{i+1})c_{i,j,k} + (\delta x_i)c_{i+1,j,k}}{\delta x_i + \delta x_{i+1}} \right)$$

$$- (\theta D_{xx})_{i+(1/2),j,k}\left(\frac{c_{i+1,j,k} - c_{i,j,k}}{\Delta x_i} \right) \tag{7}$$

Similar expressions may be derived for each of the six flux terms in Equation (2). Each of these expressions may be substituted into Equation (2) to generate algebraic equations involving the discrete concentration values associated with each block. In general, these algebraic expressions have the following form, for arbitrary block (i,j,k):

$$\varepsilon A_{i-1,j,k}^{n+1} c_{i-1,j,k}^{n+1} + \varepsilon A_{i+1,j,k}^{n+1} c_{i+1,j,k}^{n+1} + \varepsilon A_{i,j-1,k}^{n+1} c_{i,j-1,k}^{n+1} + \varepsilon A_{i,j+1,k}^{n+1} c_{i,j+1,k}^{n+1}$$

$$+ \varepsilon A_{i,j,k-1}^{n+1} c_{i,j,k-1}^{n+1} + \varepsilon A_{i,j,k+1}^{n+1} c_{i,j,k+1}^{n+1} + \varepsilon A_{i,j,k}^{n+1} c_{i,j,k}^{n+1} = F_{i,j,k}^n \tag{8}$$

where the coefficients A involve combinations of parameters and grid information. For an internal block (i,j,k) (that is, $i \neq 1, i \neq N_x, j \neq 1, j \neq N_y, k \neq 1, k \neq N_z$),

$$A_{i-1,j,k}^{n+1} = \left(\frac{(\delta x_i)(q_x)_{i-(1/2),j,k}^{n+1}}{\delta x_{i-1} + \delta x_i} + \frac{(\theta D_{xx})_{i-(1/2),j,k}^{n+1}}{\nabla x_i} \right)\delta y_j \delta z_k \Delta t$$

$$A_{i+1,j,k}^{n+1} = \left(\frac{(\delta x_i)(q_x)_{i+(1/2),j,k}^{n+1}}{\delta x_{i+1} + \delta x_i} + \frac{(\theta D_{xx})_{i+(1/2),j,k}^{n+1}}{\Delta x_i} \right)\delta y_j \delta z_k \Delta t$$

$$A_{i,j-1,k}^{n+1} = \left(\frac{(\delta y_j)(q_y)_{i,j-(1/2),k}^{n+1}}{\delta y_{j-1} + \delta y_j} + \frac{(\theta D_{yy})_{i,j-(1/2),k}^{n+1}}{\nabla y_j} \right)\delta x_i \delta z_k \Delta t$$

$$A_{i,j+1,k}^{n+1} = \left(\frac{(\delta y_j)(q_y)_{i,j+(1/2),k}^{n+1}}{\delta y_{j+1} + \delta y_j} + \frac{(\theta D_{yy})_{i,j+(1/2),k}^{n+1}}{\Delta y_j} \right) \delta x_i \delta z_k \Delta t$$

$$A_{i,j,k-1}^{n+1} = \left(\frac{(\delta z_k)(q_z)_{i,j,k-(1/2)}^{n+1}}{\delta z_{k-1} + \delta z_k} + \frac{(\theta D_{zz})_{i,j,k-(1/2)}^{n+1}}{\nabla z_k} \right) \delta x_i \delta y_j \Delta t$$

$$A_{i,j,k+1}^{n+1} = \left(\frac{(\delta z_k)(q_z)_{i,j,k+(1/2)}^{n+1}}{\delta z_{k+1} + \delta z_k} + \frac{(\theta D_{zz})_{i,j,k+(1/2)}^{n+1}}{\Delta z_k} \right) \delta x_i \delta y_j \Delta t$$

$$A_{i,j,k}^{n+1} = \theta_{i,j,k}^{n+1} \delta x_i \delta y_j \delta z_k - \varepsilon A_{i+1,j,k}^{n+1} - \varepsilon A_{i+1,j,k}^{n+1} - \varepsilon A_{i,j-1,k}^{n+1}$$

$$- \varepsilon A_{i,j+1,k}^{n+1} - \varepsilon A_{i,j,k-1}^{n+1} - \varepsilon A_{i,j,k+1}^{n+1}$$

$$F_{i,j,k} = \theta_{i,j,k}^{n} c_{i,j,k}^{n} \delta x_i \delta y_j \delta z_k + [(rxn)_{i,j,k} + (src)_{i,j,k}] \delta x_i \delta y_j \delta z_k \Delta t$$

$$- (1 - \varepsilon)[A_{i-1,j,k}^{n} + A_{i+1,j,k}^{n} + A_{i,j-1,k}^{n} + A_{i,j+1,k}^{n} + A_{i,j,k-1}^{n} + A_{i,j,k+1}^{n}]$$

$$(9)$$

where ∇ is the "backward difference operator," with $\nabla x_i \equiv x_i - x_{i-1}$. Note that the superscript on the coefficients A denotes time level, with the terms on the right side of the equation (in the term $F_{i,j,k}$) being evaluated at the "old" time level n. The reaction and source terms are placed on the right side of the equation for now; these terms are discussed in more detail later. When this discrete equation is written for a block that has one or more sides coincident with the domain boundary, the boundary condition specified along that portion of the domain boundary influences the algebraic approximation. If the mass flux is specified, then the appropriate flux component in Equation (2) is replaced by the boundary condition, which is brought to the right side of the equation (that is, it is incorporated into F), and the corresponding coefficient (A) is set to zero. If the boundary condition is a specified concentration, then c at the block boundary is known, so that c outside the boundary can be related to $c_{i,j,k}$ and the known boundary value, via Equation (4). This leads to a modification in the associated coefficients (A), as well as the right side of the equation (F). Additional discussion of boundary conditions is given in the example calculations below.

The One-Dimensional Case

To understand some of the basic properties of finite difference approximations, consider a one-dimensional problem. This system may be viewed as a rectangular tube with the direction (or one dimension) of interest being the longitudinal axis along the center of the tube. Assume the longitudinal axis is parallel to the x-axis. Let the boundary conditions at the y- and z-direction boundaries be zero flux, so that $m_y = m_z = 0$. A discretization is shown in Figure 2, with $N_x \times 1 \times 1$ blocks. Because of the zero-flux conditions in the y- and z-directions, the terms $m_y|_{y_j-(\delta y_j/2)}$, $m_y|_{y_j+(\delta y_j/2)}$, $m_z|_{z_k-(\delta z_k/2)}$ and $m_z|_{z_k+(\delta z_k/2)}$ are all zero in Equation (2). Therefore, Equation (2) becomes

$$[(c\theta)|_{i,j,k}^{n+1} - (c\theta)|_{i,j,k}^{n}]\delta x_i \delta y_j \delta z_k = (m_x|_{i-(1/2),j,k}^{n+\varepsilon} - m_x|_{i+(1/2),j,k}^{n+\varepsilon})\delta y_j \delta z_k \Delta t$$

$$+ [(rxn)_{i,j,k}^{n+\varepsilon} + (src)_{i,j,k}^{n+\varepsilon}]\delta x_i \delta y_j \delta z_k \Delta t \tag{10}$$

Division of all terms in the equation by $\delta x_i \delta y_j \delta z_k \Delta t$ leads to a more familiar one-dimensional version of the finite difference equation:

$$\frac{(\theta c)_{i,j,k}^{n+1} - (\theta c)_{i,j,k}^{n}}{\Delta t} + \frac{(m_x)_{i+(1/2),j,k}^{n+\varepsilon} - (m_x)_{i-(1/2),j,k}^{n+\varepsilon}}{\delta x_i} - (rxn)_{i,j,k}^{n+\varepsilon} = (src)_{i,j,k}^{n+\varepsilon}$$

If the reaction term is taken to be first-order with respect to c, that is,

$$(rxn) = \lambda \theta c$$

if the source/sink term is identified by $\theta\Gamma$ and if subscripts j and k are dropped (because they are equal to 1 for all blocks), then the equation becomes

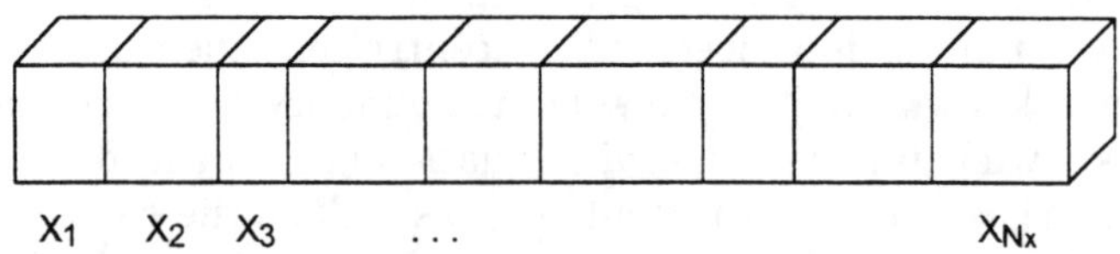

Figure 2. Discretization for a one-dimensional problem.

$$\frac{(\theta c)_i^{n+1} - (\theta c)_i^n}{\Delta t} + \varepsilon \frac{(q_x c)_{i+(1/2)}^{n+1} - (q_x c)_{i-(1/2)}^{n+1}}{\delta x_i} + (1 - \varepsilon) \frac{(q_x c)_{i+(1/2)}^{n} - (q_x c)_{i-(1/2)}^{n}}{\delta x_i}$$

$$- \frac{\varepsilon}{\delta x_i} \left[(\theta D)_{i+(1/2)} \left(\frac{c_{i+1}^{n+1} - c_i^{n+1}}{\Delta x_i} \right) - (\theta D)_{i-(1/2)} \left(\frac{c_i^{n+1} - c_{i-1}^{n+1}}{\nabla x_i} \right) \right]$$

$$- \frac{(1 - \varepsilon)}{\delta x_i} \left[(\theta D)_{i+(1/2)} \left(\frac{c_{i+1}^{n} - c_i^{n}}{\Delta x_i} \right) - (\theta D)_{i-(1/2)} \left(\frac{c_i^{n} - c_{i-1}^{n}}{\nabla x_i} \right) \right]$$

$$+ \varepsilon (\lambda \theta c)_i^{n+1} + (1 - \varepsilon)(\lambda \theta c)_i^{n} = (\theta \Gamma)_i^{n+\varepsilon}$$

For a case of saturated groundwater flow, the moisture content θ equals the porosity ϕ. For most one-dimensional flow systems, q_x is constant, which usually implies that D is constant. Also assume porosity ϕ is constant. For a case of constant node spacing (call the spacing Δx), this leads to the simplified model system:

$$\frac{c_i^{n+1} - c_i^n}{\Delta t} + \varepsilon \left[V \frac{c_{i+1}^{n+1} - c_{i-1}^{n+1}}{2\Delta x} - D \frac{c_{i+1}^{n+1} - 2c_i^{n+1} + c_{i-1}^{n+1}}{(\Delta x)^2} + \lambda c_i^{n+1} \right]$$

$$+ (1 - \varepsilon) \left[V \frac{c_{i+1}^{n} - c_{i-1}^{n}}{2\Delta x} - D \frac{c_{i+1}^{n} - 2c_i^{n} + c_{i-1}^{n}}{(\Delta x)^2} + \lambda c_i^{n} \right] = \Gamma_i^{n+\varepsilon}$$

Algebraic rearrangement of this equation leads to the following form:

$$c_i^{n+1} + \varepsilon \left(\frac{1}{2} \right) \left(\frac{V\Delta t}{\Delta x} \right) (c_{i+1}^{n+1} - c_{i-1}^{n+1}) - \varepsilon \left(\frac{D\Delta t}{(\Delta x)^2} \right) (c_{i+1}^{n+1} - 2c_i^{n+1} + c_{i-1}^{n+1}) + \varepsilon \lambda \Delta t c_i^{n+1}$$

$$= c_i^n - (1 - \varepsilon) \left(\frac{1}{2} \right) \left(\frac{V\Delta t}{\Delta x} \right) (c_{i+1}^{n} - c_{i-1}^{n}) + (1 - \varepsilon) \left(\frac{D\Delta t}{(\Delta x)^2} \right) (c_{i+1}^{n} - 2c_i^{n} + c_{i-1}^{n})$$

$$- (1 - \varepsilon)\lambda \Delta t c_i^n + \Gamma_i^{n+\varepsilon} \tag{11}$$

In Equation (11), three dimensionless groups are present: the dimensionless velocity $\mathbb{V} \equiv V\Delta t/\Delta x$, the dimensionless diffusion $\mathbb{D} \equiv D\Delta t/(\Delta x)^2$, and the dimensionless reaction rate $\Lambda \equiv \lambda \Delta t$. These dimensionless groups, or ratios of any two of them, are very important to understand the behavior of certain numerical solutions. In numerical analysis, the dimensionless velocity V is called the Courant Number, and the ratio of V to D is called the Grid Peclet Number (Pe). Ratios of any two of the dimensionless groups provide measures of the relative importance of one physical process to another, when measured over the length scale of one grid block (Δx) and the time scale of one time step (Δt). For example, the Grid Peclet Number measures the relative importance of advection to diffusion.

An Example Calculation

As an example, consider the following one-dimensional transport problem:

$$R\frac{\partial c}{\partial t} + V\frac{\partial c}{\partial x} - D\frac{\partial^2 c}{\partial x^2} + \lambda c = 0, \quad 0 < x < l, \quad t > 0$$

$$c(0,t) = C_0 \quad t > 0$$

$$\frac{\partial c}{\partial x}(l,t) = 0 \quad t > 0 \tag{12}$$

$$c(x,0) = 0, \quad 0 < x < l$$

Let the spatial domain be discretized using N blocks, each of length Δx, and let a constant time step Δt be used to march through time. The specific location of each grid block is sometimes dependent on the specified boundary conditions. For example, at the left boundary ($x = 0$), the boundary condition is a fixed value of concentration. As such, the left-most block is often placed so that $x = 0$ corresponds to the center of the block. If $x = 0$ corresponds to the left edge of the first block, then the node at the center of this block is at $x_1 = \delta x_1/2$. In this case, an extrapolation to the boundary location ($x = 0$) is required. At the right boundary, the diffusive flux is specified as a boundary condition. In this case, the right-most block is placed so that its right boundary is located at $x = l$. A discretization that places the first node at $x = 0$ and uses 10 grid cells (blocks) is shown in Figure 3.

For this grid, there are 10 unknowns (one value of concentration in each of the 10 cells.). Therefore, 10 algebraic equations are required. The equations are obtained in the following way. In the first block, the concentration is known from the boundary condition, so this trivial identity is used as one equation:

$$c_1^{n+1} = C_0$$

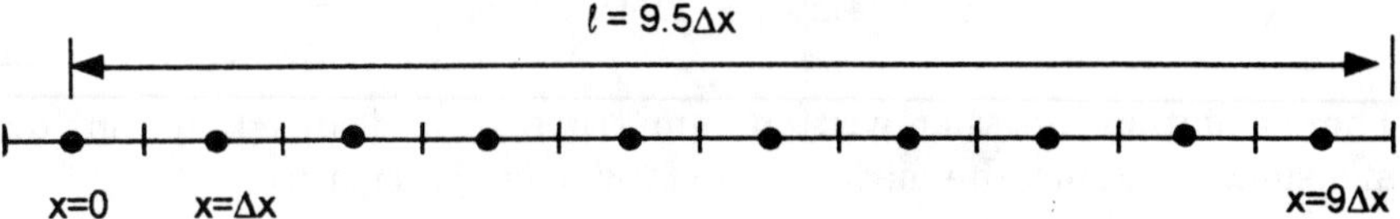

Figure 3. Finite difference discretization for the example problem.

This equation holds for all discrete time levels n. For blocks 2 through 9 (that is, for all interior blocks), the general finite difference approximation, Equation (11), is written, for $i = 2,3,\ldots,9$,

$$c_i^{n+1} + \varepsilon\left(\frac{1}{2}\right)\left(\frac{V\Delta t}{\Delta x}\right)(c_{i+1}^{n+1} - c_{i-1}^{n+1}) - \varepsilon\left(\frac{D\Delta t}{(\Delta x)^2}\right)(c_{i+1}^{n+1} - 2c_i^{n+1} + c_{i-1}^{n+1}) + (\lambda\Delta t)c_i^{n+1}$$

$$= c_i^n - (1-\varepsilon)\left(\frac{1}{2}\right)\left(\frac{V\Delta t}{\Delta x}\right)(c_{i+1}^n - c_{i-1}^n)$$

$$+ (1-\varepsilon)\left(\frac{D\Delta t}{(\Delta x)^2}\right)(c_{i+1}^n - 2c_i^n + c_{i-1}^n) + (\lambda\Delta t)c_i^n + \Gamma_i^{n+\varepsilon}$$

Consider the case where $R = 1, V = 1, D = 0.1, \lambda = 0, l = 9.5, \Delta x = 1, \Delta t = 0.2$, and $\varepsilon = 1.0$. Then, the equation for each interior node reduces to

$$(-0.12)c_{i-1}^{n+1} + (1.04)c_i^{n+1} + (0.08)c_{i+1}^{n+1} = c_i^n$$

which has the form of Equation (8). The equation associated with the right-most node ($i = N$) is complicated somewhat by the right-side boundary condition involving $\partial c/\partial x$. A typical approach is to use a "domain extension" concept wherein an extra block is added at the right end of the domain, giving rise to an additional discrete concentration value c_{N+1}. In this case, the finite difference equation is written for $i = N$, and an additional equation is written as an approximation to the boundary derivative (that is, $\partial c/\partial x\,|_{N+1/2}$). These equations are

$$(-0.12)c_{N-1}^{n+1} + (1.04)c_N^{n+1} + (0.08)c_{N+1}^{n+1} = c_N^n$$

$$\frac{c_{N+1}^{n+1} - c_N^{n+1}}{2\Delta x} = 0$$

These two equations can be combined to eliminate the nonphysical concentration c_{N+1} to yield the final equation:

$$(-0.04)c_{N-1}^{n+1} + (1.04)c_N^{n+1} = c_N^n$$

These equations are then written in matrix form and solved. For one-dimensional systems, the matrix has a tri-diagonal structure:

$$
\begin{bmatrix}
1 & 0 & 0 & 0 & \cdots & 0 \\
-0.12 & 1.04 & 0.08 & 0 & \cdots & 0 \\
0 & -0.12 & 1.04 & 0.08 & & \\
& \ddots & \ddots & \ddots & & \\
0 & \cdots & 0 & -0.12 & 1.04 & 0.08 \\
0 & \cdots & 0 & 0 & -0.04 & 1.04
\end{bmatrix}
\bullet
\begin{bmatrix}
c_1^{n+1} \\
c_2^{n+1} \\
c_3^{n+1} \\
\vdots \\
c_{N-1}^{n+1} \\
c_N^{n+1}
\end{bmatrix}
=
\begin{bmatrix}
C_0 \\
c_2^n \\
c_3^n \\
\vdots \\
c_{N-1}^n \\
c_N^n
\end{bmatrix}
\tag{13}
$$

This equation is used to march through time, beginning with $n = 0$. This means that the initial information on the right side of the equation (c_i^0) is obtained from the initial condition. For this example, the initial condition is $c(x,0) = 0$, so all entries on the right side are zero except the first one, which is equal to the left boundary concentration, C_0. The matrix is then solved for $\{c_i^1\}_{i=1}^N$. Given the solution at time step $n = 1$, this information defines the right-side vector and the solution at time level $n = 2$ is computed. The procedure continues forward in time until the user decides to end the calculation. Figure 4 shows the numerical solution at time $t = 5.0$ ($n = 25$); the analytical solution is also shown for comparison. Notice that the numerical solution exhibits some oscillations upstream of the concentration front. In addition, the front itself is less steep, or more diffuse, in the numerical solution as compared to the analytical solution. Both of these effects—oscillations and numerical diffusion—are nonphysical and result from the numerical approximation.

Additional example results are shown in Figures 5 through 7. In Figure 5, the same parameters are used as for the results of Figure 4, except the time-weighting parameter ε is set to 0.5 instead of 1.0. Notice that the solution exhibits much more pronounced oscillations, although the front is somewhat steeper. A second finite difference solution, for which the boundary $x = 0$ corresponds to the left edge of the first block (as opposed to the center of the block), is also shown in Figure 5. Notice that the modified boundary condition implementation does not affect the solution significantly. Figures 6 and 7 show solutions that use more grid blocks (smaller Δx), thereby reducing the Grid Peclet Number, Pe. Overall, as the figures indicate, numerical oscillations are more pro-

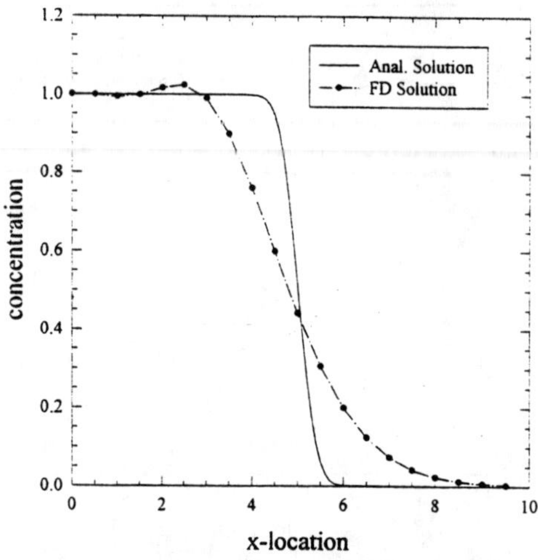

Figure 4. Results for example problem.

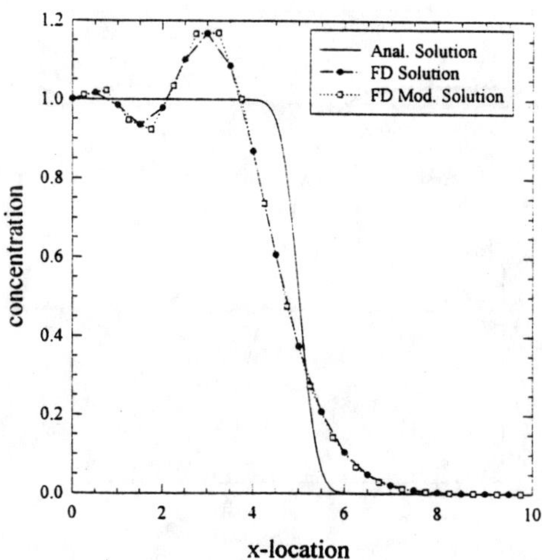

Figure 5. Results for example problem, including results using modified boundary representation.

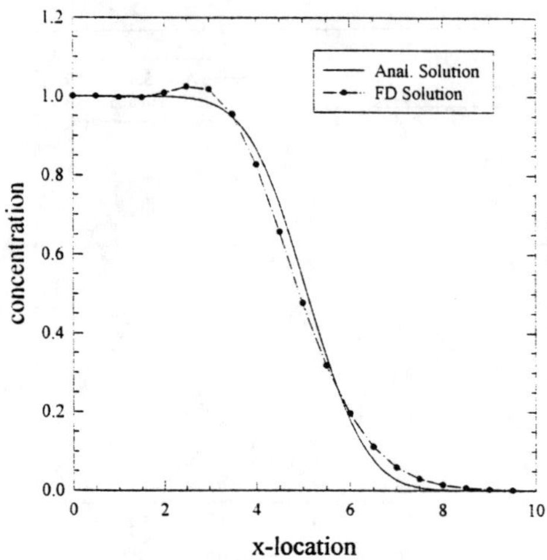

CONCENTRATION AT TIME T=5
Pe = 5, Cu = 0.5, ε = 0.5

Figure 6. Results for example problem.

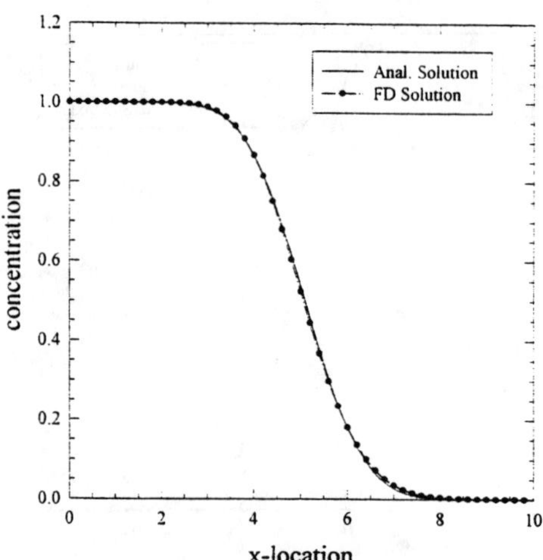

CONCENTRATION AT TIME T=5
Pe = 2, Cu = 0.625, ε = 0.5

Figure 7. Results for example problem.

72

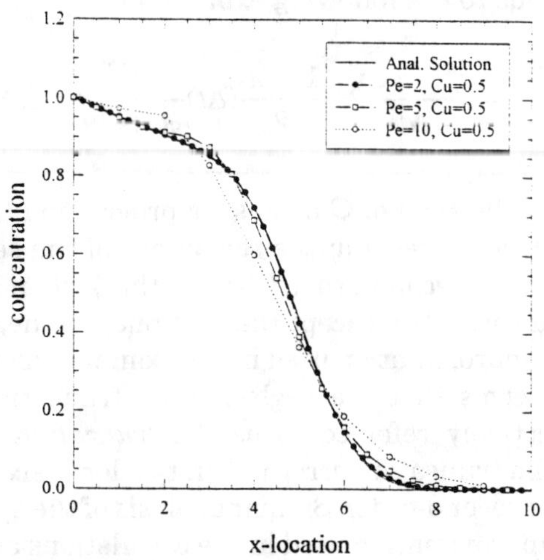

Figure 8. Results for example problem, including nonzero reaction term.

nounced when the Grid Peclet Number is larger. A similar dependence on the Grid Peclet Number applies to the case of nonzero reactions, as indicated in Figure 8.

Analysis of the Approximations

The previous set of example calculations demonstrates the basic structure of numerical calculations and also shows some of the common problems that numerical solutions exhibit when solving the transport equation. The simplest way to gain some insight into this behavior is by Taylor Series Analysis. For example, let the concentrations at node i and time levels n and $n + 1$ be expanded in Taylor series in terms of c and its derivatives evaluated at node i and time $n + \varepsilon$:

$$c_i^n = c_i^{n+\varepsilon} + \varepsilon(\Delta t)\frac{\partial c}{\partial t}\bigg|_i^{n+\varepsilon} + \frac{(\varepsilon\Delta t)^2}{2}\frac{\partial^2 c}{\partial t^2}\bigg|_i^{n+\varepsilon} + \ldots + R^M$$

$$\hspace{10cm} (14)$$

$$c_i^{n+1} = c_i^{n+\varepsilon} + (1-\varepsilon)(\Delta t)\frac{\partial c}{\partial t}\bigg|_i^{n+\varepsilon} + \frac{[(1-\varepsilon)\Delta t]^2}{2}\frac{\partial^2 c}{\partial t^2}\bigg|_i^{n+\varepsilon} + \ldots + R^M$$

where R^M denotes the Taylor remainder term, written in terms of $(\Delta t)^M$ (see, for example, Leithold, 1972). Subtraction of these two equations and division by Δt leads to the following expression:

$$\left.\frac{\partial c}{\partial t}\right|_i^{n+\varepsilon} = \frac{c_i^{n+1} - c_i^n}{\Delta t} + \frac{1-2\varepsilon}{2}(\Delta t)\left.\frac{\partial^2 c}{\partial t^2}\right|_i^{n+\varepsilon} + \mathcal{O}[(\Delta t)^2] \tag{15}$$

In Equation (15), the symbol $\mathcal{O}$ means "of order," and the term $\mathcal{O}[(\Delta t)^2]$ means that the leading term in the remainder of the series is "of order delta t squared," or "second order." Notice that, when all terms on the right side of Equation (15), except the first one, are neglected, then the finite difference approximation used in the example calculations above is recovered. The terms that are neglected, by truncation of the Taylor series, are collectively referred to as the *truncation error* (TE). For Equation (15), the truncation error is "first-order" except when $\varepsilon = 0.5$, in which case it is second-order. Similar analysis of the spatial derivatives shows that the approximation used in the calculations above is $\mathcal{O}[(\Delta x)^2]$.

To understand the increased numerical diffusion present when $\varepsilon = 1$, as opposed to $\varepsilon = 0.5$, truncation error analysis is useful. For $\varepsilon = 0.5$, the approximation is $\mathcal{O}[(\Delta t)^2]$ while for the case of $\varepsilon = 1.0$, the leading term is given by

$$\text{TE}_{\text{leading}} = \frac{-(\Delta t)}{2}\left.\frac{\partial^2 c}{\partial t^2}\right|_i^{n+\varepsilon}$$

From the governing equation [Equation (12) with $\lambda = 0$],

$$\frac{\partial c}{\partial t} = -V\frac{\partial c}{\partial x} + D\frac{\partial^2 c}{\partial x^2}$$

Subsequent differentiation with respect to t leads to

$$\frac{\partial^2 c}{\partial t^2} = -V\frac{\partial}{\partial t}\frac{\partial c}{\partial x} + D\frac{\partial}{\partial t}\frac{\partial^2 c}{\partial x^2}$$

$$= -V\frac{\partial}{\partial x}\frac{\partial c}{\partial t} + D\frac{\partial^2}{\partial x^2}\frac{\partial c}{\partial t}$$

$$= \left(-V\frac{\partial}{\partial x} + D\frac{\partial^2}{\partial x^2}\right)\left(\frac{\partial c}{\partial t}\right)$$

$$= \left(-V\frac{\partial}{\partial x} + D\frac{\partial^2}{\partial x^2}\right)\left(-V\frac{\partial c}{\partial x} + D\frac{\partial^2 c}{\partial x^2}\right)$$

$$= V^2\frac{\partial^2 c}{\partial x^2} - 2VD\frac{\partial^3 c}{\partial x^3} + D^2\frac{\partial^4 c}{\partial x^4} \qquad (16)$$

For cases of advection domination, the term $V^2(\partial^2 c/\partial x^2)$ is important and leads to numerical diffusion, which is seen in the following expression for the finite difference approximation:

$$\frac{c_i^{n+1} - c_i^n}{\Delta t} = \left.\frac{\partial c}{\partial t}\right|_i^{n+\varepsilon} - \frac{(\Delta t)}{2}\left.\frac{\partial^2 c}{\partial t^2}\right|_i^{n+\varepsilon} + \mathcal{O}[(\Delta t)^2]$$

$$= \left.\frac{\partial c}{\partial t}\right|_i^{n+\varepsilon} - \frac{(\Delta t)}{2}\left[V^2\frac{\partial^2 c}{\partial x^2} - 2VD\frac{\partial^3 c}{\partial x^3} + D^2\frac{\partial^4 c}{\partial x^4}\right] + \mathcal{O}[(\Delta t)^2]$$

$$(17)$$

Because diffusion terms correspond to second-order derivatives in space, the "numerical diffusion" coefficient introduced by this time-derivative approximation is $V^2(\Delta t)/2$.

While truncation-error analysis based on Taylor series expansions provides some insight into the origin of numerical diffusion, it does little to explain the nonphysical oscillations observed in the numerical solutions. Fourier analysis is a technique that explains both numerical diffusion and oscillations in the numerical solutions. In the Fourier approach, both the analytical solution and the numerical solution are decomposed into their respective Fourier representations [that is, in Fourier series of sines and cosines—see Celia and Gray (1992)]. The phase and amplitude changes of each Fourier component are calculated for both the analytical and the numerical solutions. Errors in phase propagation between the numerical and analytical solutions lead to oscillations in the numerical solution, while errors in amplitude may lead to numerical diffusion. While Fourier analysis is a powerful tool in numerical analysis, its mathematical presentation is beyond the scope of this chapter. The interested reader is referred to Chapter 4 of Celia and Gray (1992).

A final important note of caution involves the concept of stability. It turns out that, for the case of $\varepsilon < 0.5$, there is a restriction on the time-step size Δt. If the time step is chosen to be too large, the numerical solution may grow without bound, making the numerical solution unstable. In terms of Fourier analysis, this situation occurs when at least one of the Fourier components has an amplitude that grows with time. So a stability condition for the numerical solution may be written as a requirement that the magnitude of each Fourier coefficient must not increase as time increases. Imposition of this requirement in the mathematical implementation of Fourier analysis allows specific bounds on Δt to be derived for a given numerical approximation. Again, see Celia and Gray (1992) for details of the mathematical developments and applications of Fourier analysis.

Based on decades of research focusing in numerical methods for the contaminant transport equation, there are simple rules that can be followed to prevent the difficulties discussed above. For example, when solving Equation (12) using the finite difference approximation developed above, the solution will not exhibit oscillations as long as the Grid Peclet Number is not greater than 2, with the Courant Number not exceeding 1 [see Daus et al. (1985) for an excellent analysis of oscillation limits]. In addition, stability can be guaranteed whenever $\varepsilon \geq 0.5$.

ALTERNATIVE NUMERICAL APPROACHES

Finite Element and Other Weighted-Residual Methods

Finite element methods are fundamentally different than finite difference methods in the way that the approximating equations are derived. However, the final result is analogous to that of the finite difference approach in that algebraic equations are derived with the unknowns being discrete nodal values of the unknown. On rectangular grids, finite element approximations can often be viewed as specific cases of finite difference approximations. The major advantage of finite element methods is the relative ease with which approximations may be derived on general (nonrectangular) domains. While the detailed development of finite element approximations is beyond the scope of this chapter, a brief introduction is provided to facilitate later discussions.

In the finite element method, an explicit functional representation of the approximate solution is written in which only a finite number of terms are unknown (this means that the approximation is in a finite-dimensional function space). This approximation is called the *trial function*. If a discretization is chosen with N discrete (node) points at which the approximate solution is sought, the trial function $\hat{c}$ is usually written in terms of these nodal approximations as

$$\hat{c}(\mathbf{x}, t) = \sum_{j=1}^{N} C_j(t) \varphi_j(\mathbf{x}) \tag{18}$$

In this equation, C_j is the approximate solution at discrete (node) point j, and $\varphi_j(\mathbf{x})$ is a "basis function" associated with node j. The basis function is usually chosen to be a piecewise polynomial, with a standard choice being the so-called piecewise linear functions shown in Figure 9. Because $\hat{c}$ is an approximation to the true solution c, $\hat{c}$ will, in general, not satisfy the governing equation. The amount by which $\hat{c}$ fails to satisfy the governing equation is referred to as the residual $R(\mathbf{x},t)$. The objective of "weighted residual" methods is to minimize the residual $R(\mathbf{x},t)$. To achieve this and to simultaneously derive a set of algebraic equations in the unknowns $\{C_j\}$, a set of N test functions or weight functions, $\{w_i(\mathbf{x})\}_{i=1}^{N}$, is defined. In the finite element method, the weight functions are chosen to be the same as the basis functions, $w_i(\mathbf{x}) = \varphi_i(\mathbf{x})$. Other methods of weighted residuals use other choices of weight functions (see Celia and Gray, 1992). Given a choice for weight functions, the algebraic equations are derived by forcing the weighted average of the residual to be zero, where the "weighting" comes from multiplication by the weight functions. Mathematically, this is written as

$$\int_{\Omega} R(\mathbf{x}, t) w_i(\mathbf{x}) d\mathbf{x} = 0, \quad i = 1, 2, \ldots, N \tag{19}$$

Calculation of the resulting integrals and implementation of boundary conditions leads to a matrix equation with the vector of unknowns being $\{C_j\}_{j=1}^{N}$. Standard time-stepping methods are then used to solve for the discrete unknowns at each time level.

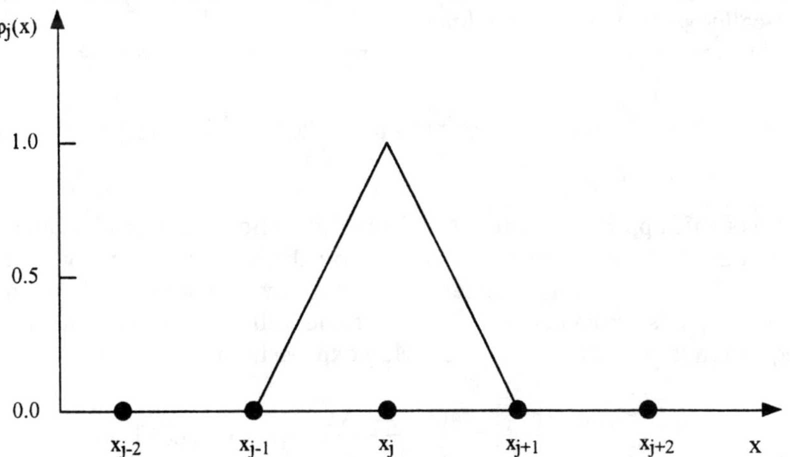

Figure 9. Piecewise linear basis function.

Upstream-Weighting Methods

When oscillatory behavior is analyzed, the approximation that is found to be the cause is that associated with the first spatial derivative (the advection term, $\nabla \cdot \mathbf{q}c$, or in the simplified case, $V\partial c/\partial x$). The approximation used in the preceding calculations is the so-called *centered approximation* for the advection term:

$$\left.\frac{\partial(q_x c)}{\partial x}\right|_i^{n+1} \cong \frac{(q_x c)_{i+(1/2)}^{n+1} - (q_x c)_{i-(1/2)}^{n+1}}{\delta x_i}$$

or for the simplified one-dimensional case with constant node spacing,

$$\left.V\frac{\partial c}{\partial x}\right|_i^{n+1} \cong V\frac{c_{i+1}^{n+1} - c_{i-1}^{n+1}}{2\Delta x} \tag{20}$$

This approximation is second-order in space, $\mathcal{O}[(\Delta x)^2]$ (this is analogous to the second-order-in-time approximation for $\partial c/\partial t\,|^{n+\varepsilon}$ when $\varepsilon = 0.5$; this choice of ε makes the approximation "centered in time" between time levels n and $n + 1$). Because the approximation in Equation (20) is centered in space, it weights equally information from "upstream" and from "downstream" of node i. However, the process of advection is inherently one-sided, in that information moves from upstream to downstream, based on the velocity V. Therefore, numerical approximations to this term might reasonably be developed that weight upstream information more heavily than downstream information. This is the basic idea of so-called *upstream-weighted methods*.

The simplest upstream-weighted approximation is given by

$$\left.V\frac{\partial c}{\partial x}\right|_i \cong V\frac{c_i - c_{i-1}}{\Delta x} \tag{21}$$

Use of this approximation in the finite difference equations has an effect similar to that of increasing ε from 0.5 to 1.0: numerical diffusion is added to the system. Simple truncation error analysis explains this behavior. When c_{i-1} is expanded in a Taylor series about the point x_i and the equation is rearranged, the following expression results:

$$V\frac{c_i - c_{i-1}}{\Delta x} = \left.V\frac{\partial c}{\partial x}\right|_i - \frac{V\Delta x}{2}\left.\frac{\partial^2 c}{\partial x^2}\right|_i + \mathcal{O}[(\Delta x)^2] \tag{22}$$

The "numerical diffusion" coefficient associated with this approximation is seen to be $V\Delta x/2$. Similar to the variably-weighted time approximation, spatial upstream weighting is often made variable by introduction of a parameter α, such that

$$V\left.\frac{\partial c}{\partial x}\right|_i \cong V\left[\alpha\frac{c_i - c_{i-1}}{\Delta x} + (1-\alpha)\frac{c_{i+1} - c_{i-1}}{2\Delta x}\right] \tag{23}$$

For the full approximation, including both the time and space derivatives, the numerical diffusion coefficient is then

$$D_{num} = \alpha\frac{V\Delta x}{2} + \frac{1-2\varepsilon}{2}V^2(\Delta t) \tag{24}$$

The literature is replete with different upstream-weighting techniques. These include methods based on finite difference approximations, finite element approximations, and other more specialized approaches. Almost all of them have in common the introduction of some kind of numerical diffusion to control unwanted numerical oscillations when the Grid Peclet Number becomes too large.

Characteristic Methods

The idea of treating each of the terms in the governing equation with a different type of numerical approximation is attractive because the physical processes that underlie each term have distinctly different behaviors. As stated earlier, advection is a process that has definite directional bias. To appreciate some of the mathematical properties of the advection part of the governing equation, consider the case of pure advection in one spatial dimension. The governing equation simplifies to

$$\frac{\partial c}{\partial t} + V\frac{\partial c}{\partial x} = 0 \tag{25}$$

This equation has a particularly simple solution, which may be derived as follows. Instead of writing Equation (25) using partial derivatives in x and t, which implies a stationary (or Eulerian) frame of reference, consider rewriting the equation using a time derivative defined with respect to a moving coordinate system. Such a time derivative is often written as the total derivative with respect to time, d/dt, with $d/dt \equiv \partial/\partial t + U\partial/\partial x$, where U is the velocity of the reference frame. In a physical system, when the velocity U corresponds to the velocity of the material under study, the total derivative is often referred to as the *material derivative* and denoted by D/Dt. For the case of advection, the appropriate material velocity is the flow velocity V. By definition of the material derivative, Equation (25) may be rewritten as

$$\frac{\partial c}{\partial t} + V \frac{\partial c}{\partial x} = \frac{Dc}{Dt} = 0$$

This equation states that when moving with characteristic velocity V (which means $Dx/Dt = V$), the time derivative of concentration, viewed from that reference frame, is zero. This means that, along the space-time curves defined by $Dx/Dt = V$, c is constant. The curves defined by $Dx/Dt = V$ are known as *characteristic curves* for the advection equation, Equation (25). In addition, the frame of reference that moves with characteristic velocity is referred to as a Lagrangian reference frame.

Similar arguments apply to the case of advection and reaction. In this case, the governing equation is

$$\frac{\partial c}{\partial t} + V \frac{\partial c}{\partial x} + \lambda c = 0 \tag{26}$$

Definition of the material derivative $Dx/Dt = V$ allows this equation to be rewritten as

$$\frac{Dc}{Dt} + \lambda c = 0$$

This means that, along each characteristic curve defined by $Dx/Dt = V$, the concentration c changes exponentially in time, with exponent $-\lambda t$. If V or λ is nonconstant or if λ is nonlinear, then the solution becomes somewhat more complicated, but the overall concept remains the same.

These observations about solution behavior along the characteristic curves lead to the possibility of numerical approximations that take advantage of that behavior. Consider the advection-only case. Equation (25) implies that c is constant along the characteristics. Therefore, if finite difference approximations are written along the characteristic directions, good numerical behavior might be expected because the truncation errors would be very small. This is because the truncation error expressions involve higher-order derivatives, and if c is constant, then its derivatives are zero. A typical numerical implementation of this concept would use the following approximation:

$$\left.\frac{Dc}{Dt}\right|_i^{n+1} \cong \frac{c_i^{n+1} - c_{i*}^n}{\Delta t}$$

where subscript i^* corresponds to the spatial location at time t^n through which passes the characteristic curve that also passes through point x_i at time t^{n+1} (see Figure 10). The point x_{i*} is sometimes referred to as the "foot of the characteristic" point. If c^n is known at the discrete node points x_i, then c_{i*}^n may be determined by spatial interpolation. For the characteristic curve shown in Figure 10, piecewise linear interpolation between nodal values leads to the following approximation:

$$\left.\frac{Dc}{Dt}\right|_i^{n+1} \cong \frac{c_i^{n+1} - [(1-\alpha)c_i^n + \alpha c_{i-1}^n]}{\Delta t}$$

$$= \frac{c_i^{n+1} - c_i^n}{\Delta t} + \frac{\alpha}{\Delta t}(c_i^n - c_{i-1}^n)$$

$$= \frac{c_i^{n+1} - c_i^n}{\Delta t} + V\frac{(c_i^n - c_{i-1}^n)}{\Delta x} \tag{27}$$

In this expression, use was made of the fact that $\alpha = V\Delta t/\Delta x$ ($=V$, the Courant Number), and the Courant Number was assumed to be less than or equal to one. Equation (27) demonstrates that the characteristic approximation for this case is identical to an upstream-weighted difference approximation. Notice also that the spatial derivative is evaluated at time t^n, as opposed to earlier examples in which the evaluations were at the later time level t^{n+1}. These two observations mean that this approximation generally has some numerical diffusion associated with it and that, when the Courant Number is an integer (for example, $V = 1$), the approximation is exact (this can be seen from truncation error analysis).

The errors that arise in the approximation of Equation (27) are due to the interpolation at the old time level t^n. The numerical errors can be reduced by using larger time steps (larger V) and/or by use of more accurate interpolation formulas. However, there is no way to eliminate these errors in general. Notice also that, when reactions are involved, then c is no longer constant along the characteristics, and additional truncation errors arise because of this. The characteristic algorithm then may take the form

$$\left.\left(\frac{Dc}{Dt} + \lambda c\right)\right|_i^{n+1} \cong \frac{c_i^{n+1} - c_{i*}^n}{\Delta t} + \lambda c_i^{n+1} \tag{28}$$

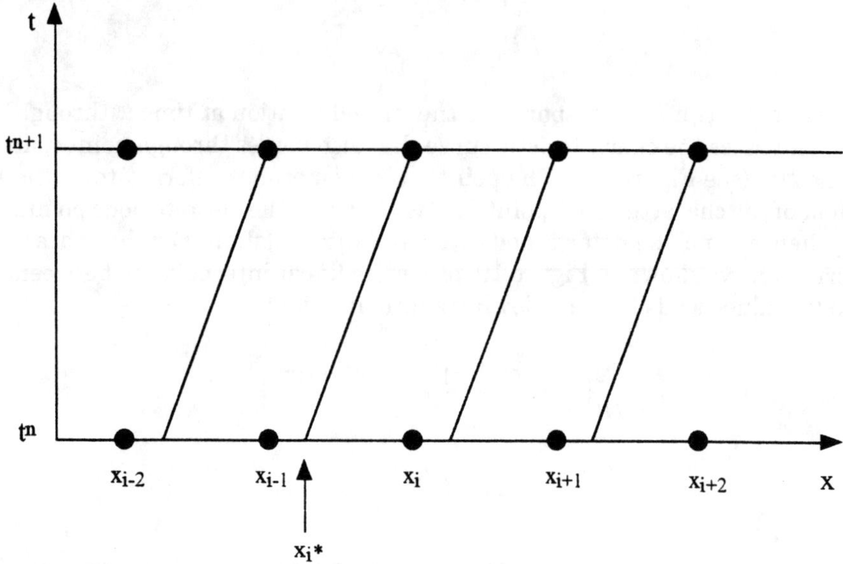

Figure 10. Characteristic curves for the advection problem, showing the "foot of the characteristic" point.

where the reaction term is evaluated at the new time level t^{n+1} (it might also be evaluated using c_i^n or using an ε-weighting between n and $n +$ 1). While characteristic methods suffer from interpolation errors at the "foot of the characteristic" points, their advantages usually outweigh these drawbacks, such that characteristic methods are usually superior to their Eulerian-based counterparts.

Operator Splitting Methods

Characteristic methods are often used within a more general procedure referred to as operator splitting. In the operator splitting technique, each of the spatial operators—advection, diffusion, and reaction—are treated separately using numerical approximations appropriate for each. The individual solutions are combined to provide the final solution at the new time level. In operator splitting methods, the advection operator is usually treated with a characteristic method, the diffusion operator is treated with a standard finite difference or finite element approximation, and the reaction operator is treated with a solver appropriate for ordinary differential equations. While this approach is similar to that shown in Equations (27) and (28), the operator splitting technique is usually considered to be somewhat more general. For a discussion of operator splitting methods, see Valocchi et al. (1993) and Chiang et al. (1991).

Treatment of Nonlinearities

A distinguishing feature of biodegradation reactions is their nonlinear nature, which means that the coefficient λ is a function of the unknown concentration c. For example, in Equation (28), if λ is a function of c, then the equation becomes

$$\left(\frac{Dc}{Dt} + \lambda c\right)\Bigg|_i^{n+1} \cong \frac{c_i^{n+1} - c_{i*}^n}{\Delta t} + \lambda_i^{n+1} c_i^{n+1} \tag{29}$$

Because c_i^{n+1} is unknown, $\lambda_i^{n+1} \equiv \lambda(c_i^{n+1})$ is also unknown. To derive a system of linear algebraic equations, the coefficient λ_i^{n+1} must be approximated so that the only unknown in the reaction term is the (linear) concentration that multiplies λ. Approximation of the unknown coefficient using some known information is referred to as *linearization*. The simplest linearization is to evaluate the coefficient using the value of concentration from the previous time level, such that $\lambda_i^{n+1} \cong \lambda(c_{i*}^n)$. If the numerical approximation is not a characteristic method but is instead Eulerian, the approximation would be $\lambda_i^{n+1} \cong \lambda(c_i^n)$. In some applications, an improved estimation is provided by projecting forward to time t^{n+1} using values from time levels t^n and t^{n-1} (appropriate formulas may be derived from Taylor-series analysis). Finally, if the nonlinearities are strong, an iterative procedure may be adopted wherein the coefficient is continually updated within the time step. If the advection-reaction equation is taken as an example, with a characteristic method such as Equation (29) used as the numerical approximation, a simple iterative procedure would be

$$\frac{c_i^{n+1,k+1} - c_{i*}^n}{\Delta t} + \lambda_i^{n+1,k} c_i^{n+1,k+1} = 0 \tag{30}$$

In this equation, superscript n represents time level, while superscript k represents iteration level. For a given n, calculations proceed by successively updating k until the iteration in k converges under a chosen measure, such as $\Sigma_i |c_i^{n+1,k+1} - c_i^{n+1,k}| < \zeta$, where ζ is a specified error tolerance.

NUMERICAL IMPLEMENTATION

A number of numerical simulation studies involving biodegradation have been reported in the literature over the last decade. These studies encompass a broad range of numerical approximations, although the most recent papers have focused on operator splitting methods in a finite element context. Of note in the recent studies are the papers by Chiang et al. (1991), Valocchi et al. (1993), and Wood et al. (1994). Each of these papers uses a version of the operator splitting technique, with a characteristic method for the advection part of the calculation. Chiang et al. (1991) used their numerical simulator to study the effects of variable hydraulic conductivity and variable linear sorption, while Wood et al. (1994) used a numerical formulation analogous to that of Chiang et al. (1991) to model a two-dimensional, intermediate-scale laboratory experiment involving a two-layer porous media system. Valocchi et al. (1993) used their operator splitting–based numerical solutions to study the interactions between biodegradation and adsorption in one-dimensional test problems. The paper by Valocchi et al. (1993) provides an excellent description of the operator splitting concept, as well as several operator splitting formulations. Wang et al. (1995) used a Eulerian-Lagrangian Localized Adjoint Method (ELLAM) to solve the biodegradation transport equations. This method is another variant of the general characteristic method presented above, where the development of the numerical approximation is based on analysis of the adjoint operator associated with the governing partial differential equation. Whereas operator splitting methods treat all operators separately, the ELLAM approach treats the reaction term(s) in conjunction with one of the other spatial operators, usually advection. This is analogous to the approximation of Equation (28). Other studies that use numerical approximations based on characteristic methods include Widdowson et al. (1988) and Dawson and Wheeler (1992).

Several earlier studies used Eulerian approximations to solve the governing equations. For example, Celia et al. (1989) developed a numerical solution procedure based on the so-called Optimal Test Function (OTF) algorithm. Like the ELLAM formulation, OTF derives the approximating equations by focusing the analysis on the adjoint operator associated with the governing partial differential equation. The end result is a method (OTF) that resembles an upstream-weighted approximation but has certain optimal properties. As shown in later analyses [see, for example, Bouloutas and Celia (1991)], the OTF method behaves similarly to an upstream-weighted finite difference approximation. Kindred and Celia (1989) used the OTF simulator to study a wide range of aerobic and anaerobic biodegradation problems. The anaerobic problems included those with microbial consortia. One of the examples presented in the paper involved aerobic biodegradation combined with nitrate-reducing metabolism. A model of the system was formulated to include transport equations for an organic substrate, oxygen, nitrate (as an electron acceptor), and assimilatory nitrogen. These four transport equations were coupled to two growth equations for bacteria, one for the aerobic biomass, and one for the nitrate-reducing biomass. Various growth inhibition factors were used to control biomass growth. The resulting six equations [see Equations (19) through (24) in the paper by Kindred and Celia (1989)] were solved using the OTF method. Figures 11a and 11b show a typical result, where the system has constant velocity in the positive x-direction, and the inflow boundary conditions are the concentration values shown on the graphs at the location $x = 0$. In this system, aerobic degradation occurs until oxygen is depleted, which corresponds approximately to the first 20 meters of the domain. Once the oxygen is depleted, nitrate-reducing bacteria take over and continue the degradation until the nitrate is depleted. The biodegradation then ceases. This behavior is reflected in the graphs, as is the inhibition of nitrate-reduction by the presence of oxygen (no nitrate reduction occurs until essentially all of the oxygen is depleted). The biomass distribution in space also reflects this behavior. For other examples like these, see the original reference.

One of the great advantages of numerical simulations is that they may be used to generate realizations of physical systems and perform "numerical experiments" by which certain scientific questions may be ad-

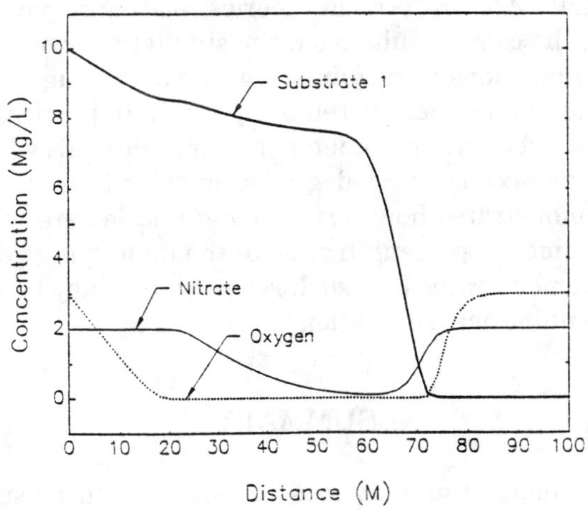

Figure 11a. Profiles of oxygen, nitrate, and a substrate [from Kindred and Celia (1989)].

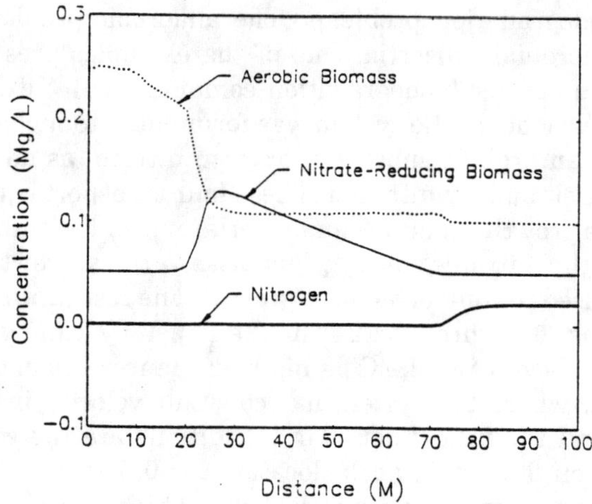

Figure 11b. Spatial distribution of biomass [from Kindred and Celia (1989)].

dressed. Macquarrie et al. (1990a, 1990b) used a specialized numerical approximation technique to develop a simulator for aerobic degradation in groundwater. This numerical method, referred to as the principal direction technique, uses approximations to the transport equation written with respect to a coordinate system that aligns with the groundwater flow field. The motivation for this is somewhat analogous to the motivation for characteristic methods, in that the importance of the velocity field is recognized when approximating contaminant transport. Simulations were carried out in realistic permeability fields, including small-scale random variability, to test certain theories relating to parameter definitions and length scale. Results from the simulations were used to demonstrate the limitations to certain stochastic averaging processes, especially when the plume size was relatively small. In particular, definition of average, or "effective," parameters becomes difficult when the plume size is shrinking because of biodegradation. This is an important study because it demonstrates how certain length scales are related (length scale of the plume versus length scale of the underlying geologic formation), and it demonstrates how such relationships may be studied using well-designed numerical simulations.

SUMMARY

The use of numerical simulators is now standard in subsurface hydrology. These simulators provide unique opportunities to study many aspects of contaminant transport problems, including biodegradation. Because the problem of contaminant transport and biodegradation is complicated, often involving a set of many coupled partial differential equations, an understanding of basic numerical principles is necessary to use these simulators effectively. This chapter has presented the basic concepts that underlie numerical approximations and numerical simulators. Methodologies have been outlined to the extent that methods in current use can be understood, at least in a general concept. Simple examples have been provided to demonstrate common numerical problems, and examples from the literature have been discussed briefly to provide a context for the current capabilities and applications.

REFERENCES

Bouloutas, E. T. and M. A. Celia. 1991. "An Improved Cubic Petrov-Galerkin Method for Simulation of Transient Advection-Diffusion Processes in Rectangularly Decomposable Domains," *Comp. Meth. Appl. Mech. Eng.*, 92:289–308.

Celia, M. A., J. S. Kindred and I. Herrera. 1989. "Contaminant Transport and Biodegradation: 1. A Numerical Model for Reactive Transport in Porous Media," *Water Resources Research*, 25(6):1141–1148.

Celia, M. A. and W. G. Gray. 1992. *Numerical Methods for Differential Equations*. Prentice Hall.

Chiang, C. Y., C. N. Dawson and M. F. Wheeler. 1991. "Modeling of In-Situ Biorestoration of Organic Compounds in Groundwater," *Transport in Porous Media*, 6:667–702.

Daus, A. D., E. O. Frind and E. A. Sudicky. 1985. "Comparative Error Analysis in Finite element Formulations of the Advection-Dispersion Equation," *Advances in Water Resources*, 8:86–95.

Dawson, C. N. and M. F. Wheeler. 1992. "Time-Splitting Methods for Advection-Diffusion-Reaction Equations Arising in Contaminant Transport," *Proc. ICIAM '91*, R. O'Malley, ed., Philadelphia: SIAM, pp. 71–82.

Gray, W. G. and G. F. Pinder. 1976. "Some Article on Fourier Analysis," *Water Resources Research*.

Kindred, J. S. and M. A. Celia. 1989. "Contaminant Transport and Biodegradation: 2. Conceptual Model and Test Simulations," *Water Resources Research*, 25(6):1149–1159.

Leithold, L. 1972. *The Calculus with Analytic Geometry*. New York: Harper and Row.

MacQuarrie, K. T. B., E. A. Sudicky and E. O. Frind. 1990a. "Simulation of Biodegradable Organic Contaminants in Groundwater 1. Numerical Formulation in Principal Directions," *Water Resources Research*, 26(2):207–222.

MacQuarrie, K. T. B., E. A. Sudicky and E. O. Frind. 1990b. "Simulation of Biodegradable Organic Contaminants in Groundwater 2. Plume Behavior in Uniform and Random Fields," *Water Resources Research*, 26(2):223–239.

Valocchi, A. J., J. E. Odencrantz and B. E. Rittmann. 1993. "Computational Studies of the Transport of Reactive Solutes: Interaction between Adsorption and Biotransformation," *Advances in Hydro-Science and -Engineering, Volume 1,* S. S. Y. Wang, ed., The University of Mississippi, pp. 1845–1852.

Wang, H. R. E. Ewing and M. A. Celia. 1995. "Eulerian-Lagrangian Localized Adjoint Methods for Reactive Transport with Biodegradation," *Numerical Methods for Partial Differential Equations,* 11:229–254.

Widdowson, M. A., F. J. Molz and L. D. Benefield. 1988. "A Numerical Transport Model for Oxygen- and Nitrate-Based Respiration Linked to Substrate and Nutrient Availablity in Porous Media," *Water Resources Research,* 24(9): 1553–1565.

Wood, B. D., C. N. Dawson, J. E. Szecsody and G. P. Streile. 1994. "Modeling Contaminant Transport and Biodegradation in a Layered Porous Media System," *Water Resources Research,* 30(6):1833–1845.

Modeling of Cometabolism for the in situ Biodegradation of Trichloroethylene and Other Chlorinated Aliphatic Hydrocarbons

LEWIS SEMPRINI
Department of Civil Engineering
Oregon State University
Corvallis, OR 97331, USA

ROGER L. ELY
Department of Civil Engineering
University of Idaho
Moscow, ID 83201, USA

MARGARET M. LANG
Department of Civil Engineering
Humbolt State University
Arcata, CA 95521, USA

INTRODUCTION

Remediation of contaminated aquifers continues to be a major focus of regulatory activity and an area of scientific and technical challenge for environmental professionals. Improper storage and waste management practices have left a legacy of contaminated soil and aquifers, threatening drinking water supplies in many areas (Fetter, 1993). Chlorinated ethenes, particularly trichloroethylene (TCE), are among the most frequently detected groundwater contaminants (Westrick et al., 1984; Barbash and Roberts, 1986). In addition to being a suspected carcinogen (Fan, 1988), TCE has been shown to undergo reductive dechlorination to vinyl chloride (VC), a known human carcinogen, under anaerobic conditions commonly observed in aquifers (Vogel and McCarty, 1985). Groundwater remediation for chlorinated ethene contamination is usually attempted with pump-and-treat systems that extract contaminated groundwater for treatment aboveground. Treatment is generally some combination of air stripping, adsorption onto activated carbon, thermal destruction, or biological degradation. In situ bioremediation is being suggested as an alternative or enhancement to pump-and-treat remediation (National Research Council, 1994).

CAHs in general do not serve as substrates for aerobic growth (McCarty and Semprini, 1993). Aerobic microorganisms transform chlorinated aliphatic hydrocarbons (CAHs) through cometabolic degradation, i.e., fortuitous degradation of a compound by nonspecific enzymes (oxygenases) that the microorganisms produce to metabolize their primary electron donor. Laboratory investigations (Wilson and Wilson, 1985; Fogel et al., 1986; Nelson et al., 1986, 1988; Little et al., 1988; Fox et al., 1990; Henry and Grbić-Galić, 1990; Folsom et al., 1990; Rasche et al., 1991; Hopkins et al., 1993; Wackett et al., 1989; Fan and Scow, 1993) have identified a number of aerobic microorganisms capable of transforming chlorinated ethenes aerobically. Aerobic cometabolism requires the addition of oxygen as the electron acceptor and an appropriate electron donor (growth substrate). Methane has been shown to be an effective electron donor for the cometabolic transformation of chlorinated ethenes in both the laboratory (Wilson and Wilson, 1985; Fogel et al., 1986; Little et al., 1988; Fox et al., 1990; Henry and Grbić-Galić, 1990) and in the field (Roberts et al., 1990; Semprini et al., 1990). Additional effective electron donors have also been identified, including phenol, toluene, ammonia, propane, and others (Nelson et al., 1986; 1988; Wackett et al., 1989; Folsom et al., 1990; Rasche et al., 1991; Hopkins et al., 1993; Fan and Scow, 1993; McCarty and Hopkins, 1995). Methane, phenol, and toluene have been demonstrated to effectively promote cometabolic transformation of chlorinated ethenes in pilot-scale field experiments, with phenol and toluene exhibiting more efficient TCE degradation than methane (Semprini et al., 1990; Hopkins et al., 1993; Hopkins and McCarty, 1995).

Modeling is an important component in the design of systems for in situ aerobic remediation of CAHs. This is particularly important for effective remediation by aerobic cometabolism due to the complexity of the cometabolic process itself and the difficulties associated with the mixing of substrates and contaminant(s) in the heterogeneous subsurface environment. This chapter will address some of the complexities associated with the cometabolic transformation process. Results of laboratory studies will be reviewed that provide a basis for kinetic models for cometabolism. Model simulations of controlled field studies of in situ biodegradation of CAHs will be presented. The chapter will conclude with model simulations for designing in situ bioremediation of TCE that optimizes for different nutrient delivery schemes for a two-well recirculation system.

KINETIC MODELS FOR AEROBIC COMETABOLISM OF CAHs

Most enzymes are highly specific in the reactions they catalyze and the substrates they accept; that is, they have quite narrow substrate ranges. However, some, principally monooxygenase and dioxygenase enzyme systems present in several aerobic microbial species, possess uncommonly broad substrate specificity, which allows them to catalyze reactions with many compounds other than their growth substrate. Cometabolic remediation approaches attempt to take advantage of the capabilities of such enzyme systems to degrade recalcitrant pollutants.

Dalton and Stirling (1982) defined cometabolism as "the transformation of a nongrowth substrate in the obligate presence of a growth substrate or another transformable compound." It is a key concept that, while microorganisms capable of cometabolism may transform various nongrowth substrates in addition to the growth substrate, the process is not sustainable in the absence of the growth substrate (or some other compound capable of supporting the microorganisms). Dalton and Stirling proposed that unsustainable "cometabolic" reactions carried out in the absence of the growth substrate be termed "biotransformations" to distinguish them from cometabolism. Because the term *cometabolism* is applied to both situations in most of the recent literature, we will do likewise; however, the concept of sustainability is important and should not be overlooked.

Bacterial Cometabolism of CAHs

Several types of bacteria are currently known to be capable of aerobic cometabolism of CAHs. For example, trichloroethylene (TCE) can be degraded cometabolically by bacteria grown on ammonia (Arciero et al., 1989; Rasche et al., 1991), cumene (Dabrock et al., 1992), 2,4-diphenoxyacetic acid (Harker and Kim, 1990), isoprene (Ewers et al., 1990), methane (Oldenhuis et al., 1991), phenol (Folsom et al., 1990), propane (Wackett et al., 1989), propylene (Ensign et al., 1992), and toluene (Wackett and Gibson, 1988). All of these use broad-specificity monooxygenase or dioxygenase enzymes to initiate oxidation of the growth substrate. In all cases, the same enzyme is responsible for initial oxidation of TCE as well (for example, see Figure 1). Nearly all C_1 and C_2 CAHs have been shown amenable to aerobic cometabolism by some or all of the bacterial types mentioned above. Only fully chlorinated compounds, such as carbon tetrachloride and tetrachloroethylene, appear resistant to aerobic cometabolism.

Substrate Interactions and Other Effects

Since the same enzyme catalyzes the initial oxidation of both the growth substrate and the nongrowth substrate, competition for the enzyme can occur, reducing the rate of growth substrate oxidation. Similarly, reduction in the rate of nongrowth substrate oxidation can occur when the growth substrate is present. The decrease in the reaction rate will depend on the relative affinity of each of the substrates for the enzyme. A nongrowth substrate with high enzyme affinity will significantly decrease the rate at which the growth substrate is oxidized, while a nongrowth substrate with low enzyme affinity will have much less effect. In addition to competitive inhibition, unproductive binding of a nongrowth substrate conceivably could cause uncompetitive inhibition or noncompetitive inhibition of growth substrate oxidation, as defined by Dixon and Webb (1979).

With several CAHs, another very significant effect is enzyme inactivation, also referred to as product toxicity, resulting from oxidation of the nongrowth substrate. While the exact nature and/or mechanism of enzyme inactivation may vary with different microbial species and various CAHs, it is generally thought to result from reaction intermediates and products that are highly reactive and that inactivate the oxygenase enzyme by covalently binding to it and changing its structure. In experiments with ammonia-oxidizing bacteria, *Nitrosomonas europaea*, Rasche

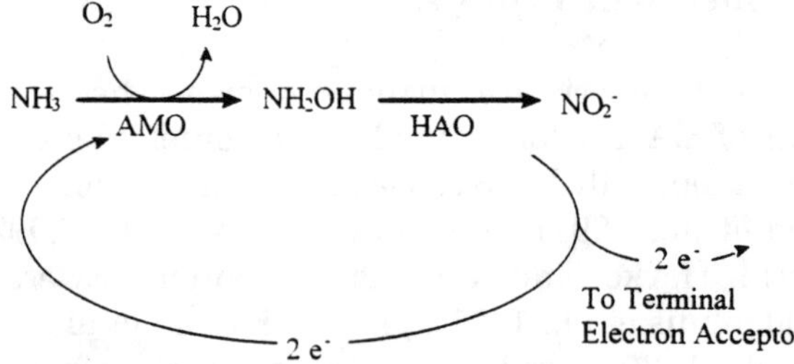

Figure 1. Reactions catalyzed by *Nitrosomonas europaea* in converting ammonia to nitrite. AMO is ammonia monooxygenase. HAO is hydroxyamine oxidoreductase. AMO catalyzes TCE oxidation as well. (Adapted from Keener and Arp, 1993.)

et al. (1991) found that 15 of the 17 C_1 and C_2 alkanes and alkenes tested were biodegradable and that 12 of them caused substantial enzyme inactivation, suggesting that the phenomenon may be fairly common. For TCE oxidation by methane-oxidizing bacteria, Little et al. (1988) and Fox et al. (1990) proposed that TCE was oxidized to TCE-epoxide, which then broke down spontaneously to dichloroacetic acid, glyoxylic acid, and formate, via acyl chloride intermediates. Evidence suggested that the inactivating compound was a diffusible, acyl chloride, such as glyoxyl chloride, formyl chloride, dichloroacetyl chloride, or dichlorocarbene. Green and Dalton (1989) showed loss of prosthetic iron during inactivation of methane monooxygenase (MMO) by TCE and 1,1-dichloroethene (1,1-DCE), suggesting a mechanism similar to heme alkylation known to occur in cytochrome P-450 enzymes (see, e.g., Ortiz de Montellano and Reich, 1986). With TCE oxidation by toluene dioxygenase from recombinant *E. coli* strains (originally from *Pseudomonas putida* F1), Li and Wackett (1992) found formate and glyoxylic acid as the major reaction products. They proposed that formate was formed from both carbon atoms and that significant TCE-epoxide was not produced, in contrast with monooxygenase-mediated reactions.

A third effect involves the supply of endogenous reductant, for example, NADH in methane-oxidizing bacteria. With bacteria using monooxygenase enzyme systems—such as methane-oxidizing or ammonia-oxidizing organisms—one oxygen atom from an O_2 molecule is inserted into the growth or nongrowth substrate to oxidize it while simultaneously reducing the O atom. This involves a two-electron transfer, for example, oxidizing C^{4-} in methane to C^{2-} in methanol while reducing the O^0 to O^{2-}. The other O atom in the O_2 molecule is reduced concurrently to form H_2O, requiring two electrons that must be provided from other reactions. In ammonia-oxidizing bacteria, these two electrons are provided, via an unidentified electron transport protein, from the subsequent oxidation of hydroxylamine to nitrite (Keener and Arp, 1993) (Figure 1). In methane-oxidizing bacteria, they are provided, via NADH, from subsequent oxidation of methanol to formaldehyde, formaldehyde to formate, and/or formate to CO_2 (Bedard and Knowles, 1989) (Figure 2). Therefore, in vivo regeneration of the reductant depends on oxidation of compounds that are products of growth substrate oxidation. If growth substrate oxidation slows or stops, reductant needed to support oxidation of growth substrate (and nongrowth substrate) will become less available and may become depleted, even though ample active enzyme may remain. Formate addition has been shown to increase oxidation of nongrowth substrates by methane-oxidizing bacteria and phenol-oxidizing bacteria in the absence of the growth substrate (see, e.g., Oldenhuis et al., 1989; Alvarez-Cohen

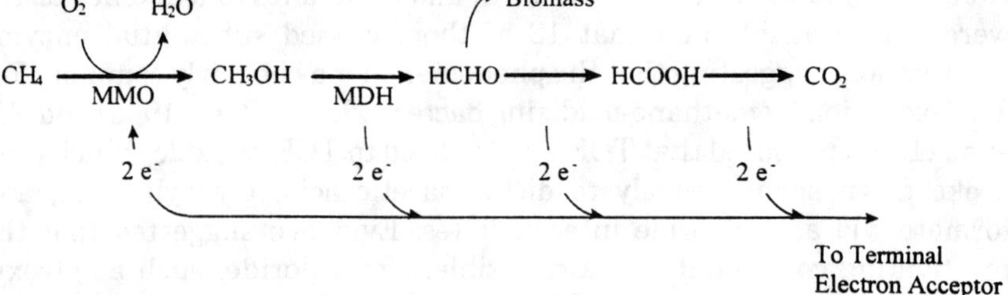

Figure 2. Reactions catalyzed by methane-oxidizing bacteria in converting methane to CO_2. MMO is methane monooxygenase. MDH is methanol dehydrogenase. (Adapted from Bedard and Knowles, 1989.)

and McCarty, 1991; Semprini et al., 1991; Hopkins et al., 1993; Chang and Alvarez-Cohen, 1995).

Another factor that can become important is natural enzyme turnover and the regulation of enzyme synthesis. For example, synthesis of toluene monooxygenase in *Pseudomonas cepacia* G4 does not occur in the absence of toluene or phenol (Shields et al., 1989), except in a recently isolated strain (Shields and Reagin, 1992), and soluble methane monooxygenase synthesis in *Methylococcus capsulatus* apparently requires the presence of methane under copper-limited conditions (Bedard and Knowles, 1989). Also, evidence suggests that ammonia monooxygenase synthesis in *N. europaea* appears to require NH_3 both as an inducer and as a nitrogen source for protein synthesis (Hyman and Arp, 1995). Since a loss of enzyme occurs in vitro due to natural enzyme turnover, activity of an induced enzyme will decrease with time in the absence of the growth substrate (or other acceptable inducer). This factor must be considered when interpreting laboratory results or planning in situ approaches in which the growth substrate is absent.

Kinetic Models

Modeling of competitive inhibition in cometabolism is well established (see, e.g., Folsom et al., 1990; Strand et al., 1990; Broholm et al., 1992, Chang and Alvarez-Cohen, 1995) and may be accomplished using a modified Michaelis-Menten enzyme kinetics expression. For example, the equations

$$\frac{1}{X}\frac{dS}{dt} = -\frac{kS}{K_m\left(1+\dfrac{I}{K_I}\right)+S} \tag{1}$$

and

$$\frac{1}{X}\frac{dI}{dt} = -\frac{k_I I}{K_I\left(1 + \dfrac{S}{K_m}\right) + I} \tag{2}$$

describe the specific rates of growth substrate and nongrowth substrate depletion, respectively, in a system in which competitive inhibition is the only effect requiring consideration. In these equations, S is the growth substrate concentration [mass/volume, or M/L^3], I is the nongrowth substrate concentration [M/L^3], X is the concentration of microorganisms engaged in cometabolism [M/L^3], k is the maximum specific utilization rate of the growth substrate [reciprocal of time, or T^{-1}], k_I is the maximum specific utilization rate of the nongrowth substrate [T^{-1}], K_m is the enzyme half-saturation constant of the growth substrate [M/L^3], and K_I is the enzyme half-saturation constant of the nongrowth substrate [M/L^3]. These equations assume that growth substrate and nongrowth concentrations are controlling the reaction rates, i.e., that all other reactants are present in excess and that enzyme concentration is constant. If only one substrate is present, for example, if $I = 0$, Equation (2) disappears and Equation (1) reverts to a Michaelis-Menten enzyme kinetics expression.

Modeling of enzyme inactivation is not as well established, perhaps partially because experimental and modeling approaches have not yet been standardized. The first model for enzyme inactivation resulting from product toxicity was published by Oldenhuis et al. (1991) who used formate-amended (20 mM) *Methylosinus trichosporium* OB3b whole cell cultures to degrade TCE and other CAHs in the absence of methane. They could not determine an inhibition constant (K_I) for TCE because of "deviation from Michaelis-Menten kinetics" but found loss of enzyme activity to be proportional to the amount of TCE degraded. Their enzyme inactivation model was based on TCE degradation activity:

$$V_{\max,t} = V_{\max,0} - p(S_a - S_t) \tag{3}$$

"where $V_{\max,0}$ and $V_{\max,t}$ represent the activity of the cells at time points 0 and t, respectively; S_a is the total amount of substrate added; S_t is the amount of substrate left at time t; and p is the inactivation constant, which means the amount of activity (nanomoles minute^{-1} milligram^{-1}) inactivated per amount of substrate converted (micromoles)." They estimated p to equal 0.48. However, because the experiments were conducted in the absence of methane, it is not clear how much activity loss could

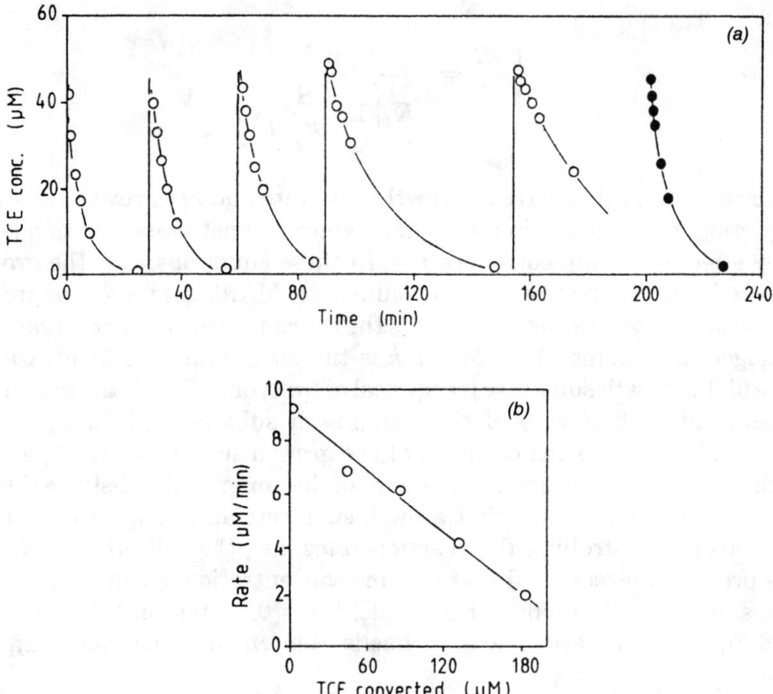

Figure 3. (a) Effect of repeated addition of 40 to 45 μM TCE on the degradation rate of TCE (O) by *M. trichosporium* OB3b cells (0.11 mg/ml^{-1}). As a control, degradation was also measured with cells to which TCE was added for the first time after 200 min (●). (b) Degradation rate versus amount of TCE converted. [Reprinted from Oldenhuis et al. (1991) with permission from the American Society for Microbiology.]

have resulted from enzyme turnover. The paper describes experiments in which a culture was dosed repeatedly to approximately the same initial TCE concentration and the maximum TCE depletion rate, i.e., the rate immediately after injecting TCE, was estimated for each TCE injection (see Figure 3). The rates thus obtained were then plotted against the cumulative amount of TCE transformed to estimate activity loss as a function of TCE transformed. Also, from the units given for activity (nmol/min-mg-mol), it is clear that the authors were referring to specific activity rather than to total activity. Using terms as defined for Equation (2), the model may be restated as

$$\left(\frac{1}{X}\frac{dI}{dt}\right)_{n,0} = \left(\frac{1}{X}\frac{dI}{dt}\right)_0 - p(I_a - I_{n,0}) \tag{4}$$

where n denotes the nth injection of TCE and I_a is the cumulative amount of TCE added. Important features of this model are that (1) it considers decreases in culture specific activity (which, according to traditional enzyme kinetics modeling approaches, would be due to decreases in active enzyme levels), and (2) it shows that decreases in culture specific activity are proportional to the amount of TCE transformed.

A second model, employing a different conceptual approach, was presented by Alvarez-Cohen and McCarty (1991b). Using a mixed, methane-oxidizing culture, previously estimated to contain 20% methane-oxidizing cells (Alvarez-Cohen and McCarty, 1991a), they examined transformation of TCE and chloroform, separately and together, in the absence of methane, both with and without amended formate, by repeatedly injecting the CAH to approximately the same initial concentration until culture activity declined to near zero. The total amount of CAH transformed in this process was divided by the total cell concentration, X, to obtain the ratio of total CAH transformed per unit of biomass, which was termed the transformation capacity. The effect was modeled essentially as a reverse Monod growth expression where, instead of growth, a decrease in "active biomass" was addressed. The Monod expression was presented as

$$-\frac{dS_i}{dt} = \frac{k_i X S_i}{K_{si} + S_i} \tag{5}$$

"where S_i is the concentration of cometabolized contaminant i at time t (milligrams per liter), X is the active microbial concentration at time t (milligrams per liter), k_i is the maximum rate of contaminant i transformation (milligrams of i per milligram [dry weight] of cells per day), and K_{si} is the half velocity constant for i (milligrams per liter)" (Alvarez-Cohen and McCarty, 1991b). Instead of modeling inactivation from product toxicity as a decrease in specific activity, it was modeled as a loss of active biomass:

$$X = X_0 - \frac{1}{T_{ci}}(S_{0i} - S_i) \tag{6}$$

"where X_0 is the initial active microbial concentration (milligrams per liter), S_{0i} is the initial concentration of cometabolized contaminant i (milligrams per liter), and T_{ci} is the transformation capacity for cometabolized contaminant i (milligrams of i per milligram of cells)." [Note the similarity of form between Equation (6) and Equation (3).] Equations (5) and (6) were then combined to yield the expression

$$-\frac{dS_i}{dt} = \frac{k_i S_i}{K_{si} + S_i}\left[X_0 - \frac{1}{T_{ci}}(S_{0i} - S_i)\right] \tag{7}$$

Their results indicate that, in the absence of methane, about 40% of culture activity was lost in 4 hours without TCE versus about 67% activity loss in 4.5 hours with TCE present (see Table 1) (Alvarez-Cohen and McCarty, 1991a). Activity loss in the absence of TCE was shown to be greater when the culture was aerated than when it was not exposed to oxygen. They speculated that cell decay due to endogenous respiration or predation could have been higher in the presence of oxygen. Because the observed activity loss was about an order of magnitude greater than expected typical decay values, they further speculated that perhaps active oxygen species generated by MMO in the absence of a reduced substrate (e.g., CH_4 or TCE) could react with MMO, thereby inactivating the enzyme. They also noted much greater TCE oxidation rates and apparent transformation capacities in formate amended cultures, suggesting that the cells were reductant limited in the absence of methane.

Criddle (1993) extended the Alvarez-Cohen model to include "loss of activity in resting cells due to endogenous decay; loss of activity in resting cells caused by transformation of nongrowth substrate; an increase in the rate and extent of cometabolism in the presence of growth substrates; an increase in the rate and extent of cometabolism in the presence of energy

Table 1. *Methane-oxidizing activity losses in cells exposed to TCE and in cells not exposed to TCE (after Alvarez-Cohen and McCarty, 1991a).*

Experimental Condition	Exposure Time (h)	TCE Transformed (μg)	Formate Addition (mM)	Methane Consumption (mg CH_4/d-mg cells)	Percent Activity Loss
Fresh cells	0	0	0	0.48	
	0	0	0	0.54	
Shaken with TCE	5.0	2110	0	0.06	88
	4.5	1700	0	0.17	67
	4.7	2900	20	0.04	90
	6.6	1600	40	0.04	92
Shaken with air	4.0	0	0	0.30	41
	4.5	0	0	0.31	39
	4.3	0	20	0.33	35
	6.5	0	20	0.32	37
Unshaken	4.0	0	0	0.45	14
	6.5	0	0	0.45	14
	7.0	0	20	0.42	19

substrates; the potential for competitive inhibition; reduced cell yield in the presence of nongrowth substrates; and decreased specific growth rates in the presence of nongrowth substrate." As an extension of the Alvarez-Cohen model, it is also based on the active biomass concept and also assumes constant culture specific activity. However, by incorporating growth substrate/nongrowth substrate competition for enzyme; cellular decay (which presumably would include natural enzyme turnover); and cometabolism-associated, deleterious effects of cell growth, Criddle's model addresses more the complexities of cometabolism.

The models presented by Oldenhuis et al. (1991), Alvarez-Cohen and McCarty (1991b), and Criddle (1993) possess significant fundamental differences. In Monod kinetics modeling, total culture activity is calculated by multiplying a culture's specific activity, i.e., the activity per unit of biomass, by the total amount of active biomass. For convenience, it is generally assumed that all of the biomass is active and X is stated as milligrams of total suspended solids (dry weight) per liter, milligrams of volatile suspended solids (dry weight) per liter, or milligrams of cell protein per liter. Therefore, we typically model culture activity as

$$\text{Total Activity} = \text{Specific Activity} \cdot \text{Total Biomass} \qquad (8)$$

which is merely a verbal restatement of Equation (5). However, specific activity—in the limiting case, the activity per cell—cannot be measured directly. Instead, culture total activity and total biomass are measured, and specific activity is calculated. Changes in total activity may then be accounted for as changes in active biomass, X, assuming de facto that the specific activity is constant, or as changes in specific activity, with X being constant as measured by one of the techniques mentioned. However, because most methods of measuring X do not evaluate activity, active biomass is generally difficult to assess. Apart from this practical issue, a conceptual issue arises as well. In the Oldenhuis model, the activity of a single cell (or unit of cells) may decrease without complete inactivation of the cell. In the Alvarez-Cohen model, the activity of a single cell cannot change. The cell (or unit of cells) is either active (presumably alive and fully functioning) or inactive (presumably dead).

To examine the Oldenhuis model further, consider that specific activity may be represented as

$$\text{Specific Activity} = \frac{kS}{K_m + S} \qquad (9)$$

Since Oldenhuis et al. (1991) measured activity repeatedly at approximately the same substrate (TCE) concentration, S may be considered a constant in their model. K_m is a property of the particular enzyme and substrate and is constant also. Therefore, changes in culture-specific activity observed in the experiments of Oldenhuis et al. can only be due to changes in k. In Michaelis-Menten or Briggs-Haldane enzyme kinetics derivations,

$$k = k_2 \frac{E_T}{X} \qquad (\text{or } V_{\max} = kX = k_2 E_T) \tag{10}$$

where k_2 is the first-order reaction rate constant for the conversion of enzyme-substrate complex (ES) to product (P) and free enzyme (E), and the ratio E_T/X is the total amount of active enzyme per unit of biomass. Because k_2 is a constant, specific activity can change only if the amount of active enzyme per unit of biomass changes. Therefore, changes in specific activity, as presented by Oldenhuis et al., reflect changes in the amount of active enzyme per unit of cells.

In that both Oldenhuis et al. (1991) and Alvarez-Cohen and McCarty (1991b) conducted their experiments in the absence of growth substrate, neither considered or modeled implications of having a growth substrate present, such as competition between growth substrate and nongrowth substrate or the potential for bacteria to respond to enzyme inactivation by synthesizing an additional enzyme. In addition, neither showed the relation of their model to basic enzyme kinetics by presenting a derivation of the model from fundamental principles.

A model presented by Ely et al. (1995a) attempted to incorporate these phenomena. For a case of competitive inhibition, the Ely model is based on two enzyme-substrate reaction equations:

$$E + S \leftrightarrow ES \rightarrow E + P_1$$

$$(\text{plus some amount of new enzyme, } E_{\text{new}}) \tag{11}$$

and

$$E + I \leftrightarrow EI \rightarrow E\left(1 - \frac{E'}{E}\right) + E' + P_2 \tag{12}$$

where E is free enzyme, S is the growth substrate, P_1 is the product of growth substrate transformation, I is the nongrowth substrate, P_2 is the product of nongrowth substrate transformation, E_{new} is newly synthesized enzyme, and E' is inactivated enzyme. Equation (12) indicates that some quantity of enzyme is inactivated in transforming I to P_2, while Equation (11) indicates that some quantity of new enzyme, related to the amount of growth substrate transformed, is synthesized by the bacteria in order to recover lost activity. Ely et al. stated this as

$$E'_{new} = g(P_1) \tag{13}$$

and

$$E' = h(P_2) \tag{14}$$

where $g(P_1)$ is some function g of P_1, i.e., the amount of growth substrate oxidized to product, and $h(P_2)$ is some function h of P_2, i.e., the amount of nongrowth substrate oxidized to product. Assuming a quasi-steady-state condition (i.e., that $d[ES]/dt = 0$ and $d[EI]/dt = 0$) and using an enzyme mass balance ($E_0 - E' + E_{new} = E + ES + EI$), Ely et al. derived equations for the rates of growth substrate and nongrowth substrate transformation:

$$\frac{1}{X}\frac{dS}{dt} = -\frac{(k - k_{inact}P_2 + k_{rec}P_1)S}{K_m\left(1 + \dfrac{I}{K_I}\right) + S} \tag{15}$$

and

$$\frac{1}{X}\frac{dI}{dt} = -\frac{\dfrac{k_I}{k}(k - k_{inact}P_2 + k_{rec}P_1)I}{K_I\left(1 + \dfrac{S}{K_m}\right) + I} \tag{16}$$

where E_0 is the total amount of enzyme in the absence of enzyme inactivation and recovery, k_{inact} is an inactivation coefficient relating specific activity loss to the amount of nongrowth substrate transformed $[L^3M^{-1}T^{-1}]$, k_{rec} is a recovery coefficient relating recovery of activity to the amount of growth substrate transformed $[L^3M^{-1}T^{-1}]$, and k_I is the maximum specific nongrowth substrate transformation rate $[NM^{-1}T^{-1}$, where N is moles]. Other terms are as defined earlier. This model is similar to the Oldenhuis model in that it is based on changes in culture-specific activity rather than on changes in active biomass. However, it differs from both the Oldenhuis and the Alvarez-Cohen models in that it combines the effects of competitive inhibition, enzyme inactivation (and, presumably, normal enzyme turnover), and recovery of activity. In addition, Ely et al. (1995a) showed that growth substrate and nongrowth substrate transformation rates depend on each other and that they may be related by a dimensionless constant that they called the first-order reaction rates ratio:

$$\frac{dS}{dt} = \mathcal{E}\,\frac{S}{I}\,\frac{dI}{dt} \quad \text{or} \quad \frac{dS}{dI} = \mathcal{E}\,\frac{S}{I} \tag{17}$$

where $\mathcal{E}$ is the first-order reaction rates ratio [equal to $(k/K_m)/(k_I/K_I)$].

To test the model, Ely et al. (1995b) used pure cultures of ammonia-oxidizing bacteria, *Nitrosomonas europaea*, to degrade TCE cometabolically in the presence of ammonia. Results of a typical experiment are shown in Figure 4. After establishing a quasi-steady-state condition in a small bioreactor, the system was perturbed by injecting a pulse of TCE, and system response was monitored and modeled in real time. In results from 13 experiments, they estimated k at about 830 nmol NO_2^-/min·mg protein, k_I at about 11.0 nmol TCE/min·mg protein, K_I at about 10.7 μM TCE, k_{inact} at about 0.069 mmol NO_2^-·L/mmol TCE·min·mg protein, and k_{rec} at about 0.00026 mmol NO_2^-·L/mmol NO_2·min·mg protein. Their results indicated that, for the bacteria used, the maximum TCE transformation rate was about 80 times slower than that of ammonia; TCE had about four times greater affinity than ammonia for AMO (based on a literature K_m value of 40 μM for ammonia), $\mathcal{E}$ was equal to about 20, and transformation of 1 mmol of TCE caused ammonia-oxidizing activity to decrease 0.069 mmol/min (per mg/l of cell protein) whereas oxidation of 1 mmol of NH_3 to NO_2^- supported the recovery of only 0.00026 mmol/min (per mg/l of cell protein) of ammonia-oxidizing activity. This suggests that

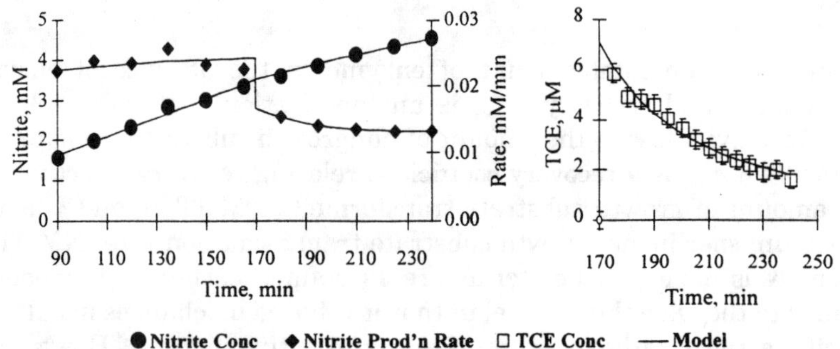

Figure 4. Sample TCE degradation experiment results and example model fits (TCE was injected at time = 170 min) (after Ely et al., 1995b).

oxidation of about 265 mmol of NH_3 would be required in order for the bacteria to recoup the activity lost in oxidizing 1 mmol of TCE. Based on Equation (17) and a presumed NH_3 concentration equal to the K_m (40 μM), this indicates that TCE at a concentration greater than about 3 μM would cause enzyme inactivation to occur at a rate too great for recovery processes to keep pace; the cells would eventually become completely inactivated. Conversely, if ammonia concentration was greater than 40 μM and/or TCE concentration was less than 3 μM, cellular recovery processes would be expected to keep pace with enzyme inactivation effects, rendering the process sustainable for an indefinite time period.

MODEL SIMULATIONS OF THE RESULTS FROM THE MOFFETT FIELD EXPERIMENTS

Pilot scale studies of in situ cometabolism of TCE and other CAHs has been evaluated at the Moffett Naval Air Station for methane-utilizing bacteria (Roberts et al., 1990; Semprini et al., 1991, 1993) and phenol-utilizing bacteria (Hopkins et al., 1993; Hopkins and McCarty, 1995). These studies have illustrated in situ many of the complexities associated with cometabolism observed in the laboratory studies. For example, in situ CAH removal rates were compound-specific, which suggests different rates of cometabolism. The ability to cometabolically degrade the contaminants was strongly associated with the utilization of the growth substrates. Termination of growth substrate addition resulted in the loss of cometabolic activity within a few days. Competitive inhibition was also observed between the growth substrate and the CAH. The addition of a noncompeting energy source such as formate was observed to temporarily enhance CAH cometabolism, by eliminating competitive inhibition and by providing a source of reducing power for the CAH transformation. However, the addition of the cometabolic growth substrate was required to sustain CAH cometabolism. The CAH concentration was also observed to be an important factor. The study of Hopkins et al. (1993) with phenol-utilizing bacteria showed the extent of removal of TCE decreased at concentrations above 500 mg/l, suggesting a deviation from first-order kinetics due to product toxicity, the concentration being close to the half-saturation value, or insufficient reducing power.

Abiotic processes were also found to be important in these studies. Rate-limited sorption and desorption were found to be a dominant process affecting the transport of CAHs through the test zone, as well as limiting the compound's availability for cometabolism (Harmon et al., 1992). For example, TCE tended to sorb more strongly and desorb more slowly than

DCE, while VC was the least strongly sorbed CAH and desorbed the fastest.

Modeling of these pilot scale studies has been performed by Semprini and McCarty (1991, 1992), Semprini et al. (1994), and Semprini (1995). The nonsteady-state model that was developed has been described in detail in those works and will not be repeated here. The 1-D numerical transport code includes terms for advection, dispersion, and rate-limited sorption in porous media. Nonsteady-state processes included microbial growth, electron donor and electron acceptor utilization, and the cometabolic transformation of the contaminants. Results from tracer studies indicated that transport from the injection well to the monitoring wells could be reasonably represented by 1-D transport. Rate-limited sorption and desorption were represented by a first-order kinetic model for the transfer of sorbed contaminants between the aquifer solids and the aqueous phase. A simple first-order decay model was used to describe microbial deactivation of the population towards the cometabolic transformation during periods when the microbial population in a given location was not growing.

The model assumes that, if a biofilm develops, it is fully penetrating based on methods described by Suidan et al. (1987). This simplifies the modeling approach considerably. Recent model development for cometabolism in methane-utilizing biofilms is described by Anderson and McCarty (1994) for completely mixed systems. Their model illustrates the complexity of models for these processes, which is probably not warranted for conditions of low substrate addition used in in situ aerobic cometabolism.

In the model of Semprini and McCarty (1991, 1992), microbial growth and electron acceptor and electron donor utilization are modeled using dual-monod kinetics. Thus, the absence of the electron donor and the electron acceptor could limit microbial growth and utilization of the electron donor and electron acceptor. Aerobic cometabolism of the CAH also required the presence of oxygen, which was also modeled using dual-monod kinetics, with terms for competitive inhibition due to the presence of the growth substrate, as illustrated in the first two equations of Table 2. The original model did not include terms for microbial deactivation using the transformation capacity model or models for enzyme deactivation discussed previously. Deactivation occurred in the absence of microbial growth using a first-order decay model for enzyme deactivation.

Dual-monod kinetics were also important in describing the process of pulsed addition of growth substrate and electron acceptor that was used in the field studies. Alternate pulsed injection of groundwater containing

Table 2. Kinetic equations used in the modified
cometabolic transformation model.

$$\frac{dC_1}{dt} = -X^* k_{C1} \left(\frac{C_1}{K_{I1} + C_1 + S \dfrac{K_{I1}}{K_m} + C_2 \dfrac{K_{I1}}{K_{I2}}} \right) \left(\frac{A}{K_{SA} + A} \right)$$

$$\frac{dC_2}{dt} = -X^* k_{C2} \left(\frac{C_2}{K_{I2} + C_2 + S \dfrac{K_{I2}}{K_m} + C_1 \dfrac{K_{I2}}{K_{I1}}} \right) \left(\frac{A}{K_{SA} + A} \right)$$

$$\frac{dS}{dt} = -Xk \left(\frac{S}{K_s + S + \dfrac{S^2}{K_{HAL}} + C_1 \dfrac{K_m}{K_{I1}} + C_2 \dfrac{K_m}{K_{I2}}} \right) \left(\frac{A}{K_{SA} + A} \right)$$

$$\frac{dX}{dt} = Y \frac{dS}{dt} \left(\frac{A}{K_{SA} + A} \right) - bX \left(\frac{A}{K_{SA} + A} \right) - T_{C1} \frac{dC_1}{dt} - T_{C2} \frac{dC_2}{dt}$$

where

C_1 and C_2 = concentrations of CAH 1 and 2
S = concentration of the growth substrate concentration
A = concentration of the electron acceptor
X = concentration of the microbial population
X^* = concentration of the microbial population active towards cometabolism
X^* = X when microbes are in the growth condition ($dx/dt \geq 0$)
X^* = decays via a first-order process when $dx/dt < 0$.

the electron donor (methane or phenol) and the electron acceptor (oxygen) was performed throughout the field studies. The numerical model captured the response to pulsing by including the appropriate boundary conditions and using a numerical finite difference method described by Leonard (1979) that eliminates numerical dispersion. The set of resulting coupled nonlinear equations was solved explicitly using a fourth-order Runge-Kutta method. Details of this method are described by Semprini and McCarty (1991).

Modeling of Methanotrophic Studies

A modeling comparison with the results from the Moffett Field studies for the methane-utilizing studies are discussed by Semprini and McCarty

(1991, 1992). Figure 5 illustrates one of the key results of that study: the competitive inhibition by methane of cometabolism of VC and *trans*-1,2-dichloroethylene (t-DCE). The initial decrease in VC and t-DCE concentrations at the S1 monitoring well results from the addition of methane in short pulses and the rapid stimulation of methane-utilizing microorganisms. The addition of methane with longer pulse cycles at 30 h resulted in the pulsed methane breakthrough and concentration oscillations of VC and t-DCE (which were not pulsed) coincident with the methane pulse, with no time lag or inhibition. While the field and simulation results strongly supported the competitive inhibition of methane on the transformation of the CAHs, rate-limited sorption and desorption was also an important process contributing to the observed response. When equilibrium sorption was assumed, the fluctuations in the aqueous phase concentration became greatly attenuated, since both the aqueous and sorbed phase concentrations were degraded simultaneously, and thus, a greater total contaminant mass was involved. When sorption and desorption was modeled as a rate-limited process, the aqueous phase concentration fluctuated greatly, while the sorbed phase did not. The importance here is that the response due to competitive inhibition was coupled to rate-limited sorption. Thus, one could greatly overestimate the extent of competitive inhibition if equilibrium sorption was assumed compared to rate-limited sorption. These results also illustrate how sorption can affect the biological response and the complexity of the combined biological and physical processes.

Rate coefficients derived from model fits to the field observation indicated that the rates of transformation of VC and t-DCE were on the order of those of methane. While VC and t-DCE had similar rate coefficients, VC was removed more quickly in the field, since it was less strongly sorbed to the aquifer solids, thus less total mass was degraded. The rate coefficients for c-DCE and TCE transformation, however, were factors of 20 and 100 lower than VC and t-DCE. These rates should be considered model dependent and can vary greatly from those reported in laboratory studies.

The second-order rate coefficient (k/K_s) of TCE that was determined in the simulation of the field results is in the range of that reported by Henry and Grbić-Galić (1989) for a culture of *Methylomonas* MM2, which was enriched from the Moffett subsurface. This MM2 culture likely expressed particulate MMO, since it was grown in the presence of copper. Cultures expressing particulate MMO have been shown to have slower TCE transformation rates than methanotrophs that express soluble MMO (Janssen et al., 1991), but faster rate of s-DCE transformation. The model-derived rates tend to support the hypothesis that, in the Moffett

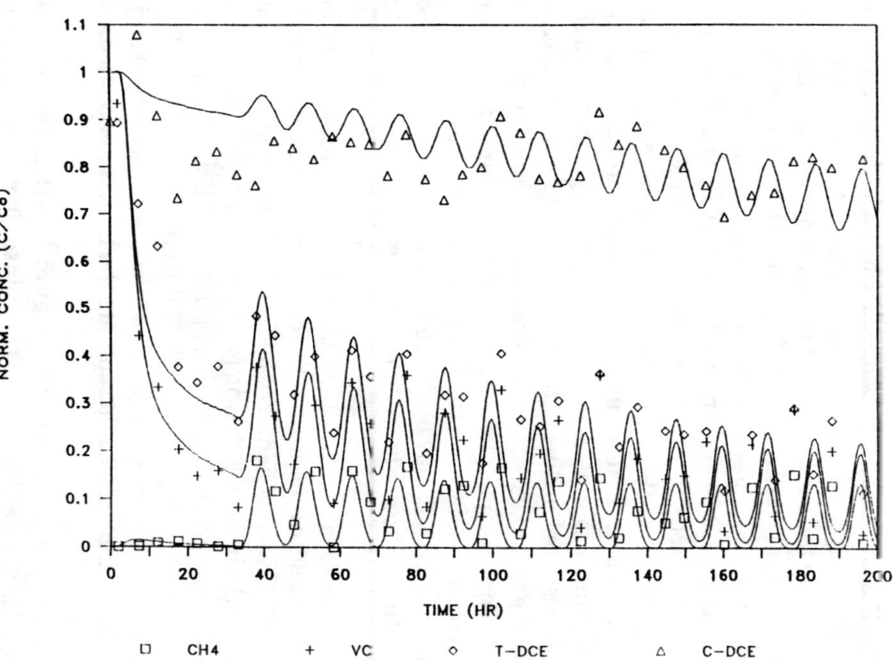

Figure 5. Field observations and model simulations of methane, VC, *trans*-DCE, and *cis*-DCE responses at well S1 (1 m from the injection well) in the Moffett Field pilot scale studies of in situ cometabolism of CAHs using methane-utilizing microorganisms. (Reprinted from Semprini and McCarty, 1992 with permission, copyright 1992, Ground Water Publishing Co.)

study, the methane utilizers that grew expressed particulate MMO, presumably due to the presence of copper in the subsurface, were likely responsible for CAH transformation.

Another important question is how quickly a cometabolic process deactivates towards CAH transformation in the absence of the growth substrate. Figure 6 provides an example from a pilot test where methane addition into the test zone was stopped while the injection of t-DCE, c-DCE, and TCE was continued. The simulation corresponds to a period of testing following the simulations of methane and DO uptake shown in Figure 11 of Semprini and McCarty (1991). Model parameters for DO and methane utilization are those reported by Semprini and McCarty (1991), and those for the transformation of c-DCE, t-DCE, and TCE are those used in the simulations in Figure 5 shown previously and reported by Semprini and McCarty (1992). For these simulations, zero time actually represents a period several hundred hours into the test after pseudo-steady-state transformation was achieved. Although these simulations and data are presented to illustrate the effects of stopping and restarting the addition of primary substrate, several other interesting observations are apparent. For example, the model tends to overpredict the degree of transformation of c-DCE and t-DCE observed in the field test. The overprediction for c-DCE may be due to the uncertainty in the fitted rate parameter obtained from Figure 5. A better match may be expected for the t-DCE data based on the good fit shown in Figure 5. The model simulations helped confirm a problem that existed in the field data shown. The field data actually represented the combined responses of t-DCE and 1,1-DCE. 1,1-DCE was present as a background contaminant in the field and coeluted with the t-DCE during gas chromatographic analysis. Thus, the normalized concentrations shown of 0.3 to 0.4 (for the time period of 0 to 100 h) most likely represent 1,1-DCE that was not likely degraded by the methane-utilizing bacteria. This coelution was later confirmed after a new GC column that could separate the DCE isomers was installed at the field site.

The main process to consider in Figure 6 is the response to the termination of methane (300 h) and the reintroduction of methane (500 h). Both the field results and the model simulations show a rapid loss of activity toward the CAH transformations upon the termination of methane addition. Upon reintroduction of methane, CAH degradation was rapidly regained. This response was reproduced reasonably well in the model simulations using the empirical approach taken in the model of Semprini and McCarty (1992) of a first-order decay in CAH cometabolism activity when the biomass was not growing. This loss of activity can represent loss of energy for the cometabolic transformation or loss of

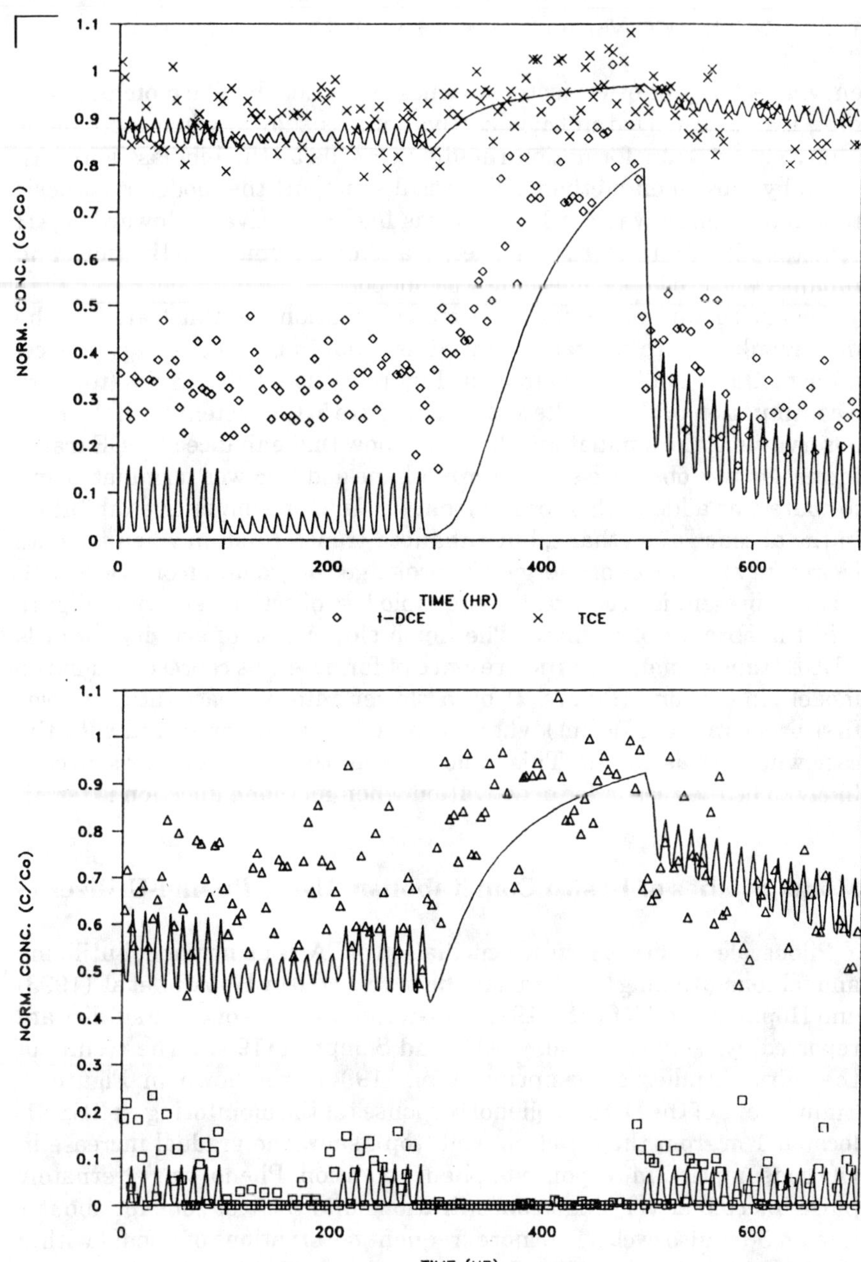

Figure 6. Field observations and model simulations of methane and c-DCE (bottom) t-DCE, and TCE (top) responses at well S1 due to the termination of methane addition (280 h) and the restarting of methane addition (500 h) in the Moffett Field pilot scale studies of in situ cometabolism of CAHs using methane-utilizing microorganisms.

enzyme activity or other factors discussed previously. The biomass, however, is not destroyed in this deactivation. As shown, when methane is reintroduced, transformation rapidly proceeds as the biomass is reactivated by growth on methane. The results support the model hypothesis that the biomass was not lost but was likely deactivated towards CAH cometabolism, due to loss of energy and/or enzyme deactivation, and rapidly recovered once methane was added.

Energy requirements for CAH transformation are illustrated in the field results where formate is added as a noncompeting energy source, after methane addition is stopped (Figure 7). The field results (top) are a continuation of the results shown in Figure 5. Consistent with the field observations, the simulations (bottom) show that enhanced t-DCE transformation was observed soon after methane addition was terminated and formate was added. This results from the lack of competitive inhibition in the absence of methane, but enhanced transformation when formate is present as a source of energy. The prolonged degradation of t-DCE with formate present is in contrast to the rapid loss of activity shown in Figure 6 in the absence of formate. The much slower loss of activity towards t-DCE transformation in the presence of formate was represented in the model simulations (Figure 7) by a slower rate of deactivation (lower first-order rate coefficient) when formate is present compared with the case when it was absent. This indicates that the loss of energy source is likely to be the reason for deactivation when methane injection is terminated.

Model Studies of in situ Cometabolism Using Phenol-Utilizers

Pilot scale studies of in situ remediation of CAHs using phenol-utilizing and toluene-utilizing bacteria have been reported by Hopkins et al. (1993) and Hopkins and McCarty (1995). Model simulations of these studies are reported by Semprini et al. (1994) and Semprini (1995). The results of the initial studies of Semprini et al. (1994) are shown in Figure 8. Simulations of the DO and phenol responses at the monitoring well SSE1, located 1 m from the injection well (top), show the gradual increase in DO consumption in response to phenol addition. Phenol was alternately pulse injected at very high concentrations into the test zone for about 1 h in an 8-h pulse cycle. The more frequent observations of phenol within the first 10 days, with most observations below the detection limit after that, suggest biostimulation of the phenol-utilizing population, as indicated in the model simulations. The model also simulates well the uptake and oscillations of DO due to the pulsed addition of phenol. Simulations supported the field observations, which showed that, upon increasing the

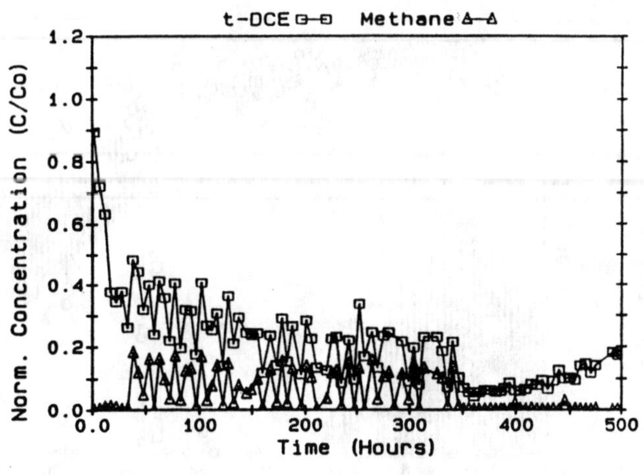

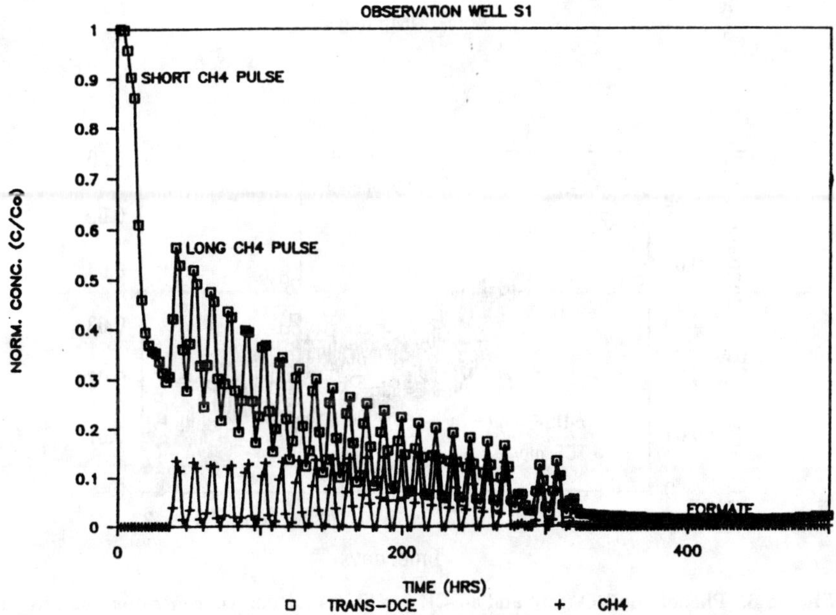

Figure 7. Field observations (top) and model simulations (bottom) of the response of *trans*-DCE to the substitution of formate for methane at 350 h in the Moffett Field pilot scale studies of in situ cometabolism of CAHs using methane-utilizing microorganisms.

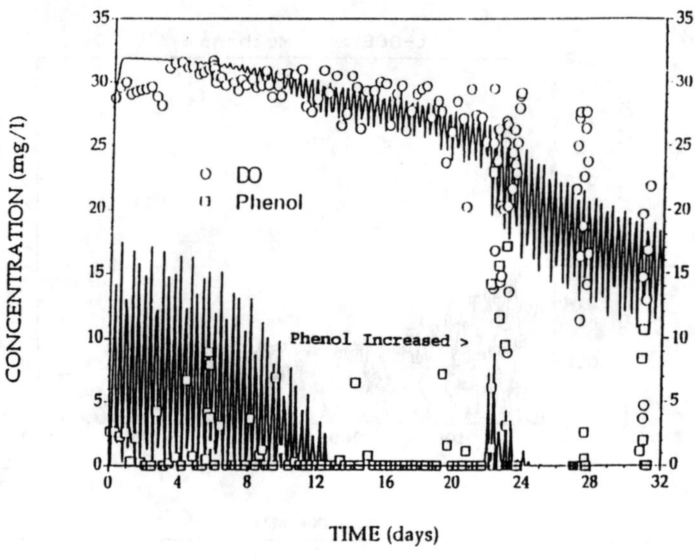

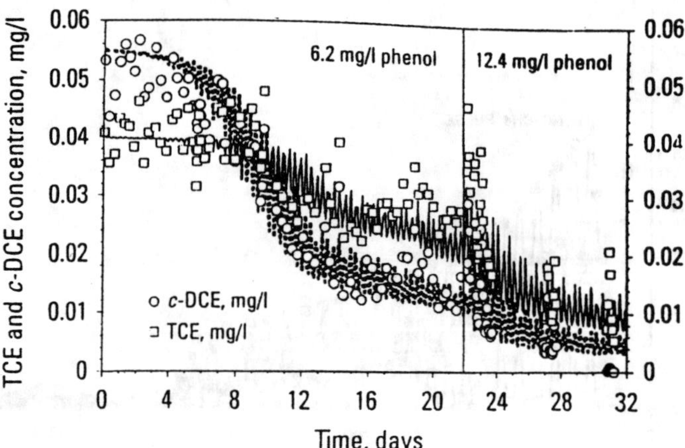

Figure 8. Phenol and DO (top) and c-DCE and TCE (bottom) concentration histories at the S1 observation well and model simulations in the Moffett Field pilot-scale studies of in situ cometabolism of CAHs using phenol-utilizing microorganisms. (Reprinted with permission from Semprini et al., 1994, copyright Lewis Publishers, an imprint of CRC Press, Boca Raton, Florida.)

112

phenol injection concentration, the DO concentration both decreased and oscillated over a greater concentration range.

The simulations of the c-DCE and TCE responses (Figure 8, bottom) are consistent with the field observations, showing decreases in the CAH concentration resulting from the biostimulation of the phenol-utilizing population. Enhanced transformation occurred after the phenol concentration was increased. The model simulations also indicated that competitive inhibition of c-DCE and TCE by phenol occurred, consistent with the field observations.

Comparison of model simulations between the methane-utilizing and the phenol-utilizing microorganisms was summarized by Semprini et al. (1994). The modeling analysis indicated that phenol was a better substrate for in situ remediation of TCE and c-DCE at Moffett Field for several reasons. Phenol addition resulted in a greater simulated biomass. This resulted both from the ability to add more phenol than methane, since stoichiometrically, less DO is required, and phenol-oxidizers have a higher yield coefficient than methane-oxidizers. The rate coefficients for the cometabolic transformation of TCE and c-DCE were also a factor of two to three greater for phenol-utilizers compared to the methane-utilizers.

The simulations presented thus far for both the methane and phenol tests used the model of Semprini and McCarty (1992), which did not include the concept of transformation capacity. For the simulations presented, the concentrations of the CAHs were low (below 250 μg/l) so that transformation product toxicity was not a dominant process. Figure 9 shows the results of later field tests with the phenol-utilizers, where the TCE concentration injected into the test zone was progressively increased from 62.5 to 1000 μg/l. The results indicate that similar extents of transformation occurred up to 500 μg/l. This indicated that the transformation was rate-limited and first-order with respect to concentration. Upon increasing the concentrations, the percent removals decreased. The results suggest transformation product toxicity was one potential reason for the decreased removal. Other factors suggested by Hopkins et al. (1993) include the TCE concentration being closer to the K_S value or insufficient reducing power (energy) to carry out the transformation.

Modifications have been made to the model described by Semprini and McCarty (1991, 1992) to include important kinetic processes that have been observed in both laboratory studies and in the field studies. The forms of the kinetic equations incorporated into the model are provided in Table 2. The model has been adapted for the simultaneous transformation of two contaminants that can inhibit the transformation of each other, as well as the utilization of the primary substrate. Cometabolic kinetics have been adapted to include transformation product toxicity in the kinetic expression for microbial growth and decay (Table 2, fourth

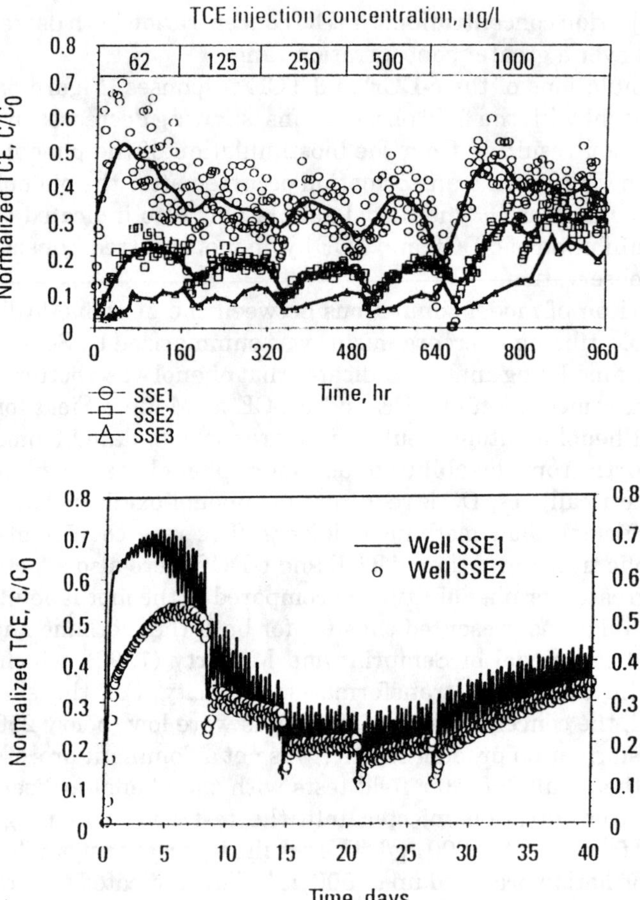

Figure 9. Normalized TCE concentrations at monitoring locations during phenol injection at a time average concentration of 12.5 mg/l with TCE concentration raised in steps from 62.5 to 1000 µg/l (top). Reprinted with permission from Hopkins et al., 1993a, copyright 1993, American Chemical Society.) Model simulations of the TCE response using a transformation capacity model (bottom).

equation). Here, terms are included for the natural decay of cells in the presence of oxygen as well as the destruction of cells resulting from the transformation of the CAHs. In the case presented, two CAHs are present and two transformation capacity terms are included. The utilization of the primary substrate and the growth on the primary substrate can also be inhibited by the CAHs and the primary substrate itself. Primary substrate inhibition is represented using Haldane kinetics described in Bailey and Ollis (1986).

An example using the inclusion of the transformation capacity concept is shown in Figure 9 (bottom) corresponding to the field observations. The basic trends are consistent with the field observations and show the increase in concentration at the monitoring well upon increasing the injected TCE concentration from 500 to 1000 μg/l. TCE concentration affects the transformation of TCE in this region, and product toxicity is one process potentially responsible for the field observations. However, incorporation of other kinetic models, such as the Oldenhuis model or that of Ely, should also be considered in future model development.

An example of competitive inhibition of a CAH on both the cometabolism of another CAH and on the utilization of the primary substrate is given by Hopkins and McCarty (1995). They present results from a demonstration at Moffett Field where the aerobic cometabolism of TCE and 1,1-DCE by phenol-utilizing microorganisms was evaluated. Concentration histories of TCE, 1,1-DCE, and phenol are shown in Figure 10. An important observation of this experiment was the strong inhibition that 1,1-DCE had on both TCE cometabolism and phenol utilization (Hopkins and McCarty, 1995). Initially, TCE and phenol were injected into the test zone at time-averaged concentrations of 250 μg/l and 12.5 mg/l, respectively. During the first 200 h of the test, effective removals of TCE and phenol were observed. Injection of 1,1-DCE at a time-averaged concentration of 130 μg/l was started at 200 h. TCE concentrations rapidly increased as 1,1-DCE concentration rose in the test zone, and biotransformation of TCE decreased from 88% to 15%. About 30% of the 1,1-DCE was biotransformed within 1 m of travel in the test zone. At 500 h the 1,1-DCE concentration was decreased to 65 μg/l. Hopkins and McCarty (1995) noted a slight improvement in TCE removal and better phenol-utilization in response. 1,1-DCE injection was terminated at 680 h. TCE biotransformation rapidly increased with removals of approximately 94%, somewhat better than when 1,1-DCE injection began. Phenol removal gradually improved. Hopkins and McCarty (1995) noted that significant product toxicity has been associated with 1,1-DCE cometabolism of methane-utilizing bacteria, and 1,1-DCE toxicity was likely observed in these in situ tests.

The model, using the kinetic equations provided in Table 2, was used to simulate the field observations. The results of model simulations of the TCE and 1,1-DCE experiment at Moffett Field are shown in Figure 11. Simulation responses were similar to those observed in the field. Strong inhibition by 1,1-DCE of TCE biotransformation and phenol utilization was required to simulate the response observed in the field. The model, for example, also indicates improvements in TCE biotransformation and phenol utilization upon the lowering of 1,1-DCE concentration, which is

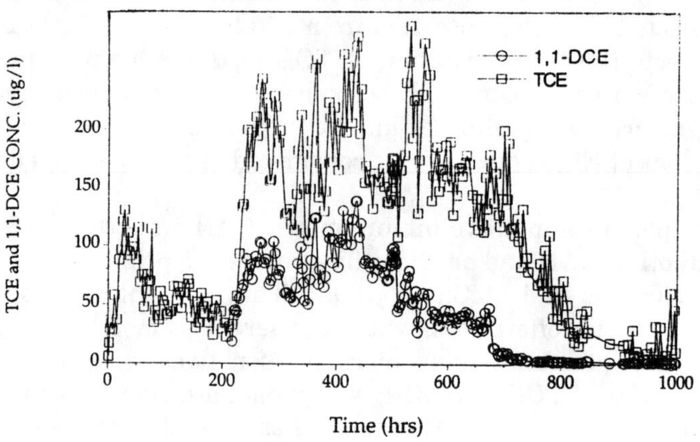

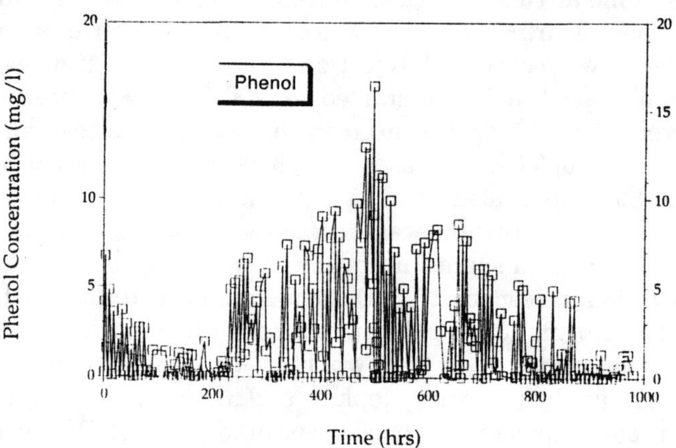

Figure 10. Concentration histories of TCE, 1,1-DCE (top) and phenol (bottom) at a monitoring well 1 m from the injection well. 1,1-DCE addition at 130 μg/l was started at 200 h and strongly inhibited TCE cometabolism and phenol utilization. 1,1-DCE injection concentration were lowered to 65 μg/l at 500 h, and injection was terminated at 680 h (adapted from Hopkins and McCarty, 1995).

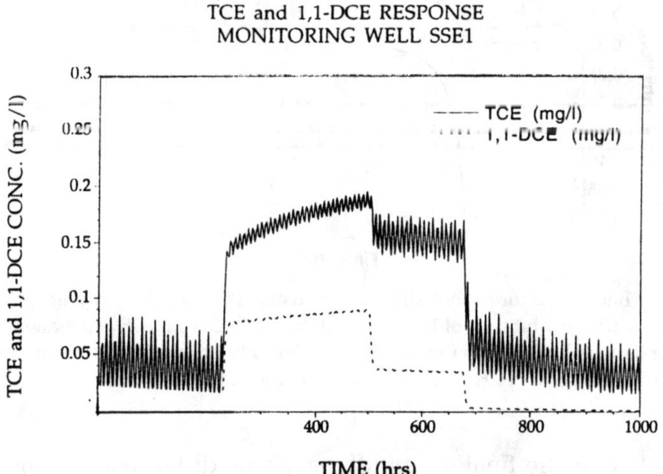

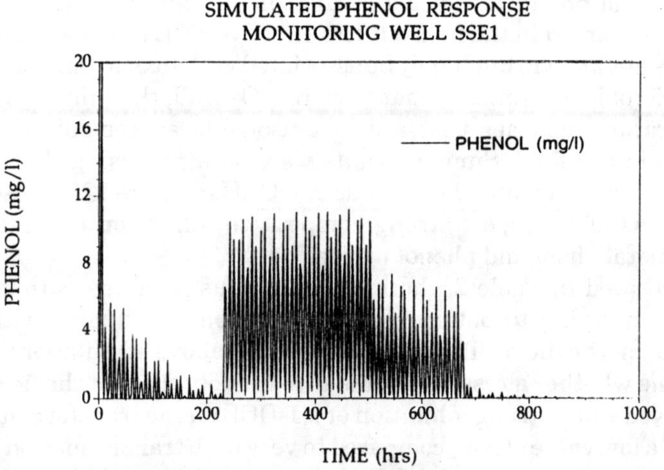

Figure 11. Model simulations of the Moffett Field results shown in Figure 10. Strong competitive inhibition of 1,1-DCE on TCE cometabolism (top) and phenol utilization (bottom) was required to simulate the response observed in the field.

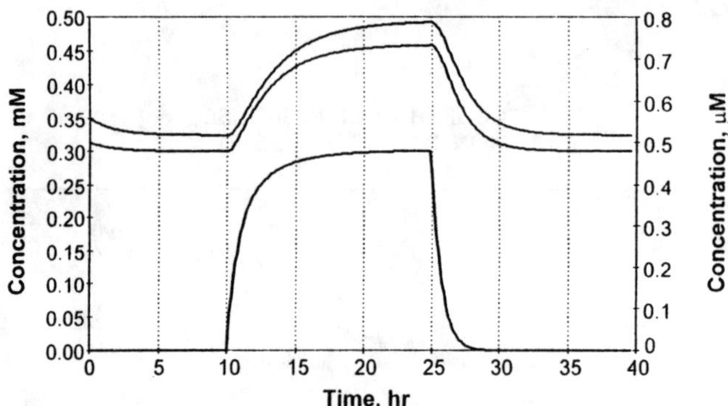

Figure 12. Batch reactor model simulations using the model of Ely et al. (1995a) to simulate competitive inhibition of 1,1-DCE (bottom line, μM) on TCE cometabolism (middle line, μM) and phenol utilization (top line, mM). The simulations require strong competitive inhibition of 1,1-DCE on TCE cometabolism and phenol utilization.

consistent with the field observations. Some differences in the simulations and the field observations are noted after 1,1-DCE injection was terminated at 680 h. The model predicts more complete phenol removal than was observed in the field but less effective TCE removal. The reason for this is not known, but it may be associated with toxicity and/or changes in the microbial community that occurred. Overall, the simulations show concentration histories that are in reasonable agreement with those obtained in the field. Similar results were obtained using the model of Ely et al. (1995a) modified to include two CAHs (Figure 12). These model simulations also support strong competitive inhibition of 1,1-DCE on TCE cometabolism and phenol utilization.

As indicated in Table 2, there are numerous parameters that can be adjusted in trying to obtain model simulations that agree with those obtained in the field. The objective of the above simulations was to determine whether a response, similar to that obtained in the field, could be achieved when strong inhibition of 1,1-DCE on the transformation was applied (a low value of K_{I2}), compared to very high transformation toxicity (high T_{C2} coefficient for 1,1-DCE). The simulations indicated strong inhibition of 1,1-DCE on the transformation could provide responses similar to that observed in the field. Thus, it is possible 1,1-DCE product toxicity is not responsible for the observed response, but rather strong competitive inhibition of TCE and phenol oxidation by 1,1-DCE. The model prediction of better phenol removal than was observed, however, suggests toxicity was also involved. Clearly, more work is required to better understand the processes involved.

MODELING OF IN SITU COMETABOLISM USING NOVEL NUTRIENT DELIVERY METHODS

Model simulations will now be presented as a tool for designing in situ bioremediation at a contaminated site. In situ bioremediation of a contaminant plume requires effective methods for the delivery of the key ingredients for cometabolic transformation—oxygen, methane or phenol, and the target contaminant in the subsurface. To guarantee the full extent of mixing of these ingredients, the pumping scenario assumed in these simulations is a two-well, fully recirculating system (Figure 13). The electron acceptor and donor are introduced in alternating pulses to minimize the presence of both at the injection point and thus minimize excess microbial growth and potential clogging at the injection well. Pulsed introduction of electron acceptor and donor (collectively referred to as substrates) results in regions of the aquifer experiencing transient periods of both, none, or only one substrate. The pulses then mix in the aquifer by the process of dispersion.

System performance will be compared over a wide range of substrate delivery policies for two different electron donors, methane and phenol, for TCE cometabolism. The effects of different microbial growth and degradation kinetics will be evaluated as well as the effects of delivery characteristics, by comparing system behavior for a very soluble, easily introduced substrate, phenol, to that for a less soluble substrate, methane. The effects of both pulse length and the ratio of electron acceptor to electron donor are illustrated. Trichloroethylene, at an initial concentra-

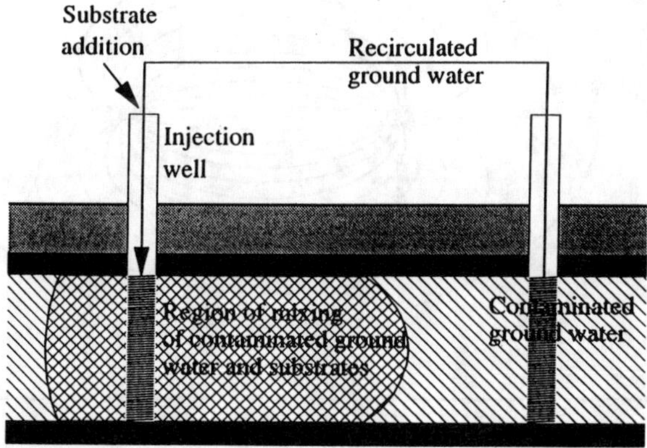

Figure 13. Two-well recirculation system used in the model simulations of nutrient delivery methods. [From Lang et al. with permission, copyright 1997, Ground Water Publishing Co.]

tion of 1 mg/l in the aquifer, is the contaminant of interest for all of the simulations presented.

Methods

The two-dimensional flow regime created by the pumping was modeled using the semi-analytical flow model, RESSQ (Javandel et al., 1984). RESSQ simulates one- and two-dimensional steady-state flow in a confined homogeneous aquifer with multiple injection and extraction wells. The application site is a relatively homogeneous, fine sand aquifer (McCarty et al., 1994) with a hydraulic conductivity of 10^{-4} m/s and porosity of 0.30. The well configuration assumed for these analyses is wells that are screened over a 2-m interval with a 10-m separation between the injection and extraction well. The injection and extraction rate was 10 gpm for all of the simulations shown.

The two-dimensional output from RESSQ is streamlines of equivalent flux established by the pumping. These streamlines define stream tubes along which a one-dimensional simulation of the bioremediation is implemented. Figure 14 shows a 2-D cross section of the stream tubes

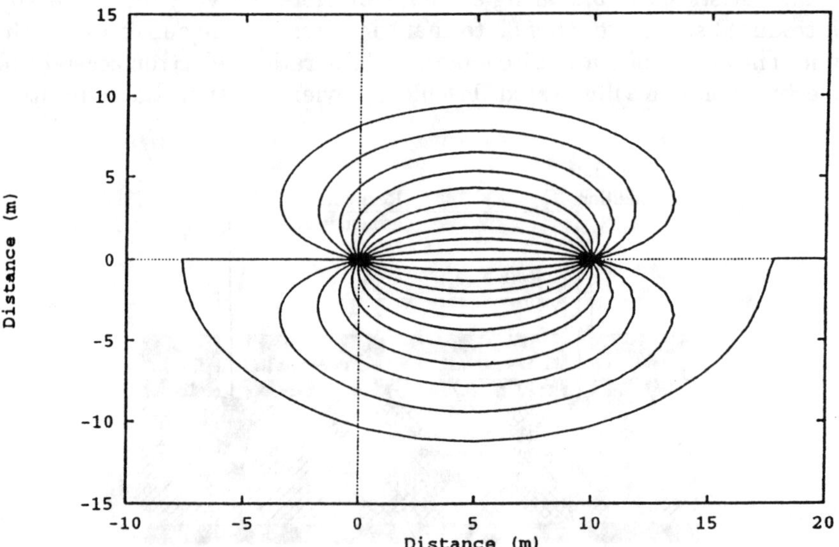

Figure 14. Capture zone developed with the extraction rate equal to the injection rate and no regional groundwater flow. [From Lang et al. with permission, copyright 1997, Ground Water Publishing Co.]

describing the flow system. Each stream tube accounts for 10% of the total flow through the system. Residence times in the stream tubes range from approximately 6 days for the innermost stream tube to 150 days for the outermost.

The in situ bioremediation model developed by Semprini and McCarty (1991, 1992) was used to simulate the cometabolic transformation and transport in each of the stream tubes. Model parameters relating to microbial growth and contaminant transformation and values determined from model simulations of previous field experiments were used in the simulations (Semprini and McCarty, 1991, 1992; Semprini et al., 1993). For these simulations, the model was modified to perform multiple one-dimensional simulations, one along each stream tube. Dispersion was permitted only in the longitudinal direction of flow. No transverse dispersion was permitted between the stream tubes. The concentration at the extraction point was calculated as a flux-weighted average of the concentrations exiting the ten stream tubes.

The performance of different substrate delivery policies (pulse length and ratio of electron acceptor to electron donor) was determined through a search method employing multiple simulations in which the two variables controlling substrate delivery, pulse length, and ratio of electron acceptor to electron donor were varied. Results are plotted as two-dimensional contour plots. The substrate delivery control variables differed in the methane-oxygen and phenol-oxygen systems. For a methane-oxygen system, the injected substrate concentrations were fixed by the gas solubility and the gas transfer efficiency. Thus, varying the pulse length was the only means of varying the mass of methane and oxygen introduced into the system. The phenol-oxygen system has three possible substrate control variables: the phenol injection concentration, the oxygen pulse length, and the phenol pulse length. In these analyses, the oxygen-to-phenol pulse length ratio was initially set at 1.5 (an arbitrary choice for which the sensitivity is addressed later), which leaves as variables the phenol pulse length and the phenol injection concentration.

System performance was evaluated by plotting contour plots of the two for two system performance criteria: the fraction of total TCE mass degraded within 200 days, with the maximum biomass concentration occurring at any time or location during the 200 day simulation (an indication of potential clogging). Experience with this system's geometry and kinetics has shown that, after 200 days, all of the groundwater has recirculated at least once, and if a steady-state biomass concentration could develop under the operative substrate delivery policy, this steady-state biomass concentration had been reached. The initial analysis of each system consisted of 25 model runs.

Methane-Oxygen Systems

The mass of oxygen and methane introduced to promote microbial growth is determined by the solubility of these gases in water and the gas transfer efficiency achieved. In an essentially closed system designed to recirculate and promote mixing of contaminated groundwater with the added substrates, the partial pressure of a gaseous substrate added in excess of its rate of utilization will accumulate in the system and decrease the ability to transfer the limited gaseous substrate into the recirculating groundwater. Thus, the additional problem of competition between dissolved gases already present in the recirculating groundwater and additional gases being introduced must also be considered in the methane-oxygen system. The simulations assumed the gas transfer characteristics measured during experimental investigation of an in-well venturi prototype (Bae et al., 1995) where approximately 50% of the oxygen's theoretical saturation concentration was achieved, after taking into account the partial pressure of a competing dissolved gas.

Figure 15 shows the normalized mass of TCE transformed using methane as the electron donor. The y-axis is the ratio of the oxygen-to-methane pulse length, and the x-axis is the methane pulse length in days. The greatest transformation occurs for a ratio of oxygen-to-methane pulse length of approximately 1.5 to 1. This rate of substrate delivery results in 2.8 grams of O_2 delivered per gram of CH_4, slightly higher than the stoichiometric requirement for methanotroph growth, 2.4 grams O_2 per gram of CH_4 (Semprini and McCarty, 1991). This ratio establishes a substrate delivery that satisfies the stoichiometric requirement of oxygen and methane for net biomass growth, i.e., the substrate requirements for biomass growth plus the oxygen demand for endogenous respiration, with neither substrate accumulating within the system. This operating policy minimizes recirculation and interference of one substrate with the injection of the other substrate and produces a steady-state biomass concentration over time. Figure 15 also shows that more frequent pulsing, i.e., shorter pulse lengths, results in better contaminant transformation at any pulse ratio but (as discussed below) at the cost of a less uniform biomass distribution and consequently higher localized biomass concentrations.

Figure 15 also illustrates the different penalties in terms of the fraction of TCE transformed when approaching the optimal substrate delivery policy from a region of excess methane or excess oxygen. In the operating region where oxygen is introduced in excess (top half of the contour plot), the gradients of the TCE transformation show broader contour lines and the penalty for moving away from the steady-state substrate delivery

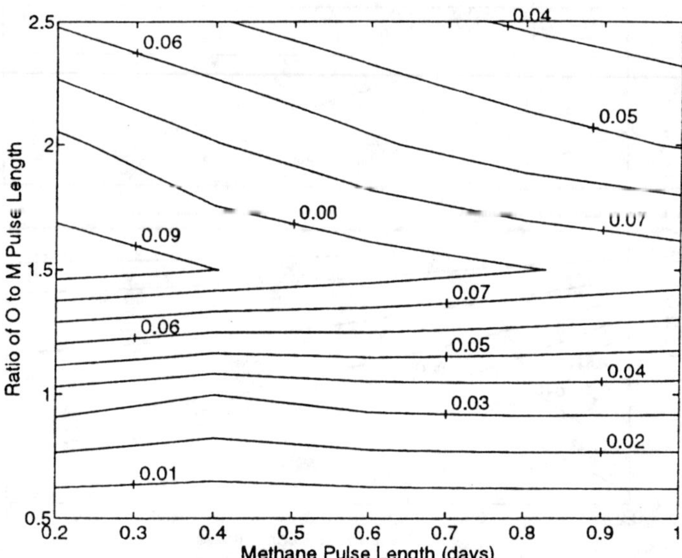

Figure 15. Normalized mass of TCE transformed after 200 days for a two-well, methane-oxygen system. [From Lang et al. with permission, copyright 1997, Ground Water Publishing Co.]

policy in the direction of increasing oxygen is mild. In the operating region where methane is in excess (bottom half of the contour plot), the penalty, as indicated by much steeper gradients in the fraction of TCE transformed, increases much more rapidly as the ratio of oxygen to methane increases. The more pronounced penalty in the region of excess methane has two causes: 1) the transformation reaction of TCE with methane monooxygenase (the active enzyme) requires oxygen, so when oxygen is depleted TCE transformation cannot occur, and 2) increased methane leads to increased competitive inhibition. When methane runs out with oxygen in excess, the microorganisms are slowly deactivating but retain some ability to transform TCE for a finite time period.

Figure 16 shows the maximum biomass concentrations at any location in the capture zone predicted to develop under this range of substrate delivery policies within the 200 days of stimulation. The concentration of high biomass around the short pulse intervals with oxygen-to-methane pulse length ratio of 1.0 illustrates the effects of the model assumption that the microbial population becomes dormant (no decay) in the absence of oxygen. In this region, methane is always present, thus, according to current model assumptions, preventing periods of net microbial death that would occur when oxygen is present but methane absent.

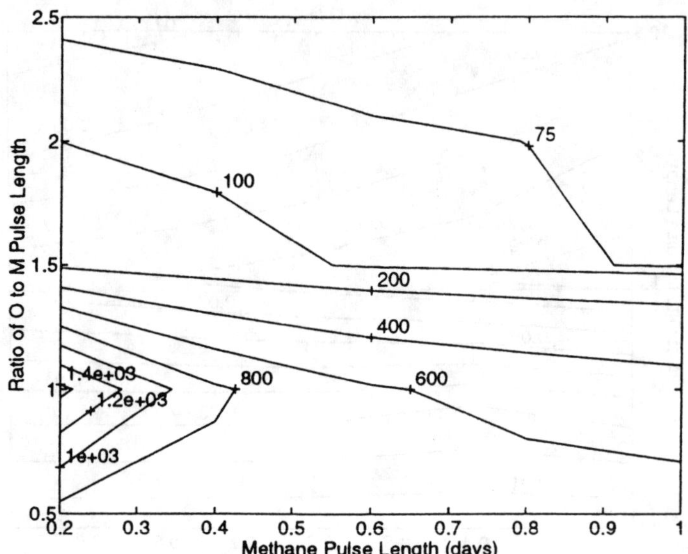

Figure 16. Maximum biomass concentration (mg/l) occurring within 200 days for a two-well, methane-oxygen system. [From Lang et al. with permission, copyright 1997, Ground Water Publishing Co.]

Clearly, from this and the previous figure, the best operating policy is the steady-state or net stoichiometric addition of substrates with a tendency to err towards providing excess oxygen. At the substrate delivery policy that achieves the best contaminant transformation, the maximum biomass concentration, 200 mg dry wt/l, is below levels that have been observed to cause clogging in soil column experiments, i.e., >550 mg dry wt/l (Taylor and Jaffe, 1990). The biomass concentrations presented in Figure 7 also show that a decrease in maximum biomass concentration can be achieved with longer pulse lengths. This decrease is more pronounced away from the steady-state substrate delivery.

Phenol-Oxygen Systems

Simulations were also performed using phenol as the electron donor to drive TCE cometabolism, based on the success of the previously discussed field studies (Hopkins et al., 1993; Hopkins and McCarty, 1995). In addition, phenol is readily soluble at the concentrations needed to stimulate active microbial populations so the phenol-oxygen systems avoid the competition for introduction of two gas phase substrates that limits the methane-oxygen system. Gas transfer efficiency still influences the system performance because the mass of oxygen transferred determines the mass of phenol that can be utilized to stimulate microbial growth and contaminant transformation. For the initial set of simulations, the oxygen pulse length was set at 1.5 times the phenol pulse length. Sensitivity to this pulse length is addressed at the end of this section.

Figure 17 shows the contour results of the fraction of the initial TCE mass transformed for the range of phenol injection concentrations and pulsing frequencies. A phenol injection concentration of approximately 23 mg/l phenol achieves the maximum TCE transformation and, again, the shorter pulse lengths perform better than the longer ones. The phenol injection concentration of 23 mg/l matches the net stoichiometric requirement for growth of phenol degraders for the mass of oxygen (oxygen concentration ~30 mg/l) being introduced. This corresponds to a net stoichiometric requirement of 1.57 grams O_2 to 1 gram of phenol based on the model parameters of Semprini et al. (1993). Comparison of oxygen plus phenol breakthrough curves at the extraction well for phenol injection concentrations of 20, 25, and 30 mg/l and phenol and oxygen pulse lengths of 0.2 and 0.3 days found that a balanced substrate delivery policy

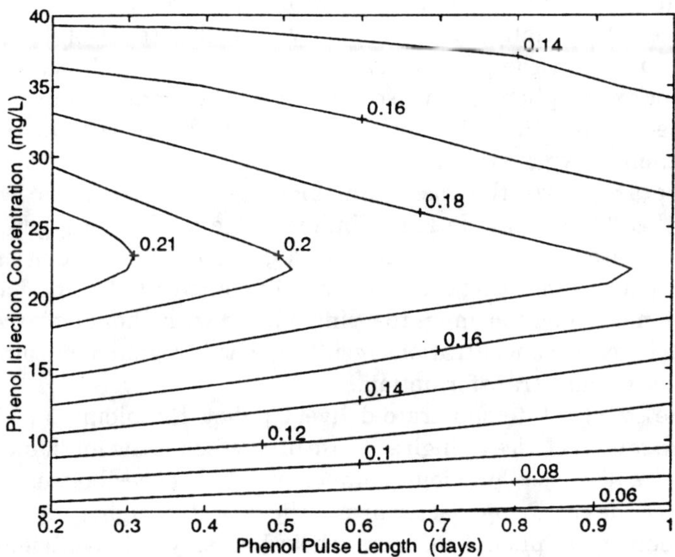

Figure 17. Normalized mass of TCE transformed after 200 days for a two-well, phenol-oxygen system. [From Lang et al. with permission, copyright 1997, Ground Water Publishing Co.]

(25 mg/l) achieved almost complete utilization of both oxygen and phenol (low values of sum of the concentrations). Operating near this region not only achieves the most efficient transformation of TCE, but near complete utilization of both added substrates minimizes disruption of discrete pulsed introduction of the substrates at the injection well. A potential problem with operating in this region, however, is that complete utilization of the added phenol (a regulated chemical) is not necessarily achieved. Balancing the requirement of complete phenol utilization and efficient contaminant degradation requires using substrate delivery policies in the region with oxygen in excess.

Substrate delivery policies introducing excess phenol are represented by the upper half of Figure 17, i.e., phenol injection concentrations exceeding 25 mg/l, while excess oxygen corresponds to the lower half of the figure. Unlike the methane-oxygen system, the penalties to contaminant transformation are slightly greater when moving away from the steady-state substrate delivery policy in the direction of excess oxygen than in the direction of excess phenol. The same processes active with excess methane, i.e., dependence of the TCE transformation reaction on the presence of oxygen and competitive inhibition, are observed with phenol as the electron donor (Hopkins et al., 1993). However, phenol and oxygen do not compete for introduction, and phenol or oxygen can be added at a constant rate independent of the concentration of the other substrate. The penalty for moving away from the steady-state substrate delivery policy in the phenol-oxygen system in either direction, excess oxygen or excess phenol, can be explained by the biomass concentrations developed, rather than the competition for substrate introduction as in the methane-oxygen system.

Figure 18 shows the maximum biomass concentrations occurring within the 200-day simulations. Similar to the methane-oxygen system, biomass is predicted to accumulate to a greater extent under substrate delivery policies injecting excess phenol. In contrast to the methane-oxygen system, oxygen continues to be introduced with phenol in excess, and the high biomass concentrations resulting in this region maintain significant contaminant transformation.

At the steady-state substrate delivery policy, the biomass concentrations are roughly twice as high as those predicted for the methane-oxygen system. As discussed previously, important differences between phenol- and methane-oxidizing bacteria are the slightly higher rate of TCE cometabolism by phenol utilizers, their higher yield coefficient of 0.8 compared with 0.5 for methane utilizers, and their lower oxygen-to-substrate requirement (Semprini et al., 1993). The higher yield and lower oxygen requirement for phenol-oxidizers may act as either a benefit to

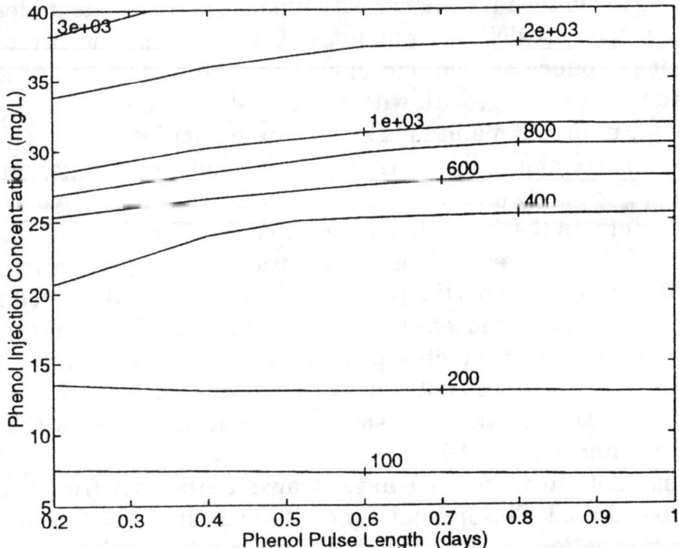

Figure 18. Maximum biomass concentration (mg/l) occurring within 200 days for a two-well, phenol-oxygen system. [From Lang et al. with permission, copyright 1997, Ground Water Publishing Co.]

remediation, i.e., greater microbial stimulation for the equivalent mass input of substrate, or an impediment, depending on the aquifer's sensitivity to clogging by bacterial growth. Biomass concentrations of 400 to 600 mg dry wt/l are in the range Taylor and Jaffe (1990) reported to cause clogging, 550 mg dry wt/l, but are much lower than the level of 3000 mg dry wt/l reported by Vandivivere and Baveye (1992).

For the analyses shown above, the ratio of the oxygen-to-phenol pulse lengths was set to a constant value of 1.5 to reduce the number of variables to two. The sensitivity of the system to the ratio of pulse lengths was also performed. When the oxygen-to-phenol pulse length ratio was changed to 1.0, the phenol injection concentration predicted to have the greatest rate of TCE transformation was approximately 15 mg/l, compared to 23 mg phenol/l predicted for an oxygen-to-phenol pulse length of 1.5 (Figure 17). The steady-state phenol injection concentration decreased to match the decrease in the mass of oxygen being introduced into the system as a result of the shorter oxygen pulse length. The maximum biomass concentration at the optimal policy was approximately the same as those predicted for the substrate delivery policy with an oxygen-to-phenol pulse length ratio of 1.5, i.e., 400 mg dry wt/l (Figure 18). This biomass concentration represents the steady-state biomass concentration that will develop given the microbial growth kinetics of phenol degraders. The substrate delivery policies that maintain this biomass concentration are those that introduce oxygen and phenol to match the requirements of phenol and oxygen for growth with additional oxygen to balance endogenous respiration and maintain a steady-state population.

The method employed to introduce gas phase substrates into the recirculating groundwater will determine the efficiency of gas transfer into the system. In the simulations presented thus far, the efficiency used was 50% of the oxygen's theoretical saturation concentration, after taking into account the partial pressure of a competing dissolved gas. In a full-scale operation, it is feasible that gas transfer efficiencies could be either lower, because competition with dissolved gases other than the introduced substrates was not considered, or higher, through the use of more efficient transfer mechanisms such as in-line mixers (Hopkins, personal communication, 1994).

The sensitivity to an increase in gas transfer efficiency from 0.5 to 0.9 was also performed. The optimal oxygen-to-methane ratio for maximum TCE transformation mass was 1.5, the same as achieved for an efficiency of 0.5. This increase in gas transfer efficiency promoted an increase in the TCE transformation from 0.09 mass TCE transformed/initial mass of TCE for an efficiency of 0.5 to greater than 0.14 for a gas transfer efficiency of 0.9. As expected, the availability of additional substrate increases the biomass growth, and thus the capacity for transformation. The mass of TCE transformed under any substrate delivery policy increases with an increase in gas transfer efficiency.

The maximum biomass concentrations predicted to occur assume the higher gas transfer efficiency. The general trends in the maximum biomass concentrations were very similar to those seen with a transfer efficiency of 0.5 (Figure 16). For most of the substrate delivery policies sampled, the maximum biomass concentrations were predicted to be almost twice as high, 200 mg/l for a gas transfer efficiency of 0.5 compared to 400 mg/l for a gas transfer efficiency of 0.9 for a substrate delivery policy of 0.30 days oxygen and 0.2 days methane. However, the maximum biomass concentrations for substrate delivery policies with methane in excess are lower than those predicted for a gas transfer efficiency of 0.5. In this region, the higher gas transfer efficiency increases the oxygen introduced, which in turn increases the endogenous respiration and slightly lowers the maximum biomass concentration predicted in the regions with methane introduced in excess.

Simulations were also performed for the phenol-oxygen system where the gas transfer efficiencies were increased from 0.5 to 0.9 for the same range of substrate delivery policies as the previous simulations. The optimal phenol injection concentration increased from approximately 23 mg/l to 42 mg/l. For the methane-oxygen system, no change in the steady-state substrate delivery policy was observed with an increase in the gas transfer efficiency, because the gas transfer efficiency influenced the introduction of both substrates. In the phenol-oxygen system, an increase in the gas transfer efficiency increases the amount of oxygen introduced into the system, and the optimal phenol injection concentration increases to match this increase in the delivery of oxygen. The predicted fractional transformation of TCE increases from 0.21 to 0.80 with the increase in substrate addition.

The maximum biomass, however, changes very little for most of the substrate delivery policies. The main difference in maximum biomass concentrations with the change in gas transfer efficiency is for those substrate delivery policies receiving phenol in excess, for example, roughly the upper half of Figure 16. The maximum biomass concentrations in this region are lower for a gas transfer efficiency of 0.9 because more oxygen is introduced, thus increasing the endogenous respiration and decreasing the biomass concentration.

REFERENCES

Alvarez-Cohen, L. and P. L. McCarty. 1991a. Effects of toxicity, aeration, and reductant supply on trichloroethylene transformation by a mixed methanotrophic culture. *Appl. Environ. Microbiol.* 57(1):228–235.

Alvarez-Cohen, L. and P. L. McCarty. 1991b. Product toxicity and cometabolic inhibition modeling of chloroform and trichloroethylene transformation by methanotrophic resting cells. *Appl. Environ. Microbiol.* 57(4):1031–1037.

Anderson, J. E. and P. L. McCarty. 1994. Model for treatment of trichloroethylene by methanotrophic biofilms. *J. Envir. Eng. ASCE.* 120(2):379–400.

Arciero, D., T. Vannelli, M. Logan, and A. B. Hooper. 1989. Degradation of trichloroethylene by the ammonia-oxidizing bacterium *Nitrosomonas europaea. Biochem. Biophys. Res. Commun.* 159(2):640–643.

Bae, J., L. Semprini, and P. L. McCarty. 1995. Down-well apparatus for adding oxygen and methane into a contaminated aquifer for bioremediation. *J. Environ. Engrg. ASCE.* 121(8):565–570.

Bailey, J. E. and D. F. Ollis, 1986. *Biochemical Engineering Fundamentals*, 2nd ed. McGraw Hill Publishing Company, NY.

Barbash, J. and P. V. Roberts. 1986. Volatile organic chemical contamination of groundwater resources in the U.S. *J. Water Poll. Cont. Fed.* 58(5):343–348.

Bedard, C. and R. Knowles. 1989. Physiology, biochemistry, and specific inhibitors of CH_4, NH_4^+, and CO oxidation by methanotrophs and nitrifiers. *Microbiol. Rev.* 53(1):68–84.

Broholm, K., T. H. Christensen, and B. K. Jensen. 1992. Modeling TCE degradation by a mixed culture of methane-oxidizing bacteria. *Wat. Res.* 29(9):1177–1185.

Chang, H. L. and L. Alvarez-Cohen. 1995. Transformation capacities of chlorinated organics by mixed cultures enriched on methane, propane, toluene, or phenol. *Biotechnol. Bioeng.* 45:440–449.

Criddle, C. S. 1993. The kinetics of cometabolism. *Biotechnol. Bioeng.* 41:1048–1056.

Dabrock, B., J. Riedel, J. Bertram, G. Gottschalk. 1992. Isopropylbenzene (cumene)—A new substrate for the isolation of trichloroethylene-degrading bacteria. *Arch. Microbiol.* 158:9–13.

Dalton, H. and D. I. Stirling. 1982. Co-metabolism. *Phil. Trans. R. Soc. Lond. B.* 297:481–496.

Dixon, M. and E. C. Webb. 1979. *Enzymes*, Academic Press, New York.

Ely, R. L., M. R. Hyman, D. J. Arp, R. B. Guenther, and K. J. Williamson. 1995a. A cometabolic kinetics model incorporating enzyme inhibition, inactivation, and recovery. II. trichloroethylene degradation experiments. *Biotechnol. Bioeng.* 46:232–245.

Ely, R. L., K. J. Williamson, R. B. Guenther, M. R. Hyman, and D. J. Arp. 1995b. A cometabolic kinetics model incorporating enzyme inhibition, inactivation, and recovery. I. model development, analysis, and testing. *Biotechnol. Bioeng.* 46:218–231.

Ensign, S. A., M. R. Hyman, and D. J. Arp. 1992. Cometabolic degradation of chlorinated alkenes by alkene monooxygenase in a propylene-grown *Xanthobacter* strain. *Appl. Environ. Microbiol.* 58(9):3038–3046.

Ewers, J., D. Freier-Schroder, and H. Knackmuss. 1990. Selection of trichloroethylene (TCE) degrading bacteria that resist inactivation by TCE. *Arch. Microbiol.* 154:410–413.

Fan, A. M. 1988. Trichloroethylene: water contamination and health risk assessment. In *Reviews of environmental contamination and toxicology*. pp. 55–92 Springer-Verlag.

Fan, S. and K. M. Scow. 1993. Biodegradation of trichloroethylene and toluene by indigenous microbial populations in soil. 1993. *Appl. Environ. Microbiol.* 59(6):1911–1918.

Fetter, C. W. 1993. *Contaminant Hydrogeology*, MacMillan, New York, New York.

Fogel, M. M., A. R. Tadeo, and S. Fogel. 1986. Biodegradation of chlorinated ethenes by a methane-utilizing mixed culture. *Appl. Environ. Microbiol.* 51(4):720–724.

Folsom, B. R., P. J. Chapman, and P. H. Pritchard. 1990. Phenol and trichloroethylene degradation by *Pseudomonas cepacia* G4: kinetics and interactions between substrates. *Appl. Environ. Microbiol.* 56:1279–1285.

Fox, B. G., J. G. Borneman, L. P. Wackett, and J. D. Lipscomb. 1990. Haloalkene oxidation by the soluble methane monooxygenase from *Methylosinus trichosporium* OB3b: methanistic and environmental implications. *Biochem.* 29(27):6419–6427.

Green, J. and H. Dalton. 1989. Substrate specificity of methane monooxygenase. *J. Biol. Chem.* 264(30):17698–17703.

Harker, A. R. and Y. Kim. 1990. Trichloroethylene degradation by two independent aromatic-degrading pathways in *Alcaligenes eutrophus* JMP134. *Appl. Environ. Microbiol.* 56(4):1179–1181.

Harmon, T. C., L. Semprini, and P. V. Roberts. 1992. Simulating Groundwater Solute Transport Using Independently Determined Sorption Parameters. *J. Environmental Engineering, ASCE.* 118(5):666–689.

Henry, S. M. and D. Grbić-Galić. 1989. Chapter 9 in *In-situ aquifer restoration of chlorinated aliphatics by methanotrophic bacteria*. Edited by P. V. Roberts, L. Semprini, G. D. Hopkins, D. Grbic-Galic, P. L. McCarty, and M. Reinhard. EPA Research Report EPA/600/S2-89/003.

Hopkins, G. D. and P. L. McCarty. 1995. Field evaluation of in situ aerobic cometabolism of trichloroethylene and three dichloroethylene isomers using phenol and toluene as the primary substrates. *Environ. Sci. Technol.* 29(6):1628–1637.

Hopkins, G. D., J. Munakata, L. Semprini, and P. L. McCarty. 1993. Trichloroethylene concentration effects on pilot field-scale in-situ groundwater bioremediation by phenol-oxidizing microorganisms. *Environ. Sci. Technol.* 27(12): 2542–2547.

Hopkins, G. D., personal communication, 1994.

Hyman, M. R. and D. J. Arp. 1995. Effects of ammonia on the de novo synthesis of polypeptides in cells of *Nitrosomonas europaea* denied ammonia as an energy source. *J. Bacteriol.* 177:4794–4979.

Javandel, I., C. Doughty, and C. F. Tsang. 1984. *Groundwater transport: Handbook of mathematical models, Water Resources Monograph 10*, American Geophysical Union, Washington, D.C.

Janssen, D. B., A. J. van den Wijngaard, J. J. van der Waarde, and R. Oldenhuis. 1991. Biochemistry and Kinetics of Aerobic Degradation of Chlorinated Aliphatic Hydrocarbons. In: *On-Site Bioreclamation: Processes for Xenobiotic and Hydrocarbon Treatment*. Eds. R. E. Hinchee and R. F. Olfenbuttel. Butterworth-Heinemann, Stoneham, Mass.

Keener, W. K. and D. J. Arp. 1993. Kinetic studies of ammonia monooxygenase inhibition in *Nitrosomonas europaea* by hydrocarbons and halogenated hydrocarbons in an optimized whole-cell assay. *Appl. Environ. Microbiol.* 59(8):2501–2510.

Lang, M. M., P. V. Roberts, and L. Semprini. "Model simulations in support of field scale design and operation of bioremediation based cometabolic degradation," *Ground Water.* 35(4):565–573.

Leonard, B. P. 1979. A stable and accurate modelizing procedure based on quadratic upstream modeling. *Comput. Methods Appl. Mech. Eng.* 19:59–98.

Li, S. and L. P. Wackett. 1992. Trichloroethylene oxidation by toluene dioxygenase. *Biochem. Biophys. Res. Comm.* 185(1):443–451.

Little, C. D., A. V. Palumbo, S. E. Herbes, M. E. Lindstrom, R. L. Tyndall, and P. J. Gilmer. 1988. Trichloroethylene biodegradation by a methane-oxidizing bacterium. *Appl. Environ. Microbiol.* 54(4):951–956.

Malachowsky, K. J., T. J. Phelps, A. B. Teboli, D. E. Minnikin, and D. C. White. 1994. Aerobic mineralization of trichloroethylene, vinyl chloride, and aromatic compounds by *Rhodococcus* species. *Appl. Environ. Microbiol.* 60(2): 542–548.

McCarty, P. L., L. Semprini, M. E. Dolan, T. C. Harmon, C. Tiedeman, and S. M. Gorelick. 1994. In situ methanotrophic bioremediation for contaminated groundwater at St. Joseph, Michigan. In: *On-Site bioreclamation*. R. E. Hinchee, R. F. Olfenbuttel, Ed., Butterworth-Heinemann, Stoneham, MA, pp. 16–40.

National Research Council. 1994. *Alternatives for Ground Water Cleanup*, National Academy Press, Washington, D.C.

Nelson, M. J. K., S. O. Montgomery, E. J. O'Neill, and P. H. Pritchard. 1986. Aerobic metabolism of trichloroethylene by a bacterial isolate. *Appl. Environ. Microbiol.* 52(2): 383–384.

Nelson, M. J. K., S. O. Montgomery, and P. H. Pritchard. 1988. Trichloroethylene metabolism by microorganisms that degrade aromatic compounds, *Appl. Environ. Microbiol.* 54(2): 604–606.

Oldenhuis, R., J. Y. Oedzes, J. J. van der Waarde, and D. B. Janssen. 1991. Kinetics of chlorinated hydrocarbon degradation by *Methylosinus trichosporium* OB3b and toxicity of trichloroethylene. *Appl. Environ. Microbiol.* 57(1):7–14.

Oldenhuis, R., R. L. J. M. Vink, D. B. Janssen, and B. Witholt. 1989. Degradation of chlorinated hydrocarbons by *Methylosinus trichosporium* OB3b expressing soluble methane monooxygenase. *Appl. Environ. Microbiol.* 55(11): 2819–2826.

Ortiz de Montellano, P. R. and N. O. Reich. 1986. Inhibition of cytochrome P-450 enzymes. In: *Cytochrome P-450: Structure, Mechanism, and Biochemistry*, P. R. Ortiz de Montellano, Ed., Plenum Press, New York.

Rasche, M. E., M. R. Hyman, and D. J. Arp. 1991. Factors limiting aliphatic chlorocarbon degradation by *Nitrosomonas europaea:* cometabolic inactivation of ammonia monooxygenase and substrate specificity. *Appl. Environ. Microbiol.* 57(10): 2986–2994.

Roberts, P. V., G. D. Hopkins, D. M. Mackay, and L. Semprini. 1990. A field evaluation of in-situ biodegradation of chlorinated ethenes: Part 1. Experimental methodology and characterization. *Ground Water.* 28(4):591–604.

Semprini, L. 1995. In situ bioremediation of chlorinated solvents. *Environmental Health Perspectives.* 103(5):101–105.

Semprini, L., G. D. Hopkins, and P. L. McCarty. 1994. A field and modeling comparison of in situ transformation of trichlorethylene by methane and phenol utilizers, In: *Bioremediation of Chlorinated and Polycyclic Aromatic Hydrocarbons*, R. E. Hinchee, A. Leeson, L. Semprini, S. K. Ong, Ed., Lewis Publishers, Chelsea, MI, pp. 248–254.

Semprini, L., G. D. Hopkins, P. V. Roberts, D. Grbić-Galić, and P. L. McCarty. 1991. A field evaluation of in-situ biodegradation of chlorinated ethenes: Part 3. Studies of competitive inhibition. *Ground Water*. 29(2):239–250.

Semprini, L. and P. L. McCarty. 1991. Comparison between model simulations and field results for in-situ biorestoration of chlorinated aliphatics: Part 1. Biostimulation of methanotrophic bacteria. *Ground Water*. 29(3):365–374.

Semprini, L. and P. L. McCarty. 1992. Comparison between model simulations and field results for in-situ biorestoration of chlorinated aliphatics: Part 2. Cometabolic transformations. *Ground Water*. 30(1):37–44.

Semprini, L., P. V. Roberts, G. D. Hopkins, and P. L. McCarty. 1990. A field evaluation of in-situ biodegradation of chlorinated ethenes: Part 2. The results of biostimulation and biotransformation experiments. *Ground Water*. 28(5):715–727.

Shields, M. S., S. O. Montgomery, P. J. Chapman, S. M. Cuskey, and P. H. Pritchard. 1989. Novel pathway of toluene catabolism in the trichloroethylene-degrading bacterium G4. *Appl. Environ. Microbiol.* 55(6):1624–1629.

Shields, M. S. and M. J. Reagin. 1992. Selection of a *Pseudomonas cepacia* strain constitutive for the degradation of trichloroethylene. *Appl. Environ. Microbiol.* 58(12):3977–3983.

Strand, S. E., M. D. Bjelland, and H. D. Stensel. 1990. Kinetics of chlorinated hydrocarbon degradation by suspended cultures of methane-oxidizing bacteria. *Res. J. Wat. Poll. Control Fed.* 62(2):124–129.

Suidan, M. T., B. E. Rittmann, and U. K. Traegner. 1987. Criteria establishing biofilm types. *Water Res.* 21(4):491–498.

Taylor, S. W. and P. R. Jaffe. 1990. Biofilm growth and the related changes in the physical properties of a porous medium, 1, Experimental investigation. *Water Resour. Res.* 26(9):2153–2160.

Vandevivere, P. and P. Baveye. 1992. Saturated hydraulic conductivity reduction caused by aerobic bacteria in sand columns. *Soil Sci. Soc. Am, J.* 56(1):1–13.

Vogel, T. M. and P. L. McCarty. 1985. Biotransformation of tetrachloroethylene to trichloroethylene, dichloroethylene, vinyl chloride, and carbon dioxide under methanogenic conditions. *Appl. Environ. Microbiol.* 49:1080–1083.

Wackett, L. P., G. A. Brusseau, S. R. Householder, and S. R. Hansen. 1989. Survey of microbial oxygenases: trichloroethylene degradation by propane-oxidizing bacteria, *Appl. Environ. Microbiol.* 55(11):2960–2964.

Wackett, L. P. and D. T. Gibson. 1988. Degradation of trichloroethylene by toluene dioxygenase in whole-cell studies with *Pseudomonas putida* F1. *Appl. Environ. Microbiol.* 54(7):703–1708.

Wackett, L. P. and S. R. Householder. 1989. Toxicity of trichloroethylene to *Pseudomonas putida* F1 is mediated by toluene dioxygenase. *Appl. Environ. Microbiol.* 55(11): 2723–2725.

Westrick, J. J., J. W. Mello, and R. F. Thomas. 1984. The groundwater supply survey, *J. Am. Water Works Assoc.* 5:52–59.

Wilson, J. T. and B. H. Wilson. 1985. Biotransformation of trichloroethylene in soil. *Appl. Environ. Microbiol.* 49:242–243.

Fundamentals and Modeling Aspects of Bioventing

C. M. TELLEZ, A. AGUILAR-AGUILA, R. G. ARNOLD AND R. Z. GUZMAN
Chemical and Environmental Engineering Department
University of Arizona
Tucson, AZ 85721, USA

INTRODUCTION

Short-chain, chlorinated aliphatics are among the most common pollutants in groundwaters and soils. Chlorinated methanes, ethanes, and ethenes have been widely used as industrial solvents without adequate consideration for their disposal and eventual fate in the environment. Their toxicity and carcinogenicity motivated the establishment of enforceable maximum contaminant limits (MCLs) for a number of compounds in the class in the low part per billion range (Westrick et al., 1984). The breadth of problems associated with alkyl halide pollutants and the levels to which remediation has been mandated necessitate the development of new, more efficient technologies for the in-place destruction of these compounds.

Physical and biological treatment strategies for the remediation of contaminated soil and groundwaters that consider the probable partitioning of chlorinated organics in multiphase environments have the greatest probability of success. Among the organic contaminants in groundwater and soils, the chlorinated aliphatics occupy a somewhat unique position in that they are characterized by both low hydrophobicity [low octanol/water partition coefficient (K_{ow})] and high volatility [large Henry's Law constant (H)] as shown in Figure 1. Using the physical models of unsaturated sediments, equilibrium considerations dictate that chlorinated aliphatics will partition, at least to some extent, into the local gas-phase, as can be observed in the partition behavior of TCE shown in Figure 2. The volatility of the chlorinated aliphatics suggests that gas-phase treatments, or methods that rely upon contaminant and nutrient transport with the bulk gas phase, may be particularly effective. Adding to the

135

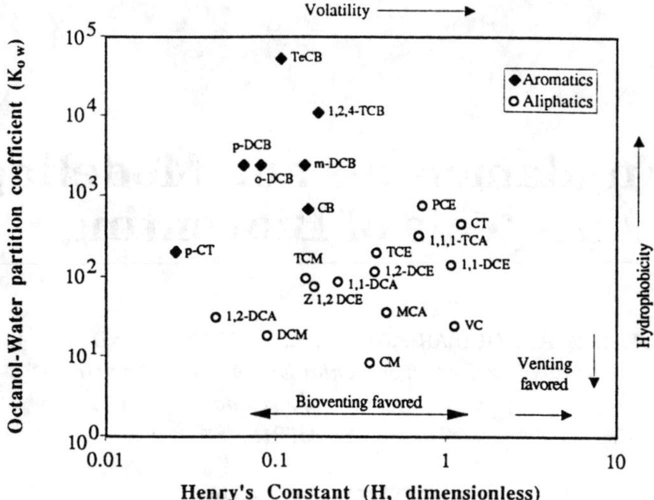

Figure 1. Partition behavior of halogenated organic compounds in terms of hydrophobicity and volatility. Based on contaminant physical properties, bioventing and soil vapor extractions (SVE) are favored when organic contaminants show high values for volatility (to promote mass transfer from the aqueous to the gas phase) and low hydrophobicity (to prevent the contaminant from adsorbing to the organic matter in the soil). Bioventing, however, can be used even when the contaminant shows low volatility, as long as it is available to the degrading microorganisms.

1,1,1-TCA	1,1,1-trichloroethane	CB	chlorobenzene
1,1-DCA	1,1-dichloroethane	1,2,4-TCB	1,2,4-trichlorobenzene
1,1-DCE	1,1-dichloroethylene	m-DCB	*m*-dichlorobenzene
1,2-DCA	1,2-dichloroethane	o-DCB	*o*-dichlorobenzene
1,2-DCE	*trans*-1,2-dichloroethylene	p-CT	*p*-chlorotoluene
CM	chloromethane	p-DCB	*p*-dichlorobenzene
CT	carbon tetrachloride	TeCB	1,2,4,5-tetrachlorobenzene
DCM	dichloromethane	TCM	chloroform
MCA	chloroethane	VC	vinyl chloride
PCE	tetrachloroethylene	Z1,2-DCE	*cis*-1,2-dichloroethylene
TCE	trichloroethylene		

attractiveness of such strategies is the relative ease with which gaseous fluids pass through the vadose zone, primarily due to the immense difference between liquid and gas viscosities.

Recently, considerable interest has been directed to the in situ treatment of unsaturated zone soils. Soil venting has been used at numerous sites to remediate low-molecular-weight, volatile contaminants. In this process, the contaminant's partition from the aqueous sorbed phases into the gas phase as air is circulated through the contaminated soil.

Bioventing is based on the same principle as soil venting or soil-vapor extraction. In addition, however, bioventing is designed to promote the biological degradation of pollutants in soil or groundwater in the remediation process. The method and its applications have been described in several publications (Norris and Matthews, 1994; Rittmann et al., 1994; Marley and Hoag, 1984; Brown et al., 1987; Batcheder et al., 1986).

Venting is aimed primarily at the restoration of sites contaminated with volatile organic chemicals (VOCs), which are suitable for transport out of soil moisture or soil surfaces in a gas stream. Simulation shows (Johnson et al., 1990) that, even under optimal conditions, heavy components of hydrocarbon mixtures (e.g., gasoline) remain in the soil after the light components have been removed.

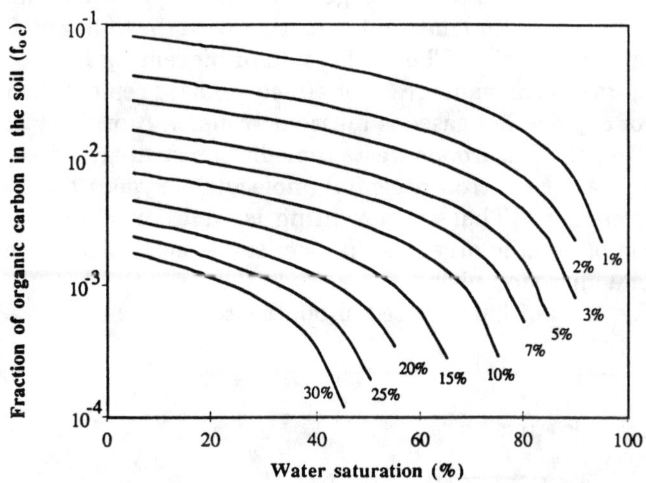

Figure 2. Trichloroethylene (TCE) distribution in the gas phase as a function of soil water saturation. Each contour represents a constant fraction of the total mass of TCE that is present in the gas phase under equilibrium conditions. TCE was assumed to distribute among organic matter and the aqueous and gaseous phases in unsaturated soil. These results were obtained based on Henry's law and on empirical relationships reported by Briggs (1981) between soil f_{oc} and the contaminant organic matter partition coefficient (K_{oc}). The behavior determines how a given contaminant partitions between the aqueous and solid phases in soil. This simulation shows that low soil f_{oc} and low moisture content promote higher TCE concentrations in the gas phase. These conditions apparently would favor venting operations (SVE and bioventing) for TCE transport and degradation. From the figure, the effect of increasing the soil moisture content, at a given f_{oc}, favors the transfer of TCE to the aqueous phase more effectively, thus making it more suitable for biodegradation. An excessively high soil water content, however, would decrease air permeability. The parameters used in the above calculation are: soil density = 1.85 g soil/cm^3 soil; soil porosity = 0.35; TCE Henry's constant = 0.392 at 25°C (dimensionless); TCE log K_{oc} = 1.93.

Bioventing combines soil venting with bioremediation, using oxygen from the venting air for aerobic biodegradation of the remaining organic contaminants. The main purpose of a bioventing system is to use the gas phase to transfer oxygen or other gaseous components (including volatile soil contaminants) to positions in the subsurface where indigenous organisms can utilize them as nutrients or electron acceptors. Elements of the bioventing process are schematically illustrated in Figure 3. Chlorinated solvents are frequently difficult to biodegrade in this manner, but they are sometimes degraded fortuitously during microbial growth on another hydrocarbon.

Bioventing has the potential to significantly reduce the concentration of a range of hydrocarbons and other organic contaminants in unsaturated soils, particularly volatile and semivolatile compounds. The volatility of the chlorinated aliphatics suggests that gas-phase treatments that rely on contaminant and nutrient transport with the bulk gas phase may be particularly effective. The application of bioventing to vadose remediation offers clear advantages over aqueous-based remediation systems in terms of oxygen and gaseous nutrient transfer. A primary advantage of bioventing as a hazardous waste remediation strategy arises from the availability of an electron acceptor (molecular oxygen) for use in biochemical reactions. That is, bioventing is well suited to promote the destruction of volatile organics that can serve as substrates for aerobic vadose zone microorganisms.

The resistance of halogenated aliphatics to aerobic biodegradation is

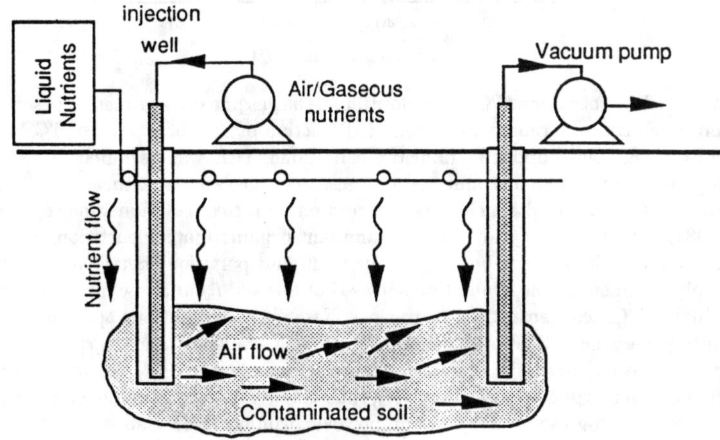

Figure 3. Schematic diagram of a bioventing system.

directly related to the degree of compound halogenation (Strand et al., 1990; Henson et al., 1989). For example, relatively few metabolic systems are able to catalyze the aerobic transformation of trichloroethene (TCE), while vinyl chloride (VC) oxidation is much more common and generally more rapid. Methanotrophic and ammonia-oxidizing bacteria are among those that can initiate aerobic transformations of TCE (Vanelli et al., 1990; Rasche et al., 1991; Arciero et al., 1989). In no case has it been shown, however, that capable bacteria themselves benefit from this activity. That is, biotransformations of TCE and other halogenated aliphatic compounds are cometabolic or, at best, fortuitous in nature. This underscores the need for a primary substrate, which must be supplied exogenously if such biotransformations are to be encouraged. For some time, it has been hypothesized that methanotrophic bacteria can be stimulated by the addition of methane and oxygen to the vadose-zone environment to accelerate the in situ, cometabolic transformation of halogenated aliphatic compounds. Nonetheless, no practical bioremediation procedure has yet been attempted in which supplemental methane was provided for that purpose. In a research/demonstration project, Stanford University researchers added methane and oxygen to a contaminated aquifer, as opposed to unsaturated sediments. However, TCE biodegradation was enhanced only modestly (McCarty et al., 1989). Semprini et al. (1994) reported that in situ rates of degradation of TCE in an aquifer were generally lower than those observed with laboratory cultures. These results emphasize the need for laboratory experiments that approach conditions encountered in the field. The record of similar investigations involving ammonia-oxidizing bacteria is also brief, and there have been no successful field trials.

These observations suggest that our current conceptual picture of cometabolic TCE transformations and mathematical models based on those concepts is incomplete. Realistic models that take into consideration field-like laboratory experiments will help optimal design parameters for large-scale bioremediation of TCE.

Unfortunately, mathematical models of bioventing processes, including cometabolic bioventing, are inadequate for most practical purposes. Relatively few such efforts are described. Johnson et al. (1990), for example, represented "bioventing" as an extension of a venting model by introducing a sink term in the transport/reaction equation to account for microbial activity. The equation so modified, because of its relative complexity, has not been satisfactorily solved. A comprehensive description of this and related efforts to represent the biochemical transformation of hazardous organics within multiphase (gas/liquid/solid) transport models is provided in material that follows.

PHYSICAL DESCRIPTION OF THE VADOSE ZONE

System Components

The most common points of entry for contaminants into the soil lie either at the surface (spills or runoffs) or a few meters below it (underground storage containers or landfill leaks). As such, most pollutants must travel through the unsaturated or vadose zone before reaching the water table. The amount of a specific contaminant, the form in which it is released to the soil environment and its physical characteristics (solubility, volatility, hydrophobicity, polarity, etc.) will contribute to its mobility in this complex physical environment. Relevant physical characteristics of the environment itself include such soil properties as porosity, degree of water saturation, hydraulic conductivity, permeability, fraction of organic carbon, particle size, and degree of homogeneity. Unsaturated soils or sediments offer a special degree of complexity and a variety of potentially interrelated determinants of pollutant and transport characteristics. A brief description of the determinants of contaminant transport characteristics in this setting follows.

Contaminants and Nutrients

Our primary focus will be on chlorinated organic compounds that are susceptible to aerobic degradation. In general, the biodegradability of such compounds is inversely related to degree of halogenation, a relationship that is equally apparent in mammalian and microbial systems (Vogel et al., 1987). Aerobic biotransformations of perchloroethylenes, for example, have not yet been observed, and microbial systems that can catalyze the transformation of TCE number only a few (Ensley, 1991). Lesser chlorinated homologs (DCE isomers and vinyl chloride) are more appropriate targets for aerobic transformation processes. Although a degree of stereospecificity is apparent in the biodegradation of haloaliphatic and haloaromatic isomers, such effects play a secondary role to degree of halogenation as determinants of expected biodegradation rates. In a sense, trichloroethylene is among the greatest anthropogenic challenges that have been overcome by aerobic biochemical systems. Aerobic biotranformations of TCE remain sufficiently special to motivate reviews such as that of Ensley (1991), which provided kinetic data and systems descriptions. Due to the obvious environmental relevance of TCE and its (imperfect) resistance to aerobic biochemical transformations, it is afforded a degree of prominence in this chapter.

A primary substrate is an absolute requirement for cometabolic transformation of TCE. Methane, propane, ammonia, and toluene or phenol have been used as primary substrate, each for a specific type of degrader. None of these systems operates in the absence of molecular oxygen, which is a cosubstrate in necessary oxygenase reactions and serves as a terminal electron acceptor for microbial respiration.

The general concept in bioventing is that volatile and semivolatile contaminants can be transformed through the vadose zone within a motile gas phase. Since air is used for this purpose, there is an ample supply of oxygen for aerobic decomposition of biodegradable contaminants. Other nutrients such as nitrogen (usually ammonia or nitrate-N) and phosphorus (phosphate) are sometimes required to promote bacterial growth in efficient bioventing. However, phosphorus must be pounded in water, and there are few examples in which gaseous ammonia and molecular nitrogen have successfully been used to stimulate microbial growth and activity (Chu et al., 1995).

Phases of the System

The vadose zone is a physiologically complex region due to the presence of multiple phases and accompanying interfaces. Gaseous, aqueous and solid phases are always present. In addition, specific applications may depend primarily on the existence of moisture or natural organic matter (NOM), an organic phase, nonaqueous-phase liquids (NAPLs), and biotic components. This description is shown schematically in Figure 4.

It is reasonable to expect that the relative ease of bulk transport of the gas phase or the liquid phase depends on sediment void volume, pore size

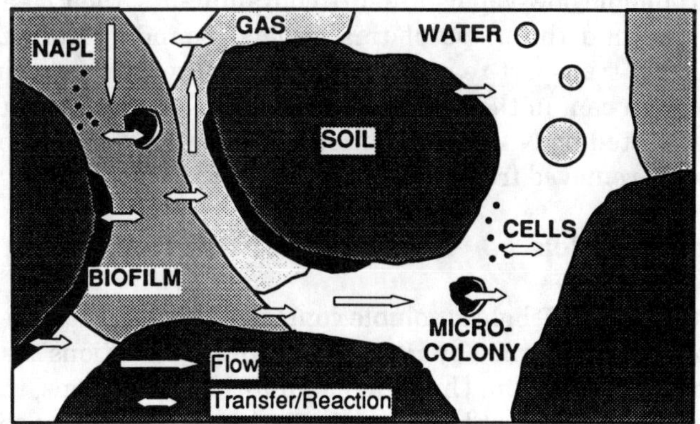

Figure 4. Components and interactions in a vadose zone system.

distribution, and water content. On relatively dry soils, water should be confined by capillarity to the smaller void spaces, making the gas the continuous phase and facilitating gas transport in response to a gradient in potential. At high water content, gases are confined to the large void spaces as bubbles, and the aqueous phase is continuous. Convective transport of the gas phase should not then, be expected.

It is common to treat NOM as a distinct, homogenous phase for representation of contaminant partitioning. The extent of that phase in a particular sediment is then estimated via measurement of the sediment fraction organic carbon (f_{oc}) using defined procedures.

Microorganisms have been represented as existing in uniformly distributed biofilms (sometimes only a few microorganisms in thickness) on solid surfaces, in randomly distributed colonies that consist of some representative number of cells, or uniformly distributed in the aqueous phase. None of these physical representations has been clearly established as superior to the others in specific modeling applications, perhaps because none faithfully represents the actual distribution of active microbes in the vadose zone.

Contaminants that are slightly soluble in water may exist as NAPLs in the vadose zone. The behavior of that phase is determined in part by NAPL density and interfacial tensions among phases. Even when the existence of a separate phase is not apparent on a bulk scale, slightly soluble contaminants may exist in an essentially pure form in small pore spaces, displacing water based on more favorable surface energy considerations. Slow pollutant desorption from these NAPL-filled pores continuously contaminates the bulk or convective phase, impeding the kinetics of engineered groundwater remediation activities. Although such effects may be of great practical importance, particularly among relatively hydrophobic, low-vapor-pressure contaminants, their treatment is essentially beyond the scope of this review. Boundary conditions for multiple-phase transport-reaction models that are presented in a subsequent section can, in theory, be adjusted to reflect contaminant availability, as affected by NAPL dissolution, recovery from micropores that are physically removed from the primary convective zones.

Partitioning Models

The distribution of slightly soluble contaminants between the gas and aqueous phases is governed by Henry's law under conditions that permit attainment of equilibrium. Under nonequilibrium conditions, the rate of intraphase mass transport between phases is proportional to the difference between the actual concentration in a phase and the equilibrium solute concentration. Consequently, contaminant partitioning and/or the direction and rate of net transport between the gas and liquid phases are sensitive to the Henry's law constant.

Intermittent partitioning between the liquid phase and an organic phase such as NOM is essentially modeled in the same way; that is, under equilibrium conditions, concentrations in adjacent aqueous and organic phases are proportional. The constant of proportionality, or the partition coefficient, is strongly affected by contaminant hydrophobicity, as indicated by octanol/water partitioning or a suitable surrogate parameter. In this approach, NOM in soil or sediment is treated as a homogenous phase, in the same manner as a nonaqueous-phase liquid. It follows that contaminant partitioning between liquid and organic phases is strongly affected by f_{oc} and the octanol/water partition coefficient of the contaminant. This is shown graphically in Figure 5. It is worth noting, however, that octanol/water partition coefficients are measured at an aqueous-phase pH of 1.0, and reported values should not be used to anticipate the distribution of organics that behave as weak acids.

In sediments of low organic content, contaminant partitioning between the liquid and solid phases is governed by adsorption at the solid-liquid interface. Both isotherms are commonly used to represent heterogeneous sorbate distributions under equilibrium conditions. The biodegradability of adsorbed organics has been questioned although sorption could, in theory, concentrate microbial substrates to concentrations that would support the growth of attached bacteria.

From the foregoing discussion, it is evident that volatility and hydrophobicity direct the distribution of contaminants among multiple phases in the vadose zone. Compound volatility and low hydrophobicity (as evidenced by relatively high Henry's constant and low octanol/water partitioning coefficient) favor partitioning in the gas phase, which is shown in Figure 5. The possibility of convective transport within the gas phase and remediation based on bioventing strategies are favorable for such compounds. Needless to say, these are not the only factors in selection of a remediation strategy, however; contaminant degradability, for example, should not be overlooked in preparing such a strategy.

Soil Properties

COMPOSITION

Soil characteristics are derived from its composition and the interaction among components, which creates soil structure. Since soils are formed by weathering of rocks and minerals, it is expected that their composition will be similar to that of the parent rock. Soil structure could range from

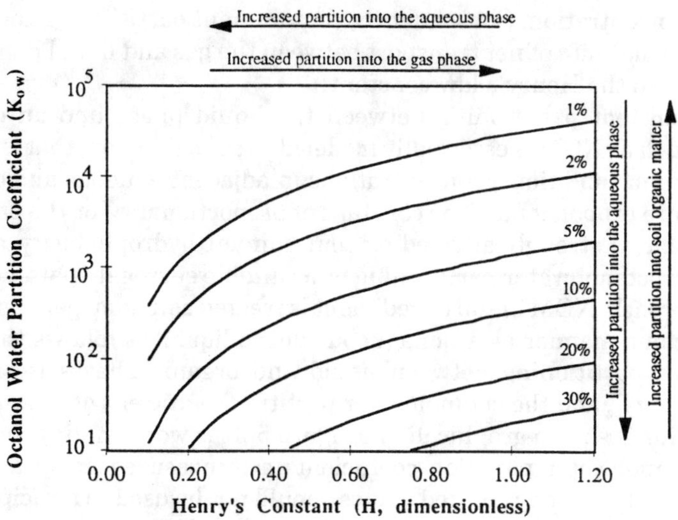

Figure 5. Distribution chart of contaminants in the gas phase in terms of hydrophobicity and volatility. Each contour represents a constant fraction of the total mass of contaminant that is present in the gas phase under equilibrium conditions. The contaminant was assumed to distribute among organic matter and the aqueous and gaseous phases in unsaturated soil. Arrows on the sides of the plot indicate the phase into which a given contaminant would preferentially partition based on its relative hydrophobicity (K_{ow}) and volatility (H_c) for given soil properties. In general, venting operations would be favored when most of the contaminant is in the gas phase. In analogy to data presented in Figure 1, the fact that a specific contaminant does not fall under the bioventing-favored section of the plot does not mean that bioventing cannot be used as a feasible degradation method. This simply implies that, under the same circumstances, longer biodegradation times will be required. The simulation was based on the following soil properties: density = 1.85 g soil/cm^3 soil; porosity = 0.35; 30% water saturation and f_{oc} = 0.005 g of organic carbon per gram of soil.

soils with low bulk density and high porosity to more compact, low-permeability soils (Yong et al., 1992).

The inorganic fraction of the soil consists primarily of alumina silicates. These vary from highly crystalline to amorphous. Organic components are the result of microbial degradation of higher organisms or their excretions and the microorganisms themselves. They range from plant tissue, in various degrees of decomposition, to highly humidified and very stable material. The presence of organic material in the surface of soil particles frequently determines the surface characteristics and, therefore, the type and strength of adsorbed contaminants and bacteria.

A direct relationship between the soil organic content and the total amount of pollutant absorbed per unit of soil mass has been described by Karickhoff et al. (1979) and Mackay and Roberts (1985).

PERMEABILITY

The permeability k, is defined as the fluid transmission capability of the soil, and it is entirely dependent on the soil characteristics. It refers to the ease with which a fluid (in the case of bioventing, air, and volatile contaminants) flows through the soil. Permeability coefficient, on the other hand, is a term that was used to describe hydraulic conductivity of the soil (K). In this chapter, the term *permeability* will be applied to the property of the soil as defined above and hydraulic conductivity as the proportionality constant in Darcy's law, which is given by

$$v = K \cdot \frac{\Delta h}{\Delta x} \qquad (1)$$

an empirical equation that describes the relationship between the flow of a fluid and the hydraulic gradient in porous media. Here v is the specific discharge of the fluid (with units of velocity), K is the hydraulic conductivity (m/s), and $\Delta h/\Delta x$ is the hydraulic gradient in the x-direction, or direction of flow. There is, however, a direct relationship between hydraulic conductivity (K) and permeability (k).

$$K = \frac{k \cdot \rho \cdot g}{\mu} \qquad (2)$$

where g is the acceleration of gravity and μ and ρ are the dynamic viscosity and the density of the fluid, respectively. Therefore, there is only one value for the permeability of a soil but several values for K, which is fluid-specific. The importance of knowing values such as porosity, gas, and liquid hydraulic conductivities lies in their contribution to the rate of air flow in the bioventing operation and determination of head requirements like pumping or suction.

BIOCHEMICAL TRANSFORMATIONS

A primary advantage of bioventing as a hazardous waste remediation strategy arises from the availability of molecular oxygen as electron acceptor in biochemical reactions; that is, bioventing is best suited to promote the destruction of volatile organics that can serve as substrates for aerobic vadose zone microorganisms. Lighter components of hydrocarbon fuels, such as the BTEX compounds (benzene, toluene, ethylbenzene, and xylene), satisfy the basic criteria for bioventing and are the most common targets for such operations. In order to extend the utility of bioventing operations for the destruction of volatile, but less biodegradable, compounds, it may be necessary to stimulate groups of bacteria with specific, sometimes unique metabolic capabilities or to encourage cometabolic activities by providing a suitable primary substrate in addition to molecular oxygen. In the latter case, the engineered process would more properly be called cometabolic bioventing.

The halogenated ethenes, up to, but not including, perchloroethene, offer appropriate targets for cometabolic bioventing. These are very common pollutants with exceptional volatility and reasonable hydrophilicity; however, their use as an energy source is questionable. There have been no convincing observations of trichloroethene (TCE) oxidation, for example, coupled to the generation of useful metabolic energy. Consequently, fortuitous transformations may be the only aerobic avenues available for TCE transformation, and a suitable primary substrate must be provided in order to support growth of the responsible bacterial population. Where these obstacles to (aerobic) TCE degradation are not sufficiently daunting, the enzymatic transformation of TCE has itself proven challenging, and only a few capable systems are known. These will be reviewed sequentially. In each case, the enzyme that initiates the transformation of TCE is a monooxygenase, and the transformation product, a first intermediate, is TCE epoxide (as shown schematically in Figure 6).

Methane Oxidizers

Methanotrophic bacteria include all species that are able to grow using methane as a sole source of carbon and energy (Ensley, 1991). They are virtually ubiquitous in nature but exist in very high numbers only at interfaces between anaerobic and aerobic environments, where methane and molecular oxygen may exist simultaneously. Such habitats include the waters above anaerobic sediments, especially freshwaters, in which sulfate-reducing bacteria are less capable competitors, and the surface soils of methanogenically active sanitary landfills.

Methane monooxygenases (MMO) catalyze the initial attack on methane by methanotrophic bacteria. This task is sufficiently difficult that nature tolerates an unusual lack of specificity in the substrates that MMO will oxygenate. MMO initiates the aerobic transformation of a variety of haloorganic compounds. Their characteristic inefficiency or nonspecificity makes methanotrophs extremely attractive candidate organisms for the in situ, cometabolic destruction of short-chain, halogenated aliphatic compounds.

Based on a number of biochemical characteristics, methanotrophs are classified as Type I, Type II, or Type X. Type X methanotrophs exhibit a mixture of characteristics that are usually assigned to either Type I or Type II species. Type II methanotrophs produce both a soluble MMO (sMMO) and a particulate (membrane-associated) MMO (pMMO) (Tsien and Hanson, 1992; Anthony, 1986; Oldenhuis et al., 1989). Even a minuscule amount of available Cu(II) (15 ppb) selects for pMMO at the expense of sMMO expression. Significantly, sMMO is much more adept at catalyzing cometabolic transformations, including the transformation of TCE, than is pMMO (Colby et al., 1977; Patel et al., 1982; Tsien and

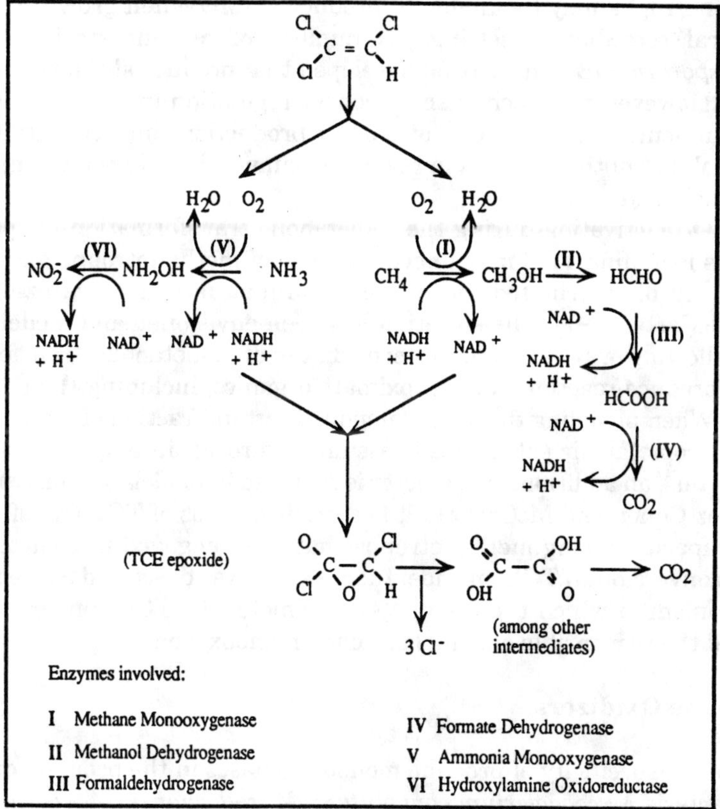

Figure 6. Cometabolic degradation of TCE by methanotrophs and nitrifiers.

Hanson, 1992). *Methylosinus trichosporium* OB3b expressing sMMO has exhibited the highest known rate of TCE biodegradation (Ensley, 1991). Unfortunately, encouragement of sMMO expression has proven problematic because copper levels in virtually any polluted environment will select for pMMO. Graham (1992) found that copper chelators, including ethylenediaminetetraacetic acid (EDTA), triethylenetetramine (Trien), and others, offer no impediment to the ability of *M. trichosporium* OB3b to acquire Cu(II) in batch cultures and can actually promote pMMO expression.

Copper appears to have a central regulatory role in methanotrophs, affecting the expression of sMMO, membrane organization, and growth rate (Green et al., 1985; Leak and Dalton, 1986; Park et al., 1991; Stanley et al., 1983). In pure cultures of *M. trichosporium* OB3b, the presence of as little as 0.2 μM Cu(II) in the growth medium eliminates sMMO synthesis. Since copper generally exceeds such levels in polluted waters, methanotrophs may be unable to produce sMMO when grown in most practical remediation settings. A number of mutant strains of *M. trichosporium* OB3b have been developed that produce sMMO constitutively. However, the mechanism of copper regulation in this system, the environmental specifications of sMMO production, and the utility or survivability of the mutants in real environmental settings all remain to be established.

MMO inactivation during the cometabolic transformation of specific targets including TCE may impede the use of methanotrophs for bioremediation of chlorinated solvents in certain situations (Alvarez-Cohen and McCarty, 1991). The situation has been envisioned and modeled in the following way: unstable intermediates of cometabolic transformations produce reactions with proximate organics, including the enzyme itself. When modeling this phenomenon, a certain fraction of transportation of cometabolic substrates is assumed to result in enzyme inactivation. Thus, an additional parameter is required in models so constructed. Alvarez-Cohen and McCarty (1991) described a loss of TCE transformation capacity among methanotrophs that were engaged in cometabolic TCE conversion to TCE epoxide. Inactivation was described as a suicide mechanism in which the product(s) of cometabolic TCE conversion attacked the responsible enzyme, methane monooxygenase.

Propane Oxidizers

The nonspecificity of propane monooxygenases in the propane-oxidizing bacteria *Mycobacterium convolutum*, *M. rodochorus*, *M. vaccae*, and perhaps others suggests that this class of bacteria can also be stimulated in situ to initiate the cometabolic destruction of TCE and less halogenated chloroethenes (Vestal and Perry, 1969; Wackett et. al., 1989). Of the species tested, *M. vaccae* JOB5 exhibited the greatest potential for cometabolic transformation of TCE and vinyl chloride. The relative convenience of providing propane as a primary substrate in the vadose zone suggests that propane oxidizers are reasonable candidate microorganisms for remediative systems based on cometabolic bioventing.

Ammonia Oxidizers

In a similar fashion, ammonia-oxidizing bacteria derive their energy for growth exclusively from the oxidation of ammonia to nitrite. Capable species are chemo(litho)autotrophs.

There are five genera in this group: *Nitrosomonas, Nitrosococcus, Nitrospira, Nitrosolobus,* and *Nitrosovibrio.* The best studied species of ammonia oxidizers is *Nitrosomonas europaea,* an obligate chemolithotroph that initiates nitrification by the reductant-dependent oxidation of ammonia to hydroxylamine using ammonia monooxygenase (AMO). Reductant for AMO-catalyzed reactions is provided by the further oxidation of hydroxylamine to nitrite by hydroxylamine oxidoreductase (Figure 6). There is evidence that the AMO of *N. europaea* can also initiate cometabolic transformation of hydrocarbons (Hyman and Wood, 1983; 1984) and haloaliphatic compounds, including TCE (Arciero et al., 1989; Vanelli et al., 1990). The situation is entirely parallel to that of the methanotrophs in that halogenated substrates are transformed by both AMO and MMO without any direct metabolic benefit. Both enzymes consume reductant in the process, which must be replenished by metabolic activities at some expense to the organisms. For this reason, a primary or energy-yielding substrate is necessary for sustained cometabolic activity.

The foregoing suggests that these ubiquitous soil bacteria may be stimulated to perform remediative functions by adding gaseous ammonia and molecular oxygen to the vadose zone. Among the advantages that are enjoyed by nitrifiers in this context is their apparent tolerance to the intermediates of TCE cometabolism. Arciero et al. (1989) reported that ammonia oxidation by *N. europaea* is not inactivated during short-term exposure to TCE.

Aromatic Oxidizers

Several species of the genus *Pseudomonas* contain enzyme systems that can undergo degradation of some aromatic hydrocarbons, such as benzene, toluene, phenol, and cresol. The initial step in aromatic degradation is performed by mono- and dioxygenases (Ensley, 1991), which have also proved to be capable of degrading TCE.

The expression of some aromatic oxygenase systems requires the induction by the presence of the compound subject to degradation. For example, *P. putida* strain F1 expresses a dioxygenase in the presence of toluene, which enables the organism to degrade TCE (Wacket and Gibson, 1988; Nelson et al., 1988). The toluene dioxygenase system doesn't have, however, the capability to degrade PCE, VC, or ethylene. Since the transformation is cometabolic, the inducer (which is often used as carbon and energy source) will act as a competitive inhibitor of cometabolic transformation.

A recombinant *Escherichia coli* containing the genes encoding for toluene dioxygenase was constructed to overcome inhibition by the inducer (Zylstra et al., 1989). Although it did not require the presence of toluene as inducer, the degradation rates were lower than with the *P. putida* strain from which the genes were cloned, possibly because of differences in cytotoxic effects between *P. putida* and *E. coli*.

Other aromatic oxygenases found to degrade TCE include *ortho-*, *para-* and *meta*-monooxygenases present in several different species, as shown in Table 1.

PROCESSES IN THE VADOSE ZONE

Factors affecting the fate of most organic contaminants, nutrients, and microorganisms involve physical, chemical, and biological processes.

Table 1. Characteristics of TCE degradation in aerobic systems (after Ensley, 1991).

Enzyme System	Inducer	TCE Degrad. Rate[a]	Species
Ammonia MO[b]	None	1	*N. europaea*
Aromatic oxygenases			
Toluene DO[b]	Phenol, toluene	1.8	*P. putida* F1
ortho-MO	Phenol, toluene, cresol	8	*P. cepacia* G4
para-MO	Toluene	1–2	*P. mendocina*
meta-MO			
Methane monooxygenase (MMO)			
Particulate MMO	Constitutive	Low	
Soluble MMO	Low Cu(II)	150	*M. thricosporium* OB3b
Propane MO	Propane	High	*M. vaccae* JOB5

[a]nmol/min-mg protein.
[b]MO = monooxygenase, DO = dioxygenase.

Physical phenomena include advection, dispersion, solubility, volatilization, and sorption. The first two refer to mass transport in a single phase, and these can take place in the liquid and/or in the gas phases. Solubility and volatilization are interface mass transfer mechanisms that take place between the gas phase and the liquid phases (either aqueous or NAPL) or between both liquids. Sorption describes another set of mass transfer mechanisms that occur between the solid phase (soil) and all other phases. Chemical processes are restricted, normally, to contaminants, and they comprise hydrolysis and oxidation/reduction reactions. Biological processes include the biodegradation of pollutants and the kinetics of both substrate consumption and cell growth mechanisms.

Physical Processes

The increasing application of SVE and bioventing demands the need to improve the understanding of the processes that limit their effectiveness. Experimental work suggests that some of the main factors affecting efficiency are limitations imposed by mass transfer in the liquid and gas phases. An effective model should account for the mass transfer processes among the constituent phases in order to describe realistically SVE and bioventing operations. A description of the main variables involved in these processes is presented in a general mass transport expression.

MASS TRANSPORT IN A SINGLE PHASE

Advection
This is the principal transport mechanism for pollutants and contaminants within the gas or within the liquid phase when the soil is saturated. In the case of the vadose zone, advection is only important when the water flow is in the vertical direction, normally during recharging. Contaminant migration during recharging, however, is not a usually modeled event, although the same type of equations that will be described later can be applied to the process. In general, advection in a bioventing operation will be dominant in terms of mass transport of nutrients and contaminants in the gas phase where dispersion effects could be considered negligible.

Dispersion
Hydrodynamic dispersion includes the combined effect of molecular diffusion and mechanical micromixing. Fick's first law is usually used to describe the process and gives the net flux of the substance diffused over a given gradient. Tucker and Nelken (1982) have used the following

modified equation to account for the diffusion of a pollutant in the porous media of the vadose zone.

$$D_C = \frac{J_C}{\Delta[C_C]} \tag{3}$$

Mixing at a micro-scale plays an important role in the value of the diffusion coefficient, especially in flow through porous media (e.g., soil) where different diameter pore sizes, pore lengths, and tortuosity heavily influence solute migration. When this type of migration of the solute is considered, in addition to molecular diffusion, the parameter is called effective, or apparent, dispersion coefficient. The following relationship has been used to determine this effective diffusivity in soil operations (Bonazountas and Kallidromitou, 1993; Harleman and Rumer, 1962):

$$D_S = \tau \cdot D_C + \alpha \cdot v^n \tag{4}$$

where τ is the tortuosity factor (generally in the range of 0.01 and 0.5), v is the pore water velocity, and α and n are empirically determined soil constants. α is generally called dispersivity and could have values as large as 100 m, while n could range between 1 and 2 (Freeze and Cherry, 1979). For most practical purposes, n has been taken as unity for granular soils.

The general equation for mass transport in a single phase (gas or liquid) that includes advection and dispersion for a component of concentration C is given by

$$\frac{\partial C}{\partial t} = D_h \frac{\partial^2 C}{\partial l^2} - v_1 \frac{\partial C}{\partial t} + S_i - R_i \tag{5}$$

Steady-state flow velocity conditions and unsteady contaminant concentrations are assumed in this equation. Chemical and biological transformations are included in this advection-dispersion equation and given by the source (S_i) and sink (R_i) terms.

MASS TRANSFER BETWEEN GAS AND LIQUID PHASES

Solubilization

The solubility in water is one of the most important properties of a contaminant to consider when determining its distribution in the soil. There are several methods to estimate the solubility of an organic contaminant in an aqueous solution, most of them are regressions based on the octanol-water partition coefficient (K_{ow}). One of the most commonly used empirical equations (Sposito, 1984; Coates et al., 1985) is described by

$$\log S = -1.12 \cdot \log K_{ow} + 7.3 - 0.15 \cdot T \tag{6}$$

When an NAPL is in direct contact with the aqueous phase, continuous contamination of the latter might occur by slow dissolution of the organic chemical. The contribution to the total aqueous concentration from this type of mass transfer is significant and, in most cases, should be included in any mathematical analysis. The above relationship is also valid here and could be incorporated as a source term (S_i) in the general transport Equation (5) applied to the aqueous phase.

Volatilization

The transport of a compound from the liquid to the gas phase may represent an important pathway for chemicals with high vapor pressures or low solubilities. Volatilization rates are a function of the compound physical properties such as solubility, vapor pressure, and structure. Soil properties and environmental conditions, including water and organic carbon contents, as well as temperature, affect the partitioning of a chemical between water soil and air.

The most common expression used to predict the extent of volatilization of a chemical is Henry's law. This empirical expression describes the relationship between equilibrium concentrations in the gas and liquid phases for a given compound, and it is usually represented by

$$C^g = H \cdot C^l \tag{7}$$

MASS TRANSFER BETWEEN SOLID AND FLUID PHASES

Sorption

Adsorption and desorption are both included in the general term sorption, which indicates the partitioning process of solutes between the soil and other phases. For organic pollutants, the most important sorption mechanism is the partitioning between water and the organic carbon content of the soil (Mackay and Roberts, 1985).

Adsorption and desorption are commonly treated as a fully reversible process, although sometimes hysteresis is observed (Bonazountas and Kallidromitou, 1993). Among the most commonly used mathematical expressions for sorption in the literature are the Langmuir and Freundlich isotherm models. The Langmuir model was developed for single-layer adsorption, assuming that all sites on the soil surface are equivalent and that there is no migration of solute molecules on the surface. This model is described by

$$\frac{dC_C^s}{dt} = k_{sw}(C_{\max C}^s - C_C^s); \quad C_{\max C}^s = \frac{M_C^{s^*} \cdot b_s \cdot C_C^l}{(1 + C_C^l)} \tag{8}$$

while the Freundlich isotherm is given by

$$C_C^s = k_f \cdot C_C^{l1/n} \tag{9}$$

For organic molecules at low concentrations, the amount of sorbed solute is proportional to the concentration in soil moisture. Therefore, in the Freundlich isotherm parameter, n in most cases is equal to 1. For organics, k_f is usually called the soil sorption coefficient (k_p) and provides a measurement of the affinity of a chemical to attach on the surface of soil particles.

Since the dominant sorption mechanism for organic molecules is the hydrophobic interactions between this molecule and the natural organic matter in the soil, k_p can be correlated with the fraction of organic carbon (f_{oc}) in the soil as

$$k_p = K_{oc} f_{oc} \tag{10}$$

The K_{oc} value (proportionality constant specific for each chemical) represents a hypothetical partition coefficient for a given chemical assuming that the soil is 100% organic carbon. This relationship is valid for soils with high organic contents (f_{oc} higher than 0.001).

Since K_{oc} values for many compounds are not known, several investigators (Chiou et al., 1982; Briggs, 1981) have developed correlations between K_{oc} and more easily available properties, like solubility (C_S) and octanol-water partition coefficient (K_{ow}). The relationships are regression equations involving several compounds of similar structure and properties and are usually expressed in a log-log form as

$$\log K_{oc} = a \cdot \log K_{ow} + b \tag{11}$$

$$\log K_{oc} = a \cdot \log C_s + b \tag{12}$$

Chemical Processes

Naturally occurring chemical reactions profoundly affect the fate of pollutants in the environment. They can be abiotic in nature or microorganism-mediated. For organic molecules, hydrolysis, elimination, and oxidation-reduction reactions are the most common type of abiotic transformations.

Hydrolysis involves the formation of a new carbon-oxygen bond, replacing generally a carbon-halogen in the original molecule. Therefore, the products of these reactions are generally alcohols. For halogenated aliphatics, the rate of hydrolysis decreases as the degree of halogenation increases, and normally higher rates are observed for compounds where bromine is the halogen compared to those of chlorinated molecules (Vogel et al., 1987).

Halogenated alkanes in water also undergo dehydrodehalogenation reactions to produce alkenes. This elimination reaction is favored for highly substituted molecules with halogens in the same carbon atom.

Some abiotic redox reactions are initiated by free radicals already present in solution, while others require transition metals to act as electron donors or acceptors. Redox reactions catalyzed by living systems are, in general, more efficient than those that occur with no microbial presence.

Biological Processes

Biodegradation processes are commonly described by Michaelis-Menten kinetics (Alvarez-Cohen and McCarty, 1991; Criddle, 1993; Landa et al., 1994). In cometabolic conversions, however, the target compound (e.g., TCE) and primary substrate compete for the same enzyme active site. A competitive inhibition model is then used to represent the kinetics of both substrate and cometabolic conversion such as

$$q_S = -q_{\max S} \frac{C_S}{C_S + K_{SS}\left(1 + \dfrac{C_C}{K_{SC}}\right)} \tag{13}$$

and

$$q_C = -q_{\max C} \frac{C_C}{C_C + K_{SC}\left(1 + \dfrac{C_S}{K_{SS}}\right)} \tag{14}$$

The parameters $q_{\max S}$ and $q_{\max C}$ are corresponding maximal rates of conversion, i.e., the rates observed in the presence of excess substrate or cometabolite, ignoring possible inhibitory roles of the hazardous target.

Since these values are proportional to cell density or enzyme concentration, they must be normalized using some measure of cell density or culture activity in order to facilitate comparison among different bacterial systems. Thus, conversion rate is coupled to cell number or culture growth rate. K_{SS} and K_{SC} are the respective half-saturation constants for enzymatic conversion of substrate and cometabolite or the concentrations at which the uninhibited reactions proceed at one-half their maximal rates. Low values of the half-maturation constants are inversely related to the affinity of the substrate or inhibitor/cometabolite for the enzyme active site.

Cell growth rate is generally proportional to the rate of substrate oxidation with a correction for cell death. The specific growth rate is thus:

$$q_X = Y_{X/S} \cdot q_S - b_X \tag{15}$$

When suicide inactivation of substrate-utilizing enzyme system accompanies cometabolic activity, a second culture decay term can be added:

$$q_S = Y_{X/S} \cdot q_S - b_X - \frac{q_C}{T_C} \tag{16}$$

where T_C is an inactivation coefficient that reflects the number of cometabolic conversions that are necessary, on average, to inactivate a single enzyme active site. This coefficient was introduced by Alvarez-Cohen and McCarty (1991) as a measure of transformation capacity in a culture of resting cells.

FATE AND TRANSPORT SYSTEMS

Mathematical Modeling

The development of a mathematical model in venting/bioventing processes involves an understanding of the interacting zone, including soil characteristics, flow and fluid patterns, and microbial systems as previously described. In this work, we try to present models that include the most relevant process variables and operating conditions. We restrict the analysis to deterministic models on aspects of fate and transport in the vadose zone under venting and/or bioventing conditions. Mathematical deterministic models for venting processes when combined with current biodegradation models have the potential to describe effectively bioventing operations. Information about stochastic models for bioremediation can be found elsewhere (Schäfer and Kinzelbach, 1992).

Several biodegradation models based on biofilms, suspended growth, and microaggregation are described, as well as modeling bioventing strategies for contaminant degradation in the field and in soil-packed columns.

A general model accounts for mass transfer between the constituent phases, advection, dispersion, biodegradation, and biomass production (Abriola, 1989). Most of the actual fate and transport models are basically modifications and additions to the general macroscopic transport equation for all the system components in a porous medium, namely a mass balance given by

{accumulation = advective flux − dispersive flux

− phase mass exchange + sink/source}

or mathematically expressed as

$$\frac{\partial}{\partial t}(\rho\varepsilon\omega_i) = \nabla\cdot(\rho\varepsilon v\omega_i) - \nabla\cdot(\rho\varepsilon D_i\cdot\nabla\omega_i) - E_i \pm G_i \qquad (17)$$

The E_i term considers mass transfer to and from other phases in the system (volatilization, solubility, sorption, etc.), while G_i is the internal source/sink component in the phase and accounts for microbial degradation. The second and third terms represent advective and diffusive fluxes of one component in the transport medium. Simplifications to this general transport equation are generally considered for most practical applications. For example, dispersive fluxes can be neglected when molecular diffusion and mechanical mixing contributions are negligible compared to those of advective flow. The following are models developed under specific conditions.

Biofilm Models

A biofilm model was developed by Chen et al. (1992) to describe the biodegradation of a dissolved hydrocarbon plume. In this model, the gas phase is assumed stationary at atmospheric pressure and without density gradients. A mass balance for bulk fluid phase transport, considering microorganism populations as fully penetrated biofilms and using Henry's law for liquid-gas partition, reduces the general mass transport equation for each component as

$$\frac{\partial}{\partial t}[(\rho^l\varepsilon^l + \rho^g\varepsilon^g H + \rho_b\rho^l K_{dc}\omega_C^l)] + \nabla\cdot(\rho^l\varepsilon^l v^l\omega_C^l)$$

$$- \nabla\cdot[(\rho^l\varepsilon^l D_C^l + \rho^g\varepsilon^g D_C^g H)\cdot\nabla\omega_C^l] = -k_C\,\frac{A_X C_X^m}{m_X^m}(\rho^l\omega_C^l - C_C)$$

$$(18)$$

The right-hand side of this equation represents the mass transfer across the biofilm boundary layer. This term for each component includes nutrients, electron acceptors, and target compounds. For contaminants, it becomes

$$k_C\,\frac{A_{Xm}C_X}{C_{Xm}}(\rho^l\omega_C^l - C_C)$$

$$= \sum_E q_{\max C}\left(\frac{C_C}{K_{SC}+C_C}\right)\left(\frac{C_E}{K_{SE}+C_E}\right)\left(\frac{C_S}{K_{SS}+C_S}\right)\prod^{CE}(C_E)C_X \quad (19)$$

For electron acceptor utilization, it is described by

$$k_E\,\frac{A_{Xm}C_X}{C_{Xm}}(\rho^l\omega_E^l - C_E)$$

$$= \sum_C Y_{E/C}q_{\max CE}\left(\frac{C_C}{K_{SC}+C_C}\right)\left(\frac{C_E}{K_{SE}+C_E}\right)\left(\frac{C_S}{K_{SS}+C_S}\right)\prod^{CE}(C_E)C_X$$

$$(20)$$

In this model, oxygen and nitrogen were used as electron acceptors. That is the model accounts for anaerobic and denitrifying populations, respectively. The growth substrate equation can be obtained as

$$k_S\,\frac{A_{Xm}C_X}{C_{Xm}}(\rho^l\omega_S^l - C_S)$$

$$= \sum_C\sum_E Y_{S/C}q_{\max C}\left(\frac{C_C}{K_{SC}+C_C}\right)\left(\frac{C_E}{K_{SE}+C_E}\right)\left(\frac{C_S}{K_{SS}+C_S}\right)\prod^{CE}(C_E)C_X$$

$$(21)$$

The substrate (S) is considered the limiting nutrient, such as any carbon and/or energy source associated with its respective electron acceptor. Biomass growth and decay of the attached biomass populations are described by

$$\frac{dC_X}{dt} = \sum_E \left\{ q_{\max C_E} Y_{X/S_E} \left(\frac{C_C}{K_{SC_E} + C_C} \right) \left(\frac{C_E}{K_{SE_C} + C_E} \right) \left(\frac{C_S}{K_{SS_E} + C_S} \right) \right.$$

$$\left. \times \prod^{CE} (C_E) - b_{SE} \right\} C_X \tag{22}$$

This equation is used for each of the bacterial groups involved in the degradation, and yield (Y) and decay (b_X) coefficients correspond to one bacteria group growing on a specific substrate and under its corresponding respiration system.

The resulting systems represent a complicated task to model. In fact, in order to obtain a reasonable mathematical description of the process, the following assumptions were made: 1) all microorganisms are attached in a biofilm; 2) microcolonies increase in biomass, but not in the size of colonies; 3) no chemical reaction occurs other than consumption or degradation; 4) degradation of each target is independent of the others; and 5) there are no interactions between microorganisms.

Suspended Growth Models

Semprini and McCarty (1992) developed a nonsteady-state model to simulate the results of field experiments where chlorinated ethenes were cometabolically degraded by methanotrophic bacteria. The model considers microbial growth, utilization of the electron donor (methane, also a substrate) and the electron acceptor (oxygen). A kinetic model for the cometabolic transformation to account for competitive inhibition by methane was used in order to match the biotransformation from the stimulation of methanotrophic bacteria in field observations. The substrates and electron acceptors are supplied as dissolved gases in groundwater through injection wells. In this model, the biomass growth and decay rates are assumed to be a function of both electron donor and acceptor, in a double limiting nutrient kinetic scheme given by

$$\frac{\partial C_X}{\partial t} = C_X q_{\max X} Y_{X/S} \left(\frac{C_S}{K_{SS} + C_S} \right) \left(\frac{C_E}{K_{SE} + C_E} \right) - b_X C_X \left(\frac{C_S}{K_{SS} + C_S} \right)$$

$$\tag{23}$$

For the electron donor transport and uptake, it becomes

$$\frac{\partial C_S}{\partial t} = \frac{D_h}{Y_{X/S}} \frac{\partial^2 C_S}{\partial C_X^2} - \frac{v}{Y_{X/S}} \frac{\partial C_S}{\partial C_X} - \frac{q_{max} x C_X}{Y_{X/S}} \left(\frac{C_S}{K_{SS} + C_S} \right) \left(\frac{C_E}{K_{SE} + C_E} \right) \tag{24}$$

and for the electron acceptor transport and uptake:

$$\frac{\partial C_E}{\partial t} = \frac{D_h}{Y_{X/E}} \frac{\partial^2 C_E}{\partial C_X^2} - \frac{v}{Y_{X/E}} \frac{\partial C_E}{\partial C_X} - \frac{q_{max} x Y_{E/S} C_X}{Y_{X/E}} \left(\frac{C_S}{K_{SS} + C_S} \right) \left(\frac{C_E}{K_{SE} + C_E} \right)$$

$$- \frac{d_E f_d C_X}{Y_{X/E}} \left(\frac{C_E}{K_{SE} + C_E} \right) \tag{25}$$

In this situation, a biomass inhibition/deactivation term can be expressed as

$$\frac{dF_a}{dt} = -b_d F_a \tag{26}$$

and a contaminant sorption term by the equation:

$$\frac{dC_C^s}{dt} = \alpha(K_d C_C - C_C^s) \tag{27}$$

while the contaminant transport and degradation is provided by

$$\frac{\partial C_C}{\partial t} = D_h \frac{\partial^2 C_C}{\partial C_X^2} - \bar{v} \frac{\partial C_C}{\partial C_X} - \frac{\rho^s}{\phi} \alpha(K_d C_C - \bar{C}_C)$$

$$- F_a C_X q_{max\,C} \left(\frac{C_C}{K_{SC} + C_C + \dfrac{C_S}{K_i}} \right) \left(\frac{C_E}{K_{SE} + C_E} \right) \tag{28}$$

This model previously developed for methanotroph growth (Semprini and McCarty, 1991) includes relevant microbial and transport processes. The model simulations allow a preliminary evaluation of in situ bioremediation scenarios, which could help determine the effectiveness to a particular application and in the design of a specific process.

Microaggregates Models

The role of aggregates and their effects on bioremediation of soil has received very little attention. Dhawan et al. (1991) developed a model to account for this physical situation under specific conditions. The CAB model, as they called it, considers cases where contaminants aggregate. The efforts to model this type of degradation is somehow complicated by the fact that the microorganisms responsible for the remediation are present in biofilm aggregates as well. The CAB model has helped elucidate the fate of contaminant under these conditions. The CAB model considers that contaminants are uniformly deposited in soil aggregates and that microorganisms are present in suspended forms and attached as micro-colonies to the solid surfaces. The model allows their growth by consumption of organic contaminant and oxygen. The analysis is performed in spherical coordinates for the aggregates. Local adsorption/desorption equilibrium is considered in the aggregate. The reaction terms for substrate, oxygen, and biomass are expressed in terms of Monod kinetics. In the model, the substrate used by microorganisms is the same contaminant C. After consideration of all the rate expressions for oxygen, substrate, and biomass growth, the fate of the contaminant is described in the system by

$$(\varepsilon^l + \rho K_{dC})\frac{\partial C_C^l}{\partial t} = \frac{D_C^l \varepsilon^l / \tau + D_C^s \rho K_{dC}}{r^2}\frac{\partial}{\partial r}\left(r^2\frac{\partial C_C^l}{\partial r}\right)$$

$$- (\varepsilon^l + \rho K_{dX})\frac{q_{\max}X}{Y_{C/X}}C_X^l\left(\frac{C_C^l}{K_{SC}+C_C^l}\right)\left(\frac{C_E^l}{K_{SE}+C_E^l}\right)$$

$$(29)$$

Transport in the aggregate is driven only by diffusion. Oxygen is required to perform the degradation, and its transport and consumption is represented by the equation

$$\varepsilon^l\frac{\partial C_E^l}{\partial t} = \frac{D_E^l \varepsilon^l}{\tau r^2}\frac{\partial}{\partial r}\left(r^2\frac{\partial C_E^l}{\partial r}\right)$$

$$- (\varepsilon^l + \rho K_{dX})\frac{q_{\max}X}{Y_{E/X}}C_X^l\left(\frac{C_S^l}{K_{SS}+C_S^l}\right)\left(\frac{C_E^l}{K_{SE}+C_E^l}\right) \qquad (30)$$

The biomass is considered present only in aggregates, follows Monod kinetics, and is represented by

$$(\varepsilon^l + \rho K_{dX}) \frac{\partial C_X^l}{\partial t} = \frac{D_X^l \varepsilon^l}{\tau r^2} \frac{\partial}{\partial r} \left(r^2 \frac{\partial C_X^l}{\partial r} \right)$$

$$+ (\varepsilon^l + \rho K_{dX}) q_{\max X} C_X^l \left(\frac{C_S^l}{K_{SS} + C_S^l} \right) \left(\frac{C_E^l}{K_{SE} + C_S^l} \right)$$

$$- (\varepsilon^l + \rho K_{dX}) b_X C_X^l \tag{31}$$

The results from simulations using this model indicate that microbiological degradation of contaminants within aggregates may play a significant role in bioremediation situations. Aggregation models might help estimate some limiting constraints in microbial growth and in bioremediation time.

VENTING MODELS

Despite their limited applicability, venting models could serve as a starting point to describe bioventing processes. A cometabolic bioventing model can be developed by including biological activity (biodegradation of the contaminant and inhibition by the carbon source, cell growth and decay, and nutrient transport) a venting model. Figure 3 schematically describes the processes to be considered in modeling bioventing.

Venting models can be classified into two different types based on the output they provide. Most of the current efforts produce a pressure distribution of gas in the soil around a vapor extraction well. They consider mostly the gas flow in the porous media and the physical radius of influence around that well. Other types of models focus on the removal of the contaminant from the soil and include mass transfer among the phases of the system, as well as other phenomena that directly affect the outcome of a remediation process (like NAPL formation).

Pressure Distribution Models

Shan et al. (1992) developed a model for steady-state gas pressure and stream function distributions for a vapor extraction well. The solution can be applied to isotropic, as well as anisotropic, homogeneous formations, the ground surface being permeable to gaseous mass transfer with the atmosphere. It is assumed that an impermeable soil layer or a water table limits gas flow below a certain depth. The model is based on the equation for gas flow in a homogeneous anisotropic porous medium. Under steady-state conditions and neglecting gravitational effects and following the analysis of Bear (1972), the model reduces to the pressure equation given by

$$\nabla \cdot (k \nabla P^2) = 0 \tag{32}$$

This equation can be directly derived from the general transport equation, assuming an isothermic ideal gas and neglecting dispersive flux and degradation of contaminants, as well as negligible interphase mass transfer.

The boundary conditions for this equation are obtained by considering atmospheric pressure at the ground level and no flux at an impermeable layer at a depth h, namely,

$$P = P_{\text{atm}} \quad \text{at} \quad z = 0 \tag{33}$$

and

$$\frac{\partial P}{\partial z} = 0 \quad \text{at} \quad z = h \tag{34}$$

The pressure solution is found first, for a continuous point sink and then, integrated along a line from the beginning of the well screen to its end, to obtain a pressure distribution for a continuous line sink. Gas streamlines are generated by direct integration of one of the Cauchy-Riemann equations in cylindrical coordinates.

The pressure distribution was used to calculate gas travel times, from any point in the soil within the radius of influence of the well to the well screen. Additionally, the model provides for the possibility of using information from gas pumping tests to calculate average gas permeabilities, in both the horizontal and vertical directions, by fitting the data to the pressure solution.

This type of model, while addressing important factors in the venting process, e.g., the presence of a line sink, instead of a point one, in the screened well and calculation of pressure distributions, lacks a more detailed description of contaminant fate present in other types of models.

Sepehr and Sambini (1993) proposed a model to calculate pressure drop and radius of influence around extraction wells in soil venting. The model is designed to account for the effect of partial penetration and partial screening of the extraction well, in addition to nonhomogeneity and anisotropy of the soil.

The governing equation in the mathematical description of the process focus on a mass balance around a finite volume element (dV), with dimensions of dx, dy, and dz, under steady-state conditions.

$$dm = d\rho(\eta dV) \tag{35}$$

Darcy's law is assumed to be valid within the element of study

$$v_{x,y,z} = -\frac{k}{\mu}\frac{dP}{d(x,y,z)} \tag{36}$$

Since $m_i = \rho v_i$ and applying the ideal gas law to substitute density as a function of pressure, the three-dimensional gas flow equation becomes

$$k_x\frac{\partial^2 P^2}{\partial x^2} + k_y\frac{\partial^2 P^2}{\partial y^2} + k_z\frac{\partial^2 P^2}{\partial z^2} = \frac{\eta\mu^g}{P}\frac{\partial P^2}{\partial t} - 2Q^g\mu^g P \tag{37}$$

In the model, the effect of soil moisture content is accounted for by introducing an effective porosity (an air-filled porosity) as a function of the effective saturation value.

The equation is solved by a successive over-relaxation method coupled to the Newton-Raphson method. The model has been applied to compute vacuum distributions around an extraction well and compares favorably with field data.

BIOVENTING MODELS

One of the first mathematical analyses applied to venting in the treatment of contaminated sites was done by Johnson et al. (1988, 1990). The model is based on a molar balance for each component in a mixture of contaminants in a given soil mass. This appears to be the first serious attempt to contemplate microbial degradation coupled to venting analysis in the cleanup of contaminated soils. The working equation can be written as

$$\frac{dM_i}{dt} = -Q^g C_i - B_i \tag{38}$$

The second term on the right-hand side of the equation (B_i) accounts for the biological or chemical degradation of component i. As in the original derivation, if this last term is set to zero, the equation describes only a venting operation.

In this analysis, a relationship between C_i and M_i at a given time in the mass balance (in moles) for each component i gives

$$M_i = \text{moles of } i, \text{ vapor phase} + \text{moles of } i, \text{ aqueous phase}$$

$$+ \text{ moles of } i, \text{ adsorbed to soil} \tag{39}$$

This model considers that the contaminant is uniformly distributed throughout the soil at all times and that local equilibrium exists between the vapor, NAPL adsorbed i in soil and aqueous phases. This latter assumption is based on laboratory venting experiments performed by Marley and Hoag (1984) and by Johnson et al. (1990). In these studies they determined the distance the vapor phase has to travel, after entering a contaminated zone, before the concentration of a contaminant i (in that vapor phase) reaches 99% of the equilibrium value.

Further assumptions include ideal gas behavior for the vapor phase, an ideal mixture behavior for the NAPL phase. However, in this analysis, the aqueous phase is considered a nonideal solution. Under these conditions, equilibrium among the phases can be described as

$$\chi_i^g P = \chi_i^n P_i^v = \gamma_i \chi_i^l P_i^v = C_i^l RT \tag{40}$$

The mole balance of component i can be expressed as

$$\chi_i^n \left\{ \frac{P_i^v \varepsilon V}{RT} + \frac{M^{ls}}{\gamma_i} + \frac{b_{si} M^s}{\gamma_i MW^l} \delta(M^{ls}) \right\} = M_i \tag{41}$$

This equation, thus, is only valid for a soil moisture content greater than zero. By definition, χ_i is the mole fraction of i in the NAPL phase and thus equal to

$$\chi_i^n = \frac{M_i^n}{M^n} \tag{42}$$

Under these conditions and with the constraint that $\Sigma \chi_i^n = 1$, the solution of the above set of equations is based on whether or not an NAPL phase is present. This is determined by calculating all the terms $\alpha_i y_i$. If all of them are less than unity, then a free organic liquid phase is not formed.

In a similar analysis, Johnson et al. (1990) considered the conditions needed to achieve steady-state gas flow in a venting operation. Depending on the soil permeability, their simulation predicts that it could take from 1 day to 1 week to reach steady state. Once steady state has been reached, the same equations can be employed to obtain estimates for pressure distribution using the radius of influence as a parameter.

Simulation of a venting operation using the above model for a gasoline spill was able to predict the successful removal of the light gasoline components. Still another contribution of Johnson et al. (1990) was the introduction of the concept of venting efficiency, defined as the ratio of mass removal for heterogeneous contaminant distribution to the mass removal rate for the case of homogeneous contamination. This concept allows one to determine the accuracy of the model predictions and, thus, provides a degree of the validity of working assumptions and its applicability to analyze results under given experimental conditions. Efficiencies as low as 8% were reported in some cases.

Falta et al. (1993) modified the Shan et al. (1992) model of gas pressure distribution to account for the partition of a single VOC between the gas and soil and between the gas and the aqueous solution. The model requires the knowledge of an initial distribution of contaminant in the soil, although it is flexible enough to accept input for a one-, two-, or three-dimensional initial contaminant distributions.

The model assumes local equilibrium in the gas-solid and gas-aqueous phase interfaces. Henry's law is applied to express the equilibrium between gas and aqueous phases. The model, however, does not account for the presence of an NAPL phase. The above assumptions are stated in a gas retardation coefficient for the specific contaminant present.

$$R_g = \frac{C_C^{l^*}}{HC_C^g} + \frac{\rho^s K_d}{H\phi C_C^g} + 1 \tag{43}$$

As reported by Karickhoff et al. (1979), adsorption to the soil by organic chemicals can be approximated as a function of the organic carbon in soil. Falta et al. (1993) used the following relationship for the sorption of a contaminant to soil:

$$K_d = K_{oc} f_{oc} \tag{44}$$

In this scheme, the soil properties were assumed constant and homogeneous throughout the soil volume under consideration. This implies that the retardation coefficient can also be assumed constant throughout the surface.

The initial step of the method consists of a coordinate transformation of the initial contaminant concentration from the r-z plane to a vapor travel time-normalized stream function plane. The pressure solution and the stream function are presented by Shan et al. (1992) and are used by Falta et al. (1993) to obtain the unretarded gas travel time (t).The advective contaminant travel time $(t_R)$is calculated afterwards by multiplying t by the retardation coefficient (R_g). The fraction of contaminated gas exiting the soil through the well can be obtained graphically from a plot of t_R versus the normalized stream function. From this composition and by knowing the total gas flow rate, the rate of removal of pollutant can be determined.

The model in this case, applied to the extraction well, allows screening over a fraction of the unsaturated zone as compared to the point sink assumption presented by Johnson et al. (1990). A drawback of this model is that it considers only the presence of a single pollutant in the soil. Multicomponent mixtures would be difficult to analyze with the existing equations, unless they share similar physical properties and a single-phase retardation coefficient can be applied to the mixture.

Recently, Gomez-Lahoz et al. (1994a) proposed a model to account for the biodegradation that occurs during soil vapor extraction (SVE). Even though, normally, gas flow rates used for SVE are much larger than those used to stimulate growth of native bacterial populations and do not contain gaseous nutrients (carbon or nitrogen sources in addition to oxygen), as is the case of bioventing, the model can be used to describe the biological degradation in a bioventing process.

Their model is based on the fact that, commonly, mass transfer limitations exist during contaminant removal in SVE processes and therefore bacterial growth could be limited by the supply of oxygen. The model describes a nonequilibrium oxygen transport process between the gas phase and the bulk aqueous phase that contains the biomass for biodegradation. Advection is regarded as the only transport mechanism of oxygen and contaminants in the gaseous phase. A lumped-parameter first-order mass transfer mechanism is used to describe this process.

The model takes into account the change in nutrient concentrations, and these are considered present only in the aqueous phase. The model is flexible enough to accommodate an indefinite number of such nutrients, although in practical applications, only the modeling of the limiting ones are of relevance. A relevant application of the model is that it accounts for the presence of a second substrate, even if this is volatile, as a competitive inhibitor of the first substrate or as a possible toxin to the biomass population.

The model is applied to a one-dimensional column geometry and divides the length of the column into a series of horizontal slabs (compartments) for analysis. The rate of change of biomass per unit volume is based on Monod kinetics and given by

$$\frac{\partial C_X^l}{\partial t} = q_{\max X} C_X^l \frac{q_{\max X/S_1} \dfrac{C_{S_1}^l}{K_{SS_1}} + q_{\max X/S_2} \dfrac{C_{S_2}^l}{K_{SS_2}}}{1 + \dfrac{C_{S_1}^l}{K_{SS_1}} + \dfrac{C_{S_2}^l}{K_{SS_2}}} \frac{C_E^l}{K_{SE} + C_E^l} \prod_{i=1}^{n} \left[\frac{C_{S_i}^l}{K_{SS_i} + C_{S_i}^l} \right]$$

$$- b_X C_X^l \tag{45}$$

In this equation, the first term represents the biomass growth by consumption of nutrients, contaminants and oxygen, while the term $-b_X C_x$ represents a death term for biomass.

The rate of change of nutrients includes a regeneration term from dead biomass, in addition to their consumption by live biomass. The rate of change in oxygen concentration in the aqueous phase includes a term for oxygen consumption for degradation of contaminants by living microorganisms. Another for the consumption of oxygen by the degradation of dead biomass and a last one as a source term that accounts for mass transfer from the gas to the liquid.

The rate of change of the first contaminant in the aqueous phase is given by

$$\frac{\partial C_{C_1}^l}{\partial t} = k_{Cc} \left[\frac{C_{C_1}^g}{H_{C_1}} \right]$$

$$- q_{\max X} C_X^l \frac{Y_{X/C_1} q_{\max X/S_1} \dfrac{C_{S_1}^l}{K_{SS_1}}}{1 + \dfrac{C_{S_1}^l}{K_{SS_1}} + \dfrac{C_{S_2}^l}{K_{SS_2}}} \frac{C_E^l}{K_{SE} + C_E^l} \prod_{i=1}^{n} \left[\frac{C_{S_i}^l}{K_{SS_i} + C_{S_i}^l} \right] \tag{46}$$

In this equation, the first term represents the mass transport of the contaminant between the vapor and liquid phases, while the second term represents biodegradation of this contaminant by production of biomass. A similar equation can be written for a second contaminant.

The rates of change for the concentrations of both contaminants and oxygen in the gas phase include a term for a lumped-parameter representation of mass transfer between the liquid and vapor phases and another term for the advective transport in the gas phase through the column.

The system of coupled nonlinear partial differential equations was approximated by a similar system of ordinary differential equations in which the variables were calculated only for a discrete set of values of the independent variables.

Simulations using this model were run using various sets of parameters, mainly changing the values of the mass transfer coefficients for contaminants between gas and liquid phases. The model yields interesting results for the effect of biological degradation of contaminants in a soil vapor extraction system. Gomez-Lahoz and Wilson (1994a) report that the theoretical cleanup time required to achieve removal percentages of 99.996% of a volatile organic compound is reduced by more than half when biodegradation occurs. Additionally, the decrease in cleanup time due to degradation becomes more important as the values for the mass transfer coefficients are smaller.

The authors report that the total amount of substrate removed by biodegradation when remediation is complete is nearly constant, even though the values of the mass transfer coefficients change over four orders of magnitude. This indicates that, under these conditions, the contaminant concentrations in the liquid phase were not a limiting factor in the biodegradation.

In general, in the model, an initial period of time is allowed for microbial growth, during which contaminant removal by biodegradation is negligible. This period (four days when nutrients are not limiting) is the time required for the microbial concentration to increase from $1\,\mu g/l$ to 1 mg/l. After this period, biodegradation increases exponentially until the substrate concentration is low enough to become a limiting factor.

Simulations with small mass transfer coefficients result in small initial periods for microbial growth, compared to the total cleanup time. In this case, under limiting oxygen transfer, the biodegradation rates increase linearly after this induction period. These results help explain observed biodegradation rates of zero order with respect to contaminant and biomass concentrations measured experimentally in the field.

Using sensitivity studies, the effect of one and two substrates was incorporated in the remediation process. The rate-limiting effects of the mass transfer coefficients of contaminant and/or oxygen between the gas and aqueous phases clearly show a strong dependency, in general, during the last part of the remediation process. At this stage, as the contaminant concentration decreases, the efficiency of the stripping is also reduced, while the biodegradation of pollutant continues. This, in fact, demonstrates that, when microbial activity is present, shorter cleanup times and higher removal levels can be accomplished (Gomez-Lahoz and Wilson, 1994b).

Similar analysis, using two substrates in the modeling scheme, provided a description of the several types of interactions and effects that could occur between two substrates in a bioremediation process. A cometabolic effect was observed when a substrate was degraded without direct benefit to the microorganism. In this case, it is possible to consider a competitive inhibitory effect among the two substrates. Results, when oxygen was considered the first substrate and a volatile contaminant as the cometabolite (secondary substrate) at low and high values of mass transfer rates, were qualitatively reasonable.

BIOVENTING OPERATIONS

In situ Studies

Several bioventing pilot tests and field studies have been performed in the past decade following the emergence of the technique in the mid-1980s. Jet fuels (JP-4 and JP-5), diesel fuels, gasoline, BTEX, polycyclic aromatic hydrocarbons, pentachlorophenol, and diesel oil are among the pollutants that have been treated by bioventing (Hoeppel et al., 1991; Bulman et al., 1993; Hinchee et al., 1995). Most of the bioventing efforts for in situ remediation have been at U.S. Air Force bases. The U.S. Air Force Center for Environmental Excellence started a program in 1992 to assess the applicability of bioventing to remediation of contaminated sites. The program was in cooperation with the Air Force Armstrong Laboratory and the U.S. Environmental Protection Agency and was named the Bioventing Initiative (Bostron, 1994). The program includes more than 135 bioventing applications and demonstrations at Air Force sites across the United States.

The first in situ bioventing operation reported in the literature occurred at Hill Air Force Base in Utah. The site was contaminated with 102,000 L of JP-4 jet fuel that was spilled in January 1985 (Dupont, 1993). Remediation did not begin until late 1988, and it started as a soil vapor extraction system. A catalytic incinerator was placed on line to treat the vented gas. During the nine months the SVE was in operation, a total of 62,600 kg of petroleum hydrocarbon was removed from the soil (Hinchee et al., 1991). In situ respiration tests performed during this SVE treatment period revealed that part of the overall contaminant removal was due to microbial activity. Since this biodegradation occurred without explicit microbial stimulation under the prevalent venting conditions, it was concluded that enhancement of biodegradation could be possible if operating conditions were modified to favor microbial growth. Posterior analyses showed that biodegradation accounted for 15 to 20% of the recovered JP-4 during the SVE operation. Changes to the existing SVE system were designed to maximize the flow path and retention time of vapors in the contaminated soil. Volatilization rates were reduced from 90–180 kg/day to less than 9 kg/day, which allowed the direct discharge of collected gas without any catalytic incineration. The modifications increased the average degradation rates for the site from 32 kg of hydrocarbon per day to greater than 45 kg/day (Dupont, 1993), corresponding to an average of 10 mg/kg/day.

During the 13 months of operation, 33,200 kg of JP-4 were removed under bioventing conditions. A total of 95,800 kg of JP-4 were removed during the remediation period that ended in November 1990 (this includes SVE and bioventing). From that amount, 53,600 kg of JP-4 were estimated to be due to biodegradation. Bioventing conditions included nutrient addition, but no significant increase in biodegradation rates was observed. However, Lee and Swindoll (1993) report that subsequent laboratory studies with soil from the Hill AFB site showed that addition of inorganic nutrients tripled overall mineralization of contaminants. Additionally, increasing moisture content from 6 to 18% raised biodegradation by 50% when inorganic nutrients were not limiting. Atlas (1995) reports that, at the end of the study at Hill AFB in 1993, 40% of the removed hydrocarbons was mineralized to carbon dioxide and water, and venting accounted for the remaining 60%.

A different in situ bioventing operation was conducted in a JP-4 contaminated site at the Tyndall Air Force Base in Florida (Miller et al., 1991). In this case, the maximum biodegradation removal rate was 20 mg/kg/day, but average (stable) biodegradation rates were approximately 5 mg/kg/day (Lee and Swindoll, 1993). Bioremediation accounted for as much as 82% of the total contaminant removal. In this soil, sufficient phosphorus and nitrogen were present to support biodegradation since further addition of these nutrients did not increase contaminant removal. Raising moisture content from 6.5–7.4% to a range of 8.5–9.8% did not enhance biodegradation either. These results suggest that nutrient availability was not the rate-limiting step at this site.

An example of bioventing in situ not related to U.S. Air Force sites is given by Bulman et al. (1993). The site to be remediated involved a spill of 50,000 L of diesel fuel at the State Railway in Marshaling Yard, Australia. Average fuel concentration in the soil was 1.5% by weight. The diesel fuel contaminated the soil over a depth of 1.5 to 3.5 m. Preliminary studies were performed with soil from this site to determine feasibility of the treatment. Experiments in shake flasks demonstrated that aerobic growth conditions with added nutrients resulted in higher rates of microbial growth and hydrocarbon biodegradation than anaerobic conditions. Further experiments with soil-packed columns under unsaturated conditions yielded a 20% hydrocarbon removal over a period of four weeks.

The in situ remediation operation included an initial six-month period of venting alone that reduced total hydrocarbon concentrations by 10% in some regions and as much as 30% in other parts of the site, to a depth of 3 m. A subsequent six-month period, over which nutrients were added, resulted in further reduction of 30% in fuel concentration to the full depth of 3.5 m. Examples of in situ bioventing are presented in Table 2.

Cost Analysis

Even though a significant number of in situ remediation operations using bioventing has been published in the literature, a surprisingly few number of them include cost analysis of the process. During remediation of a contaminated soil at the Marine Corps Air and Ground Combat Center in Twenty-nine Palms, California, Zwick et al. (1995) presented detailed projections for bioremediation of the site. The site was contaminated with gasoline, JP-5 jet fuel, and diesel fuel. An average soil TPH (total petroleum hydrocarbon) concentration of 2000 mg/kg soil was assumed for the analysis, along with a total contaminated soil volume of 14,000 yd^3. Remediation time was estimated to be four years, with a biodegradation rate of 550 mg/kg/year (average of 1.5 mg/kg/day).

The bioremediation operation described by Zwick et al. (1995) consisted of a bioventing system coupled to an irrigation system to enhance microbial activity. The total cost for complete remediation, including monitoring of hydrocarbon concentrations and operation and maintenance, was estimated to be $450,000. This results in an average cost of $31/yd^3 of soil treated (assuming a soil density of 1.85 gr/cm^3 this corresponds to $11.5/kg of contaminant treated). It is important to mention that this cost is based on the maximum biodegradation rate at one of the monitoring wells but, in general, observed biodegradation rates are not homogeneous and can change significantly from point to point within the same site.

Ramiz et al. (1995) present a cost comparison between bioventing and an SVE system. Data for the comparison is taken from a field study at the Savannah River in Aiken, South Carolina. The site was contaminated with VOCs, the principal of them being TCE. Total cost for the bioventing system, which includes a pair of injection/extraction wells, was estimated to be about $354,000. The approximate cost of the SVE system is $380,000 and includes four extraction wells, a pump-and-treat well, and a catalytic converter to oxidize vented contaminants. The bioventing process included the addition of methane (1–4%) and nutrients (nitrous oxide and triethyl phosphate). Results from the study indicated 12,096 lb of VOCs were remediated during SVE. An additional 40% of that mass was treated

Table 2. Examples of sites remediated by bioventing.

Site	Contaminant	Nutrients Added	Controlled Moisture[a]	Contaminant Removal[b]	References
Hill AFB, Utah	Jet fuel JP-4	Nutrients added, but did not have observable effects	Moisture added	10 mg/kg/day (HC)	Hinchee et al., 1991 Dupont, 1993 Lee and Swindoll, 1993
Tyndall AFB, Florida	Jet fuel JP-4	None	No	5 mg/kg/day (HC)	Lee and Swindoll, 1993
State Railway, Marshalling, Australia	Diesel fuel	Nitrogen and phosphorus	60% of water-holding capacity	8 mg/kg/day (HC)	Bulman et al., 1993
Amoco Oil petroleum storage, Michigan	Xilenes (BTEX)	None	gw and vz	1.5–3.9 mg/kg/day (HC)	Javanmardian et al., 1995
Patrick AFB, Florida	Gasoline	None	No	2.8 mg/kg/day (F)	Downey et al., 1995
Vandenberg, California	Gasoline	None	No	1.6–2.7 mg/kg/day (F)	Downey et al., 1995
MCAGCC,[c] Twentynine Palms, CA	JP-5 jet fuel	None	Moisture increased from 4 to 12% by irrigation	0.17 mg/kg/day[d] (Hex) 1.5 mg/kg/day	Zwick et al., 1995
Three wood preserving sites	PAH and PCP[e]	None	gw and vz	2.1–6.3 mg/kg/day (HC)	Gentry and Simpkin, 1995
Gas plant in Calgary, Canada	LNAPL (C_5 to C_{12})	None	No	20 mg/kg/day[f] (HC)	Moore et al., 1995

(continued)

Table 2. (continued).

Site	Contaminant	Nutrients Added	Controlled Moisture[a]	Contaminant Removal[b]	References
Waste landfill northeast USA	Benzene, phenol PAHs	None	No	22–79 mg/kg/day[g] (HC) 6.4 mg/kg/day	Heuckeroth et al., 1995
Wheeler Army Airfield and Hickam AFB, Hawaii	BTEX	None	No	0.3–0.6 mg/kg/day (HC) 4.9–13.7 mg/kg/day	Ratz et al., 1995
Natural gas production well, Traverse City, MI	BTEX, alkanes $(C_1$ to $C_{10})$	N as $NaNO_3$[h] phosphorus (TEP[i])	gw and vz	2.0 mg/kg/day (HC)	Raetz and Scharff, 1995

[a]All studies have been performed in vadose zone, except those denoted by "gw" (groundwater) and "gw and vz" (studies included both groundwater and vadose zone).
[b]Attempts have been made to standardize an indicator of the efficiency of the bioventing operation (biodegradation rate in mg of contaminant/kg of soil/year). Data refers to degradation under bioventing conditions only and not under vapor extraction when this latter technique was also employed along with bioventing. The contaminant used for the calculation of the biodegradation rate is included in parentheses. HC = hydrocarbon, Hex = hexane, and F = fuel.
[c]MCAGCC = Marine Corps Air Ground Combat Center.
[d]High and low degradation rates correspond to post- and preirrigation conditions, respectively.
[e]PAH = Polycyclic aromatic hydrocarbons; PCP = pentachlorophenol.
[f]Average in situ temperature was 8°C during the study.
[g]High degradation rates correspond to regions inside oxygen delivery areas, while the low rate is reported for a spot outside the oxygen delivery area.
[h]Nitrous oxide was also used, but it was determined that it could not be used by the microorganisms.
[i]TEP = triethyl phosphate.

under bioventing conditions. This information yields an average of $31/lb ($14/kg) of VOC remediated for the SVE system and $27/lb ($12.25/kg) of VOC for bioventing. Ramiz et al. indicate that the price could decrease to $20/lb ($9/kg) of treated VOC if microorganisms can degrade 90% more VOC mass than the SVE system, instead of only 40%.

Reported prices per unit mass of contaminant treated from the previous independent studies are remarkably close considering that costs are strongly site-dependent. A number of variables, including soil properties (moisture, homogeneity, gas permeability, etc.), the need of nutrient addition, the type of contaminant to degrade, and its degradation mechanism (cometabolism or assimilation), are among the most important factors that affect degradation time and, ultimately, the overall cost of bioremediation.

Downey et al. (1995) have made a cost comparison between bioventing and traditional SVE at three Air Force sites. In all three cases, bioventing provided the least expensive alternative for remediation. Treatment of the vented gas during an SVE operation is usually required because of the high contaminant concentrations in the stream. The most common method for disposing of this effluent is incineration. Capital, operation, and maintenance costs associated with the treatment of the recovered vapor makes SVE more expensive than bioventing. However, if the site is contaminated mainly with volatile organics, cleanup would proceed faster using SVE than bioventing. Bioventing has the added advantage of remediating sites contaminated with nonvolatile organics.

SUMMARY

Physical and biological treatment strategies for the remediation of contaminated soil and groundwaters that consider the probable partitioning of chlorinated organics in multiphase environments have the greatest probability of success. Among the organic contaminants in groundwater and soils, the chlorinated aliphatics occupy a somewhat unique position in that they are characterized by both low hydrophobicity and high volatility. Consequently, they should be relatively good candidates for treatment or gas-phase transport.

Bioventing combines soil venting with biodegradation. Compared to soil venting systems, bioventing can produce considerable reductions in remediation costs by decreasing the necessity for off-gas treatments.

Pilot- and field-scale bioventing systems have been implemented in many locations, and cometabolic applications are under consideration. Despite its promise as an in situ bioremediate strategy, mathematical representations of models in bioventing processes, including cometabolic bioventing, are inadequate for most practical purposes. Relatively few modeling efforts have been described. Future work should address the development of mathematical models that help design and operate parameters, if bioventing is to become an effective and efficient bioremediation strategy. An ideal model for cometabolic bioventing should describe the behavior of contaminants and nutrients in the gas and liquid phases. This will allow the determination of concentrations of oxygen and gaseous nutrients in the zone to be remediated and, thus, ensure the stimulation of the appropriate bacterial populations. Experimental observations support the urgent need to develop models in bioventing processes that would eventually provide direction on the manipulation of system variables. Realistic models that take into consideration field-like laboratory experiments will help find optimal design to large-scale strategies for bioremediation.

NOTATION

Variables

A = area
C = concentration
h = height
J = molar flux
l = curvilinear coordinate direction
m = mass
M = number of moles
P = pressure
P_v = vapor pressure
q = rate of production/consumption
Q = volumetric flow rate
ρ = density
r,z = radial coordinate positions
T = temperature
t = time
v = velocity
x,y,z = rectangular coordinate directions
χ = mole fraction
ε = volume fraction
μ = dynamic viscosity
θ = mass fraction

Superindex

f = interface
g = gas phase
l = liquid phase
n = NAPL phase
s = solid (soil) phase
$*$ = saturation

Subindex

C = contaminant
E = electron acceptor (in most cases oxygen)
i = component
m = microcolony or microaggregate
S = growth (limiting) substrate
X = biomass
x,y,z = rectangular coordinate directions

Constants

a,b,n = empirical parameters
b_X = decay coefficient
b_S = adsorption partition coefficient
D = diffusion coefficient
D_h = hydrodynamic dispersion coefficient
D_S = apparent dispersion coefficient
f_d = fraction of cells that is biodegradable
f_{oc} = fraction of organic carbon in the soil
g = acceleration of gravity
H = Henry's law constant
k = permeability
k_C = mass transfer coefficient
k_f = Freundlich adsorption coefficient
k_p = soil sorption coefficient
k_{sw} = Langmuir equilibrium kinetic coefficient
K = hydraulic conductivity
K_d = partition coefficient
K_{oc} = chemical-organic carbon partition coefficient
K_S = half saturation constant
q_{max} = maximum rate
R = ideal gas constant

R_g = gas retardation coefficient
S = solubility
S_w = residual aqueous phase saturation
T_C = inactivation coefficient
Y = yield coefficient
α = dispersivity
ϕ = soil porosity
γ = activity coefficient
η = porosity
τ = tortuosity

REFERENCES

Abriola, L. M. (1989). Modeling multiphase migration of organic chemicals in groundwater systems—A Review and Assessment. *Environ. Health Persp.* 83:117–143.

Alvarez-Cohen, L. and P. L. McCarty (1991). A cometabolic biotransformation model for halogenated aliphatic compounds exhibiting product toxicity. *Environ. Sci. Technol.* 25(8):1381–1387.

Anthony, C. (1986) Bacterial oxidation of methane and methanol. *Adv. Microbiol. Physiol.* 27:113–210.

Arciero, D., T. Vannelli, M. Logan, and A. B. Hooper (1989) Degradation of trichloroethylene by the ammonia-oxidizing bacterium *Nitrosomonas europaea. Biochem. Biophys. Res. Commun.* 159:640–643.

Atlas, R. M. (1995). Bioremediation. *Chem. & Engin. News.* 73(14):32–42.

Batchelder, G. V., W. A. Panzeri, and H. T. Phillips (1986). Soil ventilation for the removal of adsorbed liquid hydrocarbons in the subsurface. NWWA/API Conference on petroleum hydrocarbons and organic chemicals in ground water, Houston, TX.

Bear, J. (1972). *Dynamics of fluids in porous media.* New York, NY, Elsevier Science.

Bonazountas, M. and D. Kallidromitou (1993). Mathematical hydrocarbon fate modeling in soil systems. *Principles and practices for petroleum contaminated soils.* E. J. Calabrese and P. T. Kostecki. Chelsea, MI, Lewis Publishers, pp. 131–322.

Bostron, D. (1994). Soil bioventing: A technology comes of age, white paper, Barr Engineering Company, WWW site.

Briggs, G. G. (1981). Theoretical and experimental relationships between soil adsorption, octanol-water partition coefficients, water solubilities, bioconcentration factors and the parachor. *J. Agric. Food Chem.* 29:1050–1059.

Brown, R. A., G. E. Hoag and R. D. Norris (1987). The remediation game: Pump, dig, or treat. Water pollution control federation conference.

Bulman, T. L., M., Newland and A. Wester (1993). In situ bioventing of a diesel fuel spill. *Hydrol. Sci.* 38(4):297–308.

Chen, Y. M., L. M. Abriola, P. J. J. Alverez, P. J. Anid and T. M. Vogel (1992). Modeling transport and biodegradation of benzene and toluene in sandy aquifer material: Comparisons with experimental measurements. *Wat. Resour. Res.* 28(7):1833–1847.

Chiou, C. T., D. W. and M. Manes (1982). Partitioning of organic compounds on octanol-water system. *Environ. Sci. Technol.* 16:4–10.

Chu, K.-H., J. Vernalia and L. Alvarez-Cohen (1995). Nitrogen sources for bioremediation of TCE in unsaturated porous media. *Bioremediation of chlorinated solvents.* R. E. Hinchee, A. Leeson and L. Semprini. Columbus, Ohio, Batelle Press. 3(4).

Coates, M., I. W. Connell and D. M. Barron (1985). Aqueous solubility and octanol to water partition coefficients of aliphatics hydrocarbons. *Envir. Sci. Technol.* 19:628–632.

Colby, J., D. I. Stirling and H. Dalton. (1977), The soluble methane monooxygenase of *Methylococcus capsulatus* (Bath), its ability to oxygenate n-alkanes, ethers, and alicyclic, aromatic and heterocyclic compounds, *Biochem. J.* 165:395–402.

Criddle, C. S. (1993). The kinetics of cometabolism. *Biotechnol. Bioeng.* 41:1048-1056.

Dhawan, L. T., L. E. Erickson and P. Tuitemwong (1991). Modeling, analysis, and simulation of bioremediation of soil aggregates. *Envir. Prog.* 10(4): 251–260.

Downey, D. C., R. A. Frishmuth, R. S. Archabal, C. J. Pluhar, P. G. Blystone and R. N. Miller (1995). Using in situ bioventing to minimize soil vapor extraction costs. *In situ aeration: Air sparging, bioventing, and related remediation processes.* R. E. Hinchee, R. N., Miller and P. C. Johnson. Columbus, Ohio, Batelle Press. 3(2).

Dupont, R. R. (1993). Fundamentals of bioventing applied to fuel contaminated sites. *Envir. Prog.* 12(1):45–53.

Ensley, B. D. (1991), Biochemical diversity of trichloroethylene metabolism. *Annu. Rev. Microbiol.,* 45:283–299.

Falta, R. W., P. Karsten and D. A. Chesnut (1993). Modeling advective contaminant transport during soil vapor extraction. *Ground Wat.* 31(6):1011–1020.

Freeze, R. A. and J. A. Cherry (1979) *Groundwater.* Prentice Hall, Englewood Cliffs, NJ.

Gentry, J. L. and T. J. Simpkin (1995). Experience with bioventing at wood-preserving sites. *In situ aeration: Air sparging, bioventing, and related remediation Processes.* R. E. Hinchee, R. N. Miller and P. C. Johnson. Columbus, Ohio, Batelle Press. 3(2).

Gomez-Lahoz, R. M. and D. J. Wilson (1994a). Biodegradation phenomena during soil vapor extraction. III. *Sep. Sci. Technol.* 29(10):1275–1291.

Gomez-Lahoz, R. M. and D. J. Wilson (1994b). Biodegradation phenomena during soil vapor extraction. A high speed non equilibrium model. *Sep. Sci. Technol.* 29(4):429–463.

Graham, D. (1992) Factors affecting the utilization of methanotrophic bacteria in bioremediation. Ph.D. Thesis Env. Eng. Dept. University of Arizona.

Green, J., S. D. Prior and H. Dalton. (1985) Copper ions as inhibitors of protein C of soluble methane monooxygenase of *Methylococcus capsulatus* (Bath). *Eur. J. Biochem.* 153:137–144.

Harleman, D. R. F. and R. R. Rummer (1962). *The dynamics of salt-water intrusion in porous media.* Cambridge, MA, MIT Press.

Henson, M. J., M. Y. Yates and J. W. Cochran (1989) Metabolism of chlorinated methanes, ethanes and ethylenes by a mixed bacterial culture growing on methane. *Indust. Microbiol.,* 4:29–35.

Heuckeroth, D. M., M. F. Eberle and M. J. Rykaczewski (1995). In situ vacuum extraction/bioventing of a hazardous waste landfill. *In situ aeration: Air sparging, bioventing, and related remediation processes.* R. E. Hinchee, R. N., Miller and P. C. Johnson. Columbus, Ohio, Batelle Press. 3(2).

Hinchee, R. E., R. N. Miller and P. C. Johnson, Eds. (1995). *In situ aeration: Air sparging, bioventing, and related remediation processes.* Third International In Situ and On-Site Bioreclamation Symposium. Columbus, Ohio, Batelle Press.

Hinchee, R. E., R. N. Miller and R. R. Dupont (1991). Enhanced biodegradation of petroleum hydrocarbons: An air based in situ process. *Innovative hazardous waste treatment technologies.* H. M. Freeman and P. R. Sferra. Lancaster, Pennsylvania, Technomic Publishing Co., Inc., *Vol. 3, Biological Processes.*

Hoeppel, R. E., R. E. Hinchee and M. F. Arthur (1991). Bioventing soils contaminated with petroleum hydrocarbons. *J. Industr. Microbiol.* 8(3): 141-146.

Hyman, M. R. and P. M. Wood (1984) Ethylene oxidation by *Nitrosomonas europaea. Arch. Microbiol.* 137:155–158.

Hyman, M. R. and P. M. Wood (1983) Methane oxidation by *Nitrosomonas europaea. Biochem. J.* 212:31–37.

Javanmardian, M., J. S. Huber, C. B. Olson, W. P. Schwartz, C. A. Masin and V. J. Kremesec (1995). In situ biosparging at an Amoco site: subsurface air distribution and biostimulation. *In situ aeration: Air sparging, bioventing, and related remediation processes.* R. E. Hinchee, R. N., Miller and P. C. Johnson. Columbus, Ohio, Batelle Press. 3(2).

Johnson, P. C., M. W. Kemblowski and J. D. Colthart (1988). Practical screening models for soil venting applications. *Proceedings of petroleum hydrocarbons and organic chemicals in ground water,* Houston, TX, National Water Well Assoc.

Johnson, P. C., M. W. Kemblowski and J. Colthart (1990). Quantitative analysis for the cleanup of hydrocarbon-contaminated soils by in-situ soil venting. *Ground Wat.* 28(3):413–429.

Karickhoff, S. W., D. S. Brown and T. A. Scott (1979). Sorption of hydrophobic pollutants on natural sediments. *Wat. Res.* 13:241–248.

Landa, A. S., E. M. Sipkema, J. Weijma, A. A. C. M. Beenackers, J. Dolfing and D. B. Janssen (1994). Cometabolic degradation of trichloroethylene by *Pseudomonas cepacia* G4 in a Chemostat with Toluene as the Primary Substrate. *Appl. Env. Microbiol.* 60(9):3368–3374.

Leak, D. J. and H. Dalton. 1986. Growth yields of methanotrophs. 1. Effect of copper on the energetics of methane oxidation. *Appl. Microbiol. Biotechnol.* 23:470–476.

Lee, M. D. and C. M. Swindoll (1993). Bioventing for in situ remediation. *Hydrol. Sci.* 38(4):273–282.

Mackay, D. M. and P. V. Roberts (1985). Transport of organic contaminants in groundwater. *Environ. Sci. Technol.* 19(5):384.

Marley, M. C. and G. E. Hoag (1984). Induced soil venting for the recovery/restoration of gasoline hydrocarbons in the vadose zone. NWWA/API conference on petroleum hydrocarbons and organic chemicals in ground water.

McCarty, P. L., L. Semprini and P. V. Roberts (1989) Methodologies for evaluating the feasibility of in situ biodegradation of halogenated aliphatic ground water contamination by methanotrophs. In *Proceedings of AWMA/EPA international symposium on biosystems for pollution control* (Pittsburgh, PA: Air and Waste Management Association), pp. 69–82.

Miller, R. N., R. E. Hinchee and C. M. Vogel (1991). A field-scale investigation of petroleum hydrocarbon biodegradation in the vadose zone enhanced by soil venting at Tyndall Air Force Base, Florida. *In situ bioreclamation applications and investigations for hydrocarbon and contaminated site remediation.* R. E. Hinchee and R. F. Olfenbuttel. Butterworth-Heinemann, Stoneham, Massachusetts.

Moore, B. J., J. E. Armstrong, J. Baker and P. E. Hardisty (1995). Effects of flow rate and temperature during bioventing in cold climates. *In situ aeration: Air sparging, bioventing, and related remediation Processes.* R. E. Hinchee, R. N., Miller and P. C. Johnson. Columbus, Ohio, Batelle Press. 3(2).

Nelson, M. J. K., S.O. Montgomery, W. R. Mahaffey and P. H. Pritchard (1988). Trichloroethylene metabolism by microorganisms that degrade aromatic compounds. *Appl. Env. Microbiol.* 54:604–606.

Norris, R. D. and J. E. Matthews (1994). *Handbook of bioremediation.* Boca Raton, FL, Lewis Publishers.

Oldenhuis, R., R. L. J. M. Vink, D. B. Janssen and B. Witholt (1989) Degradation of chlorinated aliphatic hydrocarbons by *Methylosinus trichosporium* OB3b expressing soluble methane monooxygenase, *Appl. Env. Microbiol.*, (55)11: 2816–1826.

Park, S., M. L. Hanna, R. T. Taylor, and M. W. Droege. (1991) Batch cultivation of *Methylosinus trichosporium* OB3b. I. Production of soluble methane monooxygenase. *Biotechnol. Bioeng.* 38:423–433.

Patel, R. N., C. T. Hou, A. I. Laskin and A. Felix (1982) Microbial oxidation of hydrocarbons: Properties of a soluble methane monooxygenase from a facultative methane-utilizing organism, *Methylobacterium* sp. strain CRL-26. *Appl. Environ. Microbiol.* 44:1130–1137.

Raetz, R. M. and D. L. Scharff (1995). Field application and results of an engineered bioventing process. In situ aeration: Air sparging, bioventing, and related remediation processes. R. E. Hinchee, R. N., Miller and P. C. Johnson. Columbus, Ohio, Batelle Press. 3(2).

Ramiz, P. S., W. E. Showalter and S. R. Booth (1995). Cost effectiveness of in situ bioremediation at Savannah river. *Bioremediation of chlorinated solvents*. R. E. Hinchee, A. Leeson and L. Semprini. Columbus, Ohio, Batelle Press. 3(4).

Rasche, M. E., M. R. Hyman and D. J. Arp (1991). Factors limiting aliphatic chlorocarbon degradation by *Nitrosomonas europaea:* Cometabolic inactivation of ammonia monooxygenase and substrate specificity. *Appl. Environ. Microbiol,* 57(10):2986–2994.

Ratz, J. W., G. D. Pierson, K. K. Caskey and W. L. Barry (1995). In situ bioventing: Results from three pilot tests performed in Hawaii. *In situ aeration: Air sparging, bioventing, and related remediation processes.* R. E. Hinchee, R. N. Miller and P. C. Johnson. Columbus, Ohio, Batelle Press. 3(2).

Rittmann, B. E., E. Seagren, B. A. Wrenn, A. J. Valocchi, C. Ray and L. Raskin (1994). *In situ bioremediation.* Noyes Publications, Park Ridge, NJ.

Schäfer, W. and W. Kinzelbach (1992). Stochastic modeling of in situ bioremediation in heterogeneous aquifers. *J. of Cont. Hydrol.* 10:47–73.

Semprini, L. and P. L. McCarty (1991). Comparison between model simulations and field results for in-situ biorestoration of chlorinated aliphatics. Part 1. Biostimulation of methanotrophic bacteria, *Ground Water,* (29):239–250.

Semprini, L. and P. L. McCarty (1992). Comparison between model simulations and field results for in-situ biorestoration of chlorinated aliphatics: Part 2. Cometabolic Transformations. *Ground Wat.* 30(1):37–44.

Semprini, L., G. D. Hopkins, D. Grbić-Galić, P. L. McCarty and P. V. Roberts (1994). A laboratory and field evaluation of enhanced in situ bioremediation of trichloroethylene, *cis-* and *trans-*dichloroethylene, and vinyl chloride by methanotrophic bacteria. In *Bioremediation: Field experience,* Lewis Publishers, Ch. 18, pp. 383–412.

Sepehr, M. and Z. A. Sambini (1993). In situ soil remediation using vapor extraction wells, development and testing of a three-dimensional finite-difference model. *Ground Wat.* 31(3):425–436.

Shan, C., R. W. Falta and I. Javandel (1992). Analytical solutions for steady state gas flow to a soil vapor extraction well. *Wat. Resour. Res.* 28(4):1105–1120.

Sposito, G. (1984). The future of an illusion: Ion activities in soil solutions. *Soil Sci. Soc. Amer. J.* 48:531–536.

Stanley, S. H., S. D. Prior, D. J. Leak and H. Dalton. (1983) Copper stress underlies the fundamental change in intracellular location of methane mono-oxygenase in methane-oxidizing organisms: Studies in batch and continuous cultures. *Biotechnol. Lett.* 5:487–492.

Strand S. E., M. D. Bjelland and H. D. Stensel (1990) Kinetics of chlorinated hydrocarbon degradation of suspended cultures of methane-oxidizing bacteria. *J. Water Pollut. Cont. Fed.* 62:124–129.

Tsien, H. C. and R. S. Hanson (1992) Soluble methane monooxygenase component B gene probe for identification of methanotrophs that rapidly degrade trichloroethylene. *Appl. Environ. Microbiol.* 58:953–960.

Tucker, W. A. and S. W. Nelken (1982). Diffusion coefficients in air and water. *Handbook of chemical properties estimation methods.* Lyman, W. J., W. F. Reehl and D. H. Rosenblatt. New York, McGraw Hill.

Vannelli T., M. Logan, D. M. Arciero, and A. B. Hooper (1990) Degradation of halogenated aliphatic compounds by the ammonia-oxidizing bacterium *Nitrosomonas europaea, Appl. Environ. Microbiol.* 56(4):1169–1171.

Vestal, J. R. and J. J. Perry (1969). Divergent metabolic pathways for propane and propionate utilization by a soil isolate. *J. Bacteriol.* 99:216–221.

Vogel, T. M., C. S. Criddle and P. L. McCarty (1987). Transformations of halogenated aliphatic compounds. *Environ. Sci. Technol.* 21(6):722–736.

Wacket, L. P. and D. T. Gibson (1988). Degradation of toluene by toluene dioxygenase in whole cell studies with *Pseudomonas putida* F1. *Appl. Environ. Microbiol.* 54:1703–1708.

Wackett, L. P., G. A. Brusseau, S. R. Householder, and R. S. Hanson (1989). Survey of microbial oxygenases: Trichloroethylene degradation by propane-oxidizing bacteria. *Appl. Environ. Microbiol.* 55(11):2960–2964.

Westrick, J. J., J. W. Mello, and R. G. Thomas (1984). The ground-water supply survey, *J. Am. Wat. Works Assoc.* 76(5):52–59.

Yong, R. N., A. M. O. Mohamed and B. P. Warkentin (1992). *Principles of contaminant transport in soils.* Elsevier Science Publishers, Amsterdam.

Zwick, T. C., A. Leeson, R. E. Hinchee, R. E. Hoeppel and L. Bowling (1995). Soil moisture effects during bioventing in fuel-contaminated arid soils. *In situ aeration: Air sparging, bioventing, and related remediation processes.* R.E. Hinchee, R. N. Miller and P. C. Johnson. Columbus, Ohio, Batelle Press. 3(2).

Zylstra, G. J., L. P. Wacket and D. T. Gibson (1989). Trichloroethylene degradation by *Escherichia coli* containing the cloned *Pseudomonas putida* F1 toluene dioxygenase genes. *Appl. Environ. Microbiol.* 55:3162–3166.

Analysis of Bioremediation in Organic Soils

DANIEL M. WHITE AND ROBERT L. IRVINE
Department of Civil Engineering and Geological Sciences
University of Notre Dame
Notre Dame, IN 46556, USA

INTRODUCTION

Bioremediation is the task of engineers and scientists who may draw upon the resources of civil and chemical engineering, hydrology, geology, chemistry, physics, and biology. Bioremediation of contaminated soils, however, is not unlike traditional wastewater treatment. The biofilm in a trickling filter, for instance, contains many of the same organisms that are found in soil environments: aerobic, anaerobic, and facultative bacteria; fungi; algae; and protozoans (Metcalf & Eddy, Inc., 1991). Practices in wastewater treatment can be applied to soil bioremediation inasmuch as they provide a fundamental understanding of the behavior of organisms in an engineered system.

Biological treatment of contaminated soils may be accomplished in ex situ reactors, or in situ, where the ground is the reactor. In ex situ remediation, soils are treated in slurry or solid-phase systems (Lauch et al., 1992; Lewis, 1993; EPA, 1990; Troy et al., 1992; Ying et al., 1992). A soil slurry is used to increase contact between the contaminant, nutrients, organisms, and electron acceptors. A slurry phase system retards the persistence or formation of microenvironments that may host unfavorable growth conditions (e.g., low pH, nutrient, or contaminant deprivation). Solid-phase systems are designed to provide similar, but less aggressive conditions coincident with minimal energy input. Solid-phase systems commonly involve periodic tilling of a soil bed or air circulation

Please direct communication to Daniel M. White, Department of Civil and Environmental Engineering, 248 Duckering Box 755900, University of Alaska Fairbanks, Fairbanks, Alaska, 99775-5900.

through a soil heap. In situ bioremediation involves subsurface fluid circulation, bioventing, and biosparging technologies (Dineen et al., 1990; Hinchee, 1994; Piotrowski, 1991). Elimination of excavation costs is a principal driving force for in situ bioremediation. In cases where surface structures prohibit excavation, in situ treatment may be the only practical option.

Whether in situ or ex situ, all bioremediation programs require three basic analyses. First, contaminant concentration must be quantified. Second, analyses must establish that the contaminant is biodegradable in the soil environment, and third, analyses must confirm that contaminant disappearance was due to microbial degradation. Standard methods for all three analyses are well established and effective in most soils.

While most soils are principally composed of inorganic minerals (i.e., clay, silt, and sand), organic soils may contain between 5 and 60% by weight natural organic material (NOM). Natural organic material complicates all three basic analyses required of bioremediation projects. First, some compounds in NOM cannot be separated analytically from the contaminant. These compounds interfere with contaminant quantification. Second, certain NOM constituents may be readily biodegradable and serve as microbial substrates in addition to, or in place of, the contaminant. Proof that a contaminant is readily biodegradable in a mineral soil, therefore, does not mean that the same contaminant will be equally biodegradable in an organic soil environment. Third, analyses typically used to confirm that a contaminant was biologically degraded involve measurement of organism growth and carbon dioxide production. In organic soils, however, organism growth and carbon dioxide production are not strictly coupled with contaminant degradation.

In the next section of this chapter, the nature of NOM and contaminants in soil are reviewed. The following section describes the fractions of NOM that complicate contaminant quantification in organic soil. Following that, pyrolysis-GC/MS is presented as an analytical technique to help determine if a contaminant will be biodegradable in an organic soil. Finally, stable isotope analysis is presented as an analytical procedure to help confirm biological degradation of a contaminant in organic soil. Experimental results accompany the third through the fifth sections.

CONTAMINATED ORGANIC SOIL

This section contains a discussion of 1) the origin, stability, and chemical nature of NOM; 2) soil contaminants; and 3) the carbon box model as a tool for understanding the dynamics of organic material in soil.

General Consideration of Organic Soils

Organic soils are commonly classified by percent organic material (%OM) based on ashing or percent carbon (%C) based on EPA Method 415.1 (ASTM, 1995; Kopp and McKee, 1983). The %OM and %C are often equated by a factor of 1.72 (i.e., %OM = 1.72 × %C) (Dragun, 1988). For use in the present chapter, organic soils are defined as those soils containing 5–60% organic material (Lavanda et al., 1983). Organic soils are principally located near the ground surface in the "O" (i.e., organic) horizon. Although the concentration of organic material in soil generally decreases with depth, layers of organic soil may occur buried beneath surface mineral soils.

DEFINITIONS

Soil organic material (SOM) refers to all organic matter in soil (Schnitzer, 1991). Natural organic material refers to all organic compounds in soil that became part of the SOM by a natural process. Compounds introduced to soil by humans (i.e., anthropogenic) are considered contaminants. Although most organic contaminants are of natural origin, they are not considered part of the NOM.

Natural organic material is chemically divided into two categories, humic and nonhumic substances. The nonhumic substances consist of distinct biogenic molecules (e.g., carbohydrates, proteins, peptides, amino acids, nucleic acids, purines, pyrimidines, fatty acids, waxes, resins, pigments, etc.). Humic substances are derived from biogenic compounds that have undergone chemical, physical, or biological transformations. Humic substances are generally described as amorphous, dark-colored, partly aromatic, mainly hydrophilic, chemically complex, or polyelectrolyte-like materials (Schnitzer, 1991).

THE ORIGIN AND STABILITY OF NOM

Organic material accumulates in soil when production (e.g., plant and microbial growth) exceeds decomposition. The rate and extent of microbial decomposition depends on both the stability of the organic material present and on abiotic (e.g., pH, nutrient supply, electron acceptors) and environmental factors (e.g., climate, vegetation type, topography). Under biologically favorable conditions, the rate at which a compound is degraded in soil is directly related to its resistance to biological attack (i.e., stability). The rate of destruction of an organic molecule in soil is commonly expressed as turnover time, or mean residence time (MRT) (Dragun, 1988; Parton et al., 1987; Stevenson, 1994). Soil organic material may be categorized as "active," "slow," or "passive" based on MRTs of 0.14–3 years, 5–40 years, and 150–2500 years, respectively (Stevenson, 1994; Bohn et al., 1994; Parton et al., 1987). Although the compounds placed into each category are soil specific, the categories typically contain decomposing plant residues (0.14–3 years), microbial metabolites and cell wall constituents (5–40 years), and "resistant fractions" of NOM (150–2500 years) (Bohn et al., 1994). Some resistant fractions can have turnover times of greater than 3000 years (Stevenson, 1994).

CHEMICAL NATURE OF NOM

Nonhumic Organic Nitrogen

There are many forms of nitrogen (N)-containing compounds in soil. Organic N-containing compounds include proteins, peptides, amino acids, amino sugars, purines, pyrimidines, and heterocyclic ring structures. While most N-containing compounds are readily biodegradable (e.g., proteins), complexes formed in soil can stabilize them against microbial attack. For example, interactions with stabilized organic constituents such as tannins and lignins will make proteins resistant to biodegradation (Schnitzer, 1991; Stevenson, 1994).

Nonhumic Carbohydrates

Carbohydrates in soil occur primarily in the form of simple sugars, hemicellulose, and cellulose of both plant and microbial origin. Carbohydrates are the most readily degradable fraction of SOM, particularly when simple sugars represent a significant fraction of all carbohydrates (Schnitzer, 1991; Stevenson, 1994).

Nonhumic Lipid Materials

Lipid materials in soil include fatty acids, sterols, terpenes, chlorophyll, waxes, and resins. Lipid materials, or soil bitumens, can compose up to 20% of the NOM in organic soils (Braids and Miller, 1975). Certain soil bitumens (e.g., long-chain acids, alcohols, and alkanes) come from waxes that protect plant leaves (i.e., cuticle waxes), fruits, and needles; others are considered products of soil microbes (e.g., glyceride and phosphatide compounds). Soil bitumens are generally less biodegradable than the N-containing and carbohydrate fractions of NOM (Stevenson, 1966, 1994; Schnitzer, 1991).

Humic Substances

Humic substances are thought to be formed from various reactions. First, condensation reactions involve polyphenols derived from lignin and microbial metabolism with enzymatic conversion or self-condensation to

quinones. These molecules are occasionally combined with amino compounds to form nitrogen-containing polymers. The hypothetical structure
of a humic building block is illustrated in Figure 1 (adapted from Stevenson, 1994). Humic materials are generally stable against biodegradation.
The stability of humic substances is partly derived from random formation (i.e., heteropolycondensate nature), which results in physical and
steric inaccessibility to microbes. Humic substances are also resistant to
microbial degradation when complexed with polyvalent cations or sorbed
to clay minerals (Stevenson, 1994).

Contaminants

ANTHROPOGENIC CONTAMINANTS IN ORGANIC SOILS

Many contaminants are found in soil (e.g., heavy metals, organic
chemicals, oils and tars, combustible, radioactive and biologically active
materials, asbestos). Pesticides, solvents, and petroleum constituents are
common organic contaminants. Soil contamination results from land
disposal, uncontrolled dumping, accidental leakages during operation or
transportation, and stockpiling of raw materials, wastes, and finished
products. Actual contaminant concentrations in soil can range from free
product to concentrations below detection limits (Smith, 1985). Contaminants typically dissolve into the soil water or partition into hydrophobic,
aqueous exclusion zones created by NOM or inorganic particles. Compounds with limited water solubility can be present in both aqueous and
nonaqueous phases in soil. The water solubility of a compound depends
on the molecular weight and polarity of the compound, salinity of the soil
water, and the presence or absence of surface active agents (i.e., surfactants). The extent to which an organic chemical will adsorb to soil surfaces
is dependent on the molecular weight, water solubility, octanol-water
partition coefficient, and vapor pressure of the compound (Dragun, 1988).

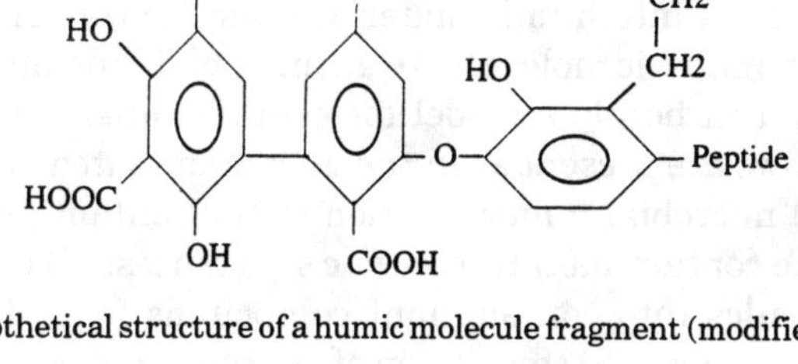

Figure 1. Hypothetical structure of a humic molecule fragment (modified from Stevenson,
1994).

Because many contaminant compounds exist naturally in organic soils (albeit in low concentrations), most soils contain organisms with the genetic capability to degrade them. A shift in the microbial population typically occurs when a contaminant is applied to soil, favoring the consortia that are more efficient at using the contaminant as a carbon and/or energy source (Dragun, 1988). If a contaminant is biodegradable and the appropriate microbes and nutrients are present, the rate of biodegradation of any contaminant is dependent on its availability to the microorganisms (i.e., bioavailability) (NRC, 1993). The soil structure, water solubility of the contaminant, genetic capability of the organisms, and environmental conditions (e.g., temperature, pH, moisture conditions) all participate in contaminant bioavailability.

Carbon Box Models

The transformation of carbon-containing compounds in soil is referred to as "carbon dynamics." Agronomists use models for carbon, nitrogen, phosphorus, and sulfur transformations in soil to describe each element's cycle in the environment. Models of carbon dynamics are used to predict changes in SOM composition due to practices in fertilizer and water management, changes in tillage and cropping strategies, and changes in crop distributions (Parton et al., 1988). Managing bioremediation in organic soils also requires an understanding of the entire ecosystem inasmuch as the nature of NOM influences the bioavailability of contaminants, nutrient turnover, and the process of diauxic growth (i.e., two phase growth).

The box model illustrated in Figure 2(a) shows the basic flow paths of carbon containing compounds between the three principal carbon reservoirs associated with an uncontaminated soil (i.e., humic substances, nonhumic substances, and the soil gas/atmosphere). The basic carbon flows (i.e., excluding translocation of water-soluble organics) illustrated in Figure 2 are: (1) microbial mineralization of humic substances; (2) microbial mineralization of nonhumic substances; (3) transformation of nonhumic materials into humic materials; and (4) conversion of humic molecules into nonhumic molecules (e.g., microbial consumption). Figure 2(b) illustrates a carbon box model for contaminated organic soil. The same carbon flows are present as in the uncontaminated model, with the addition of: (5) microbial mineralization of the contaminant; (6) transformation of the contaminant into humic substances; (7) transformation of humic molecules into contaminant compounds (e.g., degradation of plant waxes to alkanes); (8) conversion of the contaminant into nonhumic

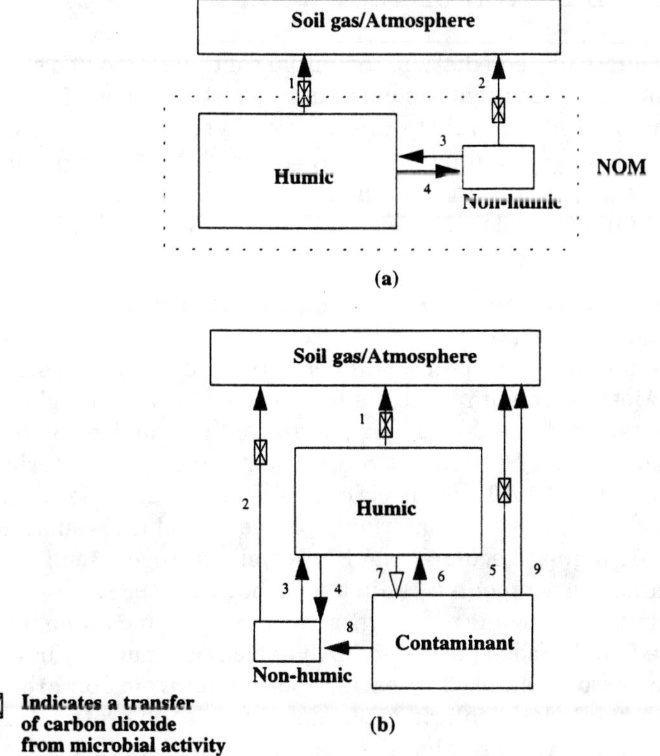

Figure 2. Carbon box models for (a) uncontaminated soil and (b) contaminated soil.

substances (e.g., microbial consumption); and (9) direct volatilization of the contaminant.

Summary

Natural organic material consists of compounds in various degrees of decomposition. Depending on the genetic capability of the soil microbes, most compounds in NOM are biologically degradable. Nonhumic substances are, by definition, recognizable biogenic compounds. Humic substances, on the other hand, are transformed, unrecognizable compounds occurring with atomic weights in the thousands of Daltons. Nonhumic substances, humic substances, and organic contaminants are intimately associated through both biological and nonbiological reactions. Carbon box models illustrate the complexity of subsurface carbon dynamics.

COMPLICATION OF TESTING PROCEDURES BY NOM

To evaluate bioremediation, the initial concentration of a soil contaminant and its subsequent disappearance must be quantified. In organic soils, the acquisition and interpretation of these data may be complex and requires an understanding of the ways in which NOM complicates testing procedures. The following section presents methods used for quantifying soil contaminants and the complexities introduced by NOM.

Solvent Extraction of Petroleum Compounds

Crude oil and other petroleum derivatives are common soil contaminants. Most petroleum products (e.g., crude oil) are complex mixtures that contain both nonpolar (e.g., n-triacontane) and polar (e.g., acetic acid) compounds (Dragun, 1988). In one report, polar materials ranged between 2.9 and 17.9 weight percent of Prudhoe Bay, South Louisiana, and Kuwait crude. Polar materials in crude oil include resins and asphaltenes (i.e., high-molecular-weight polycyclic nitrogen-, sulfur- and oxygen-containing compounds) (Gill and Robotham, 1989). Nonpolar materials in crude oil include many types of compounds, including the straight, branched- and cyclic alkanes and polynuclear aromatic hydrocarbons.

Common solvents used to extract petroleum from soil are n-hexane, benzene, methylene chloride, carbon disulfide, chloroform, and 1,1,2-trichloro-1,2,2-trifluoroethane (freon-113). Since crude oil is principally composed of nonpolar compounds, the solvents used to extract oil from soil are equally nonpolar. Since crude oil includes some polar compounds, the highly nonpolar solvents (e.g., hexane) are sometimes avoided. In Figure 3, the relative polarity of various solvents is illustrated by comparing their respective water solubilities. Since water is a polar solvent, an infinite water solubility indicates the greatest degree of polarity. Since NOM contains hydrophobic compounds (i.e., soil bitumens), the solvent used to extract crude oil from soil also extracts a fraction of NOM. A solvent will extract both those compounds from NOM with the same elemental composition as compounds in crude oil (see Table 1) and those compounds with similar degrees of polarity.

Soil Bitumens

ALIPHATIC HYDROCARBONS

Most hydrocarbons in NOM are in the form of straight-chain alkanes and are derived from plants and microorganisms (Stevenson, 1994). Some

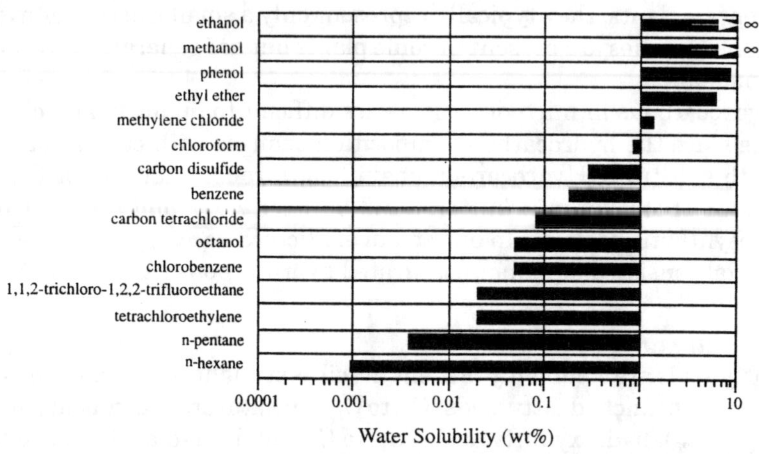

Figure 3. Solubility of various solvents in water (adapted from Riddick and Bunger, 1970).

plant waxes, such as that of the *Solandra grandiflora* contain up to 92% alkanes. The alkanes in plant waxes are primarily odd chain alkanes (n-C_{21} to n-C_{37}). Depending on the plant, n-C_{27}, n-C_{29}, n-C_{31}, or n-C_{33} may be the n-alkane component with the highest relative concentration (Tulloch, 1976; Braids and Miller, 1975; Stevenson, 1966). Alkanes in plant waxes are most likely produced through the elongation-decarboxylation mechanism. In this process, elongation occurs in two-carbon units (i.e., lipid synthesis) followed by decarboxylation of the terminal carboxyl group (Kolattukudy et al., 1976). Although branched chain alkanes are

Table 1. Compounds present in crude oil and NOM
(modified from Jorgenson et al., 1990).

Acetic acid	Heptacosane	*n*-Hentriacontane	Octanoic acid
Alkanes	Heptacosanoic acid	*n*-Heptadecane	*o*-Xylene
Benzene	Hexacosane	*n*-Hexadecane	Pentacosane
1,2-Benzofluorene	Hexadecanoic acid	*n*-Nonacosane	Pentanoic acid
Benzoic acid	Methane	*n*-Nonadecane	Perylene
Butanoic acid	Methanethiol	Nonanoic acid	Phenanthrene
Carbazole	Methanol	*n*-Octacosane	Propanoic acid
Decanoic acid	*m*-Xylene	*n*-Octadecane	*p*-Xylene
2,6-Dimethylundecane	Naphthalene	*n*-Pentadecane	Tetradecanoic acid
Eicosanoic acid	*n*-Dotriacontane	*n*-Tetracosane	Toluene
Ethanol	*n*-Docosane	*n*-Tetradecane	
Ethylbenzene	*n*-Eicosane	*n*-Triacosane	
Formic acid	*n*-Heneicosane	*n*-Triacontane	

present in plants, they typically represent only a small fraction of hydrocarbons. Alkenes are present in some plants but are generally only minor components.

Hydrocarbons in microorganisms are difficult to quantify and characterize since the hydrocarbon composition changes with culture age and growth substrate. Hydrocarbon chain length ranges between n-C_{14} and n-C_{20} for photosynthetic bacteria and between n-C_{26} and n-C_{30} for non-photosynthetic bacteria. In one strain of *Micrococcaceae*, 20% of the lipid material consisted of monounsaturated hydrocarbons (Albro, 1976).

Other Bituminous Material

Fatty acids are commonly present in soil as straight chains n-C_7 to n-C_{29}, but also as branched fatty acids (C_{12} to C_{19}), unsaturated fatty acids (n-C_{16}, n-C_{18}, n-C_{20}), hydroxy fatty acids (C_{12} to C_{16}), and α,ω-diacids (C_{15} to C_{25}). Fatty acids n-C_{26} to n-C_{38} typically come from plants and insects, while fatty acids n-C_4 to n-C_{26} are thought to come from microorganisms (Stevenson, 1994). Consistent with lipid synthesis (two carbon additions), the most prevalent fatty acids are even chain, unbranched, and saturated (though variants exist).

Waxes commonly found in soil are esters of higher aliphatic acids to higher aliphatic or cyclic alcohols (Bergmann, 1963). Consistently, fatty acids and alcohols are composed of even carbon number subunits (Stevenson, 1994). Polynuclear aromatic hydrocarbons (e.g., pyrene, chrysene, phenanthrene); alcohols (e.g., ceryl alcohol, n-alkanol C_{16} to C_{30}); sterols (e.g., β-sitosterol) and terpenes (e.g., friedelin); pigments (e.g., β-carotene, a-chlorophyll); and heterocyclic compounds (e.g., carbazole) are also present in soil bitumens (Braids and Miller, 1975).

Experimental

To illustrate the impact of the soil bitumens on crude oil analyses, organic soil from Umiat, Alaska (20% NOM w/w), was artificially contaminated with 6400 mg/kg of crude oil. The soil was then extracted with carbon disulfide and analyzed by GC/MS (White, 1995). Soil bitumens were an obvious fraction of the total organic material extracted (see Figure 4). The bitumens could not be selectively removed from the chromatogram, however, since they co-eluted with compounds from crude oil. An uncontaminated sample of the same soil was extracted with carbon disulfide in order to quantify the soil bitumens alone (see Figure 5). Most compounds were identified by mass spectrometry as long-chain n-alkanes, ketones, aldehydes, or alcohols. No fatty acids or short-chain

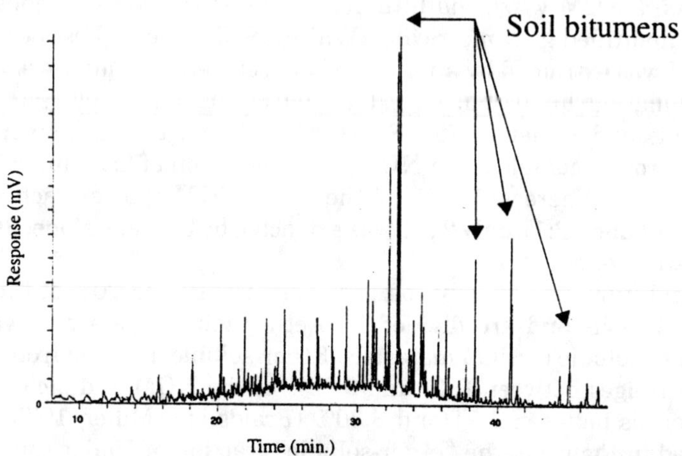

Figure 4. Chromatogram of carbon disulfide extract of soil artificially contaminated by 6400 mg/kg crude oil.

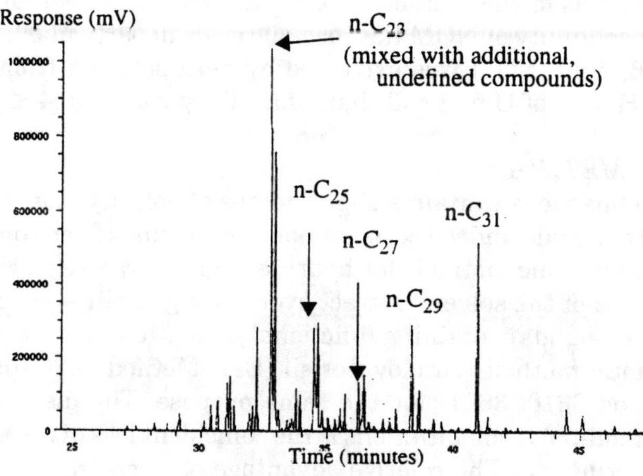

Figure 5. Chromatogram of soil bitumens.

alcohols or ketones were observed. The soil bitumens extracted amounted to roughly 5000 mg/kg (i.e., against a crude oil standard).

In experiments conducted on other organic soils from Umiat, Alaska (60–65% NOM w/w), soil bitumens were extracted by various solvents and quantified gravimetrically (White, 1995). The highest percentage of NOM was extracted by an acidified benzene:ethanol mixture, a common combination for enhanced extraction of soil bitumens (see Figure 6) (Braids and Miller, 1975). Since ethanol acts to break lipoprotein or glycoprotein complexes in NOM, the proportion of fat soluble NOM was increased. Whereas 15.5% of the soil's NOM was extracted by benzene:ethanol (2:1), only 2.2% was extracted by benzene alone, illustrating the effect of ethanol.

Soil bitumens typically compose between 10 and 20% of the NOM in organic soils and are divided into ether-soluble (fats and waxes) and alcohol-soluble (resins) materials. Ether-soluble material from peat typically ranges between 0.42 and 9.36% of the NOM and alcohol- soluble materials between 1.71 and 8.30% (Braids and Miller, 1975). As illustrated in Figure 6, the "ether-soluble" fraction of Umiat soil accounted for 6.2% of the total NOM, well within the reported range.

The EPA methods for quantification of "oil and grease" (e.g., EPA Method 413.1) require acidification of the sample to a pH below 2. While acidification aids in the extraction of oil and grease, it also can increase the solvent solubility of NOM (i.e., organic acids in particular). As shown in Figure 6, more NOM was extracted by most solvents from acidified samples (pH < 2) of Umiat soil than unacidified samples (4 ≤ pH ≤ 5).

CLEANUP METHODS

Most soil bitumens contain a slight degree of polarity (e.g., long-chain ketones). To exclude undesired, semi-polar compounds from contaminant quantification, some methods for hydrocarbon analysis (e.g., EPA 418.1) include a pass of the solvent extract over silica gel. Silica gel selectively removes compounds containing functional groups (e.g., carboxyl groups). Other cleanup methods employ florisil (EPA Method 3620) or alumina (EPA Method 3610, 3611) for the same purpose. The disadvantage to removing semi-polar soil bitumens is the coincident loss of the semi-polar fraction of crude oil. The relative advantage of removing soil bitumens from a solvent extract must be weighed against sample loss.

Summary

Many of the same compounds are found in crude oil and NOM. In addition, many compounds in crude oil and NOM behave the same way

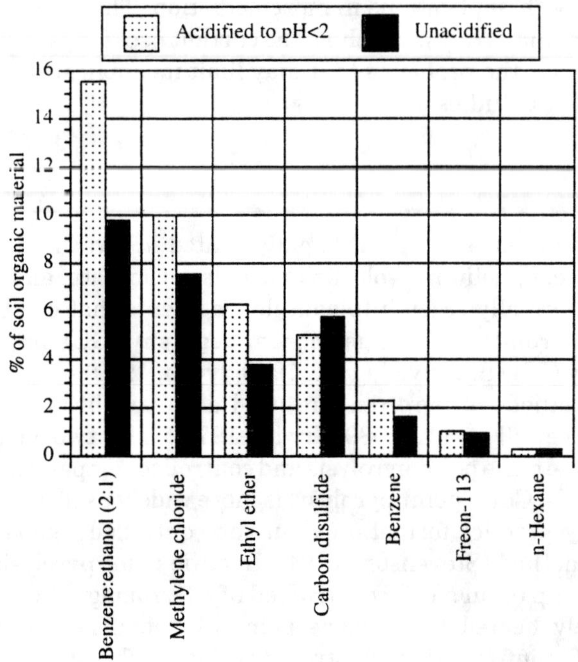

Figure 6. Mass percent of SOM soluble in various solvents and the effect of sample acidification.

when treated with an organic solvent. While most compounds from NOM contain some degree of polarity and can be removed by cleanup procedures, the removal is likely accomplished at the expense of a fraction of the contaminant desired for quantification.

PYROLYSIS-GC/MS

Natural organic material contains soft substrates, or biodegradable compounds, which compete with the contaminant as a microbial substrate. Since any compound can be a soft substrate, no one extraction method is likely to isolate all potentially biodegradable compounds in NOM. In order to identify soft substrates, therefore, a whole soil extraction method is needed. Pyrolysis is a method traditionally used in soil science to remove all volatile and nonvolatile components of NOM from soil. Using pyrolysis, all compounds (including the fragments of nonvolatile compounds) can be extracted from soil and analyzed in-line with GC/MS. The purpose of this section is to present analytical pyrolysis as a technique with applications in bioremediation. This section includes a description of the pyrolysis method, the current uses for pyrolysis in water and soil science, the problems that may limit its application, and results from two recent studies.

Introduction

During pyrolysis, a sample is rapidly heated in a vacuum or a stream of inert gas (e.g., helium). Volatile molecules evaporate, and nonvolatile molecules thermally crack into volatile fragments that can be analyzed by mass spectrometry (MS), gas chromatography (GC)/MS, or infrared spectroscopy (IR) (Bracewell et al., 1989; Irwin, 1979).

Several methods of pyrolysis are used and have been previously reviewed (Bracewell et al., 1989; Irwin, 1979). The two most common techniques are curie point pyrolysis and controlled temperature programming pyrolysis. Curie point pyrolysis is more widely used and less subject to secondary product formation than the controlled temperature programming method (Stevenson, 1994). In curie point pyrolysis, a sample is fixed in a cup or on a coil constructed of a ferromagnetic metal, which is inductively heated to its curie point (ferromagnetic limit). In the pyrolysis-GC configuration illustrated in Figure 7, a carrier gas enters the pyrex sleeve, which contains the ferromagnetic wire and carries the volatile (i.e., pyrolyzed) molecules through the injector needle directly onto a chromatographic column. The molecules are condensed on the column and then eluted and detected by one of many methods (e.g., MS).

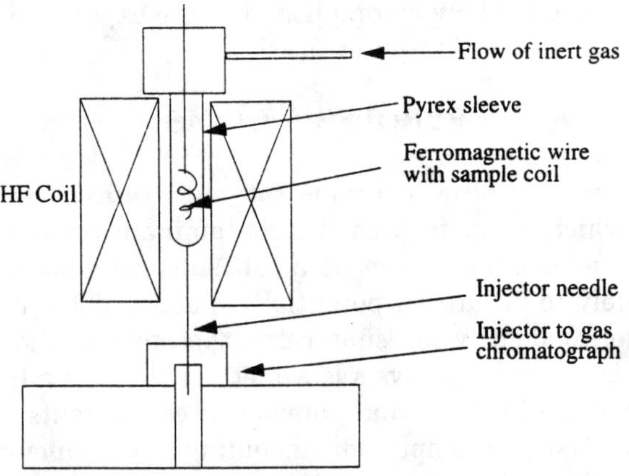

Figure 7. Diagram of a curie point pyrolysis reactor.

Controlled temperature programming pyrolysis is conducted with a resistively heated wire filament or ribbon. Liquid suspensions are applied directly to a platinum filament or ribbon. Solid samples are placed in a quartz tube around which a platinum filament is wrapped. A current is passed through the filament or ribbon, which heats and pyrolyzes the sample. Volatile compounds from the sample are carried directly into a GC column by a carrier gas (e.g., helium).

PYROLYSIS OF ENVIRONMENTAL SAMPLES

A chromatogram of pyrolyzed organic material is commonly called a fingerprint (Stevenson, 1994). Fingerprinting of organic material has enabled scientists to gain insight into complex organic mixtures in the aquatic and soil sciences (Bracewell et al., 1989). Pyrolysis has been used in the aquatic sciences to qualitatively investigate the structure, character, and origin of aquatic humic substances (Schulten et al., 1987; Abbt-Braun et al., 1989; Wilson et al., 1983; Bruchet et al., 1990; MacCarthy et al., 1985; van de Meent et al., 1980).

In soil science, fingerprints of humic substance are used to illustrate the character of NOM. Specific studies have focused on humic substance formation processes; comparisons of humic substances from different soils; correlation of climate to the nature of organic material; the effect of alkali extractants on Podzol fulvic acids; comparisons with aquatic humic substances; structural studies; and comparative analyses of desert soils, Precambrian shales, and meteorites (Bracewell, 1971; Bracewell and Robertson, 1973; Bracewell et al., 1975, 1980; Martin, 1976; Saiz-Jimenez et al., 1979; Schulten and Schnitzer, 1992; Simmonds et al., 1969; Wershaw and Bohner, 1969; Wilson et al., 1983). In one study, polluted soil and sediment samples were screened with pyrolysis-GC/MS to demonstrate that anthropogenic compounds in polluted samples evaporated and were not thermally altered with a curie point temperature of 510°C (de Leeuw et al., 1986). These data support the use of pyrolysis as a tool to monitor the rate and extent of destruction of SOM (contaminants and/or NOM) during bioremediation.

In microbiology, pyrolysis has been used in the taxonomy of microorganisms. By fingerprinting entire cells or identifying the primary components of cell walls, organisms have been differentiated or identified (Abbey et al., 1981; Bahr and Schulten, 1983; Engman et al., 1984). Variations in culture medium, age, and sample preparation (e.g., air dry, lyophilized) have an appreciable impact on an organism's chromatographic fingerprint. Some limitations may be overcome by using chemically defined media and organisms that are growth limited and in synchronous growth (Bracewell et al., 1989).

HISTORICAL BARRIERS TO QUANTITATIVE PYROLYSIS-GC/MS

Pyrolysis (usually PY-GC) has been used for quantitative analysis of onium salts, pharmaceutical drugs and some biological samples (Irwin, 1979). In environmental samples, however, pyrolysis has historically been used only as a qualitative technique. Though some studies used quantitative pyrolysis in aquatic or soil samples, the vast array of products produced and the variable effects of sample size have prevented pyrolysis from becoming a widely used quantitative technique (Bruchet et al., 1990). The pyrolytic behavior of organic material influences the "extent of cracking" (i.e., the mass ratio of pyrolysis products). Sample size affects secondary reactions that occur after the initial pyrolysis reaction and diffusion-restriction of long-chain radicals (Hancox et al., 1991). The effect of the sample size can also change with the character of the organic material pyrolyzed.

Compounds in soil pyrolyzates can be pyrolysis or evaporation products from one or more parent products. Pyrroles, for instance, can be derived from various proteinaceous compounds (e.g., proline, hydroxyproline, glutamine) and from porphyrins (Bracewell et al., 1989; Irwin, 1979). Pyridines may be derived from proteins or nucleic acids, and hydrocarbons from proteins and lipids (Irwin, 1979). In cases where compounds in pyrolyzates have only one- or two-parent molecules, the origin can sometimes be determined. To determine the contribution that each parent molecule (i.e., polyhydroxyaromatics or tyrosine) contributed to the phenol identified in pyrolyzates, a particularly noteworthy method was proposed (Bruchet et al., 1990). Since tyrosine produces equal amounts of phenol and *p*-cresol, and polyhydroxy aromatics produce predominantly phenol, the difference in response between *p*-cresol and phenol peaks should equal the phenol produced by polyhydroxy aromatics. Though this simple relation was adequate for phenol, analysis of other pyrolysis products (e.g., hydrocarbons) can be far more complex.

Experimental

Experiments were conducted by the Center for Bioengineering and Pollution Control (University of Notre Dame) in conjunction with the Institute and Testing Laboratory for Water Quality Control and Waste Management (Technical University of Munich) to determine the potential for curie point pyrolysis-GC/MS as a quantitative technique for use in bioremediation of contaminated organic soils. Included herein are two examples of the studies performed. First, in order to determine the best sample size for organic soil, a method was used in which the concentration of toluene in pyrolyzates served as an indicator of the "extent of cracking" (Kimber and Searle, 1970). Second, samples of uncontaminated soil, crude oil, and soil artificially contaminated with crude oil were qualitatively "fingerprinted" to illustrate the differences in chromatographic signatures.

MATERIALS AND METHODS

Experiment to Determine the Effect of Sample Size

Samples of organic soil from Umiat, Alaska (20% organic material, w/w) were air dried and ground to a fine powder with a mortar and pestle. Soil samples weighing between 0.5 and 5 mg (quantified using a Sartorius Microbalance) were placed in 700°C sample cups for pyrolysis. Prior to running each sample, the quartz pyrolysis sleeve was placed in a muffle furnace at 500°C for 1 h. A complete pyrolysis run with no sample was made prior to all sample runs to ensure there was no carryover from the previous sample. Pyrolysis was conducted with a Fischer Curie Point Pyrolyzer Model 0316 with the reactor temperature set at 275°C and a pyrolysis heating time of 0.1 sec and hold time of 9.9 sec. The pyrolysis reactor was mounted on a gas chromatograph, HP Series II, with an Ultra 2 column (5% Ph Me Silicon) of 50 m × 0.2 m × 0.1 μm film thickness. Chromatographic conditions were 50°C for 2 min, followed by 20°C/min to 300°C and hold for 9 min. The GC was plumed directly to an HP 5971 Series Mass Selective Detector on EI mode. The MS scanned mass units 46 to 650. The computer hardware was an HP Apollo Series 400 with HP *Chemsystem 400* software. All mass spectra were compared to the Wiley 138.1 spectral library. The column pressure was held at 137 kPa using helium as a carrier gas at 0.5 cc/min. The sample was injected with a split ratio of 100:1.

Examples for the Application to Bioremediation

Pyrolysis products of crude oil, uncontaminated soil, and soil artificially contaminated with crude oil were compared. For the first case, a dilute solution of Kuparuk crude oil (supplied by ARCO Alaska) was prepared in carbon disulfide. A 700°C pyrolysis coil was dipped into the mixture and held several minutes at room temperature for the solvent to evaporate. The sample was run at the previously described pyrolysis conditions. Chromatographic conditions were 50°C for 5 min, 10°C/min to 70°C, 5°C/min to 270°C and hold for 10 min, then 5°C/min to 300 and hold for 12 min.

Samples of uncontaminated soil were prepared in the same way as previously described for the sample size experiment and run according to the same chromatographic conditions as the crude oil. Contaminated soil samples were prepared by artificially contaminating Umiat soil with Kuparuk crude oil at an application rate of 100,000 mg/kg. The soil was homogenized by slurrying, air dried, and ground to a fine powder. Five days elapsed between contamination of the soil and analyses. The soil therefore was considered "freshly contaminated" soil. Oil that has undergone a long period of weathering would likely produce slightly different pyrolysis products due to various interactions that take place in the soil over time (e.g., aggregate formation).

RESULTS AND DISCUSSION

Experiment to Determine the Effect of Sample Size

The ratio of the chromatographic area of toluene to the total chromatographic area (i.e., the toluene ratio) is illustrated in Figure 8, as a function of total sample mass. The curve in Figure 8 was drawn by the best visual fit. The slope of the line between 0.5 and 1.2 mg of sample indicates that a change in sample mass caused a large change in relative toluene production. When the sample mass reached roughly 1.25 mg, the toluene ratio appeared to plateau at approximately 0.01. Random variation about a toluene ratio of 0.01 occurred for soil masses larger than roughly 2 mg. For samples larger than roughly 2.5 mg, residual sample (i.e., not pyrolyzed) was observed in the quartz sleeve following pyrolysis. The residual material was probably ejected from the pyrolysis probe since the probe

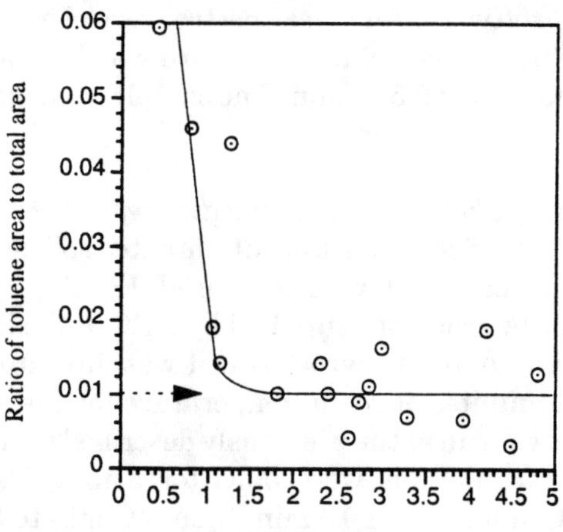

Figure 8.   Graph illustrating the effect of sample size on toluene yield.

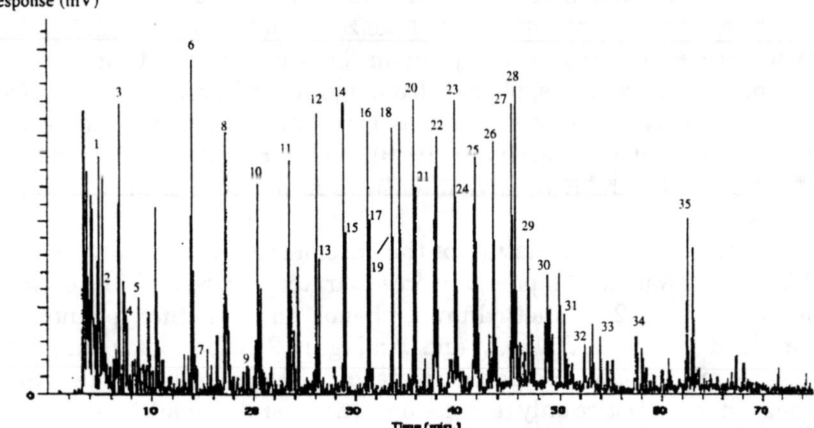

Figure 9. Chromatogram of uncontaminated Umiat soil (see Table 2 for identification of labeled compounds).

vibrated slightly during heating. Based on these results and observations, 1.8–2 mg was selected as the best sample size for Umiat soil.

Fingerprint of Uncontaminated Soil

The chromatographic fingerprint for uncontaminated soil is illustrated in Figure 9. All labeled compounds are listed in Table 2. A homologous series of *n*-alkanes and *n*-alkenes (i.e., C_{10}, C_{11}, C_{12}, . . .) appeared as the major peaks eluting between roughly 10 and 50 min. *N*-alkanes occurred with roughly equal intensities between C_{10} and C_{19}, after which the odd carbon chain was prevalent. In contrast, even chain alkenes were more prevalent than odd chain alkenes above C_{19}. Also observed as pyrolysis products were alkylbenzenes; methyl-alkyl-benzenes; and methyl-substi-

Table 2. Labeled compounds from Figure 9.

1	Benzene	11	Tridecene	21	Octadecane	31	Hexacosane
2	Pyridine	12	Tetradecene	22	Nonadecane	32	Heptacosane
3	Toluene	13	Tetradecane	23	Eicosene	33	Octacosane
4	2,5-Dimethylfuran	14	Pentadecene	24	Eicosane	34	Nonacosane
5	Methyl-pyrrole	15	Pentadecane	25	Heneicosane	35	Hentriacontane
6	Decene	16	Hexadecene	26	Docosene		
7	Phenol	17	Hexadecane	27	Triacosene		
8	Undecene	18	Heptadecene	28	Triacosane		
9	2-Methoxyphenol	19	Heptadecane	29	Tetracosene		
10	Dodecene	20	Octadecene	30	Pentacosane		

tuted benzenes, indenes, phenols, methoxyphenols, furans, pyrroles, and pyridine. Alkyl-methyl-substituted benzenes are derived from aromatic hydrocarbon precursors and proteins in soil, indenes from aromatic hydrocarbon precursors, phenols from lignins and proteins, furans from secondary polysaccharides, pyrroles from proteins, and pyridine from proteins or nucleic acids (Irwin, 1979). The same suite of products were observed with similar relative intensities in previous studies (Schulten and Schnitzer, 1992).

Due to the nonpolar nature of the chromatographic column used for this study, few polar compounds were separated very well. While pyridine, methyl pyrrole, 2,5-dimethylfuran, phenol, and 2-methoxyphenol were identified (see peaks 2, 4, 5, 7, and 9 in Figure 9), other polar compounds coeluted. Some polar compounds were identified according to their functional characteristics only (e.g., ketones, acids and alcohols) and were not labeled on the chromatogram.

Fingerprint of Crude Oil

The first 55 (of 75) min of a chromatographic fingerprint from crude oil are illustrated in Figure 10. The most obvious peaks are a homologous series of *n*-alkanes. The pristane and phytane peaks are also evident adjoining the *n*-C_{17} and *n*-C_{18} peaks, respectively. Few polar compounds (which would be expected to elute in the first 10 min from the slightly polar chromatographic column) and no alkenes were observed (see Table 3).

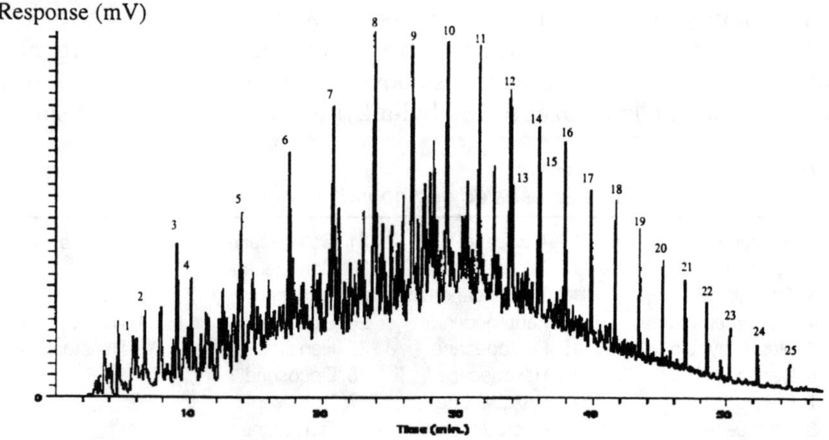

Figure 10. Chromatogram of Kuparuk crude oil (see Table 3 for identification of labeled compounds).

Table 3. Labeled compounds from Figure 10.

1 Toluene	10 Pentadecane	19 Docosane
2 Octane	11 Hexadecane	20 Triacosane
3 1,2-Dimethylbenzene	12 Heptadecane	21 Tetracosane
4 Nonane	13 Pristane	22 Pentacosane
5 Decane	14 Octadecane	23 Hexacosane
6 Undecane	15 Phytane	24 Heptacosane
7 Dodecane	16 Nonadecane	25 Octacosane
8 Tridecane	17 Eicosane	
9 Tetradecane	18 Heneicosane	

Fingerprint of Contaminated Soil

The chromatographic fingerprint for contaminated soil is illustrated in Figure 11. The most interesting feature is the distribution of *n*-alkanes and *n*-alkenes. Since *n*-alkenes are not derived from pyrolysis of crude oil, the relative fraction of *n*-alkenes to *n*-alkanes was lower than observed with uncontaminated soil. Despite the contamination, the peaks associated with the compounds exclusively derived from NOM were readily isolated (e.g., pyridine, 2,5-dimethylfuran, methyl-pyrrole, 2-methoxyphenol).

Summary

The results from these studies demonstrated that individual compounds derived from crude oil or NOM can be identified in pyrolysis-GC/MS fingerprints of contaminated soil. Since individual compounds

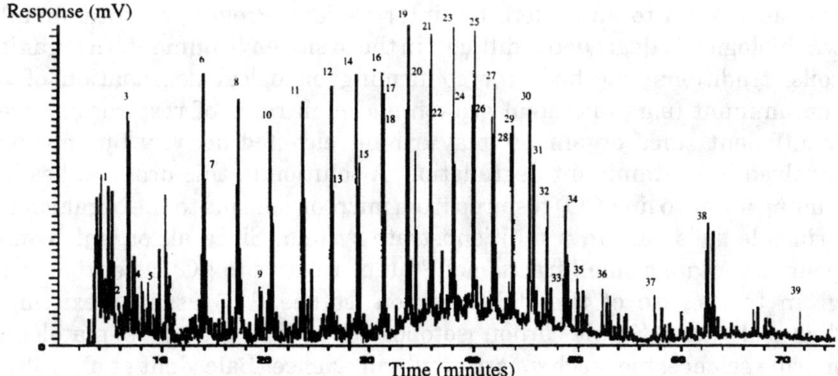

Figure 11. Chromatogram of contaminated Umiat soil (see Table 4 for identification of labeled compounds).

Table 4. Compounds labeled in Figure 11.

1 Benzene	11 Tridecene	21 Octadecane	31 Triacosane
2 Pyridine	12 Tetradecene	22 Nonadecene	32 Tetracosene
3 Toluene	13 Tetradecane	23 Nonadecane	33 Tetracosane
4 2,5-Dimethylfuran	14 Pentadecene	24 Eicosene	34 Pentacosane
5 Methyl-pyrrole	15 Pentadecane	25 Eicosane	35 Hexacosane
6 Decene	16 Hexadecene	26 Heneicosene	36 Heptacosane
7 Phenol	17 Hexadecane	27 Heneicosane	37 Nonacosane
8 Undecene	18 Heptadecene	28 Docosene	38 Hentriacontane
9 2-Methoxyphenol	19 Heptadecane	29 Docosane	39 Pentacontane
10 Dodecene	20 Octadecene	30 Triacosene	

can be isolated, chromatographic separation and quantification is possible. Through quantification of individual compounds, the relative degradation of NOM (e.g., proteins, lignins, polysaccharides) and the contaminant (e.g., alkanes, naphthalenes, phenanthrenes) may be calculated. Based on these analyses, the relative biodegradability of a contaminant can be compared to individual compounds in the NOM.

CARBON 13

The goal in bioremediation is to transform a soil contaminant into carbon dioxide (CO_2), water, and inorganic reaction products (e.g., orthophosphates). Biological degradation is typically measured by contaminant disappearance. Contaminant disappearance will also be measured, however, if the contaminant migrated off-site, was unextractable due to interactions with the mineral or organic soil fractions, or underwent a transformation to an undetected intermediate. Proving a contaminant was biologically degraded is difficult in the in situ environment. In organic soils, traditional methods for confirming biological degradation of a contaminant (e.g., microbial growth and high rates of respiration) are insufficient since organisms may exhibit elevated activity but not be involved in contaminant degradation. A solution to this problem lies in finding a way to link CO_2 respired from microorganisms to the organisms' principle substrate in a multisubstrate system. Since all organic compounds contain a small fraction of ^{13}C (i.e., relative to ^{12}C), attention was given to the use of the stable carbon isotope (^{13}C) as a preexisting, "natural" label. Stable carbon isotopes are used in geology, climatology, marine science, biogeochemistry, and soil science (Balesdent et al., 1987; Faure, 1986; Freeman and Hayes, 1992; Hayes et al., 1989; O'Brien and Stout, 1978; Saupe et al., 1989; Schell et al., 1989; Reardon et al., 1979;

Vinette, 1992). The following section includes an introduction to ^{13}C analyses and a summary of studies designed to evaluate ^{13}C analysis as a tool to confirm biodegradation of crude oil in organic soils.

Introduction

EXPRESSION OF ISOTOPES

Carbon-13 represents only a small fraction of all carbon atoms (i.e., 1.11%). The relative abundance of ^{13}C is expressed in terms of $\delta^{13}C$-values, which represent the $^{13}C/^{12}C$ ratio relative to the universal standard, Peedee Belemnite (PDB),

$$\delta^{13}C = \left(\frac{R}{R_s} - 1 \right) \times 1000$$

where

$R = {}^{13}C/{}^{12}C$ ratio in the sample
$R_s = {}^{13}C/{}^{12}C$ ratio in the standard

$\delta^{13}C$ is expressed in parts per thousand, or "per mil" (i.e., expressed using the symbol, ‰) (Craig, 1957; Faure, 1986; Hoefs, 1987). Compounds depleted in ^{13}C, relative to the standard, have negative $\delta^{13}C$-values.

ISOTOPE FRACTIONATION

In a reaction where more than one product is formed, the carbon isotope *ratio* in each product will not necessarily equal the isotope ratio in the reactant. The relative concentration of isotopes in reaction products compared to reactants is called fractionation (Hoefs, 1987). The fractionation of isotopes during a reaction is caused by an isotope effect. Although an isotope effect is not directly observable (i.e., it is a physical phenomenon), its existence is inferred to describe readily observed fractionation. Two isotope effects have been described: the isotope exchange effect and the kinetic isotope effect (Hayes, 1992). With the isotope exchange effect, carbon-carbon (C—C) bond strength controls the distribution of ^{13}C between molecules. The heavy isotope generally resides in the compound where it is most strongly bound (Biegleson, 1965). Bond strength differences arise because the heavy isotope, ^{13}C, has a smaller reserve of free energy than ^{12}C and forms bonds of lower minimum free energy. The isotope exchange effect drives the redistribution of ^{13}C between oceanic carbonates and dissolved and gaseous carbon dioxide (i.e., during the hydration of CO_2 gas) (Degens, 1969).

With the kinetic isotope effect, fractionation is governed by the *rates of reaction* involving compounds with differing isotopic composition. Because ^{12}C atoms are lighter than ^{13}C atoms, reaction rates involving ^{12}C are faster than those involving ^{13}C (i.e., isotopically light molecules are more mobile and have greater velocities than isotopically heavy molecules) (Galimov, 1981). Carbon-12 atoms therefore accumulate in compounds formed during high rate reactions.

In photosynthesis, plants take up CO_2 in a reversible diffusion process and then fix the carbon in an enzymatic carboxylation reaction. Fractionation (due to the kinetic isotope effect) occurs during both processes. In the first process, more CO_2 containing the light isotope diffuses into plant leaves due to its lighter mass and higher velocity. Accumulation of the light isotope results in dissolved intracellular CO_2 being depleted in ^{13}C relative to atmospheric CO_2. In the second step (i.e., CO_2 fixation) isotopically light CO_2 is preferentially incorporated into cell mass via carboxylation. As a result, fixed carbon is depleted relative to dissolved, intracellular CO_2. The extent of fractionation during carbon fixation depends on whether the carbon is fixed in a three-carbon molecule (phosphoglyceric acid) by the "Calvin Cycle," or a four-carbon, dicarboxylic acid, by the "Hatch-Slack Pathway." Plants using the Calvin Cycle (C3 plants) are depleted in ^{13}C relative to those using the Hatch-Slack Pathway (C4 plants). Typical isotopic depletions of fixed carbon relative to atmospheric CO_2 range between -10 and $-18‰$, and between -22 and $-32‰$ for C4 and C3 plants, respectively (Faure, 1986). The carbon in atmospheric carbon dioxide is itself depleted $-7‰$ compared to the PDB standard. As a result of fractionations over millennia, specific types of compounds generally have a characteristic carbon isotope ratio (see Figure 12).

Once carbon enters an organism, isotopes are concentrated in *specific* groups of biochemical compounds (Blair et al., 1985). In plants, for instance, the chloroform extractable lipids are the most depleted in ^{13}C, followed by total lipids, lignin, cellulose, sugars, hemicellulose, proteins, and pectin (Hoefs, 1987). Furthermore, isotopes are concentrated at certain positions within molecules themselves (Monson and Hayes, 1982). In one study, for example, acetate was isolated from a strain of *E. coli* growing on glucose (Blair et al., 1985). The $\delta^{13}C$-value of the carboxyl C (i.e., of the acetate molecules) was enriched in ^{13}C by 23.4‰, relative to the methyl C.

As a result of ^{13}C fractionation patterns within the biosphere, the $\delta^{13}C$-value of an organic compound can contain information about its origin or formation process. The $\delta^{13}C$ differential has been used to characterize carbon dynamics in petroleum geology, agriculture, and ecology.

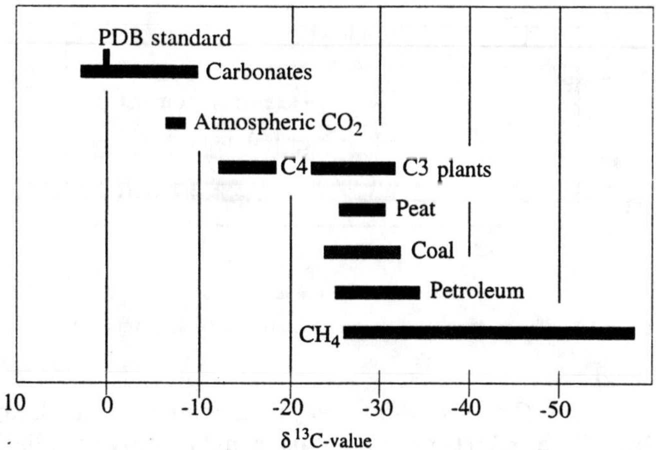

Figure 12. ^{13}C Distribution in nature (modified from Faure, 1986).

PETROLEUM GEOLOGY

Oil originates from unique geologic formations in the subsurface (i.e., the source rock). Since oil is a liquid and will migrate in the subsurface, an oil reservoir may be found far from the source rock. Correlating specific source rock to a known oil field can aid in the discovery of new oil-producing formations (i.e., by learning about the migration patterns of oil in the subsurface). The isotope-type curve method uses the difference in the carbon isotope ratios between various fractions of crude oil to correlate an oil reservoir with the organic material (i.e., kerogen) in its respective source rock. The isotopic concentration in components of crude oil follows the basic pattern:

$$\delta^{13}C_{kerogen} \geq \delta^{13}C_{asphaltenes} \geq \delta^{13}C_{heterocomponents} \geq \delta^{13}C_{aromatics} \geq \delta^{13}C_{saturates}$$

In the isotope-type curve method, the δ^{13}C-value for kerogen in various source rocks is matched with the δ^{13}C-value for each crude oil fraction (as shown in Figure 13).

AGRICULTURAL/ECOLOGICAL STUDIES

Subsurface processes involving NOM (i.e., carbon dynamics) affect soil fertility and are therefore of interest in agriculture. Carbon-13 analyses have been used in many studies attempting to describe subsurface carbon dynamics. From trends in the δ^{13}C-value of NOM with time and soil depth, for example, historical successions in surface vegetation were indicated, including transitions from native prairie to cultivated croplands (Bales-

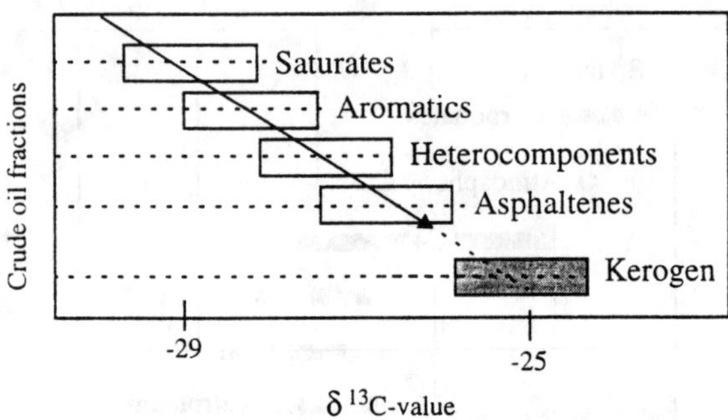

Figure 13. The isotope-type curve method (modified from Stahl, 1978).

dent et al., 1988). Other studies have investigated the landscape-scale patterns in $\delta^{13}C$, the effects of deforestation and pasture installation, and the turnover rate of NOM under various land management conditions (Balesdent et al., 1987; Desjardins et al., 1994; O'Brien and Stout, 1978; van Kessel et al., 1994).

In ecological studies, scientists found that the $\delta^{13}C$-value in the cell mass of higher organisms generally reflected that of the food source (Schell et al., 1989; Peterson and Fry, 1987). Carbon-13 analyses were used in studies to identify the food source of fish and whales in the Arctic Ocean and the migration patterns of caribou on Alaska's North Slope (Barnett, 1993; Schell, 1983; Schell et al., 1989).

Pollution Control Studies

The $\delta^{13}C$-value of a consumer's cell mass was previously correlated to the $\delta^{13}C$-value of the consumer's food source (Schell, 1983). A similar correlation for bacteria-consuming pollutants could confirm contaminant degradation. Since it is difficult to isolate microorganisms from soil, it was proposed that respired CO_2, instead of cell mass, be used to confirm biodegradation of a soil contaminant. This $^{13}CO_2$ method was previously used in attempts to prove biological degradation in field and laboratory studies (Aggarwal and Hinchee, 1991; Yocum, 1993). It was shown that CO_2 evolved from microbial respiration in soil columns and reflected the isotope ratio of the contaminant and diesel fuel, and not atmospheric CO_2 (Yocum, 1993).

Experimental

The present study was designed to determine if the isotope ratio in respired CO_2 can provide an indication of an active microbial consortium's primary substrate in a mixture of two complex substrates (i.e., NOM and crude oil).

MATERIALS AND METHODS

Closed system reactors were prepared from 50 ml septum covered vials. Each of 28 reactors received 10 ml of a phosphate buffer solution (pH = 8), 0.66 g of air-dried soil from Umiat, Alaska, and 0.5 ml of a mixed microbial consortium. The microbial consortium was isolated on trypto-case soy broth from slurry reactors prepared from the same soil contaminated with Kuparuk crude oil. The soil added to each reactor contained 10% C (w/w) by carbon, hydrogen, nitrogen (CHN) analysis (Carlo Erba CHNS-O EA 1108 Elemental Analyzer), resulting in an application rate of 66 mg of NOM-C/reactor for all reactors (NOM $\delta^{13}C$ = −28.4‰ ± 0.2‰). Half of the reactors then received 87.5 mg of crude oil, resulting in an application rate of 69 mg Oil-C/reactor (Kuparuk Crude = 78.4% C, $\delta^{13}C$ = −30.0‰ ± 0.2‰). All reactors received a headspace of pure oxygen at atmospheric pressure. Reactors containing NOM as the sole carbon source were referred to as "NOM" reactors, and reactors containing the mixture of substrates (NOM and crude oil) as "NOM-OIL" reactors.

Two reactors of each series (i.e., four total) were "killed" after 2, 4, 5, 6, 8, 9, and 10 days of incubation with 1 ml of 10 N HCl (except for Day 2 reactors, which were killed with 1 ml, 10 N NaOH to trap the CO_2 in the aqueous phase, and then treated with 1.5 ml of 10 N HCl to drive the CO_2 back into the headspace). Each reactor was subsampled to determine the mass of CO_2 produced (GOW-MAC Model 350, equipped with a thermal conductivity detector), and the remaining reactor headspace was cryogenically separated and collected. Carbon isotopes were quantified with a VG Isogas SIRA 9 triple collector mass spectrometer (Saupe et al., 1989). Separate controls were used to evaluate the affect of soil carbonates on the isotope ratio of respired CO_2.

RESULTS AND DISCUSSION

The mass of CO_2 produced in NOM reactors steadily increased in concentration during the incubation period, with the rate of production decreasing appreciably after the first four days (see Figure 14). In NOM-OIL reactors, the mass of CO_2 evolved increased during the entire reaction period. The total mass of CO_2—C produced represented less than 10% of the substrate-C in both NOM and NOM-OIL reactors.

The isotope curves (see Figure 15) indicated that the $\delta^{13}C$-value for CO_2 in NOM reactors approximated the $\delta^{13}C$-value of NOM though it became

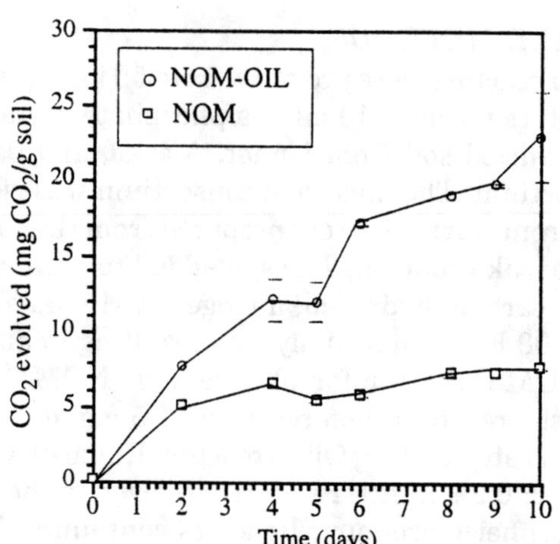

Figure 14. CO_2 production in NOM and NOM-OIL reactors.

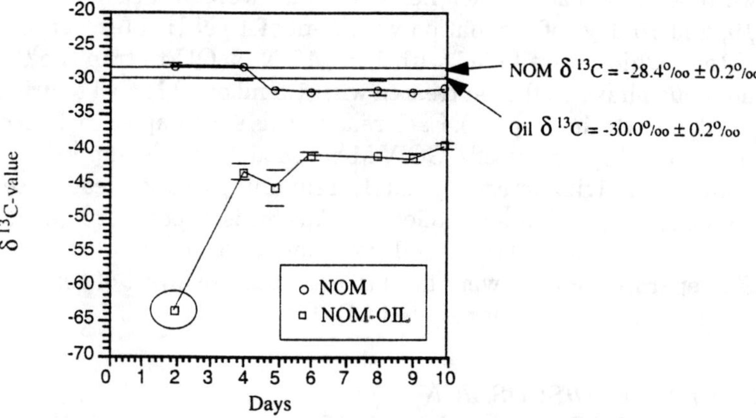

Figure 15. $\delta^{13}C$-value for CO_2 in NOM and NOM-OIL reactors.

212

slightly depleted between the 4th and 5th days of incubation. Carbon dioxide produced between days 5 and 10 in NOM reactors was depleted in ^{13}C approximately $-3.5‰$ relative to the substrate, NOM. The CO_2 in NOM-OIL reactors was appreciably depleted in ^{13}C with respect to both crude oil and NOM. The $\delta^{13}C$-value of the CO_2 increased with time from $-63‰$ to roughly $-41‰$ over five days. The large ^{13}C depletion, observed on day 2 (i.e., $33‰$ relative to crude oil) was unexpected, though depletions of similar magnitude (i.e., as high as $-24‰$) have been observed (Smejkal et al., 1971). Since the duplicate reactor for the day 2 measurement failed during incubation, the observed depletion could not be confirmed and was thus not considered. The average ^{13}C depletion for days 4 through 10 (i.e., $-10‰$ relative to crude oil and $-11.6‰$ relative to NOM) was well within the range of results reported in the literature (i.e., $0‰$ to $-24‰$) (Blair et al., 1985; Mary et al., 1992; Kaplan and Rittenberg, 1964; Stahl, 1978; Smejkal et al., 1971).

One possible explanation for the observed results is that organisms selectively degraded compounds in NOM or crude oil that were depleted in carbon-13. Although this could account for some fractionation, the distribution of isotopes between different crude oil compounds typically ranges between 2 and $4‰$, in contrast to $10‰$ as observed in NOM-OIL reactors (Stahl, 1978). Although intracellular fractionation could not be confirmed, the results indicated that an intracellular fractionation occurred in the reactors that contained crude oil and NOM.

Summary

The ratio of $^{13}C/^{12}C$ in individual compounds presents a natural label, which has been used in the sciences and has the potential for use in bioremediation. The most obvious use is to help prove biological degradation of a contaminant in soils containing a high concentration of natural organic material. In order for this method to be effective, the CO_2 respired by the active microbial consortium must be correlated with the isotopic composition of the food source. Studies indicated that the actual fractionation of ^{13}C between an organism and its respired CO_2 are not in constant proportionality at all times. In the results obtained, the microbes grown on crude oil and NOM produced CO_2 with a markedly different isotopic ratio than those grown on NOM alone (i.e., even though the isotope ratios in NOM and crude oil were similar). With further refinement, ^{13}C analysis may be an excellent technique for determining the primary substrate of subsurface microorganisms.

CONCLUSIONS

Solvent Extraction

The solvents typically used to recover crude oil from soil may extract the soil bitumen fraction of NOM. Only 1% of Umiat NOM was extracted by freon, the extraction solvent used for recovery of "oil and grease" (O + G) and "total petroleum hydrocarbons" (TPH). Although freon extracted a relatively small fraction of soil bitumens, 1% of the NOM in a highly organic soil can appreciably interfere with contaminant analysis.

Gas chromatography/mass spectrometry (GC/MS) was a useful technique for identifying and quantifying the soil bitumens in solvent extracts. In spite of the fact that GC/MS is a somewhat sophisticated technique, individual compounds from NOM and crude oil could not be separated (i.e., in many cases, the compounds from NOM and crude oil were identical).

Pyrolysis-GC/MS

Pyrolysis-GC/MS was investigated as a method for distinguishing between petroleum compounds and NOM in contaminated soil. Since pyrolysis-GC/MS is a whole soil technique, the problems associated with solvent extraction are avoided. Pyrolysis-GC/MS permitted general inspection of the organic material in uncontaminated and contaminated Umiat soil (i.e., qualitative fingerprinting). The method was particularly useful for qualitative studies in the laboratory aimed at developing an intuitive understanding of the general decomposition patterns in various fractions of NOM and crude oil.

Carbon-13 Analyses

It was observed that organisms degrading a mixture of NOM and crude oil respired carbon dioxide with an appreciably different carbon-13 concentration than organisms degrading NOM alone (i.e., even though the carbon-13 concentrations in NOM and crude oil are similar). Potentially, carbon-13 analyses can be used to describe the principle growth substrate and, possibly, the growth state of subsurface organisms, based on a soil gas analysis. In bioremediation, knowing the primary microbial substrate can be used to confirm contaminant biodegradation.

ACKNOWLEDGEMENTS

Many thanks to the members of the Technical University of Munich, namely, Dr. Peter Wilderer and Dr. Brigette Helmreich for their technical support of the Pyrolysis-GC/MS studies. Thanks also to the members of the Water Research Center at the University of Alaska, Fairbanks, namely, Dr. Donald Schell, and Norma Haubenstock for their technical support of the carbon isotope studies.

REFERENCES

Abbey, L. E. et al. 1981. "Differentiation and characterization of *Klebsiella pneumoniae* strains by pyrolysis-gas-liquid chromatography-mass spectrometry." *J. Clinical Microbiology*, 13(2):313–319.

Abbt-Braun, G. et al. 1989. "Structural investigations of aquatic humic substances by pyrolysis-field ionization mass spectrometry and pyrolysis gas chromatrography/mass spectrometry." *Water Research*, 23(12): 1579–1591.

Aggarwal, P. K. and R. E. Hinchee. 1991. "Monitoring in situ biodegradation of hydrocarbons by using stable isotopes." *Environmental Science and Technology*, 25:1178–1180.

Albro, P. W. 1976. "Bacterial Waxes." In: Kolattukudy, P. E. ed. *Chemistry and Biochemistry of Natural Waxes*. New York: Elsevier, pp. 419–445.

ASTM. 1987. "Standard test methods for moisture, ash, and organic matter of peat and other organic soils." In: *Annual Book of ASTM Standard Methods*. Philadelphia: ASTM, pp. 384–386.

Bahr, U. and H.-R. Schulten. 1983. "Pyrolysis field ionization mass spectrometry of cell wall components and bacteria cell walls." *J. Analytical and Applied Pyrolysis*, 5:27–37.

Balesdent, J. et al. 1987. "Natural ^{13}C abundances as a tracer for studies of soil organic matter dynamics." *Soil Biology and Biochemistry*, 19:25–30.

Balesdent, J. et al. 1988. "Soil organic matter in long-term field experiments as revealed by carbon-13 natural abundance." *Soil Science Society of America Journal*, 52:591–594.

Barnett, B. 1993. "Carbon and nitrogen isotope ratios of caribou tissues, vascular plants and lichens from northern Alaska." Master's Thesis. Department of Civil Engineering/Institute of Marine Science. University of Alaska, Fairbanks.

Bergmann, W. 1963. "Geochemistry of lipids." In: Berger, I. A. ed. *Organic Geochemistry*. London: Pergamon Press, pp. 504–542.

Biegleson, J. 1965. "The chemistry of isotopes." *Science*, 147:463–471.

Blair, N. et al. 1985. "Carbon isotopic fractionation in heterotrophic microbial metabolism." *Applied and Environmental Microbiology*, 50(4):996–1001.

Bohn, H. L. et al. 1994. *Soil Chemistry*. New York: John Wiley and Sons, pp. 68–261.

Bracewell, J. M. 1971. "Characterization of soils by pyrolysis combined with mass spectrometry." *Geoderma*, 6:163–168.

Bracewell, J. M. and G. W. Robertson. 1973. "Humus type discrimination using pattern recognition of the mass spectra of volatile pyrolysis products." *J. Soil Science*, 24(4):421–428.

Bracewell, J. M. et al. 1975. "Humus type discrimination from mass spectra by a simplified statistical treatment." *J. Soil Science*, 26(1):62–65.

Bracewell, J. M. et al. 1980. "Pyrolysis-mass spectrometry studies of humification in a peat and a peaty podzol." *J. Analytical and Applied Pyrolysis*, 2:53–62.

Bracewell, J. M. et al. 1989. "Thermal degradation relevant to structural studies of humic substances." In: Hayes, M. H. B. et al. eds. *Humic Substances II*. New York: John Wiley and Sons Ltd., pp. 181–222.

Braids, O. C. and R. H. Miller. 1975. "Fats, waxes and resins in soil." In: Gieseking, J. E. ed. *Soil Components: Vol. 1 Organic Components*. New York: Springer-Verlag, pp. 343–368.

Bruchet, A. et al. 1990. "Pyrolysis-GC/MS for investigating high molecular weight THM precursors and other refractory organics." *J. American Water Works Association*, September:66–74.

Buffle, J. 1990. *Complexation Reactions in Aquatic Systems*. New York: Ellis Horwood, pp. 90–120.

Craig, H. 1957. "Isotopic standards for carbon and oxygen and correction factors for mass spectrometric analysis of carbon dioxide." *Geochimica et Cosmichimica Acta*, 12:133–149.

de Leeuw, J. W. et al. 1986. "Screening of anthropogenic compounds in polluted sediments and soils by flash evaporation/pyrolysis gas chromatography-mass spectrometry." *Analytical Chemistry*, 58:1852–1857.

Degens, E. T. 1969. "Biogeochemistry of stable carbon isotopes." In: Eglinton, G. et al. eds. *Organic Geochemistry*. Berlin: Springer-Verlag, pp. 304–329.

Desjardins, T. et al. 1994. "Organic carbon and ^{13}C contents in soils and soil size fractions, and their changes due to deforestation and pasture installation in eastern Amozonia." *Geoderma*, 61:103–118.

Dineen, D. et al. 1990. "In situ biological remediation of petroleum hydrocarbons in unsaturated soils." In: Kostecki P. T and E. J. Calabrese. eds. *Petroleum Contaminated Soils*. Ann Arbor, MI: Lewis Publishers.

Dragun, J. 1988. *The Soil Chemistry of Hazardous Materials*. Silver Spring, MD: Hazardous Materials Control Research Institute, pp. 325–445.

Engman, H. et al. 1984. "Classification of bacteria by pyrolysis-capillary column gas chromatography-mass spectrometry and pattern recognition." *J. Analytical and Applied Pyrolysis*, 6:137–156.

EPA. 1990. "Slurry biodegradation." *Engineering Bulletin*, September:1–8.

Faure, G. 1986. *Principles of Isotope Geology.* New York: John Wiley and Sons, pp. 16–40.

Freeman, K. H. and J. M. Hayes. 1992. "Fractionation of carbon isotopes by phytoplankton and estimates of ancient CO_2 levels." *Global Biogeochemical Cycles,* 6(2):185–198.

Galimov, E. M. 1981. *The Biological Fractionation of Isotopes.* Orlando, FL: Academic Press, pp. 1–15.

Gill, R. A. and P. W. J. Robotham. 1989. "Composition, sources, and source identification of petroleum hydrocarbons and their residues." In: Green, J. and M. W. Trett. eds. *The Fate and Effects of Oil in Freshwater.* New York: Elsevier Applied Science, pp. 11–40.

Hancox, R. N. et al. 1991. "Sample size dependence in pyrolysis: An embarrassment or a utility." *J. Analytical and Applied Pyrolysis,* 19:333–347.

Hayes, J. M. et al. 1989. "An isotopic study of biogeochemical relationships between carbonates and organic carbon in the Greenhorn Formation." *Geochemica and Cosmichemica Acta,* 53:2961–2972.

Hayes, J. M. 1992. "Practice and principles of isotope measurements in organic geochemistry." Personal communication.

Hildebrandt, W. W. and S. B. Wilson 1992. "The use of modern on-site bioremediation systems to reduce crude oil contamination on oilfield properties." In: Kostecki, P. T. and E. J. Calabrese eds. *Hydrocarbon Contaminated Soils and Groundwater.* Boca Raton, FL: Lewis Publishers, pp. 487–501.

Hinchee, R. E. 1994. *Air Sparging for Site Remediation.* Ann Arbor, MI: Lewis Publishers.

Hoefs, J. 1987. *Stable Isotope Geochemistry.* Berlin: Springer-Verlag, pp. 1–36, 82.

Irwin, W. J. 1979. "Analytical pyrolysis—An overview." *J. Analytical and Applied Pyrolysis,* 1:89–122.

Jorgenson, M. T. et al. 1990. "Bioremediation and tundra restoration after a crude oil spill near drill site 2U, Kuparuk Oilfield, Alaska." Personal communication.

Kaplan, I. R. and S. C. Rittenberg. 1964. "Carbon isotope fractionation during metabolism of lactate by *Desulfovibrio desulfuricans.*" *J. General Microbiology,* 34:213–217.

Kimber, R. W. L. and P. L. Searle. 1970. "Pyrolysis gas chromatography of soil organic matter." *Geoderma,* 4:47–55.

Kolattukudy, P. E. et al. 1976. "Biochemistry of plant waxes." In: Kolattukudy P. E. ed. *Chemistry and Biochemistry of Natural Waxes.* New York: Elsevier, pp. 290–350.

Kopp, J. F. and G. D. McKee. 1983. *Methods for Chemical Analysis of Water and Wastes.* U.S. Environmental Protection Agency, pp. 413.1-3 to 418.1-3.

Lauch, R. P. et al. 1992. "Removal of creosote from soil by bioslurry reactors." *Environmental Progress,* 11(4):265–271.

Lavanda, A. O. et al. 1983. "Geotechnical classification of peats and organic soils." In: Jarrett, P. M. ed. *Testing of Peats and Organic Soils*. Ann Arbor, MI: American Society for Testing and Materials, pp. 37–54.

Lewis, R. F. 1993. "SITE demonstration of slurry-phase biodegradation of PAH contaminated soil." *Air and Waste*, 43(Spring):503–510.

MacCarthy, P. et al. 1985. "Pyrolysis-mass spectrometry/pattern recognition on a well-characterized suite of humic samples." *Geochimica et Cosmichimica Acta*, 49:2091–2096.

Martin, F. 1976. "Effects of extractants on analytical characteristics and pyrolysis gas chromatography of podzol fulvic acids." *Geoderma*, 15:253–265.

Mary, B. et al. 1992. "Use of ^{13}C variations at natural abundance for studying the biodegradation of root mucilage, roots and glucose in soil." *Soil Biology and Biochemistry*, 24(10):1065–1072.

Metcalf & Eddy, Inc. 1991. *Wastewater Engineering*. New York: McGraw-Hill, p. 404.

Monson, K. D. and J. M. Hayes. 1982. "Carbon isotopic fractionation in the biosynthesis of bacterial fatty acids. Ozonolysis of unsaturated fatty acids as a means of determining the intramolecular distribution of carbon isotopes." *Geochimica et Cosmichimica Acta*, 46:139–149.

NRC. 1993. *In Situ Bioremediation*. Washington, D.C.: National Academy Press, pp. 1–43.

O'Brien, B. J. and J. D. Stout. 1978. "Movement and turnover of soil organic matter as indicated by carbon isotope measurements." *Soil Biology and Biochemistry*, 10:309–317.

Parton, W. J. et al. 1987. "Analysis of factors controlling soil organic matter levels in Great Plains grasslands." *Soil Science Society of America Journal*, 51(5):1173–1179.

Parton, W. J. et al. 1988. "Dynamics of C, N, P, and S in grassland soils: A model." *Biogeochemistry*, 5:109–131.

Peterson, B. J. and B. Fry. 1987. "Stable isotopes in ecosystem studies." *Annual Reviews of Ecological Systems*, 18:293–320.

Piotrowski, M. 1991. "Bioremediation of hydrocarbon contaminated surface water, groundwater and soils: The microbial ecology approach." In: Kostecki, P. T. and E. J. Calabrese eds. *Hydrocarbon Contaminated Soils and Groundwater*. Boca Raton, FL: Lewis Publishers, pp. 203–239.

Reardon, E. J. et al. 1979. "Seasonal chemical and isotopic variations of soil CO_2 at Trout Creek, Ontario." *J. Hydrology*, 43:355–371.

Riddick, J. A. and W. B. Bunger. 1970. *Organic Solvents*. New York: Wiley-Interscience, pp. 61–551.

Saiz-Jimenez, C. et al. 1979. "Comparisons of soil organic matter and its fractions by pyrolysis mass spectrometry." *Geoderma*, 22:25–37.

Saupe, S. M. et al. 1989. "Carbon-isotope ratio gradients in western arctic zooplankton." *Marine Biology*, 103:427–432.

Schell, D. 1983. "Carbon-13 and Carbon-14 abundances in Alaskan aquatic organisms: Delayed production from peat in arctic food webs." *Science,* 219:1068–1071.

Schell, D. M. et al. 1989. "Bowhead whale (*Balaena mysticetus*) growth and feeding as estimated by $\delta^{13}C$ techniques." *Marine Biology,* 103:433–443.

Schnitzer, M. 1991. "Soil organic matter—The next 75 years." *Soil science,* 151(1):41–58.

Schulten, H.-R. et al. 1987. "Time-resolved pyrolysis field ionization mass spectrometry of humic material isolated from freshwater." *Environmental Science and Technology,* 21:349–357.

Schulten, H.-R. and M. Schnitzer. 1992. "Structural studies on soil humic acids by curie-point pyrolysis-gas chromatography/mass spectrometry." *Soil Science,* 153(3):205–224.

Simmonds, P. G. et al. 1969. "Organic analysis by pyrolysis-gas chromatography-mass spectrometry: A candidate experiment for the biological exploration of Mars." *J. Chromatographic Science,* 7:36–41.

Smejkal, V. et al. 1971. "Studies of sulfur and carbon isotope fractionation with microorganisms isolated from springs of Western Canada." *Geochimica et Cosmichimica Acta,* 35:787–800.

Smith, M. A. 1985. "Background." In: Smith, M.A. ed. *Contaminated Land, Reclamation and Treatment.* New York: Plenum Press, pp. 1–11.

Stahl, W. J. 1978. "Source-rock-crude oil correlation by isotopic type-curves." *Geochimica et Cosmichimica Acta,* 42:1573–1577.

Stevenson, F. J. 1966. "Lipids in soil." *J. American Oil Chemists Society,* 43:203–210.

Stevenson, F. J. 1994. *Humus Chemistry.* New York: John Wiley and Sons, pp. 1–211; 259–302.

Troy, M. A. et al. 1992. "Bioremediation of diesel fuel contaminated soil at a former railroad fueling yard." In: Kostecki, P. T. and E. J. Calabrese eds. *Contaminated Soils.* Boca Raton, FL: Lewis Publishers, pp. 62–80.

Tulloch, A. P. 1976. "Chemistry of waxes of higher plants." In: Kolattukudy, P. E. ed. *Chemistry and Biochemistry of Natural Waxes.* New York: Elsevier, pp. 236–290.

van de Meent, D. et al. 1980. "Chemical characterization of non-volatile organics in suspended matter and sediments of the river Rhine delta." *J. Analytical and Applied Pyrolysis,* 2:249–263.

van Kessel, C. et al. 1994. "Carbon-13 and Nitrogen-15 natural abundance in crop residues and soil organic matter." *Soil Science of America Journal,* 58:382–389.

Vinette, K. A. 1992. "Carbon and nitrogen isotope ratios in bowhead whales (*Balaena Mysticetus*) and their zooplankton prey as indicators of feeding strategy and environmental change." Master's Thesis. School of Fisheries and Ocean Sciences, University of Alaska, Fairbanks: pp. 1–147.

Wershaw, R. L. and J. Bohner, G. E. 1969. "Pyrolysis of humic and fulvic acids." *Geochimica et Cosmichimica Acta,* 33:757–762.

White, D. M. 1995. "Bioremediation of crude oil in the active layer overlying Alaska's North Slope permafrost." Ph.D. Dissertation. Department of Civil Engineering and Geological Sciences, University of Notre Dame, pp. 20–51.

Wilson, M. A. et al. 1983. "Comparison of the structures of humic substances from aquatic and terrestrial sources by pyrolysis gas chromatography-mass spectrometry." *Geochimica et Cosmichimica Acta,* 47:497–502.

Ying, A. C. et al. 1992. "Enhanced biodegradation of heavy engine oil in soil from railroad maintenance yard: Phase II field demonstration." In: Kostecki, P. T. and E. J. Calabrese eds. *Contaminated Soils.* Boca Raton, FL: Lewis Publishers, pp. 47–62.

Yocum, P. S. 1993. "Bioremediation of leachate and soil contaminated with petroleum products." Ph.D. Dissertation. Department of Civil Engineering and Geological Sciences. University of Notre Dame, pp. 179–191.

Microscale Heterogeneities in Soil Properties and Their Effects on Contaminant Sorption and Bioavailability

WALTER J. WEBER, JR., THOMAS M. YOUNG, AND ASTRID HILLERS
Department of Civil and Environmental Engineering
The University of Michigan
Ann Arbor, MI 48109-2125, USA

INTRODUCTION

The remediation of contaminated subsurface systems has been a major environmental restoration activity within the United States for more than a decade. The prohibitive costs associated with these efforts, coupled with their relatively modest effectiveness, have prompted searches for improved remediation technologies (MacDonald and Kavanaugh, 1994). In situ bioremediation has emerged as a leading alternative to more traditional "pump-and-treat" schemes. When subsurface conditions are properly controlled and suitable microorganisms are present, in situ bioremediation is capable of dramatic reductions in the subsurface concentrations of a wide range of organic contaminants under aerobic and/or anaerobic conditions. In some cases, however, bioremediation has failed to achieve established remedial targets even though both microbial population and residual contaminant levels are high enough that rapid, sustained degradation would be expected. Application failures may result when normally biodegradable materials are present in an environmental form or location that makes them unavailable or inaccessible to microorganisms. In this regard, sorption reactions between contaminant molecules and soil or aquifer materials appear to be of central importance in determining the bioavailability of a contaminant to particular microorganisms. Sorption reactions encompass a variety of interphase transfer processes ranging from absorption, in which solute is homogeneously distributed throughout the contacting phases, to adsorption, in which solute accumulates at an interface between two phases.

This chapter focuses on the nature of sorption reactions in subsurface systems and the role that such reactions may play in determining the bioavailability of organic contaminants. The chapter begins with an overview of the structure and chemical composition of soils and aquifer solids to focus attention on their intrinsic heterogeneity at the grain scale. Heterogeneity at the microscale is ubiquitous, even in apparently homogeneous subsurface domains. Macroscale, aquifer-level heterogeneities can significantly limit bioremediation performance by impeding delivery of necessary nutrients and electron acceptors to particular contaminated zones, but these problems are not addressed in this chapter. The bioavailability limitations discussed here persist even in well-mixed batch systems and are affected little by changes in the overall pattern of groundwater flow. Next, the link between sorption processes and bioavailability is explored by reviewing present knowledge about microbial utilization of sorbed contaminants, concluding that materials in this state are often not directly available to microorganisms.

The remainder of the chapter describes how variations in soil composition and structure affect sorption equilibria and rates and how this information can be applied to explain the reduced bioavailability observed for many field-contaminated soils. The fundamental hypothesis we advance is that microscale heterogeneity in soil composition and structure is a critical determinant of sorptive behavior and that these factors, in turn, control the bioavailability of hydrophobic organic contaminants. The development of a (1) basic appreciation of when and how physical-chemical processes such as desorption and mass transport control biodegradation progress and (2) efforts required to address information needs in this area are critical goals of this chapter.

AN OVERVIEW OF THE SUBSURFACE ENVIRONMENT

Soil and aquifer materials are diverse in their structure and chemical composition, with age, source materials, and environmental conditions each playing a role in establishing their characteristics. Figure 1(a) depicts an idealized subsurface cross section, highlighting some important common features of these systems. The major structural elements are aggregates formed when clay particles are bridged by polysaccharides, polyvalent cations, or humus (Oades, 1989; Stevenson, 1994). The grains range in size from a few microns to several millimeters, while an aggregate typically comprises numerous grains. A contaminant entering the subsurface will be bioavailable only if it is present at a location and in a

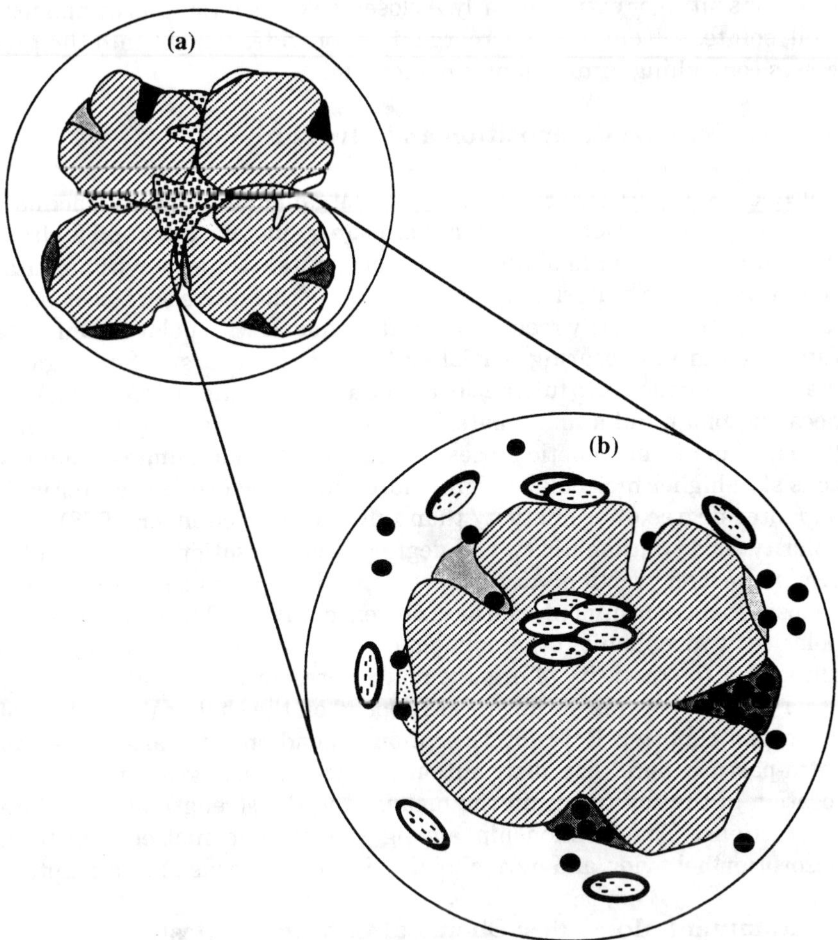

Figure 1. Schematic illustration of microscale subsurface heterogeneity at (a) the aggregate level and (b) the grain scale.

form that microorganisms can access. Solute molecules may be dissolved or adsorbed at the particle surface or may have migrated into the particle interior within micropores or soil organic matter [Figure 1(b)]. Microorganisms may also be distributed between the solid and solvent compartments, existing in either suspended or attached forms, although they are generally unable to access interior regions of soil grains. Location and accessibility of solute are important issues whether the solute serves as a primary or secondary substrate or whether a specific bacterial enzyme functions intra- or extracellularly. A closer look at each system component (soil, solute, solvent, and microorganism) provides insight into the processes controlling interactions among them.

The Soil Matrix: Composition and Structure

Typical soil particles comprise both inorganic and organic fractions. The inorganic fractions contain minerals (e.g., metal oxides and hydroxides) and clays (e.g., kaolinite, montmorillonite). Under environmental conditions, these materials typically bear a net negative surface charge, making them primarily reactive toward chemicals with polar or ionizable functional groups bearing partial or full positive charges. Soil organic matter, subdivided into fulvic acid, humic acid, and humin, is formed from decaying plant and animal materials (i.e., proteins, carbohydrates, and lipids) by biotic and abiotic processes. Humic acids and humins generally consist of higher molecular weight, more condensed organic matter with a greater degree of aromaticity than fulvic acids (Schnitzer, 1978). The polarity, acidity, aromaticity, and degree of condensation of a particular organic matter fraction vary widely, depending on its source material, depositional environment, and age (Stevenson, 1994). The sorption of low polarity, "hydrophobic" organic chemicals is dependent to a first approximation on the quantity of organic matter present in the soil, but many other salient features of the process depend on the chemical structure of the organic matter. Rates of desorption depend on intra-aggregate and intra-particle pore size distributions, concentration gradients, initial location of the solute within the matrix, and the strength of soil-solute interactions. The relationship among soil organic matter structure, desorption behavior, and bioavailability is a major focus of this chapter.

Contaminant Molecules: Chemical Characteristics

The chemical characteristics of a solute play an important role in determining its distribution in the subsurface and its susceptibility to biodegradation. Hydrophobicity, as measured by the octanol-water partition coefficient or the inverse of aqueous solubility, is central to both factors. Increasing hydrophobicity generally results in greater accumulation of the compound in the sorbed form or in the lipid fraction of biota. Both effects can lead to a reduction in substrate availability. Other characteristics relevant to predicting the fate and degradation of a compound include (1) its volatility, which determines its distribution between air and water, and (2) its charge, or polarity, which determines whether the compound sorbs primarily to the inorganic or organic soil fractions.

Uncharged solutes will sorb mainly to the organic fraction while charged or highly polar molecules will associate primarily with mineral surfaces of opposite charge or polarity.

The Solvent Phase: Saturated and Unsaturated Zones

A contaminant's persistence in the subsurface is greatly influenced by the presence or absence of water. Water sorbs strongly enough to inorganic materials that only relatively polar organic solutes compete effectively for these sites in saturated systems or at higher moisture levels in unsaturated systems. This renders organic matter the primary reactive fraction for solutes of lower polarity. In unsaturated systems at low relative humidities, sorption to mineral surfaces or polar regions of organic matter may become dominant even for nonpolar solutes (Pennell et al., 1992b; Young, 1996). At higher humidities, soil aggregates will typically be coated by a thin water film. The high surface area of this film may make sorption of hydrophobic compounds at the air/water interface an important mechanism in these systems (Pennell et al., 1992b).

Microbial community response to low moisture levels results in decreased rates and extents of biodegradation. While the permanent wilting point of higher plants is assumed to occur at a water potential of around −15 bar (Harris, 1981), bacterial activity is assumed to cease only around −80 bar (Griffin, 1972) but can be assumed to be negligible in most soils below −15 bar (Griffin, 1981). Fungi are not as sensitive to water content because of their increased ability to access dissolved substrates, nutrients, and water by forming hyphae. Bacterial movement can be severely impaired by water potentials around −1 bar (Griffin, 1972, 1981). This and other responses of bacteria to water potential fluctuations, such as increased lag periods and decreased growth rates, are highly dependent on the bacterial population. It can be assumed, though, that bacterial degradative activity at the low relative humidities described above will not be significant.

Subsurface Bacteria: Surface Attachment and Its Influence

Bacteria in subsurface systems are primarily attached to soil particles (Harvey et al., 1984). In modeling biological treatment processes involving attached bacteria, it is common to assume that they form biofilms on the surfaces of support media. This assumption is questionable for soil systems because the biomass concentration is so small that it is unlikely that a closed biofilm will exist. A simple calculation can illustrate this. Assuming a bacterial size of $1\,\mu$m $\times\ 2\,\mu$m, a biomass concentration of 10^8

bacteria per gram of aquifer sand, and a specific surface area of 1 m²/g, approximately 0.02% of the surface will be covered by bacteria. This is consistent with a common estimate that less than 0.1% of soil surface area is covered with bacteria (Griffin, 1972, 1981). Describing biomass growth using a microcolony model better represents physical reality in the subsurface (Molz et al., 1986; Baveye and Valocchi, 1989).

The main advantage of attachment in substrate-limited subsurface systems is the increased substrate transport to the cell surface resulting from bulk groundwater flow. Furthermore, high local biomass concentrations in bacterial microcolonies facilitate sharing of extracellular polymers, reducing energy requirements and allowing utilization of accumulated biomass as a food source during extended periods of starvation. Agglomeration in microcolonies also mitigates predation and convective washout (Stratton et al., 1983). The contribution of attached bacteria to total heterotrophic activity has been found to be at least four times as high as could be expected from the fraction of attached cells (van Loosdrecht et al., 1990). This indicates that attachment might enhance microbiological activity, although the experimental findings are not always consistent and depend on medium composition, type of surface, substrate, and bacterial strain (Gordon and Millero, 1985; van Loosdrecht et al., 1990). Surfaces exhibit several direct and indirect influences on microbial cells. Direct influences involve changes in membrane structure and permeability near solid surfaces (van Loosdrecht et al., 1990). Indirect influences include variation in media composition due to sorption and desorption of substrates and nutrients, changes in substrate availability, pH buffering, and water activity.

Cells can attach reversibly or irreversibly to surfaces via extracellular polymers, where the degree of reversibility depends mainly on solution composition and cell age. Reversible attachment permits exchange between attached and suspended cells, facilitating migration through porous media. Methods for quantifying migration of bacteria in saturated subsurface horizons under field conditions are not well developed and require further research. Such methods are of crucial importance for prediction of the fate and transport of biodegradable pollutants. The need for studies of the mobility of bacteria in the unsaturated zone is particularly great because little literature is available in this regard. Bacterial movement in unsaturated soils will, to a large extent, depend on the presence of continuous water-filled pathways in the soil (Griffin and Quail, 1968; Griffin, 1972, 1981). Soil matrix forces will determine the pore sizes that are either air- or water-filled at a particular moisture level. This will, in turn, depend on soil properties such as grain and pore size, surface properties, and carbon content. For at least some soil-bacteria systems, bacterial movement will be severely impaired at matric water potentials at or below −0.2 to −1 bar (Griffin, 1972, 1981).

To better understand the interaction between a bacterium and a flat solid surface, it is helpful to examine the geometry involved (Figure 2). The thickness of the adsorbed substrate layer is negligible compared to the diameter of the cell. Reversibly attached cells are located at 2–5 nm from the solid surface, and only a minimal part of the cell surface is in direct contact with the adsorbed material (van Loosdrecht et al., 1990). Consequently, direct uptake of a compound from the solid surface is not likely to be a prevailing mechanism, especially because size exclusion phenomena make a considerable fraction of the porous solid's surface inaccessible to soil bacteria. Interactions between cells and adsorbed chemicals are therefore mainly of an indirect nature through changes in the microenvironment of the cell.

In order for a compound to be taken up by a bacterium, the substrate must come into contact with the bacterial cell. Compounds sorbed onto porous materials must desorb and diffuse out of the pores and away from the particle surface to contact the cell membrane. The mechanism of uptake, either via diffusion through the membrane or active transport, will have a determining influence on bacterial kinetics and can be the limiting factor in the degradation of a substrate. Another possible mechanism of substrate degradation is exoenzymatic attack of the substrate on the surface, creating smaller fractions that can be taken up by the cell and used as carbon and energy sources. This process is generally not favorable because extra energy must be expended to produce the exoenzyme, and several reaction steps are necessary to make substrate available to the cell: exoenzyme must diffuse into the pore and sterically fit

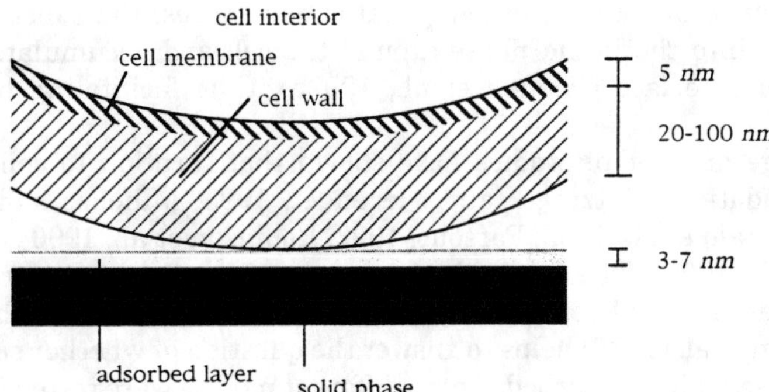

Figure 2. Schematic view of size relationships between attached cells and solid surfaces (adapted from van Loosdrecht, 1990).

the substrate in the given pore geometry, and metabolites must diffuse out of the particle to reach and be taken up by the bacterium (Criddle et al., 1991).

Combining the above information about soils, microorganisms, solutes, and solvents leads to a more detailed picture of the subsurface environment. Most microorganisms and many organic solutes are present primarily in an attached or sorbed state. Although this proximity might be expected to enhance the rate and extent of biodegradation, conflicting experimental results have been obtained, as summarized in the next section.

AVAILABILITY OF SORBED SUBSTRATE

Much controversy has evolved around the question of whether bacteria can take up substrate from solid surfaces. The discussion above illustrates that direct uptake from the solid phase is not likely to be a predominant mechanism. Substrate must be converted to an accessible form to be taken up by bacterial cells. Accessibility requires that substrate be available in a form that can bind with cellular enzymes involved in the initial stages of metabolic activity. In most systems, a substrate will have to become dissolved in the aqueous phase surrounding the cell before uptake can occur. Depending on the physical-chemical properties of the system, this will involve desorption from the solid matrix or dissolution from a nonaqueous liquid (NAPL) phase, possibly enhanced by the formation of biosurfactants. Studies of degradation from an NAPL phase are inconclusive regarding whether a solubilization step must precede uptake. Questions of dissolution and formation of biosurfactants are covered elsewhere and will not be expanded upon in this chapter (see, e.g., Stucki and Alexander, 1987; Rosenberg et al., 1992; Efroymson and Alexander, 1995). In the case of volatile substances, the substrate will partition into the liquid phase around the cell and accumulate at the air-water interface (Pennell et al., 1992b), thus facilitating bacterial uptake.

A large number of studies have shown that sorption to soils limits biodegradation (Hatzinger and Alexander, 1995; Mihelcic and Luthy, 1991; Ogram et al., 1985; Parsons, 1992; Robinson et al., 1990; Weissenfels et al., 1992). The role of long-term sorption processes in reducing bioavailability has been recently reviewed (Alexander, 1995). The study of Ogram et al. (1985) helps to answer the question of whether substrate can be used if it is adsorbed. Three different models were tested for their ability to fit mineralization data of (2,4-dichlorophenoxy) acetic acid (2,4-D) in four different soils: 1) only solution phase 2,4-D can be degraded and only by suspended bacteria, 2) sorbed bacteria degrade sorbed 2,4-D and suspended bacteria degrade dissolved 2,4-D, and 3) only dissolved 2,4-D is available and can be degraded by both suspended and attached bacteria. Only the last model adequately described the data, suggesting that sorbed substrate was unavailable. Some recent evidence indicates that this conclusion may in part be a function of the bacterial species under study. Differences in bioavailability of sorbed naphthalene were identified for two different strains of bacteria (Guerin and Boyd, 1992). While sorbed naphthalene was not readily available to one strain and the bacteria had to rely on passively desorbing substrate, the other strain appeared to increase desorption rates since the initial degradation rates in the presence of soil exceeded the rates expected from the dissolved equilibrium concentrations. Therefore, bioavailability may depend, not only on the sorbate-sorbent interactions, but also on the properties of the bacteria in the system.

Positive effects of sorption on biodegradation occur if sorption to the soil decreases otherwise toxic concentrations of a substrate below the toxicity threshold value (Earhardt and Rehm, 1985; Mörsen and Rehm, 1987). Sorption can also enhance degradation if the surface serves as a pool of locally increased substrate concentration when bulk substrate levels cannot sustain a bacterial culture. Soil surfaces and dissolved organic matter can also act to concentrate essential nutrients thus enhancing degradation (Rijnaarts et al., 1990).

Experimental results on the impact of sorption on biodegradation are available for a wide variety of soils, soil components, and model systems. Summarizing selected work from this literature, the remainder of this section reviews bioavailability results for clay minerals and humic substances, concluding with a review of modeling approaches and their implications. The inherent influence of various soil constituents on bacterial metabolism must be considered in comparing bacterial kinetics in soil-containing and soil-free systems. Consequently, some literature on the effect of soil components, especially humic material, on enzyme activity will be reviewed.

Influence of Clay Minerals

The influence of clay minerals on substrate degradation rate and extent is highly system-specific. An inverse relationship was found between the degree of adsorption of organic acids on porous hydroxyapatite and the degree of mineralization and assimilation (Gordon and Millero, 1985). An increase in sorbent-sorbate adsorption energy resulted in decreased degradation rates. Similar trends have been observed in other studies

(Knaebel et al., 1994; Wszolek and Alexander, 1979). Competition between bacteria and substrate for surface sites was indicated by comparing results of sorption experiments using either preattached cells or preadsorbed substrate (Gordon and Millero, 1985). This is consistent with results from biological activated carbon systems, where pore blockage by the biofilm imposes an extra diffusional barrier decreasing sorption rates (Earhardt and Rehm, 1989).

The influence of clay minerals on bacterial degradation processes is dependent on the clay to solution ratio, the type of clay, the principal cation associated with the clay, and the type of microorganism (Nováková, 1972; Filip, 1973). The effect of the cation exchange capacity (CEC) on biodegradation is most pronounced in unbuffered systems where the clays serve to stabilize pH by exchanging H^+ for the respective cation (Stotzky, 1966). No effect of CEC on mineralization rates was observed in buffered systems. The degradation of anionic substrates is not affected by the presence of clays because of their negative surface charge (Subba-Rao and Alexander, 1982).

Sorption within clay interlayer regions, as well as sorption to external surfaces, can play a significant role in determining bioavailability. Sorption of substrates to swelling clay components such as montmorillonite decreases degradation rates to a much larger degree than sorption onto sands or nonswelling clays. For example, degradation of linear surfactants having a sufficiently small diameter to migrate into the interlayers of montmorillonite was considerably retarded (Knaebel et al., 1994). In contrast, the degradation of dodecyl alkylbenzene sulfonate associated with montmorillonite was not significantly impeded because the diameter of the benzene side chain is too large to allow migration into the inner clay layers. Other experiments suggest similar mechanisms of size exclusion (Weber and Coble, 1968).

Influence of Humic Substances

Research on the impacts of dissolved humic materials on biodegradation processes have shown that these compounds may either enhance or suppress bacterial activity and that their influence goes beyond the scavenging of substrate by some form of association with the humic acid. The role of humic substances seems to depend primarily on their chemical structure, functionality, and molecular size, all of which determine their possible interference with enzyme activity (Butler and Ladd, 1971; Mato et al., 1972; Ladd and Butler, 1975). Binding of catabolic enzymes to the humic material due to structural similarity to certain substrates can result in reduced activity toward the target substrates. Some researchers hypothesize that humics bind enzymes as cations, and the effect can therefore be reversed by addition of cations (Ladd and Butler, 1975).

Amador and Alexander (1988) tested the effect of humic acid on the degradation of benzylamine and aromatic acids. For benzylamine, a decreased acclimation period was observed, possibly caused by an increase in substrate concentrations at the surface of the humic acid molecule. The extent of mineralization was somewhat diminished at all substrate levels, resulting from ion exchange mechanisms and binding of the substrate to humic acid. The reduction of rate and extent of degradation of the aromatic acids was attributed to the binding of phenolic degradation products to humic acid or to an inhibition of degradative enzymes by humic acid. The causes and effects of binding of certain substrates to humic material will be discussed below.

Other researchers have reported a decrease in specific substrate degraders with the addition of increased levels of humic acids (Shimp and Pfaender, 1985). Enzyme inactivation by binding to humics, resulting from structural similarity between substrate and humic acid, or nonreversible substrate sorption could explain this result.

Enhanced degradation in the presence of humic and fulvic acids has also been observed. Increased cell yields can either be attributed to partial cooxidation of humic and fulvic acids and growth on a cometabolic metabolite or to induction of enzymes by the humic material rendering substrate biodegradable that would otherwise not be attacked in a specific soil system (deHaan, 1974; Visser, 1985).

Covalent Binding to Soil Constituents

One mechanism that leads to decreased bioavailability of certain aromatic compounds in soils is covalent binding. This process is not to be confused with physical sorption processes as described in the previous section. It is included in this chapter because it may contribute to the phenomenon of reduced bioavailability over long soil-solute contact times in certain soils and for specific substrates. Binding occurs via a free radical mechanism (Bollag, 1992) induced by soil enzymes (Bollag, 1992; Sjoblad and Bollag, 1981) or inorganic surfaces (Shindo and Huang, 1984; Whelan and Sims, 1992; Dec and Bollag, 1994). The incorporation of organic pollutants into soil organic matter is similar to the humification process and has been demonstrated for several substituted phenols and pesticides (Bollag, 1992; Capriel et al., 1985). In one study, about 50% of the ^{14}C-labeled atrazine applied to soils under field conditions nine years earlier was found as a nonextractable residue, with the highest percentage of the bound form found in the humic acid fraction (Capriel et al.,

1985). A considerable amount of the bound atrazine was present as hydroxylated metabolites of the parent molecule, which suggests that incorporation within soil organic matter prevents the complete mineralization of this compound.

The participation of a free radical in these reactions limits their practical significance to molecules containing electron-withdrawing substituents that can stabilize the free radical structure (e.g., hydroxylated compounds). However, even if pollutants would not be expected to undergo direct polymerization via this mechanism, other subsurface transformations may make the reaction possible. For example, nonreactive aromatic parent compounds may be converted to their hydroxylated intermediates by biotic or abiotic processes, rendering them amenable to polymerization reactions. Hydroxylated intermediates of polyaromatic substrates could react with soil organic matter following a similar reaction pathway as 1-naphthol (Sjoblad et al., 1976). This could render some polyaromatic hydrocarbon metabolites "nonextractable" over extended contact times, consistent with the observations of Weissenfels et al. (1992). The binding induced by horseradish peroxidase has been identified by ^{13}C NMR as a covalent bond (Bortiatynski et al., 1994).

Enzymatic or mineral surface-induced coupling reactions can be accompanied by the dehalogenation of chlorinated aromatic substrates, a phenomenon demonstrated for chlorinated phenols (Dec and Bollag, 1994, 1995) and chlorinated anilines (Dec and Bollag, 1995). Dehalogenation patterns indicate that the halogen is released when the unpaired electron from the free radical is located at the halogen substituted carbon atom (Dec and Bollag, 1994). Dehalogenation has been shown to be coupled to the free-radical mechanism of oxidative coupling and cannot be enhanced independently (Dec and Bollag, 1995).

The observed polymerization of pollutants and their degradation products to soil organic matter suggests that this mechanism may contribute to decreased release of pollutants from soil matrices after extended exposure. The frequency and extent of these reactions for common subsurface pollutants under typical field conditions has not been established. Consequently, their importance in determining contaminant bioavailability remains uncertain.

Modeling Desorption/Diffusion Limited Biodegradation

Studies with purified clay and humic materials, major reactive components of soils, show that each has the ability to enhance or retard pollutant degradation, depending on conditions. Beneficial results can be expected when the pollutant's concentration is toxic to the bacterium. Sorption can lower the aqueous phase concentration below the toxicity threshold, enabling the biological degradation of the compound and shortening apparent lag phases. In other cases the reduced aqueous phase concentration, especially for compounds with low water solubilities, can depress substrate concentrations below the minimum level, S_{min}, necessary to sustain a viable steady-state population of bacteria. This minimum concentration is calculated from a mass-balance on bacteria under steady-state conditions and assuming that Monod kinetics apply (McCarty, 1984):

$$S_{min} = \frac{bK_s}{Yk - b} \tag{1}$$

where

K_s = half-saturation constant
 k = maximum rate of substrate utilization
 Y = cell yield coefficient
 b = cell decay coefficient

It has been found that Monod expressions alone cannot adequately describe biological degradation in soils in many cases (Scow et al., 1986). This should not be surprising since Monod-based models were developed for suspended, well-mixed systems with the substrate or another constituent (e.g., electron acceptor) serving as the limiting factor during the degradation process. There are a wide variety of reasons that such models may fail to describe pollutant degradation in porous media. The limiting factor (e.g., electron donors, electron acceptors, or trace nutrients) might change during the course of the degradation or across the pollutant plume due to mass transfer considerations. Even microscale variations in these variables may change the limiting factor within the local environment surrounding the cell. Accurate modeling of the degradation process requires that the limiting factor be reassessed at each microscale point. Shifting selective pressure may also alter the composition of the bacterial population spatially and temporally so that a different set of rate variables will apply. Haack et al. (1995) recently published a study attempting to characterize soil microbial communities using BIOLOG® substrate utilization patterns. Sample replicates from greenhouse soils were not reproducible, a fact that the authors attributed to spatial heterogeneity of microbial consortia in the soil. This shows that microbial communities not only vary on a macroscale, but also within quite small samples.

When whole soils are considered, the complexity observed for humic and clay systems is multiplied manyfold. To sort out the varied influences at work in the system, several models have been developed to describe the interactions among substrate, bacteria, and soil surfaces. The level of complexity of these models varies, but comparison with experimental data suggests that mass transfer is a major factor controlling pollutant degradation in porous media. Experiments performed in conjunction with these modeling efforts typically compare results from sterile and nonsterile systems to infer limiting mechanisms. A shortcoming of this type of comparison is that the diffusion rate out of a porous particle depends on the concentration gradient, which will be steeper in a nonsterile system than in a sterile system. In the limit, the aqueous phase concentration in the biologically active system will be zero. This increased driving force makes it possible for biodegradation to accelerate desorption (Criddle et al., 1991; Rijnaarts et al., 1990; Guerin and Boyd, 1992).

The failure of Monod-based models to describe biodegradation may also occur because substrate flux to an attached organism differs from that to a suspended cell. Attached cells cannot obtain substrate from all directions, as can a cell suspended in the solution; rather, the diffusive flux is restricted by neighboring bacteria and by the solid surface (Harms and Zehnder, 1994). A higher bulk concentration must be maintained to ensure the same surface concentration, resulting in an apparent increase in the half velocity constant. The system-specific K_s, for attached cells can be estimated from bacteria-specific rate coefficients and appropriate system dimensions. For cells attached to spherical particles (Harms and Zehnder, 1994):

$$(K_S)_{\text{attached}} = (K_S)_{\text{suspended}} + \frac{k_{\text{cell}} n}{2 \eta A v} \tag{2}$$

where

K_S = half-saturation constant
k_{cell} = maximum uptake rate for a specific cell
n = number of cells attached to a spherical particle
η = single collector efficiency
A = cross-sectional area of the particle
v = linear velocity of the liquid phase

Diffusional limitations to growth and degradation have been observed by others (e.g., Li and DiGiano, 1983; Scow and Alexander, 1992; Rijnaarts et al., 1990). Most mathematical models do not include corrections of this type.

The apparent inability of Monod rate expressions to describe pollutant fate data is often caused by a failure to incorporate all mass transfer resistances and to identify the limiting substrate. In unsaturated systems these issues pose particular problems because the water potential influences bacterial diversity and mobility, transport of metabolites away from the cell, and migration of electron donors, electron acceptors, and nutrients to the cell. The relevant components of water potential in the unsaturated zone are osmotic and matric forces. Drying of soils changes the osmotic potential of the cells resulting in altered growth rate, energy production, and degradation patterns. Different matric potentials will alter water availability and mass transfer around the cell. Only recently has moisture content been correlated to pollutant degradation, and a few attempts have been made to relate this to soil water potentials (Hillers and Adriaens, 1995; Holden and Firestone, 1995; Holman and Tsang, 1995; Schnell and King, 1995). It is expected that an optimum water potential for microbial respiration exists (Alexander, 1977; Howard and Howard, 1993). Reduction of the water potential below this optimum, either by osmotic or matrix forces, has been shown to result in an increased lag-phase, decreased growth rate, and cell yield (Scott, in Harris, 1981). However, the effects depend on the nutritional status of the cell (Siegele and Kolter, 1992). Current knowledge is insufficient to adequately incorporate these mechanisms in a mathematical model. Work is currently in progress in our laboratory to elucidate the prevailing mechanisms and to quantify their impacts on degradation rates.

Several models have been proposed to describe changes in apparent degradation rates in the presence of soils suspended in completely mixed batch reactor (CMBR) systems. Most models are system-specific at this point, failing to incorporate all of the phenomena discussed above and frequently making implicit assumptions. For example, bacterial growth is often neglected, which may be inaccurate in heavily polluted subsurface environments. Sorption and diffusion processes are also often described using simplifying assumptions, including the existence of only one type of sorption site or diffusional resistance or that equilibration is instantaneous. Some of the more sophisticated models are discussed as examples in the following paragraphs, but the discussion is by no means intended to be comprehensive. Models describing macroscale subsurface transport are not discussed here because these add yet another level of complexity.

Decreased rates of benzylamine degradation in the presence of montmorillonite have been observed and modeled by Miller and Alexander (1991). Although montmorillonite is an expandable clay, benzylamine is expected to sorb mostly on outer surfaces due to size exclusion. The model assumes nongrowing populations, readily desorbable organics, and that only solution-phase substrate is available to bacteria. Sorption is described by the Langmuir isotherm. Given these assumptions, the experimental data were well-described by the model. Mihelcic and Luthy (1991) found sorption and desorption of naphthalene reversible and fast compared to the rate of degradation. The maximum degradation rate was observed to be dependent on the soil-to-solution ratio. The model incorporates retarded intraaggregate diffusion.

Scow and Alexander (1992) formulated a model for substrate degradation under saturated conditions that accounts for limitations caused by diffusion and sorption to unavailable sites. Their diffusion-sorption-biodegradation model (DSB model) incorporates a Fickian diffusion term, a linear isotherm expression, and one of several possible biological rate equations. Critical assumptions include local sorption equilibrium and uniform dispersion of bacteria throughout the bulk solution phase. The model was tested in the presence and absence of synthetic aggregates, as well as in the presence of gel exclusion beads. Both the shapes of the degradation curves and the rates were found to differ in the presence of the aggregates. Among the rate equations tested, the two-compartment model (Scow et al., 1986) provided the best fit in most cases. The model cannot be easily extrapolated to new systems because each parameter comprises several more fundamental parameters. Furthermore, the model's assumption of uniformly distributed microorganisms in the bulk solution may introduce additional error if the population of attached bacteria is significant.

Rijnaarts et al. (1990) employed a sorption-retarded radial diffusion model (RDM), which considers intraaggregate diffusion as the rate-limiting mass transport step to describe mineralization data for α-hexachlorocyclohexane in soil. Assumptions of the RDM are instantaneous equilibrium within the particle, linear sorption, and that mass transport across the liquid boundary layer is not limiting. The model was found to fit abiotic desorption data well. The experimental system was noted to be far from equilibrium after ten days, and the model indicated that equilibration would require weeks to months. Porosity values calculated from the model were too small to bear physical significance, indicating the contribution of retardation mechanisms other than those built into the model (Rijnaarts et al., 1990). Furthermore, measured K_{oc} values could not account for the contaminant sorption. The authors attributed this to sorption by soil components other than organic matter but did not question the linear isotherm model. Their experiments showed that, in most cases, biodegradation was limited by intraaggregate diffusion. For the model to fit the data, the diffusional distance needed to be reduced in the presence of bacteria, suggesting that bacteria might penetrate the solid matrix to some extent. A first-order biodegradation model was also found to fit the data well, possibly because of the low solid-solution ratio in the suspended slurries. Diffusional resistances are expected to be far more important in column experiments or field applications where much higher solid-solution ratios are typical (Scow and Alexander, 1992).

The adequate modeling of sorption and biodegradation processes in natural systems requires that the complex interactions among solutes, soils, and microorganisms be well understood. As biodegradation is dependent on desorption, elucidation of sorption and desorption mechanisms and incorporation of microscale soil heterogeneities is of critical importance.

SORPTION PROCESSES AND THEIR RELATIONSHIPS TO BIOAVAILABILITY

The experimental and modeling results presented above lead to the general conclusion that sorbed substrate, especially when present in matrices that exclude entry of microorganisms, is not available for microbial utilization. Desorption equilibria and rates typically control solution-phase composition in contaminated subsurface systems and, therefore, play a central role in establishing bioavailability. In general, no degradation of a primary substrate will occur if its equilibrium aqueous-phase concentration (C_e) is less than the minimum substrate level required for microbial growth according to the Monod equation (S_{min}). Circumstances of this nature may occur in the subsurface because a substrate is sorbed to soil particles or because it is dissolved in other organic contaminants present as nonaqueous phase liquids (Efroymson and Alexander, 1995). Even if equilibrium concentrations exceed minimum substrate levels, biodegradation will be limited if desorption rates are significantly lower than microbial uptake rates. In such cases, degradation rates cannot exceed desorption rates, and if the differences are large enough, steady-state aqueous-phase solute concentrations may be too low for biodegradation to proceed. Predictions of biotransformation rate and extent must therefore be based on a thorough understanding of controlling desorption processes. This section focuses on establishing conceptual linkages among microscale soil heterogeneity, sorption processes, and bioavailability. Space constraints dictate that it can provide only a few of the equations necessary for rigorous process modeling. For a detailed overview of mathematical models on subsurface sorption phenomena and their conceptual bases, see, for example, Weber et al. (1991).

Sorption Equilibria: Isotherm Models and Energetics

The equilibrium distribution of a chemical between the solid (soil) and solvent (air or water) phases is usually described by an isotherm equation, which relates the concentration in one phase to that in the other at a particular temperature. Sorption of hydrophobic organic contaminants to soils is frequently modeled by a linear isotherm:

$$q_e = K_D C_e \tag{3}$$

in which q_e and C_e are the solid and solution-phase solute concentrations, respectively, and K_D is the distribution coefficient. This isotherm has been extensively used in fate and transport models both because it fits experimental data from many soil-solute systems reasonably well, particularly over low ranges of solute concentration, and because it greatly simplifies the overall modeling effort. The ratio of the distribution coefficient to the organic carbon content of a soil, referred to as the organic carbon-normalized sorption coefficient (K_{OC}), has been observed to be approximately constant for a particular solute over a range of soils (Lambert, 1967; Chiou et al., 1979). Fortuitously, this allows K_{OC} values to be predicted from knowledge of a soil's organic content and a solute's octanol-water partition coefficient (K_{OW}), in some cases to within a factor of two (Karickhoff et al., 1979; Chiou et al., 1983; Means et al., 1980). These findings led to the development of a conceptual model viewing sorption as the "partitioning" of solute into the macromolecular structure of soil organic matter (Chiou et al., 1985). Additional observations supporting the partitioning model for nonpolar organic sorption include the 1) low sorption observed in soils from which organic matter has been removed, 2) low sorption energies ($\sim -\Delta H_{\mathrm{soln}}$), and 3) lack of competition between solutes sorbing simultaneously. The partitioning model implies that 1) each solute molecule sorbs with the same energy regardless of solid-phase loading, 2) that there are no well defined sorption "sites," and 3) that the sorption reaction involves relatively weak, reversible chemical interactions. More recent sorption studies challenge the completeness and generality of this conceptual model.

When sorption data are collected over a wider range of solution-phase concentrations, particularly for sorption of more hydrophobic compounds on aquifer materials, the linear model is often inappropriate (Mingelgrin and Gerstl, 1983; Weber and Miller, 1988; Weber et al., 1992; Young and Weber, 1995). The Freundlich isotherm has proven useful in these cases:

$$q_e = K_F C_e^n \tag{4}$$

where K_F is the Freundlich capacity factor and n is the Freundlich exponent. When n equals one, an isotherm reduces to a linear form. This model is consistent with an exponential distribution of sorption energies as a function of solid-phase loading, with the most energetically favorable sites being occupied first, followed by successively less favorable sites. Experimental results have confirmed that sorption enthalpies vary exponentially with solid-phase concentrations (Young and Weber, 1995). Additional support for this hypothesis is provided by the observation that sorption of one organic solute may be suppressed by the presence of a second, particularly for soils with more nonlinear single-solute isotherms (McGinley et al., 1993; Pignatello, 1991). Competitive sorption is not consistent with the linear partitioning model (Chiou et al., 1983); it would, however, be predicted for site energy variations with solid loading. The wider applicability of the Freundlich model results from its ability to accommodate the effects of heterogeneities present in soils at both the aggregate and grain scales. Nonlinear isotherms have also been explained by referencing variations in sorbed-phase activity coefficients with solid loading predicted by the Flory-Huggins model of polymer solution thermodynamics (Spurlock and Biggar, 1994a, 1994b).

The distributed reactivity model (DRM) represents a more explicit way to consider subsurface heterogeneity (Weber et al., 1992; McGinley et al., 1993; Young and Weber, 1995). The model views soils and aquifer materials as a number of discrete components, each of which features a distinct sorption capacity and linearity. Where a particular soil component falls along this spectrum is largely determined by the degree of diagenetic alteration to which its organic matter fractions have been subjected. Older, more diagenetically altered organic matter can have an order of magnitude higher sorption capacity and significantly greater nonlinearity than newer, less altered organic matter. Structural changes accompanying diagenesis, including reductions in polarity and increases in aromaticity, degree of condensation, and molecular weight, appear to be responsible for the change in sorption characteristics. The greater sorption of nonpolar organic molecules by more reduced organic matter, as indicated by lower oxygen-carbon (O/C) or higher hydrogen-oxygen (H/O) ratios, provides important support for this hypothesis (Grathwohl, 1990; Garbarini and Lion, 1986; Young, 1996). The emphasis placed by the distributed reactivity model on a range of possible sorption behaviors also meshes with observations that sorption is less linear on minerals coated with small amounts of humic acids than on the same minerals with more dense humic acid coverage (Murphy et al., 1990). Mathematically, the DRM describes the overall sorption reaction as a weighted average of sorptions that occur on or within individual soil components. While this approach provides insights into sorption mechanisms, its data requirements make it unwieldy for field use. Fortunately, the Freundlich model does a good job of empirically describing the same data over reasonably wide concentration ranges.

The conceptual model of sorption resulting from the DRM is an extension and refinement of the partitioning model. The organic matter associated with soils and aquifer materials is viewed as a complex macromolecule with properties that range from an open, flexible, amorphous structure [e.g., fulvic acid shown in Figure 3(a)] to one that becomes more condensed, rigid, and glassy [e.g., kerogen as shown in Figure 3(b)] after extensive diagenetic alteration (Young and Weber, 1995). Relatively young, amorphous organic structures such as those in peat exhibit sorption isotherms that are well described by the linear partitioning model. The expanded structure, low cross-linking density, and low molecular weight of the organic macromolecules in these materials permit extensive segmental motion, resulting in sorption behavior similar to partitioning into a liquid phase. Far older organic matter, such as the kerogens present in shales, include both amorphous and glassy regions, resulting in varied sorption energies, nonlinear isotherms, higher capacities, and competition among solutes for sorption sites within the macromolecule. The condensed regions of these macromolecules feature restricted segmental motion caused by (1) the proximity of neighboring chains, (2) the higher molecular weight, (3) increased cross-linking and (4) π-π bonding between adjacent aromatic sheets. Sorption in these condensed regions will be more *adsorption*-like, involving localization of solute molecules at particular sites within the macromolecules. Almost any intermediate position between these two extremes is possible given the underlying heterogeneity in organic matter structure described earlier. For example, Figure 3(c) depicts organic matter containing both amorphous, disordered regions and condensed regions in which chain segments are more densely packed and display specific orientations relative to one another. Physical evidence for microscale heterogeneity in the structure of soil organic matter has been provided by electron diffraction (Schnitzer and Kodama, 1975) and ^{13}C-NMR (Preston and Newman, 1992).

Sorption in the heterogeneous macromolecules that comprise soil organic matter can be understood by analogy to results for synthetic

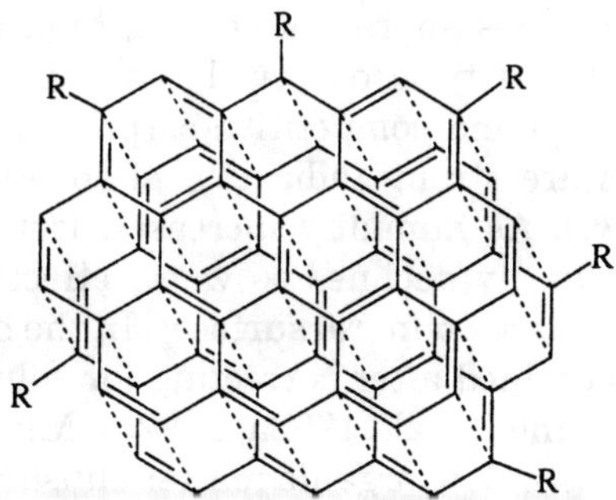

(a) Fulvic acid structure proposed by Schnitzer (1978).

(b) Partial kerogen structure proposed by Tissot and Welte (1984).

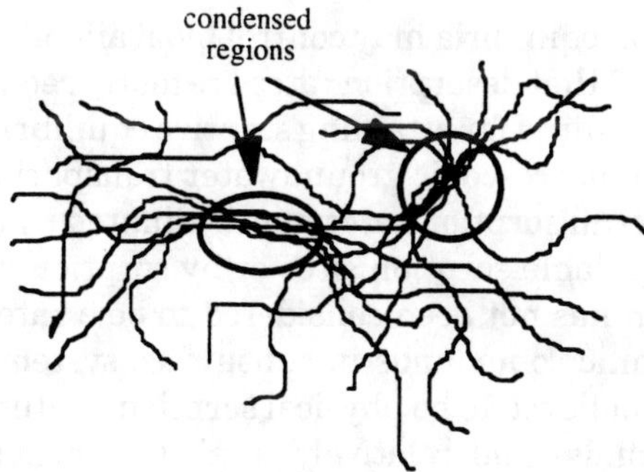

(c) The fringed micelle model for polymer structure after Stevens (1990).

Figure 3. Proposed chemical structures for natural organic matter.

polymer systems above and below their glass transition temperatures (Barrer et al., 1958). Isotherm results from these systems are linear and have low capacities when a polymer is in an amorphous form (above T_g). Glassy forms of the same polymer (below T_g) yield an order of magnitude higher sorption capacities and nonlinear isotherms. Sorption in these polymers has been described successfully by a two-site model, with Langmuir and linear isotherm models used for the glassy and amorphous regions, respectively, directly analogous to the distributed reactivity model (Koros and Paul, 1978).

The BET adsorption isotherm has been shown to fit experimental sorption data in unsaturated soils, indicating the importance of multi-layer sorption as sorbate partial pressures increase. Sorption capacity and isotherm shape are strong functions of moisture levels in unsaturated systems. At relative humidities approaching zero, high capacities and significant isotherm nonlinearity is observed. The highly exothermic sorption energies ($\sim -\Delta H_{vap}$) and competitive sorption phenomena observed for these systems are strong indicators of sorption to mineral surfaces (Rao et al., 1989). As humidity increases, isotherm linearity increases and sorption capacity declines as water effectively displaces organic solutes from high-energy mineral surfaces. In the humidity limit, a saturated isotherm is obtained after correcting for solute sorption at the air-water interface (Pennell et al., 1992a, 1992b). Minerals may also be important sorption sites in low organic matter subsurface materials such as those often found in sandy, highly productive aquifers.

Sorption Rates: Concepts and Models

Although desorption equilibria may control bioavailability, experimental data have suggested that desorption rates are more frequently limiting because time scales required for attaining sorptive equilibria often exceed those associated with macroscopic groundwater transport and biodegradation processes. Overall sorption rates may be limited by either rates of mass transport to available sorption sites or by sorption reaction rates. The sorption reaction has not been considered to be a rate-limiting step for hydrophobic organic compounds in subsurface systems because the binding mechanism, believed to be physical sorption, features rapid rates (milliseconds to seconds) and relatively weak (<5 kcal/mol) energies (Weber et al., 1992). Microscopic mass transport resistances are consequently viewed as the main reason for the slow sorption observed in some subsurface systems. Solute transport from bulk solution to sorption sites comprises several steps, including transport (1) across the solvent film surrounding an aggregate, (2) through solvent-filled regions within an aggregate, (3) through solvent-filled micropores within a grain, and (4) into macromolecular natural organic matter. These steps can occur in series or in parallel, depending on the characteristics of soil grains and aggregates. Any one or a combination of these impedances may control mass transport, in particular subsurface systems. The identification of specific rate-limiting steps allows simpler transport models to be constructed without loss of accuracy or physical significance of model parameters. The development of an understanding of the associated mechanisms and their respective roles in determining sorption rates has therefore been a major research objective.

Adsorption and desorption rate studies in CMBR systems have consistently revealed an initial fast step, lasting minutes to hours, during which sorbed-phase concentrations reach between 25 and 75% of their equilibrium values, followed by a much slower step lasting from days to months, during which equilibrium is attained (Karickhoff, 1980). The simplest approach to modeling sorption rates is as a first-order process described by a single mass transfer coefficient. This approach does not adequately describe the frequently "bi-phasic" nature of sorption rate data (Karickhoff, 1980; Miller and Weber, 1988). An improvement to the first-order model is obtained by dividing the sorption domain into two distinct types of sites, "fast" sites that reach equilibrium instantaneously or very rapidly and "slow" sites that attain equilibrium over much longer periods. Such two-site models have proven capable of fitting adsorption and desorption rate data from a wide range of different systems (Cameron and Klute, 1977; Brusseau et al., 1991; Ball and Roberts, 1991a). Several different physical explanations have been advanced to differentiate between the "sites" of two-site models. Mobile and immobile regions of groundwater flow (Cameron and Klute, 1977), differences in chemical reaction rates, and differences in microscale mass transport (Karickhoff, 1980) have each been postulated as possible causes of such observed behavior. The mathematical form of the two-site model, which is identical in each case, does not allow the true rate-limiting mechanism to be inferred from this body of work.

Microscale mass transport limitations have been accepted as the most plausible cause of sorption rate limitations and as the underlying reason for differentiation between two types of sites. The mobile/immobile water explanation has been discarded as an important explanation because nonsorbing solutes typically do not exhibit the same tailing as sorbing solutes in column experiments (Brusseau et al., 1991). Two different types of impedances have been identified as potentially limiting microscale mass transport. One relates to the migration of solute through highly tortuous, solvent-filled micropores within soil or aquifer particles.

The second involves slow diffusion through macromolecular natural organic matter phases associated with these particles. Each of these theories has garnered experimental support in particular systems but has failed to describe some aspect of sorption behavior in other systems.

Intraparticle diffusion models used extensively to describe sorption rates in various types of solids (e.g., Weber and Rumer, 1965; Weber and Crittenden, 1975) were subsequently extended to soils, sediments, and aquifer materials (Miller and Weber, 1986; Wu and Gschwend, 1986; Ball and Roberts, 1991b). In porous solids, the path that a solute must travel to reach a sorption site is lengthened because it must follow the twists and turns of fluid-filled micropores. Along the way, the solute's progress may be hindered by constrictions in pore diameter and by sorption to pore walls. To fully describe intraparticle transport rates, information is required on the tortuosity, internal porosity, pore shape, and particle size distribution. This level of detail is frequently unavailable for heterogeneous environmental sorbents. Consequently, an empirical effective diffusion coefficient (D_e) is commonly used in conjunction with a homogeneous-phase diffusion model to describe the relationship between solute flux rates and concentration gradients. Experimental sorption rate data for soils and sediments fit using this approach have yielded effective diffusion coefficient values between 10^{-9} and 10^{-11} cm²/s (Miller and Weber, 1986; Wu and Gschwend, 1986). In groundwater systems, slow advective flow rates may make it important to consider additional mass transport limiting steps. For example, sorption rate data in column systems have been found to be better fit by a dual-resistance model that couples intraparticle diffusion and film transfer impedances (Miller and Weber, 1988).

Diffusion through macromolecular natural organic matter, or intraorganic matter diffusion, is the principal alternative to intraparticle diffusion as a mechanism for microscale mass transport limitations. Sorption of hydrophobic organic compounds is known to occur primarily in the organic matter fraction of soils, and this model stresses the mass transfer resistance or impedance imparted by that organic phase. One implementation of the intraorganic matter diffusion model will be examined in more detail to illustrate the approach (Brusseau et al., 1991). In this model labile sites are presumed to reach equilibrium instantaneously (e.g., surface sites), while resistant sites approach equilibrium pursuant to a first-order rate equation. Breakthrough curve data were obtained from miscible displacement experiments for eight hydrocarbons on two low carbon aquifer materials, which were used to calculate two model parameters for each system, these parameters being the fraction of instantaneous sites and a corresponding first-order rate coefficient. The larger hydrocarbons exhibited greater instantaneous sorption fractions and smaller rate coefficients for transport to resistant sites than their smaller homologs. This positive correlation between fractions of sites that sorb instantaneously and sizes of sorbing molecules can be explained by noting that larger molecules have access to fewer sites within the organic matrix due to steric hindrance. Consequently, a larger fraction of the sorption of these compounds will occur on the surfaces of organic matter. The inverse relationship between molecular size and sorption rate constant can be explained by the slower diffusion of larger compounds through the polymer matrices.

To better understand the mass transport impedance imparted by the soil organic matter matrix, it is helpful to return to the synthetic polymer analogy. In contrast to pore diffusion, migration of small solute molecules through polymer gels occurs in a largely solvent-free environment through a network of "pores" that may exhibit little or no permanent structure. Solute must either travel along existing gaps or vacancies or exchange places with a polymer chain when the potential energy barrier to segmental motion is overcome (Pignatello, 1989). Diffusion rates will be higher for solutes of lower molecular volumes and cross-sectional areas. A particular solute will move more slowly through polymers that are more dense, more rigid, and more highly cross-linked. Below a polymer's glass transition temperature, diffusion coefficients drop dramatically as migration by place-change mechanisms requires that significant activation energy barriers be surmounted. These features act to restrict the polymer's segmental motion, forcing the solute molecules to follow more defined, possibly more tortuous paths through pores of molecular dimensions. Although rate models typically feature two distinct types of sites, in reality the microscale heterogeneity of soil organic matter means that a solute molecule will experience a range of diffusivities along its migration path as it travels between expanded and condensed regions. Transport through condensed regions is further restricted by the more energetic solute-organic matrix interactions occurring in these domains (Young and Weber, 1995) in much the same way that strong sorption energies retard movement through micropores. The large differences between solute diffusion coefficients in amorphous and glassy polymers and the similarity of these features to the expanded and condensed regions of natural organic matter have been used as the basis for a model of slow desorption of PCBs from Hudson River sediments (Carroll et al., 1994).

Differentiating between intraparticle and intraorganic matter diffusion hypotheses for explaining slow uptake is difficult because the form of the two-site model used to fit the data is the same in both cases (Ball and Roberts, 1991b; Brusseau et al., 1991). A variety of indirect approaches have been used to support a particular hypothesis, however. For example, Ball and Roberts (1991a, 1991b) studied the sorption of tetrachloroethene (PCE) and 1,2,4,5-tetrachlorobenzene (TeCB) on a sandy aquifer material over long equilibration periods using sealed glass ampoules to minimize volatilization losses. On the largest soil size fraction examined ($d = 0.3$ mm), TeCB was estimated to require approximately three years to attain an equilibrium distribution, while 30–60 days was sufficient equilibration time for pulverized samples of the same soil. Similar results were obtained for desorption rates (Steinberg et al., 1987). This appears inconsistent with the intraorganic matter diffusion model because crushing is not expected to appreciably alter the dimensions of the organic matter regions in a low carbon aquifer material, and thus, it should not affect rates of attaining equilibrium.

In contrast, effective diffusivities calculated using the intraparticle diffusion model are frequently too low to explain based on tortuosity alone. For example, effective diffusivities on the order of 10^{-17} cm^2/s have been measured for desorption of ethylene dibromide (EDB) from agricultural soils (Steinberg et al., 1987). Hindered diffusion caused by pore constrictions may play a role in the process (Farrell and Reinhard, 1994) but does not seem likely to be of general importance in soils based on the relative dimensions of typical pores and solute molecules (Brusseau et al., 1991). The intraorganic matter diffusion hypothesis has the advantage that the mass transport resistance is imparted by the phase known to be the primary sorption site for HOCs. Brusseau et al. (1991) have noted that the intraparticle diffusion hypothesis requires that the bulk of the sorbing material, organic matter, must be located in the particle interior, a requirement that is not necessarily consistent with common soil grain structures. Direct experimental support for the intraorganic matter diffusion hypothesis is limited, but examination of rate parameters for a homologous series of aromatic compounds has revealed that addition of methyl groups to a benzene side chain has a limited effect on sorption rates after the side chain reaches five carbons (Brusseau et al., 1991). This observation is analogous to those made in similar studies in synthetic polymer systems but is hard to reconcile with the intraparticle diffusion model.

One limitation of both the intraparticle and intraorganic matter diffusion models is that they have almost always been implemented using linear isotherms, often with the distribution coefficient estimated from the same fitting exercise used to obtain the rate coefficient. If the sorption isotherm is actually nonlinear and this is not reflected in model architecture, significant errors may result (Weber et al, 1992; Farrell and Reinhard, 1994). For example, if higher energy sites are distributed throughout the sorption domain, regardless of whether the domain be a pore network or an organic matter phase, the transport of solute molecules past those sites will be slower than past sites of average sorption energy (i.e., the linear isotherm model). If the pores are small relative to solute dimensions or the organic matter is condensed, then the slower rate will affect all solute molecules traveling behind the strongly sorbed molecule. Thus, the presence of a limited number of high-energy sorption sites in heterogeneous pores or organic matter phases can lead to important changes in the mass transport properties of a system. When equilibrium and rate parameters are obtained simultaneously from the same data set, the model specification error biases both values and can lead to significant mistakes when the results are used to make mechanistic inferences or transport predictions.

APPLICATION OF SORPTION PRINCIPLES TO PREDICTION OF BIOAVAILABILITY

Both sorption equilibrium and rate studies suggest that a sizable fraction of sorbed contaminant is located within tortuous micropores or within a sorbent's natural organic matter matrix. In either case, the solute is not likely to be readily accessible to microorganisms nor to exoenzymes they may produce for extracellular degradation. Consequently, desorption processes are likely to control bioavailability in soils or aquifer materials exposed to contaminants for extended periods, which includes the vast majority of subsurface contamination scenarios requiring remediation efforts. This section explores how the sorption principles reviewed above can be applied to help conceptualize bioavailability issues and implement effective bioremediation systems.

An example of the type of research that has been conducted to explore sorption-biodegradation interactions is that of Weissenfels et al. (1992). The results confirmed the biphasic nature of desorption and attributed the slow phase of PAH sorption to migration of PAHs into less accessible sites within the soil matrix. Extractability was found to decrease with contact time. For one of the soils tested, sorption was observed to reduce contaminant toxicity. In another soil, long-term sorption attributed to migration of PAHs into soil organic matter prevented biodegradation. Freshly added PAHs were degraded rapidly, but degradation was not complete, with 28% of PAHs remaining in the soil. The same bacterial population degraded PAHs completely in low organic carbon sandy soils.

Attainable remediation levels for contaminants at a site are potentially the most important variables considered when deciding to use an in situ bioremediation system. Remedial targets are frequently specified independently for the aqueous and solid phases at a site. The attainable in situ aqueous phase concentration is set by S_{min} for microorganisms making direct use of a solute as a primary substrate, while lower concentrations can be obtained in cometabolic systems. If S_{min} is less than the aqueous remedial target (e.g., maximum contaminant limits or drinking water standards), then the solid-phase cleanup goal should be reconsidered. The minimum attainable solid-phase concentration when the solute serves as a primary substrate is the solid-phase concentration that would exist in equilibrium with an aqueous-phase concentration of S_{min}. Obviously, the value predicted for the minimum solid-phase loading will depend on the isotherm model specified to describe the sorption process. For matters of convenience and unawareness of consequence, sorption data is often collected only over a limited range of aqueous concentrations (e.g., approximately one order of magnitude), and a linear model is used to fit the data. If the underlying sorption process is truly nonlinear, this procedure will result in serious underestimation of sorption at the low aqueous concentrations typical of cleanup standards. To illustrate this concept, Figure 4 depicts experimental data (Young and Weber, 1995) for phenanthrene sorption on two surface soils, Chelsea soil (6.4% organic carbon) and Webster soil (3.0% organic carbon), fit by the linear and Freundlich models. In Figure 4(a), the isotherm fits are displayed on a linear scale, and although the Freundlich fit appears superior, the difference between the two models does not seem to be significant. The discrepancy between the models would likely be less apparent if fewer data points had been collected over a narrower range of aqueous concentrations, as is more common practice. Figure 4(b) displays the same data and model fits on a logarithmic scale. Viewing the data in this manner reveals the order of magnitude errors in estimated solid-phase loadings that can result from improper specification of an isotherm model. Assuming an aqueous-phase phenanthrene standard of 1 μg/l, the linear model would predict minimum solid-phase cleanup levels of 0.71 mg/kg and 0.24 mg/kg for Chelsea and Webster soils, respectively. The Freundlich model, which is far more accurate in this concentration region, would predict equilibrium solid-phase phenanthrene loadings of 2.8 mg/kg and 2.1 mg/kg for these two soils, values between 0.6 and 1.0 orders of magnitude larger than the linear model predictions. On these soils, far less of the contaminant will be available for biodegradation than would be predicted from the linear model, and solid-phase cleanup targets may not be met even when solution-phase standards have been satisfied. This

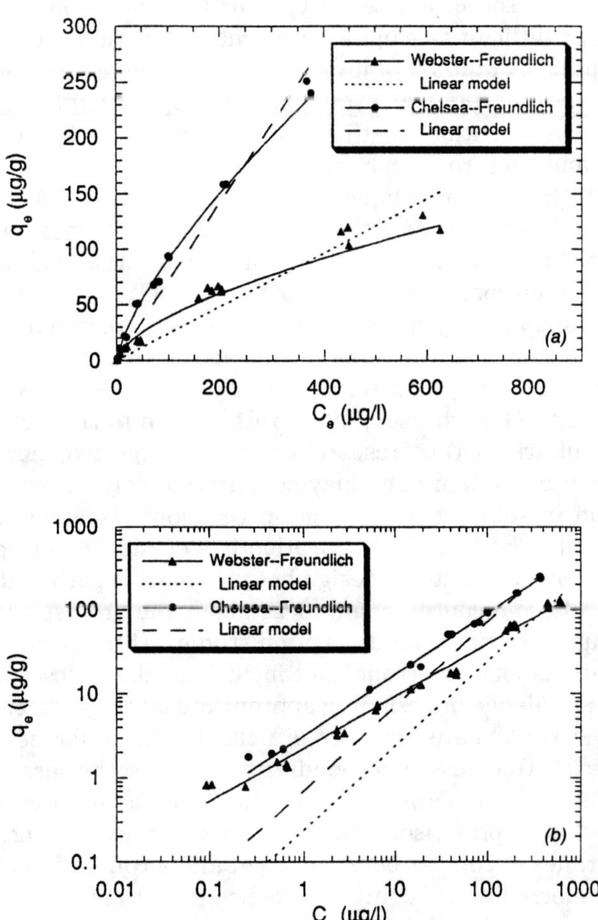

Figure 4. Phenanthrene sorption isotherms plotted on (a) a linear scale and (b) a logarithmic scale.

argument is equivalent to saying that some fraction of solute molecules have sorption energies high enough that they are not bioavailable.

Even with careful attention to isotherm model specification and a good set of adsorption isotherm data for a particular contaminant-soil system, it may still be difficult to generate accurate predictions about contaminant desorption. A number of investigators have noted that distribution coefficients for desorption are substantially larger than the corresponding adsorption partition coefficients (DiToro and Horzempa, 1982; Horzempa and DiToro, 1983; Steinberg et al., 1987; Pignatello and Huang, 1991; Schrap and Opperhuizen, 1992; Weissenfels et al., 1992). For example, desorption partition coefficients for PCBs from lake sediments were shown to be up to three times higher than values of corresponding sorption coefficients (Horzempa and DiToro, 1983). Hysteresis effects for nonpolar organic compounds in soils commonly have been ascribed to incomplete equilibration in adsorption experiments or to experimental artifacts (Gschwend and Wu, 1985; Brusseau and Rao, 1989; Scrap and Opperhuizen, 1992) rather than to true differences in reaction equilibria. Other researchers have suggested, however, that some fraction of contaminant may sorb irreversibly, removing it from participation in subsequent desorption reactions (Isaacson and Frink, 1984; Bollag and Bollag, 1990). Sorption hysteresis, real or apparent, is of great concern for remedial design because most experimental results, related models, and concepts and ideas about the process are based on sorption studies rather than desorption studies. If the resulting models and correlations lead to erroneous conclusions about desorption, engineers face a challenge in designing appropriate cleanup systems because desorption isotherms are far more difficult to obtain for field-contaminated materials than are corresponding sorption isotherms.

The problem is compounded by the fact that the difference between sorption and desorption isotherm parameters appears to increase with increasing times of contact between a soil and a contaminant, an effect commonly referred to as "aging." Evidence from the field and the laboratory indicates that sorbed-phase organic compounds become increasingly difficult to remove via flushing, solvent extraction, or biodegradation the longer the soil has been exposed to the compound (McCall and Agin, 1985; Steinberg et al., 1987; Pignatello and Huang, 1991; Pavlostathis and Jaglal, 1991; Pavlostathis and Mathavan, 1992). Studies with soils exposed to contaminants under laboratory conditions have shown a continuing increase in the partition coefficients associated with desorption reactions with increased residence times (McCall and Agin, 1985; Pavlostathis and Mathavan, 1992). Extremely slow microscale mass transport of solute to less accessible regions of soil grains is viewed as the most likely cause of this phenomenon. Support for this hypothesis can be found in both modeling and experimental results. Two-site models of desorption rate data following increased periods of aging reveal a decreasing fraction of "labile" or easily removed contaminant. Experimental results reveal that the fraction of contaminant removed by rapid extraction also declines over longer aging times, paralleling the desorption rate model results (McCall and Agin, 1985; Hatzinger and Alexander, 1995). One of the first extensive studies of aging effects on field-contaminated materials (Steinberg et al., 1987) found that soils exposed to 1,2-dibromoethane (EDB) years earlier had sorption partition coefficients of 1.5–2.1 ml/g while partition coefficients associated with the desorption reaction ranged from 170 to 300 ml/g. Freshly sorbed EDB could be completely removed by a nitrogen purge in 100 minutes, while only 5% of the native EDB could be removed under the same conditions. Pulverization of the soil samples allowed significantly more of the EDB to be desorbed by nitrogen purging (40% instead of 5%). When the desorption process was assumed to be diffusion-controlled, half lives of 23 to 31 years were calculated for the resistant EDB fractions.

A close parallel has been demonstrated between reductions in rate and extent of contaminant desorption with increased contact times and reductions in bioavailability (Steinberg et al., 1987; Scribner et al., 1992; Hatzinger and Alexander, 1995; Landrum, 1989; Umbreit et al., 1986; Alexander, 1995). A variety of organic chemicals that are readily biodegraded when first added to soil samples are slowly degraded after months or years of contact with the soils (Steinberg et al., 1987; Scribner et al., 1992; Hatzinger and Alexander, 1995). These results suggest that a useful test to distinguish between microbiological and physical-chemical constraints on biodegradation is to monitor degradation of labeled (freshly added) contaminant and aged contaminant in the same sample. Similar rates and extents of degradation indicate microbiological controls (e.g., limitations in nutrients, electron acceptors, microorganisms) while large differences in rates suggest that desorption rates or equilibria control biodegradation. Uptake of aged chemicals is also reduced in larger plant and animal species after extended soil-contaminant contact times. For example, toxicity of TCDD containing soil from an old 2,4,5-trichlorophenoxyacetic acid site was found to be correlated to the contact time (Umbreit et al., 1986). TCDD that had been exposed to the soils for years proved to be nontoxic to guinea pigs, whereas freshly added chemical in the same concentration led to acutely toxic responses. The same trend was found for the herbicide simazine and its toxic effects to sugarbeets (Scribner et al., 1992). The reduction in bioavailability with aging has been observed in different soil types, and reductions in availability continue to occur after more than a year of contact between contaminants and soil (Hatzinger and Alexander, 1995). Sonication to disrupt soil aggregates has been observed to increase biodegradation rates of aged contaminants, but not to return degradation rates to the same levels observed for freshly added contaminants (Hatzinger and Alexander, 1995).

CLOSURE

In situ bioremediation represents the most financially and environmentally attractive cleanup option for many organic-contaminated sites. Low bioavailability of these chemicals poses significant problems for the design and implementation of subsurface bioremediation systems, potentially rendering the technology ineffective. The problem is exacerbated by further reductions in bioavailability after long exposure periods. Biodegradation will proceed more slowly than predicted and may fail to reach remedial targets if bioavailability is not explicitly considered. Desorption processes appear to control bioavailability in many, if not most, subsurface systems. In almost all cases, the extent of desorption will be far less than would be predicted by applying a K_{OC}-K_{OW} correlation developed from adsorption isotherm data to obtain a linear distribution coefficient. This discrepancy results from a combination of factors, including isotherm nonlinearity, rate limitations, desorption hysteresis, and aging effects. Errors in desorption and biodegradation predictions increase as sorbed-phase concentrations decline and soil-contaminant contact times increase. Reduced bioavailability appears to result from declining desorption rates and increasingly resistant fractions of sorbed-phase contaminant with aging. Despite their practical importance, the mechanisms underlying these process behaviors are not well understood. For example, the impact of contaminant and soil characteristics on changes in bioavailability upon aging has not been systematically investigated. The development of a more complete picture of the desorption process appears to be the key to sorting out the complex interactions within soil-contaminant-microorganism systems. Microscale heterogeneity of soils and aquifer materials has often been neglected, with conceptual models featuring only one (organic matter) or two (organic matter and minerals) homogeneous compartments. Future work in this area should be guided by an improved understanding of the vast differences in composition and structure within and between soil organic matter and inorganic fractions. Only with this information will it be possible to make accurate simplifications for modeling purposes and useful a priori predictions of bioavailability.

REFERENCES

Alexander, M. *Introduction to Soil Microbiology;* John Wiley and Sons: New York, 1977.

Alexander, M. *Environ. Sci. Technol.* 1995, 29, 2713–2717.

Amador, J.; Alexander, M. *Soil Biol. Biochem.* 1988, 20, 185–191.

Ball, W. P.; Roberts, P. V. *Environ. Sci. Technol.* 1991a, 25, 1223–1236.

Ball, W. P.; Roberts, P. V. *Environ. Sci. Technol.* 1991b, 25, 1237–1249.

Barrer, R. M.; Barrie, J. A.; Slater, J. *J. Polymer Sci.* 1958, 27, 177–197.

Baveye, P.; Valocchi, A. *Water Resour. Res.* 1989, 25, 1413–1421.

Bollag, J.-M.; Bollag, W. B. *Intern. J. Environ. Anal. Chem.* 1990, 39, 147–157.

Bollag, J.-M. *Environ. Sci. Technol.* 1992, 26, 1876–1881.

Bortiatynski, J. M.; Hatcher, P. G.; Minard, R. D.; Dec, J.; Bollag, J.-M. Enzyme-Catalyzed Binding of ^{13}C-Labeled 2,4-Dichlorophenol to Humic Acid Using High Resolution ^{13}C NMR. In *American Chemical Society, Division of Fuel Science: 207th National Meeting.* 1994. San Diego, CA.

Brusseau, M. L.; Rao, P. S. C. *CRC Critical Reviews in Environmental Control 1989, 19, 33–99.*

Brusseau, M. L.; Jessup, R. E.; Rao, P. S. C. *Environ. Sci. Technol.* 1991, 25, 134–142.

Butler, J. H. A.; Ladd, J. N. *Soil Biol. Biochem.* 1971, 3, 249–257.

Cameron, D. R.; Klute, A. *Water Resour. Res.* 1977, 13, 183–188.

Capriel, P.; Haisch, A.; Khan, S. U. *J. Agric. Food Chem.* 1985, 33, 567–569.

Carroll, K. M.; Harkness, M. R.; Bracco, A. A.; Balcarcel, R. R. *Environ. Sci. Technol.* 1994, 28, 253–258.

Chiou, C. T.; Peters, L. J.; Freed, V. H. *Science* 1979, 206, 831–832.

Chiou, C. T.; Porter, P. E.; Schmedding, D. W. *Environ. Sci. Technol.* 1983, 17, 227–231.

Chiou, C. T.; Shoup, T. D.; Porter, P. E. *Organic Geochem.* 1985, 8, 9–14.

Criddle, C. S.; Alvarez, L. A.; McCarty, P. L. in *Transport Processes in Porous Media;* J. Bear and M. Y. Corapcioglu, Eds.; Kluwer Academic Publishers: Netherlands, 1991.

Dec, J.; Bollag, J.-M. *Environ. Sci. Technol.* 1994, 28, 484–490.

Dec, J.; Bollag, J.-M. *Environ. Sci. Technol.* 1995, 29, 657–663.

deHaan, H. *Freshwater Biol.* 1974, 4, 301–310.

DiToro, D. M.; Horzempa, L. M. *Environ. Sci. Technol.* 1982, 16, 594.

Earhardt, H. M.; Rehm, H. J. *Appl. Environ. Microbiol.* 1985, 21, 32–36.

Earhardt, H. M.; Rehm, H. J. *Appl. Microbiol. Biotechnol.* 1989, 30, 312–317.

Efroymson, R. A.; Alexander, M. A. *Environ. Sci. Technol.* 1995, 29, 515–521.

Farrell, J.; Reinhard, M. *Environ. Sci. Technol.* 1994, 28, 53–62.

Filip, Z. *Folia Microbiol.* 1973, 18, 56–74.

Garbarini, D. R.; Lion, L. W. *Environ. Sci. Technol.* 1986, 20, 1263–1269.

Gordon, A. S.; Millero, F. J. *Microbial Ecology* 1985, 11, 289–293.

Grathwohl, P. *Environ. Sci. Technol.* 1990, 20, 1687–1693.

Griffin, D. M. *Ecology of Soil Fungi* Chapman and Hall, Ltd.: London, 1972.

Griffin, D. M. in *Water Potential Relations in Soil Microbiology,* SSSA Spec. Pub. 9; J. F. Parr, W. R. Gardner and L. Elliot, Eds.; Soil Science Society of America: Madison, WI, 1981.

Griffin, D. M.; Quail, G. *Australian J. Biol.Sci.* 1968, 21, 579–582.

Gschwend, P. M.; Wu, S. *Environ. Sci. Technol.* 1985, 19, 90–96.

Guerin, W.; Boyd, S. *Appl. Environ. Microbiol.* 1992, 58, 1142–1152.

Haack, S. K.; Garchow, H.; Klug, M. J.; Forney, L. J. *Appl. Environ. Microbiol.* 1995, 61, 1458–1468.

Harms, H.; Zehnder, A. J. B. *Appl. Environ. Microbiol.* 1994, 60, 2736–2745.

Harris, R. F. in *Water Potential Relations in Soil Microbiology,* SSSA Spec. Pub. 9; J. F. Parr, W. R. Gardner and L. Elliot, Eds.; Soil Science Society of America: Madison, WI, 1981.

Harvey, R. W.; Smith, R. L.; George, L. *Appl. Environ. Microbiol.* 1984, 48, 1197–1202.

Hatzinger, P. B.; Alexander, M. *Environ. Sci. Technol.* 1995, 29, 537–545.

Hillers, A.; Adriaens, P. In *In situ and On-Site Bioreclamation, 3rd International Symposium;* San Diego, CA, 1995.

Holden, P. A.; Firestone, M. K. In *American Society for Microbiology Conference Proceedings;* Washington, D.C., 1995.

Holman, H. Y.; Tsang, Y. In *In situ and On-Site Bioreclamation, 3rd International Symposium;* San Diego, CA, 1995.

Horzempa, L. M.; DiToro, D. M. *Water Res.* 1983, 17, 851–859.

Howard, D. M.; Howard, P. J. A. *Soil Biol. Biochem.* 1993, 25, 1537–1546.

Isaacson, P. J.; Frink, C. R. *Environ. Sci. Technol.* 1984, 18, 43–48.

Karickhoff, S. W.; Brown, D. S.; Scott, T. A. *Water Res.* 1979, 13, 241–248.

Karickhoff, S. W. In *Contaminants and Sediments, Vol. 2;* R. A. Baker, Ed.; Ann Arbor Science: Ann Arbor, MI, 1980.

Knaebel, D. B.; Federle, D. C.; McAvoy, J. R.; Vestal, J. R. *Appl. Environ. Microbiol.* 1994, 60, 4500–4508.

Koros, W. J.; Paul, D. R. *J. Polymer Sci.* 1978, 16, 1947–1963.

Ladd, J. N.; Butler, J. H. A. In *Soil Biochemistry;* E. A. Paul and A. D. McLaren, Eds.; Vol. 4, Marcel Dekker: New York, 1975.

Lambert, S. M. *J. Agr. Food Chem.* 1967, 15, 572–576.

Landrum, P. F. *Environ. Sci. Technol.* 1989, 23, 588–595.

Li, A. Y. L.; DiGiano, F. A. *J. WPCF* 1983, 55, 392–399.

MacDonald, J. A.; Kavanaugh, M. C. *Environ. Sci. Technol.* 1994, 29, 362A–368A.

Mato, M. C.; Olmedo, M. G.; Mendez, J. *Soil Biol. Biochem.* 1972, 4, 469–473.

McCall, P. J.; Agin, G. L. *Environ.Toxicol.Chem.* 1985, 4, 37–44.

McCarty, P. L. In *Proceedings of the Second International Conference on Ground Water Quality,* Tulsa, OK, 1984.

McGinley, P. M.; Katz, L. E.; Weber, W. J., Jr. *Environ. Sci. Technol.* 1993, 27, 1524–1531.

Means, J. C.; Wood, S. G.; Hassett, J. J.; Banwart, W. L. *Environ. Sci. Technol.* 1980, 14, 1524–1528.

Mihelcic, J. R.; Luthy, R. G. *Environ. Sci. Technol.* 1991, 25, 169–177.

Miller, C. T.; Weber, W. J., Jr. *J. Contam. Hydrol.* 1986, 1, 243–261.

Miller, C. T.; Weber, W. J., Jr. *Water Res.* 1988, 22, 465–474.

Miller, M. E.; Alexander, M. *Environ. Sci. Technol.* 1991, 25, 240–245.

Mingelgrin, U.; Gerstl, Z. *J. Environ. Qual.* 1983, 12, 1–11.

Molz, F. J.; Widdowson, M. A.; Benefield, L. D. *Water Resour. Res.* 1986, 22, 1207–1216.

Mörsen, A.; Rehm, H. J. *Appl. Microbiol. Biotechnol.* 1987, 26, 283–288.

Murphy, E. M.; Zachara, J. M.; Smith, S. C. *Environ. Sci. Technol.* 1990, 24, 1507–1516.

Nováková, J. *Zentralblatt für Bakteriologie* 1972, 127, 367–372.

Oades, J. M. In *Minerals in Soil Environments,* 2nd Ed.; J. B. Dixon and S. B. Weed, Eds.; Soil Science Society of America: Madison, WI, 1989.

Ogram, A. V.; Jessup, R. E.; Ou, L. T.; Rao, P. S. C. *Appl. Environ. Microbiol.* 1985, 49, 582–587.

Parsons, J. R. *Chemosphere* 1992, 25, 1973–1980.

Pavlostathis, S. G.; Jaglal, K. *Environ. Sci. Technol.* 1991, 25, 274–279.

Pavlostathis, S. G.; Mathavan, G. N. *Environ. Sci. Technol.* 1992, 26, 532–538.

Pennell, K. D.; Rhue, R. D.; Hornsby, A. G. *J. Environ. Qual.* 1992a, 21, 419–426.

Pennell, K. D.; Rhue, R. D.; Rao, P. S. C.; Johnston, C. T. *Environ. Sci. Technol.* 1992b, 26, 756–763.

Pignatello, J. J. In *Reactions and Movement of Organic Chemicals in Soils;* B. L. Sawhney and K. Brown, Eds.; SSSA Spec. Pub. 22; Soil Science Society of America & American Society of Agronomy: Madison, WI, 1989.

Pignatello, J. J. In *Organic Substances in Sediments and Water, Vol. 1;* R. A. Baker Ed.; Lewis Publishers: Chelsea, MI, 1991.

Pignatello, J. J.; Huang, L. Q. *J. Environ. Qual.* 1991, 20, 222–228.

Preston, C. M.; Newman, R. H. *Can. J. Soil Sci.* 1992, 72.

Rao, P. S. C.; Ogwada, R. A.; Rhue, R. D. *Chemosphere* 1989, 18, 2177–2191.

Rijnaarts, H. H. M.; Bachmann, A.; Jumelet, J. C.; Zehnder, A. J. B. *Environ. Sci. Technol.* 1990, 24, 1349–1354.

Robinson, K. G.; Farmer, W. S.; Novak, J. T. *Water Res.* 1990, 24, 345–350.

Rosenberg, E.; Legmann, R.; Kushmara, A.; Taube, R.; Adler, E.; Ron, E. Z. *Biodegradation,* 1992, 3, 337–350.

Schnell, S.; King, G. M. In *American Society for Microbiology Conference Proceedings;* Washington, DC, 1995.

Schnitzer, M.; Kodama, H. *Geoderma* 1975, 13, 279–287.

Schnitzer, M. In *Soil Organic Matter;* M. Schnitzer and S. U. Khan, Eds.; Elsevier: Amsterdam, 1978.

Schrap, S. M.; Opperhuizen, A. *Chemosphere* 1992, 24, 1259–1282.

Scow, K. M.; Simkins, S.; Alexander, M. *Appl. Environ. Microbiol.* 1986, 51, 1028–1035.

Scow, K. M.; Alexander, M. *Soil Science American Journal* 1992, 56, 128–134.

Scribner, S. L.; Benzing, T. R.; Sun, S.; Boyd, S. A. *J. Environ. Qual.* 1992, 21, 115–120.

Shimp, R.; Pfaender, F. K. *Appl. Environ. Microbiol.* 1985, 49, 402–407.

Shindo, H.; Huang, P. M. *Soil Sci. Soc. Am. J.* 1984, 48, 927–934

Siegele, D. A.; Kolter, R. *J. Bacteriology* 1992, 174, 345–348.

Sjoblad, R. D.; Bollag, J.-M. *Oxidative Coupling of Aromatic Compounds by Enzymes from Soil Microorganisms;* Marcel Dekker: New York, 1981.

Sjoblad, R. D.; Minard, R. D.; Bollag, J.-M. *Pesticide Biochemistry and Physiology* 1976, 6, 457–463.

Spurlock, F.; Biggar, J. W. *Environ. Sci. Technol.* 1994a, 28, 989–995.

Spurlock, F.; Biggar, J. W. *Environ. Sci. Technol.* 1994b, 28, 996–1002.

Steinberg, S. M.; Pignatello, J. J.; Sawhney, B. L. *Environ. Sci. Technol.* 1987, 21, 1201–1208.

Stevenson, F. J. *Humus Chemistry: Genesis, Composition, Reactions;* John Wiley and Sons, Inc.: New York, 1994.

Stotzky, G. *Can. J. Microbiol.* 1966, 12, 1235–1246.

Stratton, R. G.; Namkung, E.; Rittman, B.E. *J. AWWA* 1983, 54, 463–469.

Stucki, G.; Alexander, M. *Appl. Environ. Microbiol.* 1987, 53, 292–297.

Subba-Rao, R. V.; Alexander, M. *Appl. Environ. Microbiol.* 1982, 44, 659–668.

Tissot, B. P.; Welte, D. H. *Petroleum Formation and Occurrence;* Springer-Verlag: New York, 1984.

Umbreit, T. H.; Hesse, E. J.; Gallo, M. A. *Science* 1986, 232, 497–499.

van Loosdrecht, M. C. M.; Lyklema, J.; Norde, W.; Zehnder, A. *Microbiol. Rev.* 1990, 54, 75–87.

Visser, S. A. *Soil Biol. Biochem.* 1985, 17, 457–462.

Weber, J. B.; Coble, H. D. *J. Agr. Food Chem.* 1968, 16, 475–478.

Weber, W. J., Jr.; Crittenden, J. C. *J. Wat. Pollut. Control Fed.* 1975, 47, 924–940.

Weber, W. J., Jr. ; Miller, C. T. *Water Res.* 1988, 22, 457–464.

Weber, W. J., Jr. ; McGinley, P. M.; Katz, L. E. *Water Res.* 1991, 25, 499–528.

Weber, W. J., Jr. ; McGinley, P. M.; Katz, L. E. *Environ. Sci. Technol.* 1992, 26, 1955–1962.

Weber, W. J., Jr.; Rumer, R. R., Jr. *Water Resour. Res.* 1965, 1, 361–362.

Weissenfels, W. D.; Klewer, H. J.; Langhoff, *J. Appl. Microbiol. Biotechnol.* 1992, 36, 689–696.

Whelan, G.; Sims, R. C. *Hazardous Waste and Hazardous Materials* 1992, 9, 245–265.

Wszolek, P. C.; Alexander, M. *J. Agric. Food Chem.* 1979, 27, 410–414.

Wu, S.; Gschwend, P. M. *Environ. Sci. Technol.* 1986, 20, 717–725.

Young, T. M. Ph.D. Dissertation, University of Michigan, Ann Arbor, MI, 1996.

Young, T. M.; Weber, W. J., Jr. *Environ. Sci. Technol.* 1995, 29, 92–97.

Interactions between Organic Contaminants and Soil Affecting Bioavailability

DANIEL CASSIDY AND ROBERT L. IRVINE
Department of Civil Engineering and Geological Sciences
University of Notre Dame
Notre Dame, IN 46556, USA

INTRODUCTION

The interactions between organic compounds and soil are of extreme environmental importance. Binding to the soil matrix causes the attenuation of contaminants in groundwater (i.e., a decrease in contaminant concentration in the direction of flow, which is greater than that which would result from dispersion alone). Slow release of soil-bound and nonaqueous phase liquid (NAPL) contaminants into groundwater limits the effectiveness of pump-and-treat remediation operations and renders the soil matrix a long-term source of groundwater and air contamination.

Interactions between organic compounds and soil also control the effectiveness of bioremediation technology. Rates of contaminant biodegradation in soil are typically less than those observed in a system without soil (Geerdink, 1995) because a portion of the contaminants become bound to the soil matrix or are entrapped within micropores and, in these forms, are not readily available to microorganisms. Bioavailability describes the rates of mass transfer of contaminants from a phase or environment in which the availability to microorganisms is limited or absent to a phase or environment in which microbial consumption can readily occur. Bioavailability is of extreme importance because it controls the rates of and, therefore, the cost-effectiveness of contaminant bioremediation.

Rates of biological degradation of contaminants in soil often decrease markedly when some relatively low concentration is reached. Figure 1

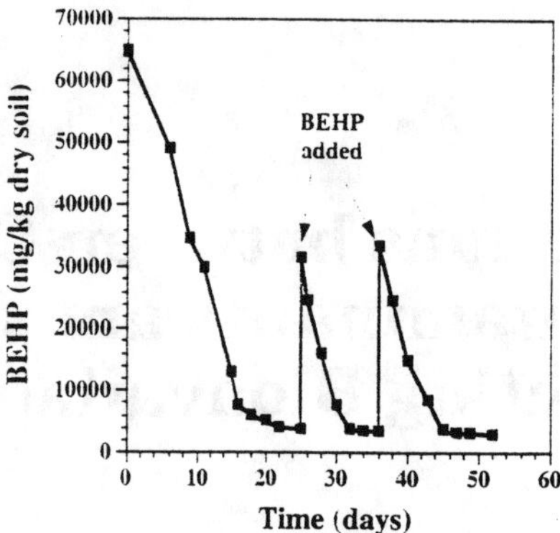

Figure 1. Biodegradation of BEHP in a silty loam took place until a concentration of 4000 mg/kg was reached. The BEHP added on days 25 and 36 was in a form more biodegradable than the residual 4000 mg/kg, resulting in greater rates of biodegradation (Cassidy, 1995).

shows the biological removal of bis-(2-ethylhexyl) phthalate (BEHP, or DEHP) from a silty loam in a slurry reactor. Rates of BEHP removal decrease drastically at a concentration of approximately 4000 mg/kg. Additions of BEHP on days 25 and 36 (without additional nutrients) resulted in increased rates of biodegradation, similar to the rates observed during the first 15 days. The contrast in rates of biodegradation at concentrations above and below 4000 mg/kg indicates that the residual BEHP in the silty loam was in a form less bioavailable than the BEHP at higher concentrations.

Although some microorganisms can grow directly on the surface of a nonaqueous phase liquid (NAPL), microbial consumption of contaminants typically occurs only in the aqueous phase (Wodzinski and Coyle, 1974). Bound or sequestered contaminants must first dissolve in the aqueous phase and then be transported to a microenvironment in which they come into direct contact with microorganisms or the appropriate extracellular enzymes in order to be transformed. A limited availability of organic molecules to microbes in soil results from (a) sorption to humic substances and mineral surfaces, (b) presence as NAPLs, (c) covalent bonding with humic substances (humification), and (d) entrapment within micropores of fine-grained soil aggregates and/or 2:1 clay minerals. The nature and reversibility of these interactions are controlled primarily by the properties of the soil and the contaminant but are also affected by the length of time during which the soil has been contaminated. Interactions of organic compounds with clay minerals can also alter soil properties such as cation exchange capacity and aggregate stability (Miehlich and Wagner, 1989; Goetz and Wiechmann, 1989), which can in turn affect rates of contaminant removal during bioremediation.

The impact of all of these factors on contaminant bioavailability in soil are discussed in this chapter. Data compiled from laboratory investigations on the biodegradation of BEHP in a silty loam and diesel fuel in a silty clay loam (Cassidy, 1995) are reviewed to illustrate the effect of these interactions on bioavailability.

SORPTION

The soil matrix offers much surface area for sorption of contaminants to occur. The active surfaces for contaminant sorption are most commonly humic substances and clay minerals but can also include amorphous Fe or Al oxides and oxyhydroxides and carbonate minerals. Some clay minerals (e.g., montmorillonite) have a surface area of up to 800 m^2 per gram, and humic substances typically have more than 1000 m^2 of surface area per gram. A typical soil is composed of 45% mineral substances (primarily silicates), 25% air, 23% water, and 7% organic substances (including humic substances and microflora and microfauna) on a volumetric basis (Schweisfurth, 1988). The mineral fraction typically comprises more than 90% of the mass of the solid matrix of most soils.

Adsorption refers to the accumulation of the contaminant on the surface itself, and *absorption* refers to the retention of the contaminant within the mass of solid. *Sorption* is used to include both adsorption and absorption. The sorptive capacity of soil can affect microbial contaminant degradation by removing contaminants from the aqueous phase, retaining microorganisms, depressing the activity of extracellular enzymes, rendering microbial inhibitors less toxic, and modifying the pH of the aqueous microenvironment adjacent to the site of sorption.

A number of factors control the sorption or partitioning of organic compounds to soil. These include the type and concentration of contaminant in the surrounding solution, the amount of humic substances present in the soil, the type and quantity of clay minerals, the age of contamination, and the pH and temperature of the surrounding solution.

Humic Substances

Humic substances are naturally occurring heterogeneous refractory organic molecules formed through the microbial degradation of plant and animal remains in soil. Although biodegradative processes lead to the production of humic substances, microbially mediated polymerization reactions are also involved in their formation (Stevenson, 1985). Humic

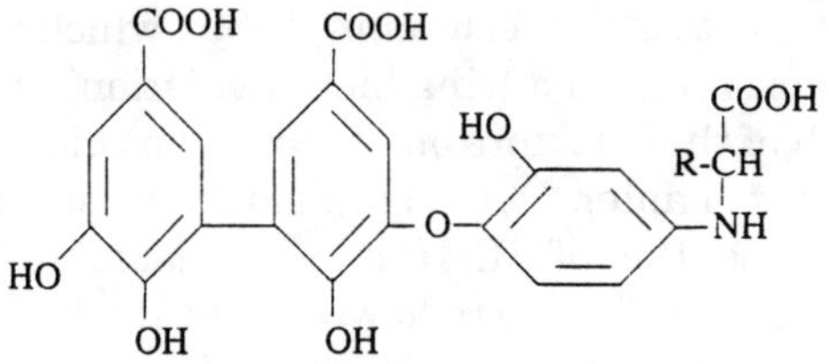

COOH   COOH

HO

COOH

R-CH

HO   O   NH

HO

OH   OH

Figure 2. Hypothetical structure for a fraction of a humic molecule (modified from Stevenson, 1994).

substances can be divided into three fractions, based on their solubility in acids and bases. Humic acids precipitate at low pH, upon acidification of an alkaline extract of soil organic matter. Fulvic acids are soluble at all pH values and remain in solution upon acidification of an alkaline extract of soil organic material. Humins are insoluble at all pH values and are therefore not extractable from soils in aqueous solutions.

The structure of humic substances is heterogeneous, but certain functional groups appear repeatedly. Figure 2 shows a hypothetical structure for a fraction of a humic molecule. Humic substances are composed primarily of carbon, oxygen, and hydrogen, with smaller amounts of nitrogen and phosphorus. The hydrophobic nature of humic substances plays a predominant role in their interaction with organic contaminants. Because of their large molecular size, humic substances adsorb readily to clay minerals, largely through van der Waals forces (Stumm and Morgan, 1981; Stevenson, 1985). Catroux and Schnitzer (1987) found that, in some soils, the humics associated with the clay fraction have a higher content of aliphatics and a lower content of aromatics and oxygen-containing functional groups (i.e., are more hydrophobic) than humic substances associated with larger particle sizes.

Humic substances have oxygen-containing functional groups (e.g., acids, ketones, alcohols) and amines, which also play a role in their chemistry. The behavior of the functional groups is affected by pH. At low pH values, the organic acid groups are protonated and therefore neutral. The amine groups, however, are positively charged. At high pH values, the acid groups are negatively charged and the amine groups have a neutral charge. At neutral pH values, the acid groups are negatively charged and the amine groups are positively charged. The high content of acid functional groups gives humic substances the ability to retain cations at neutral and high pH values. The retained cations typically can be exchanged with other cations in the soil solution. The capacity of a material to retain and exchange cations is expressed as the cation exchange capacity (CEC). Table 1 lists the CEC and the specific surface area of humic substances and other common constituents of soils. The oxygen atoms in humic substances can donate pairs of free electrons to participate in complexation reactions with organic cations and metals. Gieseking (1975), Aiken et al. (1985), and Stevenson (1985) offer a more thorough discussion of humic substances in soil.

The native organic fraction in soil sorbs many contaminants, particularly those that are strongly hydrophobic in nature. Sorption to humic substances, rather than to clay minerals, is responsible for the diminished bioavailability of many polycyclic aromatic hydrocarbons (PAHs), petroleum products, polychlorinated biphenyls (PCBs), and phthalic acid esters (Dragun, 1988). Figure 3 shows the rate and extent of biodegradation (measured by the cumulative oxygen consumption) of BEHP and diesel fuel added to whole soil, extracted humic substances, the mineral fraction (i.e., whole soil without humic substances), and a mineral salts solution. Bioavailability can be determined by the rate and extent of oxygen consumption in the reactors. The bioavailability of BEHP was limited by the presence of humic substances but completely unaffected by the mineral fraction of the silty loam. The bioavailability of diesel fuel was affected by the humic fraction (though less markedly than was BEHP) and was also somewhat affected by the mineral fraction of the silty clay loam.

The extent of sorption of an organic compound to humic substances is in direct proportion to the hydrophobicity of the compound and to the humic content of the soil. The hydrophobicity of a compound is often expressed by the octanol-water partition coefficient (K_{ow}). The K_{ow} is determined by measuring the ratio of equilibrium concentration of the test chemical in octanol (an arbitrarily chosen hydrophobic solvent) to that in water. The greater the K_{ow} value of an organic compound, the

Table 1. CEC and specific surface area for common soil constituents (from Dragun, 1988).

Soil Constituent	CEC (meq/100 g)	Specific Surface Area (m²/g)
Quartz sand	1–2	1–3
Kaolinite	3–15	5–39
Mica (illite)	10–40	100–200
Montmorillonite	80–150	700–800
Vermiculite	100–150	750–800
Humics	>200	1000–1200
Oxides and oxy-hydroxides	2–6	50–100

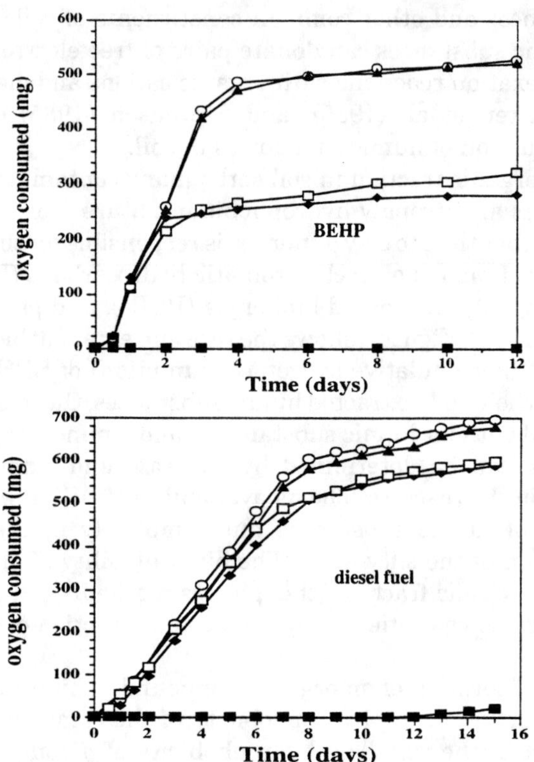

Figure 3. Oxygen consumption accompanying the biodegradation of BEHP and diesel fuel in the presence of whole soil (open squares), extracted humics (closed diamonds), the mineral fraction (closed triangles), a mineral salts solution (open circles), and for a killed control (closed squares) (Cassidy, 1995).

greater its hydrophobic nature. Values of log K_{ow} are often used as a qualitative indicator of the tendency of an organic contaminant to sorb to humic substances in soil. Table 2 lists the log K_{ow} values for some organic compounds.

Two mechanisms have been proposed to explain the nature of the retention of organic contaminants by humic substances. One mechanism is that the organic compounds are adsorbed onto the surfaces of the humic substances (Calvet, 1989). The second mechanism is that the hydrophobic nature of the humic substances dissolves or partitions the organic compound (Chiou, 1989). The term *hydrophobic bonding* is sometimes used to describe the partitioning process. It is likely that both mechanisms participate in the sorption of organic contaminants to humic substances.

Although sorption is thought to be a reversible process, the rate of release of sorbed contaminants from humic substances has been shown to control the rates of biodegradation of many contaminants in soils (Dragun, 1988; Lyman et al., 1992; Riser-Roberts, 1992; Rutherford et al., 1992). Many studies have shown that organic compounds sorb extensively to humic substances in soil and sediments and that the sorbed compounds exhibit limited bioavailability. Diminished bioavailability of organic compounds sorbed to soil humic substances has been shown for BEHP and other phthalic acid esters (Fairbanks et al., 1986; Gibbons and Alexander, 1989), phenanthrene (Manital and Alexander, 1991), and PAHs (Tiberg et al., 1992; van Afferden et al., 1992; Weissenfels et al., 1992). The humic substances in most soils are in intimate contact with clay minerals (Stevenson, 1985; Catroux and Schnitzer, 1987; Dragun, 1988; Lyman et al., 1992). A source of some confusion in the literature is whether humic substances or the clay minerals are directly responsible for controlling the bioavailability of sorbed organic contaminants in soil. Fitch and Du (1996) offered experimental evidence that humic acid competes with the interlayer region of 2:1 clays for sorption of organic compounds and can physically block access to the interlayer region.

Clay Minerals

Clay minerals also serve as solid surfaces for the sorption of organic contaminants. The bioavailability of contaminants sorbed to clay minerals (especially montmorillonite) is often the limiting factor in bioremediation of a site (Dragun, 1988; Alexander, 1994). Contaminants sorbed

Table 2. Log of octanol-water partition coefficients for some organic compounds (from Laane et al., 1987).

Chemical	log K_{ow}
Dioxane	−1.1
Acetone	0.23
Pyridine	0.71
Benzene	2.0
Toluene	2.4
Xylene	3.1
Diethyl phthalate	3.3
Diphenyl ether	4.3
Decane	5.6
Tetradecane	7.6
Dioctyl phthalate	8.8

to montmorillonite and other clay minerals often display diminished bioavailability relative to systems not containing the clay minerals (Geerdink, 1995). Figure 4 shows the rate and extent of biodegradation of added BEHP and diesel fuel (as measured by oxygen consumption) in the presence of several common soil constituents. The bioavailability of BEHP (determined by the extent of oxygen consumption) is controlled by humic substances but is unaffected by the mineral constituents tested. The bioavailability of diesel fuel, on the other hand, is influenced by the presence of sodium and calcium montmorillonite as well as humic substances. Kaolinite and quartz sand apparently play no role in controlling the bioavailability of BEHP or diesel fuel.

The clay mineral fraction varies considerably from soil to soil. Clay

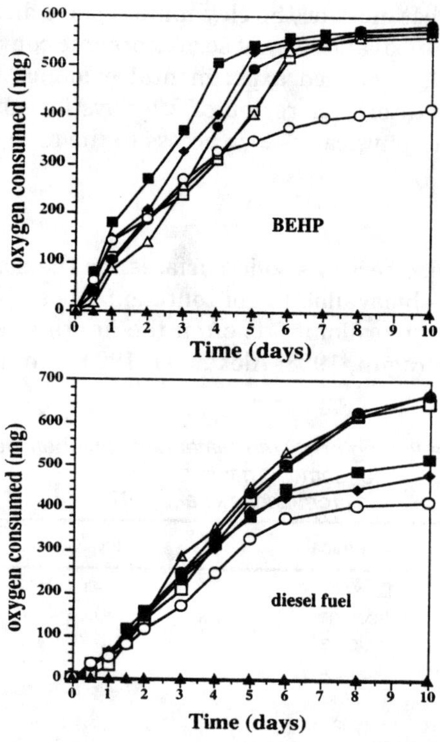

Figure 4. Bioavailability of BEHP and diesel fuel (determined as O_2 consumption) in presence of mineral salts (open squares), quartz sand (closed circles), kaolinite (open triangles), sodium and calcium montmorillonite (closed diamonds and squares, respectively), extracted humic substances (open circles), and a killed control (closed triangles) (Cassidy, 1995).

minerals play a predominant role in sorption of organic contaminants in most soils because of their high specific surface area compared to other soil minerals (Table 1). Clay minerals (phyllosilicates) are layered silicates composed of alternating layers of SiO_4 tetrahedral sheets and aluminum oxyhydroxide octahedral sheets. Typical clay minerals found in soils are comprised of units of tetrahedral-octahedral layers (1:1 clays) or of tetrahedral-octahedral-tetrahedral layers (2:1 clays). Figure 5 illustrates 1:1 and 2:1 structures and lists the names of common examples represented by each. The relative abundance of a particular clay mineral depends primarily on the parent rock from which the soil developed. Kaolinite, montmorillonite, and micas are the most commonly occurring clay minerals in soils (Gieseking, 1975; Hillel, 1980).

If the octahedral sheets contained only Al^{3+} and the tetrahedral sheets contained only Si^{4+}, the clays would be electrically neutral. However, when clays are formed, Fe^{2+} and Mg^{2+} often substitute for Al^{3+} in the octahedral layer. Sometimes (though less frequently) Al^{3+} substitutes for Si^{4+} in the tetrahedral layer. These processes are termed isomorphous substitution because they do not change the lattice structure and because they result in a net negative charge on the surface of the clay minerals. Isomorphous substitution is characteristic of montmorillonite, vermiculite, and other 2:1 clays but occurs only rarely in kaolinite. The negative charges on the surfaces of clay minerals give them a high CEC. Montmorillonite exhibits stacking of the individual 2:1 sheet structures with water and exchangeable cations in the interlamellar spaces (Figure 5). The thickness of the interlamellar space is less than one nanometer (10^{-9} m) but changes somewhat, depending on the type of cation present and the dielectric constant of the water in the interlamellar space. For this reason, montmorillonite is often referred to as an "expandable" or a "swelling" clay. The narrow openings of the interlamellar spaces of montmorillonite impose severe mass transfer limitations on contaminants sorbed to those surfaces.

Organic cations bind ionically to montmorillonite and other clay minerals with negatively charged surfaces by replacing the exchangeable cations. Nonionic, hydrophobic organics also sorb extensively to clay minerals, especially montmorillonite (Dragun, 1988; Lyman et al., 1992). Examples of hydrophobic organics that sorb extensively to montmorillonite include naphthalene (Mihelcic and Luthy, 1990), hexachlorohexane (Rijnaarts et al., 1990), diesel fuel and kerosene (Brown and Thomas, 1984), polycyclic aromatic hydrocarbons (PAHs) (Buchanan, 1964), and phthalic acid esters, including BEHP (Fairbanks et al., 1986). Extensive sorption onto clay surfaces does not necessarily hinder bioavailability, as long as the rate of desorption is equal to or greater than the rate of

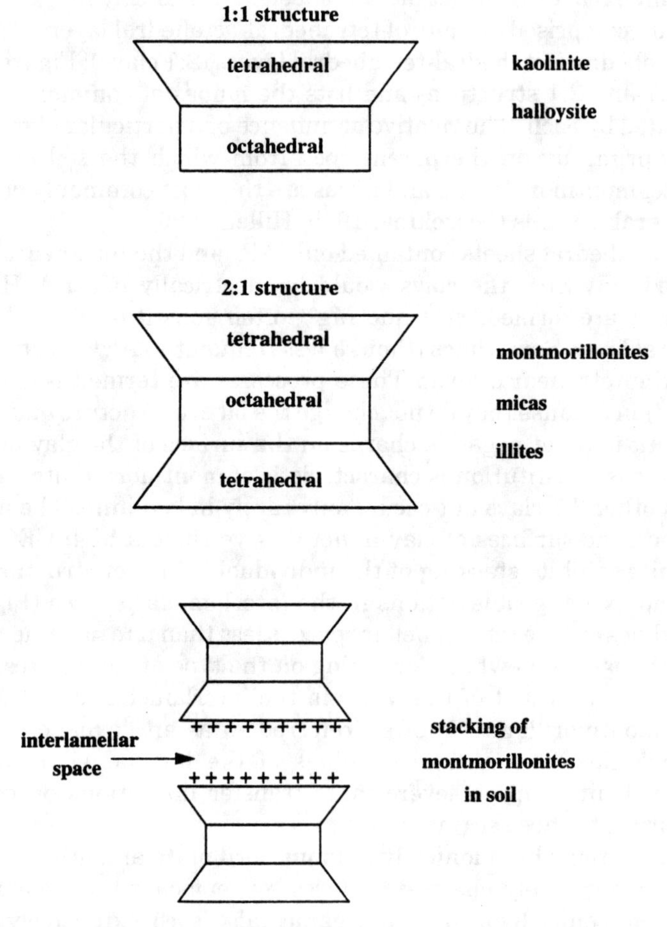

Figure 5. The structure and names of common 1:1 and 2:1 clay minerals in soil and the stacked structure of montmorillonites.

268

contaminant biodegradation. This is apparently the case for BEHP (Figure 2).

Hydrophobic organics are at a lower energy state when sorbed to solids (even those that are not hydrophobic) than in aqueous solution (Stumm and Morgan, 1981). This is the primary driving force for the sorption of hydrophobic organic contaminants to the surfaces of montmorillonite and other clay minerals. The less water-soluble an organic compound is, the greater will be its propensity to sorb to solid surfaces. Sorption of hydrophobic organics to clay minerals is maintained predominantly by van der Waals forces (Stumm and Morgan, 1981; Dragun, 1988; Lyman et al., 1992). The larger the organic molecule, the greater the influence of van der Waals attraction forces, and the greater the extent of sorption. Organic cations (e.g., amino acids) held ionically as exchangeable cations can sorb hydrophobic organics and serve as a bridge between the contaminant and the clay mineral surface.

NONAQUEOUS PHASE LIQUIDS

Many organic compounds that contaminate soil exist, not in the aqueous phase or sorbed to solids, but rather in liquids that are immiscible with water. Because microorganisms require an aqueous environment to function, biodegradation can occur only on the peripheries of the NAPL body that are in contact with soil water. This drastically limits the bioavailability of contaminants present in soil as NAPLs. The NAPL represents a long-term source of groundwater and air pollution because the contaminants present in the NAPL will continue to replace that which is biodegraded or transported away from the site in water or air. If the NAPL contaminant has a density greater than water, it will move downward and collect at the bottom of the aquifer. Such contaminants are called dense nonaqueous phase liquids (DNAPLs).

Remediation efforts have shown that NAPLs are quite persistent. It is often the case that an NAPL consists of several hydrophobic compounds. Under these conditions, many of the compounds in the NAPL may be biodegraded simultaneously (Foght et al., 1990), or some may be transformed only after more readily metabolized compounds are degraded (Oberbremer and Müller-Hurtig, 1989). Laboratory investigations have demonstrated that a compound is more resistant to biodegradation when present in an NAPL and that the same hydrophobic compound present in different NAPLs may be biodegraded at markedly different rates (Efroymson and Alexander, 1991). Biodegradation of one compound present in an NAPL may be inhibited by the presence of another compound in the NAPL (Inoue and Horikoshi, 1991).

Laboratory studies indicate that bacteria use only that fraction of a sparingly soluble compound that is in aqueous solution. Wodzinski and Johnson (1968) showed that the growth rates of PAH degrading bacteria increased progressively when grown on anthracene, phenanthrene, and naphthalene (i.e., as the solubility of the substrate increased). Other studies have shown that the growth rate of certain bacteria on sparingly soluble substrates is the same when the substrate is available only in dissolved form as when it is present in excess of its solubility (Wodzinski and Coyle, 1974; Wodzinski and Larocca, 1977). Another indication of the dependence of microbial growth on the presence of the substrate in the aqueous phase is the results from a study in which bacteria were found to grow readily on the dissolved portion of hydrophobic substrates but decreased drastically as soon as the dissolved concentration became nondetectable (Thomas et al., 1986; Stucki and Alexander, 1987).

Growth would be much limited if microorganisms had to rely on spontaneous partitioning of hydrophobic compounds to the aqueous phase. Some microorganisms grow directly on the surface of an NAPL (Rosenberg and Rosenberg, 1985; Efroymsom and Alexander, 1991). Many species that grow on sparingly water-soluble compounds produce surface-active or emulsifying agents, which convert the NAPLs to droplets 1 μm or smaller (Einsele et al., 1975). The increased surface area of the emulsified NAPL gives it a higher apparent or "pseudo" solubility, thus resulting in enhanced rates of biodegradation. Laboratory studies have shown that the rate of apparent solubilization caused by biosurfactants is enough to account for the substrate utilization and growth rates (Singer and Finnerty, 1984; Goswami and Singh, 1991). The biosurfactants that have been chemically characterized are proteins, polysaccharides, protein-polysaccharide complexes, and glycolipids (Rosenberg, 1986).

Figure 6 shows the results of a study in which surface tension, emulsification capacity, and oxygen uptake rates (a surrogate measure for microbial activity) were monitored during growth on BEHP in a mineral salts solution. Biosurfactant production during growth on BEHP is indicated by a reduction in surface tension and an increase in emulsification capacity.

The NAPL serves as a long-term reservoir of contaminants in soil. As biodegradation proceeds at the periphery of an NAPL, some inorganic nutrient (e.g., oxygen, nitrogen, and phosphorus) is likely to become the reactant that limits the rate of contaminant biodegradation. Once the inorganic nutrient is depleted by biological reactions, it must diffuse from other parts of the soil matrix or the atmosphere. Bioslurry treatment of soils mixes contaminated soil in water, which provides energy to aid in

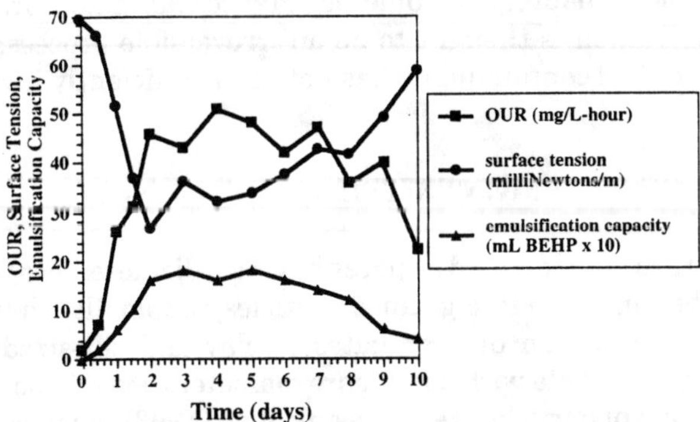

Figure 6. Oxygen uptake rate (OUR), surface tension, and emulsification capacity measured during microbial consumption of BEHP in a mineral salts solution. OUR is a surrogate measure for biological activity (Cassidy, 1995).

emulsification of the NAPL and replenishes the inorganic nutrients present at the NAPL-water interface. This is one reason for the better performance of bioslurry remediation compared with static biological reactors (e.g., solid-phase bioremediation).

HUMIFICATION

Contaminant molecules, or their metabolic intermediates, can also undergo microbially mediated covalent bonding reactions with themselves or with humic substances in soil (Stotzky, 1972; Bollag and Loll, 1983; Aiken et al., 1985; Stevenson, 1985; Kaestner et al., 1992). This polymerization process results in the production of more humic substances and is called humification. The microbiology of humification reactions is not well understood. It is thought, however, that one reason for these polymerization reactions is an attempt to reduce the toxicity of the contaminant or its metabolic products to the microorganisms involved (Bollag and Loll, 1983).

Humification has been shown to occur with PAHs (Qiu and McFarland, 1991; Kaestner et al., 1992) and with petroleum compounds (Stevenson, 1985). In instances where considerable humification takes place, it is difficult to distinguish a reduction in extractable contaminants due to biodegradation from incorporation into humic substances. In some cases, hydrolysis of the soil with strong acids or bases is necessary to yield the parent molecules or related metabolites, indicating that the contaminants were incorporated in the humic substances and not merely sorbed to the soil matrix (Singh and Agarwal, 1992). Humification strongly controls the bioavailability of some organic contaminants in soil (e.g., PAHs). Humification is thought to be an irreversible process; however, the fate of humified contaminants has not been sufficiently investigated.

ENTRAPMENT WITHIN MICROPORES

Micropores are often defined as pores having a diameter less than $1\,\mu$m. The higher the content of fine-grained particles, the smaller the pore sizes in a soil. The formation of aggregates of clay and silt-sized particles creates localized pockets with small pore diameters, even in soils with an even distribution of particle sizes. Hassink et al. (1993) reported that over 30% of the pore volume of some soils has pores with diameters smaller than 0.2 μm. The majority of pores in fine-grained aggregates are less than 1 μm. The interlamellar spaces of montmorillonite and other 2:1 clay minerals are less than 1 nanometer in thickness. Most soil microorganisms have a diameter ranging from 1–10 μm, which renders micropores inaccessible to microbial cells. The greater the content of montmorillonite and other 2:1 clay minerals in a soil, the greater will be the fraction of the total pore space consisting of micropores.

The small openings of micropores impose severe mass transfer resistances to the biodegradation of contaminants sorbed to the surfaces of these pores (Stotzky, 1972; Hattori and Hattori, 1976). Contaminants present within micropores must desorb from the pore surfaces and diffuse to the exterior of the aggregates or the stacked structure of the 2:1 clay minerals to be biodegraded. The desorbed molecules are subject to further sorption as they diffuse through the tortuous and narrow micropores. Extracellular enzymes and biosurfactants are also subject to being sorbed to the interlamellar surfaces, which strongly reduces their effectiveness.

Clay and silt particles aggregate naturally in soils (Hillel, 1980; Kemper and Rosenau, 1982; Blume, 1992). Humic substances are in intimate contact with clay minerals and serve as a glue to stabilize aggregates. Because hydrophobic organics readily adsorb to clay minerals and partition into humic substances, aggregates have the potential to concentrate organic contaminants in soil. Table 3 shows the results of the soil analyses performed on the bulk soil and aggregates isolated from the bulk soil. The CEC of the aggregates was greater than the bulk soil, which is to be expected from the higher clay content. The total organic content and the humic content of the aggregates were also greater than the bulk soil. The concentrations of BEHP and TPH were approximately two times greater in the aggregates than in the bulk soil.

Several studies have shown that the mass transfer resistances imposed

Table 3. Properties of silty loam contaminated with BEHP and petroleum compounds (from Cassidy, 1995).

Parameter	Bulk Soil	Aggregates
Bulk density (g/cm^3)	1.65	1.9
Percent clay (<2 microns)	12.0	30.1
Carbonate content (%)	28	12
CEC (meq/100 g)	10.3	15.0
Total organic content (%)	5.2	14.6
Humic content (%)	3.1	7.0
BEHP (g/kg soil sieved to 2 mm)	42.6	92.3
TPH (g/kg soil sieved to 2 mm)	17.7	34.4

by soil aggregates limit the bioavailability of hydrophobic organics in soil (Steinberg et al., 1987; Mihelcic and Luthy, 1990; Rijnaarts et al., 1990; Battaglia et al., 1991; Mueller et al., 1991a, 1991b; Scow and Alexander, 1992; Chung et al., 1993). Other studies have shown that reduction in aggregate size increases the bioavailability of contaminants in slurry and solid-phase (static pile) systems (Riijnaarts et al., 1990; Cassidy, 1995; Irvine and Cassidy, 1995). Figure 7 shows the effect of disaggregation (through mixing) on the biodegradation of BEHP in solid-phase bioreactors. Mixing provided disaggregation, which markedly enhanced the rate and extent of BEHP biodegradation relative to the well-aggregated, unmixed soil.

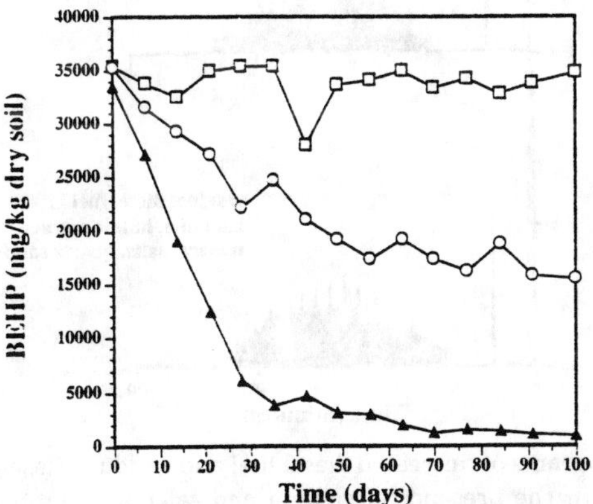

Figure 7. The effect of disaggregation on rate and extent of biodegradation of BEHP in aerobic solid-phase reactors. Soil that was mixed to provide disaggregation (closed triangles) shows a markedly greater rate and extent of BEHP biodegradation than soil that was not mixed (open circles) and a killed control (open squares) (Cassidy, 1995).

Results from several studies have shown that the presence of soil causes an increase in the relative ratio of linear to branched alkanes remaining after biodegradation (Wang and Bartha, 1990; Kleijntjens, 1991; Yocum, 1993; Geerdink, 1995). Linear alkanes are biodegraded more readily than branched alkanes (Morgan and Watkinson, 1989), and if both were equally bioavailable, the residual hydrocarbons after biodegradation would consist predominantly of branched alkanes. In the presence of soil, however, the relative amount of linear alkanes in the residual contamination is often considerably higher than in systems without soil. This is illustrated in Figure 8, which shows chromatograms of untreated diesel

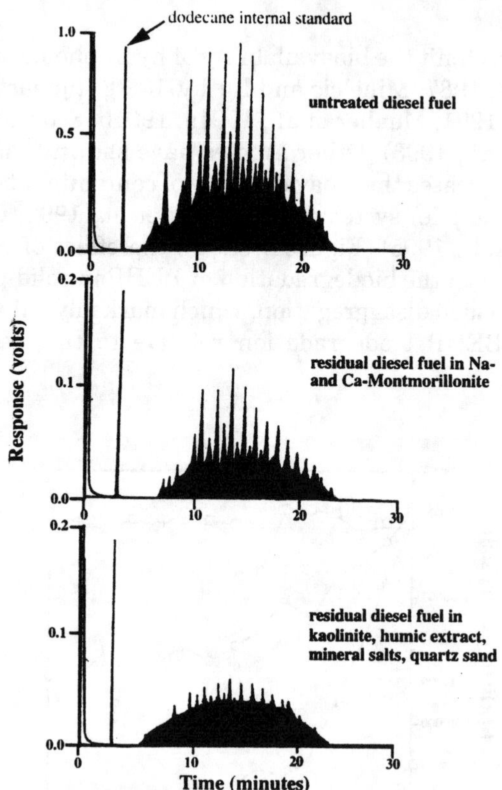

Figure 8. Chromatograms of untreated diesel fuel and residual diesel fuel after ten days' biological treatment in the presence of sodium and calcium montmorillonite, kaolinite, extracted humic substances, quartz sand, and a mineral salts solution. The sharp peaks ("grass") represent straight chained alkanes. The "hump" represents the branched alkanes, or unresolved complex mixture (UCM) (Cassidy, 1995).

fuel and residual diesel fuel after ten days of biological treatment in the presence of several common soil constituents. The sharp peaks in the chromatograms represent straight alkanes, and the "hump" represents the unresolved complex mixture (UCM), which is comprised predominantly of branched alkanes. Chromatograms from sodium and calcium montmorillonite showed that a considerable amount of linear alkanes that remained were not biodegraded. In contrast, the chromatograms from kaolinite, humic extract, quartz sand, and mineral salts showed a diesel fuel residue consisting almost entirely of branched alkanes.

The high relative content of linear alkanes after biodegradation in the presence of some soil constituents may be due to the slow diffusion of diesel fuel entrapped in aggregates into the bulk solution during biodegradation—the aggregates serving as a source of linear alkanes that have already been consumed in the bulk solution (Geerdink, 1995). Another possibility is that linear alkanes have a greater mobility into the interlamellar spaces of montmorillonite. Sorption to the interlamellar surfaces may accumulate linear alkanes, which then slowly diffuse out of the interlamellar spaces of the 2:1 clay structure into the bulk solution during biodegradation.

AGE OF CONTAMINATION

The longer some contaminants remain in soil, the more resistant they become to desorption and biodegradation. This persistent fraction of contaminants is sometimes called aged or bound residues. A greater resistance to the desorption of residual contamination in soil has been reported for aldrin-dieldrin (Elgar, 1975), trichloroethylene (Pavlostathis and Mathavan, 1992), simazine (Scribner et al., 1992), and the fumigant EDB (Steinberg et al., 1987). The resistance to desorption of compounds present in soil for long periods of time has been correlated with increased resistance to biodegradation (Scribner et al., 1992; Steinberg et al., 1987).

Figure 9 shows the results of experiments with the biodegradation of BEHP after contamination ages ranging from 1–6 months in sandy soils with two different humic contents. The soil with a low humic content showed no noticeable decrease in BEHP biodegradation with increasing age of contamination. In contrast, the sandy soil with a high humic content showed a marked reduction in the rate and extent of BEHP biodegradation with increasing age of contamination. The results in Figure 9 show that a reduction in contaminant bioavailability can be observed over a period of months.

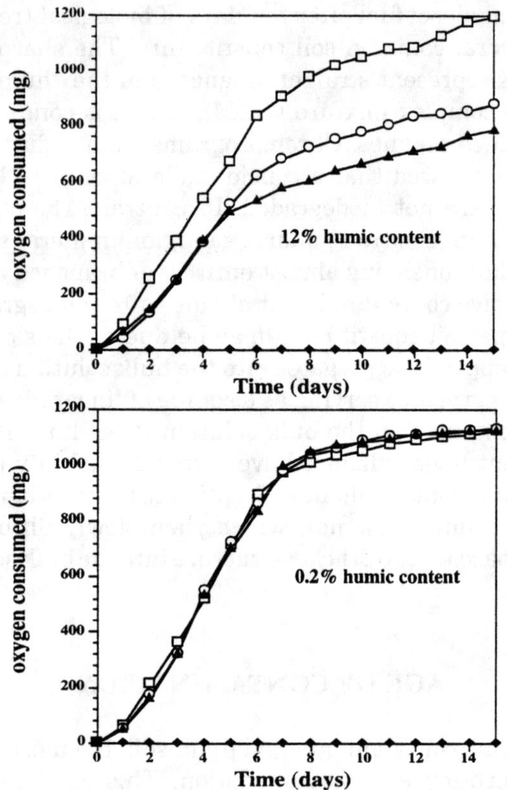

Figure 9. Biodegradation (measured as oxygen consumption) of BEHP in a sandy soil with a humic content of 12% and a sandy soil with a humic content of 0.2% for contamination times of 1 month (open squares), 4 months (open circles), 6 months (closed triangles), and a killed control (closed diamonds) (Cassidy, 1995).

EFFECTS OF ORGANIC CONTAMINANTS ON SOIL PROPERTIES

The presence of organic contaminants can change some chemical and physical properties of a soil, which can in turn affect contaminant bioavailability. Miehlich and Wagner (1989) showed that artificially contaminating four different soils with diesel fuel and lubricating oil reduced the CEC of all the soils tested. The greater the amount of hydrocarbons added, the greater was the observed decrease in CEC. This effect is thought to be due to adsorption of the contaminants to the surfaces of clay minerals, which blocks potential cation exchange sites.

In a similar study with the same four soils, Goetz and Wiechmann (1989) observed that, as the amount of added diesel fuel and lubricating oil was increased, the soil aggregates showed a decrease in compressibility and an increase in stability. The compressibility of an aggregate is the amount of compressive strain that takes place under a given stress. Stability refers to the resistance of an aggregate to disruptive forces (e.g., shear and compression). These effects can also be attributed to adsorption and partitioning of hydrophobic organics to clay minerals and humic substances present in the aggregates. The significance of an increase in aggregate stability due to the presence of organic contaminants is that more energy is required during biological treatment to disrupt aggregates in order to increase contaminant bioavailability.

An increase in the presence of hydrophobic organics in the interlamellar spaces of expandable 2:1 clays (e.g., montmorillonite) results in a decrease in the dielectric constant of the solution, causing the interlamellar spacing to decrease (Dragun, 1988; Blume, 1992). If hydrocarbons are introduced to the soil environment, the interlamellar spacing of 2:1 clays will decrease (Dragun, 1988; Brown and Thomas, 1984). A decrease in the interlamellar spacing of 2:1 clays causes a reduction in volume of materials containing these clays. Shrinkage cracks result that cause an increase in bulk hydraulic conductivity. This phenomenon is partly responsible for the premature leaking of liners composed of clay minerals. Brown and Thomas (1984) showed that treatment of montmorillonite with diesel fuel and kerosene increased the hydraulic conductivity of the clays by approximately 1500 times and 4000 times, respectively.

SUMMARY

Interactions between organic contaminants and the soil matrix control the fate and transport of contaminants in the subsurface and the effectiveness of in situ and on-site biological remediation techniques. Rates of contaminant biodegradation in soil are typically less than those observed in aqueous systems because a portion of the contaminants becomes bound to or sequestered in the soil matrix. The bound and sequestered contaminants are not in a form that is readily available to microorganisms. Limited bioavailability of organic contaminants in soil results from (a) sorption to humic substances and mineral surfaces, (b) presence as NAPLs, (c) covalent bonding with humic substances (humification), and (d) entrapment within micropores of fine-grained soil aggregates and/or 2:1 clay minerals (e.g., montmorillonite, vermiculite).

The rate at which a soil-bound contaminant becomes bioavailable depends on the nature of the contaminant and the soil. As the hydrophobic nature of a contaminant increases, so does its tendency to sorb onto the mineral and humic fraction of soil and to exist as a NAPL. A high clay content (especially 2:1 clays, such as montmorillonite) provides a large surface area for sorption of contaminants and increases the stability of fine-grained aggregates in soil. Aggregates have a higher humic content than the soil as a whole and tend to concentrate hydrophobic organic compounds. Interlamellar spaces of 2:1 clay minerals and small pore spaces in aggregates impose a severe restriction on the mass transfer of contaminants and other biological reactants (e.g., oxygen, nitrogen, and phosphorus). Humic substances in soil provide a large surface area and a hydrophobic sink for contaminant sorption and increase the stability of fine-grained aggregates. Compounds sorbed to humic substances can become involved in humification reactions, which can irreversibly polymerize contaminants to form more humic substances. Contaminant bioavailability can become increasingly reduced with increasing age of contamination.

REFERENCES

Aiken, G. R., D. M. McKnight, R. L. Wershaw, and P. MacCarthy (Eds.). 1985. "An Introduction to Humic Substances in Soil, Sediment and Water." In: *Humic Substances in Soil Sediment and Water.* John Wiley and Sons, New York.

Alexander, M. 1994. *Biodegradation and Bioremediation.* Academic Press, San Diego, CA.

Battaglia, A., D. J. Morgan, D. G. Linz, and T. D. Hayes. 1991. "Application of the GRI Accelerated Protocol to Contaminated Soils from Manufactured Gas Plant Sites." In: P. T. Kostecki, E. J. Calabrese, and M. Bonazountas (eds.), *Hydrocarbon Contaminated Soils.* Lewis Publishers, Chelsea, MI.

Blume, H. P. 1992. *Handbuch des Bodenschutzes: Bodenökologie und -Belastung; Vorbeugende und Abwehrende Schutzmassnahmen,* Auflage 2, Ecomed Verlasgesellschaft, GmbH, Landsberg, Germany.

Bollag, J. M., and M. J. Loll. 1983. "Incorporation of Xenobiotics into Soil Humus." *Experientia,* 39:1221–1231.

Brown, K. W. and J. C. Thomas. 1984. "Conductivity of Three Commercially Available Clays to Petroleum Products and Organic Solvents." *Hazardous Waste,* 1:545–553.

Buchanan, P. N. 1964. "Effect of Temperature and Adsorbed Water on Permeability and Consolidation of Sodium and Calcium Montmorillonite." Ph.D. Dissertation, Texas A&M, College Station, TX.

Calvet, R. 1989. *Environmental Health Perspective,* 83:145–177.

Cassidy, D. P. 1995. "Bioremediation of Soils Contaminated with Hydrophobic

Compounds Using Slurry and Solid Phase Techniques." Doctoral Dissertation. University of Notre Dame, Notre Dame, Indiana.

Catroux, G. and M. Schnitzer. 1987. "Chemical, Spectroscopic and Biological Characteristics of the Organic Matter in Particle Size Fractions Separated from an Aquisol." *Soil Sci. Soc. Am. Jour.*, 51(5):1200–1206.

Chiou, C. T. 1989. In: *Reactions and Movement of Organic Chemicals in Soils.* (B. L. Sawhney and K. Brown, eds.), Soil Science Society of America, Madison, WI, pp. 1–29.

Chung, G. Y., J. McCoyb, and K. M. Scow. 1993. "Criteria to Assess when Biodegradation Is Kinetically Limited by Intraparticle Diffusion and Sorption." *Biotechnol. and Bioeng.*, 41:625–632.

Dragun, J. 1988. *The Soil Chemistry of Hazardous Materials.* Hazardous Materials Control Resources Institute. Silver Springs, Maryland.

Efroymson, R. A., and M. Alexander. 1991. *Appl. Environ. Microbiol.*, 57:1441–1447.

Einsele, A., H. Schneider, and A. Fiechter. 1975. *Journ. Ferment. Technol.*, 53:241–243.

Elgar, K. E. 1975. *Environ. Quality and Safety*, 3:250–257.

Fairbanks, B. C., G. A. O'Connor, and S. E. Smith. 1986. "Fate of Di-(2-ethylhexyl) Phthalate in Three Sludge Amended Soils." *Journ. of Environmental Quality*, 18:215–229.

Fitch, A. and J. Du. 1996. "Solute Transport in Clay Media: Effect of Humic Acid." *Environmental Science & Technology*, 30(1):12–15.

Foght, J. M., Fedorak, P. M., and D. W. S. Westlake. 1990. *Canadian Journal of Microbiology*, 36:169–175.

Geerdink, M. J. 1995. "Kinetics of the Microbial Degradation of Oil in Soil Slurry Reactors." Doctoral Dissertation, Technical University of Delft, The Netherlands.

Gibbons, J. A. and M. Alexander. 1989. "Microbial Degradation of Sparingly Soluble Organic Chemicals: Phthalate Esters." *Environ. Toxicol. Chem.*, 8:283–291.

Gieseking, J. E. (editor). 1975. *Soil Components. Vol. 1: Organic Components, Vol. 2: Inorganic Components.* Springer-Verlag, New York.

Goetz, D. and H. Wiechman. 1989. "Einfluss von Oelkontaminationen auf bodenphysikalische und -mechanische Eigenschaften von kontaminierten Standorten." Sonderforschungsbereich 188 der DFG-Reinigung kontaminierter Boedon: Arbeits- und Ergebnisbericht 1989–90–91; Technische Universitaet Hamburg-Harburg. Project D2, pp. 355–390.

Goswami, P. C., and H. D. Singh. 1991. *Biotechnol. Bioeng.*, 37:1–11.

Greenland, D. J., and M. H. B. Hayes (eds.). 1981. *The Chemistry of Soil Processes.* John Wiley and Sons, New York.

Hassink, J. L., A. Bouwman, K. B. Zwart, and L. Brussard. 1993. *Soil Biol. Biochem.*, 25:47–55.

Hattori, T., and R. Hattori. 1976. "The Physical Environment of Soil Microbiology: An Attempt to Extend Principles of Microbiology to Soil Microorganisms." *Critical Reviews in Microbiology*, 4:423–462.

Hillel, D. 1980. *Fundamentals of Soil Physics*. Academic Press, Inc., Harcourt Brace Jovanovich Publishers. Orlando, Florida.

Inoue, A. and K. J. Horikoshi. 1991. *Journ. Ferment. Bioeng.*, 71:194–196.

Irvine, R. L. and D. P. Cassidy. 1995. "Periodically Operated Bioreactors for the Treatment of Soils and Leachates." In: *Biological Unit Processes for the Treatment of Soils and Leachates* (R. B. Hinchee, G. D. Sayles, and R. S. Skeen, editors). pp. 289–298. Battelle Press, Columbus, Ohio.

Jaffe, P. R., and R. A. Ferrara. 1983. "Desorption Kinetics in Modeling of Toxic Chemicals." *Journal of Environmental Engineering*, 109(40):859–867.

Kaestner, M., G. Schaefer, M. Breuer, and B. Mahro. 1992. "Microbial Degradation of PAH in Soil-Mineralization, Biotransformation and Formation of Bound Residues." pp. 281–283. In: *Preprints from the International Symposium on Soil Decontamination Using Biological Processes*. DECHEMA. Karlsruhe, Germany, Dec. 6–9.

Kemper, W. D. and R. C. Rosenau. 1982. "Aggregate Stability and Size Distribution." In: A. Klute (editor), *Methods of Soil Analysis, Part 1: Physical and Mineralogical Methods*. Soil Science Society of America, Madison, WI.

Kleijntjens, R. H. 1991. "Biotechnological Slurry Process for the Decontamination of Excavated Polluted Soils." Ph.D. Dissertation, Delft University of Technology, Delft, the Netherlands.

Laane, C., Boeren, S., Hilhorst, R., and C. Veeger. 1987. In: *Biocatalysis in Organic Media* (C. Laane, J. Tramper, and M. D. Lilly, eds.), pp. 65–84. Elsevier, Amsterdam.

Lyman, J., P. J. Reidy, and B. Levy. 1992. *Mobility and Degradation of Organic Contaminants in Subsurface Environments*. C. K. Smoley, Inc., Chelsea, Michigan.

Manital, V. B., and S. Alexander. 1991. "Factors Affecting the Microbial Degradation of Phenanthrene in Soil." *Appl. Microbiol. Biotechnol.*, 35:401–405.

Miehlich, G. and A. Wagner. 1989. "Veraenderung bodenchemischer Eigenschaften durch Oelverunreinigung und-dekontamination." Sonderforschungsbereich 188 der DFG-Reinigung kontaminierter Boedon: Arbeitsund Ergebnisbereicht 1989–90–91; Technische Universitaet Hamburg-Harburg. Project D1, pp. 325–354.

Mihelcic, J. R. and R. G. Luthy. 1990. "Sorption and Microbial Degradation of Naphthalene in Soil-Water Suspensions under Denitrifying Conditions." *Environ. Sci. & Technol.*, 25(1):169–177.

Morgan, P. and R. J. Watkinson, 1989. "Hydrocarbon Degradation in Soils and Methods for Soil Biotreatment." *CRC Critical Review Biotechnol.*, 8:305–333.

Mueller, J. G., S. E. Lantz, B. O. Blattmnan, and P. J. Chapman. 1991a. "Bench-Scale Evaluation of Alternative Biological Treatment Processes for the Remediation of Pentachlorophenol- and Creosote-Contaminated Materials: Slurry Phase Bioremediation." *Environ. Sci. & Technol.*, 25:1045–1055.

Mueller, J. G., S. E. Lantz, B. O. Blattmnan, and P. J. Chapman. 1991b. "Bench-Scale Evaluation of Alternative Biological Treatment Processes for the Remediation of Pentachlorophenol- and Creosote-Contaminated Materials: Solid Phase Bioremediation." *Environ. Sci. & Technol.*, 25:1055–1065.

Oberbremer, A. and R. Müller-Hurtig. 1989. *Appl. Microbiol. Biotechnol.*, 31: 582–586.

Pavlostathis, S. G. and G. N. Mathavan. 1992. *Environ. Sci. Technol.*, 26:532–538.

Qiu, X., and M. J. McFarland. 1991. Bound Residue Formation in PAH Contaminated Soil Composting Using *Phanerochaete chrysosporium*." *Haz. Waste & Haz. Mat.*, 8:115–126.

Rijnaarts, H. M., A. Bachmann, J. C. Jumelet, and A. J. B. Zehnder. 1990. "Effect of Desorption and Intraparticle Mass Transfer of the Aerobic Biomineralization of Hexachlorohexane in a Contaminated Calcareous Soil." *Environ. Sci. & Technol.*, 24(9):1349–1354.

Riser-Roberts, E. 1992. *Bioremediation of Petroleum Contaminated Sites.* CRC Press, Inc., Boca Raton, FL.

Rosenberg, E. 1986. *CRC Crit. Rev. Biotechnol.*, 3:109–132.

Rosenberg, M. and E. Rosenberg. 1985. *Oil Petrochem. Pollut.*, 2:155–162.

Rutherford, D. W., C. T. Chiou, and D. E. Elke. 1992. "Influence of Soil Organic Matter Composition on Partition of Organic Compounds." *Environ. Sci. & Technol.*, 26:336–340.

Schweisfurth, R. 1988. *Angewandete Mikrobiologie der Kohlenwasserstoffe in Industrie und Umwelt*, volume 164, Ehnigen Expert-Verlag.

Scow, K. A. and M. Alexander. 1992. "Effect of Diffusion on the Kinetics of Biodegradation: Experimental Results with Synthetic Aggregates." *Soil Sci. Soc. Am. Jour.*, 56:128–134.

Scribner, S. L., Benzing, T. R., Sun, S., and S. A. Boyd. 1992. *Journ. Environ. Qual.*, 21:115–120.

Singer, M. and W. R. Finnerty. 1984. In: *Petroleum Microbiology* (R. M. Atlas ed.) pp. 1–59. Macmillan, New York.

Singh, D. K. and H. C. Agarwal. 1992. *Journ. Agric. Food Chem.*, 40:1713–1716.

Steinberg, S. M., Pignatello, J. J., and B. L. Sawhney. 1987. *Environ. Science Technol.*, 21:1201–1208.

Stevenson, F. J. 1985. "Geochemistry of Soil Humic Substances." In: *Humic Substances in Soil, Sediment and Water.* John Wiley and Sons, New York.

Stevenson, F. J. 1994. *Humus Chemistry: Genesis, Composition, Reactions.* Wiley and Sons, New York, NY.

Stotzky, G. 1972. "Activity, Ecology, and Population Dynamics of Microorganisms in Soil," *CRC Critical Reviews in Microbiology,* November 1972, pp. 59–137.

Stucki, G. and M. Alexander. 1987. *Appl. Environ. Microbiol.,* 53:292–297.

Stumm, W. and J. J. Morgan. 1981. *Aquatic Chemistry.* John Wiley and Sons, New York.

Tchobanoglous, G. and E. D. Schroeder. 1985. *Water Quality.* Addison Wesley Publishers, USA.

Thomas, J. M., Yordy, J. R., Amador, J. A., and M. Alexander. 1986. *Appl. Environ. Microbiol.,* 52:290–296.

Tiberg, E., W. Fischer, A. Mürsen, H. Knoblauch, U. Genz, E. Kaun, and R. Adam. 1992. "Studies of the Microbial Degradation of Hydrocarbons (Mineral Oil and PAHs) in Different Contaminated Soils." pp. 387–391. In: Preprints from the *International Symposium on Soil Decontamination Using Biological Processes.* DECHEMA. Karlsruhe, Germany, Dec. 6–9.

van Afferden, M., M. Beyer, and J. Klein. 1992. "Significance of Bioavailability for the Microbial Remediation of PAH-Contaminated Soils." pp. 605–610. In: Preprints from the *International Symposium on Soil Decontamination Using Biological Processes.* DECHEMA. Karlsruhe, Germany, Dec. 6–9.

Wang, X. and R. Bartha. 1990. "Effects of Bioremediation on Residues: Activity and Toxicity in Soil Contaminated by Fuel Spills." *Soil Biol. Biochem.,* 22(4):501–505.

Weissenfels, W. D., H. J. Klewer, and H. J. Langhoff. 1992. "Adsorption of Polycyclic Aromatic Hydrocarbons (PAHs) by Soil Particles: Influence on Biodegradability and Biotoxicity." *Applied Environ. Biotechnol.,* 36:689–696.

Wodzinski, R. S. and J. E. Coyle. 1974. "Physical State of Phenanthrene for Utilization by Bacteria," *Applied Microbial Technology,* V. 27, pp. 1081–1084.

Wodzinski, R. S. and M. J. Johnson. 1968. *Appl. Microbiol.,* 16:1886–1891.

Wodzinski, R. S. and D. Larocca. 1977. *Appl. Environ. Microbiol.,* 33:660–665.

Yocum, P., 1993. "Bioremediation of Leachate and Soil Contaminated with Petroleum Products." Doctoral Dissertation, University of Notre Dame, Notre Dame, IN.

Microorganisms and Their Role in Soil

Ludwig Davids, Hans-Curt Flemming, and Peter A. Wilderer
Institute of Water Quality and Waste Management
Technical University of Munich
Am Coulombwall
D-85748 Garching, Germany

INTRODUCTION

Soil is undoubtedly the most complex of all microbial habitats. Microorganisms play a significant role in soil formation, which occurs by the action and interaction of five factors: (1) soil organisms, (2) climate, (3) topography, (4) parent material, and (5) time. The physical and chemical breakdown of rocks to fine particles with large surface areas and the accompanying release of plant nutrients initiate the soil-forming process.

Several major nutrients that are deficient in the early stages of the process are carbon, nitrogen, phosphorus, and sulfur; therefore, the initial colonizers of soil parent material are usually organisms capable of photosynthesizing, nitrogen fixing and releasing phosphorus and sulfur from insoluble forms. These are predominantly the cyanobacteria, also known as blue-green algae. Many of the rock colonizing microorganisms are involved in microbial weathering of stone. There they live in biofilms (Krumbein, 1988). As an interfacial process, weathering—similar to metal corrosion—depends on physical factors such as light, temperature, abrasion, and chemical factors such as pH, E_h, and the concentration of salts, O_2 and CO_2. Biofilms will influence these and represent an important factor in solubilization of metal ions and decomposition of minerals (Eckhardt, 1985). Microorganisms colonize the pores of natural stone (as well as those of concrete), and form endolithic colonies. The microbial penetration rate of sandstone can be as high as 0.47 cm/h (Jennemann et al., 1985). Microorganisms are involved in the process of turning mineral components of rock into bioavailable material. This is a prerequisite for

283

the development of plants. After higher vegetation has become established, a continuum of soil processes produces the dynamic mixture of living and dead cells, soil organic matter (SOM), and mineral particles in sufficiently small sizes, to permit the intimate colloidal interactions characteristic of soil.

A modern definition of soil relates to the earth's surface layer exploited by plant roots. Microbial growth occurs to much greater depth; former assumptions that deeper layers of the underground are sterile had to be abandoned. Paul and Clark (1989) point out that living organisms are found in oil wells, amidst rocks where they have been protected from surface environments for millions of years. Active denitrification occurs in subsoils much below the rooting depth if a source of carbon is percolated downward together with nitrate. There appear to be no known natural areas on the earth's surface, with the exception of active volcanoes, where microbial life is absent. Microbes are present in the Gobi desert, where diurnal temperature fluctuations can attain 50°C, at polar rocks at −50°C (Atlas and Bartha, 1987) and in sulfur-containing wells on the sea floor at 115°C (Stetter et al., 1985).

The soil microbiologist was primarily concerned with those areas of the earth's surface that, when undisturbed, constitute major terrestrial ecosystems, and when managed provide arable soils or other resources. Soil organisms show their greatest diversity of species and usually their largest populations in productive soils. In the last decades, soil microbiology was increasingly concerned with the effects of pollutants to soil microflora and various methods the biological sanitation of polluted soils with the help of microorganisms.

The size of the microbial biomass usually shows a direct correlation with the amount of plant growth (the primary productivity) and soil organic matter (SOM) levels (Paul and Clark, 1989). Good discussions of the microbial environment have been given by McLaren and Skujins (1968), Lynch and Poole (1979), Focht and Martin (1979), and Chen and Avnimelech (1986).

SOIL AS A MICROBIAL HABITAT

Soil Structure

A prominent feature of soil is its heterogeneity, both in space and time. Soil consists of mineral particles of various sizes, shapes and chemical characteristics, together with plant roots, the living population, and an organic matter component in various stages of decomposition. Soil gases, pore water, and dissolved minerals complete the soil habitat. An understanding of the soil structure requires knowledge of spatial arrangements. These relate to both the size and shape of the components. The relative dimensions of components of the soil matrix range from 2 mm or greater for macroaggregates to fractions of less than a micrometer for bacteria and colloidal particles. Enzymatic and other molecular reactions occur at size dimensions at least orders of magnitude smaller.

Soil aggregation is one of the most important factors controlling microbial activity and SOM turnover. Aggregate formation is initiated when microflora and roots produce filaments and polysaccharides that combine with clays to form organic matter-mineral-complexes. Soil structure is created when physical forces (drying, shrink-swell, freeze-thaw, root growth, animal movement, and compaction) mold the soil into aggregates. Figure 1 shows schematically the arrangement of microorganisms within an aggregate. Although this diagram is simplified and size relationships somewhat altered, it shows how pore sizes are affected by distances between particles and how organic matter may be protected. Organic matter in portions of the aggregate will not be mineralized if it is physically separated from the microorganisms and their enzymes.

Most organisms exist on the outside of aggregates and in the small pore spaces between them; relatively few reside within the aggregate. Those that do usually remain there from the time of its formaton to the time of its disruption. Electron micrographs and calculations of microbial density indicate that microorganisms occupy less than 1% of the total available pore space. Pore neck sizes determine accessibility to pores by organisms according to their body sizes. Occupancy is also affected by the water content of the pore. Pores as small as a few micrometers in diameter and, soaked with water are generally suitable for bacteria. Fungi invade somewhat larger pores. Pore size limits the availability of the fauna to move around; in turn, this affects grazing on the soil microflora. Microaggregates (25–250 μm) are subunits of macroaggregates (>2 mm), the interstices of which form macropores. The macropores provide refuges for the microfauna, protecting them from the larger predatory organisms.

Soil aggregates and their constituent clays influence the interaction of enzymes with their substrates. The clay particle with its large external and internal surface aras is capable of adsorbing enzymes such as urease and protease. Enzymes adsorbed to clay or intertwined with humate constituents are protected from hydrolysis by other enzymes. This is due to the structure of the extracellular polymer substances (EPS), known as slime, which bacteria excrete when immobilized. Adsorption alone makes the catalytic site of an enzyme less available; thus, immobilization in a hydrated matrix such as EPS is an ecological advantage.

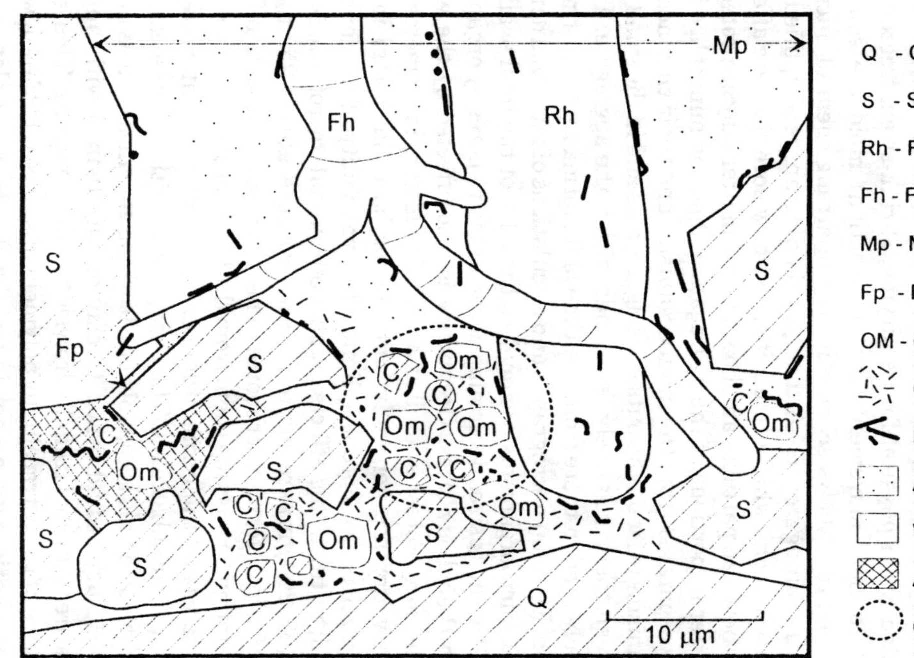

Figure 1. Section through a soil crumb showing microhabitats and patchy distribution of bacterial microcolonies. The section also shows the occurrence and air within the spaces.

Soil Atmosphere

The major gases in the soil atmosphere are those found in the atmosphere, namely nitrogen, oxygen and carbon dioxide, and methane. Gases arising from biological activity, such as nitrogen oxides, may at times be present. Because of their high reactivity with soil components and their susceptibility to biological activity, they are usually transitory. However, they can play an important role in maintaining temporal steady-state conditions under high biological activity. In well-aerated soils, the oxygen content seldom falls below 18–20%, and CO_2 seldom rises above 1–2%. Given a clay texture, however, and high moisture content coupled with high microbial activity, the CO_2 content of the soil atmosphere may reach as high as 10%.

The extent of aeration of soil can be inferred from the water content. Mineral particulates usually have an average density of 2.6 g cm^{-3}. The bulk density of surface soil generally ranges from 0.9–1.3 g cm^{-3}. Therefore, mineral soils usually consist of pores of assorted sizes, which account for 50–60% of the total volume. The water content of a soil at -0.01 megapascals (MPa) ranges from 15–30% for sandy loams and from 40–45% for clays. The difference between the total pore space and the water content represents the air-filled space. A minimum air-filled pore space of 10% of the total volume is commonly considered necessary for adequate aeration. Clay soils with 45% water may not have enough air spaces for aeration if their bulk density is 1.3 and total pore space is only 50% (Paul and Clark, 1989).

It has been determined that the change from aerobic to anaerobic metabolism occurs at O_2 concentrations of less than 1%. The overall aeration of soil is not as important as that of the individual crumbs and aggregates, however. Calculations show that water-saturated soil crumbs with a radius larger than 3 mm contain no O_2 in the center. The fact that anaerobic processes such as denitrification and sulfate reduction occur in many soils indicates that anaerobic microsites must be common. Further evidence that anaerobic microsites exist can be deduced from the common occurrence of anaerobic bacteria such as *clostridia* in the upper layers of soil. Several studies have shown that the population of anaerobic bacteria in the upper few centimeters of soil can be as much as ten times their number at greater depths (Tiedje et al., 1984). Aerobic bacteria play a precursor role in producing an environment for anaerobes. During their initial growth within a microsite oxygen is consumed, thus, allowing anaerobes to develop. This has been shown in Figure 2 recently by microelectrode measurements of biofilms.

Some soil organisms have become adapted on a geological time scale to

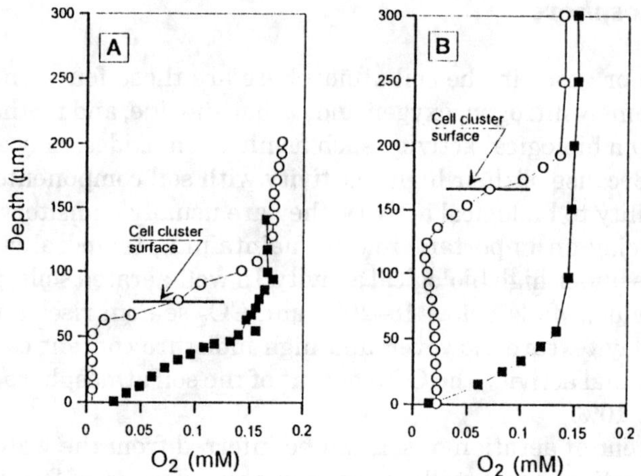

Figure 2. Oxygen profiles and mapping in a microbial biofilm (after de Beer et al., 1994).

the enhanced level of CO_2 that exists in the soil atmosphere. Certain species of fungi show a preference for a soil depth of 10–20 cm. Nitrifying bacteria also prefer CO_2 levels greater than those found aboveground. With experimentally controlled atmospheres in closed containers, little nitrification occurs with CO_2 at the atmospheric level of 0035% compared to that occurring at the 0.07–0.23% level commonly present in soil (Wong-Chong and Loehr, 1978).

Soil Water

Soil water affects not only the moisture available to organisms, but also the soil aeration status, the nature and amount of soluble materials, the osmotic pressure, and the pH of soil solution. The presently accepted terminology for soil water characteristics is based on the concept of matrix and osmotic potentials. Matrix potentials are attributed to the attraction of water to solid surfaces. Since this reduces the free energy of water, matrix potentials are negative. Because water is a solvent, solutes in the soil also reduce the free energy of water and create another negative potential, the osmotic potential. The concept of soil water potential relates to basic definitions of the free energy involved (Stumm and Morgan, 1981) and is expressed in pascals, usually given in megapascals (MPa).

Soil water has commonly been described as existing in three forms: gravitational, capillary, and hygroscopic. Gravitational water is drawn through the soil by gravitational forces. This will occur after irrigation or heavy rain. Immediately after gravitational water has drained away, the soil water is at field capacity. The micro-, or capillary, pores are, however, still filled with water that is available for plant and microbial growth. The matrix potential of this water will be −0.01 to −0.03 MPa. As water is lost from soil, either by evaporation or transpiration, plants will not be able to obtain enough to remain turgid both night and day and will be said to be wilted. Hygroscopic soil water is variously defined as that adsorbed by a dry soil from an atmosphere of high relative humidity, as that remaining in soil after air drying, or as that held by a soil when in equilibrium with a specified relative humidity (RH) and temperature, usually 98% RH and 25°C. These definitions are now obsolete, as is the term pF (logarithm of the soil water tension expressed in centimeter height of a column of water). The description of soil water availability relative to its matrix potential is a much more useful concept because it can be reproducibly measured and is scientifically based (Paul and Clark, 1989). The combined matric and osmotic components of soil water determine the stress against which an organism must work to obtain water. Generally, microbial activity in soil is optimal at −0.01 MPa and decreases as the soil becomes either waterlogged near zero water potential or more droughty at large, negative water potentials (Figure 3).

Fungi are generally more tolerant of more negative water potentials (greater water stress) than are bacteria. Table 1 shows differences, among microorganisms, in their upper tolerance levels to moisture potential. The nitrifiers, as typified by *Nitrosomonas*, are less tolerant of stress than are the ammonifiers, such as *Clostridium* and *Penicillium*. Ammonia may accumulate in droughty soil because the nitrifiers cannot operate at water

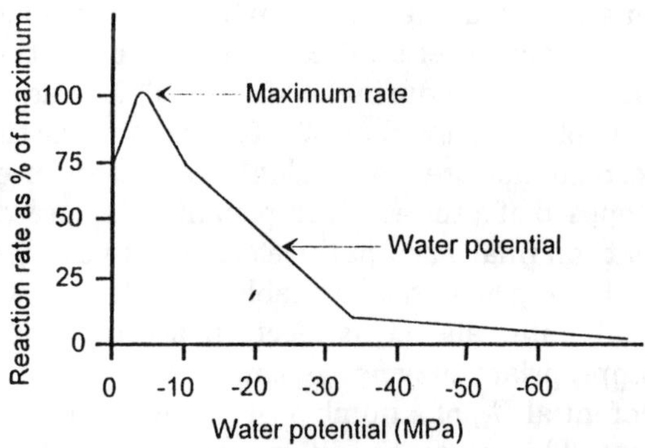

Figure 3. Relative microbial reaction rates at various moisture stresses (after Paul and Clark, 1989). Because sands and clays have different retention characteristics, the curve would occur at much lower water contents in a sandy soil than in a clay.

Table 1. Upper tolerance levels to water potential controlled by solute concentration (after Paul and Clark, 1989).

Designation of Water Potential			Solution Concentration		
MPa	Bar	A_w	NaCl (w/v)	Sucrose (w/v)	Organisms
−1.5	−15	0.99	2.0	17	*Rhizobium, Nitrosomonas*
−10	−100	0.93	12.3	52	*Clostridium, Mucor*
−25	−250	0.83	25.3	70	*Micrococcus, Penicillium*
−65	−650	0.62	—	83	*Xeromyces, Saccharomyces*

potentials at which the ammonifiers such as *Penicillium* are still active. The enhanced nitrate content sometimes observed in the surface layer of a droughty soil may not be due to nitrification but to upward movement of capillary water carrying nitrate. The evaporation of the water component at the soil surface leaves the nitrate behind. This effect is well known from other habitats such as sandstone, from where the term *salpeter* arises: it is the Latin expression for *sal petrae*, salt of stone, produced by nitrifiers within the pore matrix. As the process leads to lower pH, the carbonatic material which glues the sand pebbles together will be dissolved. Thus, the process results in microbial weathering of sandstone (Bock et al., 1989).

Redox Potential

Reduction-oxidation reactions are of major significance in explaining both soil chemical and biological phenomena (Stumm and Morgan, 1981). Life obtains energy from the oxidation of reduced materials; i.e., it removes electrons from either organic or inorganic substrates to capture the energy that is available during oxidation. This is accomplished in a series of steps involving a number of intermediate reactions. Electrons from reduced compounds are moved along respiratory or electron transport chains composed of a series of components. The process depends on a compound that can finally accept the electrons. In the aerobic pathway, this is oxygen. If oxygen is not available, nitrate, ferric iron, manganese(II), and sulphate can act as electron acceptors if the organism possesses the appropriate enzyme system.

The redox potential E_h, of a number of important biological reactions is given in Table 2. The greater the difference between electron potentials between donor and final acceptor, the greater is the potential for energy capture by biological systems.

Soil pH

Many of the soils of the world are affected by excess acidity, a problem exacerbated by heavy fertilization with certain nutrients such as ammonia (which is nitrified under acid production) and by acid rain. Biological nitrogen fixation also creates acidity in that H^+ is produced during the fixation process. The concept of pH values at a specific site must be related to the size of the organism and the multiplicity of enzymes at the microbial level. A bacterial cell contains about 1000 enzymes; many of these are pH-dependent and associated with cell wall components. The pH optimum of enzymes is affected by sorption phenomena. In the soil matrix, adsorption of enzymes to the soil humates shifts their pH optima to higher values. Paul and Clark (1989) point out that the soil microbiologist must be satisfied with the descriptive pH obtained by the traditional soil-paste measurement, which involves the addition of a $CaCl_2$ solution and measurement of the solution pH with an appropriate electrode. However, the obtained value only represents an overall pH and does not allow the detection of pH gradients in the microenvironments where they may be different and of considerable local importance.

Soil Temperature

Temperature affects not only the physiological reaction rates of cells, but also most of the physicochemical characteristics of the environment; examples include soil volume, pressure, redox potentials, diffusion, Brownian motion, viscosity, surface tension, and water structure. The

Table 2. Redox pairs arranged in order from the strongest reductants (negative potentials) to the strongest oxidants (positive potentials) (after Stumm and Morgan, 1981).

Redox Pair	E_h (volts)
CO_2/CH_2O	-0.43
N_2/NH_4^+	-0.35
CO_2/acetate	-0.28
SO_4^{2-}/H_2S	-0.22
Fumarate/succinate	$+0.03$
Nitrate/nitrite	$+0.42$
MnO_2/Mn^{2+}	$+0.48$
Cytochrome$_3$(ox)/cytochrome$_3$(red)	$+0.55$
Nitrite/nitrogen	$+0.74$
Fe^{3+}/Fe^{2+}	$+0.76$
$1/2O_2/H_2O$	$+0.82$

activities of microbial cells are governed by the laws of thermodynamics as are those of other organisms. It is therefore not surprising that changes in soil temperature have marked effects on microbial activity.

The rate of a chemical reaction is a direct function of temperature and generally obeys the relationship originally described by Arrhenius:

$$k = Ae^{-E/Rt}$$

where k is the reaction velocity, A the frequency with which molecules collide, E the activation energy of the reaction, R the gas constant, and t the temperature in degrees Kelvin (K). For two different temperatures, t_1 and t_2, this can be rewritten as

$$\frac{k_2}{k_1} = \frac{\Delta E}{R}\left(\frac{1}{t_2} - \frac{1}{t_1}\right)$$

At moderate temperatures, a plot of the reaction constant versus the reciprocal of the temperature in degrees Kelvin will yield a straight line with a slope of $-E$. This slope represents the energy hump (activation energy) that must be overcome for a reaction to proceed. Figure 4 shows the effect of temperature and pH on the rate constants for ammonia oxidation by *Nitrosomonas* and nitrite oxidation by *Nitrobacter.* Extremes of temperature cause a sharp fall-off at both ends of the curve. The abrupt fall in growth rate at high temperatures is caused by the thermal denaturation of proteins and alterations in the permeability of membranes. The maximum temperature of growth is the temperature at which these destructive forces become overwhelming. This temperature is usually only a few Kelvin higher than the temperature at which the growth rate is maximal. The interaction of temperature response with pH, as shown in Figure 4, explains why *Nitrobacter,* which oxidizes nitrite, is so much more sensitive to environmental conditions than *Nitrosomonas;* the result of this disturbance will be an accumulation of nitrite (Wong-Chong and Loehr, 1979).

Most measurements relating microbial activity to temperature show growth stopping at 0°C. Some psychrophilic bacteria are capable of growth below the freezing point (Atlas and Bartha, 1987), provided that the osmotic concentrations of the ambient solution or of the organism's cytoplasmatic constituents are sufficiently high to prevent the cell interior from freezing. A generalized temperature response curve for microbial activity is shown in Figure 5. Individual species differ in their optimum temperature, but the general shape of temperature response curves is quite similar for many organisms.

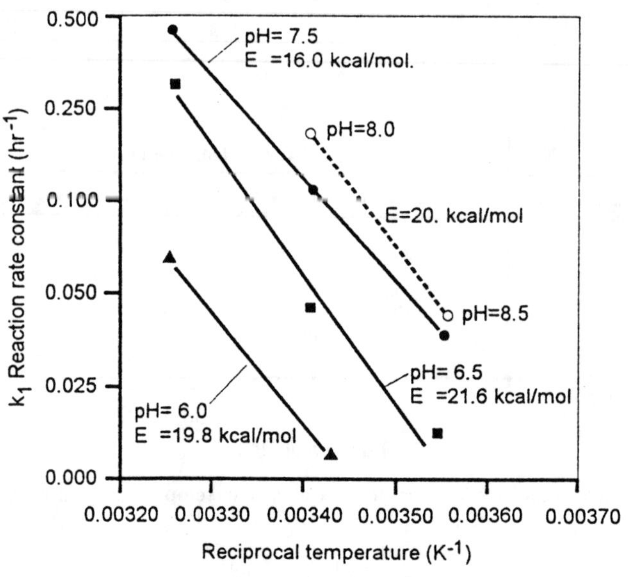

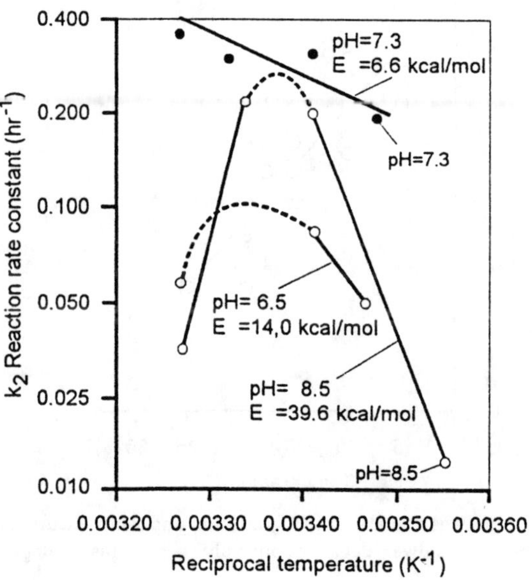

Figure 4. Combined effect of pH and temperature on reaction rate constants (h^{-1}), ammonia oxidation (k_1) by *Nitrosomonas* (top graph), and nitrite oxidation (k_2) by *Nitrobacter* (bottom graph) (after Wong-Chong and Loehr, 1978).

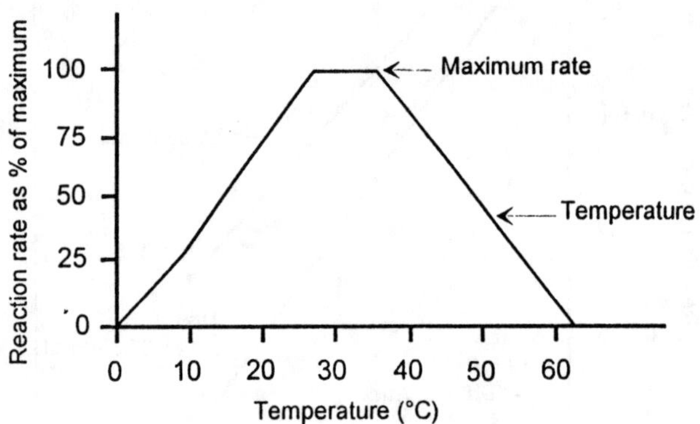

Figure 5. Relative microbial reaction rates at various temperatures (after Paul and Clark, 1989).

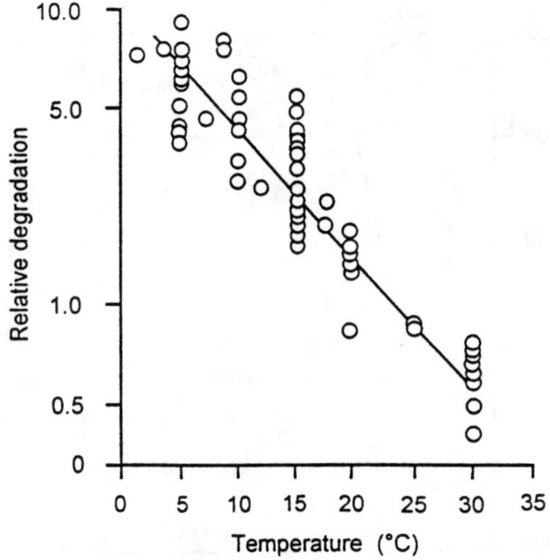

Figure 6. Relationship between relative biodegradation and temperature (reference temperature 25°C). Values > 1, slower degradation; values < 1, faster degradation (after Domsch, 1992).

Table 3. Effects of stress conditions on microorganisms and biochemical adaptations induced (after Paul and Clark, 1989).

Stress	Effect on Cells	Organisms Involved	Biochemical Adaptions or Responses
Heat	Denaturation of enzymes	Thermophiles	Synthesis of heart-stable proteins
Cold	Decrease in membrane fluidity	Psychrophiles	Production of more unsaturated fatty acids
Water potential	Dehydration and inhibition of enzymes' activity	Osmophiles, Halophiles, Xerophiles	Compensating, solute accumulations in enzyme adaptions
Acidity	Protein denaturation, enzyme inhibition	Acidophiles	Proton exclusion, adaptation in surface appendages
Anaerobiosis	Alternation of metabolic pathway	Anaerobes, Microaerophiles	Use of alternate electron sinks, fermenation

The effect of temperature can be demonstrated by the biodegradation of herbicides. 25°C was the reference temperature (relative biodegradation = 1.0). Values greater than one show the increase, values smaller than one the decrease of the turnover time (compiled by Domsch, 1992). In Figure 6, the relative biodegradation is plotted against time. It shows quantitatively how metabolism slows down at lower temperatures.

Interactions of Environmental Factors

It is difficult to interpret interactions involving temperature, moisture, soil pH, soil aeration, redox potential, and soil type. In nature, stress factors very rarely act independently. Only computer modeling techniques may attempt to describe all the interactions. A summary of microbial acclimatization to a number of stress factors is given in Table 3.

SOIL MICROORGANISMS

Bacteria

Bacteria are the most numerous among the microorganisms. Indeed, they are the most common living organisms on the face of the earth. In order to sustain life functions, they need energy sources, carbon sources, electron donors, and electron acceptors.

Ecological theory that examines the density of a species with respect to its food supply involves the concept of r- and K-selection. A species adapted to living under conditions of a bountiful energy source is designated as K-selected. It would be exposed to selection pressures different from those affecting an r-selected organism living in uncrowded, but possibly physically restricting, environments. The r-selected organisms exposed to flushes of substrates in an otherwise uncrowded environment would place a premium on high growth increase per unit of food rather than on competitiveness, as would be the case of a K-selected organism. Most soil organisms growing on flushes of substrate are r-selected, whereas rhizosphere organisms and plant pathogens are usually K-selected. Table 4 gives a survey of some important soil bacteria.

Members of the genus *Arthrobacter* are believed to be the numerically predominant bacteria in soil, as determined by the plate count method. Some estimates place their number as high as 40% of the total plate count population. *Arthrobacter* spp. are slow growing (they form small colonies on agar surfaces) and are poor competitors in the early stage of residue decomposition during which the easily decomposable materials (sugars, amino acids) are rapidly attacked by other genera.

Three genera, *Streptomyces, Pseudomonas,* and *Bacillus,* vie for the runner-up position to *Arthrobacter* as commonly occurring bacteria in soil. Any one of the three may at various times account for 5–20% of the total bacterial count as determined by plate counting. Within the bacterial kingdom, the genus *Streptomyces* falls within the order Actinomycales. The main features of five genera common in soil are shown in Table 5.

Members of the genus *Pseudomonas* occur, not only in soil, but also in

Table 4. Some important soil bacteria (compiled by Gisi, 1990).

Genus	Main Features
Micromonospora	Filaments do not grow above medium; single spores produced in and on surface of medium; colonies rather slow growing on most media
Nocardia	Filaments unstable, fragmenting into bacteria-like units; filaments usually not growing above medium and spores rarely produced
Streptomyces	Long chains of spores formed on filaments growing above the medium; species very numerous in soil and many produce antibiotics
Streptosporangium	Spores formed in sporangia or in chains on the filaments above the medium; colony appearance similar to *Streptomyces*
Thermoactinomyces	Single spores formed on filaments above and within the medium; spores heat resistant; all species thermophilic

Table 5. Major features of some genera of actinomycetes found in soil (after Paul and Clark, 1989).

Important Bacteria in Soil		
Procaryotes		
Eubacteria		
Purpurbacteria (Gram-negative bacteria)	Aerobic cocci	*Acetobacter, Acinetobacter, Agrobacterium, Azotobacter, Beijerinckia, Chromobacterium*
	Aerobic rods	*Alcaligenes, Nitrobacter, Nitrosomonas, Pseudomonas, Rhiziobium, Thiobacillus, Xanthomonas*
	Anaerobic	*Azospirillum, Spirillum, Desulfovibrio*
Enterobacteriaceae	Facult. anaerobic rods	*Enterobacter, Erwinia, Escherichia, Klebsiella, Proteus, Serratia*
Myxobacteriaceae	Aerobic rods	*(Aeromonas, Vibrio), Chondromyces, Cytophaga, Myxococcus*
Flavobacteriaceae	Aerobic rods	*Achromobacter, Flavobacterium*
Gram-positive bacteria		
Lactobacteriaceae	Aerobic rods	*Lactobacillus*
	Anaerobic cocci	*Streptococcus*
Corynebacteriaceae	Aerobic	*Arthrobacter, Cellulomonas, Corynebacterium*
Mycobacteriaceae	Aerobic	*Mycobacterium*
Actinomycetaceae Streptomycetaceae	Aerobic	*Actinomyces (S), Frankia, Microminospora (S), Nocardia, Streptomyces (S), Thermoactinomyces (S)*
Micrococcaceae	Aerobic	*Micrococcus*
	Anaerobic	*Sarcina (S)*
Bacillaceae	Aerob. rods	*Bacillus (S)*
	Anaerobic rods	*Clostridium (S)*
Cyanobacteria	Aerobic	*Anabaena (P), Gloeobacter (P), Nostoc (P), Oscillatoria (P)*
Archaebacteria	Anaerobic rods	*Methanobacterium*

S (Spores), P (phototroph).

fresh and marine waters. As a group, they attack a wide variety of organic substrates, including sugars and amino acids, alcohols and aldose sugars, hydrocarbons, oil, humic acids, and many of the synthetic pesticides.

Sporulating bacilli are mostly vigorous organotrophs, and their metabolism is either strictly respiratory, strictly fermentative, or both. Several species produce lytic enzymes and antibiotics of the polypeptide class that are destructive to other bacteria. The toxin produced by *Bacillus thuringiensis* is pathogenic to some insect larvae and is widely used as a biological control agent. Temperature tolerance in the genus ranges from about $-5°C$ to 75°C, tolerance to acidity from pH 2 to 8, and salt tolerance as high as 25% NaCl. *Clostridium* is also a sporogenic genus. Most species are strict anaerobes, but a few are microaerophilic. The genus is of economic importance; its species are used commercially for the production of alcohols and commercial solvents. Several species, such as *C. butyricum* and *C. pasteuricum,* are known to fix nitrogen. The genus is widely distributed in soils, marine and freshwater sediments, manures, and also in animal intestinal tracts. Two well-known pathogens are *C. tetani* and *C. botulinum,* the spores of which can persist in soil for extended periods.

Cyanobacteria are photosynthetic procaryotes containing chlorophyll and also phycobiliprotein pigments such as phycocyanin. They exist in unicellular, colonial, and filamentous forms. Cyanobacteria are ubiquitous in their distribution, occurring in saline and fresh waters, in soil, and on bare rocks and sand. On soil parent materials, they are important as primary colonizers, either alone or as symbionts of fungi in lichens. In some ecosystems, cyanobacteria are of great significance because of their ability to fix nitrogen.

The growth rates of bacteria in soil may differ quite strongly. Williams (1985) has collected examples of estimates of bacterial growth rates (Table 6).

Fungi

The fungi embrace eucaryotic organisms, variously referred to as molds, mildews, rusts, smuts, yeasts, mushrooms, and puffballs. Of the soil organisms, the fungi as a group are the organotrophs primarily responsible for the decomposition of organic residues, even though in plate counts they

Table 6. Examples of estimates of bacterial growth rates in soil (after Williams, 1985).

Bacterial Populations in	Generation Time (h)	References
Grassland soil	1,200	Babiuk and Paul (1970)
Deciduous woodland soil	97.7–244.8	Gray et al. (1973)
Deciduous woodland soil	480	Gray (1976)
Tundra soil	17.7–92.9	Gray (1976)
Peat	39	Clarhom and Rosswall (1980)
Pine forest humus	66	Clarhom and Rosswall (1980)
Pine forest mineral soil	55	Clarhom and Rosswall (1980)

are outnumbered by bacteria. Features characterizing eight fungal classes commonly present in soil are summarized in Table 7.

SLIME MOLDS

The Myxomycetes are the true slime molds. They form an acellular creeping plasmodium that suggests that they should be assigned to the animal kingdom. They are animal-like in their feeding, plasmodial form but fungus-like in their reproductive structures and spore formation. The group is widely distributed in soil, especially in association with decaying vegetation in cool, moist sites.

The Acrasiomycetes are the cellular slime molds. The unit of structure is a uninucleate amoeba that feeds by engulfing bacteria. Single cells characteristically aggregate into a pseudoplasmodium in which the cells do not fuse but behave as a mobile communal unit. The pseudoplasmodium changes into a fruiting structure, sporocarp, that bears asexual spores. Acrasiomycetes are found on decaying plant materials in moist environments.

FLAGELLATE FUNGI

Oomycetes can be found in water and soil, and many are highly destructive plant pathogens. They are unique in that they produce biflagellate asexual motile spores, termed zoospores. *Pythium* and *Phytophthora* are genera commonly found.

The Chytridiomycetes are especially prevalent in aquatic habitats but also commonly occur in soil. Their production of polarly uniflagellated motile zoospores distinguishes them from other fungi. Some members are parasitic on algae, higher plants, or insect larvae. Among the genera commonly found in soil are *Allomyces* and *Rhizophydium*.

SUGAR FUNGI

The Zygomycetes are called the sugar fungi because of their fermentations of diverse carbohydrate substrates. They usually produce a well-developed mycelium of coenocytic hyphae, asexual sporangiospores, and thick-walled resting zygospores. The group is mostly saprobic, but some are phytopathogenic, some parasitic on other fungi, and some produce animal-trapping mechanisms. The Mucorales (the largest order of Zygomycetes) are important economically; individual species are used for the commercial production of alcohols and organic acids, such as lactic, citric, oxalic, or fumaric acid.

HIGHER FUNGI

The Ascomycetes and the Basidiomycetes are sometimes called the

Table 7. Characteristics of fungi associated with soils and plants (after Paul and Clark, 1989).

Group	Class	Features	Representative Soil Genera
Slime molds	Acrasiomycetes	Uninucleate myxamoeba, pseudoplasmodium sporocarp bears asexual spores; found as bacterial feeders and decaying vegetation	*Dictylostelium*
	Myxomycetes	True slime molds; acellular creeping plasmodium, which is animal-like but produces fungal-like reproductive structures	*Physarum*
Flagellate fungi	Oomycetes	Produce biflagellate oospores within a sporangium; reproduction by game tangy	*Pythium, Plasmopara, Phytophthora, Saprolegnia*
	Chytridiomycetes	Aquatic habitats; polarly uniflagellate motile zoospores; some parasitics	*Allomyces, Rhizophydium*
Sugar fungi	Zygomycetes	Nonseptate coenocytic hypha; internal sporangiospores and thick-walled zygospores; mostly saprobic; some phytopathogenic or parasitic and other fungi	*Mucor, Mycotypha, Rhizopus, Zygorhyncus*
Higher fungi	Ascomycetes	Sac or ascus containing ascospores formed from karyogamy and meiosis, e.g., sexual stages, unicellular yeasts, or septate hyphae	*Endothia, Ceratocystus, Claviceps, Saccharomyces*
	Basidomycetes	Septate hyphae; sexual spores, produced by meiosis, held externally on basidium; many mushrooms, rusts, etc.	*Agaricus, Poria, Boletus, Fomes*
Fungi imperfecti	Deuteromycetes	Septate hyphae reproduce only by asexual conidia; if sexual stage found, usually ascomycete	*Aspergillus, Trichoderma, Penicillium, Helminthosporium, Fusarium, Arthrobotrys*
	(Mycelia Sterila)	Subgroups of Deuteromycetes; no conidia produced; reproduce by hyphal fragmentation	*Rhizoctonia*

"higher fungi." The former are distinguished by other fungi by the formation of a sac or ascus within which the ascospores are formed following sexual reproduction. Many species are saprophytic in soil, with the mycelium remaining underground but at intervals sending up very large fruiting bodies. The Ascomycetes have a wide range of impacts on people. Many are parasitic on plants, causing root rots, corn ear rots, brown rots of stone fruits, and powdery mildews. The ergot fungus (*Claviceps purpurea*) invades fruiting structures of grasses and produces alkaloids toxic to humans and other animals.

The Basidiomycetes include a wide assortment of fungi, notably mushrooms, puffballs, stinkhorns, shelf and bracket fungi, bird's-nest fungi, jelly fungi, smuts, and rusts. Particularly damaging are the smuts and stem rusts of cereals, diseases of forest and shade trees, and the rotting of lumber and of wooden structures. The Basidiomycetes are extremely vigorous decomposers of woody material, including standing and felled timber, tree stumps, and lumber products. Fungi decomposing cellulose but not lignin are causal agents of wood brown rots, so named because the partly oxidized dark-colored lignin remains. White rots are caused by fungi destroying both cellulose and lignin. These organisms have recently drawn attention because of their ability to degrade some organic pollutants.

Algae

The eucaryotic algae have been given names such as pond scum, water moss, seaweed, and red tide. Algae are the simplest of the chlorophyllus eucaryotes, distinguishable from all other green plants by sexual characteristics. In unicellular algae, the entire organism may function as a gamete (sexually reproductive cell). In multicellular algae, gametes are produced in either unicellular or multicellular gametangia. In asexual reproduction, algae produce flagellated and/or nonmotile spores.

Algae are unquestionably the most widely distributed of all green plants. They are predominantly aquatic. Terrestrial forms occur on and in the soil; on rocks, mud, and sand; on snowfields and buildings; and attached to plants and animals. Subsurface soil samples kept moist and under illumination commonly develop algal blooms. Most algal units that are found below ground are dormant forms, waiting for light, but some are known to be facultative organotrophs.

Lichens

Lichens are symbiotic associations of a fungus and an alga, with the two so intergrown as to form a single thallus, an undifferentiated body. The fungal member is usually an ascomycete and less often a basidiomycete or a deuteromycete. The association permits the survival of the two symbionts in harsh environments in which neither could live separately.

Lichens are found worldwide on inhospitable rock, soil, and other surfaces such as fenceposts, rooftops, gravestones, and tree barks. They grow slowly and become very old, persisting for centuries. The lichens are commonly among the first colonizers of bare rocks and soil parent materials and thus are important in the early stage of pedogenesis. In general, lichens are not ascribed major roles in the cycling of carbon and minerals in highly productive ecosystems. Ahmadjian and Hale (1973) have offered an in-depth discussion of the morphology, physiology, and taxonomy of lichens.

Protozoa

Protozoa are unicellular animals, most of which are microscopic in size but some attaining macroscopic dimensions. The group is greatly diverse in morphology and feeding habits but does show communality in that all require water envelopment for metabolic activity. A resume of the protozoa occurring in soil or in association with other soil organisms has been prepared by Stout et al. (1982) and is compiled in Table 8.

Free-living protozoa feed on dissolved organic substances and on other organisms. The soil ciliates depend primarily on bacteria as food; some feed additionally on yeasts and other protozoa, and even on small metazoa such as rotifers. Amoeba feed on bacteria, other protozoa, yeasts, fungal spores, and algae.

Soil protozoa have an effect on the structure and the functioning of microbial communities. The rise in bacterial numbers commonly observed following addition of fresh residues to soil is almost always followed by a rise of the protozoan numbers. Selective feeding by protozoa may alter the composition of the bacterial population. It is assumed that protozoan grazing keeps the population young, hence more efficient in residue decomposition.

Equilibria of Soil Microbial Populations

The more biologically complex or diverse a system is, the more "biologically buffered" it is. The microbial populations therein consist of active and inactive propagules. Many of the inactive organisms are viable, but as the conditions (biotic and abiotic) are altered, inactive forms may

Table 8. Principal free-living soil protozoan groups (after Stout et al., 1982).

Small flagellates	Ubiquitous, e.g., *Oikomonas* (one flagellum), *Bodo* (two flagella, one trailing)
Naked amoebae	Ubiquitous Small monopodal: *Vahlkampfia, Naegleria, Hartmannella* (including parasitic species) Multipodal: *Nuclearia* Pellicle layer: *Thecamoeba* Slime molds: *Dictyostelium*

Ciliates	Reflect soil structure, moisture, and aeration
	Trichostomes: almost ubiquitous, e.g., *Colpoda* (oval, flat body with indented cytostome), *Leptopharynx* (small, falt, with coarse cilia in furrows)
	Small edaphic ciliates
	  Chilodenella, humped, with cilia restricted to ventral surface
	  Cyrtolophosis, ovoid, with cytostome in depression in anterior end
	Larger and more complex hypotrichs: *Uroleptus, Keronopsis, Gonostomum*
	Sessile peritrichs, e.g., *Vorticella striata*
	Predatory species: *Spathidium, Bresslaua*
	Histophage: *Tetrahymena rostrata* (facultative parasite)
	Suctoria, sessile adult with ciliated larva, e.g., *Podophyra*

Testacea	Shape of the test commonly reflects the soil moisture regime Globose or hemispherical tests, e.g., *Cyclopyxis, Phryganella* Oval, wlth subterminal aperture, e.g., *Trinema, Corythion, Centropyxis* High-vaulted tests, normally associated with forest litters and mosses, e.g., *Nebela, Difflugia* Flattened, planoconvex (in profile) tests, normally in open structures, e.g., litter (*Arceila, Microchlamys*)

commence active growth and metabolism. Thus, by replacing a previously functioning population inhibited by the newly imposed environmental conditions, the continuity of biological function is preserved, albeit the actual microbial species that catalyze the process are different. For example, nitrogen fixation catalyzed by *Azotobacter* in aerobic soil is precluded when the soil is flooded and anaerobic conditions are established. However, the nitrogen fixation process may continue but is catalyzed by anaerobic *Clostridium* spp. (Hubbell, 1986).

There are numerous other examples of the continuation of a process in a soil system, which is mediated by different microbial populations in response to alterations in soil conditions that are either conducive or inimical to the growth activity of one or another component of the soil microflora. Here, the concept of microbial infallibility (Alexander, 1965) proves valid. It means that, for each organic compound, there exists a microbial population capable of mineralizing it. For many organic compounds, there is no such thing as a single microbial population, which totally mineralizes it. For such compounds, the metabolic end products of one organism-substrate combination may become the substrate of other microbes as soil conditions change. However, for chemically synthesized molecules, many of which are toxic and highly resistant to biodegradation, the situation is different. In cases where biodegradation pathways for such a compound are nonexistent or inordinately slow, avenues of escape from the adverse and cumulative effects of environmental pollution, such as cometabolism and nonbiological degradation, may not be effective.

There have been several attempts to classify the types of interactions that occur among bacterial species and between bacteria and eucaryotic organisms (Bull and Slater, 1982). These systems can facilitate the description and categorization of the many interactions that occur in microbial communities. One of the more widely used schemes, originally proposed by Odum (1953), is based on the effects of one species on the population size of the second species. The effects are rated as positive (+), neutral (0), or negative (−), depending on whether the population size of the affected species increases, remains the same, or decreases. The effects of a given species 1 on species 2 and vice versa are arranged in a binary matrix that contains all the possible combinations. As Odum (1971) points out, considering population interactions in this manner avoids the confusion that results when only terms or definitions are employed. Table 9 summarizes the possible effects. There are numerous examples for all

Table 9. Analysis of population interactions (after McInverney, 1986).

Type of Interactions	Species[a]		General Nature of Interactions
	1	2	
Neutralism	0	0	Neither population affects the other
Competition	−	−	Inhibition when resource is short supply
Amensalism	−	0	Population 1 inhibited, 2 not affected
Parasitism	+	−	Population 1, the parasite, usually smaller than 2, the host
Predation	+	−	Population 1, the predator, usually larger than 2, the prey
Commensalism	+	0	Population 1 benefits, while 2 is unaffected
Protocooperation	+	+	Interaction favorable to both but not obligatory
Mutualism	+	+	Interaction favorable to both and obligatory

[a]0, no significant interaction; +, growth, survival, or other populations attribute benefited (positive term to growth equation); −, population growth or other attribute inhibited (negative term added to growth equation).

kinds of interactions, as listed in Table 9. For further details, the reader is referred to the review of McInverney (1986).

SOIL MICROORGANISMS IN GEOCHEMICAL CYCLES

Carbon Cycling and Soil Organic Matter

"Microorganisms are nature's garbage disposal agents" (Paul and Clark, 1989). They convert the carbon in organic materials into CO_2 and thereby complete the biological carbon cycle that was initiated during photosynthesis. The burning of fuels, extensive forest fires, and intensive soil cultivation have increased atmospheric CO_2 to levels that could have major climatic effects. Global carbon cycles are closely tied to biological productivity and soil organic matter (SOM) turnover.

The largest fraction of all organic carbon entering the soil is that contributed by plant residues. Plants contain 15–30% cellulose, 10–30% hemicellulose, 5–10% lignin, and 2–15% protein. Soluble substances such as sugars, amino sugars, organic acids, and amino acids constitute 10% of the dry weight. They are readily leached from plant residues and are quickly utilized by soil organisms. It is difficult to separate the decomposition of plant residues from that of SOM. By definition, fine identifiable residues and microbial bodies are part of the SOM fraction. Reviews on plant residue decomposition are given by Stevenson (1982a) and Tate (1987). Moisture and temperature interactions are known to control the accumulation of SOM. Under moist conditions, the rate of carbon accumulation through photosynthesis is larger than the opposite process of decomposition. A summary of SOM contents from 3600 well-drained soil profiles sampled to a depth of 1 meter and representing the major moisture-temperature interaction zones of the world shows highest carbon accumulation in the wet rain forests of all temperature zones.

CARBOHYDRATES

Cellulose is the most abundant constituent part of plant residues, often being associated with hemicellulose and lignin. It occurs in a semicrystalline state with a molecular weight of 10^6 and is composed of glucose units with $\beta(1,4)$ linkages. Decomposition occurs via cellulases from a variety of bacteria, including *Pseudomonas, Chromobacterium, Bacillus, Clostridium, Streptomyces,* and *Cytophaga,* and fungi such as *Trichoderma, Chaetomium,* and *Penicillium.* The heterogeneous group of compounds collectively called hemicelluloses are various polymers of hexoses, pentoses, and sometimes uronic acids. Commonly occurring monomers of the group include xylose and mannose. In the pure state, hemicelluloses are easily decomposed. In nature, they are frequently complexed with other substances that may make the breakdown performed by enzymes known as pectinases more difficult.

Soil carbohydrates are not an integral part of the humic acid core. They may be attached as peripheral side chains but, more often, occur as free polysaccharides. Although readily degradable in the free state, carbohydrates account for 15% of the soil carbon. Adsorption to clays and interaction with polycationic materials such as Fe, Al, and Cu, have been found to greatly increase the resistance of polysaccharides to microbial attack.

Lignin structure is based on the phenyl propanoid unit, which consists of an aromatic ring and a three-carbon side chain. Formed by polycondensation, lignin, like aromatic SOM, is not the result of a specific enzyme reaction but a chemical reaction involving phenols and free radicals. Therefore, the material does not show a specific order. It is formed as an encrusting material on the cellulose and hemicellulose matrix. The decomposition is primarily attributed to fungi. The color of the decayed substrate is indicative of the mode of attack. White-rot fungi are the most active lignin-degrading microorganisms; however, they only can do so if oxygen is present in adequate amounts. Brown-rot fungi degrade the polysaccharides associated with the lignin and remove the methyl groups, subgroups, and $R—O—CH_3$ side chains. This leaves phenols behind, which on oxidation turn brown. The turnover rate of aromatics in nature is a major factor in determining SOM dynamics.

ORGANIC NITROGEN

The organic nitrogen content of cultivated soil generally parallels that of organic carbon, with the carbon/nitrogen ratio being 10 (Stevenson, 1982b). Of the total soil nitrogen, at least one-third remains structurally unidentified. Half or more of the total nitrogen occurs in amino acids and amino sugars. Because they are rich in nitrogen, amino acids comprise about 20% of the soil carbon but 30–40% of the soil nitrogen. Amino acids and proteins in the free state are readily degraded. However, microbially produced organic nitrogen is stabilized in soil. This is due to a relatively slow turnover of the microbial biomass itself and to stabilization of the organic nitrogen by association with resistant organics or within soil aggregates.

CELL WALLS OF ORGANISMS

The cell walls of fungi and bacteria are the major precursors to SOM formation. Decomposition of fungal cell walls is brought about by genera such as *Streptomycetes*, *Pseudomonas*, *Bacillus*, and *Clostridium*. Fungi such as *Mortierella* are important under more acidic conditions. Chitin, one of the contributors of amino sugars in soil, is a major component of fungal cell walls where it is associated with a number of other fibrous carbohydrate constituents.

Cell walls of bacteria have a rigid layer composed of chains of N-acetyl-glucosamine and N-acetylmuraminic acid. These chains are joined to each other by a limited number of amino acids linked through peptide bonds. The thick wall of gram-positive bacteria contains peptidoglycans linked to other wall constituents that include a variety of polysaccharide and polyphosphate molecules joined through phosphoester linkages. Gram-negative bacteria have more complex, thinner cell walls. Microbial growth on ^{14}C-labeled substrate in soil has been found to lead to a buildup of residual ^{14}C-containing cell wall materials. These materials are still a major component of the readily decomposable or soil-active fraction.

FORMATION OF SOIL ORGANIC MATTER AND HUMIFICATION

Soil organic matter is constituted of decomposing residues, by-products formed by organisms responsible for the decomposition of the residues, the microorganisms themselves, and the more resistant soil humates. Residues of plants differ in phenolic content and in the proportion of lignin to cellulose and protein. Microbial attack of carbohydrates and proteins results in the production of microbial products and, depending on the carbon to nitrogen to sulfur to phosphorus ratios, in the production of sulfate and ammonia as well as carbon dioxide (Figure 7).

Recalcitrant SOM slowly turns into the major characteristic of soils, which is humus—the leftovers of decomposition of readily biodegradable material. As organic matter is mineralized, the portion remaining becomes increasingly resistant to microbial attack. Also, this material becomes chemically altered due to microbiological and abiotic processes. These remains, usually termed humic substances, have an increasing turnover time as function of age and, therefore, tend to accumulate, comprising a large fraction of the organic material of soils, sediment, and water. Under certain environmental conditions, some of these materials will never mineralize but will be incorporated into sediments. They then undergo further metamorphoses through abiotic processes to become the organic components of sedimentary rocks (kerogen) and fossil fuels (petroleum, lignite, coal).

In terrestrial soils, the process of humification has been well studied. In temperate forest soils, the initial transformations can be followed in the vertical zonation with discrete layers corresponding to the annual litter falls. This transformation is, first of all, characterized by a decrease

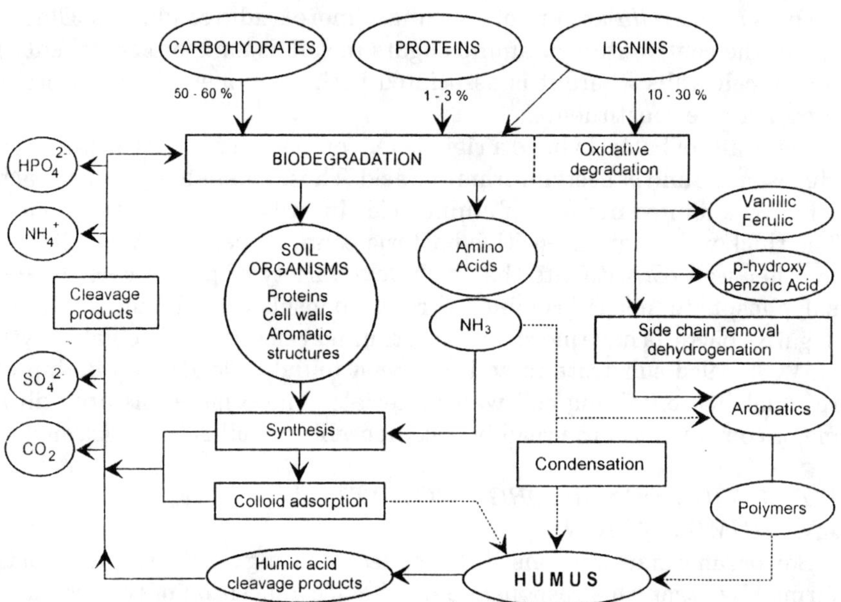

Figure 7. Degradation of plant residues and formation of soil organic matter (modified after Paul and Clark, 1989).

in the more easily decomposable constituents (e.g., carbohydrates, including cellulose, lipids, and proteins) and a relative increase in the more resistant constituents (e.g. lignin, cork substances, and resins).

This material is then transformed into humic substances through microbial, as well as nonbiological, processes. The core of humic substances is made up of aromatic rings; these derive from lignin residues, phenols, and quinones synthesized by microorganisms and later polymerized together with nitrogenous compounds to form humic substances.

The microbial degradation of humic substances is difficult to study since it is a very slow process, the precise chemical composition of the substrate is not known, and the decomposition may be restricted to various attached compounds (e.g., amino acids), whereas the aromatic core is left intact. Figure 8 gives a tentative picture of humic acids containing several kinds of functional groups and aromatic rings (after Stevenson, 1982). It is believed that the degradation of humic substances in soils is largely due to basidiomycetes and ascomycetes (Fenchel and Blackburn, 1979).

Under anaerobic conditions, in water-logged soils and swamps, the

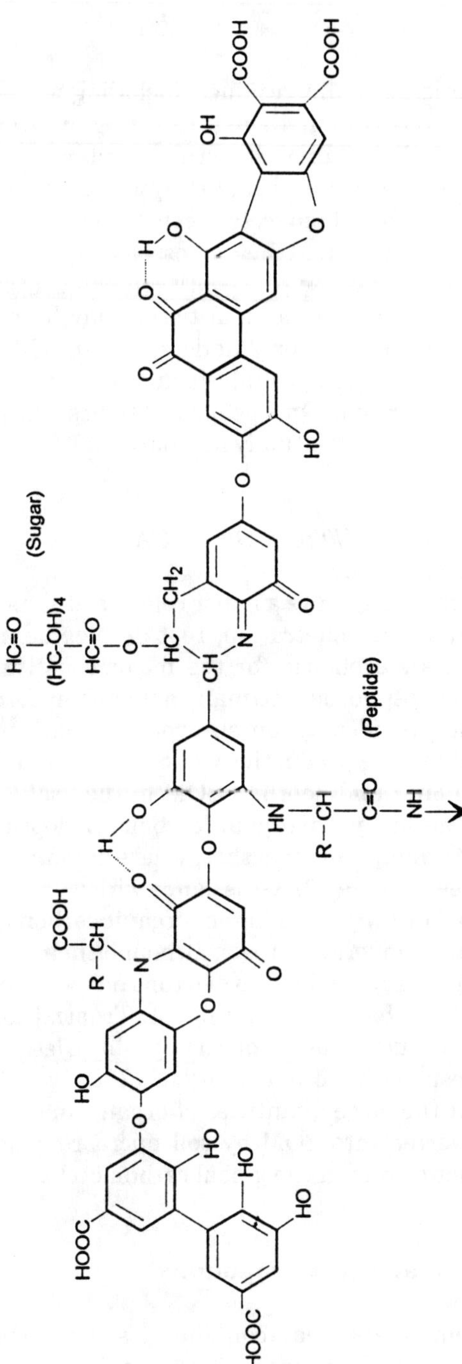

Figure 8. Proposed type structure for humic acid (after Stevenson, 1982).

mineralization of resistant plant residues, including waxes, resins, cork substances, and lignin, is very inefficient resulting in low pH and peats, which often preserve the original structure of plant tissue. Aromatic rings, such as are found in lignin, seem to require oxygen for biologically catalyzed cleavage. Also basidiomycetes seem to be the most important primary decomposers of lignin and they are essentially aerobic organisms. Finally, apart from having a low pH, peat seems to contain substances inhibitory to bacteria, and even animal tissue may be preserved. The nature of these substances is poorly understood. In addition to humic substances, peat contains lignin, cellulose, and bitumen, which consists of waxes, paraffins, and resins. On a geological time scale, peat may turn into lignite (brown coal) and eventually into hard coal through abiological processes.

QUANTITY AND DISTRIBUTION OF ORGANIC MATTER IN SOILS

Soil organic matter (SOM) plays a major role in soil structure and thus has great impact on water penetration, root development, and erosion resistance. It is also a storehouse for the major constituents, such as nitrogen, sulfur, and phosphorus, and many minor elements. Also, it gives color to the soil and provides cation absorption capacity. The greatest single factor controlling the productivity of both cultivated and uncultivated soils is the amount and depth of SOM in the profile. The highest accumulation, 700 tons $ha^{-1}y^{-1}$ of organic carbon to a depth of 1 m, occurs in highly productive swamps and marshes, where decomposition is inhibited by a lack of oxygen (Table 10). Grasslands with wet-dry seasons, and especially if stabilized by Ca^{2+}, have higher organic carbon levels than the equivalent boreal and temperate forests, although their carbon input, as shown by the plant primary production of 6 tons $ha^{-1}y^{-1}$ is on the average only half of the 12 tons $ha^{-1}y^{-1}$ in the forest. Tropical forests with an annual average net production of 19 tons $ha^{-1}y^{-1}$ have less organic carbon than temperate forests with a productivity of 12 $ha^{-1}y^{-1}$. Table 10 gives an impression about the large quantities of organic material, which are mineralized or converted into SOM by soil microorganisms and, thus, reveals their paramount role in the global carbon cycle.

Cycling of Nitrogen and Its Compounds

Nitrogen is an element essential to all life. In soil, sediment, and fresh and salt water, nitrogen exists in both inorganic and organic forms. Microbially important inorganic forms include ammonia and ammonium

*Table 10. Global distribution of plant biomass and soil organic carbon
(after Paul and Clark, 1989).*

Ecosystem	Area (ha × 10^8)	Net Primary Production (Mg carbon ha^{-1}year^{-1})	Plant Biomass (Mg carbon ha^{-1})	Soil Organic Carbon (Mg carbon ha^{-1})
Tropical forest	25	19	19	100
Temperature forest[a]	24	12	12	135
Shrubland and savanna	23	8	2	50
Temperature grassland	9	6	0.7	190
Tundra	8	1	0.3	220
Desert scrub	18	1	0.01	60
Rocks and desert	24	0.03	0.5	1
Cultivated	14	6	7	130
Swamp and marsh	2	30		700
	147			

[a]Deciduous plus boreal forest.

ion, nitrite, nitrate, and gaseous oxides of nitrogen. Organic nitrogen compounds include humic and fulvic acids, proteins, peptides, and amino acids, purines, pyrimidines, other amines, and amides. Inorganic nitrogen compounds exist in nature either as gases in the atmosphere and dissolved in water or as compounds in aqueous solution. Exceptions are small deposits of nitrates of sodium, potassium, calcium, magnesium, or ammonium known as guano or cave, playa or caliche nitrates (Lewis, 1965). These nitrate deposits were apparently formed by bacteriological transformations of organic nitrogen fixed by nitrogen-fixing bacteria, including cyanobacteria and, in some cases, from organic nitrogen in animal droppings, such as those of birds or bats, into nitrate via ammonia (Ericksen, 1983). Organic nitrogen in nature may exist dissolved in an aqueous phase, or in a solid state, in the latter case usually in polymers (e.g., certain proteins like keratin). Insofar as is known, naturally occurring organic nitrogen is usually metabolizable by microbes as far as it is bioavailable (Ehrlich, 1991).

AMMONIFICATION

Most plants derive the nitrogen that they assimilate from soil. In most instances, nitrogen occurs in the form of nitrate. The nitrate supply in soil depends on recycling of spent organic nitrogen (plant excretions and remains). The first step in this recycling is ammonification, in which organic nitrogen is transformed into ammonia. Ammonification is always an essential first step when an amino compound such as an amino acid serves as an energy source. Ammonia is also formed as a result of hydrolysis of urea catalyzed by the enzyme urease:

$$NH_2CONH_2 + 2H_2O \rightarrow 2NH_4^+ + CO_3^{2-}$$

NITRIFICATION

Since the ammonia produced in ammonification in aqueous systems at neutral pH exists as the positively charged ammonium ion, NH_4^+, due to protonation, and since ammonium ion is adsorbed by clays, it is important that it be converted into an anionic nitrogen species that is not adsorbed by clays and is thus more readily utilizable by plants (Ehrlich, 1991). The nitrifying bacteria can be divided into two groups: the first includes those bacteria that oxidize ammonia to nitrous acid, e.g., *Nitrosomonas, Nitrocystis*. The second group oxidizes nitrite to nitrate, e.g. *Nitrobacter, Nitrococcus*. They are all aerobes. Representatives are found in soil, fresh water, and seawater (for a further characterization see Alexander, 1977). Nitrification is a process typical for the interaction of microorganisms of different kinds in microconsortia, immobilized in biofilms on surfaces (de Boer et al., 1991).

Ammonia oxidation by the ammonia oxidizers involves hydroxylamine as an intermediate (see review by Wood, 1988). The formation of hydroxylamine involves an oxygenase:

$$NH_3 + 0.5O_2 \rightarrow NH_2OH$$

This reaction is endogenic and depends on the subsequent enzymatic oxidation of hydroxylamine in order to proceed in the direction as written above. The overall reaction of oxidation of hydroxylamine to HNO_2 can be summarized as

$$NH_2OH + O_2 \rightarrow HNO_2 + H_2O$$

but actually involves some intermediate steps (Hooper, 1984). The process leads to an acidification, because the nitric acid dissociates, releasing protons. This can enhance weathering processes (Eckhardt, 1985). Ammonia oxidizers can also form some NO and N_2O in side reactions (Knowles, 1985). This is an important observation because it means that biogenically formed N_2O and NO are not solely the result of denitrification. The nitrite oxidizers convert nitrite to nitrate:

$$NO_2^- + 0.5O_2 \rightarrow NO_3^-$$

They generate useful energy by coupling the process chemiosmotically to

ATP generation (Aleem and Sewell, 1984; Wood, 1988). NH_4^+ may also be converted heterotrophically to NO_3^- although this process is probably of minor importance in nature in most instances.

DENITRIFICATION

Nitrate, nitrite, and nitrous and nitric oxides can serve as electron acceptors in microbial respiration, usually under anaerobic conditions. The transformation of nitrate to nitrite is called dissimilatory reduction, and the reduction of nitrate to nitric oxide (NO), nitrous oxide (N_2O), and/or dinitrogen is called denitrification. Assimilatory nitrate reduction is the first step in a process in which nitrate is reduced to ammonia for the purpose of assimilation. Only as much nitrate is consumed in this process as is needed for assimilation. This is not a form of respiration and is performed by many organisms that cannot use nitrate for respiration. All the nitrate respiratory processes have been found to operate to varying degrees in terrestrial, freshwater, and marine environments and represent an important part of the nitrogen cycle favoured by anaerobic conditions. As explained earlier, micoorganisms need an electron acceptor as end product of respiration to maintain their turnover. If oxygen is not available, other compounds can be reduced, including nitrate. The following half-reaction describes nitrate reduction:

$$NO_3^- + 2H^+ + 2e^- \rightarrow NO_2^- + H_2O$$

The electron donor may be one of a variety of organic metabolites or reduced sulfur in a form such as H_2S or S^0. The enzyme-catalyzing the reaction is called nitrate reductase and is an iron-molybdo-protein. It is not only capable of catalyzing nitrate reduction, but may also catalyze the reduction of ferric to ferrous iron. Nitrate can inhibit ferric iron reduction by nitrate reductase (Ottow, 1969).

Nitrite may be reduced to dinitrogen by the following series of half-reactions, with organic metabolites or reduced sulfur acting as electron donor:

$$NO_2^- + 2H^+ + e^- \rightarrow NO + H_2O$$

$$2NO + 2H^+ + 2e^- \rightarrow N_2O + H_2O$$

$$N_2O + 2H^+ + 2e^- \rightarrow N_2 + H_2O$$

The reduction of nitrite to ammonia may be summarized by the equation

$$NO_2^- + 7H^+ + 6e^- \rightarrow NH_3 + 2H_2O$$

The electron donor may be one of a variety of organic metabolites. For a more complete discussion of denitrification, the reader is referred to a monograph (Payne, 1981) and a review article by the same author (Payne, 1983).

NITROGEN FIXATION

If nature had not provided for a microbial mechanism—nitrogen fixation—to reverse the effect of microbial depletion of fixed nitrogen from soil or water as volatile nitrogen oxides or dinitrogen by denitrification, life on earth would not have continued for long after the process of denitrification first evolved. Nitrogen fixation is dependent on a special enzyme, nitrogenase, which is found only in procaryotic organisms, including aerobic and anaerobic photosynthetic and nonphotosynthetic eubacteria and archaebacteriae. Nitrogenase catalyzes the reaction

$$N_2 + 6H^+ + 6e^- \rightarrow 2NH_3$$

The reducing power needed for dinitrogen reduction is in the form of reduced ferredoxin and, in heterotrophs, may come from a reaction in which pyruvate is oxidatively decarboxylated. In phototrophs, the reduced ferredoxin is produced as part of the photophosphorylation mechanism. Nitrogen fixation is a very energy-intensive reaction that consumes as many as 16 moles of ATP per mole of dinitrogen becoming reduced to ammonia (Newton and Burgess, 1983). Nitrogen fixation may proceed symbiotically or nonsymbiotically. Symbiotic nitrogen fixation requires that the nitrogen-fixing bacterium associates with a specific plant, e.g., legumes, several nonleguminous angiosperms, the water fern *Azolla*, fungi (certain lichens), or in rare cases, with an animal host in order to carry out nitrogen fixation. In nonsymbiotic nitrogen fixation, the active organisms are free-living in soil or water and fix nitrogen if fixed nitrogen is limiting. Their nitrogenase is not distinctly different from that of symbiotic nitrogen fixers. The capacity for nonsymbiotic nitrogen fixation is widespread among procaryotes. For a more detailed discussion of nitrogen fixation, the reader is referred to Alexander (1984).

Owing to their special capacities for transforming inorganic nitrogen compounds that plants and animals lack, microbes and especially procaryotes and certain fungi play a central role in the nitrogen cycle (Figure 9). The direction of the transformations in the cycle is determined, not only by environmental conditions, especially the availability of oxygen, but also by the supply of particular nitrogen compounds. Fixed nitrogen is frequently a growth-limiting factor in the marine environment but not in fresh water, where phosphate is more likely to limit productivity. Fixed

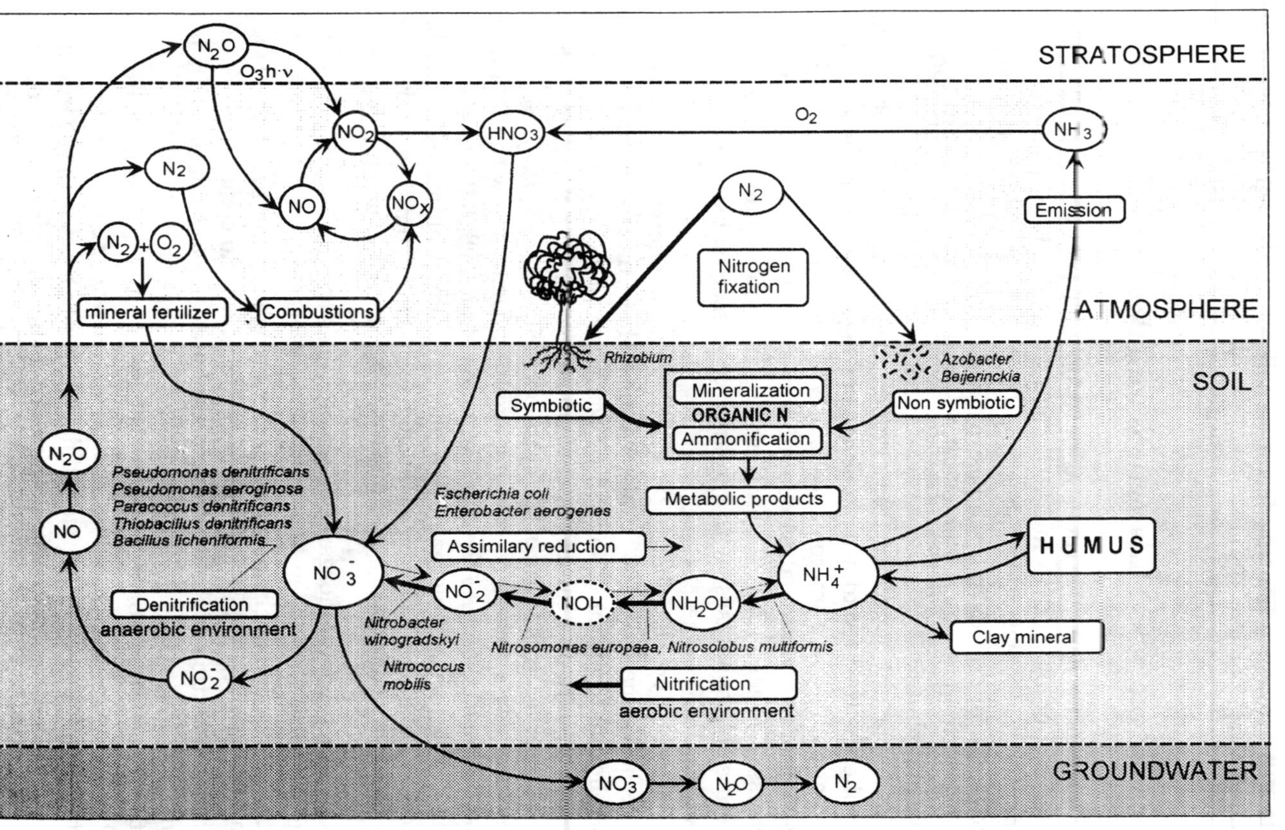

Figure 9. The nitrogen cycle.

nitrogen may be a limiting factor in soil, especially agriculturally exploited soils (Ehrlich, 1991).

Geomicrobiological Interactions with Phosphorus

Phosphorus, being a structural and functional component of all organisms, is fundamental to life. It is found universally in such vital cell constituents as nucleotides, phosphoproteins, phospholipids, and teichoic and teichuronic acids in gram-positive bacteria. The availability of phosphorus is, in many cases, the growth-limiting factor for both microorganisms and plants.

An important source of free organic phosphorus compounds in the biosphere is the breakdown of animal and vegetable matter, as well as its extraction by living microbial cells and by animals. In soil, as much as 70–80% of the microbial population may be able to participate in this process (Ehrlich, 1991). Upon their death, these organisms secrete or liberate phosphatases with greater or lesser substrate specificity (Skujins, 1967).

Inorganic pyrophosphate (PP) can serve as an energy source to some bacteria (Varma et al., 1983). Like pyrophosphate, intracellular, inorganic polyphosphate granules formed by some microbial cells are a form of metaphosphate and can represent an energy storage as well as a phosphate reserve. In the case of the cyanobacterium *Anabaena cylindrica,* it may also play a role as detoxifying agent by combining with aluminum ions that are taken into the cell (Petterson et al., 1985).

Insoluble forms of inorganic phosphorus (calcium, aluminum, and iron phosphates) may be solubilized through microbial action. The mechanism by which the microbes accomplish this solubilization varies. It may be

- the production of inorganic or organic acids that attack the insoluble phosphates
- the production of chelators, such as 2-ketogluconate (Banik and Dey, 1983), citrate, oxalate and lactate, all of which can complex the cationic portion of the insoluble phosphate salts and thus force their dissociation
- the reduction of iron in ferric phosphate (strengite) to ferrous iron by dissimilatory iron reduction in sediment-water systems
- the production of hydrogen sulfide, which can react with the iron in iron phosphate and precipitate it as iron sulfide, thereby liberating the phosphate

Microorganisms can cause fixation or immobilization of phosphate, either by promoting the formation of inorganic precipitates or by assimilation into organic cell constituents or intracellular polyphosphate granules.

Sulfur Transformations in Soil

Sulfur belongs to the vital elements and is an integral part of proteins and, thus, represents an essential nutrient for all living systems. Plants and microorganisms can produce and utilize a variety of forms of sulfur. It is harmful to plants in soils if it occurs in excess. Acid rain containing sulphate and nitrate, has great potential for acidifying lakes and for damaging forests. Table 11 shows the distribution of sulfur in some representative world soils (Chae and Lowe, 1980; Paul and Clark, 1988); most in agricultural soils range from 20–2000 μg g^{-1}. Inorganic sulfur exists in nature in a number of oxidation states, ranging from +6 in sulfate to −2 in sulfide.

Organic sulfur in soils is a component of SOM and can be classified into two major types. Carbon-bonded sulfur occurs in amino acids such as cysteine, cystine, and methionine; in cofactors such as biotin, thiamine, and coenzyme A; in iron-sulfur proteins (ferredoxins); and in lipoic acids. In the second type of soil, organic sulfur is attached to the organic matrix via —O— or sometimes via —N— bonds. Collectively, they may be termed as sulfate esters.

Table 11. Amount and distribution of sulfur in some world soils (after Chae and Lowe, 1980; Paul and Clark, 1988).

| | | Total Sulfur (metric tons) | | | |
| | | | | Organic | |
Location	Type of Soil	Total Sulfur (μg g^{-1})	Inorganic, Reducible	Carbon Bonded	Ester Sulfate
Saskatchewan	Agricultural	88–760	0.5–13	29–59	41–71
British Columbia	Grassland	286–928	ND	31–61	39–69
	Forest	162–2,328	ND	20–47	53–80
	Organic	1,122–30,430	ND	28–75	25–72
	Agricultural	214–438	2	18–45	55–82
Iowa	Agricultural	57–618	2–8	43–60	7–18
Carolinas	Tidal marsh	3,000–35,000	—	—	—
Hawaii	Volcanic ash	180–2,200	6–50	50–94	50–94
Eastern Australia	Agricultural	38–545	4–13	10–70	24–76
Nigeria	Agricultural	25–177	4–20	80–96	80–96
Brazil	Agricultural	43–398	5–23	20–65	24–59

Biomass sulfur represents 2–3% of the total organic sulfur in soils. Microbial sulfur is measured by lysing the cells with chloroform and measuring the sulfur released to a $CaCl_2$ or $NaHCO_3$ extractant. The majority occurs in amino acids. Fungi have been reported to store intracellular sulfur as choline sulfate, but little else is known about the microbial storage of ester sulfur compounds. Forms of sulfur such as phenyl sulfates and elemental sulfur have been found in a variety of soil organisms. Elemental sulfur has been identified in the sporocarps of ectomycorrhizal fungi, as well as in other self-inhibited and dormant structures. Elemental sulfur is deposited by some of the bacteria capable of utilizing H_2S as reducing agent during photosynthesis (Ehrlich, 1991).

The carbon to sulfur (C:S) ratio in SOM is not as consistent as the carbon to nitrogen (C:N) ratio. Major differences are found due to type of parent material, leaching, and sulfur inputs. Mollisols of the prairies have C:N:S ratios of 90:8:1, and lufisols and spotosols can range up to 200:12:1. A general worldwide ratio could be stated as 130:10:1.3 (Paul and Clark, 1989). Figure 10 shows the interrelations of sulfur transformations in soil.

MINERALIZATION OF ORGANIC SULFUR

The transformation of C—O—S, C—N—S, and R—C—S compounds originating from plants or microorganisms can proceed through both aerobic and anaerobic pathways (Figure 11). The hydrolysis of ester sulfates occurs by the splitting of the O—S bond by sulfatase enzymes:

$$R\!-\!O\!-\!SO_3^- + H_2O \rightarrow ROH + H^+ + SO_4^-$$

The mineralization of amino acids such as cysteine can occur anaerobically via cysteine desulfhydrase or serine sulfhydrase. It results in the release of hydrogen sulfide, which can react abiotically with other soil constituents such as iron to form insoluble black iron sulfide.

Oxidation of inorganic sulfur components in soils, waters, and sediments can be carried out by a diverse group of microorganisms. These can be subdivided into those growing at neutral pH and those that live at acidic pH values. The latter can also use ferrous iron as an electron donor, thus, closely coupling sulfur and iron transformations.

An important group among the sulfur oxidizers are the thiobacilli. These are chemolithotrophic bacteria that oxidize hydrogen sulfide, elemental S, thiosulfate, and polythionite. Some representatives live in very acidic conditions, such as *Thiobacillus thiooxidans* or *Thiobacillus ferrooxidans,* who are both strictly aerobic and show their optimum in a range between pH 1.5 and 5.0. These organisms also live in biofilms on

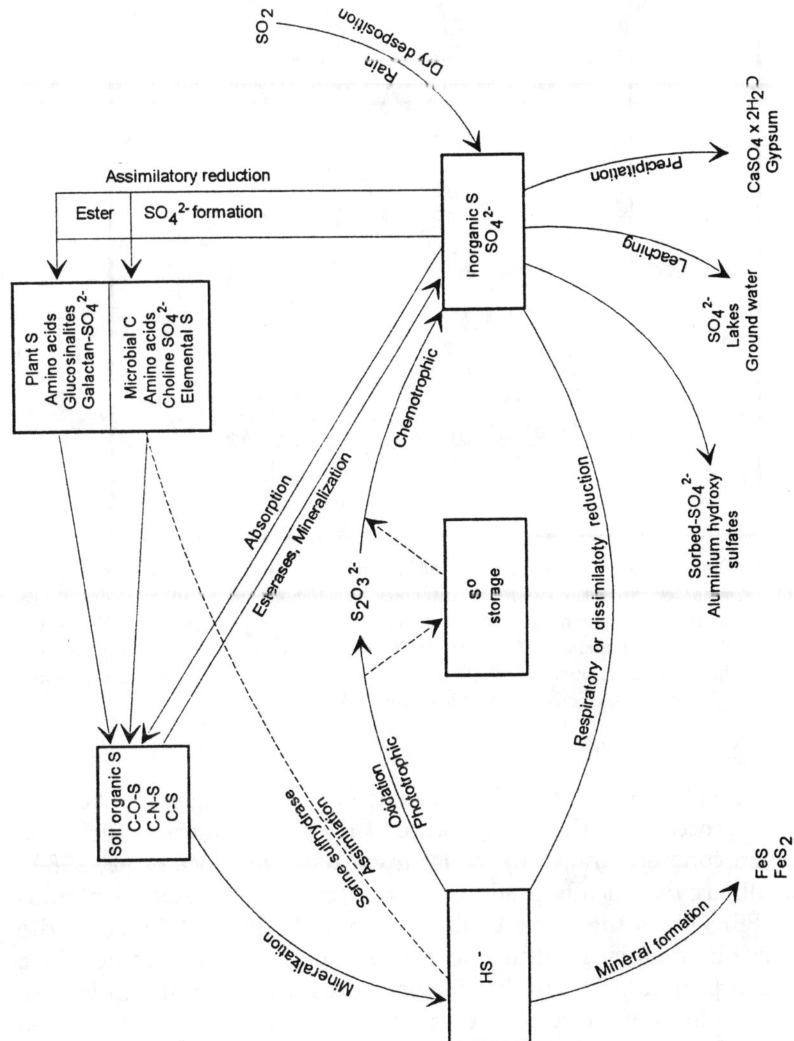

Figure 10. Sulfur (S) transformations in nature. Elemental sulfur is shown as a storage product and the possibility of sulfate sorption in certain soils is included (modified after Paul and Clark, 1989).

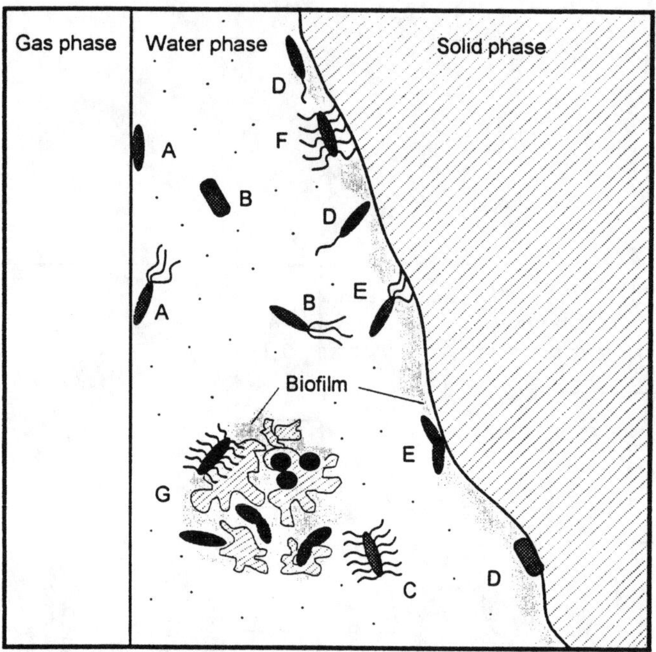

Figure 11. Diagrammatic representation of bacteria in the aqueous phase of soils or sediments. A = bacterium at the gas-liquid interface, B = bacterium in bulk aqueous phase, C = bacterium in aqueous phase where colloidal material is present, D = bacterium reversibly sorbed at solid surface, E = bacterium irreversibly sorbed at solid surface, F = bacterium as in E but enveloped in colloidal material, G = coflocculation of bacteria and particle of comparable size (modified after Marshall, 1980).

the gaseous side of sewer pipelines. Provided with hydrogen sulfide from anaerobic processes in the sewage water, they may produce sulfuric acid directly on concrete and, thus, cause great damage (Bock et al., 1989). Thiobacilli are technically used for the reclamation of metals (Hutchins et al., 1986). One of the oldest techniques exploits the oxidation of pyrite by thiobacilli, producing sulfuric acid. The technology is well established, cheap, and particularly suitable for poor ores. However, it can lead to substantial environmental problems when the leachate is not processed anymore. Slagheaps of overburden are still leached, and the trickling water contaminates soil, groundwater, and surface waters. A recent example is the remobilization of uranium from almost exhausted ores in the former GDR by leaching microorganisms (Flemming, 1993).

The reduction of inorganic forms of sulfur is also performed microbially. It is known as respiratory sulfate reduction and usually mediated by anaerobic, organotrophic organisms that use organic acids, alcohols, and often H_2 as electron donors (Swank and Fitzgerald, 1984). These organisms are responsible for sulfide formation in waterlogged soils and sediments. Sulfate-reducing bacteria are found over an extensive range of pH and salt concentrations, in saline lakes, evaporation beds, deep sea sediments, and oil wells. The organisms tolerate heavy metals and dissolved sulfide concentrations up to 2% (v/v). Although sulfate reducing bacteria are largely organotrophic, in that most of the carbon fixed is derived from organic matter, some organic molecules, such as low-molecular-weight fatty acids are inhibitory to growth. In such cases, only certain organic substrates, including lactic and pyruvic acids, are utilized as electron donors. In some cases, in which H_2 can act as an electron source, CO_2 can be fixed, although small concentrations of complex organic compounds are required for growth. When *Desulfovibrio* utilizes H_2, the following reactions occur:

$$S_2O_3^{2-} + 4H_2 \rightarrow 2HS^- + 3H_2O$$

$$S_4O_6^{2-} + 9H_2 \rightarrow 2HS^- + 2H_2S + 6H_2O$$

Although most sulfate reducers require ammonia as a nitrogen source, certain isolates reduce N_2 to NH_3. Therefore, they can grow in a carbonate and sulfate containing medium under atmospheres of H_2 and N_2. Sulfur reduction in the geological past has led to the high concentrations of reduced sulfur in oil and coal fields. Unless removed during combustion, this sulfur leads to major pollution problems.

THE ROLE OF INTERFACES

The water phase adjacent to the solid components in soils provides the major medium for microbial growth and development within such ecosystems. There is no doubt, however, that the nature and organization of the solid phase has profound effects on the ecology of microorganisms. The various interactions between microorganisms and solid components are summarized in Figure 11. All or some of these conditions may exist within any finite site in a soil or sediment.

With the exception of the plant root zone (rhizosphere) or following seasonal addition of organic matter (crop debris, leaf litter, fertilizing), most soils are in a state of chronic nutrient deficiency with respect to microbial growth. Extremely small bacteria are common in soils, which suggests the development of dwarf forms (Marshall, 1980). However, some authors point out that overall assessments of the rates of energy input to the soil mass and the biomass of bacteria maintained within suggests that soil is a grossly oligotrophic environment (Williams, 1985). Organism A sould benefit from nutrients accumulated at the gas-liquid interface (Kjelleberg, 1985), provided that the gas phase is suitable for the particular organism. The gas phase may be air or, in restricted microenvironments and in waterlogged conditions where oxygen is almost or completely absent, other gases such as carbon dioxide, hydrogen, and methane. This latter condition is the norm in anaerobic soils where gases accumulate. Microorganisms in the bulk aqueous phase (organism B) should be nutrient deficient and, possibly, present as dwarf form. If colloidal materials such as clays or organic matter are suspended in the aqueous phase, these may adhere to microbial surfaces (organism C) and alter nutrient input, metabolite output, buffering, and other conditions that may stimulate or inhibit growth. Reversible or irreversible sorption to the solid surface (organisms D and E) should allow the microorganisms to benefit from any nutrients accumulating at or near the solid-liquid interface (Stotzky, 1985). Firm adhesion would provide a selective advantage only where high shear forces exist near the surface. In soils containing high levels of colloidal material, particularly clay (Filip, 1979; Stotzky and Burns, 1982), some microorganisms may be so effectively enveloped by the colloidal material (organism F) that normal metabolism is inhibited. Figure 12 shows a scanning electron micrograph of fungal hyphae on a soil crumb covered with mineral particles.

Organic matter may accumulate under such conditions because of the reduced rate of microbial degradation. Coflocculation of microorganisms and particulates of comparable size (organism G) may lead to sedimentation and certainly modifies the behaviour of the microorganisms significantly (Marshall, 1980; Stotzky, 1985).

Microbial Utilization of Particulate Material

A considerable proportion of all nutrients occurs in particulate form. If those insoluble materials shall be available for microbial cell metabolism, they have to be broken down into small molecules that can be transported across the cell wall. This is usually performed by extracellular enzymes. Such enzymes are either bound to the outer cell wall or to other surfaces, integrated into the extracellular polymeric substances (EPS) matrix, or dissolved in pore water. Most of the exoenzymes investigated are hydrolytic enzymes (Chróst, 1991). They are produced in abundant quantities. This indicates that they are of great importance for the cell. Table 12 summarizes some of the enzymes found in soil and the reactions they catalyze. Soil enzymes are often entrapped in soil organic and

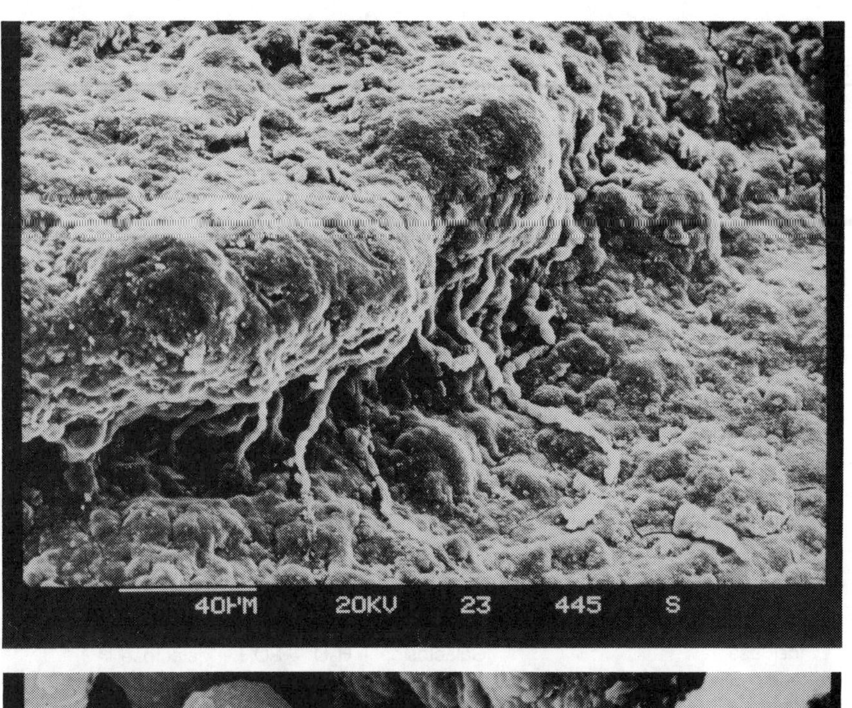
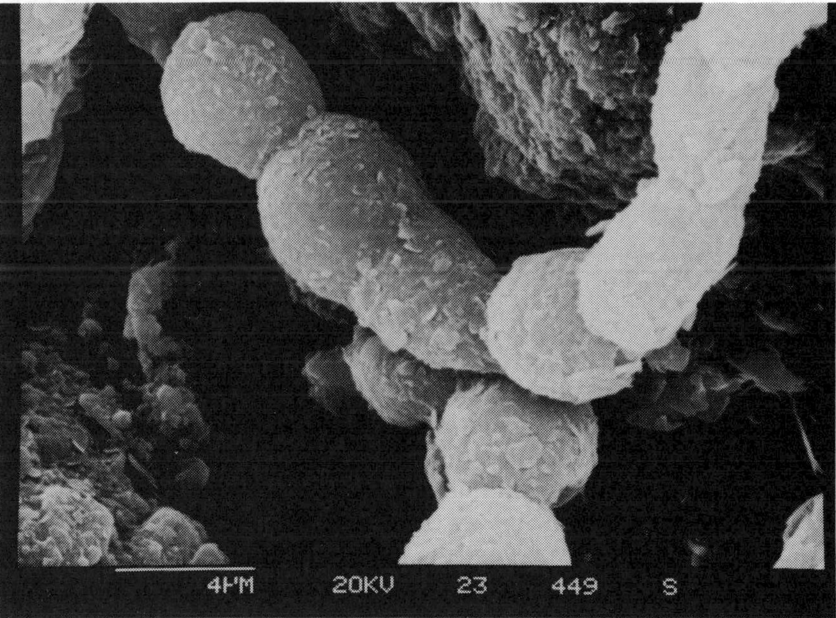

Figure 12. SEM photograph of fungal hyphae, covered with mineral particles.

Table 12. Some soil enzymes found in soil and the reactions they catalyze (Paul and Clark, 1989).

Enzymes	Reaction Catalyzed
Oxidoreductases	
Catalase	$2H_2O_2 \rightarrow 2H_2O + O_2$
Catechol oxidase (tyrosinase)	o-Diphenol + $1/2O_2 \rightarrow o$-quinone + H_2O
Dehydrogenase	$XH_2 + A \rightarrow X + AH_2$
Diphenyl oxidase	p-Diphenol + $O_2 \rightarrow p$-quinone + H_2O
Glucol oxidase	Glucose + $O_2 \rightarrow$ gluconic acid + H_2O
Peroxidase and polyphenol oxidase	$A + H_2O_2 \rightarrow$ oxidized A + H_2O
Transferases	
Transaminase	$R_1{-}R_2{-}CH{-}N^+H_1 + R_1R_4{-}CO \rightarrow R_1R_4{-}CH{-}N^+H_3 + R_1R_2CO$
Hydrolases	
Acetylesterase	Acetic ester + $H_2O \rightarrow$ alcohol + acetic acid
α- and β-Amylase	Hydrolysis of $\beta(1{\rightarrow}4)$glucosidic bonds
Asparaginase	Asparagine + $H_2O \rightarrow$ aspartate + NH_3
Cellulase	Hydrolysis of $\beta(1{\rightarrow}4)$glucan bonds
Deaminase	Carboxylic acid amide + $H_2O \rightarrow$ carboxylic acid + NH_3
α- and β-Galactosidase	Galactoside + $H_2O \rightarrow$ ROH + galactose
α- and β-Glucosidase	Glucoside + $H_2O \rightarrow$ ROH + glucose
Lipase	Triglyceride + $3H_2O \rightarrow$ glycerol + 3 fatty acids
Metaphosphatase	Metaphosphate $\rightarrow$ orthophosphate
Nucleotidase	Dephosphorylation of nucleotides
Phosphatase	Phosphate ester + $H_2O \rightarrow$ ROH + phosphate
Phytase	Inositol hexaphosphate + $6H_2O \rightarrow$ inositol + 6-phosphate
Protease	Proteins $\rightarrow$ peptides + amino acids
Pyrophosphatase	Pyrophosphate + $H_2O \rightarrow$ 2-orthophosphate
Urease	Urea $\rightarrow 2NH_3 + CO_2$

inorganic colloids. Therefore, the soil has a large background of extracellular enzymes not directly associated with the microbial biomass.

A lot of soluble organic materials are bound to clays. This decreases their availability as nutrient or energy sources for microorganisms and, furthermore, has a marked influence on the activity, ecology, and population dynamics in soil (Stotzky, 1980). Surface interactions between organics and clays may prevent the migration of organics through soil and may explain, in part, the lower concentration of organic matter in deeper horizons. On the other hand, the association of microorganisms with clay may enhance their activity; this was already known by Söhngen in 1915. Klaas et al. (1994) found that the degradation of pesticides such as chlortolurone, isoproturone, and metoxurone which are poorly degradable, was improved significantly in the presence of montmorillonite particles. However, adhesion to clay does not enhance the activity of microorganisms in all cases, as shown with *Histoplasma capsulatum.*

The decomposition of sorbed or solid substates presents several interesting features (Nedwell and Gray, 1987):

1. They are usually small in quantity at any one point, so that rapid microbial growth will cease quickly when the growth-limiting nutrient is exhausted.
2. For rapid exploitation, penetration of the substrate is desirable; for slow exploitation, gradual erosion of the surface is sufficient.
3. Individual substrate particles may be separated by considerable distances (in microbial terms), although individual substrate particles may be ingested or moved in other ways through animal activity.
4. In soils with a clay content greater than 12–14%, aggregation of particles will occur; aggregate formation may entrap both substrate and microorganism, effectively isolating them from other parts of the soil.
5. Humified materials are often deposited in particular horizons in a soil profile; organisms exploiting these substrates may need to adhere to the soil particles to avoid being separated and washed out of the soil.

Thus, organisms inhabiting soil are often found to have developed one or more of the following characteristics: (a) filamentous/mycelial/cord habit or small colonies of unicells, (b) resistant wall structures, (c) dormant propagules, or shut-down cells, (d) adhesive mechanisms, and (e) dimorphic or pleomorphic potential. Nedwell and Gray (1987) proposed a series of growth patterns of microorganisms in soil that incorporated this diversity: nonmigratory unicells, migratory unicells, plasmodia, substrate-restricted hyphae, locally spreading hyphae, mycelial strands/rhizomorphs, and diffuse spreading hyphae.

Mineral Nutrients and Decomposition

An important aspect of decomposition is that mineral nutrients (e.g,. N, P, K) found in dead organic matter are released in an inorganic form, available for the primary producers. General chemical analyses may only reveal the principal presence of these elements, but tell nothing about their availability. Microbial mechanisms such as complexing and leaching are of fundamental importance (Eckhart, 1985; Krumbein, 1988). Phosphorus can be growth-limiting; even under conditions where copious amounts of phosphorus are present, it may be in a form that is not accessible to the microorganisms.

However, the relationship between decomposition and mineral nutrients is more complex. It has long been known that when nutrient-poor organic substrates such as pure cellulose or straw are added to soils, the immediate effect may be an immobilization of mineral nutrients. Bacteria have a relatively high content of essential elements relative to plant tissue. They assimilate dissolved inorganic nutrients during the uptake of organic substrates. Bacteria utilize organic substrates for the synthesis of new cell material and for their energy metabolism. Assuming that C_b and N_b are the amounts of carbon and nitrogen in one bacterial cell to produce biomass with the carbon content C_b organic carbon where C_r is the respired carbon and C_b/C_a is the growth yield. In order for the substrate to satisfy the need for nitrogen, its C:N ratio should therefore not exceed

$$(C_b + C_r)/N_b = (C_b/N_b)/(C_b/C_a)$$

i.e., the C:N ratio of the bacteria cytoplasm divided by the growth yield. If the substrate has a higher C:N value, a net immobilization of dissolved nitrogen will take place. Conversely, if the substrate has a lower C:N value, a net mineralization of nitrogen will take place (Figure 13).

Growth yields of aerobic bacteria vary according to the substrate, growth rate, and other factors. In the literature, values between 20 and 80% have been reported for different strains growing on different kinds of substrates. Agronomic experience indicates that net mineralization will already take place when the substrate C:N ratio is lower than 20 (Fenchel and Blackburn, 1979). The immobilization of mineral nutrients by decomposer microorganisms has several important ecological implications. Plant-derived, detrital particles often have a content of mineral nutrients that is too low to serve as a food source for animals. The microbial growth on the surface of such particles will enrich them with mineral nutrients and, thus, increase the nutritive value for detritus feeders and browsing animals (Fenchel and Blackburn, 1979). Another important aspect is that the immobilization of nutrients will lead to their retention in soils outside the growing season of plants, whereas inorganic nutrients, especially nitrate, normally tend to be washed out. It is an important question whether, or to what extent, the availability of mineral nutrients limits the rate of mineralization in natural ecosystems. There are examples to show that the addition of, for example, inorganic N to natural waters increases the mineralization rate of cellulose in leaf litter, but its importance is not completely understood in quantitative terms in

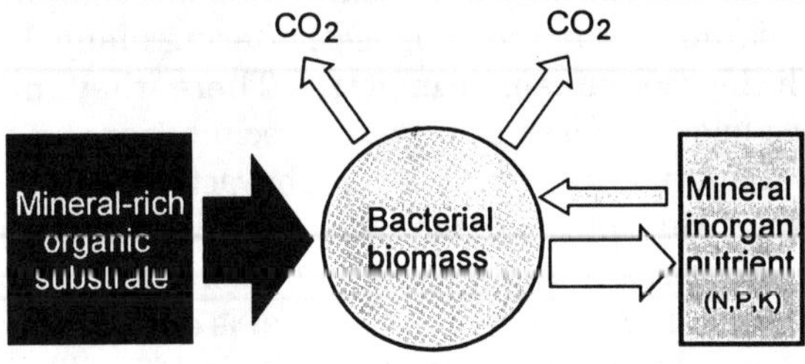

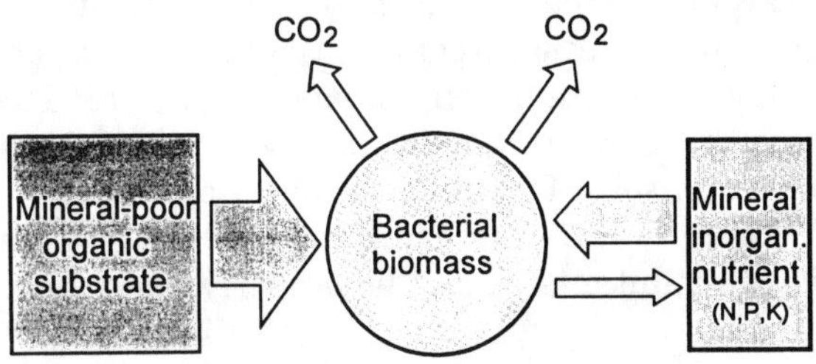

Figure 13. A schematic presentation of mineral uptake and release of mineral nutrients by bacteria decomposing mineral-rich and mineral-poor substrates, respectively (modified after Fenchel and Blackburn, 1979).

any kind of ecosystem. It is possible that, just as the availability of N and P in many systems limits the primary production, these elements also control the mineralization rate. Thus, immobilization and release of nutrients may, in part, control the relation between decomposition and production. The large steady-state pool of dead organic matter, characterizing many ecosystems, may then reflect the fact that plant tissues have lower nutrient contents than do the decomposer organisms and the primary producers and decomposers compete for essential nutrients. The large pool of dead organic matter may, however, reflect the low rate at which certain structural plant polymers can be hydrolyzed even under optimal conditions and, thus, explain the fact that decomposition lags behind production (Fenchel and Blackburn, 1979).

CONCLUSIONS

The majority of microorganisms on earth live in soils, preferably in biofilms, embedded in their polymeric network of polysaccharides, proteins, and other metabolic macromolecules. Soil is a complex matrix and includes solid, liquid, and gaseous phases, offering many different interfaces on which microorganisms can settle. There, they influence interfacial processes such as dissolution and precipitation, weathering, and the formation of humus. Soil microflora is characterized by heterogeneity in space and time. Multiple functional relationships exist among various species of microorganisms and between microorganisms and higher organisms. The degradation of organic material is one of the major roles of the soil microflora in maintaining an ecological equilibrium between production and decomposition of biological matter; it is performed mainly by bacteria and fungi. Soil microorganisms are involved in the geochemical cycles of carbon, nitrogen, phosphorus, sulfur, and many metals. Chemical factors such as mineral composition, pH, redox potential, buffer capacity, and metabolic inhibition, as well as physical factors such as diffusion, temperature, water tension, and viscosity, rule the kinetics of biotransformation in soils. Habitats can differ profoundly in microscale dimensions. Without consideration of these limits, overall data of biological degradation of pollutants cannot be properly understood.

REFERENCES

Ahmadjian, V. and Hale, M. E. (eds.) (1973): *The Lichens*. Acad. Press, New York.

Aleem, M. I. H. and Sewell, D. L. (1984): Oxidoreductase systems in *Nitrobacter agilis*. In: Strohl, W. R. and Tuovinen, O. H. (eds.): *Microbial chemoautotrophy*. Ohio State Univ. Press, Columbus, 185–210.

Alexander, M. (1965): Biodegradation: Problems of molecular recalcitrance and microbial infallibility. *Adv. Appl. Microbiol.* 7, 35–80.

Alexander, M. (1977): *Introduction to soil microbiology*. 2nd ed., John Wiley, New York.

Alexander, M. (ed.) (1984): *Biological nitrogen fixation. Ecology, Technology and Physiology*. Plenum Press, New York.

Atlas, R. M. and R. Bartha (1987): *Microbial ecology: Fundamentals and applications*. Benjamin/Cummings, Menlo Park.

Babiuk, L. A. and Paul, E. A. (1970): The use of fluorescein isothiocyanate in the determination of bacterial biomass of a grassland soil. *Can. J. Microbiol.* 16, 57–62.

Banik, S. and Dey, B. K. (1983): Phosphate solubilizing potentiality of the microorganisms capable of utilizing aluminum phosphate as a sole phosphate source. *Z. Mikrobiol.* 138, 17–23.

Bock, E., B. Ahlers and C. Meyer (1989): Biogene Korrosion von Beton und Natursteinen durch Salpetersäure bildende Bakterien. *Bauphysik* 11, 141–144.

Brown, A. D. (1990): Microorganisms and their habitats. In: Brown, A. D.: *Microbial water stress physiology*. John Wiley, New York; 18–38.

Bull, A. T. and Slater, J. H. (1982): Microbial interactions and community structure. In: Bull, A. T. and Slater, J. H. (eds.): *Microbial interactions and communities*. Academic Press, London; 13–44.

Chae, Y. and Lowe, L. E. (1980): Distribution of lipid S and total lipids in soils of British Columbia. *Can. J. Soil Sci.* 60, 633–640.

Chen, Y. and Avnimelech, Y. (1986): *The role of organic matter in modern agriculture*. Martinus Nijhoff Publ., The Hague.

Chrost, R. J. (1991): *Microbial enzymes in aquatic environments*. Springer, New York, Berlin.

Clarholm, M. and Rosswall, T. (1980): Biomass and turnover of bacteria in a forest soil and a peat. *Soil Biol. Biochem.* 12, 49–57.

de Beer, D., Stoodley, P., Roe, F. and Lewandowski, Z. (1994): Effects of biofilm structures on oxygen distribution and mass transport. *Biotechnol. Bioeng.* 43, 1131–1138.

de Boer, W., Klein Gunnewiek, P. J. A., Veenhuis, M., Bock, E. and Laanbroeck, H. J. (1991): Nitrification at low pH by aggregated chemolithotrophic bacteria. *Appl. Environ. Microbiol.* 57, 3600–3604.

Domsch, R. (1992): *Abbau von Pestiziden im Boden*. VCH Verlag, Weinheim.

Eckhardt, F. E. W. (1985): Solubilization, transport and deposition of mineral cations by microorganisms—Efficient rock weathering agents. In: J. I. Drever (ed.): *The chemistry of weathering*. D. Reidel Publ.; 161–173.

Ehrlich, H. L. (1991): *Geomicrobiology*. Marcel Dekker, New York and Basel.

Ericksen, G. E. (1983): The Chilean nitrate deposits. *Am. Scientist* 71, 366–374.

Fenchel, T. and Blackburn, T. H. (1979): *Bacteria and mineral cycling*. Academic Press, London.

Filip, Z. (1979): Wechselwirkungen von Mikroorganismen und Tonmineralien— eine Übersicht. *Z. Pflanzenernaehr. Bodendk.* 142, 375–386.

Flemming, H.-C. (1993): Biofilms and environmental protection. *Water Sci. Technol.* 27, 1–10.

Focht, D. D. and Martin, J. P. (1979): Microbiological and biochemical aspects of semiarid agricultural soils. *Ecol. Stud.* 34, 119–147.

Gisi, U. (1990): *Bodenökologie*. Thieme Verlag, Stuttgart; 304 pp.

Gray, T. R. G. (1976): Survival of vegetative microbes in soil. In: Gray, T. R. G. and Postgate, J. R: (eds.): *The survival of vegetative microbes*. Cambridge Univ. Press, Cambridge; 327–364.

Gray, T. R. G., Hissett, R. and Duxbury, T. (1973): Bacterial populations of litter and soil in a deciduous woodland. II: numbers, biomass and growth rates. *Revue d'Ecologie et Biologie du Sol* 11, 15–26.

Hill, T. C. J., McPherson, E. F., Harris, J. A. and Birch, P. (1993): Microbial biomass estimated by phospholipid phosphate in soils with diverse microbial communities. *Soil Biol. Biochem.* 25, 1779–1786.

Hooper, A. B. (1984): Ammonia oxidation and energy transduction in the nitrifying bacteria. In: Strohl, W. R. and Tuovinen, O. H. (eds.): *Microbial chemoautotrophy.* Ohio State Univ. Press, Columbus; 133–167.

Hubbell, D. H. (1986): Soil microbiology and autecological studies. In: Tate, R. L. (ed.): *Microbial autecology.* John Wiley, New York; 233–248.

Hutchins, S. R., Davidson, M. S., Brierley, J. A. and Brierley, C. L. (1986): Microorganisms in reclamation of metals. *Ann. Rev. Microbiol.* 40, 311–336.

Jenneman, G. E., McInerney, M. J. and Knapp, R. M. (1985): Microbial penetration through nutrient-saturated berea sandstone. *Appl. Environ. Microbiol.* 50, 383–391.

Kjelleberg, S. (1985): Mechanisms of bacterial adhesion at gas-liquid interfaces. In: D. C. Savage and M. M. Fletcher (eds.): *Bacterial adhesion.* Plenum Press, New York; 163–194.

Klaas, N., Ruck, W. and Flemming, H.-C. (1994): Sorption von Pestiziden an Tonmineralien und der Einfluß der Mikroorganismen. Poster, Jahrestagung der Fachgruppe Wasserchemie in der GDCH, Coburg, Mai 1994.

Knowles, R. (1985): Microbial transformation as sources and sinks of nitrogen oxides. In: Caldwell, D. E., Brierley, J. A. and Brierley, C. L. (eds.): *Planetary ecology.* Van Nostrand Reinhold, New York; 411–426.

Krumbein, W. E. (1988): Microbial interactions with mineral materials. In: D. R. Houghton, R. N. Smith and H. O. W. Eggins (eds.): *Biodeterioration 7.* Elsevier Appl. Sci., London; 78–100.

Lewis, R. W. (1965): Nitrogen. In: *Mineral facts and problems. Bull. 630,* U.S. Bureau of Mines, Washington, D.C., 621–629.

Lynch, J. M. and Poole, N. J. (1979): *Microbial ecology: a conceptual approach.* John Wiley, New York.

Marshall, K. C. (1980): Adsorption of microorganisms to soils and sediments. In: Bitton, G. and Marshall, K. C. (eds.): *Adsorption of microorganisms to surfaces.* John Wiley, New York; 317–329.

McInverney, M. J. (1986): Transient and persistent associations among procaryotes. In: J. S. Pointdexter and E. R. Leadbetter (eds.): *Bacteria in nature. Vol. 2: Methods and special applications in bacterial ecology.* Plenum Press, New York; 293–338.

McLaren, A. D. and Skujins, J. (1968): The physical environment of microorganisms in soil. In: Gray, T. R. G. and Parkinson, D. (eds.): *The ecology of soil bacteria.* Univ. of Toronto Press, Toronto; 3–24.

Morisaki, H., Kasahara, Y. and Hattori, T. (1989): The role of interfaces in microhabitats. In: Hattori, T., Ishida, Y., Maruyamara, Y., Morita, R. Y. and Uchida, A. (eds.): *Recent advances in microbial ecology.* Jap. Sci. Soc. Press, Tokyo; 123–127.

Nedwell, D. B. and Gray, T. R. G. (1987): Soils and sediments as matrices for microbial growth. In: Fletcher, M. M., Gray, T. R. G. and Jones, J. G. (eds.): *Ecology of microbial communities.* Cambridge Univ. Press, Cambridge; 21–54.

Newton, W. E. and Burgess, B. K. (1983): Nitrogen fixation: its scope and importance. In: Mueller, A. and Newton, W. E. (eds.): *Nitrogen fixation. The chemical-biochemical-genetic interface.* Plenum Press, New York; 1–19.

Odum, E. P. (1953): *Fundamentals of ecology.* W. B. Saunders, Philadelphia

Odum, E. P. (1971): *Fundamentals of ecology.* 3rd ed., W. B. Saunders, Philadelphia.

Ottow, J. C. G. (1969): Der Einfluss von Nitrat, Chlorat, Sulfat, Eisenoxydform und Wachstumsbedingungen auf das Ausmaß der bakteriellen Eisenreduktion. *Z. Pflanzenernähr. Düngung, Bodenkunde* 124, 238–253.

Paul, E. A. and Clark, F. E. (1989): *Soil microbiology and biochemistry.* Academic Press, San Diego; 273 pp.

Payne, W. J. (1981): *Denitrification.* John Wiley, New York.

Payne, W. J. (1983): Bacterial denitrification: asset or defect. *BioScience* 33, 319–325.

Petterson, A., Kunst, L., Bergman, B. and Roomans, G. M. (1985): Accumulation of aluminium by Anabaena cylindrica into polyphosphate granules and cell walls: An X-ray energy-dispersive microanalysis study. *J. Gen. Microbiol.* 131, 2545–2548.

Sand, W. and Bock, E. (1991): Biodeterioration of ceramic materials by biogenic acids. *Int. Biodet.* 27, 175–183.

Söhngen, N. L. (1915): Einfluß von Kolloiden auf mikrobiologische Prozesse. *Centralbl. f. Bakt. Abt. II Bd.* 26, 621–647.

Skujins, J. J. (1967): Enzymes in soil. In: McLaren, A. D. and Peterson, G. H. (eds.): *Soil Biochemistry.* Marcel Dekker, New York; 371–414.

Stetter, K. O., Segerer, A., Zillig, W., Huber, G., Fiala, G. Huber, R., König, H. (1985): Extremely thermophilic sulfur-metabolizing archebacteria. *System. Appl. Microbiol.* 7, 393–397.

Stevenson, F. J. (1982): *Humus chemistry: Genesis, composition, reactions.* John Wiley, New York.

Stotzky, G. (1980): Surface interactions between clay minerals and microbes, viruses and soluble organics, and the probable importance of these interactions to the ecology of microbes in soil. In: Berkeley, R. C. W., Lynch, J. M., Melling, J., Rutter, P. R. and Vincent, B. (eds.): *Microbial adhesion to surfaces.* Ellis Horwood, Chichester; 231–247.

Stotzky, G. (1985): Mechanisms of adhesion to clays, with reference to soil systems. In: Savage, D. C. and Fletcher, M. M. (eds.): *Bacterial adhesion.* Plenum Press, New York and London; 195–253.

Stotzky, G. and R. G. Burns, (1982): The soil environment: Clay—humus—microbe interactions. In: R. G. Burns and J. H. Slater (eds.): *Experimental Microbial Ecology.* Blackwell Publ., Oxford; 105–133.

Stout, J. D:, Bamforth, S. A. and Lousier, J. D. (1982): *Protozoa. Agron. Monogr. 9.*

Stumm, W. and Morgan, J. J. (1981): *Aquatic chemistry: An introduction emphasizing chemical equilibria in natural waters.* 2nd ed., John Wiley, New York.

Swank, W. T. and Fitzgerald, J. W. (1984): Microbial transformations of SO_4^- in forest soils. *Science* 223, 182–184.

Tate, R. L. III (1987): *Soil organic matter. Biological and ecological effectors.* John Wiley, New York.

Tiedje, J. M., Sextone, A. J., Parkin, T. B., Revsbech, N. P. and Shelton, D. R. (1984): Anaerobic processes in soil. *In. Dev. Plant Soil Sci.* 2, 197–212.

Williams, S. T. (1985): Oligotrophy in soils: Fact or fiction? In: Fletcher, M. M. and Floodgate, G. D. (eds.): *Bacteria in their natural environments.* Acad. Press London, 81–110.

Wong-Chong, G. M. and Loehr, R. C. (1978): Kinetics of microbial nitrite nitrogen oxidation. *Water Res.* 12, 605–609.

Wood, P. (1988): Chemolithotrophy. In: Antony, C. (ed.): *Bacterial energy transduction.* Acad. Press, London; 183–230.

Varma, A. K., Rigsby, W. and Jordan, D. C. (1983): A new inorganic pyrophosphate utilizing bacterium from a stagnant lake. *Can. J. Microbiol.* 29, 1470–1474.

Mechanisms of Organic Pollutant Transformation and Degradation by Microorganisms

MICHAEL D. AITKEN

Department of Environmental Sciences and Engineering
School of Public Health
University of North Carolina
Chapel Hill, NC 27599-7400, USA

Now I am terrified at the Earth, it is that calm and patient,
It grows such sweet things out of such corruptions,
It turns harmless and stainless on its axis, with such endless succession of
*   diseas'd corpses,*
It distills such exquisite winds out of such infused fetor,
It renews with such unwitting looks its prodigal, annual, sumptuous crops,
It gives such divine materials to men, and accepts such leavings from them at
*   last.*

—Walt Whitman

INTRODUCTION

Scientific discoveries are often followed by periods of public enthusiasm and, even among scientists, leaps of faith that are not justified by existing knowledge. In contrast, the business of refining and developing scientific principles proceeds in increments well after the public ardor has faded. Certainly, the naive perception, held a decade ago, that microorganisms could be reconstructed to solve a wide range of our most difficult pollution problems has been tempered by a realization that the simplest forms of genetic engineering may not represent much of an improvement over Nature. While we must continue to innovate and to apply untried concepts at an appropriate scale if we are to advance, we cannot afford to spend too much time or money fitting square pegs into

333

round holes. If we are to mimic natural processes to our advantage, we must look carefully at the mechanisms that have survived evolutionary pressures and, more importantly, learn from them. As Arthur Busch (1971) once noted, "Concepts must be applicable before they can be applied."

The ultimate objective of bioremediation is to gain control over the biological activity in a given system. On the premise that such control is best achieved through knowledge, the intent in this chapter is to review some of the mechanisms by which microorganisms are known to respond to and transform chemicals. The word *mechanism* can be interpreted in different ways and at various levels of detail, but I use it to describe the processes and characteristics that determine an organism's ability to transform or degrade a particular chemical in a given situation. Process engineering relies on an ability to manipulate key system variables to produce desired outcomes, and these outcomes become much more predictable if we know which variables to manipulate and in which direction to adjust them. Various strategies have been proposed to enhance biodegradation, but without a mechanistic understanding of cause and effect, it is impossible to distinguish between appropriate and inappropriate situations in which to apply them. In addition to reviewing mechanisms of pollutant transformation and degradation by microorganisms, I therefore attempt in this chapter to match these strategies with situations in which they might be practicable. The subsequent engineering challenges associated with implementing a conceptual design are significant but are beyond the scope of this chapter.

Most field bioremediation projects to date have involved the removal of contaminants that are readily biodegradable by indigenous microorganisms (Bourquin, 1993). Such contaminants usually serve as major growth substrates for the microbial community, so it is relatively straightforward to stimulate biodegradation in these situations. There are, however, several mechanisms unrelated to growth by which microorganisms can mediate pollutant transformation. Some microbes also have evolved mechanisms to increase the availability of chemicals whose low solubility in water would make them otherwise unsuitable growth substrates. Understanding these mechanisms is a key step toward developing effective approaches to bioremediation of pollutants generally regarded as "recalcitrant." In addition, more realistic models of biodegradation can be built as more mechanistic detail is incorporated (although the level of detail required will vary with the application of the model). Finally, as we learn more about microbial mechanisms, we will be able to broaden the range of chemicals for which bioremediation can be applied in the field.

TRANSFORMATION AND ITS CONSEQUENCES

The term *transformation* is distinguished from *degradation* to indicate that chemicals can be modified by microbial activity in a manner that does not necessarily lead to a reduction in ecological or human health risks. Gibson (1993) has made a similar distinction between these terms, implying that degradation represents a series of catabolic transformations that lead to products that can enter intermediary or central metabolic pathways in microorganisms.

Any change in a chemical's structure, whether major or minor, is considered to be a transformation. Transformation of some chemicals to nonmineral products has been observed to result in increases in toxicity or genotoxicity relative to the parent compound (Park et al., 1988; Wang et al., 1990; Donnelly et al., 1992; Aitken et al., 1994; Massey et al., 1994). For this reason, the need to consider toxicity endpoints during remediation projects has been recognized for a number of years (Athey et al., 1989; Sims et al., 1990). Increases in toxicity resulting from biotransformation might only be transient during bioremediation, due to the accumulation and subsequent transformation or degradation of toxic products (Wang et al., 1990; Donnelly et al., 1992). For hydrophobic chemicals, the mobility of biotransformation products in soil also might be greater than that of the parent compound (Sims et al., 1990); the desirability of an increase in mobility depends on the toxicity of the product, its potential for increasing exposure risks, or its potential to increase the product's bioavailability to organisms capable of transforming it further.

For particularly recalcitrant compounds, a single transformation step may be sufficient to trigger a series of events that lead to ultimate mineralization, assimilation, or detoxification. These subsequent reactions may be carried out by different organisms than the one that performed the initial transformation step(s). Such interactions among organisms in a consortium are difficult to demonstrate, but some recent examples relevant to pollutant biodegradation can be found in the literature (Jiménez et al., 1991; Uchiyama et al., 1992; Singh et al., 1993). The mere *potential* for subsequent metabolism of a transformation product implies that mixed cultures of microorganisms are, in general, preferable to pure cultures when attempting to maximize biodegradation. Hamer (1994) has noted that we increasingly have forsaken research involving microbial consortia while focusing on subcellular details. Details are exceedingly important in understanding how chemicals are transformed or degraded, but there are many situations in which we may focus too much on the subcellular details and not enough on intercellular or extracellular mechanisms. Tiedje (1993) suggests that research questions should shift from molecular details to environmental interactions as the number of organisms able to degrade a particular compound increases in frequency.

Transformation and degradation mechanisms are categorized by metabolic processes in the following sections of this chapter. The first level of distinction is between growth-related and nongrowth metabolism. Nongrowth metabolism is further subdivided into uncatalyzed and catalyzed reactions, and the catalyzed reactions are grouped according to whether or not they directly involve enzymes. Finally, enzyme-catalyzed reactions are classified as those involving enzymes associated with primary metabolism and those associated with secondary metabolism. Examples are provided for each type of mechanism. Transformation and degradation reactions are organized this way to emphasize that optimizing the expression of each type of mechanism in the field, or even in the lab, will require different strategies.

DEGRADATION OF GROWTH SUBSTRATES

Regardless of the mechanism by which a chemical is transformed by microorganisms, the only way to sustain transformation over extended periods of time is to ensure that the responsible organisms remain active, which in most cases, implies that those organisms must grow. Growth of organisms, communities, or consortia on the pollutants themselves is the most desirable situation in terms of bioremediation. Many microorganisms capable of growing on xenobiotics as sole carbon and/or energy sources have been isolated from various environments. Palleroni (1994) has even suggested that environmental contamination may actually be contributing to a global increase in bacterial diversity.

Growth of microorganisms can occur under aerobic conditions, in the presence of alternative electron acceptors such as nitrate, sulfate, or iron or under fermentative or methanogenic conditions. For a given chemical, however, microbes able to grow on that compound may not be found under all of these conditions. Anaerobic growth in the absence of exogenous electron acceptors would be desirable for in situ bioremediation, since many of the contaminated zones in the subsurface are likely to be anaerobic. However, limited availability of oxygen often is responsible for the persistence of contaminants in the subsurface (Lee et al., 1988), illustrating that anaerobic growth, if it occurs at all, may be far slower than we would prefer. It also is worth looking for organisms able to grow in the presence of alternative electron acceptors such as nitrate or sulfate, because nitrate and sulfate are very soluble in water and therefore would be easier to distribute in the subsurface than is oxygen. It is important to note, however, that even though facultative organisms are able to grow on some substrates using nitrate as electron acceptor, the initial transformation steps leading to biodegradation of some chemicals by these organisms often require molecular oxygen (Hutchins, 1991; Mikesell et al., 1993). It has been suggested, in fact, that the addition of nitrate and limiting concentrations of oxygen might actually increase the amount of oxygen available for such oxygenase reactions (Mikesell et al., 1993). The role of iron-reducing bacteria in subsurface transformation of pollutants is only beginning to be explored (Lovley et al., 1989, 1993; Albrechtsen and Christensen, 1994), so that generalizations cannot yet be made.

Growth implies, not only that the growth substrate is transformed, but that the organism has metabolic pathways adapted to the utilization of the substrate. Such specificity for a chemical ensures that the organism can respond to the presence of that chemical. A response is not guaranteed in all situations, however. For example, if a potential growth substrate is only a minor component within a mixture of substrates, growth on the substrate of interest is subject to the rules of microbial selection, as well as the rules of metabolic regulation. Organisms able to grow on the more dominant carbon and energy sources might compete for inorganic nutrients, electron acceptors, or space in such a way that the growth of the desired organisms is precluded or inhibited. Furthermore, the presence of alternative substrates might repress the expression of pathways for the transformation or degradation of the substrate of interest (see Hollender et al., 1994, for a recent example). Strategies designed to select for the appropriate organisms and/or to artificially induce the required transformation pathways would be necessary in these situations. One such strategy is to add carbon sources that are selective growth substrates and that also will induce the desired metabolic activity (Ogunseitan et al., 1991; Ogunseitan and Olson, 1993; Colbert et al., 1993a, 1993b). The same strategy also can be used to enhance the survivability of introduced organisms or to express nongrowth mechanisms of transformation, as discussed below.

A pollutant that can serve as a growth substrate for some microorganisms can still be resistant to biodegradation if there are no indigenous species able to degrade it. Inoculation of a system with nonindigenous species (sometimes referred to as "bioaugmentation") might help in these cases but will not necessarily lead to increases in biodegradation. Inoculated organisms must not only be present, but must also be (or become) adapted to the environmental conditions (temperature, pH, the presence of other chemicals, and other factors) (Tiedje, 1993). Effects of competition from other species would be mitigated if the pollutant(s) of concern is the major carbon or energy source in the system. Nevertheless, as Parkinson (1993) succinctly stated, "Life in the microbial world is no picnic."

The concentration of a given pollutant can also dictate whether it will be utilized as a growth substrate. Many chemicals are known to exhibit substrate inhibition at high concentrations, with phenolic compounds representing the most widely studied class of inhibitory substrates (Allsop et al., 1990). Substrate inhibition refers to the inhibition of growth or substrate utilization by the substrate itself and is distinct from inhibition that might be caused by nongrowth compounds or the inhibition manifested by competing substrates. Substrate inhibition can be overcome in reactor systems by appropriate design techniques or operating strategies (Suidan et al., 1988; Hill and Robinson, 1989; Allsop et al., 1990; Aitken, 1993). For in situ applications, there are far fewer options to control the reaction environment, so that strategies to overcome inhibition mechanisms may be difficult to implement. In general, design and operating strategies to minimize the effects of substrate inhibition are based on controlling substrate concentrations at noninhibitory levels.

Growth of microorganisms can also be difficult to achieve at extremely low concentrations of potential growth substrates. Unfortunately, it cannot be assumed a priori that concentrations low enough to preclude microbial growth will also be low enough to be of limited ecological or human health concern. Threshold concentrations for growth on a particular substrate should be quite low in most cases but should also be related to the energy available from that substrate and the efficiency with which that energy is used to synthesize biomass. The addition of selective growth substrates might be necessary to achieve meaningful transformation or degradation rates when the concentration of a pollutant is too low to sustain growth. The possibility that other factors, such as inorganic nutrient or electron acceptor concentrations, might limit growth should also be considered (Tiedje, 1993).

Competition among growth substrates for enzymes within a metabolic pathway also can lead to reduced rates of biodegradation for a given chemical. Pollutants of biogenic origin, particularly the hydrocarbons [aliphatic, monoaromatic, and polycyclic aromatic hydrocarbons (PAH)], occur in polluted environments as mixtures of related compounds. Many bacteria able to degrade these compounds have relatively broad substrate ranges. Such metabolically diverse organisms may have evolved pathways that permit the transformation of related substrates by a common set of enzymes. For example, the initial steps in the degradation of naphthalene and phenanthrene by *Pseudomonas putida* OUS82 were shown to involve the same enzymes, although the pathways for further degradation of the initial transformation products diverged (Kiyohara et al., 1994). Until recently, the implications of broad, but overlapping, metabolic capabilities with respect to biodegradation rates have not been articulated. Classic expressions for competitive inhibition of one substrate by another have now been used to describe the degradation of simple mixtures of monoaromatic hydrocarbons (Chang et al., 1993) and low-molecular-weight PAH (Stringfellow and Aitken, 1995). Competitive processes are also relevant to the transformation of nongrowth substrates, as discussed below. Detailed kinetic analyses of competitive metabolism in complex systems, particularly over extended time periods, have not been carried out.

Finally, a distinction must be made between chemicals that *cannot* serve as growth substrates because there are no known pathways for their degradation and those chemicals that *do not* serve as growth substrates because of their limited bioavailability. This might seem like an arbitrary distinction, but there is an important class of chemicals for which the latter case is relevant. High-molecular-weight PAH (those with four or more rings) are biogenic and known to be transformable by microorganisms. Only a limited number of bacteria can grow on four-ring PAH (Mueller et al., 1990; Weissenfels et al., 1990; Walter et al., 1991; Boldrin et al., 1993), and no organism is known to grow on PAH with five or more rings. Some bacteria can, however, mineralize high-molecular-weight PAH even though they are not growth substrates for those organisms (Heitkamp and Cerniglia, 1988, 1989; Heitkamp et al., 1988; Mahaffey et al., 1988; Grosser et al., 1991). These findings are puzzling because even partial mineralization implies a reasonable level of specificity and normally is expected to provide enough energy to sustain growth and to be accompanied by assimilation of some of the metabolites. If the dissolution rate for such poorly soluble compounds is the rate-limiting factor for biodegradation (Volkering et al., 1993), then it is possible that the rate at which the energy made available from high-molecular-weight PAH metabolism is too slow to sustain activity beyond that required for cell maintenance. It also has been shown clearly that limited bioavailability can preclude the degradation of otherwise readily biodegradable substrates in contaminated soils (Weissenfels et al., 1992; Erickson et al., 1993).

To summarize this section, there are five significant factors that can influence the biodegradability of pollutants able to serve as growth substrates: (1) whether the substrate is a major or minor component within a mixture, (2) whether there are indigenous degraders present, (3) whether other growth requirements (electron acceptors, environmental conditions, inorganic nutrient availability) are sufficient to achieve growth, (4) the concentration of the substrate, and (5) the availability of the substrate. Strategies to ensure that these compounds *are* degraded must be based on a recognition that one or more of these factors controls biodegradation in any situation.

NONGROWTH METABOLISM

Many organic pollutants can be transformed or degraded by microorganisms in a manner that does not lead to growth. The most significant distinction between growth substrates and those that are transformed by nongrowth mechanisms is that organisms carrying out nongrowth transformation processes are not selected by the presence of the pollutant itself (Tiedje, 1993). This simple fact creates an entirely new set of challenges in ensuring that the desired transformation reactions occur in a given situation. First, one of the greatest selection pressures for survival and proliferation of the required microorganisms (the presence of a significant carbon and/or energy source) is uncoupled from the presence of the pollutant. Additionally, the required metabolic activity is also not likely to be induced by the pollutant itself, so that alternative means of expressing that activity must be used. Finally, and perhaps most significantly, traditional methods of searching for or isolating organisms with the desired activity will not work. For all we know, every chemical that has or can be synthesized might be transformable by some microorganism somewhere [an early version of this optimistic view was expressed by Gale over 40 years ago as the "principle of microbial infallibility" (Stanier et al., 1979)]; the problem in testing this hypothesis for any given chemical is that there currently is no systematic way to find such activity. For the time being, we must continue to rely on observation of the environment to identify novel activities that, in turn, may lead to the discovery of previously unknown or unrecognized microbial capabilities. As an example, the dechlorination of polychlorinated biphenyls (PCBs) in Hudson River sediments had been observed before it was clearly established that dechlorination was microbially mediated (Alder et al., 1993).

The transformation of organic pollutants by nongrowth mechanisms is important for two reasons. First, there are a number of important pollutants for which there is no known organism able to use them as growth substrates (Table 1). Second, transformations unrelated to growth are likely to take place in the environment whether we want them to or not. These transformations can lead to the formation of products that may continue to be of concern and that can have different properties than the parent compound, as discussed previously.

Table 1. Some chemicals that have not been
documented as growth substrates for any
known microorganism.

Highly chlorinated biphenyls
PAH with five or more rings
2,3,7,8-Tetrachlorodibenzo-*p*-dioxin
Tetrachloroethylene
Trichloroethylene

Organism Groups

The ability to use xenobiotic pollutants as growth substrates is (to our knowledge) the almost exclusive dominion of bacteria, while in addition to bacteria, some fungi can grow on hydrocarbons and other biogenic pollutants. The transformation of organic pollutants under nongrowth conditions can, however, theoretically be carried out by virtually every class of organism. In the microbial world, these classes include bacteria, fungi, algae, protozoa, and microinvertebrates, but most of our knowledge and research activity remains focused on bacterial mechanisms of nongrowth metabolism. Fungi in particular are known to be important decomposer organisms in the terrestrial environment, but except for the substantial amount of research conducted in the past decade on *Phanerochaete chrysosporium* and other white-rot fungi, fungal metabolism of organic pollutants continues to be a relatively unexplored area.

Work conducted by Cerniglia and coworkers (Cerniglia and Heitkamp, 1989; Cerniglia et al., 1990; Pothuluri et al., 1990; Sutherland et al., 1991; Pothuluri et al., 1992, 1993) on PAH transformations by fungi clearly illustrated that there can be distinct differences between eukaryotic (fungal) and prokaryotic (bacterial) mechanisms. Not only are ring hydroxylation products different in fungi, but initial oxidative transformations lead to conjugates that are similar to conjugates produced as a result of mammalian detoxification mechanisms (Cerniglia and Heitkamp, 1989). In some cases, the stereospecificity of fungal metabolites can even differ from that of mammalian metabolites, and this can influence the genotoxicity of fungal metabolites compared to that of the mammalian analogues (Cerniglia et al., 1990; Pothuluri et al., 1992). There are a few other recent examples of biotransformations involving yeasts and filamentous fungi other than *P. chrysosporium* (Griffiths et al., 1992; Katayama-Hirayama et al., 1992; Polnisch et al., 1992; Field et al., 1992; Hofrichter et al., 1993; Jones et al., 1993, 1994). Studies on xenobiotic transformation by *P. chrysosporium* are summarized below.

Relatively little information is available on pollutant transformation by indigenous fungi in contaminated soils or sediments, but two recent studies indicated that bacteria appeared to play a much more significant role than fungi in situ (Yarden et al., 1990; MacGillivray and Shiaris, 1994). Field work has been conducted using inoculated fungal cultures for remediation of contaminated soil (Lamar and Dietrich, 1990; Davis et al., 1993; Lamar and Evans, 1993). Various bench-scale studies have also been described in which pure fungal cultures (the majority involving *P. chrysosporium*) have been used to treat individual contaminants or actual wastewaters (Joyce et al., 1984; Anselmo et al., 1989; Kühn and Pretorius, 1989; Yin et al., 1989a, 1989b; Kirkpatrick et al., 1990; Lewandowski et al., 1990; Prouty, 1990; Hamdi et al., 1991; Lin et al., 1991; Armenante et al., 1992; Cammarota and Sant'Anna, 1992; Sublette et al., 1992; Venkatadri et al., 1992; Alleman et al., 1995). To my knowledge, extended full-scale application of fungal processes has not been used either for soil remediation or in reactors for treatment of wastewater or contaminated groundwater. In addition, I am not aware of work in which strategies to select for fungal members of indigenous microbial communities have been attempted.

Little research has been conducted on the ability of phototrophic microorganisms (algae, cyanobacteria, and the anaerobic purple and green bacteria) to transform organic pollutants. One of the more comprehensive studies was conducted by Zepp (1983), and the limited literature prior to 1986 was the subject of a review (Mouchet, 1986). More recent examples include the transformation of heterocyclic aromatic compounds by an anaerobic phototroph (Sasikala et al., 1994), the transformation of lindane by filamentous cyanobacteria (Kuritz and Wolk, 1994), and the degradation of aromatic substrates by purple nonsulfur bacteria (Shoreit and Shabeb, 1994). Phototrophic transformation of pollutants is the subject of another chapter in this volume.

As a final note on the transformation of organic pollutants by various classes of microorganisms, there has been interest recently in the potential capabilities of rhizosphere microbial communities (i.e., those microorganisms associated with the root zone of plants) (Anderson et al., 1993). Whether transformation mechanisms unique to rhizosphere microorganisms will be discovered remains to be seen, but undoubtedly, much more will be learned about these microbial communities over the next decade.

Uncatalyzed Transformations

While most of what we know about organic pollutant transformation by microorganisms involves growing microbial cells or enzymatic activity in nongrowing cells, the possibility exists that some mechanisms of transformation can be carried out in an uncatalyzed manner. This possibility is raised only to point out that uncatalyzed reactions, if they occur, will lead to stoichiometric consumption of one or more components of the microbial culture. Such stoichiometric consumption would have direct implications for the extent to which those components must be renewed to sustain the transformation activity. One example is the direct reductive dehalogenation of haloacetophenones by NADH (Tanner and Stein, 1988), one of the major redox cofactors found in all cells. The extent to which reductants such as NADH react directly (i.e., in an uncatalyzed manner) with halogenated pollutants in vivo is unknown; a number of reductive dehalogenation reactions have also been observed to be catalyzed by metal-containing cofactors, which are discussed further in a later section.

Another possible reaction that falls into the uncatalyzed category is the reaction between a substrate and an enzyme that leads to suicide inactivation of the enzyme. Suicide, or mechanism-based, inactivation occurs when the product of an enzymatic reaction interacts with the enzyme itself, leading to an irreversible loss of activity. The stoichiometric consumption of an enzyme this way is obviously a very expensive mechanism (in terms of the investment in biosynthesis) for pollutant transformation but perhaps can be significant as a mechanism of long-term losses of pollutants in the environment. Suicide inactivation of enzymes actually can be considered to be a consequence of catalytic activity as well, since it often occurs at less than a 100 percent frequency of catalytic turnovers by the enzyme and therefore is discussed in more detail below.

A Note on Cometabolism

Cometabolism is a term that has become part of the working vocabulary in the literature on biodegradation and biotransformation, perhaps irreversibly so. For such a seemingly innocuous word, it has been the subject of extensive analysis. I interpret the controversy over its use to derive from a preference of some scientists to impart precise meanings to technical terms, and *cometabolism* apparently has a very imprecise meaning (Gibson, 1993). Its popularity might be due to the fact that it is a simple, single word; sounds like something we think we understand (metabolism); and easily leads to derivative words (cometabolize, cometabolic, cometabolite; this latter word introduces one problem, since a cometabolite, which is the substance cometabolized, is not a metabolite at all but is a substrate).

Dalton and Stirling (1982) wrote a review on cometabolism and attempted to provide a narrower definition than had been used prior to that: "the transformation of a non-growth substrate in the obligate presence of a growth substrate or another transformable compound." However, even the examples of cometabolism given by Dalton and Stirling (1982) can be achieved by resting cells in the absence of growth substrates, leading back to the original broad definition of Horvath (1972): that cometabolism does not infer the presence or absence of a growth substrate. Gibson (1993), as a result, appropriately questions why the prefix "co-" is used in cases where a growth substrate is not required. He further reiterates early concerns that processes described as cometabolism do not represent novel metabolic events—most of what has been termed cometabolism does, after all, involve enzyme-catalyzed transformations—and quotes Dagley in emphasizing the real reason we should be interested in these phenomena at all: "The disappearance of a compound from a system may require a complex analysis before we can understand precisely what has happened. To say that its disappearance is due to the 'phenomenon of cometabolism' is about as helpful as saying that Beethoven's fifth symphony is due to the 'phenomenon of sound' " (Gibson, 1993).

Terms other than cometabolism have also been used to describe the transformation of nongrowth substrates: "fortuitous" and "adventitious" metabolism are examples. Both of these terms imply that the transformation reaction is not intended (i.e., the transformation capability is not induced or selected as a result of the presence of the substrate) by the organism carrying it out. Two examples can be given, however, of nongrowth transformations that are probably fully intended by microorganisms. One example is the detoxification of organic (or inorganic) pollutants, such as the transformation of PAH by fungi as described above. A second example is the degradation of lignin by wood-rotting fungi. Lignin, the second largest terrestrial source of organic carbon, does not serve as a sole carbon source for any microorganism; instead, it is degraded under conditions of secondary (nongrowth) metabolism (Kirk, 1984). Certainly, the ability to elicit a metabolic capability to transform lignin and to utilize its degradation products offers substantial selective advantages to the organisms involved. Higgins et al. (1980) also suggest that the ability of methane monooxygenases to catalyze the oxidation of a wide range of organic compounds, including halogenated aliphatics, is not entirely accidental and probably is of survival value to methanotrophic bacteria.

Since it is clear that there are distinctions among the various processes for which we have used the term *cometabolism,* it seems that we should focus instead on these distinctions. Particularly important issues are: How long can a particular transformation capability be sustained in the absence of a growth or energy substrate? Is the turnover of cellular cofactors necessary to sustain the activity? Are suicide products formed as a result of the transformation? How much energy is used per unit of transformed substrate? Is any part of the original substrate assimilated into biomass? What is necessary to express the transformation activity? These questions are addressed below.

Nonenzyme Catalyzed Transformations

The reductive dehalogenation of halogenated organic compounds is the most relevant example of reactions that are catalyzed by biological catalysts other than enzymes. Iron porphyrins, which normally are found as prosthetic groups in heme-containing enzymes, have been observed in their reduced [Fe(II)] form to catalyze the reductive dechlorination of a variety of compounds (Klecka and Gonsior, 1984; Marks et al., 1989; Baxter, 1990; Gantzer and Wackett, 1991; Marks and Maule, 1992; Schanke and Wackett, 1992). Two other metal cofactors, the nickel-containing coenzyme F_{430} found in methanogenic bacteria (Krone et al., 1989; Gantzer and Wackett, 1991) and the cobalt-containing cyanocobalamin (vitamin B_{12}) (Marks et al., 1989; Gantzer and Wackett, 1991; Assaf-Anid et al., 1992; Marks and Maule, 1992; Schanke and Wackett, 1992; Assaf-Anid et al., 1994; Becker and Freedman, 1994; Bosma et al., 1994) have also been observed to catalyze reductive dehalogenation reactions. Gantzer and Wackett (1991) have compared the rates of reductive dehalogenation for all three cofactors with carbon tetrachloride and a series of chlorinated ethylenes.

Reductive dehalogenation reactions have been known to occur under anaerobic conditions for 30 years (Mohn and Tiedje, 1992). These reactions have been observed in the environment, in anaerobic reactors, and in pure cultures (Mohn and Tiedje, 1992). The extent to which reductive dehalogenations are catalyzed in vivo by the reduced metal cofactors identified above, or by enzymes, is largely unknown (Wackett, 1994). Although the catalysis of reductive dehalogenation reactions has been observed in cell-free extracts (Mohn and Tiedje, 1992), few studies have attempted to evaluate the relative role of enzymes versus nonprotein catalysts. One reason for this is that, like most catalytic activity derived from cells, it is difficult to isolate the activity of interest. Egli et al. (1990) inferred that the reductive transformation of chlorinated methanes was not an enzymatic process in two different anaerobic microorganisms because the activity was retained in heat-treated cell extracts. Stromeyer et al. (1992) separated the nonprotein components of cell extracts from *Acetobacterium woodii* into several fractions that exhibited reductive dehalogenation capability. Only one of these fractions contained vitamin B_{12}, but the protein fractions from the same cell extracts correlated poorly to the dehalogenation activity. Conversely, DeWeerd and Suflita (1990) found that reductive dehalogenation activity was proportional to the protein content in cell extracts of *Desulfomonile tiedjei*. The ability of chlorinated compounds to serve as electron acceptors in growth-related processes (Mohn and Tiedje, 1992; Holliger et al., 1992; Cole et al., 1994) and the inducibility of this activity (Cole et al., 1994), suggest that dehalogenation in these cases is an enzyme-catalyzed process. Reductive dechlorination of 2,4,6-trichlorophenol appeared to be related to growth in another recent study (Madsen and Aamand, 1992).

It is important to determine whether reductive dehalogenations are enzyme-catalyzed or not, because it is difficult to stimulate a particular biological activity in the environment or in an engineered system if we do not understand the basis of the activity. For example, if the activity is primarily associated with reduced metal cofactors, then methods to stimulate the synthesis of these cofactors should be evaluated (a relevant question might be whether the metal itself is a limiting factor in the environment or in a reactor). As proteins, enzymes are direct gene products (usually one gene, but at most a few genes) whereas cofactors are synthesized via a series of enzymes that are themselves products of a series of genes. This distinction can have implications for our ability to artificially induce the desired activity, to incorporate the activity into other organisms via recombinant techniques, or for horizontal transfer of the activity in the environment or in a reactor. It is also not possible to tell yet whether the electron donors (i.e., the expendable reductants) required for reductive dehalogenations might be different for reactions catalyzed by enzymes versus those catalyzed by metal cofactors.

Natural and engineered anaerobic systems always contain heterogeneous communities of microorganisms. Given our inability to identify reductive dehalogenation mechanisms even in pure cultures, it is that much more difficult to elucidate such mechanisms in complex systems. For example, site-specific factors seemed to affect dechlorination patterns of polychlorinated biphenyls (PCBs) in sediments from three different sites (Quensen et al., 1990; Williams, 1994), but these factors remain unknown. One approach to studying complex systems has been to use selective methods of inhibiting specific organism groups to determine which one(s) might be responsible for the dehalogenation activity; such approaches have been used to implicate methanogens, nonmethanogens, sulfate reducers, and spore-forming anaerobes (not all necessarily mutually exclusive) in different studies (Fathepure and Boyd, 1988; Freedman and Gossett,1989; Kuhn et al., 1990; de Bruin et al., 1992; Madsen and

Aamand, 1992; Mohn and Kennedy, 1992; Ye et al., 1992; Perkins et al., 1994). Although these approaches do not identify the specific subcellular mechanisms responsible for dehalogenation activity, they are a good place to start. Without sufficient information on the major mechanisms of reductive dehalogenation in complex systems, however, we have a limited range of options for optimizing that activity. Activity associated with sulfate reducers would be straightforward to stimulate, but our only recourse for other anaerobes might, for now, be in the selection of organic substrates or other electron donors (e.g., H_2) added to the system. A variety of organic substrates have been used to stimulate reductive dehalogenation, and in some cases, the extent or rate of dehalogenation depended on the substrate used (Nies and Vogel, 1990; Gibson and Sewell, 1992; Holliger et al., 1992). It is not usually clear, however, whether different substrates vary in their potential to stimulate the production of direct electron donors (i.e., the ultimate reductants in the dehalogenation reactions) or in their ability to select for different populations within the anaerobic consortium.

Enzyme-Catalyzed Transformations

Both xenobiotic and biogenic pollutants have been observed to be transformed in enzymatic reactions that do not lead to growth of the microorganisms involved. When the enzymes responsible for these transformations are synthesized as a result of primary metabolism (i.e., during the growth of the organism on some other substrate), then the addition of an appropriate growth substrate will at least ensure that the desired enzyme(s) will be produced. When the required enzymes are only expressed as a result of secondary metabolism, however, then a much more complex situation arises. Nongrowth transformations catalyzed by enzymes of primary and secondary metabolism are discussed separately below.

ENZYMES OF PRIMARY METABOLISM

The best, and furthest advanced, example of pollutant transformation in this category is the oxidation of chlorinated aliphatic solvents by methanotrophic bacteria. The original observations that methane monooxygenases from methanotrophic bacteria could catalyze the oxidation of various halogenated substrates were made over 15 years ago (see Dalton and Stirling, 1982). This information seemed to remain obscure to the environmental research community until Wilson and Wilson (1985) reported on the aerobic degradation of trichloroethylene in soil columns exposed to natural gas. Other laboratory investigations on the oxidation of several chlorinated organic solvents followed (e.g., Fogel et al., 1986; Strand and Shippert, 1986), and quantitative studies on rate and stoichiometric parameters as well as reactor applications emerged several years later (Oldenhuis et al., 1989; Strandberg et al., 1989; Tsien et al., 1989; Broholm et al., 1990; Strand et al., 1990, 1991; Alvarez-Cohen and McCarty, 1991a, 1991b, 1991c, 1991d; Arvin, 1991; Oldenhuis et al., 1991; Bilbo et al., 1992; Broholm et al., 1992; Speitel and Leonard, 1992; McFarland et al., 1992; Fennell et al., 1993). Small-scale, but comprehensive, field studies were conducted beginning in the late 1980s (Semprini et al., 1990, 1991; Semprini and McCarty, 1991, 1992).

More recent work has indicated that organisms able to grow on aromatic substrates such as toluene and phenol are able to catalyze the transformation of chlorinated aliphatic compounds as well (Nelson et al., 1987, 1988; Wackett and Gibson, 1988; Shields et al., 1989; Wackett and Householder, 1989; Zylstra et al., 1989; Folsom et al., 1990; Harker and Kim, 1990; Nelson et al., 1990; Folsom and Chapman, 1991; Shields et al., 1991; Fan and Scow, 1993; Hopkins et al., 1993; Landa et al., 1994; Malachowsky et al., 1994; Mu and Scow, 1994). Results from much of this work parallel the findings of research on methanotrophic transformations of chlorinated compounds and therefore are not discussed further here.

Methane monooxygenases (MMOs) are intracellular enzymes that catalyze the first step in the biodegradation of methane by methanotrophic bacteria: the oxidation of methane to methanol using molecular oxygen as the oxidant. Some MMOs are also able to catalyze the oxidation of a variety of other compounds besides methane, including several chlorinated aliphatic compounds, but these other compounds are not able to support the growth of methanotrophic organisms (Dalton and Stirling, 1982). It is the ability of MMO to catalyze the oxidation of chlorinated organic solvents that has generated the most interest in applications of methanotrophic bacteria for bioremediation.

Several key concepts have emerged on both quantifying and optimizing the transformation of chlorinated hydrocarbons by methanotrophs. First, it became apparent that the growth substrate for methanotrophs—methane—would compete with the nongrowth substrate(s) for oxidation by MMO (Strand and Shippert, 1986; Broholm et al., 1990). The competition between methane and other substrates can be described by classic competitive inhibition kinetics for enzymes (Broholm et al., 1990, 1992; Alvarez-Cohen and McCarty, 1991b; Semprini et al., 1991; Criddle, 1993; Speitel et al., 1993; Anderson and McCarty, 1994). Since a growth substrate is required to sustain a culture's long-term capability to transform a nongrowth substrate, competition between substrates presents an obvious challenge in optimizing nongrowth transformations. Reactor designs have been developed to overcome the effects of competitive inhibition (Alvarez-Cohen and McCarty, 1991d; McFarland et al., 1992; Speitel and Leonard, 1992), based on a recognition that growth does not necessarily have to be continuous. In essence these reactors separate, either in space or time, the growth of a methanotrophic culture on methane and exposure of the culture to the halogenated nongrowth substrates. Simulation models incorporating mechanistic details can be used to help optimize the reactor conditions for growth and subsequent transformation of nongrowth substrates in these reactors (Alvarez-Cohen and McCarty, 1991d).

Another challenge in applying methanotrophic cultures to transform halogenated organic compounds is that there is a finite amount of nongrowth substrate that can be transformed per unit of active biomass. The existence of a finite transformation capacity for nongrowth substrates was first articulated by Alvarez-Cohen and McCarty (1991a, 1991c) and leads to an important stoichiometric link between the amount of growth substrate ultimately required and the transformation of a given mass of nongrowth substrate. There are at least two factors that lead to the loss of a methanotrophic culture's capability to transform nongrowth substrates. Understanding the effect of the first of these factors requires some insight into the biochemistry of methane oxidation by methanotrophs. MMO requires NADH as a reducing cofactor, as shown in Figure 1; the NADH used in this reaction is often referred to as "reducing power." In the complete metabolism of methane, the NAD^+ generated during the oxidation of methane is converted back to NADH in either of two subsequent enzymatic steps: the oxidation of formaldehyde to formate and the oxidation of formate to carbon dioxide (Figure 1). NADH is consumed stoichiometrically when nongrowth substrates are transformed by MMO and are not metabolized further, so that other cellular processes have to make up for the loss in reducing power. Endogenous metabolism of storage compounds can provide some of the reducing power (Henry and Grbić-Galić, 1991a; Henrysson and McCarty, 1993), but the addition of formate to the culture has been used for this purpose and has been

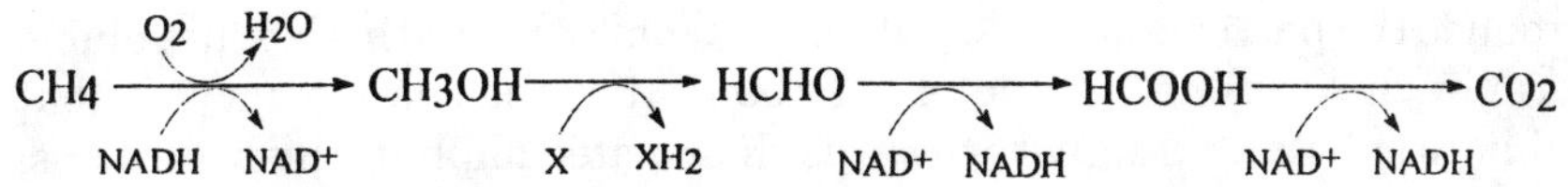

$$CH_4 \longrightarrow CH_3OH \longrightarrow HCHO \longrightarrow HCOOH \longrightarrow CO_2$$

Figure 1. Pathway for methane mineralization by methanotrophs. Note regeneration of NADH in the last two steps. X = reducible prosthetic group of methanol dehydrogenase (after Dalton and Leak, 1985).

observed to enhance the effective transformation capacity for halogenated compounds (Oldenhuis et al., 1989; Alvarez-Cohen and McCarty, 1991a, 1991b; Henry and Grbić-Galić, 1991a; Semprini et al., 1991; McFarland et al., 1992; Speitel and Leonard, 1992).

The toxicity of transformation products also decreases the capacity to transform nongrowth substrates via MMO in methanotrophic cultures. Toxic effects of chlorinated aliphatic hydrocarbon transformations have been observed by several investigators (Oldenhuis et al., 1989, 1991; Broholm et al., 1990; Alvarez-Cohen, 1991a, 1991b, 1991c; Henry and Grbić-Galić, 1991a; Speitel and McLay, 1993). Experimental evidence indicates that the toxicity is manifested at least in part by reaction of the transformation products with MMO itself, inactivating the enzyme. Fox et al. (1990) obtained direct evidence for in vitro inactivation of the soluble MMO isolated from *Methylosinus trichosporium* OB3b during the oxidation of trichloroethylene (TCE). The products of TCE oxidation have been shown to lead to the formation of protein adducts with MMO in vitro (Fox et al., 1990) and in vivo (Oldenhuis et al., 1991), thus causing suicide inactivation of MMO.

Henry and Grbić-Galić (1991b) described an indirect effect of transformation products on nongrowth substrate transformation. Carbon monoxide was observed as a product of TCE oxidation by a methanotroph and was itself susceptible to oxidation via MMO. Carbon monoxide therefore influenced the rate and extent of TCE transformation by competing with TCE and by consuming some NADH that otherwise could have been used for TCE oxidation, respectively.

Another important concept that has emerged from studies on chlorinated solvent transformations by methanotrophs is that not all methanotrophic cultures have the same capability to carry out these transformations (Broholm et al., 1993). Methanotrophic bacteria have been grouped into several taxonomic types (Type I, Type II, Type X) based on morphological and physiological traits (Dalton and Leak, 1985), although the extent to which chlorinated solvents and other nongrowth compounds can be transformed by individual species may not necessarily correlate to the taxonomic grouping (Koh et al., 1993). Methanotrophs produce two types of MMO: a membrane-associated or particulate enzyme (pMMO) and a water soluble enzyme (sMMO). All methanotrophs appear to produce the particulate MMO, while only some can synthesize the soluble form of the enzyme (Dalton et al., 1990). It is the soluble enzyme that exhibits substantial activity towards chlorinated aliphatic hydrocarbons, but pMMOs have also been shown to oxidize TCE at relatively low rates (DiSpirito et al., 1992). The synthesis of sMMO is regulated by the level of copper in the growth medium, so that sMMO is only synthesized in response to copper limitation. Several investigators have shown that the transformation of chlorinated hydrocarbons in methanotrophic cultures correlates to low copper concentrations (Tsien et al., 1989; Newman and Wackett, 1991; Oldenhuis et al., 1991; Fennell et al., 1993; Koh et al., 1993).

In principle, pure cultures with superior transformation capabilities could be used in aboveground reactors, but few practical applications of biological treatment processes involve pure cultures. In situ bioremediation applications would rely on stimulating native methanotrophs (Semprini et al., 1990, 1991; Semprini and McCarty, 1991, 1992), which are ubiquitous in the terrestrial environment. Therefore, it is important to be aware that stimulation of a mixed microbial community with methane is likely to enhance the growth of a mixture of methanotrophs with variable abilities to transform nongrowth substrates. The implication of this fact is that specific rates of transformation of nongrowth substrates (i.e., rates per unit biomass) will always be lower than the maximum rate measured for any member of the microbial community, unless further selection of that member from the community can be achieved. One such additional selective pressure might be to control the available copper concentration in the medium, although this obviously would be difficult to accomplish in the subsurface.

In summary, there are several important lessons to be learned from the knowledge we have gained on the transformation of nongrowth substrates by methanotrophic bacteria. First, very little of what has been learned about applying and optimizing the process would have been possible if the biochemistry, microbial physiology, and microbial ecology of methane oxidation had not already been well studied. In other words, sound science is prerequisite to sound process engineering. Second, it is clear that, when the enzymes responsible for transformation of nongrowth substrates are associated with primary (growth) metabolism, the possibility for competition between growth and nongrowth substrates must always be considered. Additionally, the in vivo chemistry of the enzyme itself must be understood if the limitations of the process are to be defined. The reactivity and potential toxicity of transformation products need to be considered as well. Finally, it must be recognized that the use of a growth substrate alone may not be sufficient to ensure that the desired members of a microbial community are selected predominantly over other members of the community.

ENZYMES OF SECONDARY METABOLISM

Many bacteria and fungi begin to secrete metabolites into the medium at the onset of the stationary phase in batch culture. Such metabolites are referred to as secondary metabolites, and the culture is said to be undergoing secondary metabolism or idiophasic metabolism. Secondary metabolism is associated with processes not essential to growth but depends, to a great extent, on the medium used for growth (Bennett and Bentley, 1989; Brock et al., 1994). Secondary metabolites are specific to the organism producing them and represent a far more diverse range of chemicals than those found intracellularly during primary metabolism (growth) (Bennett and Bentley, 1989; Brock et al., 1994). A number of metabolic pathways become induced (or de-repressed) during the stationary phase in batch culture, and relatively little is known about the regulation of these pathways (Kolter et al., 1993). It has been noted, however, that shifts in metabolism associated with the onset of the stationary phase are not subject to "on-off" control but, instead, increase in significance as growth rate declines; in other words, slow continuous growth can overlap with processes normally associated with secondary metabolism in batch culture (Bu'Lock, 1975; Demain, 1986; Bennett and Bentley, 1989; Kolter et al., 1993).

The influence of secondary metabolism on pollutant transformation and degradation has not been studied extensively for most microorganisms and chemicals of interest. There is, however, one case that illustrates the potential benefit of secondary metabolic processes in biotransformation and biodegradation: the transformation of a wide range of pollutants by *Phanerochaete chrysosporium* and other wood-rotting fungi. In 1985, Bumpus and coworkers (Bumpus et al., 1985) reported on the ability of *P. chrysosporium* to mineralize several pollutants regarded to be highly recalcitrant in the environment. Since then, a large body of literature has emerged on the biotransformation and biodegradation capabilities of *P. chrysosporium* and related species of white-rot fungi (Hammel, 1992; Barr and Aust, 1994). Other chapters in this volume focus specifically on these capabilities, which are summarized briefly below.

The degradation of lignin, the major structural component of wood, is carried out by white-rot fungi as a secondary metabolic process (Kirk, 1984). A wide range of organic pollutants, particularly aromatic compounds, has been observed to be transformed or degraded by *P. chrysosporium* under secondary metabolic conditions as well (Bumpus et al., 1985; Mileski et al., 1988; Ryan and Bumpus, 1989; Valli and Gold, 1991; Hammel et al., 1991; Valli et al., 1992; Spadaro et al., 1992; Pasti-Grigsby et al., 1992; Joshi and Gold, 1993; Armenante et al., 1994). Lignin degradation in *P. chrysosporium* is correlated with the production of two types of extracellular peroxidases: lignin peroxidase and manganese peroxidase (Cai and Tien, 1993). The direct or indirect involvement of these lignin-degrading enzymes in pollutant transformations has been shown for several classes of compounds (Hammel et al., 1986; Hammel and Tardone, 1988; Mileski et al., 1988; Valli and Gold, 1991; Valli et al., 1992; Pasti-Grigsby et al., 1992; Joshi and Gold, 1993; Aitken et al., 1994; Moen and Hammel, 1994; Chung and Aust, 1995), although some transformation reactions clearly do not involve these enzymes or can be achieved during primary metabolism (Sutherland et al., 1991; Hammel et al., 1992; Dhawale et al., 1992; Thomas et al., 1992; Yadav and Reddy, 1993; Mougin et al., 1994; Yadav et al., 1995). Valli and Gold (1991) illustrated how the extracellular peroxidases can interact with intracellular enzymes in the complete mineralization of 2,4-dichlorophenol, and similar mechanisms involving both intracellular and extracellular enzymes have been suggested for the mineralization of 2,4-dinitrotoluene (Valli et al., 1992) and trichlorophenols (Joshi and Gold, 1993; Armenante et al., 1994).

Secondary metabolism has been studied in great detail for industrially important secondary metabolites such as antibiotics (Brock et al., 1994). Industrial fermentations, however, involve very well characterized microorganisms grown under highly controlled conditions in batch processes. It is an entirely different matter to control the expression of secondary metabolism in reactors used to treat complex waste mixtures or in situ in contaminated soil. In general, conditions that permit intermittent growth, followed by extended periods of starvation, should lead to the periodic onset of secondary metabolism by the microorganisms present in a reactor or in a field environment. As with any transformation activity not induced by the presence of the pollutant itself, growth of the microorganism(s) with the desired transformation activity must be achieved with selective growth substrates. One potential problem is that secondary metabolism by pure cultures often is achieved with very common growth substrates, which would not be selective at all in mixed cultures. For wood-rotting fungi, the use of lignocellulosic substrates clearly is one way to select for this group of organisms (Lamar and Glaser, 1994).

For potential secondary metabolic reactions other than those carried out by wood-rotting fungi, we have the same problem in searching for the activity of interest as with other nongrowth processes: there is no systematic way to identify such activity in environmental samples. Even when a unique activity seems to be observed in the environment or in environmental samples in the laboratory, the source of the activity might elude detection if the transformation reaction is the result of a secondary metabolic process. Secondary metabolism might be indicated if the activity of interest correlates with the production of what appear to be secondary metabolites.

STRATEGIES

There is a common set of ingredients necessary to ensure biotransformation or biodegradation of organic pollutants by any of the mechanisms described in the above sections. First, one or more organism(s) with the required activity must be *present* in the system of interest or else must be *added* to the system. Second, the required activity must be *expressed* before any reaction can be expected to take place. And third, once expressed, the required activity also must be *manifested*; in other words, other processes (e.g., competing reactions or mass transfer limitations) must not interfere extensively with the required activity. Various strategies have been proposed to enhance biodegradation, but the need to match these strategies with specific transformation or degradation mechanisms is not always articulated. Furthermore, the suggestion to implement such strategies is not often accompanied by recognition of the factor(s) limiting pollutant transformation or degradation in the first place, whether it be the *presence* of microorganisms or the *expression* or *manifestation* of the activity of interest.

Microbial Strategies

Although the purpose of this section is to outline some of the approaches proposed to enhance bioremediation or biological treatment processes, it should be recognized that microorganisms have themselves evolved several strategies to respond to organic chemicals. These microbial strategies are mentioned here only to illustrate that many of our own approaches attempt to mimic the microbial processes.

The ability of microorganisms to adapt to changing conditions, including the introduction of xenobiotic chemicals to their environment, has been well established. Mutation and gene transfer represent powerful mechanisms by which such adaptation occurs, but the extent to which they have led to the selection of microbes able to degrade xenobiotics in the environment is difficult to ascertain (van der Meer et al., 1992). Furthermore, it is not possible to predict, for any given situation, the time scale over which adaptive responses will occur. Our current approaches to pollution control involving genetic manipulation of microorganisms represent efforts to speed up or tailor the kinds of genetic changes that might occur naturally in the environment over long enough periods of time.

One of the biggest problems encountered in the bioremediation of contaminated soil is that many pollutants are hydrophobic and, therefore, are associated with nonaqueous phases; such nonaqueous phases can include mineral surfaces, soil organic matter, or nonaqueous phase liquids. Rates of biodegradation of hydrophobic compounds can be limited by the rate at which they dissolve (Volkering et al., 1992, 1993) or the rate at which they desorb (Robinson et al., 1990; Weissenfels et al., 1992; Harms and Zehnder, 1995) from the nonaqueous phases. The uptake of insoluble or sorptive substrates from the aqueous phase indirectly leads to increases in mass transfer processes by maximizing the concentration gradient against which dissolution or desorption take place. Once aqueous phase concentrations are reduced to near-zero (defined as concentrations well below half-saturation coefficients for microbial activity), however, more direct mechanisms to improve bioavailability are required.

Microorganisms have evolved mechanisms of enhancing the availability of water-insoluble chemicals, the best studied examples of which are the aliphatic hydrocarbons. These mechanisms have been studied primarily with respect to growth on or biodegradation of a bulk hydrocarbon phase. Little is known about mechanisms by which microbes might gain access to sorbed chemicals other than by attachment to and growth on the sorptive surface (van Loosdrecht et al., 1990). For chemicals that are sorbed in intraparticle pore spaces that may not be physically accessible to bacteria, no direct mechanism of enhancing desorption has been described in the literature.

Organisms that grow on aliphatic hydrocarbons have developed two strategies for improving hydrocarbon bioavailability: (1) specific mechanisms of adhesion to the hydrocarbon-water interface and (2) emulsification or solubilization of the hydrocarbon (Hommel, 1990; Rosenberg, 1991; Rosenberg et al., 1992; Georgiou et al., 1992). Both adhesion at the hydrocarbon-water interface and emulsification or solubilization of hydrocarbons have been associated with the production of extracellular biosurfactants. The role of biosurfactants in adhesion may be to serve as an effective bridge between the cell surface and the hydrocarbon surface (Gerson, 1993), but surfactants added exogenously might just as easily hinder the attachment of microbes to hydrocarbons (Gerson, 1993; Ortega-Calvo and Alexander, 1994; Zhang and Miller, 1994). Furthermore, detachment and redevelopment of adhesive cell surface properties have been suggested to be essential in the life cycle of bacteria able to adhere to the hydrocarbon surface (Rosenberg et al., 1992). Such reversible mechanisms of attachment might be difficult to mimic with synthetic surfactants or even biosurfactants added exogenously.

Emulsification of a hydrocarbon by a biosurfactant has been proposed as a means of increasing the effective surface area over which hydrocarbon dissolution or bacterial attachment takes place (Rosenberg et al., 1992). The extent to which emulsification might take place in environmental systems (as opposed to in a reactor) is open to question—it is not clear that the high surfactant concentrations or mechanical energy input required to achieve substantial emulsification can occur in the environment, particularly the microenvironments associated with subsurface microbial communities. It is likely that solubilization of hydrocarbons in biosurfactant micelles is more relevant in subsurface systems.

Micellar solubilization of a hydrocarbon can increase its total concentration in the liquid phase, which is a good place for stimulation of microbial activity to occur. However, unless the microorganism has direct access to the hydrocarbon inside the micelles, the partitioning of hydrocarbon between the micellar and aqueous phases might still influence its bioavailability and hence the rate of biodegradation (Grimberg and Aitken, 1995). Direct mechanisms for microbial uptake of micellized hydrocarbons remain speculative. Hommel and Ratledge (1993) have suggested the possibilities of direct cell-to-micelle or cell-to-emulsified droplet contact (thus precluding the need for mass transfer via the aqueous phase) or the existence of specific channels in the microbial cell envelope, through which micelles would enter and transfer hydrocarbon to the interior of the cell. Even if indirect mass transfer from the micelle to the aqueous phase to the microbial cell surface were required, the transfer of hydrocarbon between micelles and the aqueous phase is likely to be considerably faster than the processes of dissolution from a nonaqueous phase liquid or desorption from the interior of a porous aggregate.

Biosurfactants are known to be produced during growth on aliphatic hydrocarbons, and some pure cultures will produce biosurfactants when grown on polar substrates such as sugars (Hommel, 1990; Rosenberg, 1991; Rosenberg et al., 1992; Georgiou et al., 1992). It seems, then, that the production of some biosurfactants is inducible by the hydrophobic substrate (a desirable situation for hydrophobic pollutants other than aliphatic hydrocarbons) and that some biosurfactants appear to be similar to other secondary metabolites in that they are synthesized under general growth-limiting conditions (Syldatk and Wagner, 1987). Production of nonspecific biosurfactants under secondary metabolic conditions might be helpful in the absence of microbes that can produce biosurfactants during growth on the hydrophobic pollutant of concern, unless highly specific interactions between the cell and the surfactant (such as those described above) are required for surfactants to have a significant impact on biodegradation.

Human Interventions

The various strategies that have been suggested to improve bioremediation can be summarized as efforts to (1) control the reaction environment to preclude limitations in nutrient supply or for aerobic processes, oxygen supply, and to optimize other environmental conditions (e.g., pH); (2) control the reaction environment to preclude inhibitory or toxic concentration effects; (3) select for and enrich a system with indigenous microorganisms containing the desired metabolic capabilities (including the addition of alternative electron acceptors or selective growth substrates); (4) add specialized microorganisms to a system; and (5) enhance mass transfer kinetics. Some of these strategies are well established or have advanced to the field level and are covered elsewhere in this volume. I therefore elaborate on only two of them, enhancement of mass transfer and the addition of specialized microorganisms, because the efficacy of these two approaches is perhaps the least proven.

MASS TRANSFER ENHANCEMENT

Mass transfer processes (dissolution or desorption) can be improved in aboveground reactors by providing a reasonable degree of mixing. For contaminated soils, one way to do so would be to use slurry reactors for treatment. The use of slurry reactors is described in other chapters in this volume and, therefore, is not considered further here. An alternative means of enhancing mass transfer, either in situ or in aboveground reactors, is to add agents that can influence dissolution or desorption processes. Polar organic solvents, for example, can influence mass transfer processes, but the concentrations that might be necessary to have a significant impact on mass transfer are also likely to add a tremendous organic load to the already contaminated system; as a result, solvent-based processes are not likely to be coupled to biodegradation. Surfactants, on the other hand, may achieve significant improvements in mass transfer at reasonably low concentrations. Additionally, coupling biodegradation to surfactant-based soil flushing or soil washing may permit reuse of the surfactant in continuous or semi-continuous processes [unless the surfactant biodegrades at a rate similar to that of the pollutant(s)].

The suggested approaches to improving the bioavailability of hydrophobic chemicals with surfactants are to add surfactants to the aqueous phase directly or to add biosurfactant-producing microorganisms to the system (Georgiou et al., 1992; Müller-Hurtig et al., 1993; Finnerty, 1994; Rouse et al., 1994). However, the simple addition of surfactants, either biological or synthetic, in experimental systems containing hydrophobic pollutants has led to results ranging from inhibition to no effect to stimulation of biodegradation (reviewed in Rouse et al., 1994, and summarized in Liu et al., 1995). Little of this prior work on the effects of surfactants on biodegradation has first considered the influence of the surfactant(s) on the mass transfer processes themselves. Obtaining such knowledge is of fundamental importance before the effects on biodegradation can be predicted or controlled. Also, specific interactions between surfactant monomers or micelles and the microbial cells, such as those described above for biosurfactants (also see Rouse et al., 1994), may play an important role in determining the efficacy of surfactant addition for bioremediation.

SPECIALIZED MICROORGANISMS

The use of genetic engineering has probably captured the attention of the public more than any other strategy proposed to improve the biodegradation of organic pollutants. From a naive perspective, it is logical to believe that, since we have tremendous problems with chemicals in the environment that don't seem to biodegrade easily and since we have the technology to create microorganisms with novel metabolic capabilities, then we should be able to construct microorganisms able to biodegrade these recalcitrant pollutants. Such logic belies the fact that, in most cases, the existence of a microorganism with an appropriate metabolic capability is not the factor limiting biotransformation of a chemical in the environment (although more extensive biodegradation of the same chemical might be a different story). Recombinant or mutant organisms face the same challenges of expressing and manifesting the required activity as do indigenous microorganisms, with the added challenge that nonindigenous organisms must first survive in their new environment. In fact, Lindow et al. (1989) have stated that "it is presently difficult or impossible to transfer all the genetic determinants enabling a bacterium to survive in a habitat to which it is not already adapted." As a result, they suggest that the best approach is to modify indigenous organisms isolated from the environment of interest. Another possibility, though, is the transfer of introduced genes to indigenous organisms (Nüßlein et al., 1992).

Genetic approaches to improving bioremediation potential fall into three categories: pathway engineering, modifying the regulation of gene expression, and protein engineering. In pathway engineering, components from different metabolic pathways (i.e., genes coding for enzymes from different pathways) are patched together in a single organism to create a complete pathway for the degradation of one or more chemicals that previously could only be transformed partially (Ensley, 1994). This technique has been used to permit the simultaneous degradation of chloro- and methylaromatics by a single organism (Rojo et al., 1987), the extensive degradation of halogenated aliphatic compounds (Wackett et al., 1994; Hur et al., 1994), the degradation of both chlorobenzoate and toluene under denitrifying conditions (Coschigano et al., 1994), and the utilization of trinitrotoluene as the sole carbon and nitrogen source (Ensley, 1994). In a similar manner, the substrate range of a single enzyme, a multicomponent aromatic dioxygenase (i.e., an enzyme comprising more than one protein subunit), was enhanced by linking the genes for different subunits of dioxygenases obtained from different organisms (Furukawa et al., 1993). The modified dioxygenase also was observed to catalyze the oxidation of trichloroethylene at rates substantially higher than the rate of either of the native dioxygenases (Furukawa et al., 1994).

Construction of novel pathways to expand the substrate range of a single organism may be of questionable benefit. If two or more chemicals in a mixture can each serve as growth substrates for *different* organisms in a mixed culture, it is reasonable to ask why having this capability in a single organism provides an advantage. In fact, Tiedje (1993) has suggested that inoculation with nonnative organisms that are apparently superior to naturally occurring species is fruitless when native degraders are present. One possible advantage of an "enhanced" microorganism would be to preclude the accumulation of dead-end products from incomplete metabolism, but it has not been documented that this is an actual concern that would limit the feasibility of bioremediation at any site. Furthermore, the composition of chemical mixtures in contaminated environments is usually site-specific, so that an organism designed to degrade a certain mixture may be relevant to a limited number of sites. Construction of pathways that permit the growth of an organism on an otherwise nongrowth substrate can, however, have substantial benefit. Modified organisms designed to allow growth on substrates under denitrifying conditions could also be useful for in situ bioremediation applications.

Manipulation of the manner in which the expression of genes is regulated is of particular importance for the transformation of nongrowth substrates. As discussed previously, nongrowth substrates are not likely to induce the expression of genes required for their transformation, so that alternative means of induction are required. Genes responsible for the transformation of xenobiotics can be transferred to other organisms and simultaneously linked to regulatory genes that are induced by growth substrates that do not competitively inhibit the desired transformation activity. For example, genes encoding for the soluble methane monooxygenase from *M. trichosporium* OB3b were cloned into a *Pseudomonas* strain in such a way that MMO could be expressed during growth on glucose (Jahng and Wood, 1994). Degradation of chlorinated hydrocarbons by the recombinant *Pseudomonas* was not susceptible to competitive inhibition by the growth substrate, nor was sMMO expression affected by the concentration of copper in the medium. Genes with the required transformation activity can also be introduced to organisms while being linked to constitutive promoters [i.e., regions of DNA that are always "on," so that expression of the associated gene(s) is neither inducible nor repressible]. Organisms that constitutively express genes for the transformation of nongrowth substrates must still be selected for after they are introduced to a mixed culture environment. Lajoie et al. (1994) described a novel approach in which genes encoding for the degradation of PCBs were linked to a constitutive promoter and introduced in a *Pseudomonas* strain able to use a surfactant as a growth substrate; the surfactant was suggested to be both a selective growth substrate and of potential benefit in improving the bioavailability of PCBs. Even if a modified organism can be enriched in the contaminated environment, one of the major challenges associated with the use of recombinant organisms for the transformation of nongrowth substrates is in ensuring that the required genes are not lost under conditions in which those genes provide no selective advantage to the microorganism. Constitutive genes associated with plasmids might be especially susceptible to instability (Lajoie et al., 1994).

Protein engineering involves an intentional alteration in a gene that results in an alteration of the protein coded for by that gene, usually an enzyme of interest (Blundell et al., 1989; Hodgson, 1990). The goal of protein engineering for application purposes is to create a modified enzyme that would retain its catalytic activity against the substrate(s) of interest but would be more stable or more catalytically active than the native enzyme or would have an expanded substrate range compared to the native enzyme (Hodgson, 1990; Janssen and Schanstra, 1994). For example, the substrate range of a multicomponent dioxygenase catalyzing the oxidation of chlorinated biphenyls was broadened through the use of protein engineering techniques (Erickson and Mondello, 1993). Much of the work on using protein engineering to enhance stability has been directed at enhanced thermal stability (Janssen and Schanstra, 1994), but for biotransforming enzymes, it is probably more important to focus on improving catalytic stability—that is, at reducing the susceptibility of the transforming enzymes to suicide inactivation reactions. Such improved catalytic stability would lead to corresponding increases in transformation capacities of the organisms containing those enzymes, which would have direct economic benefits. Critical limitations of protein engineering at present are that we still have a relatively limited understanding of the relationship between protein structure and function

(Blundell et al., 1989), that there are few environmentally relevant proteins for which the three-dimensional structure is known anyway, and that catalytic mechanisms are not well understood for most enzymes of environmental interest (Janssen and Schanstra, 1994).

Despite significant limitations in ultimate field implementation, genetically modified organisms might represent the best long-term strategy for biodegradation of highly recalcitrant classes of pollutants. However, one of the greatest difficulties with genetic approaches to bioremediation is that, at present, significant investments in both time and money are required for what amount to relatively limited solutions to relatively narrow problems. It would be unfortunate if we invested too many of our resources in genetically manipulating the limited set of biotransformation mechanisms about which we have sufficiently detailed knowledge, at the expense of increasing our knowledge of the biochemistry, physiology, and ecology of a wider range of microbial transformation processes.

As a final note on the use of "specialized" microorganisms, it may be possible to design reactors to select or enrich for organisms with a desired activity from within the indigenous microbial community. Effluent from the enrichment reactor can then be used to inoculate the system (e.g., reinjection of the effluent into the subsurface to enhance in situ activity). Enrichment could be accomplished using known inducing substrates that might be less toxic and/or more water soluble than the target substrates (Babcock et al., 1992; Babcock and Stenstrom, 1993) or using the target compounds themselves in a biologically active carbon bed that would preclude the release of the target compounds into the effluent (Weber and Corseuil, 1994).

Matching Strategies to Mechanisms and Situations

An algorithm to help identify the critical issues in evaluating the feasibility of bioremediation for a given situation is presented in Figures 2 through 5. In this algorithm, bioremediation problems are divided into three major types: those requiring the biodegradation of growth substrates that are the predominant carbon sources in the system (Figure 3), those requiring the biodegradation of potential growth substrates that are only minor constituents in complex mixtures (Figure 4), and those requiring the transformation or degradation of nongrowth substrates (Figure 5). Strategies used to achieve bioremediation in any of these situations must consider general issues such as bioavailability, concentration-dependent effects such as substrate inhibition, whether biodegradation can occur under anaerobic conditions, and whether other factors such as the presence of high concentrations of heavy metals might inhibit

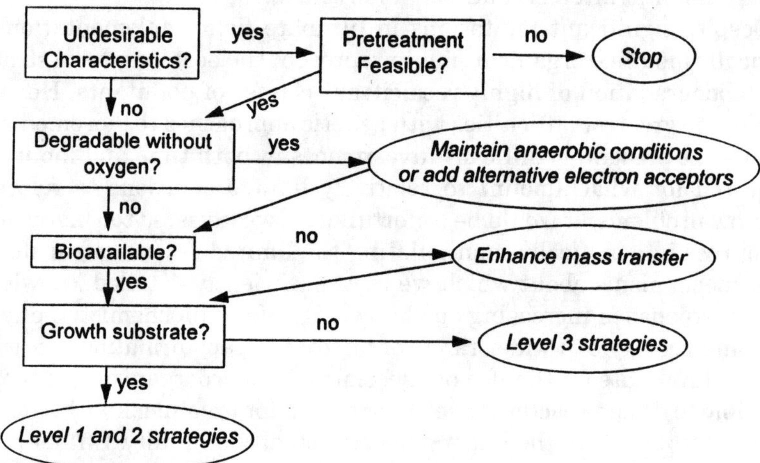

Figure 2. Preliminary assessment of strategies required for bioremediation.

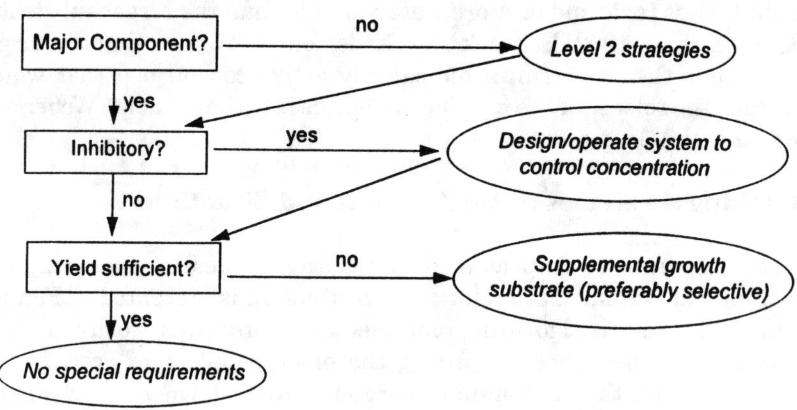

Figure 3. Level 1 strategies for bioremediation of growth substrates that are major biodegradable components of waste mixture.

362

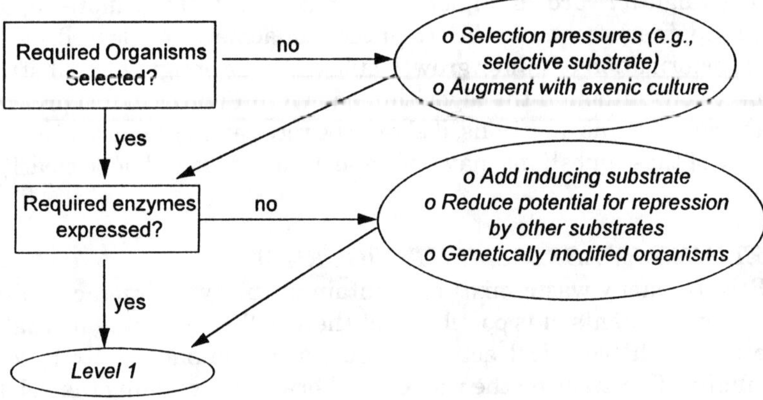

Figure 4. Level 2 strategies for bioremediation of growth substrates that are not major components of waste mixture.

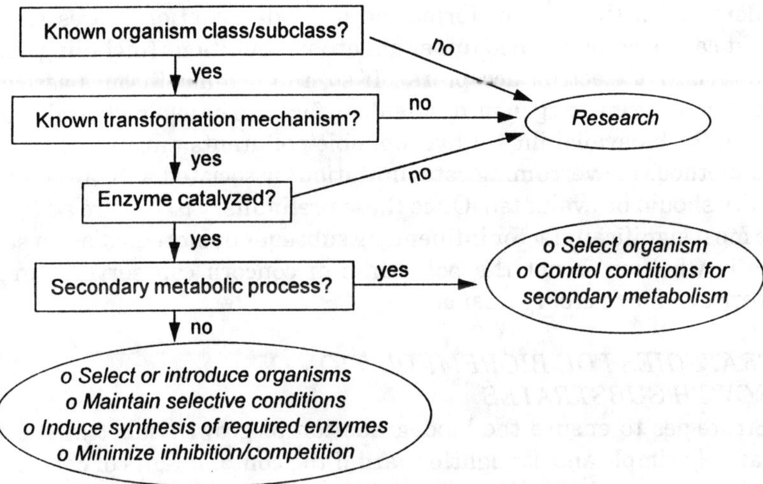

Figure 5. Level 3 strategies for bioremediation of nongrowth substrates.

or preclude the use of biological treatment (Figure 2). Once these issues are confronted, the strategies required to achieve the biodegradation of growth substrates that are major carbon sources are the simplest and most straightforward to implement, while progressively more sophisticated approaches would be required to achieve biodegradation or biotransformation of minor growth substrates or nongrowth substrates, respectively. Although the algorithm shown in Figures 2–5 is presented as a series of yes/no questions, it should be recognized that actual answers for any of these questions may not be so easily answered or so clearly yes or no.

PRELIMINARY ASSESSMENT (FIGURE 2)

Because many waste mixtures contain a variety of organic and inorganic contaminants, it is possible that there will be constituents that can interfere with biological activity. Thus, a preliminary analysis of the feasibility of pretreating the waste or otherwise overcoming the deleterious effects of these constituents should be the first step towards establishing the feasibility of bioremediation. If it is not feasible to overcome these effects, then there is no need to consider bioremediation further.

For in situ bioremediation in particular, it is clearly desirable to understand if the biotransformation or biodegradation processes of interest can be accomplished under anaerobic conditions (including the use of alternative electron acceptors). If so, the technical issues associated with transferring oxygen into the subsurface can be avoided. As discussed above, the bioavailability of hydrophobic pollutants must be considered, and methods of overcoming rate limitations associated with low bioavailability should be evaluated. Once these preliminary issues are addressed, the most significant factor influencing subsequent bioremediation strategies is whether or not the pollutants of concern can serve as growth substrates for microorganisms.

STRATEGIES FOR BIOREMEDIATION OF GROWTH SUBSTRATES

Strategies to ensure the biodegradation of growth substrates can be relatively simple and straightforward if the contaminants of concern are major carbon and energy sources within a mixture or may require more sophisticated approaches if they are minor components. These strategies are identified as "Level 1" and "Level 2" in Figures 3 and 4, respectively. Even though one or more chemicals in a mixture can serve as major carbon and energy sources for indigenous microorganisms, many such chemicals can exhibit concentration-dependent inhibitory effects. Also, for very dilute waste mixtures, it is possible that even the most dominant growth substrates would not be present in high enough concentration (or made available at a high enough rate) to sustain meaningful populations of degrader organisms. In such cases, adding supplemental growth substrates might enhance biodegradation rates (except in the case of low bioavailability, where other approaches would be warranted). As a rule, it is almost always likely to be preferable to add growth substrates that are selective for the microbial populations or communities of interest.

Selection is, of course, the key to optimizing the biodegradation of potential growth substrates that may only be minor components of a mixture. If the appropriate organisms are present in very low numbers, then selection pressures (such as the addition of selective carbon and energy sources) must be applied. If the appropriate organisms are not present in the indigenous community at all, then the waste matrix must be augmented with cultures derived from other sources as described above. Expression of the appropriate biodegradative enzymes is still an issue even if the appropriate organisms are present, and it is likely that there are many situations in which such enzymes must be artificially induced. If something is known about repressive effects of other chemicals in the matrix, then extending remediation times beyond the point at which such repressive substances are degraded should help (note: *repression* refers to the "turning off" of the synthesis of an enzyme or enzymatic pathway within an organism, which can occur in the presence of a variety of otherwise innocuous chemicals and under a variety of conditions, as opposed to inhibition or other toxic effects of a chemical). One of the appropriate situations in which it may be preferable to add genetically modified organisms is for the biodegradation of chemicals for which indigenous degraders do not exist or for which induction of the required degradative enzymes is not likely to occur under the extant conditions.

STRATEGIES FOR BIOREMEDIATION OF NONGROWTH SUBSTRATES (FIGURE 5)

We are fortunate to have learned much about the biotransformation of several widespread pollutants which no known organism can utilize as growth substrates. In these situations, we still need to apply our knowledge of the transformation mechanisms by selecting or introducing the necessary organisms, maintaining selective conditions, ensuring that the appropriate enzymes are expressed, and minimizing inhibition or competition from growth substrates. It should be clear, however, that for nongrowth substrates we have relatively limited options to accomplish biotransformation.

CONCLUSIONS

The objective of this chapter was to convey that biotransformation and biodegradation mechanisms represent a diverse array of microbial and biochemical activities and often involve complex interactions between microorganisms or between microorganisms and their immediate environment. The example of nongrowth metabolism of chlorinated hydrocarbons by methanotrophic bacteria illustrates how scientific discoveries can be translated eventually into practical application and, conversely, how an application can be accelerated by paying attention to mechanistic details. For that example, the time frame over which we have gone from original discovery to application at the field scale represents, hopefully, a near worst-case. The time scale for development and application of newly discovered biotransformation capabilities hopefully will decrease as more scientists become aware of the potential environmental implications of their work and as more engineers discover the value of exploring the basic scientific literature. It is important, however, to target efforts at defining the potential limitations of a process early in its development, either to redirect research efforts toward missing information required for technical feasibility evaluations or to help define the scope of future development activities. It is also important to realize that economic feasibility cannot be assessed fully until technical feasibility is established. For processes that eventually may be judged to be economically infeasible, it is important to define the most costly component(s) of the process so that possible alternatives can be considered.

ACKNOWLEDGEMENTS

This chapter is based on presentations made in 1992 at the American Institute of Chemical Engineers Summer National Meeting and at the University of Notre Dame. I thank Lisa Alvarez-Cohen (University of California–Berkeley) for erudite and entertaining discussions on "cometabolism" and for helping to make distinctions among nongrowth substrates.

REFERENCES

Aitken, M. D. (1993) Batch biological treatment of inhibitory substrates, *J. Environ. Eng.*, 119:855–870.

Aitken, M. D., I. J. Massey, T. Chen and P. E. Heck (1994) Characterization of reaction products from the enzyme catalyzed oxidation of phenolic pollutants, *Water Res.*, 28:1879–1889.

Albrechtsen, H.-J. and T. H. Christensen (1994) Evidence for microbial iron reduction in a landfill leachate-polluted aquifer (Vejen, Denmark), *Appl. Environ. Microbiol.*, 60:3920–3925.

Alder, A. C., M. M. Häggblom, S. R. Oppenheimer and L. Y. Young (1993) Reductive dechlorination of polychlorinated biphenyls in anaerobic sediments, *Environ. Sci. Technol.*, 27:530–538.

Alleman, B. C., B. E. Logan and R. L. Gilbertson (1995) Degradation of pentachlorophenol by fixed films of white rot fungi in rotating tube bioreactors, *Water Res.*, 29:61–67.

Allsop, P. J., M. Moo-Young and G. R. Sullivan (1990) The dynamics and control of substrate inhibition in activated sludge, *Crit. Rev. Environ. Control*, 20:115–167.

Alvarez-Cohen, L. and P. L. McCarty (1991a) Effects of toxicity, aeration, and reductant supply on trichloroethylene transformation by a mixed methanotrophic culture, *Appl. Environ. Microbiol.*, 57:228–235.

Alvarez-Cohen, L. and P. L. McCarty (1991b) Product toxicity and cometabolic competitive inhibition modeling of chloroform and trichloroethylene transformation by methanotrophic resting cells, *Appl. Environ. Microbiol.*, 57:1031–1037.

Alvarez-Cohen, L. and P. L. McCarty (1991c) A cometabolic biotransformation model for halogenated aliphatic compounds exhibiting product toxicity, *Environ. Sci. Technol.*, 25:1381–1386.

Alvarez-Cohen, L. and P. L. McCarty (1991d) Two-stage dispersed-growth treatment of halogenated aliphatic compounds by cometabolism, *Environ. Sci. Technol.*, 25:1387–1393.

Anderson, J. E. and P. L. McCarty (1994) Model for treatment of trichloroethylene by methanotrophic biofilms, *J. Environ. Eng.*, 120:379–400.

Anderson, T. A., E. A. Guthrie and B. T. Walton (1993) Bioremediation of the rhizosphere, *Environ. Sci. Technol.*, 27:2630–2636.

Anselmo, A. M., J. M. S. Cabral and J. M. Novais (1989) The adsorption of *Fusarium flocciferum* spores on celite particles and their use in the degradation of phenol, *Appl. Microbiol. Biotechnol.*, 31:200–203.

Armenante, P. M., G. Lewandowski and I. U. Haq (1992) Mineralization of 2-chlorophenol by *P. chrysosporium* using different reactor designs, *Hazard. Waste Hazard. Mater.*, 9:213–229.

Armenante, P. M., N. Pal and G. Lewandowski (1994) Role of mycelium and extracellular protein in the biodegradation of 2,4,6-trichlorophenol by *Phanerochaete chrysosporium*, *Appl. Environ. Microbiol.*, 60:1711–1718.

Arvin, E. (1991) Biodegradation kinetics of chlorinated aliphatic hydrocarbons with methane oxidizing bacteria in an aerobic fixed biofilm reactor, *Water Res.*, 25:873–881.

Assaf-Anid, N., K. F. Hayes and T. M. Vogel (1994) Reductive dechlorination of carbon tetrachloride by cobalamin(II) in the presence of dithiothreitol:

Mechanistic study, effect of redox potential and pH, *Environ. Sci. Technol.*, 28:246–252.

Assaf-Anid, N., L. Nies and T. M. Vogel (1992) Reductive dechlorination of a polychlorinated biphenyl congener and hexachlorobenzene by vitamin B_{12}, *Appl. Environ. Microbiol.*, 58:1057–1060.

Athey, L. A., J. M. Thomas, W. E. Miller and J. Q. Word (1989) Evaluation of bioassays for designing sediment cleanup strategies at a wood treatment site, *Environ. Toxicol. Chem.*, 8:223–230.

Babcock, R. W., Jr., K. S. Ro, C.-C. Hsieh and M. K. Stenstrom (1992) Development of an off-line enricher-reactor process for activated sludge degradation of hazardous wastes, *Water Environ. Res.*, 64:782–791.

Babcock, R. W. and M. K. Stenstrom (1993) Use of inducer compounds in the enricher-reactor process for degradation of 1-naphthylamine wastes, *Water Environ. Res.*, 65:26–33.

Barr, D. P. and S. D. Aust (1994) Mechanisms white rot fungi use to degrade pollutants, *Environ. Sci. Technol.*, 28:78A–87A.

Baxter, R. M. (1990) Reductive dechlorination of certain chlorinated organic compounds by reduced hematin compared with their behavior in the environment, *Chemosphere*, 21:451–458.

Becker, J. G. and D. L. Freedman (1994) Use of cyanocobalamin to enhance anaerobic biodegradation of chloroform, *Environ. Sci. Technol.*, 28:1942–1949.

Bennett, J. W. and R. Bentley (1989) What's in a name?—Microbial secondary metabolism, *Adv. Appl. Microbiol.*, 34:1–28.

Bilbo, C. M., E. Arvin, H. Holst and H. Spliid (1992) Modeling the growth of methane-oxidizing bacteria in a fixed biofilm, *Water Res.*, 26:301–309.

Blundell, T. L., G. Elliott, S. P. Gardner, T. Hubbard, S. Islam, M. Johnson, D. Mantafounis, P. Murray-Rust, J. Overington, J. E. Pitts, A. Sali, B. L. Sibanda, J. Singh, M. J. E. Sternberg, M. J. Sutcliffe, J. M. Thornton and P. Travers (1989) Protein engineering and design, *Phil. Trans. R. Soc. Lond.*, B324:447–460.

Boldrin, B., A. Thiem and C. Fritzsche (1993) Degradation of phenanthrene, fluorene, fluoranthene and pyrene by a *Mycobacterium* sp., *Appl. Environ. Microbiol.*, 59:1927–1930.

Bosma, T. N. P., F. H. M. Cottaar, M. A. Posthumus, C. J. Teunis, A. van Veldhuizen, G. Schraa and A. J. B. Zehnder (1994) Comparison of reductive dechlorination of hexachloro-1,3-butadiene in Rhine sediment and model systems with hydroxocobalamin, *Environ. Sci. Technol.*, 28:1124–1128.

Bourquin, A. W. (1993) Guest editorial, *Biodegradation*, 4:205–206.

Brock, T. D., M. T. Madigan, J. M. Martinko and J. Parker (1994) *Biology of Microorganisms*, 7th ed., Prentice Hall, Englewood Cliffs, N.J.

Broholm, K., B. K. Jensen, T. H. Christensen and L. Olsen (1990) Toxicity of

1,1,1-trichloroethane and trichloroethene on a mixed culture of methane-oxidizing bacteria, *Appl. Environ. Microbiol.*, 56:2488–2493.

Broholm, K., T. H. Christensen and B. K. Jensen (1992) Modelling TCE degradation by a mixed culture of methane-oxidizing bacteria, *Water Res.*, 26:1177–1185.

Broholm, K., T. H. Christensen and B. K. Jensen (1993) Different abilities of eight mixed cultures of methane-oxidizing bacteria to degrade TCE, *Water Res.*, 27:215–224.

Bu'Lock, J. D. (1975) Secondary metabolism in fungi and the relationships to growth and development, pp. 33–58 in *The Filamentous Fungi*, v. 1, ed. J. E. Smith and D. R. Berry, Edward Arnold Pub., London.

Bumpus, J. A., M. Tien, D. Wright and S. D. Aust (1985) Oxidation of persistent environmental pollutants by a white rot fungus, *Science*, 228:1434–1436.

Busch, A. W. (1971) *Aerobic Biological Treatment of Waste Waters*, Oligodynamics Press, Houston.

Cai, D. and M. Tien (1993) Lignin-degrading peroxidases of *Phanerochaete chrysosporium*, *J. Biotechnol.*, 30:79–90.

Cammarota, M. C. and G. L. Sant'Anna, Jr. (1992) Decolorization of kraft bleach plant E_1 stage effluent in a fungal bioreactor, *Environ. Technol.*, 13:65–71.

Cerniglia, C. E. and M. A. Heitkamp (1989) Microbial degradation of polycyclic aromatic hydrocarbons (PAH) in the aquatic environment, pp. 41–68 in *Metabolism of Polycyclic Aromatic Hydrocarbons in the Environment*, ed. U. Varanasi, CRC Press, Boca Raton, Fla.

Cerniglia, C. E., W. L. Campbell, P. P. Fu, J. P. Freeman and F. E. Evans (1990) Stereoselective fungal metabolism of methylated anthracenes, *Appl. Environ. Microbiol.*, 56:661–668.

Chang, M.-K., T. C. Voice and C. S. Criddle (1993) Kinetics of competitive inhibition and cometabolism in the biodegradation of benzene, toluene, and *p*-xylene by two *Pseudomonas* isolates, *Biotechnol. Bioeng.*, 41:1057–1065.

Chung, N. and S. D. Aust (1995) Veratryl alcohol-mediated indirect oxidation of phenol by lignin peroxidase, *Arch. Biochem. Biophys.*, 316:733–737.

Colbert, S. F., Isakeit, T., Ferri, M., Weinhold, A. R., Hendson, M. and Schroth, M. N. (1993a) Use of an exotic carbon source to selectively increase metabolic activity and growth of *Pseudomonas putida* in soil, *Appl. Environ. Microbiol.* 59:2056–2063.

Colbert, S. F., M. N. Schroth, A. R. Weinhold and M. Hendson (1993b) Enhancement of population densities of *Pseudomonas putida* PpG7 in agricultural ecosystems by selective feeding with the carbon source salicylate, *Appl. Environ. Microbiol.*, 59:2064–2070.

Cole, J. R., A. L. Cascarelli, W. W. Mohn and J. M. Tiedje (1994) Isolation and characterization of a novel bacterium growing via reductive dehalogenation of 2-chlorophenol, *Appl. Environ. Microbiol.*, 60:3536–3542.

Coschigano, P. W., M. M. Häggblom and L. Y. Young (1994) Metabolism of both

4-chlorobenzoate and toluene under denitrifying conditions by a constructed bacterial strain, *Appl. Environ. Microbiol.*, 60:989–995.

Criddle, C. S. (1993) The kinetics of cometabolism, *Biotechnol Bioeng.*, 41:1048–1056.

Dalton, H. and D. J. Leak (1985) Methane oxidation by microorganisms, pp. 173–200 in *Microbial Gas Metabolism: Mechanistic, Metabolic and Biotechnological Aspects*, ed. R. K. Poole, Academic Press, London.

Dalton, H., D. D. S. Smith and S. J. Pilkington (1990) Towards a unified mechanism of biological methane oxidation, *FEMS Microbiol. Rev.*, 87:201–208.

Dalton, H. and D. I. Stirling (1982) Co-metabolism, *Phil Trans. R. Soc. Lond.* B297:481–496.

Davis, M. W., J. A. Glaser, J. W. Evans and R. T. Lamar (1993) Field evaluation of the lignin-degrading fungus *Phanerochaete sordida* to treat creosote-contaminated soil, *Environ. Sci. Technol.*, 27:2572–2576.

de Bruin, W. P., M. J. J. Kotterman, M. A. Posthumus, G. Schraa and A. J. B. Zehnder (1992) Complete biological reductive transformation of tetrachloroethene to ethane, *Appl. Environ. Microbiol.*, 58:1996–2000.

Demain, A. L. (1986) Regulation of secondary metabolism in fungi, *Pure Appl. Chem.*, 58:219–226.

DeWeerd, K. A. and J. M. Suflita (1990) Anaerobic aryl reductive dehalogenation of halobenzoates by cell extracts of "*Desulfominile tiedji*," *Appl. Environ. Microbiol.*, 56:2999–3005.

Dhawale, S. W., S. S. Dhawale and D. Dean-Ross (1992) Degradation of phenanthrene by *Phanerochaete chrysosporium* occurs under ligninolytic as well as nonligninolytic conditions, *Appl. Environ. Microbiol.*, 58:3000–3006.

DiSpirito, A. A., J. Gulledge, A. K. Shiemke, J. C. Murrell, M. E. Lidstrom and C. L. Krema (1992) Trichloroethylene oxidation by the membrane-associated methane monoxygenase in type I, type II and type X methanotrophs, *Biodegradation*, 2:151–164.

Donnelly, K. C., C. S. Anderson, J. C. Thomas, K. W. Brown, D. J. Manek and S. H. Safe (1992) Bacterial mutagenicity of soil extracts from a bioremediation facility treating wood-preserving waste, *J. Haz. Mater.*, 30:71–81.

Egli, C., S. Stromeyer, A. M. Cook and T. Leisinger (1990) Transformation of tetra-and trichloromethane to CO_2 by anaerobic bacteria is a non-enzymic process, *FEMS Microbiol. Lett.*, 68:207–212.

Ensley, B. D. (1994) Designing pathways for environmental purposes, *Curr. Opin. Biotechnol*, 5:249–252.

Erickson, B. D. and F. J. Mondello (1993) Enhanced biodegradation of polychlorinated biphenyls after site-directed mutagenesis of a biphenyl dioxygenase gene, *Appl. Environ. Microbiol.*, 59:3858–3862.

Erickson, D. C., R. C. Loehr and E. F. Neuhauser (1993) PAH loss during bioremediation of manufactured gas plant site soils, *Water Res.*, 27:911–919.

Fan, S. and K. M. Scow (1993) Biodegradation of trichloroethylene and toluene by indigenous microbial populations in soil, *Appl. Environ. Microbiol.*, 59:1911–1918.

Fathepure, B. Z. and S. A. Boyd (1988) Reductive dechlorination of perchloroethylene and the role of methanogens, *FEMS Microbiol. Lett.*, 49:149–156.

Fennell, D. E., Y. M. Nelson, S. E. Underhill, T. E. White and W. J. Jewell (1993) TCE degradation in a methanotrophic attached-film bioreactor, *Biotechnol. Bioeng.*, 42:859–872.

Field, J. A., E. de Jong, G. Feijoo Costa and J. A. M. de Bont (1992) Biodegradation of polycyclic aromatic hydrocarbons by new isolates of white rot fungi, *Appl. Environ. Microbiol.*, 58:2219–2226.

Finnerty, W. R. (1994) Biosurfactants in environmental biotechnology, *Curr. Opin. Biotechnol.*, 5:291–295.

Fogel, M. M., A. R. Taddeo and S. Fogel (1986) Biodegradation of chlorinated ethenes by a methane-utilizing mixed culture, *Appl. Environ. Microbiol.*, 51:720–724.

Folsom, B. R. and P. J. Chapman (1991) Performance characterization of a model bioreactor for the biodegradation of trichloroethylene by *Pseudomonas cepacia* G4, *Appl. Environ. Microbiol.*, 57:1602–1608.

Folsom, B. R., P. J. Chapman and P. H. Pritchard (1990) Phenol and trichloroethylene degradation by *Pseudomonas cepacia* G4: Kinetics and interactions between substrates, *Appl. Environ. Microbiol.*, 56:1279–1285.

Fox, B. G., J. G. Borneman, L. P. Wackett and J. D. Lipscomb (1990) Haloalkene oxidation by the soluble methane monoxygenase from *Methylosinus trichosporium* OB3b: mechanistic and environmental implications, *Biochemistry*, 29:6419–6427.

Freedman, D. L. and J. M. Gossett (1989) Biological reductive dechlorination of tetrachloroethylene and trichloroethylene to ethylene under methanogenic conditions, *Appl. Environ. Microbiol.*, 55:2144–2151.

Furukawa, K., J. Hirose, A. Suyama, T. Zaiki and S. Hayashida (1993) Gene components responsible for discrete substrate specificity in the metabolism of biphenyl (*bph* operon) and toluene (*tod* operon), *J. Bacteriol.*, 175:5224–5232.

Furukawa, K., J. Hirose, S. Hayashida and K. Nakamura (1994) Efficient degradation of trichloroethylene by a hybrid aromatic ring dioxygenase, *J. Bacteriol.*, 176:2121–2123.

Gantzer, C. J. and L. P. Wackett (1991) Reductive dechlorination catalyzed by bacterial transition-metal coenzymes, *Environ. Sci. Technol.*, 25:715–722.

Georgiou, G., S. C. Lin and M. M. Sharma (1992) Surface-active compounds from microorganisms, *Bio/Technology*, 10:60–65.

Gerson, D. F. (1993) The biophysics of microbial surfactants: Growth on insoluble substrates, pp. 269–286 in *Biosurfactants: Production, Properties, Applications*, ed. N. Kosaric, Marcel Dekker, New York.

Gibson, D. T. (1993) Biodegradation, biotransformation and the Belmont, *J. Ind. Microbiol.*, 12:1–12.

Gibson, S. A. and G. W. Sewell (1992) Stimulation of reductive dechlorination of tetrachlorethene in anaerobic aquifer microcosms by addition of short-chain organic acids or alcohols, *Appl. Environ. Microbiol.*, 58:1392–1393.

Griffiths, D. A., D. E. Brown and S. G. Jezequel (1992) Biotransformation of warfarin by the fungus *Beauveria bassiana*, *Appl. Microbiol. Biotechnol.*, 37:169–175.

Grimberg, S. J. and M. D. Aitken (1995) Biodegradation kinetics of phenanthrene solubilized in surfactant micelles, pp. 59–66 in *Microbial Processes for Bioremediation*, ed. R. E. Hinchee, C. M. Vogel and F. J. Brockman, Battelle Press, Columbus, OH.

Grosser, R. J., D. Warshawsky and J. R. Vestal (1991) Indigenous and enhanced mineralization of pyrene, benzo[a]pyrene, and carbazole in soils, *Appl. Environ. Microbiol.*, 57:3462–3469.

Hamdi, M., H. Bouhamed and R. Ellouz (1991) Optimization of the fermentation of olive mill wastewaters by *Aspergillus niger*, *Appl. Microbiol. Biotechnol.*, 36:285–288.

Hamer, G. (1994) Transient state behaviour of microbial consortia in aerobic biotreatment processes, *Extended Abstracts*, American Institute of Chemical Engineers Annual Meeting, San Francisco, Calif., p. 523.

Hammel, K. E. (1992) Oxidation of aromatic pollutants by lignin-degrading fungi and their extracellular peroxidases, pp. 41–60 in *Metal Ions in Biological Systems*, vol. 28, *Degradation of Environmental Pollutants by Microorganisms and Their Metalloenzymes*, ed. H. Sigel and A. Sigel, Marcel Dekker, New York.

Hammel, K. E., B. Green and W. Z. Gai (1991) Ring fission of anthracene by a eukaryote, *Proc. Natl. Acad. Sci. USA*, 88:10605–10608.

Hammel, K. E., B. Kalyanaraman and T. K. Kirk (1986) Oxidation of polycyclic aromatic hydrocarbons and dibenzo[p]-dioxins by *Phanerochaete chrysosporium* ligninase, *J. Biol Chem.*, 261:16948–16952.

Hammel, K. E., W. Z. Gai, B. Green and M. A. Moen (1992) Oxidative degradation of phenanthrene by the ligninolytic fungus *Phanerochaete chrysosporium*, *Appl. Environ. Microbiol.*, 58:1832–1838.

Hammel, K. E. and P. J. Tardone (1988) The oxidative dechlorination of polychlorinated phenols is catalyzed by extracellular fungal lignin peroxidase, *Biochemistry*, 27:6563–6568.

Harker, A. R. and Y. Kim (1990) Trichloroethylene degradation by two independent aromatic-degrading pathways in *Alcaligenes eutrophus* JMP134, *Appl. Environ. Microbiol.*, 56:1179–1181.

Harms, H. and A. J. B. Zehnder (1995) Bioavailability of sorbed 3-chlorodibenzofuran, *Appl. Environ. Microbiol.*, 61:27–33.

Heitkamp, M. A. and C. E. Cerniglia (1988) Mineralization of polycyclic aromatic hydrocarbons by a bacterium isolated from sediment below an oil field, *Appl. Environ. Microbiol.*, 54:1612–1614.

Heitkamp, M. A. and C. E. Cerniglia (1989) Polycyclic aromatic hydrocarbon degradation by a *Mycobacterium* sp. in microcosms containing sediment and water from a pristine ecosystem, *Appl. Environ. Microbiol.*, 55:1968–1973.

Heitkamp, M. A., W. Franklin and C. E. Cerniglia (1988) Microbial metabolism of polycyclic aromatic hydrocarbons: Isolation and characterization of a pyrene-degrading bacterium, *Appl. Environ. Microbiol.*, 54:2549–2555.

Henry, S. M. and D. Grbic-Galic (1991a) Influence of endogenous and exogenous electron donors and trichloroethylene oxidation toxicity on trichloroethylene oxidation by methanotrophic cultures from a groundwater aquifer, *Appl. Environ. Microbiol.*, 57:236–244.

Henry, S. M. and D. Grbić-Galić (1991b) Inhibition of trichloroethylene oxidation by the transformation intermediate carbon monoxide, *Appl. Environ. Microbiol.*, 57:1770–1776.

Henrysson, T. and P. L. McCarty (1993) Influence of the endogenous storage lipid poly-β-hydroxybutyrate on the reducing power availability during cometabolism of trichloroethylene and naphthalene by resting methanotrophic mixed cultures, *Appl. Environ. Microbiol.*, 59:1602–1606.

Higgins, I. J., D. J. Best and R. C. Hammond (1980) New findings in methane-utilizing bacteria highlight their importance in the biosphere and their commercial potential, *Nature*, 286:561–564.

Hill, G. A. and C. W. Robinson (1989) Minimum tank volumes for CFST bioreactors in series, *Can. J. Chem. Eng.*, 67:818–824.

Hodgson, J. (1990) Protein design: Rules, empiricism and nature, *Bio/Technology*, 8:1245–1247.

Hofrichter, M., T. Günther and W. Fritsche (1993) Metabolism of phenol, chloro- and nitrophenols by the *Penicillium* strain Bi 7/2 isolated from a contaminated soil, *Biodegradation*, 3:415–421.

Hollender, J., W. Dott and J. Hopp (1994) Regulation of chloro- and methylphenol degradation in *Comamonas testosteroni* JH5, *Appl. Environ. Microbiol.*, 60:2330–2338.

Holliger, C., G. Schraa, A. J. M. Stams and A. J. B. Zehnder (1992) Enrichment and properties of an anaerobic mixed culture reductively dechlorinating 1,2,3-trichlorobenzene to 1,3-dichlorobenzene, *Appl. Environ. Microbiol.*, 58:1636–1644.

Hommel, R. K. (1990) Formation and physiological role of biosurfactants produced by hydrocarbon-utilizing microorganisms, *Biodegradation*, 1:107–119.

Hommel, R. K. and C. Ratledge (1993) Biosynthetic mechanisms of low molecular weight surfactants and their precursor molecules, pp. 3–63 in *Biosurfactants: Production, Properties, Applications*, ed. N. Kosaric, Marcel Dekker, New York.

Hopkins, G. D., J. Munakata, L. Semprini and P. L. McCarty (1993) Trichloroethylene concentration effects on pilot field-scale in-situ groundwater bioremediation by phenol-oxidizing microorganisms, *Environ. Sci. Technol.*, 27:2542–2547.

Horvath (1972) Microbial co-metabolism and the degradation of organic compounds in nature, *Bacteriol. Rev.*, 36: 146-155.

Hur, H.-G., M. J. Sadowsky and L. P. Wackett (1994) Metabolism of chlorofluorocarbons and polybrominated compounds by *Pseudomonas putida* G786-(pHG-2) via an engineered metabolic pathway, *Appl. Environ. Microbiol.*, 60:4148–4154.

Hutchins, S. R. (1991) Biodegradation of monoaromatic hydrocarbons by aquifer microorganisms using oxygen, nitrate, or nitrous oxide as the terminal electron acceptor, *Appl. Environ. Microbiol.*, 57:2403–2407.

Jahng, D. and T. K. Wood (1994) Trichloroethylene and chloroform degradation by a recombinant pseudomonad expressing soluble methane monooxygenase from *Methylosinus trichosporium* OB3b, *Appl. Environ. Microbiol.*, 60: 2473–2482.

Janssen, D. B. and J. P. Schanstra (1994) Engineering proteins for environmental applications, *Curr. Opin. Biotechnol.*, 5:253–259.

Jiménez, L., A. Breen, N. Thomas, T. W. Federle and G. S. Sayler (1991) Mineralization of linear alkylbenzene sulfonate by a four-member aerobic bacterial consortium, *Appl. Environ. Microbiol.*, 57:1566–1569.

Jones, K. H., P. W. Trudgill and D. J. Hopper (1993) Metabolism of *p*-cresol by the fungus *Aspergillus fumigatus*, *Appl. Environ. Microbiol.*, 59:1125–1130.

Jones, K. H., P. W. Trudgill and D. J. Hopper (1994) 4-Ethylphenol metabolism by *Aspergillus fumigatus*, *Appl. Environ. Microbiol.*, 60:1978–1983.

Joshi, D. K. and M. H. Gold (1993) Degradation of 2,4,5-trichlorophenol by the lignin-degrading basidiomycete *Phanerochaete chrysosporium*, *Appl. Environ. Microbiol.*, 59:1779–1785.

Joyce, T. W., H.-M. Chang, A. G. Campbell, Jr., E. D. Gerrard and T. K. Kirk (1984) A continuous biological process to decolorize bleach plant effluents, *Biotech. Advs.*, 2:301–308.

Katayama-Hirayama, K., S. Tobita and K. Hirayama (1992) Aromatic degradation in yeast *Rhodotorula rubra*, *Water Sci. Technol.*, 26:773–781.

Kirk, T. K. (1984) Degradation of lignin, pp. 399–437 in *Microbial Degradation of Organic Compounds*, ed. D. T. Gibson, Marcel Dekker, New York.

Kirkpatrick, N., I. D. Reid, E. Ziomek and M. G. Paice (1990) Biological bleaching of hardwood kraft pulp using *Trametes* (*Coriolus*) *versicolor* immobilized in polyurethane foam, *Appl. Microbiol. Biotechnol.*, 33:105–108.

Kiyohara, H., S. Torigoe, N. Kaida, T. Asaki, T. Iida, H. Hayashi and N. Takizawa (1994) Cloning and characterization of a chromosomal gene cluster, *pah*, that encodes the upper pathway for phenanthrene and naphthalene utilization by *Pseudomonas putida* OUS82, *J. Bacteriol.*, 176:2439–2443.

Klecka, G. M. and S. J. Gonsior (1984) Reductive dechlorination of chlorinated methanes and ethanes by reduced iron(II) porphyrins, *Chemosphere*, 13:391–402.

Koh, S.-C., J. P. Bowman and G. S. Sayler (1993) Soluble methane monooxygenase production and trichloroethylene degradation by a type I methanotroph, *Methylomonas methanica* 68-1, *Appl. Environ. Microbiol.*, 59:960–967.

Kolter, R., D. A. Siegele and A. Tormo (1993) The stationary phase of the bacterial life cycle, *Annu. Rev. Microbiol.*, 47:855–874.

Krone, U., K. Laufer, R. K. Thauer and H. P. C. Hogenkamp (1989) Coenzyme F_{430} as a possible catalyst for the reductive dehalogenation of chlorinated C_1 hydrocarbons in methanogenic bacteria, *Biochemistry*, 28:10061–10065.

Kühn, A. L. and W. A. Pretorius (1989) Fungal purification of an industrial effluent containing volatile fatty acids by means of a crossflow-microscreen technique, *Water Sci. Technol.*, 21:221–229.

Kuhn, E. P., G. T. Townsend and J. M. Suflita (1990) Effect of sulfate and organic carbon supplements on reductive dehalogenation of chloroanilines in anaerobic aquifer slurries, *Appl. Environ. Microbiol.*, 56:2630–2637.

Kuritz, T. and C. P. Wolk (1995) Use of filamentous cyanobacteria for biodegradation of organic pollutants, *Appl. Environ. Microbiol.*, 61:234–238.

Lajoie, C. A., A. C. Layton and G. S. Sayler (1994) Cometabolic oxidation of polychlorinated biphenyls in soil with a surfactant-based field application vector, *Appl. Environ. Microbiol.*, 60:2826–2833.

Lamar, R. T. and D. M. Dietrich (1990) In situ depletion of pentachlorophenol from contaminated soil by *Phanerochaete* spp., *Appl. Environ. Microbiol.*, 56:3093–3100.

Lamar, R. T. and J. A. Glaser (1993) Field evaluations of the remediation of soils contaminated with wood-preserving chemicals using lignin-degrading fungi, pp. 239–247 in *Bioremediation of Chlorinated and Polycyclic Aromatic Hydrocarbon Compounds*, ed. R. E. Hinchee, A. Leeson, L. Semprini and S. K. Ong, Lewis Pub., Boca Raton, Fla.

Lamar, R. T. and J. W. Evans (1993) Solid-phase treatment of a pentachlorophenol-contaminated soil using lignin-degrading fungi, *Environ. Sci. Technol.*, 27:2566–2571.

Landa, A. S., E. M. Sipkema, J. Weijma, A. A. C. M. Beenackers, J. Dolfing and D. B. Janssen (1994) Cometabolic degradation of trichloroethylene by *Pseudomonas cepacia* G4 in a chemostat with toluene as the primary substrate, *Appl. Environ. Microbiol.*, 60:3368–3374.

Lee, M. D., J. M. Thomas, R. C. Borden, P. B. Bedient, C. H. Ward and J. T. Wilson (1988) Biorestoration of aquifers contaminated with organic compounds, *CRC Crit. Rev. Environ. Control*, 18:29–89.

Lewandowski, G. A., P. M. Armenante and D. Pak (1990) Reactor design for hazardous waste treatment using a white rot fungus, *Water Res.*, 24:75–82.

Lin, J.-E., H. Y. Wang and R. F. Hickey (1991) Use of coimmobilized biological systems to degrade toxic organic compounds, *Biotechnol Bioeng.*, 38: 273–279.

Lindow, S. E., N. J. Panopoulos and B. L. McFarland (1989) Genetic engineering of bacteria from managed and natural habitats, *Science*, 244:1300–1307.

Liu, Z., A. M. Jacobson and R. G. Luthy (1995) Biodegradation of naphthalene in aqueous nonionic surfactant systems, *Appl. Environ. Microbiol.*, 61:145–151.

Lovley, D. R., M. J. Baedekker, D. J. Lonergan, I. M. Cozzarelli, E. J. P. Phillips and D. I. Siegel (1989) Oxidation of aromatic contaminants coupled to microbial iron reduction, *Nature*, 339:297–300.

Lovley, D. R., S. J. Giovannoni, D. C. White, J. E. Champine, E. J. Phillips, Y. A. Gorby and S. Goodwin (1993) *Geobacter metallireducens* gen. nov. sp. nov., a microorganism capable of coupling the complete oxidation of organic compounds to the reduction of iron and other metals, *Arch. Microbiol.*, 159:336–344.

MacGillivray, A. R. and M. P. Shiaris (1994) Relative role of eukaryotic and prokaryotic microorganisms in phenanthrene transformation in coastal sediments, *Appl. Environ. Microbiol.*, 60:1154–1159.

Madsen, T. and J. Aamand (1992) Anaerobic transformation and toxicity of trichlorophenols in a stable enrichment culture, *Appl. Environ. Microbiol.*, 58:557–561.

Mahaffey, W. R., D. T. Gibson and C. E. Cerniglia (1988) Bacterial oxidation of chemical carcinogens: Formation of polycyclic aromatic acids from benz[*a*]anthracene, *Appl. Environ. Microbiol.*, 54:2415–2423.

Malachowsky, K. J., T. J. Phelps, A. B. Teboli, D. E. Minnikin and D. C. White (1994) Aerobic mineralization of trichloroethylene, vinyl chloride, and aromatic compounds by *Rhodococcus* species, *Appl. Environ. Microbiol.*, 60:542–548.

Marks, T. S. and A. Maule (1992) The use of immobilized porphyrins and corrins to dehalogenate organochlorine pollutants, *Appl. Microbiol. Biotechnol.*, 38:413–416.

Marks, T. S., J. D. Allpress and A. Maule (1989) Dehalogenation of lindane by a variety of porphyrins and corrins, *Appl. Environ. Microbiol.*, 55:1258–1261.

Massey, I. J., M. D. Aitken, L. M. Ball and P. E. Heck (1994) Mutagenicity screening of reaction products from the enzyme-catalyzed oxidation of phenolic pollutants, *Environ. Toxicol. Chem.*, 13:1734–1752.

McFarland, M. J., C. M. Vogel and J. C. Spain (1992) Methanotrophic cometabolism of trichloroethylene (TCE) in a two stage bioreactor system, *Water Res.*, 26:259–265.

Mikesell, M. D., J. J. Kukor and R. H. Olsen (1993) Metabolic diversity of aromatic hydrocarbon-degrading bacteria from a petroleum-contaminated aquifer, *Biodegradation*, 4:249–259.

Mileski, G. J., J. A. Bumpus, M. A. Jurek and S. D. Aust (1988) Biodegradation of pentachlorophenol by the white rot fungus *Phanerochaete chrysosporium*, *Appl. Environ. Microbiol.*, 54:2885–2889.

Moen, M. A. and K. E. Hammel (1994) Lipid peroxidation by the manganese peroxidase of *Phanerochaete chrysosporium* is the basis for phenanthrene oxidation by the intact fungus, *Appl. Environ. Microbiol.*, 60:1956–1961.

Mohn, W. W. and J. M. Tiedje (1992) Microbial reductive dehalogenation, *Microbiol. Rev.* 56:482–507.

Mohn, W. W. and K. J. Kennedy (1992) Reductive dehalogenation of chlorophenols by *Desulfomonile tiedjei* DCB-1, *Appl. Environ. Microbiol.*, 58:1367–1370.

Mouchet, P. (1986) Algae reactions to mineral and organic micropollutants, ecological consequences and possibilities for industrial-scale application: A review, *Water Res.*, 399–412.

Mougin, C., C. Laugero, M. Asther, J. Dubroca, P. Frasse and M. Asther (1994) Biotransformation of the herbicide atrazine by the white rot fungus *Phanerochaete chrysosporium*, *Appl. Environ. Microbiol.*, 60:705–708.

Mu, D. Y. and K. M. Scow (1994) Effect of trichloroethylene (TCE) and toluene concentrations on TCE and toluene biodegradation and the population density of TCE and toluene degraders in soil, *Appl. Environ. Microbiol.*, 60:2661–2665.

Mueller, J. G., P. J. Chapman, B. O. Blattmann and P. H. Pritchard (1990) Isolation and characterization of a fluoranthene-utilizing strain of *Pseudomonas paucimobilis*, *Appl. Environ. Microbiol.*, 56:1079–1086.

Müller-Hurtig, R., R. Blaszczyk, F. Wagner and N. Kosaric (1993) Biosurfactants for environmental control, pp. 447–469 in *Biosurfactants: Production, Properties, Applications*, ed. N. Kosaric, Marcel Dekker, New York.

Nelson, M. J., J. V. Kinsella and T. Montoya (1990) In situ biodegradation of TCE contaminated groundwater, *Environ. Prog.*, 9:190–196.

Nelson, M. J. K., S. O. Montgomery and P. H. Pritchard (1988) Trichloroethylene metabolism by microorganisms that degrade aromatic compounds, *Appl. Environ. Microbiol.*, 54:604–606.

Nelson, M. J. K., S. O. Montgomery, W. R. Mahaffey and P. H. Pritchard (1987) Biodegradation of trichloroethylene and involvement of an aromatic biodegradative pathway, *Appl. Environ. Microbiol.*, 53:949–954.

Newman, L. M. and L. P. Wackett (1991) Fate of 2,2,2-trichloroacetaldehyde (chloral hydrate) produced during trichloroethylene oxidation by methanotrophs, *Appl. Environ. Microbiol.*, 57:2399–2402.

Nies, L. and T. M. Vogel (1990) Effects of organic substrates on dechlorination of Aroclor 1242 in anaerobic sediments, *Appl. Environ. Microbiol.*, 56:2612–2617.

Nüßlein, K., D. Maris, K. Timmis and D. F. Dwyer (1992) Expression and transfer of engineered catabolic pathways harbored by *Pseudomonas* spp. introduced into activated sludge microcosms, *Appl. Environ. Microbiol.*, 58:3380–3386.

Ogunseitan, O. A., Delgado, I. L., Tsai, Y.-L. and Olson, B. H. (1991) Effect of

2-hydroxybenzoate on the maintenance of naphthalene-degrading pseudomonads in seeded and unseeded soil, *Appl. Environ. Microbiol.* 57:2873–2879.

Ogunseitan, O. A. and B. H. Olson (1993) Effect of 2-hydroxybenzoate on the rate of naphthalene mineralization in soil, *Appl. Microbiol. Biotechnol.*, 38: 799–807.

Oldenhuis, R., J. Y. Oedzes, J. J. van der Waarde and D. B. Janssen (1991) Kinetics of chlorinated hydrocarbon degradation by *Methylosinus trichosporium* OB3b and toxicity of trichloroethylene, *Appl. Environ. Microbiol.*, 57:7–14.

Oldenhuis, R., R. L. J. M. Vink, D. B. Janssen and B. Witholt (1989) Degradation of chlorinated aliphatic hydrocarbons by *Methylosinus trichosporium* OB3b expressing soluble methane monoxygenase, *Appl. Environ. Microbiol.*, 55:2819–2826.

Ortega-Calvo, J.-J. and M. Alexander (1994) Roles of bacterial attachment and spontaneous partitioning in the biodegradation of naphthalene initially present in nonaqueous-phase liquids, *Appl. Environ. Microbiol.*, 60:2643–2646.

Palleroni, N. J. (1994) Some reflections on bacterial diversity, *ASM News*, 60:537–540.

Park, K. S., R. C. Sims, W. Doucette and J. E. Matthews (1988) Biological transformation and detoxification of 7,12-dimethylbenz[*a*]anthracene in soil systems, *J. Water Pollut. Control Fed.*, 60:1822–1825.

Parkinson, J. S. (1993) Signal transduction schemes of bacteria, *Cell*, 73:857–871.

Pasti-Grigsby, M. B., A. Paszczynski, S. Goszczynski, D. L. Crawford and R. L. Crawford (1992) Influence of aromatic substitution patterns on azo dye degradability by *Streptomyces* spp. and *Phanerochaete chrysosporium*, *Appl. Environ. Microbiol.*, 58:3605–3613.

Perkins, P. S., S. J. Komisar, J. A. Puhakka and J. F. Ferguson (1994) Effects of electron donors and inhibitors on reductive dechlorination of 2,4,6-trichlorophenol, *Water Res.*, 28:2101–2107.

Polnisch, E., H. Kneifel, H. Franzke and K. H. Hofmann (1992) Degradation and dehalogenation of monochlorophenols by the phenol-assimilating yeast *Candida maltosa*, *Biodegradation*, 2:193–199.

Pothuluri, J. V., J. P. Freeman, F. E. Evans and C. E. Cerniglia (1993) Biotransformation of fluorene by the fungus *Cunninghamella elegans*, *Appl. Environ. Microbiol.*, 59:1977–1980.

Pothuluri, J. V., J. P. Freeman, F. E. Evans and C. E. Cerniglia (1990) Fungal transformation of fluoranthene, *Appl. Environ. Microbiol.*, 56:2974–2983.

Pothuluri, J. V., R. H. Heflich, P. P. Fu and C. E. Cerniglia (1992) Fungal metabolism and detoxification of fluoranthene, *Appl. Environ. Microbiol.*, 58:937–941.

Prouty, A. L. (1990) Bench-scale development and evaluation of a fungal bioreactor for color removal from bleach effluents, *Appl. Microbiol. Biotechnol.*, 32:490–493.

Quensen, J. F. III, S. A. Boyd and J. M. Tiedje (1990) Dechlorination of four commercial polychlorinated biphenyl mixtures (Aroclors) by anaerobic microorganisms from sediments, *Appl. Environ. Microbiol.*, 56:2360–2369.

Robinson, K. G., W. S. Farmer and J. T. Novak (1990) Availability of sorbed toluene in soils for biodegradation by acclimated bacteria, *Water Res.*, 24:345–350.

Rojo, F., D. H. Pieper, K.-H. Engesser, H.-J. Knackmuss and K. N. Timmis (1987) Assemblage of ortho cleavage route for simultaneous degradation of chloro- and methylaromatics, *Science*, 238:1395–1398.

Rosenberg,, M. (1991) Basic and applied aspects of microbial adhesion at the hydrocarbon:water interface, *Crit. Rev. Microbiol.*, 18:159–173.

Rosenberg, E., R. Legmann, A. Kushmaro, R. Taube, E. Adler and E. Z. Ron (1992) Petroleum bioremediation—A multiphase problem, *Biodegradation*, 3:337–350.

Rouse, J. D., D. A. Sabatini, J. M. Suflita and J. H. Harwell (1994) Influence of surfactants on microbial degradation of organic compounds, *Crit. Rev. Environ. Sci. Technol.*, 24:325–370.

Ryan, T. P. and J. A. Bumpus (1989) Biodegradation of 2,4,5-trichloro-phenoxyacetic acid in liquid culture and in soil by the white rot fungus *Phanerochaete chrysosporium*, *Appl. Microbiol. Biotechnol.*, 31:302–307.

Sasikala, C., C. V. Raman and P. R. Rao (1994) Photometabolism of heterocyclic aromatic compounds by *Rhodopseudomonas palustris* OU 11, *Appl. Environ. Microbiol.*, 60:2187–2190.

Schanke, C. A. and L. P. Wackett (1992) Environmental reductive elimination reactions of polychlorinated ethanes mimicked by transition-metal coenzymes, *Environ. Sci. Technol.*, 26:830–833.

Semprini, L. and P. L. McCarty (1991) Comparison between model simulations and field results for in-situ biorestoration of chlorinated aliphatics: Part 1. Biostimulation of methanotrophic bacteria, *Ground Water*, 29:365–374.

Semprini, L. and P. L. McCarty (1992) Comparison between model simulations and field results for in-situ biorestoration of chlorinated aliphatics: Part 2. Cometabolic transformations, *Ground Water*, 30:37–44.

Semprini, L., P. V. Roberts, G. D. Hopkins and P. L. McCarty (1990) A field evaluation of in-situ biodegradation of chlorinated ethenes: Part 2. Results of biostimulation and biotransformation experiments, *Ground Water*, 28:715–727.

Semprini, L., G. D. Hopkins, P. V. Roberts, D. Grbić-Galić and P. L. McCarty (1991) A field evaluation of in-situ biodegradation of chlorinated ethenes: Part 3. Studies of competitive inhibition, *Ground Water*, 29:239–250.

Shields, M. S., S. O. Montgomery, P. J. Chapman, S. M. Cuskey and P. H. Pritchard (1989) Novel pathway of toluene catabolism in the trichloroethylene-degrading bacterium G4, *Appl. Environ. Microbiol.*, 55:1624–1629.

Shields, M. S., S. O. Montgomery, S. M. Cuskey, P. J. Chapman and P. H. Pritchard (1991) Mutants of *Pseudomonas cepacia* G4 defective in catabolism of aromatic compounds and trichloroethylene, *Appl. Environ. Microbiol.*, 57:1935–1941.

Shoreit, A. A. M. and M. S. A. Shabeb (1994) Utilization of aromatic compounds by phototrophic purple nonsulfur bacteria, *Biodegradation*, 5:71–76.

Sims, J. L., R. C. Sims and J. E. Matthews (1990) Approach to bioremediation of contaminated soil, *Hazard. Waste Hazard. Mater.*, 7:117–149.

Singh, N., A. Sahoo, D. Misra, V. R. Rao and N. Sethunathan (1993) Synergistic interaction between two bacterial isolates in the degradation of carbofuran, *Biodegradation*, 4:115–123.

Spadaro, J. T., M. H. Gold and V. Renganathan (1992) Degradation of azo dyes by the lignin-degrading fungus *Phanerochaete chrysosporium*, *Appl. Environ. Microbiol.*, 58:2397–2401.

Speitel, G. E. and D. S. McLay (1993) Biofilm reactors for treatment of gas streams containing chlorinated solvents, *J. Environ. Eng.*, 119:658–678.

Speitel, G. E., Jr. and J. M. Leonard (1992) A sequencing biofilm reactor for the treatment of chlorinated solvents using methanotrophs, *Water Environ. Res.*, 64:712–719.

Speitel, G. E., R. C. Thompson and D. Weissman (1993) Biodegradation kinetics of *Methylosinus trichosporium* OB3b at low concentrations of chloroform in the presence and absence of enzyme competition by methane, *Water Res.*, 27:15–24.

Stanier, R.Y., E. A. Adelberg, J. L. Ingraham and M. L. Wheelis (1979) *Introduction to the Microbial World*, Prentice-Hall, Englewood Cliffs, N.J.

Strand, S. E. and L. Shippert (1986) Oxidation of chloroform in an aerobic soil exposed to natural gas, *Appl. Environ. Microbiol.*, 52:203–205.

Strand, S. E., J. V. Wodrich and H. D. Stensel (1991) Biodegradation of chlorinated solvents in a sparged, methanotrophic biofilm reactor, *Res. J. Water Pollut. Control Fed.*, 63:859–867.

Strand, S. E., M. D. Bjelland and H. D. Stensel (1990) Kinetics of chlorinated hydrocarbon degradation by suspended cultures of methane-oxidizing bacteria, *Res. J. Water Pollut. Control Fed.*, 62:124–129.

Strandberg, G. W., T. L. Donaldson and L. L. Farr (1989) Degradation of trichloroethylene and *trans*-1,2-dichloroethylene by a methanotrophic consortium in a fixed-film, packed-bed bioreactor, *Environ. Sci. Technol.*, 23:1422–1425.

Stringfellow, W. T. and M. D. Aitken (1995) Competitive metabolism of naphthalene, methylnaphthalenes and fluorene by phenanthrene-degrading pseudomonads, *Appl. Environ. Microbiol.* 61:357–362.

Stromeyer, S. A., K. Stumpf, A. M. Cook and T. Leisinger (1992) Anaerobic degradation of tetrachloromethane by *Acetobacterium woodii*: Separation of dechlorinative activities in cell extracts and roles for vitamin B_{12} and other factors, *Biodegradation*, 3:113–123.

Sublette, K. L., E. V. Ganapathy and S. Schwartz (1992) Degradation of munition wastes by *Phanerochaete chrysosporium*, *Appl. Biochem. Biotechnol.*, 34/35:709–723.

Suidan, M. T., I. N. Najm, J. T. Pfeffer and Y. T. Wang (1988) Anaerobic biodegradation of phenol: Inhibition kinetics and system stability, *J. Environ. Eng.*, 114:1359–1375.

Sutherland, J. B., A. L. Selby, J. P. Freeman, F. E. Evans and C. E. Cerniglia (1991) Metabolism of phenanthrene by *Phanerochaete chrysosporium*, *Appl. Environ. Microbiol.*, 57:3310–3316.

Syldatk, C. and F. Wagner (1987) Production of biosurfactants, pp. 89–120 in *Biosurfactants and Biotechnology*, ed. N. Kosaric, W. L. Cairns and N. C. C. Gray, Marcel Dekker, New York.

Tanner, D. D. and A. R. Stein (1988) On the mechanism of reduction by reduced nicotinamide adenine dinucleotide dependent alcohol dehydrogenase. α-Halo ketones as mechanistic probes, *J. Org. Chem.*, 53:1642–1646.

Thomas, D. R., K. S. Carswell and G. Georgiou (1992) Mineralization of biphenyl and PCBs by the white rot fungus *Phanerochaete chrysosporium*, *Biotechnol Bioeng.*, 40:1395–1402.

Tiedje, J. M. (1993) Bioremediation from an ecological perspective, pp. 110–120 in *In Situ Bioremediation: When Does It Work?* Committee on in situ Bioremediation, Water Science and Technology Board, National Research Council, National Academy Press, Washington, D.C.

Tsien, H.-C., G. A. Brusseau, R. S. Hanson and L. P. Wackett (1989) Biodegradation of trichloroethylene by *Methylosinus trichosporium* OB3b, *Appl. Environ. Microbiol.*, 55:3155–3161.

Uchiyama, H., T. Nakajima, O. Yagi and T. Nakahara (1992) Role of heterotrophic bacteria in complete mineralization of trichloroethylene by *Methylocystis* sp. strain M, *Appl. Environ. Microbiol.*, 58:3067–3071.

Valli, K. and M. H. Gold (1991) Degradation of 2,4-dichlorophenol by the lignin-degrading fungus *Phanerochaete chrysosporium*, *J. Bacteriol.*, 173:345–352.

Valli, K., B. J. Brock, D. K. Joshi and M. H. Gold (1992) Degradation of 2,4-dinitrotoluene by the lignin-degrading fungus *Phanerochaete chrysosporium*, *Appl. Environ. Microbiol.*, 58:221–228.

van der Meer, J. R., W. M. de Vos, S. Harayama and A. J. B. Zehnder (1992) Molecular mechanisms of genetic adaptation to xenobiotic compounds, *Microbiol. Rev.*, 56:677–694.

van Loosdrecht, M. C. M., J. Lyklema, W. Norde and A. J. B. Zehnder (1990) Influence of interfaces on microbial activity, *Microbiol. Rev.*, 54:75–87.

Venkatadri, R., S.-P. Tsai, N. Vukanic and L. B. Hein (1992) Use of a biofilm membrane reactor for the production of lignin peroxidase and treatment of pentachlorophenol by *Phanerochaete chrysosporium*, *Hazard. Waste Hazard. Mater.*, 9:231–243.

Volkering, F., A. M. Breure and J. G. van Andel (1993) Effect of micro-organisms on the bioavailability and biodegradation of crystalline naphthalene, *Appl. Microbiol. Biotechnol.*, 40:535–540.

Volkering, F., A. M. Breure, A. Sterkenburg and J. G. van Andel (1992) Microbial degradation of polycyclic aromatic hydrocarbons: Effect of substrate availability on bacterial growth kinetics, *Appl. Microbiol. Biotechnol.*, 36:548–552.

Wackett, L. P. (1994) Dehalogenation in environmental biotechnology, *Curr. Opin. Biotechnol.*, 5:260–265.

Wackett, L. P. and D. T. Gibson (1988) Degradation of trichloroethylene by toluene dioxygenase in whole-cell studies with *Pseudomonas putida* F1, *Appl. Environ. Microbiol.*, 54:1703–1708.

Wackett, L. P. and S. R. Householder (1989) Toxicity of trichloroethylene to *Pseudomonas putida* F1 is mediated by toluene dioxygenase, *Appl. Environ. Microbiol.*, 55:2723–2725.

Wackett, L. P., M. J. Sadowsky, L. N. Newman, H.-G. Hur and S. Li (1994) Metabolism of polyhalogenated compounds by a genetically engineered bacterium, *Nature*, 368:627–629.

Walter, U., M. Beyer, J. Klein and H.-J. Rehm (1991) Degradation of pyrene by *Rhodococcus* sp. UW1, *Appl. Microbiol. Biotechnol.*, 34:671–676.

Wang, X., X. Yu and R. Bartha (1990) Effect of bioremediation on polycyclic aromatic hydrocarbon residues in soil, *Environ. Sci. Technol.*, 24:1086–1089.

Weber, W. J. and H. X. Corseuil (1994) Inoculation of contaminated subsurface soils with enriched indigenous microbes to enhance bioremediation rates, *Water Res.*, 28:1407–1414.

Weissenfels, W. D., H.-J. Klewer and J. Langhoff (1992) Adsorption of polycyclic aromatic hydrocarbons (PAHs) by soil particles: influence on biodegradability and biotoxicity, *Appl. Microbiol. Biotechnol.*, 36:689–696.

Weissenfels, W. D., M. Beyer and J. Klein (1990) Degradation of phenanthrene, fluorene and fluoranthene by pure bacterial cultures, *Appl. Microbiol. Biotechnol.*, 32:479–484.

Williams, W. A. (1994) Microbial reductive dechlorination of trichlorobiphenyls in anaerobic sediment slurries, *Environ. Sci. Technol.*, 28:630–635.

Wilson, J. T. and B. H. Wilson (1985) Biotransformation of trichloroethylene in soil, *Appl. Environ. Microbiol.*, 49:242–243.

Yadav, J. S. and C. A. Reddy (1993) Degradation of benzene, toluene, ethylbenzene and xylenes (BTEX) by the lignin-degrading basidiomycete *Phanerochaete chrysosporium*, *Appl. Environ. Microbiol.*, 59:756–762.

Yadav, J. S., R. E. Wallace and C. A. Reddy (1995) Mineralization of mono- and dichlorobenzenes and simultaneous degradation of chloro- and methyl-substituted benzenes by the white rot fungus *Phanerochaete chrysosporium*, *Appl. Environ. Microbiol.*, 61:677–680.

Yarden, O., R. Salomon, J. Katan and N. Aharonson (1990) Involvement of fungi and bacteria in enhanced and nonenhanced biodegradation of carbendazim and other benzimidazole compounds in soil, *Can. J. Microbiol.*, 36:15–23.

Ye, D., J. F. Quensen III, J. M. Tiedje and S. A. Boyd (1992) Anaerobic dechlorination of polychlorobiphenyls (Aroclor 1242) by pasteurized and ethanol-treated microorganisms from sediments, *Appl. Environ. Microbiol.*, 58:1110–1114.

Yin, C. F., T. W. Joyce and H.-M. Chang (1989a) Kinetics of bleach plant effluent decolorization by *Phanerochaete chrysosporium*, *J. Biotechnol.*, 10:67–76.

Yin, C.-F., T. W. Joyce and H.-M. Chang (1989b) Role of glucose in fungal decolorization of wood pulp bleaching effluents, *J. Biotechnol.*, 10:77–84.

Zepp, R. G. and P. F. Schlotzhauer (1983) Influence of algae on photolysis rates of chemical in water, *Environ. Sci. Technol.*, 17:462–468.

Zhang, Y. and R. M. Miller (1994) Effect of a *Pseudomonas* rhamnolipid biosurfactant on cell hydrophobicity and biodegradation of octadecane, *Appl. Environ. Microbiol.*, 60:2101–2106.

Zylstra, G., L. P. Wackett and D. T. Gibson (1989) Trichloroethylene degradation by *Escherichia coli* containing the cloned *Pseudomonas putida* F1 toluene dioxygenase genes, *Appl. Environ. Microbiol.*, 55:3162–3166.

Molecular Probes and Biosensors in Bioremediation and Site Assessment

GARY S. SAYLER,*,**,† UDAYAKUMAR MATRUBUTHAM,† FU-MIN MENN,†
WADE H. JOHNSTON,*,† AND RAYMOND D. STAPLETON, JR.**,†
University of Tennessee
Knoxville, TN 37916, USA

INTRODUCTION

Bioremediation is a process utilizing living organisms to destroy, transform, and in some cases, sequester environmental pollutants. In the broadest sense, all forms of biological waste treatment can be referred to as bioremediation, including such conventional processes as activated sludge and composting for sewage and sludge treatment, respectively. However, in the usual context, it is generally associated with destruction by biodegradation and biotransformation of toxic chemicals at sites contaminated as a result of industrial activities. While not universal in its application, there is a growing awareness of the utility and ability of bioremediation to complement other physical-chemical technologies for site remediation.

There are some significant differences between conventional waste treatment and bioremediation. Conventional waste treatment is generally directed toward the stabilization of organic materials that are putrescible. These technologies involve the controlled degradation of wastes to prevent the production of nuisance odors and/or the deterioration of water supplies by uncontrolled biodegradation in the environment. As such, conventional treatment generally involves the biodegradation of materials that are a problem due to the fact that they are readily biodegradable. In contrast, industrial contamination tends to be a prob-

*Department of Microbiology.
**Graduate Program in Ecology.
†The Center for Environmental Biotechnology.

lem due to the human or animal toxicity of the contaminants, which may or may not be readily degradable. This difficulty is compounded by the fact that previous spills or disposal practices have resulted in the presence of contaminants in environments not conducive to any currently available technologies for detoxification or destruction. Stimulation of competent microbial populations in these environments may be constrained by mass-transfer of nutrients and contaminants, resulting in low active biomass concentrations and patchy distribution. In some environments, low initial concentrations of competent strains may be present due to the limited distribution of the specialized genotypes capable of degrading the more recalcitrant contaminants. For these reasons, bioremediation and biological waste treatment, as well as fundamental environmental biodegradation research, can benefit from the continued development of more specific, and potentially quantitative, molecular approaches for measuring the abundance, activity, and dynamics of specific degradative microbial populations.

Efficient bioremediation is dependent on achieving adequate population density, metabolic capability, and physiological activity of the microorganisms at the contaminated site (Sayler et al., 1982, 1983). Absolute quantification of these microbiological parameters under field conditions is necessary for documenting that contaminant loss is resulting from biodegradation, that rates of degradation have been maximized to the extent possible, and that these rates are sufficiently high and uniform throughout the site to achieve effective treatment (Sayler et al., 1985; Jain et al., 1988). Recently, different approaches employing novel molecular techniques have been developed to quantify microbial populations and degradative activities. The availability of comprehensive molecular genetic information and the applicability of sensitive biological techniques for microbial ecology have enabled development of such molecular techniques.

Molecular techniques such as direct extraction of nucleic acids from environmental samples and nucleic acid hybridization using specific probes for a biodegradation gene or gene message have become somewhat routine research procedures in assessing the bioremediating potential of a contaminated site (Sayler and Layton, 1990b; Barkey and Sayler, 1988; Ogram and Sayler, 1988; Selenska and Klingmuller, 1992; Sayler, 1991). Based on site contaminants and nucleic acid sequence databases, an array of nucleic acid probes can be designed, with a certain degree of bias, to evaluate a variety of microbiological and biodegradative parameters. The probes may be employed to determine (1) the overall genetic diversity, (2) the dominant and active gene pool, and (3) the density and frequency of specific gene line required to degrade a target compound at a site.

In recent years, the potential for noninvasive and in situ remediation assessment strategies involving biosensors and biomonitors has gained significant attention. A biosensor or biomonitor is an analytical device consisting of a biological element that produces an environmental signal corresponding to the presence or absence of a contaminant or a microbial response. In bioremediation and site assessment, biosensors can be utilized to monitor (1) the bioavailability of a specific compound, (2) the degradation and disappearance of the specific compound, (3) the physiological activities of specific bacterial groups capable of degrading the compound, and (4) to assess the optimal bioremediation conditions by manipulating environmental factors. This chapter attempts to compile the majority of contemporary molecular diagnostic techniques with utility in bioremediation and site assessment.

PERSPECTIVE ON MICROBIOLOGICAL ASSESSMENT IN BIOREMEDIATION

Feasibility of Biodegradation Testing

Bioremediation typically begins by making a survey of the contaminated site to determine if biological remediation is possible. This is accomplished using treatability assays under laboratory conditions (Table 1). In open laboratory systems such as pan studies, contaminant biodegradation is determined by monitoring the pollutant loss over time. In closed systems, such as flasks and column assays, contaminant biodegradation is determined by monitoring pollutant loss over time, by metabolite accumulation, or by adding ^{14}C-labeled pollutant and assaying for the evolution of $^{14}CO_2$ over time concomitant with pollutant loss. Each assay may be designed to mimic a particular type of bioremediation strategy such as solid-phase, slurry-phase, liquid-phase, or in situ treatment. These treatability studies may be simple in that contaminant disappearance is monitored, or they may be complex in that contaminant mineralization is studied along with several chemical, physical, and microbial parameters concomitantly (Nelson et al., 1991). The degree of simplicity in a treatability study is dependent on the prior biochemical knowledge of the biodegradation of the pollutant. It has become apparent and deemed mandatory that attempts at mass balance are included in these studies (Unterman, 1991). With the proper use of controls, these treatibility assays provide evidence that microbiological biodegradation of the pollutant is possible.

Treatability studies are limited in that they are rarely a realistic

Table 1. Types of treatability tests.

Treatability Test	Treatment Simulation	Indication
Flask studies	Liquid-phase	• General test for biodegradation in water or soil
	Slurry-phase	• Simulate soils or water in bioreactors • Proper use is for rapid screening of samples • Do not mimic field conditions
Pan studies	Solid-phase landfarming	• Used to test whether soils/sludges may be treated without slurrying • Testing for appropriate amendments and physical mixing
Column studies	in situ	• Valuable for use in determining soil flushing vs. biodegradation • Identification of physical constraints of nutrient delivery • Identification of chemical constraints such as precipitation and clogging • Statistical variation is large between column replicates

Source: Nelson et al., 1991.

surrogate for the genuine environment that is to be remediated. Treatability studies are usually conducted in the laboratory under somewhat controlled conditions. The treatments may be amended with nutrients, pH buffered, oxygenated, etc. These amendments are usually intended to optimize the biodegradation process but may lead to results that are too optimistic to be achieved in the environment. Treatability studies with no amendments have been shown to overestimate the in situ biodegradation rates in deep aquifers (Chapelle and Lovley, 1990).

Despite the limitations of treatability studies, they do provide valuable information. Treatability studies do demonstrate that in situ biodegradation may occur if the proper conditions are met. If a site is newly contaminated, treatability studies may give the minimum time necessary for the indigenous microbiota to acclimate for contaminant biodegradation (Wilson and Jones, 1993). Transformation products may be monitored, and the possibility of producing compounds more toxic and or more mobile may be determined (Park et al., 1988). Perhaps most importantly, treatability studies will demonstrate whether, under the most ideal conditions or those that can be reasonably provided, a regulatory minimum concentration of pollutant can be obtained (Nelson et al., 1991).

Microbiological Characterization— Culturability and the Pure Culture

Contaminated sites have been typically characterized microbiologically by removing a soil or water sample and attempting to culture microorganisms capable of biodegrading the pollutant of interest. The most prevalent method of doing this has been through the use of enrichment culture. An environmental sample is incubated with or without the use of additional pollutant substrate and microbes are screened for degradation of the compound of interest. This type of assay is biased toward microbes that can be cultured—generally heterotrophic bacteria that are capable of growth on rich media—from the sample in that these cultivable microorganisms may not be the most significant degraders in the sample.

Another method of obtaining microbial isolates that can degrade pollutants is through the use of agar contaminant spray-plate tests (Kiyohara et al., 1982). A diluted environmental sample is spread plated onto agar plates, and the pollutant, which is dissolved in a volatile solvent, is sprayed directly onto the plate to give a thin film over the surface of the agar. The isolated colonies that can degrade the pollutant forms a clearing zone around the colony. This method has been used to examine the polynuclear aromatic hydrocarbon (PAH) utilization patterns from PAH-contaminated soils (Kästner et al., 1994). This assay does not demonstrate mineralization but, rather, solubilization and utilization of the test substrate. Thus, this assay is an indicator of transformation of the test compound(s). However, spray plating is a good method of isolating microbes capable of at least partially degrading the test compound(s) of interest. The agar spray plate method also suffers from culture bias. It has been generally accepted that, in a typical sample, only 1–10% of the bacteria are culturable, making it impossible to determine if the culturable microorganisms are the most abundant and functionally active for degrading the pollutant of interest. Even in light of these limitations, pure culture isolates form the fundamental foundation of the biochemical and genetic knowledge in the current database on biodegradation.

Environmental and Contaminant Complexity

Extrapolation of treatability studies to the field has always been a problem. The most obvious reason for this difficulty is the geological and hydrological heterogeneity of soil and subsurface environments. Also many polluted environments are contaminated with multiple compounds, making bioremediation more difficult since adaptation may take longer for some compounds. There are several factors that may be more pronounced in situ than in a treatability study. The use of nutrient supplements is usually more difficult in an actual environmental situation. The addition of oxygen, other electron acceptors, micronutrients, etc., are more problematic due to heterogeneity in hydraulic conductivity in subsurface environments. The porosity of the subsurface may be low and subject to plugging with biomass, making flow through the contaminated zone or plume more difficult or impossible. Soil properties such as particle size and organic matter have a broad impact on the rate and extent of biodegradation in soils (Manilal and Alexander, 1991).

Microorganisms degrade pollutants either as a primary source of carbon and energy or by cometabolism (cooxidation) (Horvath, 1972). During cometabolism, the degradative enzyme(s) degrades the pollutant even though the compound is not a normal substrate for growth. Examples include the degradation of trichloroethylene by the toluene dioxygenase (*tod*) system for toluene oxidation and the methane monooxygenase (*mmo*) for the methane oxidation. The cometabolized compound is not an inducer for the degradative system. The degradation of high-molecular-weight PAHs also seems to be a cometabolic process (Keck et al., 1989).

Many pollutant mixtures include compounds of regulatory concern that have a very low solubility in aqueous environments. This translates into situations where the pollutant is not bioavailable. Also contributing to bioavailability is the partitioning of the pollutant onto soil and organic particles and the complexation with organic matter. In these situations, the pollutants are usually unavailable for catabolism by the microbiota in that environment, and this may be the rate-limiting factor for biodegradation. In some cases, the pollutants are in very low concentrations. This may be due to bioavailability or a truly low pollutant concentration. It has been proposed that a threshold concentration may exist at which microbes will not metabolize a compound. This proposed threshold concentration, $[S]_{min}$, has been discussed (Alexander, 1985, Boethling and Alexander, 1979). The chief consequence of $[S]_{min}$ is that there will be a residual concentration of pollutant that may be above the regulatory minimum.

LEVEL OF BIOCHEMICAL AND MOLECULAR UNDERSTANDING OF BIODEGRADATION

Microbial degradation of environmental pollutants have been studied extensively on biochemical and molecular biological levels. Most pollutants are not of biosynthetic origin but are derived from either pyrolysis of organic materials, e.g., aromatic hydrocarbons (Gibson, 1977), or man-made chemicals, e.g., chlorinated compounds. Several excellent reviews have been written during the last decade that describe microbial biodegradative capacity and mechanisms (Gibson and Subramanian, 1984; Gunsalus, 1985; Timmis et al., 1985; Weightman et al., 1985; Dagley, 1986; Frantz and Chakrabarty, 1986; Rochkind-Dubinsky et al., 1987; Reineke and Knackmuss, 1988; Commandeur and Parsons, 1990; Smith, 1990; Chaudhry and Chapalamadugu, 1991; Cerniglia, 1992; van der Meer et al., 1992; Furukawa, 1994; Janssen et al., 1994; Layton et al., 1994a; Singleton, 1994; Williams and Sayers, 1994). Since that time, a significant number of catabolic genes have been isolated and characterized from various bacteria and have been reviewed previously (Sayler and Blackburn, 1989; Sayler et al., 1990a; Wallace and Sayler, 1992). An updated list of recent studies on bacterial genes involved in degradation of environmental pollutants is given in Table 2. This table illustrates a wide variety of microorganisms participating in degradation of a broad number of environmental pollutants. The soil bacterium, *Pseudomonas*, still plays a major role in the biodegradation task among the known bacteria. In this section, the discussion will be focused primarily on aerobic processes, although anaerobic metabolism is a significant, if less well defined, mechanism used by bacteria to degrade environmental contaminants.

Toluene is a major product of the petroleum industry and is used extensively in the synthesis of a wide variety of chemicals and is present in fuel hydrocarbon mixtures. As a result, toluene is widespread and can be detected in soil, groundwater, and the atmosphere of urban environments as a contaminant. Biodegradation of toluene by bacteria under aerobic processes has been studied extensively in the past (Gibson and Subramanian, 1984b), and five different pathways have been demonstrated in the bacteria. The five aerobic pathways for the degradation of toluene by bacteria are illustrated in Figure 1. *Pseudomonas putida* mt-2 contains the TOL plasmid that initiates the pathway by monooxygenation at the methyl group to yield benzyl alcohol (Worsey and Williams, 1975). The initial reaction by *Pseudomonas cepacia* G4 (Shields et al., 1989), *Pseudomonas pickettii* PKO1 (Kaphammer et al., 1991), and *Pseudomonas mendocina* KR (Whited and Gibson, 1991) is a monohydroxylation at the *ortho-*, *meta-*, and *para-*positions, respectively. Cresols are the products formed by hydroxylation. Another pathway, *Pseudomonas putida* F1, initiates the reaction by a dioxygenase and forms *cis*-toluene dihydrodiol (Gibson et al., 1970). It is important to note that all substrates are converted to catechols, which are substrates for enzymatic cleavage of the aromatic nucleus. Both atoms of molecular oxygen are required in all pathways at the initial oxidation step and the cleavage of the aromatic ring.

Table 2. Environmental pollutants for which the microbiological and molecular basis of biodegradation is well established.

Pollutant(s)	Gene	Plasmid/Chromosome Encoded	Strain	Reference
Alkane (C_6–C_{10})	*alk*	Plasmid (OCT)	*Pseudomonas putida*	(Fennewald et al., 1979; Harder and Kunz, 1986)
Alkane (up to C_8)	*amo*	—	*Nitrosomonas europaea*	(Hyman et al., 1988)
Alkene (up to C_5)	*amo*	—	*Nitrosomonas europaea*	(Hyman et al., 1988)
Aniline	*amo*	—	*Nitrosomonas europaea*	(McTavish et al., 1993; Keener and Arp, 1994)
Anthracene	*nah*	Plasmid (pKA1)	*Pseudomonas fluorescens 5R*	(Menn et al., 1993; Sanseverino et al., 1993a)
Benzene	n.d.	Plasmid (pWW174)	*Acinetobacter calcoaceticus RJE74*	(Winstanley et al., 1987)
	n.d.	Chromosome	*Bacillus stearothermophilus BR325*	(Natarajan et al., 1994)
	amo	—	*Nitrosomonas europaea*	(Hyman et al., 1985)
	bed	Plasmid (pHMT112)	*Pseudomonas putida ML2*	(Tan and Fong, 1993; Tan et al., 1993)
Benzoate	*ben*	Chromosome	*Acinetobacter calcoaceticus*	(Neidle et al., 1987)
	cat	Chromosome	*Acinetobacter calcoaceticus*	(Shanley et al., 1986; Neidle et al., 1988)
	pca	Chromosome	*Pseudomonas putida*	(Doten et al., 1987; Parales and Harwood, 1993)
Benzoate(3-Cl-)	*cba*	Plasmid (pBRC60)	*Alcaligenes* sp. BR60	(Nakatsu and Wyndham, 1993)
	tfdCDE	Chromosome/Plasmid (pJP4)	*Alcaligenes eutrophus* JMP134	(Don et al., 1985b; Perkins et al., 1990)
	clc	Plasmid (pAC27)	*Pseudomonas putida*	(Frantz and Chakrabarty, 1987; Coco et al., 1993)
Benzoate(4-Cl-)	n.d.	Plasmid (pASU1)	*Arthrobacter* sp. SU	(Schmitz et al., 1992)
	n.d.	Chromosome	*Pseudomonas* sp. CBS3	(Scholten et al., 1991)
Biphenyl/4-CB	*bph*	Chromosome	*Alcaligenes eutrophus* A5	(Springael et al., 1993)
	bph	Plasmid (pWW100)	*Pseudomonas* sp. CB406	(Lloyd-Jones et al., 1994)

Table 2. *(continued).*

Pollutant(s)	Gene	Plasmid/Chromosome Encoded	Strain	Reference
Carbon mon-	*amo*	—	*Nitrosomonas europaea*	(Jones and Morita, 1983)
oxide 2,4-D	*tfd*	Plasmid (pJP4)	*Alcaligenes eutrophus* JMP134	(Don et al., 1985a; Perkins et al., 1990)
	n.d.	Plasmid (pKA2)	*Alcaligenes paradoxus* 2811P	(Ka and Tiedje, 1994)
	n.d.	Plasmid (pKA4)	*Alcaligenes pickettii* 712	(Ka and Tiedje, 1994)
Dibenzo-*p*-dioxin	*dbf*	—	*Sphingomonas* sp. RW1	(Happe et al., 1993)
Dibenzofuran	*dbf*	—	*Sphingomonas* sp. RW1	(Happe et al., 1993)
Dibenzothiphene	*dox*	Plasmid	*Pseudomonas* sp. C18	(Denome et al., 1993)
	n.d.	Plasmid	*Pseudomonas alcaligenes* DBT2	(Montcello et al., 1985)
DCE(1,2-)	*dhl*	Plasmid (pXAU1)	*Xanthobacter autotrophicus* GJ10	(Tardif et al., 1991; Pries et al., 1994)
DDT	*bph*	Chromosome	*Alcaligenes eutrophus* A5	(Nadeau et al., 1994)
Ethylbenzene	*amo*	—	*Nitrosomonas europaea*	(McTavish et al., 1993; Keener and Arp, 1994)
	tbu	Chromosome	*Pseudomonas pickettii* PKO1	(Olsen et al., 1994)
Furan	*thd*	Chromosome	*Escherichia coli* NAR30	(Alam et al., 1990)
Naphthalene	*dox*	Plasmid	*Pseudomonas* sp. C18	(Denome et al., 1993)
	nah	Plasmid (pKA1)	*Pseudomonas fluorescens* 5R	(King et al., 1990)
	bph	Chromosome	*Pseudomonas paucimobilis* Q1	(Kuhm et al., 1991)
	nah	Plasmid (Nah7)	*Pseudomonas putida* PpG7	(Dunn and Gunsalus, 1973)
	nah	Plasmid (pDTG1)	*Pseudomonas putida* NCIB9816-4	(Serdar and Gibson, 1989)
	pah	Chromosome	*Pseudomonas putida* OUS82	(Kiyohara et al., 1994)
Octane	*alk*	Plasmid (OCT)	*Pseudomonas oleovorans*	(Chakrabarty et al., 1973)
Phenanthrene	*dox*	Plasmid	*Pseudomonas* sp. C18	(Denome et al., 1993)
	nah	Plasmid (pKA1)	*Pseudomonas fluorescens* 5R	(Menn et al., 1993; Sarseverino et al., 1993a)
	pah	Chromosome	*Pseudomonas putida* OUS82	(Kiyohara et al., 1994)

(continued)

Table 2. (continued).

Pollutant(s)	Gene	Plasmid/Chromosome Encoded	Strain	Reference
Phenol	phl	Chromosome	Alcaligenes eutrophus JMP134	(Pieper et al., 1989; Kim et al., 1994)
	amo	—	Nitrosomonas europaea	(Hyman et al., 1985)
	dmp	Plasmid (pVI150)	Pseudomonas sp. CF600	(Shingler et al., 1992; Powlowski and Shingler, 1994)
	tom	—	Pseudomonas cepacia G4	(Nelson et al., 1987, 1988)
	tbu	Chromosome	Pseudomonas pickettii PKO1	(Kukor and Olsen, 1990)
PCB	bph	Chromosome	Alcaligenes eutrophus H850	(Yates and Mondello, 1989)
	bph	Chromosome	Arthrobacter sp. M5	(Peloquin and Greer, 1993)
	bph	Chromosome	Pseudomonas sp. ENV307	(Sharma et al., 1991)
	bph	Chromosome	Pseudomonas sp. KKS102	(Kimbara et al., 1989; Kikuchi et al., 1994a, 1994b)
	bph	Chromosome	Pseudomonas sp. LB400	(Mondello, 1989; Erickson and Mondello, 1992)
	bph	Chromosome	Pseudomonas paucimobilis Q1	(Taira et al., 1988)
	bph	Chromosome	Pseudomonas pseudoalcaligenes KF707	(Taira et al., 1992)
	bph	Chromosome	Pseudomonas putida KF715	(Hayase et al., 1990)
	cbp	Chromosome	Pseudomonas putida OU83	(Khan and Walia, 1991)
	bph	Chromosome	Pseudomonas testosteroni B-356	(Ahmad et al., 1990, 1991)
	bph	Chromosome	Rhodococcus globerulus P6	(Asturias et al., 1994)

Table 2. (continued).

Pollutant(s)	Gene	Plasmid/Chromosome Encoded	Strain	Reference
Protocatechuate	pca	Chromosome	Acinetobacter calcoaceticus	(Doten et al., 1987)
Styrene	amo	—	Nitrosomonas europaea	(McTavish et al., 1993; Keener and Arp, 1994)
Thiophenes	thd	Chromosome	Escherichia coli NAR30	(Alam et al., 1990)
Toluate (m-)	n.d.	Chromosome/Plasmid (pTDN1)	Pseudomonas putida UCC22	(Saint et al., 1990)
Toluene	amo	—	Nitrosomonas europaea	(McTavish et al., 1993; Keener and Arp, 1994)
	tom	—	Pseudomonas cepacia G4	(Shields et al., 1991)
	tmo	Chromosome	Pseudomonas mendocina KR1	(Yen et al., 1991)
	tbu	Chromosome	Pseudomonas pickettii PKO1	(Olsen et al., 1994)
	tod	Chromosome	Pseudomonas putida F1	(Zylstra et al., 1988)
	n.d.	Plasmid (pDK1)	Pseudomonas putida HS1	(Kunz and Chapman, 1981)
	xyl	Plasmid (TOL pWWO)	Pseudomonas putida mt-2	(Williams and Murray, 1974)
TCB (1,2,4-)	tcb	Plasmid (pP51)	Pseudomonas sp. P51	(van der Meer et al., 1991)
TCE	phl/tfd	Chromosome/Plasmid (pJP4)	Alcaligenes eutrophus JMP 134	(Harker and Kim, 1990; Kim et al., 1994)
	mmo	Chromosome	Methylococcus capsulatus (Bath)	(Green and Dalton, 1989)
	mmo	Chromosome	Methylosinus trichosporium OB3b	(Oldenhuis et al., 1989; Tsien et al., 1989)
	n.d.	—	Methylomonas GJ6	(Oldenhuis and Janssen, 1993)

(continued)

Table 2. (continued).

Pollutant(s)	Gene	Plasmid/Chromosome Encoded	Strain	Reference
TCE (continued)	*tomA*	—	*Pseudomonas cepacia* G4	(Nelson et al., 1987; Shields et al., 1991; Shields and Reagin, 1992)
	tmo	Chromosome	*Pseudomonas mendocina* KR1	(Winter et al., 1989)
	tbuABC	Chromosome	*Pseudomonas pickettii* PKO1	(Leahy and Olsen, 1994; Olsen et al., 1994)
	todC1C2BA	Chromosome	*Pseudomonas putida* F1	(Wackett and Gibson, 1988; Zylstra et al., 1989; Li and Wackett, 1992)
	n.d.	Plasmid (pBD2)	*Rhodococcus erythropolis* BD2	(Dabrock et al., 1994)
TMB (1,2,4-)	n.d.	Plasmid (pDK1)	*Pseudomonas putida* HS1	(Kunz and Chapman, 1981)
Xylene	n.d.	Plasmid (XYL)	*Pseudomonas* Pxy	(Friello et al., 1976)
Xylene (o-)	*tbu*	Chromosome	*Pseudomonas pickettii* PKO1	(Olsen et al., 1994)
Xylene (m-)	n.d.	Plasmid (pKJ1)	*Pseudomonas* sp. TAB	(Yano and Nishi, 1980)
	tbu	Chromosome	*Pseudomonas pickettii* PKO1	(Olsen et al., 1994)
	xyl	Plasmid (TOL)	*Pseudomonas putida* mt-2	(Worsey and Williams, 1975)
	n.d.	Plasmid (pDK1)	*Pseudomonas putida* HS1	(Kunz and Chapman, 1981)
Xylene (p-)	*amo*	—	*Nitrosomonas europaea*	(McTavish et al., 1993; Keener and Arp, 1994)
	n.d.	Plasmid (pKJ1)	*Pseudomonas* sp. TAB	(Yano and Nishi, 1980)
	tbu	Chromosome	*Pseudomonas pickettii* PKO1	(Olsen et al., 1994)
	xyl	Plasmid (TOL)	*Pseudomonas putida* mt-2	(Worsey, 1975)
	n.d.	Plasmid (pDK1)	*Pseudomonas putida* HS1	(Kunz and Chapman, 1981)

n.d.: Not determined. 4-CB: 4-chlorobiphenyl. 2,4-D: 2,4-dichlorophenoxyacetic acid. DCE: 1,2-dichloroethane. DDT: 1,1,1-trichloro-2,2-bis(4-chlorophenyl)ethane. PCB: polychlorinated biphenyl. TCB: 1,2,4-trichlorobenzene. TCE: trichloroethylene; cometabolism. TMB: 1,2,4-trimethylbenzene.

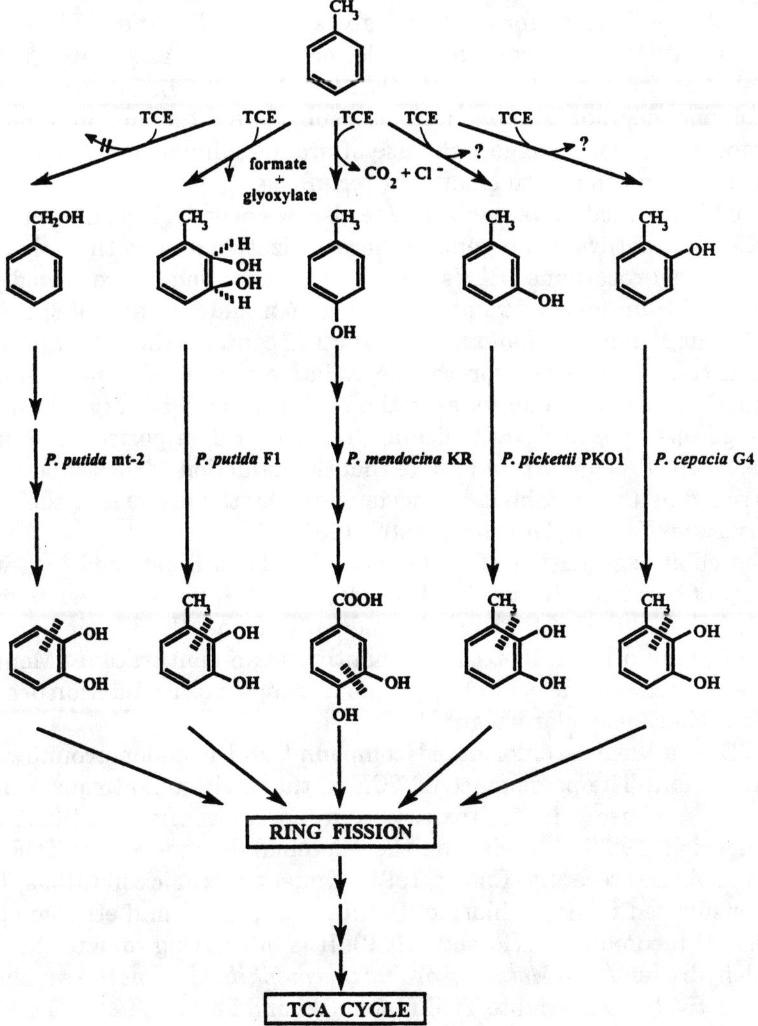

Figure 1. Toluene degradation pathways used by soil bacteria. The initial enzymes, in F1, KR1, PKO1, and G4, in the pathways also involve cooxidation of TCE (see text).

As indicated by Figure 1, four of the five initial steps in toluene oxidation also promote the cometabolic degradation of trichloroethylene (TCE). Both the dioxygenase and monooxygenases appear to cometabolically oxidize TCE to formate and glyoxylate in *P. putida* F1 (Li and Wackett, 1992) or to carbon dioxide, chloride ion, and water-soluble metabolites in *P. mendocina* KR1 (Winter et al., 1989). The inducibility of toluene degradation by aromatic substrates, toluene and phenol (monooxygenation), suggest the use of aromatic inducers to drive TCE remediation as an in situ treatment technology.

The biodegradation of naphthalene also has been well studied on both biochemical pathway and genetic characterization among the polycyclic aromatic hydrocarbons (PAHs) and has been thoroughly reviewed previously (Gibson and Subramanian, 1984; Yen and Serdar, 1988; Schell, 1990). The strain *Pseudomonas putida* PpG7 contains the NAH7 plasmid that encodes the genes for the degradative pathway of naphthalene (Figure 2). The catabolic genes in the NAH7 plasmid are organized into two operons, *nah* and *sal*, which are controlled by a positive regulator gene, *nahR*. It is important to note that the induction of these operons is controlled by the metabolite, salicylate, and by the product of the regulatory gene (Yen and Gunsalus, 1982, 1985).

Microbial degradation of other PAHs has been illustrated in a wide variety of bacteria (Gibson, 1984; Dagley, 1986). It has also been demonstrated that the naphthalene plasmids, pKA1 and NAH7, mediate the catabolism of other PAHs, i.e., phenanthrene and anthracene (Menn et al., 1993; Sanseverino et al., 1993a), and perhaps initial oxidation of some other higher molecular weight PAHs.

TCE is a volatile chlorinated compound and a major groundwater contaminant. The persistence of TCE in the environment causes more concern because of its toxicity and carcinogenicity (Miller and Guengerich, 1983). A review on the microbial degradation of TCE has been published recently (Ensley, 1991). Under anaerobic conditions, TCE is transformed to vinyl chloride (Suflita et al., 1982) and ethylene ultimately (Freedman and Gossett, 1989). It is interesting to note that the CO dehydrogenase in *Methanosarcina thermophila* also has the capability to reductively dechlorinate TCE (Jablonski and Ferry, 1992). TCE is not only oxidized aerobically by toluene oxygenases and soluble methane monooxygenase (see Table 2), but also can be degraded by either a recombinant strain *P. putida* F1/pSMMO20 (Jahng and Wood, 1994) or a hybrid aromatic dioxygenase (*todC1bphA2bphA3bphA4*) that hosts in *E. coli* (Furukawa et al., 1994). In addition, ammonia monooxygenase in *Nitrosomonas europaea* (Arciero et al., 1989; Vannelli et al., 1990), alkene monooxygenase in *Xanthobacter* strain Py2 (Ensign et al., 1992), propane

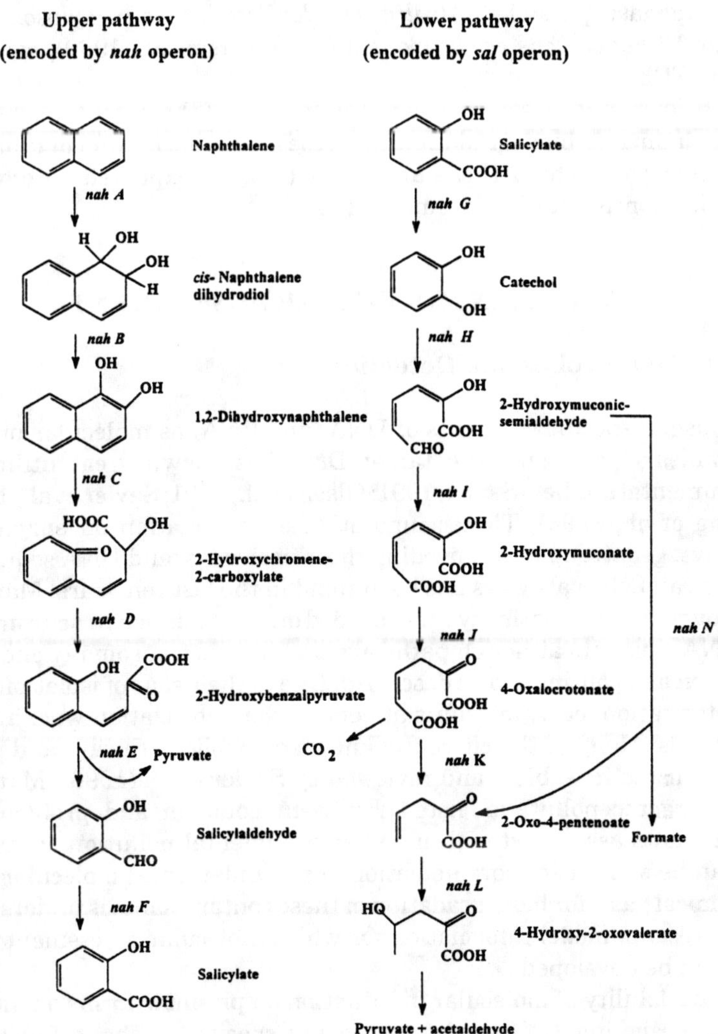

Figure 2. Naphthalene degradation pathway used by *P. putida* PpG7 encoded by the NAH7 plasmid. The oxidation of other higher molecular weight PAHs by the upper pathway has been demonstrated.

monooxygenase in *Mycobacterium vaccae* JOB-5 (Wackett et al., 1989; Vanderberg and Perry, 1994) and isoprene-utilizing bacteria (propene monooxygenase possible), *Alcaligenes denitrificans* ssp. *xylosoxidans* JE75 and *Rhodococcus erythropolis* JE77 (Ewers et al., 1990), can also degrade TCE.

The brief examples are given as an indication of the potential depth of understanding of biodegradation processes. The basic information derived from research in these areas can then be exploited to develop molecular approaches in bioremediation.

ADVANCES IN MOLECULAR ASSESSMENT

Nucleic Acid Probes and Detection

The use of specific segments of DNA (and RNA) as molecular probes of organisms possessing the target DNA has shown great utility in environmental studies (Sayler, 1991; Olson et al., 1991; Sayler et al., 1985; Fleming et al., 1993). The accumulation of information on enzymatic pathways, genetic operons encoding these pathways, and DNA sequence data for catabolic pathways has been rapid in the last ten years. Many of the known catabolic pathways are encoded on plasmids or extrachromosomal DNA elements although pathways that are chromosomally encoded have been and continue to be described. Due to their ease of isolation and characterization, early attempts at genetic characterization were aimed at plasmids. Many of the currently known catabolic pathways and plasmids are listed in Table 2 and reviewed by Sayler et al. (1990). Many of the 35 organic pollutants listed represent common and problematic contaminants associated with many environmental pollution scenarios that can be subject to bioremediation. The fundamental molecular and biochemical basis for biodegradation of these contaminants is understood and represents model information for which molecular assessment protocols can be developed.

The availability of molecular information on plasmids forms a foundation for designing DNA probes for many specific catabolic functions involved in biodegradation of hazardous pollutants. Once a catabolic function or group of functions has been localized on a plasmid (or chromosome), that information can be used to design such specific probes. This information may include a detailed restriction map or also DNA sequence data. If only restriction site data are known, specific fragments may be used as a probe. If DNA sequence data are also known, very specific probes may be developed using polymerase chain reaction technology as described in a later section. Such use of DNA probes aids in microbial process monitoring for wastestreams and bioremediation (Sayler, 1991). It is stressed that DNA probing is one of many microbial monitoring technologies available (Jain et al., 1988).

The nature of DNA:DNA and DNA:RNA interactions lends itself to the production of very specific probes. The major limitation of nucleic acid probing is in the basic knowledge of pathways in terms of the genes involved. At the present, the biochemistry and genetic knowledge of biodegradative pathways and organisms are extensive, although limited to the systems listed in Table 2 and Sayler (1991). A few of these genetically characterized systems are considered model systems, and their relevance to the open environment is questioned (Sayler et al., 1988). There may be other biodegradative determinants that are equally or more important. One example is the NAH7 plasmid for naphthalene biodegradation. There is now evidence that there are naphthalene bio-degradative pathway(s) that are not based on the naphthalene dioxy-genase (*nah*A). Ahn et al. (1994) have discovered a *Pseudomonas pauci-mobilis* that does not hybridize with the *nah*A gene probe but did mineralize naphthalene. Thus, gene probing may underestimate the total biodegradative potential of a particular environment.

Some typical analytical results derived from field investigations using a variety of gene probes are summarized in Table 3. Research was conducted at the Westinghouse Savannah River (SRS) TCE integrated field demonstration site to quantify the occurrence and distribution of soluble methane monooxygenase (sMMO) producing methanotrophs thought to be responsible for TCE in situ bioremediation (Gregory, 1995). Besides sMMO, the other genotypes investigated included methanol dehydrogenase (MDH) and toluene dioxygenase (*tod*C1C2). Field treat-ments at SRS were a pre-air test, air test, 1% CH_4 injection, 4% CH_4 injection, and a 4% CH_4 + gaseous N and P to stimulate methanotrophic sMMO activity in deep subsurface. In general, the density of type II methanotrophs was low at the limits of gene probe detection, making quantification difficult. Samples demonstrating positive hybridization signals were divided by the total number of samples probed to produce a percent positive value for each probe. At the SRS, the percentage sample frequency showing positive hybridization actually declined with 1% and 4% CH_4 injections until additional nutrients supporting growth were also injected into the subsurface.

Studies on PAH degradative populations in six coal tar–contaminated soils and a creosote-contaminated site (Sanseverino et al., 1993b) dem-onstrated that these soils were highly enriched in PAH degraders based on naphthalene dioxygenase (*nah*A) gene frequency. Estimated *nah*A

Table 3. Application of molecular diagnostics to field studies.

Case Scenario	Gene Probes	Results	Reference
TCE			
DOE SRS: subsurface injection of CH_4 for the stimulation of methanotrophs			
Post air	*sMMO*	42% positive samples	(Gregory et al., 1995)
	MDH	6% positive samples	
	todC1C2	46% positive samples	
4% CH_4	*sMMO*	6% positive samples	
	MDH	21% positive samples	
	todC1C2	13% positive samples	
4% CH_4 + N and P	*sMMO*	28% positive samples	
	MDH	51% positive samples	
	todC1C2	0% positive samples	
PAH			
MGP soils; aerobic remediation of near surface soils			
Uncontaminated soils			(Sanseverino et al., 1993b)
Etowah soil: colony hybridization	*nahA*	$<3.0 \times 10^5$ cfu/gram soil	
Pamlico soil: colony hybridization	*nahA*	$<3.0 \times 10^5$ cfu/gram soil	
Coal tar–contaminated			
Soil A	*nahA*	0.8–2.8×10^8 bacteria/gram soil	
Soil C	*nahA*	0.6–2.4×10^8 bacteria/gram soil	
Soil D	*nahA*	0.2–1.3×10^7 bacteria/gram soil	
Soil E	*nahA*	1.0–3.6×10^8 bacteria/gram soil	
Soil N	*nahA*	1.0–5.6×10^7 bacteria/gram soil	
Creosote-contaminated			
Soil G	*nahA*	0.3–1.1×10^{10} bacteria/gram soil	

Table 3. *(continued).*

Case Scenario	Gene Probes	Results	Reference
PCB			
Bioremediation potential in contaminated soils			
New England soil: DNA extraction	*bphBC*	below detection limit	(Layton et al., 1994b)
colony hybridization	*bphBC*	9×10^2 cfu/gram soil	
enriched: DNA extraction	*bphBC*	5×10^{10} sequences/gram soil	
colony hybridization	*bphBC*	2×10^9 cfu/gram soil	
TVA soil DNA extraction	*bphBC*	below detection limit	
colony hybridization	*bphBC*	below detection limit	
Mixed Hydrocarbons			
CAFB; Baseline analysis for potential to atten-uate introduced groundwater contaminants			
Upgradient	*alkB*	$3.13–6.72 \times 10^6$ bacteria/gram soil	(Stapleton and Sayler, 1995)
	nahA	$3.72–7.83 \times 10^6$ bacteria/gram soil	
	nahH/xylE	$2.69–6.90 \times 10^6$ bacteria/gram soil	
	todC1C2	$6.25–13.9 \times 10^6$ bacteria/gram soil	
	xylA	$1.25–3.90 \times 10^6$ bacteria/gram soil	
Downgradient	*alkB*	$3.37–8.46 \times 10^6$ bacteria/gram soil	
	nahA	$3.58–7.10 \times 10^6$ bacteria/gram soil	
	nahH/xylE	$2.99–7.25 \times 10^6$ bacteria/gram soil	
	todC1C2	$5.82–15.2 \times 10^6$ bacteria/gram soil	
	xylA	$1.66–3.64 \times 10^6$ bacteria/gram soil	

populations densities ranged from 10^7 to 10^{10} cells g^{-1} soil based on colony hybridization results as compared to 10^5 cells g^{-1} soil in uncontaminated farm soil. These dramatic differences were magnified by direct DNA extraction (Ogram et al., 1987) and probing that demonstrated greater than 10^{11} *nah*A gene copies g^{-1} soil in the highly contaminated creosote soil.

Studies of groundwater aquifer material at Columbus Air Force Base, which had previously been exposed to an intentional injection of hydrocarbons including benzene, ethyl benzene, *p*-xylene, and *o*-dichlorobenzene, during the macrodispersion experiment (MADE 2) were also conducted (Stapleton and Sayler, 1995). This study demonstrated the disappearance of up to 94% of the contaminants over a 15-month period (Boggs et al., 1993). Microbiological degradation of the hydrocarbons was proposed as the mechanism of disappearance, but no data exist to support this hypothesis. As a follow-up, the Natural Attenuation Study (NAT) has been initiated to include microbiological aspects in the fate of hydrocarbons in the subsurface. The initial site characterization of the native microbial community showed that bacteria possessing the genotype to degrade the contaminant mixture to be used in the NAT study exist. DNA hybridization studies with the gene probes *alkB* (alkane monooxygenase), *nahA, nahH/xylE* (catechol 2,3-dioxygenase), *todC1C2*, and *xylA* (xylene monooxygenase) showed that the study site maintains these degradative populations at approximately 10^6 bacteria/gram of sample even in pristine control samples. This initial characterization will be used to compare future microbial community responses after the hydrocarbons have been released at the study site.

In the case of PCB gene diagnostics, studies were undertaken to determine if electrical power substation soil (TVA) had genetic capacity for indigenous PCB biodegradation (Layton et al., 1994b). In the case of the TVA soils examined, virtually no organisms could be detected as to being genetically competent for PCB degradation even after enrichment with 2-chlorobiphenyl. In contrast, industrial soils contaminated with PCB were shown to maintain 10^2 PCB biodegrading organisms g^{-1} soil and after enrichment with 2-chlorobiphenyl, these organisms could be enhanced to 10^9–10^{10} g^{-1} soil as measured by colony hybridization or DNA extraction and probing, respectively.

Unfortunately, biodegradation cannot be monitored by a single nucleic acid probe. The use of a wide array of gene probes representing the various possible biochemical pathways expected to be active at the site of interest will provide the best possible results. It is interesting to note that many degradative pathways "funnel" intermediates to certain enzymes that show striking similarities. An example of this is seen in both the function and nucleotide sequence of the aromatic ring cleavage enzyme catechol 2,3-oxygenase (Bartilson and Shingler, 1989).

Polymerase Chain Reaction (PCR)

PCR was first introduced in 1985 and has been a major technological improvement in molecular biology (Saiki et al., 1985; Mullis and Faloona, 1987). PCR allows for in vitro replication (amplification) of specific DNA fragments through an enzymatic process involving DNA polymerase that allows for discrete cycles of replication, serving to effectively double the amount of the specific fragment of interest each cycle. After 20 amplification cycles, the target sequence can be amplified a million times (2^{20}) assuming 100% efficiency. Biologists have been quick to apply this powerful technology to environmental research (Steffan and Atlas, 1991).

The PCR process is illustrated in Figure 3 and begins by denaturing or melting the double-stranded target or template DNA to produce single-stranded DNA. Primers consisting of 20–25 base length oligonucleotides specific for the sequence bounding the target sequence (both strands) are annealed to the template DNA. Some sequence knowledge about the target DNA is required in order to design the primers. The primers are designed to provide a free 3′ hydroxyl facing the target sequence. In the presence of reaction buffer, $MgCl_2$, deoxynucleotide triphosphates, and heat-stable polymerase, the target DNA is copied via a program of temperature changes. This process begins at the 3′-OH of each of the primers and extends across the target. This process constitutes one PCR cycle. Discrete copies of the original target DNA are effectively doubled each cycle. The specifics concerning concentration of reagents, temperature, and cycle time length are described in detail elsewhere (Innis and Gelfand, 1990; Saiki, 1992). PCR has also been adapted for use in RNA analysis.

One of the most obvious applications of PCR is the detection of rare sequences in environmental matrices (Steffan and Atlas, 1991). DNA must first be extracted from the matrix and purified before PCR can be applied. There are two approaches for recovering DNA from an environmental matrix. One method involves the extraction of cells from the sample followed by cell lysis and recovery of the DNA (Torsvik, 1980). The second method involves lysing the cells within the matrix and then purifying the DNA from the mixture (Ogram et al., 1987; Johnston et al., 1995). This procedure often yields DNA of high enough quality to PCR without further purification, as is seen in Figure 4. Extensive purification will not remove all impurities in a sample and PCR may be adversely affected. Chelators of ions such as Mg^{+2}, will have a dramatic effect on

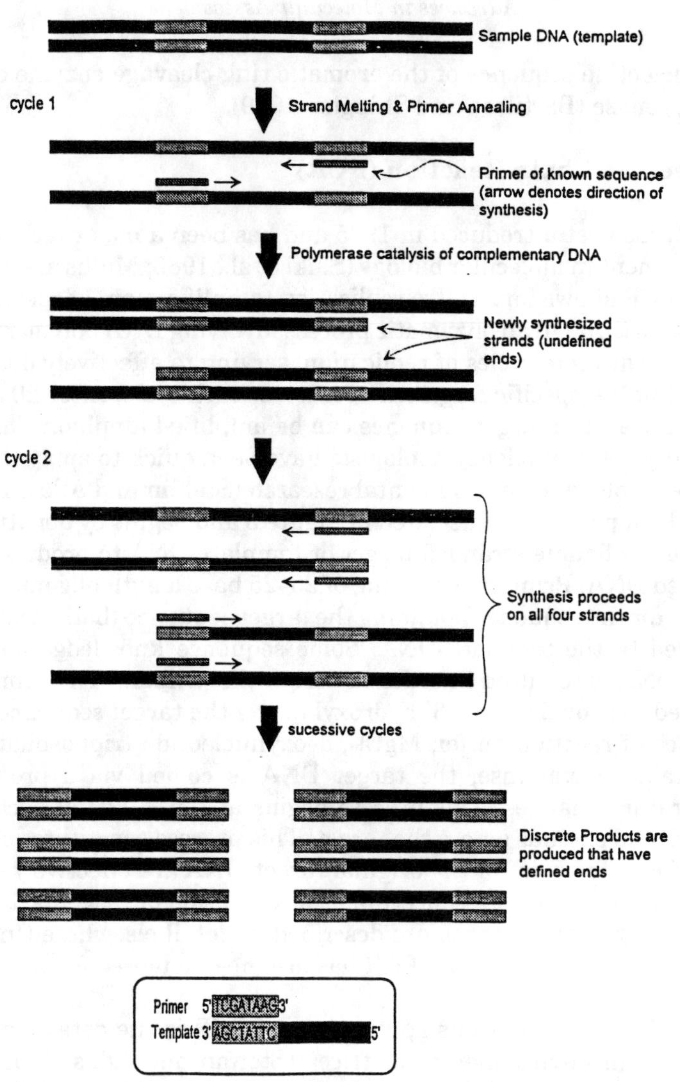

Figure 3. Schematic of PCR amplification cycles and the resulting amplified products from early cycles. Target DNA is melted or denatured by heating to 92–95°C to produce single-stranded DNA that acts as template. Primers designed as complementary to the DNA regions flanking the target region are annealed at a specific temperature. Heat-tolerant polymerase then synthesizes two new strands of DNA, thus producing a copy of the original target sequence. Both the copy and the original template can serve as template for the second round of replication. After four cycles, discrete copies of the target are produced and are replicated exponentially. The nondiscrete or variable length fragments are produced linearly. The different intensities of the bands are due to unequal amplification at priming sites (Bassam et al., 1992). (Adapted from Arnheim and Levenson, 1990.)

Figure 4. PCR of 16S ribosomal DNA sequences from (A) sludge and (B) soil. Panel (A) shows amplification of the 1500 base pair 16S fragment from DNA extracted from industrial sludges. Panel (B) shows amplification of the 1500 base pair fragment from soil. For both the industrial sludges and the soils, the DNA was extracted by the method of Johnston (1995). PCR (negative) controls were included and shown in each panel to the right of the 1-kb standard ladder.

the PCR process since the activity of DNA polymerase requires this ion. Conditions for PCR will have to be ascertained for each environmental milieu. Another obstacle in using PCR to monitor microbial populations is bacterial evolution. As bacteria existing in nature are exposed to increasing types and amounts of anthropogenic contaminants, the existing degradative pathways must evolve and adapt to the changing environmental conditions. This evolution will be seen at the DNA sequence level, making the PCR detection more difficult. This may be accounted for by altering the stringency of primer annealing (i.e., Mg^{+2} concentration and temperature) and incorporating primer degeneracy, but amplification may be difficult.

For the purpose of bioremediation, the occurrence of a degrading population at a concentration of less than 10^5 cfu/g of soil is considered insignificant for effective and timely removal of a pollutant. The limit of detection for direct DNA extraction from soils and sediments has been determined to range from 10^5 to 10^6 (Applegate et al., 1995). The direct extraction procedure requires a 50-g sample, and the extracted DNA is slot blotted and probed with ^{32}P-labeled DNA (Ogram et al., 1987). The use of PCR may allow detection of genotypes that are difficult to detect with traditional DNA extraction techniques.

DNA extraction and PCR have a great utility in feasibility studies.

Samples from contaminated sites may be screened for specific degrader populations to assess the potential for bioremediation at that site. Additionally, these techniques may be used to monitor degrader populations over the course of a bioremediation process as a monitor of process efficacy. PCR will also be a very useful tool for monitoring genetically modified organisms (GMOs) in bioremediation efforts if such organisms become approved for field use. Part of the monitoring process for risk assessment may involve monitoring levels of GMOs, which may decline below the limits of detection of other monitoring protocols.

The PCR process has been adapted for quantitative use. Quantitation of DNA sequences based on PCR products is difficult since the efficiency of the reaction seldom approaches 100% thus making back calculation of the starting concentration of target impossible. Several methods have been developed to make quantitative PCR realistic (Gilliland et al., 1990; Wiesner et al., 1992). Competitive PCR is one quantitative PCR technique (Gilliland et al., 1990). The target DNA is coamplified in the presence of a known quantity of competitive DNA. This competitive DNA is an internal standard that amplifies with the same kinetics as the target DNA. The target DNA is quantitated by titrating against an elution series of competitor DNA.

The second method for quantitative PCR, exponential PCR, is based on the theoretical doubling of each strand of DNA per cycle (Wiesner et al., 1992). As already indicated, the amplification efficiency is less than 100%. This procedure determines the actual number of target molecules of the PCR reaction by measuring the concentration of product accumulation in consecutive cycles and determining the amplification efficiency. The equation describing product accumulation,

$$\log N_n = \log \textit{efficiency} \cdot n + \log N_0$$

where

N_n = DNA concentration at cycle n
N_0 = the target concentration at cycle zero

can be analyzed by linear regression and the molar concentration of target at cycle zero, N_0, estimated.

Regardless of the quantitative PCR method used, strict quality assurance protocols must be observed at all times. The extreme sensitivity of the PCR process means that one contaminating molecule of previously amplified target DNA may invalidate the quantitative PCR process. The use of proper controls and procedures are critical to monitor this possibility.

An interesting application of PCR technology that has gained interest for use as a monitoring tool utilizes molecular knowledge of the bacterial chromosome. Repetitive DNA sequences found on the chromosome of bacteria are conserved in the bacterial genome. They include the 35- to 40-bp repetitive extragenic palindromic (REP) sequence (Gilson et al., 1984), the 124- to 127-bp enterobacterial repetitive intergenic consensus (ERIC) sequence (Hulton et al., 1991), and the newly discovered 154-bp BOX element (Martin et al., 1992). These three sequences have the capacity to form stem-loop structures and may play a significant role in the organization of the bacterial genome. Since the genome organization is speculated to be dependent on selection, the distribution of REP, ERIC, and BOX elements may be indicative of the structure and evolution of the bacterial genome. This assumption has led to the use of the distribution of the repetitive DNA sequences for the fingerprinting of bacterial genomes (Versalovic et al., 1991). Oligonucleotide primers are designed to include consensus sequences in the REP, ERIC, or BOX elements and used with PCR to generate DNA products of varying lengths that may be resolved by electrophoresis. Figure 5 shows the use of ERIC primers to generate banding patterns for a variety of eubacteria (Versalovic et al., 1991). This computer-generated version of a gel in their paper is intended to illustrate the utility of the technology. Bacterial species and strain-specific fingerprint patterns were developed using amplification of ERIC sequences of genomic DNA (Versalovic et al., 1991). The use of ERIC primers generated banding complexity sufficient to show species differences but not strain differences in some cases, the exception being the differences between lab strains and pathogenic isolates. It was possible with certain species of bacteria to generate species-specific banding (Figure 5, indicated by arrows). A REP-PCR banding generation resulted in patterns sufficient to fingerprint specific strains (too complex to reproduce). REP-PCR has been applied to soil isolates capable of biodegrading the herbicide 2,4-dichlorophenoxyacetate (Ka et al., 1994). The technique was used for grouping individual strains from the site with good success.

A similar fingerprinting application of PCR, known as Randomly Amplified Polymorphic DNA (RAPD), uses RAPD markers detected by PCR amplification of small inverted repeats located throughout the bacterial genome (Hadrys et al., 1992). The amplification protocol differs from REP-PCR, in that only one short oligonucleotide is used and no prior knowledge of the genome is needed. There is a high probability that, when a short oligonucleotide is used, there will be priming sites close to one another. The technique then effectively scans the genome for small inverted repeats, thus amplifying the intervening DNA. The PCR prod-

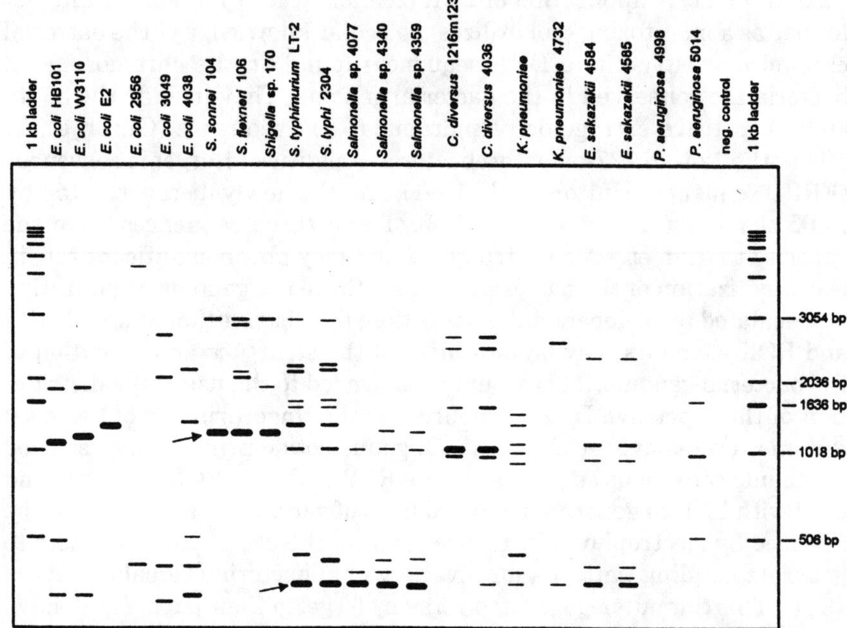

Figure 5. PCR fingerprinting of enterobacterial genomic DNA using ERIC primers for amplification. The negative control contains no template DNA. Examples of species-specific bands are indicated by arrows. (Adapted from Versalovic et al., 1991.)

ucts are resolved on agarose gels, and the polymorphisms are determined. It has been proposed to use this technique on mixed genomic samples of bacteria from environmental matrices. Empirical data to demonstrate the utility of this application in mixed culture situations are lacking.

REP-PCR and RAPD-PCR suffer from many of the same limitations and difficulties (Hadrys et al., 1992). In both processes, the design of the oligonucleotide primer is critical since the size of the primer will determine the degree of specificity of genome scanning. The PCR reaction conditions are particularly limiting. Slight changes in reaction conditions tend to alter the reproducibility of the assay. Additionally, if the techniques are applied to environmentally derived DNA, contaminants in the DNA extract will make intrasite comparisons difficult, if not impossible. Contaminating substances may render the PCR process nonreproducible. Comigration of PCR products may make gel interpretation difficult. Both of these techniques are relatively new, and very little empirical data exist to demonstrate their utility for analyzing mixed culture DNA from the environment.

BIOREPORTERS AND BIOSENSORS IN MONITORING BIOREMEDIATION

Bioavailability and biodegradation of pollutants are important issues in predicting the bioremediation potential of contaminated environments. Ways to continuously monitor these two processes have been difficult. While bioavailability and biodegradation of compounds can influence specific gene expression, the two tasks have often been detected ex situ using laboratory techniques. However, these techniques may be impractical for real-time monitoring of the two processes. Genetic manipulation of biodegradative operons has resulted in the creation of easily detectable phenotypes (reporters) that can be useful in the continuous monitoring of bioavailability and biodegradation. Such altered genes and the bacteria harboring them are called reporters and bioreporters, respectively.

Bioluminescent Bioreporters

Bioluminescent genes, *luxCDABE*, are classical examples of reporter genes used in the genetic manipulation of other genes (Shaw et al., 1988). Luciferase enzyme encoded by the *lux* genes is responsible for the bioluminescent phenotype. Genetic modification of a biodegradative gene promoter by fusion of the promoterless *lux* genes results in a bioluminescent bioreporter. When expressed, the bioreporter gene produces light that can be easily detected and continuously monitored with optical devices. Bioluminescent bioreporters constructed in this manner have been investigated in laboratory conditions for the monitoring of expression of biodegradative genes.

Molecular genetic studies on the luminous marine bacteria *Vibrio fischeri* paved the way for the development of bioluminescent bioreporters. To date, there are a handful of bioluminescent bioreporters capable of detecting specific xenobiotics and heavy metals (Table 4). The bioluminescence reaction per se is employed commercially in the assessment of environmental pollutants. MICROTOX® is one such commercial toxicity kit developed using the luminescent bacteria *Photobacterium phosphoreum*. The bioassay measures the amount of light produced per unit time when the bacteria are incubated in samples containing toxins (Beckman Instruments, 1978; de Zwart and Sloof, 1983). Toxic compounds that are interfering with the NADPH and ATP requiring luciferase reaction are nonspecifically detected.

The first catabolic bioreporters, *Pseudomonas fluorescens* HK44 and *Pseudomonas* sp. RB1351, were constructed employing *lux* genes to

Table 4. Biosensors of xenobiotics and heavy metals.

Type/Brand or Generic Term	Detection Method	Purpose	Microorganism/Plasmid	Reference
Specific Biosensors				
nah-lux	Bioluminescence	Detect naphthalene and salicylate	*Pseudomonas fluorescens*, pUTK21; *Pseudomonas* sp., pUTK9	(King et al., 1990; Burlage et al., 1990)
bph-lux	Bioluminescence	Detect PCBs	*Alcaligenes* sp. A5	(Springael et al., 1991)
xyl-lux	Bioluminescence	Detect xylene	*Pseudomonas putida* pUTK24	(Burlage et al., 1992)
tod-lux	Bioluminescence	Detect toluene and TCE degradation	*Pseudomonas putida* B2	(Applegate et al., 1995)
pvd-inaZ	Ice nucleation	Detect iron	*P. fluorescens, P. syringae*	(Loper and Lindow, 1994)
mer-lux	Bioluminescence	Bioavailable mercuric ions	*E. coli*; pRB28, pOS14, pOS15	(Selifonova et al., 1993)
cup-lux	Bioluminescence	Detect copper	*Alcaligenes eutrophus* DS185; AE984	(Corbisier et al., 1992, 1993a)
czc-lux	Bioluminescence	Detect zinc	*A. eutrophus*	(Corbisier et al., 1992)
cnr-lux	Bioluminescence	Detect nickel and cobalt		(Corbisier et al., 1992)
chr-lux	Bioluminescence	Detect chromium		(Corbisier et al., 1992)

Table 4. (continued).

Type/Brand or Generic Term	Detection Method	Purpose	Microorganism/Plasmid	Reference
Specific Biosensors (*continued*)				
thl-lux	Bioluminescence	Detect thallium		(Corbis er et al., 1992)
cad-lux	Bioluminescence	Detect arsenic and cadmium	*Staphylococcus aureus*; pl258	(Corbis er et al., 1993b)
Nonspecific Biosensors				
RODTOX®	Respiration	Rapid BOD	Native microbial community	(Van Rolleghem et al., 1990)
MICROTOX®	Bioluminescence	Toxicity Test	*Photobacterium phosphoreum*	(Beckman Instruments, 1978; de Zwart and Sloof, 1983)
TOXI-Chromotest	Inhibition of de novo synthesis	Antibiotics and toxins test	*Escherichia coli*	(Environmental Biodetection Products Inc., Ontario, Canada)
MetPad™	Enzyme inhibition, β-galactosidase activity	Heavy metals: Cd, Cu, Pb, Hg, Ni and Zn ions	*E. coli*	(Bitton et al., 1992; de Vevey et al., 1993)
Ames test	Mutagenesis	Genotoxicity	*Salmonella typhimurium*	(Ames, 1971, 1972; Maron and Ames, 1983)
Mutatox™	Mutagenesis/ Bioluminescence	Genotoxicity	*Vibrio fischeri* M169	(Microbics reference D006)

detect the bioavailability and degradation of naphthalene (King et al., 1990; Burlage et al., 1990). These bacteria produce light in the presence of either naphthalene or its catabolic intermediate, salicylate. The strain HK44 harbors a NAH7-like plasmid encoding the degradation of naphthalene to salicylate, the latter being mineralized through chromosomally encoded pathway. The position of the *lux* genes in this NAH7-like plasmid, pUTK21, is shown in Figure 6.

Pseudomonas fluorescens HK44, was found to produce a linear bioluminescence response ($r^2 = 0.99$) at a concentration range of 72 ppb to 3.25 ppm of naphthalene in liquid culture studies (Heitzer et al., 1992). It even was able to detect 45 ppb of naphthalene, which may not be the lower limit (Heitzer et al., 1992). The bacteria demonstrated its sensitivity and applicability in environmental conditions at low pollutant concentrations. In the studies conducted in bioreactors with perturbed naphthalene feed, the strain displayed interrupted bioluminescence response (King et al., 1990). The interruption reflected the rapidity and inducibility of the luciferase enzyme system, which is dependent on the availability

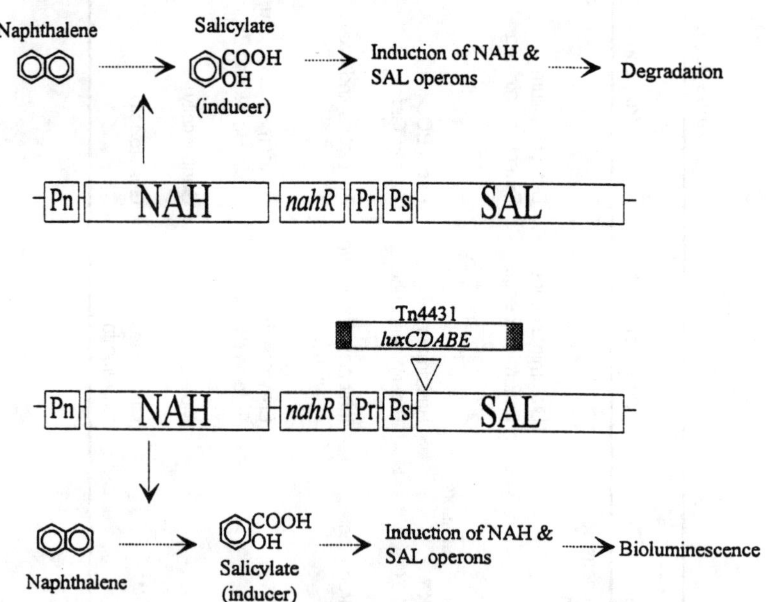

Figure 6. Organization of naphthalene degradative operons and the *lux* genes. (a) Genes and induction of naphthalene degradation on the NAH7-like plasmid of the wild-type bacteria; (b) genes and induction of bioluminescence on the mutagenized NAH7-like plasmid, pUTK21, of *Pseudomonas fluorescens* HK44.

and concentration of naphthalene. The bioluminescent strain was later demonstrated to have a specific response to naphthalene in complex matrices such as soil and soil slurry (Heitzer et al., 1992). These studies ascertained the potential scope of application of HK44 in environmental systems requiring in situ or on-line monitoring of naphthalene bioavailability and biodegradation.

Bioluminescent Biosensors in Environmental Monitoring

The bioluminescent bacteria are potential candidates for biosenor technology. In conjunction with sensitive, powerful, light-amplifying, and light-collecting devices, the bacteria can be deployed in the (1) rapid detection of gene expression, (2) monitoring of bioavailability and biodegradation of pollutants, (3) investigation of factors affecting environmental processes, and (4) study of the fate and impact of genetically modified microorganisms released into the environment.

In a broad sense, a biosensor can be defined as an apparatus consisting of a biological sensing component, connected or integrated to a physico-chemical transducer (van der Lelie et al., 1993). The biological component may be a protein, a nucleic acid, a whole cell, a tissue, or an organism. The transducer may be any electrical or chemical element that produces amperometric, potentiometric, conductimetric outputs or optical changes like light emission, reflection, absorption, and fluorescence. Recently, with the use of the bioreporter bacteria *P. fluorescens* HK44, a bioluminescent biosensor was developed and investigated for potential environmental applications (Heitzer et al., 1994).

Naphthalene Biosensor *Pseudomonas fluorescens* HK44

Results of investigations with HK44 as an on-line biosensor of naphthalene were encouraging because this biosensor produced specific and reliable results (Heitzer et al., 1994). The biosensor design consisted of a probe (the tip of a liquid light pipe with immobilized HK44) set on-line in effluent streams. The effluents were artificially spiked aqueous naphthalene solutions and realistic PAH-contaminated solutions (soil extracts) with other compounds as well. The setup also had ports for periodic sampling of the effluents. The probe was interfaced to a personal computer for continuous data acquisition. In each of the tested conditions, the bioluminescence response of the biosensor was correlated with the amount of naphthalene present in the effluent stream. With realistic contaminants such as jet fuel JP-4 and manufactured gas plant soil (MGP) leachates, the bioluminescence response was rapid and specific to the bioavailable naphthalene. These observations have ascertained the utility of the biosensor for contaminants from the environment. However, as a technical limitation, unknown toxic compounds present in jet fuel effluents were speculated to have dramatically inhibited the bioluminescence response, as the magnitude of light was lower than that produced with the MGP soil leachates (Heitzer et al., 1994).

Biosensor technology can be extended to field situations if properly implemented. The bioreporter bacteria may be contained in sensor modules and installed beneath the ground surface to monitor contaminant bioavailability and biodegradation in groundwater. The design of the module should allow groundwater to flow through it and have provisions to feed nutrients and other growth factors required for the biosensor bacteria. Necessarily, bacterial containment should be an important feature of the module to avoid ecological risk due to bacterial release. A prototype module, as shown in Figure 7, has been proposed and is currently being investigated.

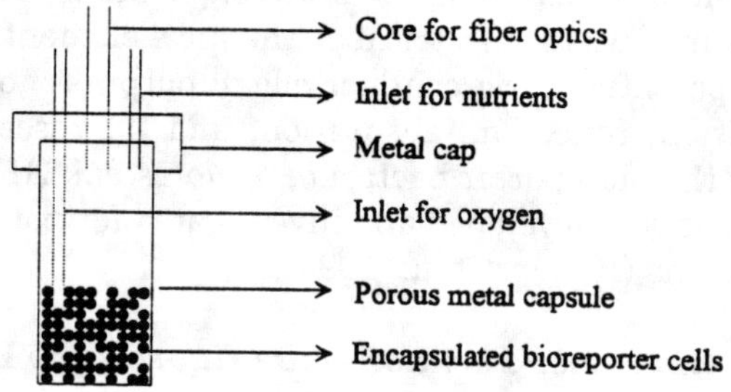

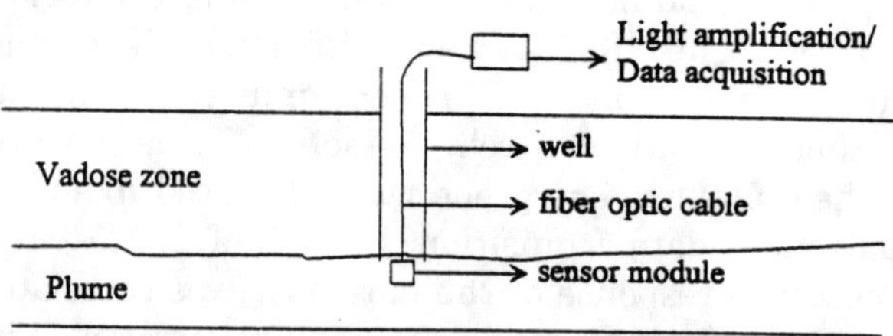

Figure 7. Cross section of a field biosensor module (top panel) and a schematic representation of its potential application (bottom panel).

When employed in the above-mentioned situation, the biosensor bacteria should be able to withstand prolonged incubation under stressed conditions and simultaneously generate stable bioluminescent responses. The naphthalene biosensor HK44 has been found to possess these characteristics, based on a recent investigation (Matrubutham et al., 1995). Alginate-encapsulated HK44 demonstrated robustness in survival and consistency in bioluminescence response over a period of 5 weeks when incubated in groundwater. The response was significantly more stable in groundwater with pH 6 than in any other condition used in the study. More investigations are required to confirm the stability and steady state of the bioluminescence reaction when the contaminant is the only carbon and energy source for the bacteria.

Field biosensors are, in many aspects, surrogates of the native bacterial community. When employed in contaminated scenarios, they reflect the physiological processes and demands of the bacterial population actually involved in biodegradation. Besides, such an on-line or in situ biosensor will assist in the proper manipulation of environmental factors facilitating bioremediation.

Biosensors of Other Xenobiotics

Bioluminescent biosensors similar to the naphthalene biosensor have been proposed and are under trial. One such biosensor, *Pseudomonas putida* B2 for toluene, is of special interest because of its dual purpose based on the circumstance in which it is employed. For instance, the biosensor can be a specific sensor of toluene just as HK44 for naphthalene or can be a biosensor of TCE biodegradation as toluene is a cometabolite of TCE. In the latter scenario, though it is not specific for the compound, the biosensor is specific for a biodegradation process. Such biosensors may be called process or system biosensors. Scope for the development of such biosensors is broad and promising. Table 4 shows some of the potential biosensors that are either being investigated or in application.

Biosensors of Heavy Metals

Development of bioluminescent biosensors for heavy metal detection have mushroomed in recent years. Bacterial genes encoding heavy metal resistance and/or their regulatory genes have been employed in the construction of specific biosensors. Most metal resistance determinants are inducible by metals in bacteria, and this phenomenon benefited in vivo use of these genes with an instantaneously assayable phenotype.

A mercury biosensor, *mer-lux*, has been demonstrated to detect nano-molar concentrations of bioavailable Hg(II) present in natural waters supplemented with mercury (Selifonova et al., 1993). Interestingly, a *cup-lux* biosensor has been shown to detect an insoluble form of copper, which cannot be detected with the commercial chemical system "Mercko-quant" (Corbisier et al., 1993a). This additional advantage of bacterial biosensors versus chemical detection techniques has increased the sensi-tivity of detecting bioavailable copper. The biosensor also was found useful in detecting copper in complex mixtures like cuproslodowskite, a uranium-copper mineral (Corbisier et al., 1993a).

Bioluminescent reporter technology provides a promising novel tech-nique to understand the performance of bacteria in complex environ-mental conditions and to biomonitor the presence and fate of various environmental contaminants. As discussed, biosensors, along with other bioanalytical techniques, will be valuable in assessing contaminated sites and will essentially facilitate the optimization of environmental condi-tions for microbial degradation of toxic chemicals in the environment.

DISCUSSION

The continued generation of new information on the biochemistry, genetics, and molecular biology of microbial biodegradative pathways should fuel and expand opportunities for molecular environmental analy-sis. However, the information alone cannot result in routine and practical molecular applications. Several fundamental factors involving contempo-rary environmental practices and practitioners will influence whether molecular approaches become a part of a standard protocol in predicting and measuring the performance of natural biodegradation or bioreme-diation.

Among the questions concerning the use of molecular technologies is the foremost question of whether new relevant site or process information is developed that could not be obtained with existing conventional meth-ods. In answering the question affirmatively, it must be recognized that there are significant limitations on the robustness, sensitivity, and speci-ficity of most conventional assessment protocols. The majority of conven-tional microbiological tests, to either quantify degradative organisms or biodegradative activity of a sample, are highly biased towards the labo-ratory environment and are not particularly relevant to the site environ-ment. Molecular methods should ultimately be more specific, subject to little influence of laboratory processing and more amenable to standard-izable statistically representative field analysis. In those cases where known mechanisms of biodegradation are the result of cometabolism, the advantages should be greater for molecular site assessment over that of cultivation-based microbial enumeration.

A critical evaluation of molecular assessment technology must also include the fact that currently available hybridization/probe methods, as well as bioreporter/sensor technology, are largely based on information derived from laboratory investigation. There is the argument that other organisms, not recovered from the environment, may play at least as an important or greater role in realized biodegradation or bioremediation. This is certainly a legitimate argument, but it should also be acknowledged that there is the expectation for a degree of unity in the biochemistry and genetics showed by organisms in natural habitats. However, there is no question that the biodiversity of organisms and biodegradation pathways are likely to be expanded as discovered by further research. It is also hypothesized that a major portion of this diversity will be variations of existent central themes based on our current knowledge. Where unique biochemistry and genetics exist, the information can be rapidly integrated into a molecular assessment paradigm. Arguably, this serves as a major driving force to continue exploration of biodegradative organisms and mechanisms.

Recent advances in 16S rRNA analysis and research in mRNA recovery from environmental samples may also lead to furthering our understanding of biodiversity of microbial communities engaged in biodegradation. Technical application of hierarchical 16S rRNA gene probing can provide a reasonably unbiased estimate of the representativeness of the microbial population in a given site. The potential ability to sample environmentally transcribed DNA as mRNA can also provide an unbiased measure of whether or not laboratory target genes represent those functionally important genes in the environment. Recent adaptation of PCR-based techniques for both DNA and RNA analyses will significantly increase the detection and quantitation limits of relevant microbial population and genetic activities.

Currently, there is a significant amount of general disagreement as to whether or not any microbiological or molecular monitoring tool is absolutely necessary in preliminary site assessment or process monitoring. To some extent, this argument derives from the disciplinary perspective of the bioremediation practitioner, the degradation of the target contaminants, the hydrogeochemical characteristics of the site, costs and/or regulatory oversight of the remediation process, and the degree to which the process is passive or actively engineered. Given that current laboratory-based assessments are extremely site-specific, unpredictable relative to field performance, and of uncertain dynamics and ultimate endpoints relative to the field, it would seem that there is strong rationale for more than a perfunctory biological assessment.

Assuming that molecular site assessment and monitoring approaches are desirable, there remains the fact that these approaches are of questionable practical application in routine use. While many probe applications are available, standardization of methods for cross-site and inter-laboratory comparisons have not been made. Besides, the methods require more reduction to practices in terms of automated sample processing, data quantification, and interpretation. The gap between research development and demonstration and actual use, while not large, still remains an impediment.

However, what we can expect from the current technology follows:

1. Quantitative and qualitative processes of known genes responsible for biodegradation
2. Gene dynamics with respect to natural or experimental processes
3. Assessment of sufficient abundance and genetic potential to promote degradation
4. Relative proportion of degradative genotypes as an indication of process effectiveness or past experimental history
5. Establishment of pattern and time in genetic drift due to evolutionary processes as a representation of specific site scenario
6. Specific assessment and quantification of contaminant bioavailability and degradation based on cellular and molecular activities
7. Real-time monitoring of physiological events to facilitate constant and appropriate fabrication of processes leading to effective biodegradation
8. Generation of case-specific and comprehensive data bases enabling futuristic integrated approaches.

ACKNOWLEDGEMENTS

We thank C. Lajoie for technical review of the manuscript and M. James and K. Berry for the preparation of the manuscript.

This work was supported by the U.S. Air Force Office of Scientific Research (contract numbers F49620-92-J0147, F49620-91-I-0222, and F08637-94-R-6046) and U.S. Department of Energy (project number DE-FG05-91ER61193).

REFERENCES

Ahmad D., Masse R. and Sylvestre M. 1990. "Cloning and expression of genes involved in 4-chlorobiphenyl transformation by *Pseudomonas testosteroni*:

Homology to polychlorobiphenyl-degrading genes in other bacteria." *Gene.* 86:53–61.

Ahmad D., Sylvestre M. and Sondossi M. 1991. "Subcloning of *bph* genes from *Pseudomonas testosteroni* B356 in *Pseudomonas putida* and *Escherichia coli*: Evidence for dehalogenation during initial attack on chlorobiphenyls." *J. Bacteriol.* 57:2880–2887.

Ahn Y., A. Sonesson and Sayler G. S. 1994. *94th General Meeting, Amer. Soc. Microbiol.* May 23–27. Las Vegas, NV. p. 420.

Alam K. Y., Worland M. J. and Clark D. P. 1990. "Analysis and molecular cloning of genes involved in thiophene and furan oxidation by *E. coli.*" *Appl. Biochem. Biotechnol.* 24–25:843–856.

Alexander M. 1985. "Biodegradation of organic chemicals." *Environ. Sci. Technol.* 18:106–111.

Ames B. N. 1971. *Chemical mutagens, principles and methods for their detection.* New York. Plenum Press. pp. 267–282.

Ames B. N. 1972. *Mutagenic effects of environmental contaminants.* New York. Academic Press. pp. 57–66.

Applegate B. M., Matrubutham U., Sanseverino J. and Sayler G. S. 1995. *Molecular microbial ecology manual.* Dordrecht, The Netherlands. Kluwer Academic Publishers.

Arciero D., Vannelli T., Logan M. and Hooper A. B. 1989. "Degradation of trichloroethylene by the ammonia-oxidizing bacterium *Nitrosomonas europaea.*" *Biochem. Biophys. Res. Commun.* 159:640–643.

Arnheim, N. and Levenson, C. H. 1990. "Polymerase chain reaction." *Chem. Engineering News* 1:36–47.

Asturias J. A., Eltis L. D., Prucha M and Timmis K. N. 1994. "Analysis of three 2,3-dihydroxybiphenyl 1,2-dioxygenases found in *Rhodococcus globerulus* P6. Identification of a new family of extradiol dioxygenases." *J. Biol. Chem.* 269:7807–7815.

Barkay T. and Sayler G. S. 1988. *Aquatic toxicology and hazard assessment.* American Society for Testing and Materials, pp. 29–36.

Bartilson M. and Shingler V. 1989. "Nucleotide sequence and expression of the catechol 2,3-dioxygenase-encoding gene of phenol-catabolizing *Pseudomonas* CF600." *Gene* 85:233–238.

Bassam, B. J., Caetano-Anolles G., and Gresshoff P. M. 1992. "DNA amplification fingerprinting of bacteria." *Appl. Microb. Biotechnol.* 38:70–76.

Beckman Instruments. 1978. "Microtox model 2055 toxicity analyser system." Carlsbad, CA, USA.

Bitton G., Koopman B. and Agami O. 1992. "MetPad™: A bioassay kit for the rapid assessment of heavy metal toxicity in waste water." *Water Environm. Res.* 64:834–836.

Boethling R. S. and Alexander M. 1979. "Effect of concentration of organic chemicals on their biodegradation by natural microbial communities." *Appl. Environ. Microbiol.* 37:1211–1216.

Boggs J. M, Beard L. M., Waldrop W. R., Stauffer T. B., MacIntyre W. G. and Antworth C. P. 1993. "Transport of tritium and foru organic compounds during a natural-gradient experiment (MADE 2)." EPRI Report TR-101998. Electric Power Research Institute, Palo Alto, CA.

Burlage R. S., Heitzer A. and Sayler G. S. 1992. "Bioluminescence: A versatile bioreporter for monitoring bacterial activity." *BFE* 9:704–709.

Burlage R. S., Sayler G. S. and Larimer F. 1990. "Montoring of naphthalene catabolism by bioluminescence with nah-lux transcriptional fusions." *J. Bacteriol.* 172:4749–4757.

Cerniglia C. E. 1992. "Biodegradation of polycyclic aromatic hydrocarbons." *Biodegradation* 3:351–368.

Chakrabarty A. M., Chou G. and Gunsalus I. C. 1973. "Genetic regulation of octane dissimilation plasmid in *Pseudomonas*." *Proc. Natl. Acad. Sci. USA* 70:1137–1140.

Chapelle G. H. and Lovley D. R. 1990. "Rates of microbial metabolism in deep coastal plain aquifers." *Appl. Environ. Microbiol.* 56:1865–1874.

Chaudhry G. R. and Chapalamadugu S. 1991. "Biodegradation of halogenated organic compounds." *Microbiol. Rev.* 55:59–79.

Coco W. M., Rothmel R. K., Henikoff S. and Chakrabarty A. M. 1993. "Nucleotide sequence and initial functional characterization of the *clcR* gene encoding a *LysR* family activator of the *clcABD* chlorocatechol operon in *Pseudomonad putida*." *J. Bacteriol.* 175:417–427.

Commandeur L. C. M and Parsons J. R. 1990. "Degradation of halogenated aromatic compounds." *Biodegradation* 1:207–220.

Corbisier P., Diels L., Nuyts G., Baeyens W. and Mergeay M. 1992. *92nd General Meeting Amer. Soc. Microbiol.*, May 26–30. New Orleans, LA. p. 202.

Corbisier P., Diels L., van der Lelie D. and Mergeay M. 1993a. *Proceedings of the Sixth International Symposium, toxicity assessment and on-line monitoring.* May 10–14. Berlin, Germany. p. 58.

Corbisier P., Ji G., Nuyts G., Mergeay M. and Silver S. 1993b. "*LuxAB* fusions with the arsenic and cadmium resistance operons of *Staphylococcus aureus* plasmid pI258." *FEMS Microbiol. Lett.* 110:231–238.

Dabrock B., Kebeler M., Averhoff B. and Gottschalk G. 1994. "Identification and characterization of a transmissible linear plasmid from *Rhodococcus erythropolis* BD2 that encodes isopropylbenzene and trichloroethylene catabolism." *Appl. Environ. Microbiol.* 60:853–860.

Dagley S. 1986. *The biology of* Pseudomonas. New York. Academic Press, Inc., pp. 527–555.

De Vevey E., Bitton G., Rossel D., Ramos L. D., Munguia G. L. and Tarradellas J. 1993. "Concentration and bioavailability of heavy metals in sediments in Lake Yojoa (Honduras)." *Bull. Environ. Contam. Toxicol.* 50:253–259.

de Zwart D. and Sloof W. 1983. "The Microtox as an alternative assay in the acute toxicity assessment of water pollutants." *Aquat. Toxicol.* 4:129–138.

Denome S. A., Stanley D. C., Olson E. S. and Young K. D. 1993. "Metabolism of dibenzothiophene and naphthalene in *Pseudomonas* strains: Complete DNA sequence of an upper naphthalene catabolic pathway." *J. Bacteriol.* 175: 6890–6901.

Don R. H. and Pemberton J. M. 1985. "Genetic and physical map of the 2,4-dichlorophenoxyacetic acid-degradative plasmid pJP4." *J. Bacteriol.* 161:466–468.

Don R. H., Wieghtman A. J., Knackmuss H. H. and Timmis T. N. 1985. "Transposon mutagenesis and cloning analysis of the pathways for degradation of 2,4-dichlorophenoxyacetic acid and 3-chlorobenzoate in *Alcaligenes eutrophus* JMP134(pJP4)." *J. Bacteriol.* 161:85–90.

Doten R. C., Ngai K.-.L, Mitchell D. J. and Ornston L. N. 1987. "Cloning and genetic organization of the *pca* gene cluster from *Acinetobacter calcoaceticus.*" *J. Bacteriol.* 169:3168–3174.

Dunn N. W. and Gunsalus I. C. 1973. "Transmissible plasmid coding early enzymes of naphthalene oxidation in *Pseudomonas putida.*" *J. Bacteriol.* 114:974–979.

Ensign S. A., Hyman M. R. and Arp D. J. 1992. "Cometabolic degradation of chlorinated alkenes by alkene monooxygenase in a propylene-grown *Xanthobacter* strain." *Appl. Environ. Microbiol.* 58:3038–3046.

Ensley B. D. 1991. "Biochemical diversity of trichloroethylene metabolism." *Annu. Rev. Microbiol.* 45:283–299.

Erickson B. D. and Mondello F. J. 1992. "Nucleotide sequencing and transcriptional mapping of the genes encoding biphenyl dioxygenase, a multicomponent polychlorinated-biphenyl-degrading enzyme in *Pseudomonas* strain LB400." *J. Bacteriol.* 174:2903–2912.

Ewers J., Freier-Schröder D. and Knackmuss H.-J. 1990. "Selection of trichloroethylene (TCE) degrading bacteria that resist inactivation by TCE." *Arch. Microbiol.* 154:410–413.

Fennewald M., Benson S., Oppici M. and Shapiro J. 1979. "Insertion element analysis and mapping of the *Pseudomonas* plasmid *alk* regulation." *J. Bacteriol.* 139:940–952.

Fleming J., Sanseverino J. and Sayler G. S. 1993. "Quantitative relationship between naphthalene catabolic frequency and expression in predicting PAH degradation in soils at town gas manufacturing sites." *Env. Sci. Technol.* 27:1068–1074.

Frantz B. and Chakrabarty A. M. 1986. *The biology of* Pseudomonas. New York. Academic Press, Inc. pp. 295–323.

Frantz B. and Chakrabarty A. M. 1987. "Organization and nucleotide sequence determination of a gene cluster involved in 3-chlorocatechol degradation." *Proc. Natl. Acad. Sci. USA* 84:4460–4464.

Freedman D. L. and Gossett. 1989. "Biological reductive dechlorination of tetrachloroethylene and trichloroethylene to ethylene under methanogenic conditions." *Appl. Environ. Microbiol.* 55:2144–2151.

Friello D. A., Mylroie J. R., Gibson D. T., Rogers J. E. and Chakrabarty A. M. 1976. "XYL, a nonconjugative xylene degradative plasmid in *Pseudomonas* Pxy." *J. Bacteriol.* 127:1217–1224.

Furukawa K. 1994. "Molecular genetics and evolutionary relationship of PCB-degrading bacteria." *Biodegradation* 5:289–300.

Furukawa K., Hirose J., Hayashida S. and Nakamura K. 1994. "Efficient degradation of trichloroethylene by a hybrid aromatic ring dioxygenase." *J. Bacteriol.* 176:2121–2123.

Gibson D. T. 1977. *Fate and effects of petroleum hydrcarbons in marine organisms and ecosystems.* New York, Pergamon Press. pp. 36–46.

Gibson D. T. (ed.) 1984. *Microbial degradation of organic compounds.* New York. Marcel Dekker, Inc.

Gibson D. T., Hensley M., Yoshioka H. and Mabry T. J. 1970. "Formation of (+)-*cis*-2,3-dihydroxy-1-methylcyclohexa-4,6-diene from toluene by *Pseudomonas putida.*" *Biochemistry* 9:1626–1630.

Gibson D. T. and Subramanian V. 1984. *Microbial degradation of organic compounds.* New York, Marcel Dekker, Inc. pp.181–252.

Gilliland G., Perrin S., Blanchard K. and Bunn H. G. 1990. "Analysis of cytokine mRNA and DNA: Detection and quantitation by competitive polymerase chain reaction." *Proc. Natl. Acad. Sci. USA* 87:2725–2729.

Gilson E, Clément J. M., Brutlag D. and Hofnung M. 1984. "A family of dispersed repetitive extragenic palindromic DNA sequences in *E. coli.*" *EMBO J.* 3:1417–1421.

Green J. and Dalton H. 1989. "Substrate specificity of soluble methane monooxygenase." *J. Biol. Chem.* 264:17698–17703.

Gregory I. R., Jimenez L., Zhang D., Fleming J., Pfiffner S., Bowman J., White D. C. and Sayler G. S. 1995. "Co-metabolic gene distribution and expression in a trichloroethylene-contaminated site undergoing in situ bioremediation." *Appl. Environ. Microbiol.* In Review.

Gunsalus I. C. 1985. *Plasmid in bacteria.* New York. Plenum Press. pp. 687–706.

Hadrys H., Balick M. and Schierwater B. 1992. "Applications of random amplified polymorphic DNA (RAPD) in molecular ecology." *Mol. Ecol.* 1:55–63.

Happe B., Eltis L. D., Poth H., Hedderich R. and Timmis K. N. 1993. "Characterization of 2,2′,3-trihydroxybiphenyl dioxygenase, an estradiol dioxygenase from the dibenzofuran- and dibenzo-*p*-dioxin-degrading bacterium *Sphingomonas* sp. strain RW1." *J. Bacteriol.* 175:7313–7320.

Harder P. A. and Kunz D. A. 1986. "Characterization of the OCT plasmid encoding alkane oxidation and mercury resistance in *Pseudomonas putida.*" *J. Bacteriol.* 165:650–653.

Harker A. R. and Kim Y. 1990. "Trichloroethylene degradation by two independent aromatic-degrading pathways in *Alcaligenes eutrophus* JMP134." *Appl. Environ. Microbiol.* 56:1179–1181.

Hayase N., Taira K. and Furukawa K. 1990. "*Pseudomonas putida* KF715 *bphABCD* operon encoding biphenyl and polychlorinated biphenyl degradation: Cloning, analysis, and expression in soil bacteria." *J. Bacteriol.* 172:1160–1164.

Heitzer A., Malachowsky K., Thonnard J. E., Bienkowski P. R., White D. C. and Sayler G. S. 1994. "Optical biosensor for environmental on-line monitoring of naphthalene and salicylate bioavailability with an immobilized bioluminescent catabolic reporter bacterium." *Appl. Environ. Microbiol.* 60:1487–1494.

Heitzer A., Webb O. F., Thonnard J. E. and Sayler G. S. 1992. "Specific quantitative assessment of naphthalene and salicylate bioavailability using a bioluminescent catabolic reporter bacterium." *Appl. Environ. Microbiol.* 58: 1839–1846.

Horvath R. S. 1972. "Microbial co-metabolism and the degradation of organic compounds in nature." *Microbiol. Rev.* 36:146–155.

Hulton C. S. J., Higgins C. F. and Sharp P. M. 1991. "ERIC sequences: A novel family of repetitive elements in the genomes of *Escherichia coli*, *Salmonella typhimurium* and other enterobacteria." *Mol. Microbiol.* 5:825–834.

Hyman M. R., Murton I. B. and Arp D. J. 1988. "Interaction of ammonia monooxygenase from *Nitrosomonas europaea* with alkanes, alkenes, and alkynes." *Appl. Environ. Microbiol.* 54:3187–3190.

Hyman M. R., Sansome-Smith A. W., Shears J. H. and Wood P. M. 1985. "A kinetic study of benzene oxidation to phenol by whole cells of *Nitrosomonas europaea* and evidence for the further oxidation of phenol to hydroquinone." *Arch. Microbiol.* 43:302–306.

Innis M. A. and Gelfand D. H. 1990. *PCR protocols: A guide to methods and applications*. New York, Academic Press, Inc. pp. 3–12.

Jablonski P. E. and Ferry J. G. 1992. "Reductive dechlorination of trichloroethylene by the CO-reduced CO dehydrogenase enzyme complex from *Methanosarcina thermophila*." *FEMS Microbiol. Lett.* 96:55–60.

Jahng D. and Wood T. K. 1994. "Trichloroethylene and chloroform degradation by a recombinant Pseudomonad expressing soluble methane monooxygenase from *Methylosinus trichosporium* OB3b." *Appl. Environ. Microbiol.* 60: 2473–2482.

Jain R. K., Burlage R. S. and Sayler G. S. 1988. "Methods for detecting recombinant DNA in the environment." *CRC Crit. Rev. Microbiol.* 8:33–84.

Janssen D. B., van der Ploeg J. R. and Pries F. 1994. "Genetics and biochemistry of 1,2-dichloroethane degradation." *Biodegradation* 5:249–257.

Johnston W. H., Stapleton R. D. and Sayler G. S. 1995. "Direct extraction of microbial DNA from soils and sediments," In: (A. D. L. Akkermans, J. D. van Elsas, and F. J. de Bruijn, eds.) *Molecular microbial ecology manual.* Dordrecht, The Netherlands. Kluwer Academic Publishers.

Jones R. D. and Morita R. Y. 1983. "Carbon monoxide oxidation by chemolithotrophic ammonium oxidizers." *Can. J. Microbiol.* 29:1545–1551.

Ka J. O., Holben W. E. and Tiedje J. M. 1994. "Genetic and phenotypic diversity of 2,4-dichlorophenoxyacetic acid (2,4-D)-degrading bacteria isolated from 2,4-D-treated field plots." *Appl. Environ. Microbiol.* 60:1106–1115.

Ka J. O. and Tiedje J. M. 1994. "Integration and excision of a 2,4-dichlorophenoxyacetic acid-degradative plasmid in *Alcaligenes paradoxus* and evidence of its natural intergeneric transfer." *J. Bacteriol.* 176:5284–5289.

Kaphammer B. J., Kukor J. J. and Olsen R. H. 1991. *Biodeterioration and biodegradation 8.* London, Elsevier Applied Science. pp. 571–572.

Kästner M., Breuer-Jammali M. and Mahro B. 1994. "Enumeration and characterization of the soil microflora from hydrocarbon-contaminated soil sites able to mineralize polycyclic aromatic hydrocarbons (PAH)." *Appl. Microbiol. Biotechnol.* 41:267–273.

Keck J., Sims R. C., Coover M. P., Park K. S. and Symons B. 1989. "Evidence for cooxidation of polynuclear aromatic hydrocarbons in soil." *Water Res.* 23:1467–1476.

Keener W. K. and Arp D. J. 1994. "Transformations of aromatic compounds by *Nitrosomonas europaea.*" *Appl. Environ. Microbiol.* 60:1914–1920.

Khan A. A. and Walia S. K. 1991. "Expression, localization, and functional analysis of polychlorinated biphenyl degradation genes *cbpABCD* of *Pseudomonas putida.*" *Appl. Environ. Microbiol.* 57:1325–1332.

Kikuchi Y., Nagata Y., Hinata M., Kimbara M., Fukuda M., Yano K. and Takagi M. 1994a. "Identification of the *bphA4* gene encoding ferredoxin reductase involved in biphenyl and polychlorinated biphenyl degradation in *Pseudomonas* sp. strain KKS102." *J. Bacteriol.* 176:1689–1694.

Kikuchi Y., Yasukochi Y., Nagata Y., Fukuda M. and Takagi M. 1994b. "Nucleotide sequence and functional analysis of the *meta*-cleavage pathway involved in biphenyl and polychlorinated biphenyl degradation in *Pseudomonas* sp. strain KKS102." *J. Bacteriol.* 176:4269–4276.

Kim Y., Ayoubi T. and Harker A. 1994. *94th General Meeting Amer. Soc. Microbiol.,* May 23–27. Las Vegas, NV. p. 307.

Kimbara K. T., Hashimoto T., Fukuda M., Koana T., Takagi M., Oishi M. and Yano K. 1989. "Cloning and sequencing of two tandem genes involved in degradation of 2,3-dihydroxybiphenyl to benzoic acid in the polychlorinated biphenyl-degrading soil bacterium *Pseudomonas* sp. strain KKS102." *J. Bacteriol.* 171:2740–2747.

King J. M. H., DiGrazia P. M., Applegate B. M., Burlage R., Sanceverino J., Dunbar P., Larimer F. and Sayler G. S. 1990. "Rapid, sensitive bioluminescent reporter technology for naphthalene exposure and biodegradation." *Science* 249:778–781.

Kiyohara H. K., Nagao and K. Yana. 1982. "Rapid screen for bacteria degrading water insoluble, solid hydrocarbons on agar plates." *Appl. Environ. Microbiol.* 43:454–457.

Kiyohara H., Torigoe S., Kaida N., Asaki T., Iida T., Hayashi H. and Takizawa N. 1994. "Cloning and characterization of a chromosomal gene cluster, *pah*, that encodes the upper pathway for phenanthrene and naphthalene utilization by *Pseudomonas putida* OUS82." *J. Bacteriol.* 176:2439–2443.

Kuhm A. E., Stolz A. and Knackmuss H.-J. 1991. "Metabolism of naphthalene by the biphenyl-degrading bacterium *Pseudomonas paucimobilis* Q1." *Biodegradation* 2:115–120.

Kukor J. J. and Olsen R. H. 1990. "Molecular cloning, characterization, and regulation of a *Pseudomonas pickettii* PKO1 gene encoding phenol hydroxylase and expression of the gene in *Pseudomonas aeruginosa* PAO1c." *J. Bacteriol.* 8:4624–4630.

Kunz D. A. and Chapman P. J. 1981. "Isolation and characterization of spontaneously occurring TOL plasmid mutants of *Pseudomonas putida* HS1." *J. Bacteriol.* 146:952–964.

Layton A. C., Lajoie C. A., Easter J. P., Jernigan R., Beck M. J. and Sayler G. S. 1994a. *Recombinant DNA technology II, Annals of the New York Academy of Sciences Vol 721.* The New York Academy of Sciences, New York, New York. pp. 407–422.

Layton A. C., Lajoie C. A., Easter J. P., Jernigan R., Sanseverino J., and Sayler G. S. 1994b. "Molecular diagnostics and chemical analysis for assessing biodegradation of polychlorinated biphenyls in contaminated soils." *J. Indust. Microbiol.* 13:392–401.

Leahy J. G. and Olson R. H. 1994. *94th General Meeting, Amer. Soc. Microbiol.,* May 23–27. Las Vegas, NV. p. 307.

Li S. and Wackett L. P. 1992. "Trichloroethylene oxidation by toluene dioxygenase." *Biochem. Biophys. Res. Commun.* 185:443–451.

Lloyd-Jones G., de Jong C., Ogden R. C., Duetz W. A. and Williams P. A. 1994. "Recombination of the *bph* (biphenyl) catabolic genes from plasmid pWW100 and their deletion during growth on benzoate." *Appl. Environ. Microbiol.* 60:691–696.

Loper J. E. and Lindow S. E. 1994. "A biological sensor for iron available to bacteria in their habitats on plant surfaces." *Appl. Environ. Microbiol.* 60:1934–1941.

Manilal V. B. and Alexander M. 1991. "Factors affecting the microbial degradation of phenanthrene in soil." *Appl. Microbiol. Biotechnol.* 35:401–405.

Maron D. M. and Ames B. N. 1983. "Revised methods for the *Salmonella* mutagenicity test." *Mutation Res.* 113:173–215.

Martin B., Humbert O., Camara M., Guenzi E., Walker J., Mitchell T., Andrew P.,

Prudhomme M., Alloing G., Hakenbect R., Morrison D. A., Boulnois G. J. and Claverys J. P. 1992. "A highly conserved repeated DNA element located in the chromosome of *Streptococcus pneumoniae.*" *Nucleic Acids Res.* 20:3479–3483.

Matrubutham U., Thonnard J. E. and Sayler G. S. 1995. "Bioluminescence induction response of the biosensor bacterium *Pseudomonas fluorescens* HK44 under stress." In preparation.

McTavish H., Fuchs J. A. and Hooper A. B. 1993. "Sequence of the gene coding for ammonia monooxygenase in *Nitrosomonas europaea.*" *J. Bacteriol.* 175:2436–2444.

Menn F.-M., Applegate B. M. and Sayler G. S. 1993. "NAH-plasmid-mediated catabolism of anthracene and phenanthrene to naphthoic acids." *Appl. Environ. Microbiol.* 59:1938–1942.

Microbics reference #D006: The mode of action of genotoxic agents in the restoration of light in the Mutatox™ system. Microbics Corporation, 2232 Rutherford Rd, Carlsbad, CA 92008.

Miller R. E. and Guengerich F. P. 1983. "Metabolism of trichloroethylene in isolated hepatocytes, microsome, and reconstituted enzyme systems containing cytochrome P-450." *Cancer Res.* 43:1145–1152.

Mondello F. J. 1989. "Cloning and expression in *Escherichia coli* of *Pseudomonas* strain LB400 genes encoding polychlorinated biphenyl degradation." *J. Bacteriol.* 171:1725–1732.

Monticello D. J., Bakker D. and Finnerty W. R. 1985. "Plasmid-mediated degradation of dibenzothiophene by *Pseudomonas* species." *Appl. Environ. Microbiol.* 49:756–760.

Mullis K. B., and Faloona F., 1987. "Specific synthesis of DNA in vitro via a polymerase chain reaction." *Methods Enzymol.* 155:335–350.

Nadeau L. J., Menn F.-M., Breen A. and Sayler G. S. 1994. "Aerobic degradation of 1,1,1-trichloro-2,2-bis(4-chlorophenyl)ethane (DDT) by *Alcaligenes eutrophus* A5." *Appl. Environ. Microbiol.* 60:51–55.

Nakatsu C. H. and Wyndham R. C. 1993. "Cloning and expression of the transposable chlorobenzoate-3,4-dioxygenase genes of *Alcaligenes* sp. strain BR60." *Appl. Environ. Microbiol.* 59:3625–3633.

Natarajan M. R., Lu Z. and Oriel P. 1994. "Cloning and expression of a pathway for benzene and toluene from *Bacillus stearothermophilus.*" *Biodegradation.* 5:77–82.

Neidle E. L., Hartnett C., Bonitz S. and Ornston L. N. 1988. "DNA sequence of the *Acinetobacter calcoaceticus* catechol 1,2-dioxygenase I structural gene *catA*: evidence for evolutionary divergence of intradiol dioxygenases by acquistion of DNA sequence repetitions." *J. Bacteriol.* 170:4874–4880.

Neidle E. L., Shapiro M. K. and Ornston L. N. 1987. "Cloning and expression in *Escherichia coli* of *Acinetobacter calcoaceticus* genes for benzoate degradation." *J. Bacteriol.* 169:5496–5503.

Nelson M. J. K., Compeau G., Maziarz T. and Mahaffey W. R. 1991. *Bioremediation: Field experience.* Lewis Publishers, Boca Raton, FL. pp. 59–78.

Nelson M. J. K., Montgomery S. O. Mahaffey W. R. and Pritchard P. H. 1987. "Biodegradation of trichloroethylene and involvement of an aromatic biodegradative pathway." *Appl. Environ. Microbiol.* 53:949–954.

Nelson M. J. K., Montgomery S. O. and Pritchard P. H. 1988. "Trichloroethylene metabolism by microorganisms that degrade aromatic compounds." *Appl. Environ. Microbiol.* 54:604–606.

Ogram A. and Sayler G. S. 1988. "The use of gene probes in the rapid analysis of natural microbial communities." *J. Ind. Microbiol.* 3:281–292.

Ogram A., Sayler G. S. and Barkay T. 1987. "The extraction and purification of microbial DNA from sediments." *J. Microbiol. Methods* 7:57–66.

Oldenhuis R. and Janssen D. B. 1993. *Microbial growth on C_1 compounds.* Andover, UK. Intercept Ltd. pp. 121–133.

Oldenhuis R., Vink R. L. J. M., Janssen D. B. and Witholt B. 1989. "Degradation of chlorinated aliphatic hydrocarbons by *Methylosinus trichosporium* OB3b expressing soluble methane monooxygenase." *Appl. Environ. Microbiol.* 55:2819–2826.

Olsen R. H., Kukor J. J. and Kaphammer B. 1994. "A novel toluene-3-monooxygenase pathway cloned from *Pseudomonas pickettii* PKO1." *J. Bacteriol.* 176:3749–3756.

Olson, B. H. 1991. "Tracking and using genes in the environment." *Environ. Sci. Technol.* 25:604–611.

Parales R. E. and Harwood C. S. 1993. "Regulation of the *pcaIJ* genes for aromatic acid degradation in *Pseudomonas putida*." *J. Bacteriol.* 175:5829–5838.

Park K. S., Sims R., Doucette W. J. and Matthews J. E. 1988. "Biological transformation and detoxification of 7,12-dimethylbenz(a)anthracene in soil systems." *J. Wat. Poll. Control Fed.* 60:1822–1825.

Peloquin L. and Greer C. W. 1993. "Cloning and expression of the polychlorinated biphenyl-degrading gene cluster from *Arthrobacter* M5 and comparison to analogous genes from Gram-negative bacteria." *Gene* 125:35–40.

Perkins E. J., Gordon M. P., Caceres O. and Lurquin P. F. 1990. "Organization and sequence analysis of the 2,4-dichlorophenol hydroxylase and dichlorocatechol oxidative operons of plasmid pJP4." *J. Bacteriol.* 172:2351–2359.

Pieper D. H., Engesser K.-H. and Knackmuss H.-J. 1989. "Regulation of catabolic pathways of phenoxyacetic acids and phenols in *Alcaligenes eutrophus* JMP134." *Arch. Microbiol.* 151:365–371.

Powlowski J. and Shingler V. 1994. "Genetics and biochemistry of phenol degradation by *Pseudomonas* sp. CF600." *Biodegradation* 5:219–236.

Pries F., van der Ploeg J. R., van den Wijngaard A. J., Bos R. and Janssen D. B. 1994. *Bioremediation of chlorinated and polycyclic aromatic hydrocarbon compounds.* Boca Raton, FL. Lewis Publishers, pp. 259–265.

Reineke W. and Knackmuss H. J. 1988. "Microbial degradation of haloaromatics." *Annu. Rev. Microbiol.* 42:263–287.

Rochkind-Dubinsky M. L. Sayler G. S. and Blackburn J. W. 1987. *Microbiological decomposition of chlorinated aromatic compounds.* New York. Marcel Dekker, Inc.

Saiki R. K. 1992. *PCR technology: Principles and applications for DNA amplification.* New York. W. H. Freeman and Company, pp.7–16.

Saiki R. K., Schart S., Faloona F., Mullis K. B., Horn G. T., Erlich H. A. and Arnheim N. 1985. "Enzymatic amplification of ß-globin genomic sequences and restriction site analysis for diagnosis of sickle cell anemia." *Science* 230:1350–1354.

Saint C. P., McClure N. C. and Venables W. A. 1990. "Physical map of the aromatic amine and *m*-toluate catabolic plasmid pTDN1 in *Pseudomonas putida*: Location of a unique *meta*-cleavage pathway." *J. Gen. Microbiol.* 136: 615–625.

Sanseverino J., Applegate B. M., King J. M. H. and Sayler G. S. 1993a. "Plasmid-mediated mineralization of naphthalene, phenanthrene and anthracene." *Appl. Environ. Microbiol.* 59:1931–1937.

Sanseverino J., Werner C., Fleming J. T., Applegate B. M., King J. M. H and Sayler G. S. 1993b. "Molecular diagnostics of polycyclic aromatic hydrocarbon biodegradation in manufactured gas plant soils." *Biodegradation.* 4:303–321.

Sayler G. S. 1991. "Contribution of molecular biology to bioremediation." *J. Haz. Mat.* 28:13–27.

Sayler G. S. and Blackburn J. W. 1989. *Biotreatment of agricultural wastewater.* Boca Raton, FL. CRC Press, Inc., pp. 53–71.

Sayler G. S., Blackburn J. W. and Donaldson T. A. 1988. "Environmental Biotechnology of Hazardous Wastes," *Proceedings of the NSF Workshop,* Gatlinburg, TN, ORNA/TM0853, NTIS, Washington, DC.

Sayler G. S., Hooper S. W., Layton A. C. and King J. M. H. 1990. "Catabolic plasmids of environmental and ecological significance." *Microb. Ecol.* 19:1–20.

Sayler G. S. and Layton A. C. 1990. "Environmental application of nucleic acid hybridization." *Ann. Rev. Microbiol.* 44:625–648.

Sayler G. S., Perkins R. E., Sherrill T. W., Perkins B. K., Reid M. C., Shields M. S., Kong H. L. and Davis J. W. 1983. "Microcosm and experimental pond evaluation of microbial community response to synthetic oil contamination in freshwater sediments." *Appl. Environ. Microbiol.* 46:211–219.

Sayler G. S., Sherrill T. W., Perkins R. E., Mallory L. M., Shiaris M. P. and Pederson. 1982. "Impact of coal-coking effluent on sediment microbial communities: A multivariate approach." *Appl. Environ. Microbiol.* 44:1118–1129.

Sayler G. S., Shields M. S., Tedford E. T., Breen A., Hooper S. W., Sirotkin K. M. and Davis J. W. 1985. "Application of DNA-DNA colony hybridization to the detection of catabolic genotypes in environmental samples." *Appl. Environ. Microbiol.* 49:1295–1303.

Schell M. A. 1990. Pseudomonas: *Biotransformations, pathogenesis, and evolving biotechnology.* Washington, D.C. American Society for Microbiology. pp. 165–176.

Schmitz A., Gartemann K. H., Fiedler J., Grund E. and Eichenlaub R. 1992. "Cloning and sequence analysis of genes for dehalogenation of 4-chlorobenzoate from *Arthrobacter* sp. strain SU." *Appl. Environ. Microbiol.* 58:4068–4071.

Scholten J. D., Chang K.-H., Babbitt P. C., Charest H., Sylvestre M. and Dunaway-Mariano D. 1991. "Novel enzymic hydrolytic dehalogenation of a chlorinated aromatic." *Science.* 253:182–185.

Selenska S. and Klingmuller W. 1992. Direct recovery and molecular analysis of DNA and RNA from soil. *Microbial Releases.* 1:41–46.

Selifonova O., Burlage R. and Barkay T. 1993. "Bioluminescent sensors for the detection of bioavailable Hg(II) in the environment." *Appl. Environ. Microbiol.* 59:3083–3090.

Serdar C. M. and Gibson D. T. 1989. "Isolation and characterization of altered plasmids in mutant strains of *Pseudomonas putida* NCIB9816." *Biochem. Biophys. Res. Commun.* 164:764–771.

Shanley M. S., Neidle E. L., Parales R. E. and Ornston L. N. 1986. "Cloning and expression of *Acinetobacter calcoaceticus catBCDE* genes in *Pseudomonas putida* and *Escherichia coli.*" *J. Bacteriol.* 165:557–563.

Sharma A., Chunn C. D., Rothmel R. K. and Unterman R. 1991. *91st General Meeting, Amer. Soc. Microbiol.,* May 5–9. Dallas, TX. p. 284.

Shaw J. J., Settles L. G. and Kado C. I. 1988. "Transposon *Tn4431* mutagenesis of *Xanthomonas campestris* pv. *campestris*: Characterization of a nonpathogenic mutant and cloning of locus for pathogenicity." *Mol. Plant-Microbe Interact.* 1:39–45.

Shields M. S., Montgomery S. O., Chapman P. J., Cuskey S. M. and Pritchard P. H. 1989. "Novel pathway for toluene catabolism in the trichloroethylene degrading bacterium G4." *Appl. Environ. Microbiol.* 55:1624–1629.

Shields M. S., Montgomery S. O., Cuskey S. M., Chapman P. J. and Pritchard P. H. 1991. "Mutants of *Pseudomonas cepacia* G4 defective in catabolism of aromatic compounds and trichloroethylene." *Appl. Environ. Microbiol.* 57:1935–1941.

Shields M. S. and Reagin M. J. 1992. "Selection of a *Pseudomonas cepacia* strain constitutive for the degradation of trichloroethylene." *Appl. Environ. Microbiol.* 58:3977–3983.

Shingler V., Powlowski J. and Marklund U. 1992. "Nucleotide sequence and functional analysis of the complete phenol/3,4-dimethylphenol catabolic pathway of *Pseudomonas* sp. strain CF600." *J. Bacteriol.* 174:711–724.

Singleton I. 1994. "Microbial metabolism of xenobiotics: Fundamental and applied research." *J. Chem. Tech. Biotechnol.* 59:9–23.

Smith M. R. 1990. "The biodegradation of aromatic hydrocarbons by bacteria." *Biodegradation* 1:191–206.

Springael D., Baeyens W., Hannes L., De Wilde K. and Mergeay M. 1991. "Use of *Tn4431* containing the promotor-less *lux* genes of *Vibrio fischeri* to detect genes involved in PCB degradation of *Alcaligenes* sp. A5." *Arch. Int. Physiol. Biochim.* 99:77.

Springael D., Kreps S. and Mergeay M. 1993. "Identification of a catabolic transposon, Tn*4371*, carrying biphenyl and 4-chlorobiphenyl degradation genes in *Alcaligenes eutrophus* A5." *J. Bacteriol.* 175:1674–1681.

Stapleton R. D. Jr. and Sayler G. S. 1995. "Biodegradative potential for natural attenuation of hydrocarbons in a shallow, unconfined aquifer field test site." *Proceedings of the 88th Annual Meeting of the Air & Waste Management Association,* San Antonio, Texas.

Steffan R. J. and Atlas R. M. 1991. "Polymerase chain reaction: Applications in environmental microbiology." *Ann. Rev. Microbiol.* 45:137–161.

Suflita J. M., Horowitz A., Shelton D. R. and Tiedje J. M. 1982. "Dechlorination: A novel pathway for the anaerobic biodegradation of haloaromatic compounds." *Science* 218:1115–1117.

Taira K., Hayase N., Arimura N., Yamashita S., Miyazaki T. and Furukawa K. 1988. "Cloning and nucleotide sequence of 2,3-dihydroxybiphenyl dioxygenase gene from the PCB-degrading strain *Pseudomonas paucimobilis* Q1." *Biochemistry* 27:3990–3996.

Taira K., Hirose J., Hayashida S. and Furukawa K. 1992. "Analysis of *bph* operon from the polychlorinated biphenyl-degrading strain of *Pseudomonas pseudoalcaligenes* KF707." *J. Bacteriol.* 267:4844–4853.

Tan H.-M. and Fong K. P. Y. 1993. "Molecular analysis of the plasmid-borne *bed* gene cluster from *Pseudomonas putida* ML2 and cloning of the *cis*-benzene dihydrodiol dehydrogenase." *Can. J. Microbiol.* 39:357–362.

Tan H.-M., Tang H.-V., Joannou C. L., Abdel-Wahab N. H., and Mason J. R. 1993. "The *Pseudomonas putida* ML2 plasmid-encoded genes for benzene dioxygenase are unsual in codon usage and low in G+C content." *Gene* 130:33–39.

Tardif G., Greer C. W., Labbe D. and Lau P. C. K. 1991. "Involvement of a large plasmid in the degradation of 1,2-dichloroethane by *Xanthobacter autotrophicus.*" *Appl. Environ. Microbiol.* 57:1853–1857.

Timmis K. N., Lehrbach P. R., Harayama S., Don R. H., Mermod N., Bas S., Leppik R., Weightman A. J., Reineke W., and Knackmuss H. J. 1985. *Plasmid in bacteria.* New York. Plenum Press. pp. 719–739.

Torsvik V. L. 1980. "Isolation of bacterial DNA from soil." *Soil Biol. Biochem.* 10:7–12.

Tsien H.-C., Brusseau G. A., Hanson R. S. and Wackett L. P. 1989. "Biodegradation of trichloroethylene by *Methylosinus trichosporium* OB3b." *Appl. Environ. Microbiol.* 55:3155–3161.

Unterman R. 1991. *Environmental biotechnology for waste treatment.* New York. Plenum Press. pp. 159–162.

van der Lelie D., Corbisier P., Baeyans W., Wuertz S., Diels L. and Mergeay M. 1993. *EERO-VLAB Workshop on Environmental Technology.* Nov. 21–25. Grobbondonk, Bolgium. pp. 67–71.

van der Meer J. R., de Vos W., Harayama S. and Zehnder A. J. B. 1992. "Molecular mechanisms of genetics adaption to xenobiotics compounds." *Microbiol. Rev.* 56:677–694.

van der Meer J. R., van Neerven A. R. W., de Vries E. J., de Vos W. M. and Zehnder A. J. B. 1991. "Cloning and characterization of plasmid-encoded genes for the degradation of 1,2-dichloro-, 1,4-dichloro-, and 1,2,4-trichlorobenzene of *Pseudomonas* sp. strain P51." *J. Bacteriol.* 173:6–15.

Van Rolleghem P., Drues D. and Verstraete W. 1990. *Proceedings of 5th European Congress on Biotechnology,* July 8–13, 1990, Copenhagen, Denmark. pp. 161–164.

Vanderberg L. A. and Perry J. J. 1994. "Dehalogenation by *Mycobacterium vaccae* JOB-5: Role of the propane monooxygenase." *Can. J. Microbiol.* 40:169–172.

Vannelli T., Logan M., Arciero D. and Hooper A. B. 1990. "Degradation of halogenated aliphatic compounds by the ammonia-oxidizing bacterium, *Nitrosomonas europaea.*" *Appl. Environ. Microbiol.* 56:1169–1171.

Versalovic J., Koeuth T. and Lupski J. R. 1991. "Distribution of repetitive DNA sequences in eubacteria and application to fingerprinting of bacterial genomes." *Nuc. Acids Res.* 19:6823–6831.

Wackett L. P., Brusseau G. A., Householder S. R. and Hanson R. S. 1989. "Survey of microbial oxygenases: Trichloroethylene degradation by propane-oxidizing bacteria." *Appl. Environ. Microbiol.* 55:2960–2964.

Wackett L. P. and Gibson D. T. 1988. "Degradation of trichloroethylene by toluene dioxygenase in whole-cell studies with *Pseudomonas putida* F1." *Appl. Environ. Microbiol.* 54:1703–1708.

Wallace W. H. and Sayler G. S. 1992. *Encyclopedia of Microbiology. Vol. 1.* New York. Academic Press, Inc. pp. 417–430.

Weightman A. J., Reineke W. and Knackmuss H. J. 1985. *Plasmid in bacteria.* New York. Plenum Press. pp. 719–739.

Whited G. and Gibson D. T. 1991. "Toluene-4-monooxygenase, a three-component enzyme system that catalyzes the oxidation of toluene to *p*-cresol in *Pseudomonas mendocina* KR1." *J. Bacteriol.* 173:3010–3016.

Wiesner R. J., Rüegg J. C. and Morano I. 1992. "Counting target molecules by exponential polymerase chain reaction: Copy number of mitochondrial DNA in rat tissues." *Biochem. Biophys. Res. Comm.* 183:553–559.

Williams P. A. and Murray K. 1974. "Metabolism of benzoate and the methylbenzoates by *Pseudomonas putida* (*arvilla*) mt-2: Evidence for the existence of new TOL plasmids." *J. Bacteriol.* 120:416–423.

Williams P. A. and Sayers J. R. 1994. "The evolution of pathways for aromatic hydrocarbon oxidation in *Pseudomonas*." *Biodegradation* 5:195–217.

Wilson S. C. and Jones K. C. 1993. "Bioremediation of soil contaminated with polynuclear aromatic hydrocarbons (PAHs): A review." *Environ. Poll.* 81:229–249.

Winstanley C., Taylor S. C. and Williams P. A. 1987. "pWW174: A large plasmid from *Acinetobacter calcoaceticus* encoding benzene catabolism by the β-ketoadipate pathway." *Mol. Microbiol.* 1:219–227.

Winter R. B., Yen K.-M. and Ensley B. D. 1989. "Efficient degradation of trichloroethylene by a recombinant *Escherichia coli*." *Bio/Technology* 7:282–285.

Worsey M. J. and Williams P. A. 1975. "Metabolism of toluene and xylenes by *Pseudomonas putida* (*arvilla*) mt-2: Evidence for a new function of the TOL plasmid." *J. Bacteriol.* 124:7–13.

Yano K. and Nishi T. 1980. "pKJ1, a naturally occurring conjugative plasmid coding for toluene degradation and resistance to streptomycin and sulfonamides." *J. Bacteriol.* 143:552–560.

Yates J. R. and Mondello F. J. 1989. "Sequence similarities in the encoding polychlorinated biphenyl degradation by *Pseudomonas* strain LB400 and *Alcaligenes eutrophus* H850." *J. Bacteriol.* 171:1733–1735.

Yen K.-M. and Gunsalus I. C. 1982. "Plasmid gene organization: Naphthalene/salicylate oxidation." *Proc. Natl. Acad. Sci. USA.* 79:874–878.

Yen K.-M. and Gunsalus I. C. 1985. "Regulation of naphthalene catabolic genes of plasmid NAH7." *J. Bacteriol.* 162:1008-1013.

Yen K.-M., Karl M. R., Blatt L. M., Simon M. J., Winter R. B., Fausset P. R., Lu H.-S., Harcourt A. A. and Chen K.-K. 1991. "Cloning and characterization of a *Pseudomonas mendocina* KR1 gene cluster encoding toluene-4-monooxygenase." *J. Bacteriol.* 173:5315–5327.

Yen K.-M. and Serdar C. M. 1988. "Genetics of naphthalene catabolism in pseudomonads." *Crit. Rev. Microbiol.* 15:247–268.

Zylstra G. J., McCombie W. R., Gibson D. T. and Finette B. 1988. "Toluene degradation by *Pseudomonas putida* F1: Genetic organization of the *tod* operon." *Appl. Environ. Microbiol.* 54:1498–1503.

Zylstra G. J., Wackett L. P. and Gibson D. T. 1989. "Trichloroethylene degradation by *Escherichia coli* containing the cloned *Pseudomonas putida* F1 toluene dioxygenase genes." *Appl. Environ. Microbiol.* 55:3162–3166.

Designing Metabolic Pathways for the Biodegradation of Halogenated Compounds

Lawrence P. Wackett

Department of Biochemistry and
Institute for Advanced Studies in Biological Process Technology
University of Minnesota
St. Paul, MN 55108, USA

BIODEGRADATION, EVOLUTION, AND HALOGENATED ORGANIC COMPOUNDS

Bioavailable organic matter on the earth is largely recycled by microorganisms. This is accomplished by the action of catabolic enzymes. This diversity of enzymes, biosynthesized for the metabolism of almost innumerable natural product organic molecules, has arisen over millions of years. The evolutionary forces of mutation, gene transfer, and selection produce and maintain catabolic enzymes in microbial populations.

When toxic chemicals are not readily metabolized by microbial enzymes, they persist in the environment and are then called pollutants. The failure of a compound to be metabolized is a function of its great chemical stability and its nonresemblance to other metabolizable compounds. One class of chemicals that combines both of these features is certain halogenated organic compounds. For this reason and because of their toxicity to mammals, halogenated organic compounds comprise the largest single class of Environmental Protection Agency Priority Pollutants (Leisinger, 1983).

There are known to be many natural product organohalides; a recent compilation listed over 1500 (Gribble, 1992). Uniquely synthetic organohalides comprise a much larger set, and some present unique challenges to microbial metabolism. Some representative members of each set are

435

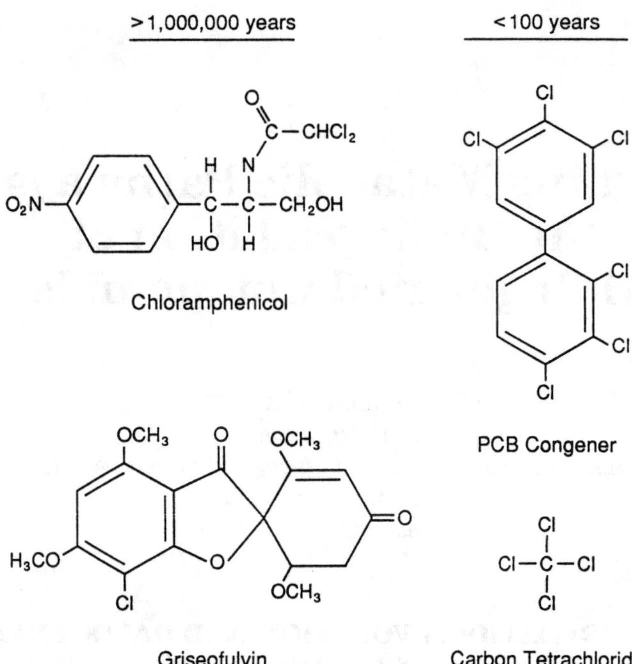

Figure 1. Halogenated compounds in the environment. Those on the left are natural products and have likely been biosynthesized for millions of years. On the right are compounds of anthropogenic origin.

shown in Figure 1. In general, the halogenated natural products would be more readily metabolized by bacteria than the synthetic compounds shown in Figure 1. This is likely due to the longer time that the natural products have been found in nature. Additionally, some synthetic compounds were highly chlorinated for the purpose of resisting biodegradation and, hence, persisting in the environment. For example, herbicides and insecticides must kill target organisms and remain in the environment to exert that effect over weeks and months. After some period of usage, some pesticides have shown toxic effects on mammals. Then environmental persistence becomes a negative feature.

Many synthetic chlorinated compounds have been in the environment for only the last few decades, a relatively short period on an evolutionary time scale. This is likely one of the reasons that few microorganisms have been isolated in pure culture to metabolize polyhalogenated compounds. One exception is a *Flavobacterium* sp. that grows on pentachlorophenol as a sole source of carbon and energy. Details of this metabolism have been elucidated by the laboratories of Ron Crawford and Cindy Orser

(Steiert et al., 1987; Orser et al., 1993a, 1993b). These studies indicate that multiple mechanisms are required to cleave all the carbon-halogen bonds of a highly chlorinated compound. Thus, more than one type of enzyme must be expressed simultaneously to create a new metabolic pathway to take a complex, highly chlorinated starting compound to simple molecules that can enter preexisting metabolic pathways. Each individual enzyme must evolve, and the new genes must be properly regulated to make the new dehalogenating metabolic pathway function properly. Thus, it might take decades for the right series of events to construct a new pathway for the metabolism of a highly chlorinated polychlorinated biphenyl (PCB), for example.

THE POTENTIAL OF BIOTECHNOLOGY

New metabolic pathways can also be created in the laboratory using biotechnological methods. These constructions start with genes that are available and encode for proteins of known function. The knowledge base that these methods use is currently extensive and is growing at a very rapid rate. As an example, it is estimated that, by the year 2000, over 1 million gene sequences will be known. The number of three-dimensional protein structures that will be elucidated by X-ray crystallography or NMR cannot keep pace with the explosion in information at the gene level. However, it must be considered that a large group of investigators are working to understand how a protein amino acid sequence determines its three-dimensional structure. The solution to this problem will increase the knowledge base by orders of magnitude. Over 1 million protein structures will then become known.

A greater knowledge of protein structure also presages a greater ability to change protein function in known ways. Genetic manipulations are now in widespread use to change genes selectively or in large-scale, random fashion to produce new proteins (Janssen and Schanstra, 1994). Groups are engineering proteins for new substrate specificities and, in some cases, new catalytic functions. A striking example of the latter is the widespread interest in catalytic antibodies (Benkovik, 1992). While antibodies are made by nature as binding proteins, insights into enzyme catalytic mechanisms have made it possible to recruit these proteins for use as catalysts.

Protein engineering and related fields of study have found notable applications in the production of industrial enzymes and pharmaceutical products. An example of the former is the reengineering of bacterial proteases used in laundry detergents, for enhanced stability at the elevated temperatures used in washing machines (Pantoliano et al.,

1987). Recently, biotechnological methods have been applied to enzymes that are important in hazardous waste metabolism (Timmis et al., 1994). They are applied in two general ways. One is to learn more about the genes and enzymes operative in naturally evolved biodegradation. The other is to design new metabolic pathways and engineer enzymes for altered specificity.

The present chapter pertains to metabolic pathway engineering but with a new variation. One enzyme in the engineered pathway was used to catalyze a reaction that is completely different from what it was designed to do by natural evolution. The other enzyme was made to metabolize substrates with structures that appear quite different from those that it is well known to metabolize. As the compiled information on gene structure and enzyme structure and mechanism increases, it will become increasingly possible to use biodegradative enzymes in creative new ways to metabolize chemicals that are poorly degraded by natural microorganisms. The studies described below address the metabolism of environmentally persistent polyhalogenated compounds by engineering new metabolic pathways in the laboratory.

METABOLISM OF POLYHALOGENATED COMPOUNDS BY A GENETICALLY ENGINEERED METABOLIC PATHWAY

Basic Knowledge of Gene Structures and Enzyme Mechanisms Related to Biodehalogenation

The engineering sciences build their disciplines on the basic sciences. For example, every chemical engineer receives training in thermodynamics, chemical kinetics, and chemical structure and bonding. This knowledge is essential for developing a sound applied science. This idea is also true for genetic and metabolic engineering. These new fields were built on a solid foundation of knowledge on gene structure and enzyme structure/mechanism, respectively.

To successfully engineer pathways for metabolism of halogenated compounds, it must be first considered how nature has evolved pathways to metabolize selected haloorganic compounds. Several crucial lessons emerge from this study. First, the metabolism of polyhalogenated chemicals typically requires more than one type of enzymatic mechanism to completely remove all halogen substituents (Wackett, 1991). Second, nature evolves genes encoding new enzymes by recruiting them from other genes that function in encoding enzymes in completely different metabolic pathways. An example of the latter concept is the recruitment of genes involved in fatty acid metabolism for new functions in catalyzing hydrolytic dechlorination of *p*-chlorobenzoic acid (Babbitt et al., 1992) (Figure 2). Bacterial cells oxidize fatty acids after first derivatizing them as thioesters using coenzyme A. In parallel fashion, *Pseudomonas* sp. CBS-3, which metabolizes *p*-chlorobenzoic acid, first converts that compound to a thioester coenzyme A derivative. This increases the reactivity of the *para*-chlorine substituent to nucleophilic displacement reactions, thus explaining the metabolic logic of recruiting a step from fatty acid oxidative metabolism. Then, a hydrolytic reaction eliminates the chlorine substituent to make *p*-hydroxybenzoic acid, an intermediate metabolized by many *Pseudomonas* strains. This step also recruits a gene from fatty acid metabolism. A hydrolase, similar to one that adds water to an aliphatic double bond of a fatty acid ester, adds water to an aromatic ring to displace a chlorine substituent at the *para* carbon atom.

In the example above and other instances, the gene recruited must encode an enzyme with a catalytic function that is versatile enough to work in a new way. In this context, we and others have explored biological metallocofactors that undergo oxidation/reduction for the potential to catalyze reductive dechlorination reactions. This reaction is formally the input of two electrons into the carbon-halogen bond to make a corresponding C—H bond or an alkene in the case of the substrate having a readily eliminatable halogen on the adjacent carbon (Figure 3).

Reductive dehalogenation reactions are very important in the dehalogenation of polychlorinated compounds, and few pure cultures of bacteria have been isolated from nature that catalyze these reactions (Wackett et al., 1994; Holliger, 1992). Thus, the usual course of isolating single bacterial cultures, purifying their reductive dehalogenases, and delineating the mechanism of the enzyme in vitro has not been readily accomplished. However, several redox active metallomacrocycles that bacteria biosynthesize can be purified and studied in isolation. Those shown in Figure 4 are demonstrated to catalyze reductive dehalogenation reactions in vitro in the presence of a suitable electron donor system. These cofactors are involved in diverse biochemical processes and may or may not be involved in environmental reductive dehalogenation reaction. However, they can be studied to give insights into the chemistry of the reactions and yield information with the potential for use in biotechnological applications.

The most versatile and well-studied of the metallomacrocycles in Figure 4 are the iron porphoryins, or heme, cofactor. Proteins bind heme for many functions (Table 1). These include electron transfer reactions, oxygenation reactions, sulfuroxyanion reduction, and oxygen reduction (Wackett et al., 1989a). The latter biological functions, reductive chemis-

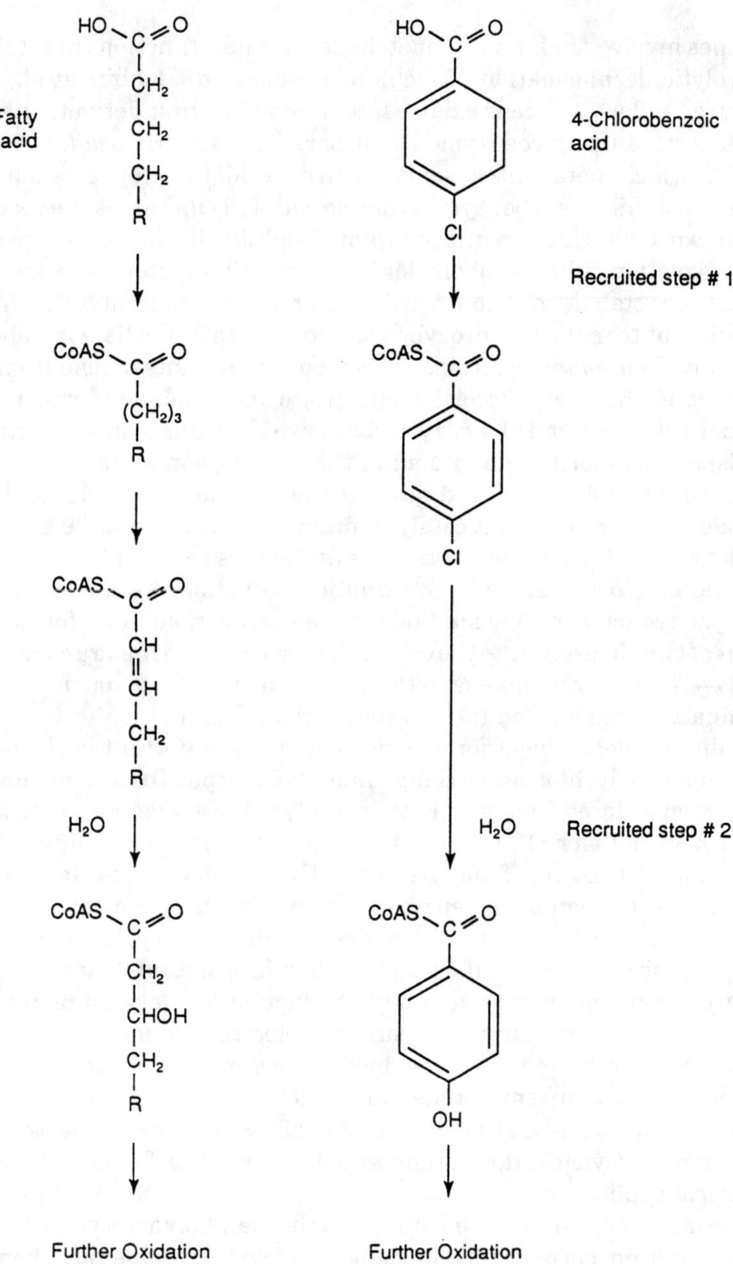

Figure 2. Parallels between metabolism of fatty acids and *p*-chlorobenzoic acid.

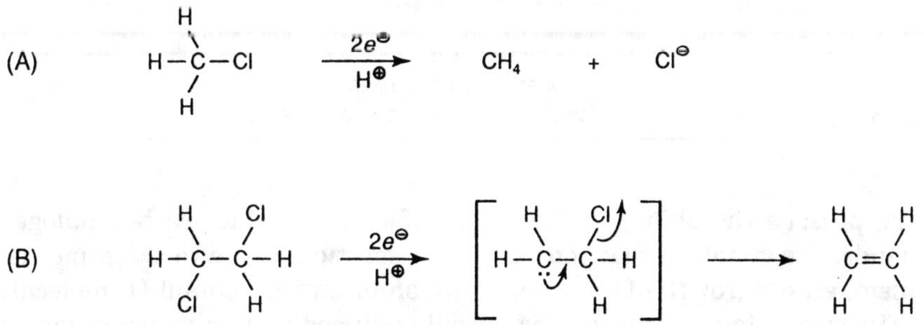
(A)
H
H—C—Cl
H
2e⊖
H⊕
CH4
+
Cl⊖
(B)
H Cl
H—C—C—H
Cl H
2e⊖
H⊕
H Cl
H—C—C—H
H
H H
C=C
H H

Figure 3. General mechanism of reductive dehalogenation of chloroalkanes leading to alkane or alkene products.

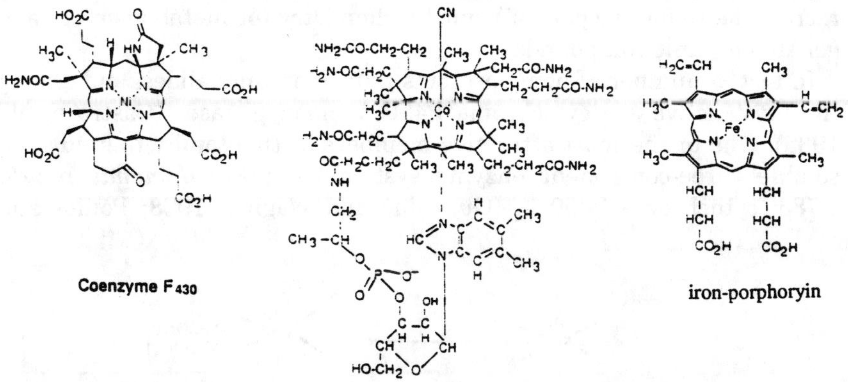
Coenzyme F430
Vitamin B12
iron-porphoryin

Figure 4. Redox active metallomacrocycles biosynthesized by bacteria.

Table 1. Examples of important heme proteins in bacteria.

Electron transfer cytochromes
Cytochrome oxidase
Sulfite reductase
Hydroxylamine oxidoreductase
Catalase
p-Cresol hydroxylase
L-tryptophan 2,3-dioxygenase
Cytochrome P450 monooxygenases

try, presage the ability of the heme cofactor to reduce carbon-halogen bonds. Oxygenation reactions are, on inspection of the reaction mechanism, an outgrowth of enzyme reduction of an iron-bound O_2 molecule (Figure 5). Molecular oxygen is initially reduced and, in monooxygenase reactions, ultimately cleaved to release the reduced product water and generate a reactive iron-bound monoatomic oxygen species. The oxygen atom reacts with organic substrates bound by the enzyme to make oxygenated products. But the potential exists, in the absence of O_2, to recruit the reductive part of heme biochemistry for metabolism of halogenated organic compounds.

In fact, a number of heme proteins reduce organohalides, and among the most active are cytochrome P450 monooxygenases (Castro et al., 1985). One of the most attractive members of this family to study is a soluble three-component enzyme system from *Pseudomonas putida* G786, cytochrome P450$_{CAM}$ (Gunsalus and Wagner, 1978; Poulos and

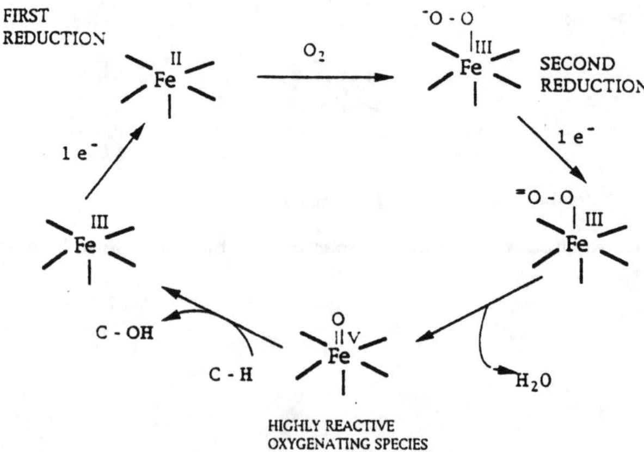

Figure 5. Oxygen reduction and activation by a protein bound metal center.

Raag, 1992). A fully active system in vitro requires NADH, a flavoprotein oxidoreductase, an iron-sulfur protein denoted putidaredoxin, and the heme-containing cytochrome P450 protein (Figure 6). It is called cytochrome P450 because a sulfur ligand to the iron gives the protein an absorbance maximum near 450 nm when the Fe(II)-atom binds carbon monoxide. The attractiveness of cytochrome $P450_{CAM}$ as a model protein for catalyzing reductive dehalogenation derives from the available extensive knowledge that dates back to 1959 (Bradshaw et al., 1959). Irwin Gunsalus and his coworkers have extensively characterized this enzyme system, making it the paradigm for understanding monooxygenase reaction mechanisms. The gene encoding cytochrome $P450_{CAM}$ has been cloned and sequenced (Unger et al., 1986), and numerous site-directed mutants have been prepared by Steven Sligar and his coworkers (Sligar et al., 1991). The X-ray structure of the protein was solved by Thomas Poulos and collaborators in 1985 (Poulos et al., 1985). The X-ray coordinates of the protein have been used in molecular dynamics simulations to better understand how substrates are bound and react with the protein (Paulsen and Ornstein, 1992; Paulsen et al., 1993).

The wealth of available information on cytochrome $P450_{CAM}$ can be used for studying and manipulating its reactions with halogenated organic compounds. The enzyme can act as an oxygenase or a reductive dehalogenase, and those reactivities are modulated by the oxygen tension (Figure 7). In either case, substrate binding to form an enzyme-substrate complex is the prerequisite for reaction. Halogenated substrate binding by cytochrome $P450_{CAM}$ was recently investigated (Li and Wackett, 1993) and indicated that halogenated substrates can be tightly bound (dissociation constants, K_D, as low as 0.7 μM). This and other studies indicate that the protein is reasonably nondiscriminating; 16 substrates have been shown to be reduced at this date (Table 2).

The products of reductive dehalogenation of polyhalogenated compounds still contain halogen substituents and must be further metabolized, likely by other mechanisms, for complete detoxification. A further step for metabolizing chloroalkene intermediates (Figure 3) was sug-

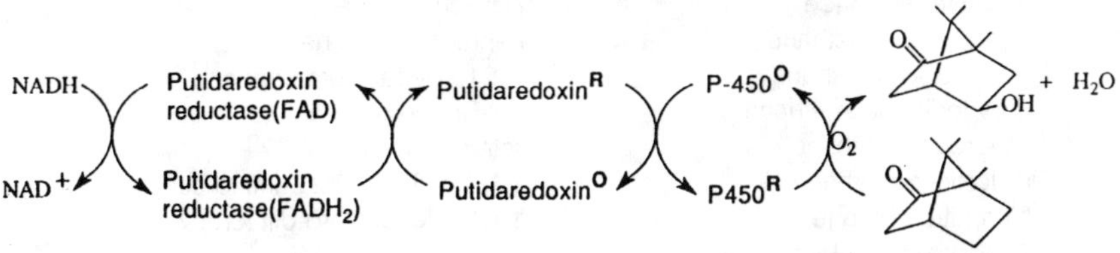

NADH — Putidaredoxin reductase(FAD) — Putidaredoxin^R — P-450° — OH + H_2O

NAD$^+$ — Putidaredoxin reductase($FADH_2$) — Putidaredoxin^O — P450^R — O_2

Figure 6. The cytochrome $P450_{CAM}$ monooxygenase system and reaction catalyzed.

$$C - H \xrightarrow[2\ e^-,\ 2\ H^+]{O_2} C - OH + H_2O$$

$$C - Cl \xrightarrow[2\ e^-,\ 2\ H^+]{} C - H + HCl\text{-}$$

Figure 7. The heme protein, cytochrome P450$_{CAM}$, catalyzes oxygenation (top) and reductive dehalogenation (bottom) reactions, modulated by substrate and oxygen tension.

gested by other studies in our laboratory on chloroalkene oxidation (Li and Wackett, 1992). This indicated the reactions might be coupled, and a representative reaction sequence is shown in Figure 8. In fact, a long list of oxygenases have now been shown to oxidize trichloroethylene (TCE) (Table 3). These studies have been driven by the widespread prevalence of TCE as a groundwater pollutant. Despite extensive study, no bacteria have been conclusively established to grow on TCE as a carbon and energy source. So the oxygenases listed in Table 3 represent broad-specificity oxygenases that were previously studied with other substrates but fortuitously oxidize TCE.

Toluene dioxygenase (Table 3) is a multicomponent enzyme system (Figure 9) that oxidizes over 40 different substrates (Gibson et al., 1990). In vitro activity requires NADH, a flavoprotein oxidoreductase, an iron-sulfur protein, and a nonheme iron protein that dioxygenates toluene. The same complete system is required for TCE oxidation (Li and Wackett, 1992). Expression of toluene dioxygenase in vivo requires the expression of four genes denoted *todABC₁C₂*. Expression of these genes in *E. coli* confers on the recombinant strain the ability to oxidize TCE (Zylstra et al., 1989). Elucidation of the products from TCE oxygenation required the use of [^{14}C]-TCE and purified protein components in vitro. The products were determined to be simple organic acids, and these could be analyzed by high-pressure liquid chromatography (HPLC) fitted with an

Table 2. List of known substrates of cytochrome P450$_{CAM}$ undergoing reductive dehalogation.

Carbon tetrachloride	Hexachloroethane
Bromotrichloromethane	Pentachloroethane
Fluorotrichloromethane	1,1,1,2-Tetrachloroethane
Dibromodichloromethane	1,2-Dibromotetrachloroethane
Trichloronitromethane	Tetrabromoethane
Dichloronitromethane	1,1,1-Trichloro-2,2,2-trifluoroethane
Chloronitromethane	1,1,1,2-Tetrachloro-difluoroethane
1,2-Dibromo-3-chloropropane	

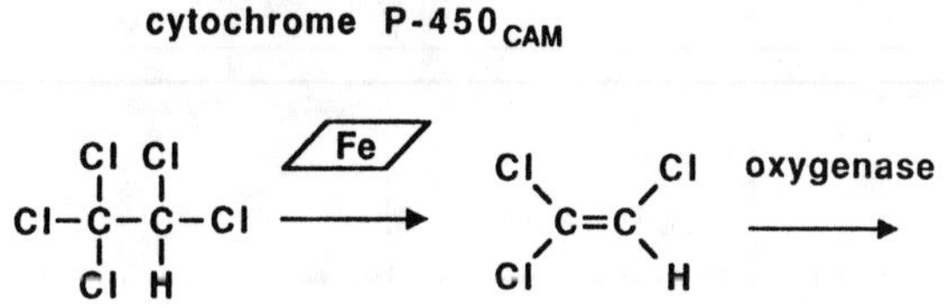

cytochrome P-450$_{CAM}$

Figure 8. Coupled reductive and oxidative dehalogenation reactions with the starting compound pentachloroethane.

organic acids column (Figure 10). The products are glyoxylate and formate. Both compounds are nontoxic and are readily metabolized by many bacteria.

Recruiting Enzymes to Design New Pathways for Metabolizing Polyhalogenated Compounds

The basic studies described above set the stage for using recombinant DNA methodologies to express cytochrome P450 and toluene dioxygenase in the same cell and create a metabolic pathway for complete dehalogenation of pentachloroethane and other compounds (Figure 11). We searched unsuccessfully in nature for such a strain (Wackett et al., 1994). The general strategy was to transfer the *todABC$_1$C$_2$* genes into *Pseudomonas*

Table 3. *Organisms and their oxygenases implicated in the biological oxidation of trichloroethylene.*

Enzyme	Organism	Reference
Dioxygenase		
Toluene dioxygenase	*Pseudomonas putida* F1	Wackett and Gibson, 1988
Monooxygenases		
Soluble methane mono-oxygenase	*Methylosinus trichosporium* OB3b (and other methano-trophs)	Oldenhuis et al., 1989 Tsien et al., 1989
Particulate methane mono-oxygenase	*Methylocystis parvus* OBBP	DiSpirito et al., 1992
Toluene 2-monooxygenase	*P. cepacia* G4	Shields et al., 1991
Toluene 4-monooxygenase	*P. mendocina*	Winter et al., 1989
Phenol hydroxylase	*Alcaligenes eutrophus* JMP 134	Harker and Kim, 1990
Ammonia monooxygenase	*Nitrosomonas europea*	Arciero et al., 1989 Rasche et al., 1991
Isoprene monooxygenase	*Rhodococcus erythropolis*	Ewers et al., 1990
Propane monooxygenase	*Mycobacterium vacea* JOB 5	Wackett et al., 1989
Alkene monooxygenase	*Xanthobacter* Py2	Ensign et al., 1992

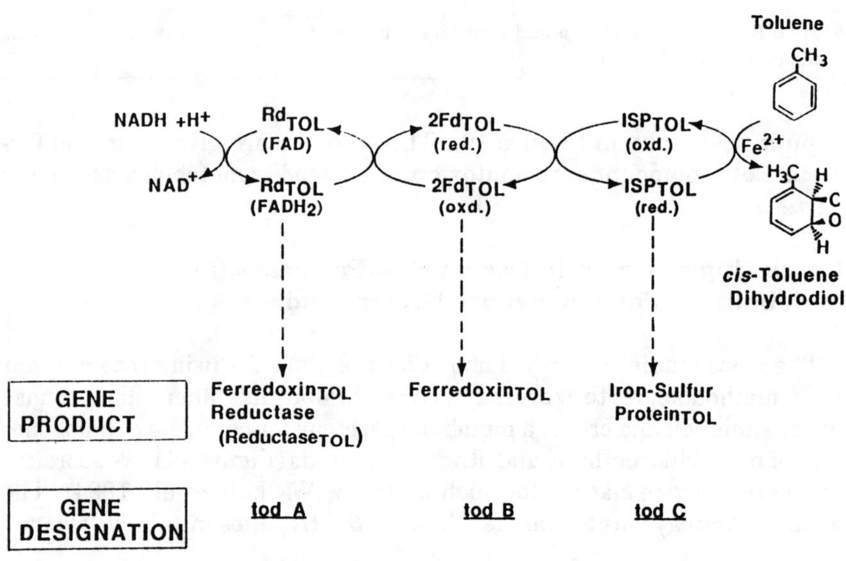

Figure 9. The toluene dioxygenase enzyme system—proteins and genes.

Figure 10. Products of trichloroethylene oxidation by toluene dioxygenase determined by high-pressure liquid chromatography.

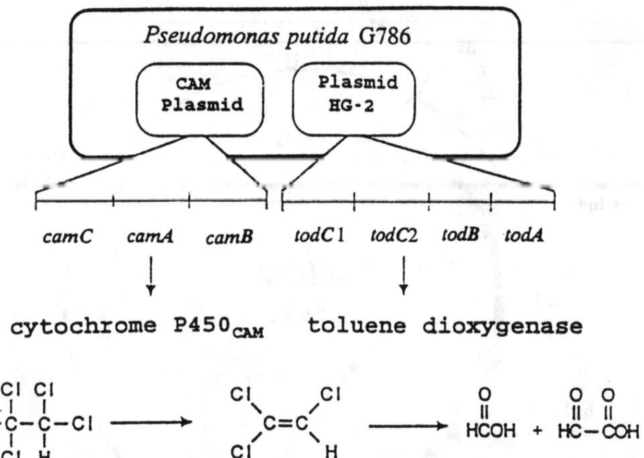

Figure 11. Construction of a recombinant bacterium containing *cam* and *tod* genes and expressing both enzyme activities against pentachloroethane as the starting substrate.

putida G786 that already contained the CAM plasmid encoding cyto-chrome $P450_{CAM}$ (Zylstra et al., 1988). The *tod* genes had been cloned by David Gibson and coworkers (1990) and were provided to us on an *E. coli* plasmid that would not replicate in *Pseudomonas* species. After transfer to a suitable broad host range vector, the recombinant plasmid (pHG-1) was put into *P. putida* G786 by triparental mating. The resultant strain expressed cytochrome $P450_{CAM}$ and toluene dioxygenase but was geneti-cally unstable. The recombinant strain constitutively expressed high levels of toluene dioxygenase during growth, and this was negatively selected against. To overcome this problem, a regulatory element ($lacI^Q$) was put onto the same plasmid containing toluene dioxygenase. The resulting plasmid, pHG-2 (Figure 12), allowed metabolic control of tolu-ene dioxygenase expression with the gratuitous inducer β-isopropyl-thiogalactoside (IPTG). Plasmid pHG-2 was mated into *P. putida* G786 by triparental mating. Toluene dioxygenase by *P. putida* G786 was ex-pressed at a low level in the absence of inducer and at a high level when IPTG was added after significant growth of the cells had occurred.

Pentachloroethane was used as a model substrate for determining how well the recombinant metabolic pathway would function in vivo (Wackett et al., 1994). To follow the metabolism in discrete steps, it was decided to run the first reaction under anaerobic conditions. This is optimum for the cytochrome $P450_{CAM}$-dependent reductive dehalogenation and pre-cludes oxygenation of TCE by toluene dioxygenase. This allowed a dem-onstration that the recombinant strain stoichiometrically transformed

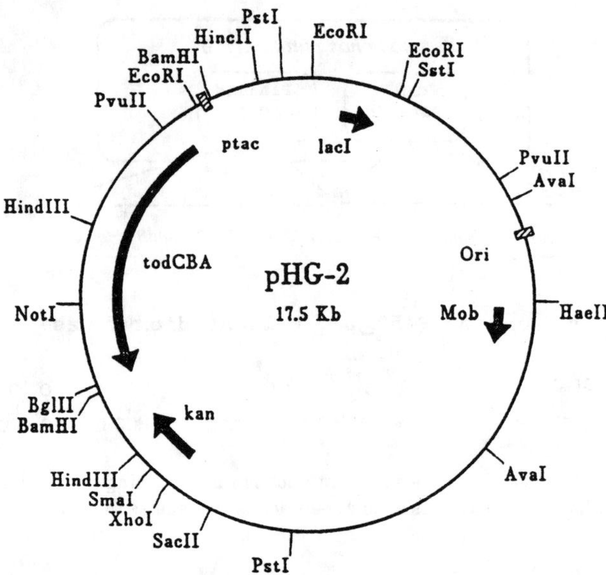

Figure 12. Plasmid pHG-2.

pentachloroethane to TCE. Then oxygen was admitted to the culture, and the accumulated TCE was further metabolized. In separate experiments, [^{14}C]-TCE was added and shown to be metabolized to CO_2 and glyoxylate in a 2:1 ratio. Purified toluene dioxygenase in vitro oxidizes TCE to formate and glyoxylate in a 2:1 ratio so we presume the cells oxidized formate to CO_2.

In nature, many catabolic enzymes have broad substrate specificity, and this is important for the biodegradation of multiple structurally related compounds. For example, toluene dioxygenase oxidizes dozens of singly and doubly substituted benzenoid compounds. It catalyzes the first metabolic step during the growth of *Pseudomonas putida* F1 on toluene, ethylbenzene, *n*-propylbenzene, and phenol. In our pathway designed to metabolize polyhalogenated compounds, the recruited catabolic enzymes also show broad specificity with halogenated aliphatic compounds. This is advantageous for the biodegradation of structural analogous halogenated compounds, some of which might be present at the same site in the environment. *P. putida* G786 (pHG-2) is known to metabolize several compounds, in addition to pentachloroethane, by the sequential action of toluene dioxygenase and cytochrome P450$_{CAM}$ (Hur et al., 1994) (Table 4). A less chlorinated compound, 1,1,1,2-tetrachloroethane, is metabolized by cytochrome P450$_{CAM}$ to yield 1,1-dichloroethene. Previously, 1,1-dichloroethene was shown to be oxidized by toluene dioxygenase, but the product(s) of the reaction was not identified (Wackett and Gibson, 1988).

Perhaps of greatest interest, *P. putida* G786 (pHG-2) metabolizes certain chlorofluorocarbons (Hur et al., 1994) (Table 4). This class of compounds has attracted widespread attention due to their implication in stratospheric ozone depletion (Molina and Rowland, 1974). Oxalate, one of the end products formed via metabolism of two different chlorofluoroethanes, is known to be oxidized to CO_2 by soil *Pseudomonas* species. Additionally, a polybrominated compound is metabolized by *P. putida* G786 (pHG-2).

It is also important to point out the compounds that *P. putida* G786 (pHG-2) fails to metabolize. Cytochrome P450$_{CAM}$ is not detectably active in the reductive dechlorination of 1,1,2,2-tetrachloroethane and 1,1,1-trichloroethane (Logan et al., 1993). Under aerobic conditions, the compounds are oxidized to dichloroacetate and 1,1,1-trichloroethanol, respectively, but the oxygenation reactions are largely uncoupled from NADH oxidation (Lefever and Wackett, 1994). Thus, these reactions would largely deplete NADH in vivo and be detrimental to the cell. Also, the end products may not be further metabolized by *P. putida* G786 (pHG-2). Cytochrome P450$_{CAM}$ is poorly, if at all, active in the reductive dehalogenation of aromatic compounds (Li and Wackett, unpublished data). It is not a particularly good catalyst for the oxygenation of aromatic compounds (Fruetal et al., 1992). In another example, cytochrome P450$_{CAM}$ transforms 1,1,2-trichloroethane to several products, including vinyl chloride

Table 4. *Compounds degraded by consecutive reactions catalyzed by cytochrome P450$_{CAM}$ and toluene dioxygenase.*

Substrate	Intermediate(s)	End Product
Pentachloroethane	Trichloroethene	Formate, Glyoxylate
1,1,1,2-Tetrachloroethane	1,1-Dichloroethene	ND[a]
1,1,1-Trichloro-2,2,2-trifluoroethane	1,1-Dichloro-2,2-difluoroethene 1,1-Dichloro-2,2,2-trifluoroethane	Oxalate ND
1,1,1,2-Tetrachloro-2,2-difluoroethane	1,1-Dichloro-2,2-difluoroethene 1,1,2-Trichloro-2,2-difluoroethane	Oxalate ND
Tetrabromoethane	*cis*-1,2-dibromoethene *trans*-1,2-dibromoethene	ND

[a]ND = not determined.

(Castro and Belser, 1990). Vinyl chloride is not oxidized by toluene dioxygenase (Wackett and Gibson, 1988), and it is known to be a mutagen and a human carcinogen (Maltoni and Lefemine, 1974).

Further Developments in Enzyme and Metabolic Engineering

The shortcomings of *P. putida* G786 (pHG-2) are clear and could potentially be addressed by further metabolic and protein engineering. Some of these shortcomings are listed immediately below and are discussed in subsequent paragraphs.

1. It is not practical to cycle between anaerobic and aerobic conditions on a large-scale system, and *P. putida* G786 (pHG-2) only functions optimally under these conditions.
2. Genetic controls on cytochrome $P450_{CAM}$ and toluene dioxygenase require impractical induction regimes using camphor and IPTG, respectively.
3. Some environmentally important compounds, such as 1,1,1-trichloroethane (discussed above), are not appreciably metabolized by *P. putida* G786 (pHG-2).
4. The metabolic pathways described above do not provide sufficient carbon and energy to allow the growth of the recombinant strain, necessitating the addition of other sustaining carbon sources.

Some possible solutions to these problems are discussed below.

The effect of oxygen tension on metabolism has previously been considered (Wackett et al., 1994). At subambient oxygen concentrations, both reductive and oxygenative dechlorination reactions are operative. However, reductive dechlorination is significantly inhibited above 5% O_2, and oxygenative dechlorination falls off significantly below 5% O_2. Thus, both reactions occur at 5% O_2, but it could prove difficult to poise the system at this degree of aerobicity. Improvements could be accomplished by decreasing the oxygen sensitivity of reductive dechlorination by cytochrome $P450_{CAM}$ or enhancing oxygenative dechlorination at low O_2 tensions. The former might be realized by protein engineering of cytochrome $P450_{CAM}$. The latter goal could be solved by using oxygenases with a lower K_m for O_2. As an example of the latter, *Pseudomonas* strains have been identified that oxygenate aromatic substrates under microaerophilic conditions, and some of these are known to oxidize trichloroethylene (Kukor et al., 1994).

Improved genetic control of cytochrome $P450_{CAM}$ and toluene dioxygenase in *P. putida* G786 (pHG-2) could be designed in several ways. Few well-characterized regulatory systems are known to be responsive to specific halogenated compounds (LaRoche and Leisinger, 1991) and those are not matched to the enzymatic dehalogenation specificities displayed by *P. putida* G786 (pHG-2). In this context, induction by suitable environmental conditions might be more desirable (Matin, 1994). Constitutive expression would be metabolically wasteful and make for selective pressure against maintenance of the cytochrome $P450_{CAM}$ and toluene dioxygenase genes.

The failure of the enzymes to metabolize compounds of particular environmental concern could be addressed by protein engineering to alter the substrate specificity of the enzymes. This is not trivial to accomplish, but successes in directed protein engineering to alter specificity are increasing in frequency (Janssen and Schanstra, 1994). It is typically accomplished with enzymes, for which a knowledge of X-ray structure and extensive mechanistic details are available. Cytochrome $P450_{CAM}$ falls into this category. The X-ray structure has been solved for native enzyme, camphor-bound enzyme, and with various alternative substrate bound. Structures have been solved for several active site mutant cytochrome $P450_{CAM}$ proteins. Additionally, molecular dynamics simulations have been performed for numerous substrates (Paulsen et al., 1993), including halogenated aliphatic compounds (Paulsen and Ornstein, 1994). With this knowledge base in hand, reasonable efforts can be made to redesign cytochrome $P450_{CAM}$ for more desirable substrate specificity. It should also be considered that there are many other ongoing efforts to modify protein reactivities, use antibody molecules as catalysts, and solve the protein folding problem. Progress in these areas will positively impact efforts to engineer catabolic enzymes for enhanced biodegradation of environmental pollutants.

Reengineered enzymes will operate most efficiently in vivo when incorporated into well-designed metabolic pathways that confer a survival advantage on the host bacterium. Such a system also offers the potential for further improvements via additional rounds of natural mutation and selection. Successful selection for new variants can be accomplished using chlorinated substrates that serve as carbon and/or energy sources for the bacterium. The most highly chlorinated compounds such as carbon tetrachloride and hexachloroethane will not be suitable for purposes of growth selection. However, it is conceivable to substitute other functional substituents, such as methyl groups, in place of chlorine atoms to provide pollutant-mimicking substrates of sufficient metabolic potential energy to support an organism's growth. Ideally, the metabolic intermediates produced should funnel into constitutively expressed trunk metabolic pathways. Selection for improved metabolism would be accomplished by scoring for more rapid bacterial growth using standard methods of batch or continuous culture. In one recent example, mutant enzymes of 1,2-dihaloalkane dehalogenase with altered substrate specificity were obtained by selection for bacterial growth on 1-chlorohexane (Pries et al., 1994).

The engineering of metabolic pathways for novel catabolism has been described most notably with aromatic substrates. In one example, an organism able to grow on a wider range of substituted benzoic acids was obtained by genetic engineering (Rojo et al., 1987). Recently, mineralization of 2,4,6-trinitrotoluene (TNT) was accomplished by introducing the TOL plasmid genes into a *Pseudomonas* strain identified to remove the nitro substituents from TNT (Duque et al., 1993). The state of knowledge of bacterial aromatic ring metabolism is more advanced than the general knowledge of organohalide metabolism. However, the latter field is developing quickly. Also, recent advances in protein and genetic engineering presage rapid progress in metabolic engineering for the biodegradation of halogenated compounds.

ACKNOWLEDGEMENTS

I wish to thank the coworkers who contributed to the original work described here. Financial support was provided by National Institutes of Health GM 41235 and Environmental Protection Agency cooperative agreement CR820771-01-0. I thank Ms. Bonnie Allen for assistance in preparing the manuscript.

REFERENCES

Arciero, D., T. Vannelli, M. Logan and A. B. Hooper (1989) Degradation of trichloroethylene by the ammonia-oxidizing bacterium *Nitrosomonas europeae. Biochem. Biophys. Res. Commun.* 159:640–643.

Babbitt, P. C., G. L. Kenyon, B. M. Martin, H. Charest, M. Sylvestre, J. D. Scholten, K.-H. Chang, P.-H. Liang and D. Dunaway-Marino (1992) Ancestry of the 4-chlorobenzoate dehalogenase: Analysis of amino acid sequence identities among families of acyl: Adenyl ligases, enoyl-CoA hydratases/isomerases, and acyl-CoA thioesterases. *Biochemistry* 31:5594–5604.

Benkovic, S. J. (1992) Catalytic antibodies. *Ann. Rev. Biochem.* 61:29–54.

Bradshaw, W. H., H. E. Conrad, E. J. Corey, I. C. Gunsalus and D. Lednicer (1959) Microbiological degradation of (+)-camphor. *J. Am. Chem. Soc.* 81:5507.

Castro, C. E. and N. O. Belser (1990) Biodehalogenation: Oxidative and reductive metabolism of 1,1,2-trichloroethane by *Pseudomonas putida*—Biogeneration of vinyl chloride. *Environ. Toxicol. Chem.* 9:707–714.

Castro, C. E., R. S. Wade and N. O. Belser (1985) Biodehalogenation: Reactions of cytochrome P450 with polyhalomethanes. *Biochemistry* 24:204–210.

DiSpirito, A. A., J. Gulledge, A. K. Schiemke, J. C. Murrell, M. E. Lidstrom and C. L. Crema. (1992) Trichloroethylene oxidation by the membrane-associated methane monooxygenase in Type I, Type II, and Type X methanotrophs. *Biodegradation* 2:151–164.

Duque, E., A. Haidor, F. Godsy and J. L. Ramos (1993) Construction of a *Pseudomonas* hybrid strain that mineralizes 2,4,6-trinitrotoluene. *Appl. Environ. Miorobiol.* 175:2278–83.

Ensign, S. A., M. R. Hyman and D. J. Arp (1992) Cometabolic degradation of chlorinated alkenes by alkene monooxygenase in a propylene-grown Xanthobacter strain. *Appl. Environ. Microbiol.* 58:3038–3046.

Ewers, J., D. Freier-Schröder and H-J. Knackmuss (1990) Selection of trichloroethene (TCE) degrading bacteria that resist inactivation by TCE. *Arch. Microbiol.* 154:410–413.

Fruetel, J. A., J. R. Collins, D. L. Camper, G. H. Loew and P. R. Ortiz de Montellano (1992) Calculated and experimental absolute stereochemistry of the styrene and β-methylsytrene epoxides formed by cytochrome $P450_{CAM}$. *J. Am. Chem. Soc.* 114:6987–6993.

Gibson, D. T., G. J. Zylstra and S. Chauhan (1990) Biotransformations catalyzed by toluene dioxygenase from *Pseudomonas putida* F1. In S. Silver, A. M. Chakrabarty, B. Iglewski and S. Kaplan, eds. Pseudomonas: *Biotransformation, Pathogenesis and Emerging Biotechnology*. ASM Press, Washington, D.C., pp. 121–132.

Gribble, G. W. (1992) Naturally occurring organohalogen compounds—A survey. *J. Natl. Prod.* 55:1353–1395.

Gunsalus, I. C. and G. C. Wagner (1978) Bacterial $P450_{CAM}$ methylene monooxygenase components: Cytochrome *m*, putidaredoxin and putidaredoxin reductase. ed. S. Fleischer and L. Packer in *Methods in Enzymology, Vol. LII*, Academic Press, New York, pp. 166–188.

Harker, A. R. and Y. Kim (1990) Trichloroethylene degradation by two independent aromatic-degrading pathways in *Alcaligenes eutrophus* JMP134. *Appl. Environ. Microbiol.* 56:1179–1181.

Holliger, C. (1992) Reductive dehalogenation by anaerobic bacteria. Ph.D. thesis, Wageningen University, Netherlands.

Hur, H-G., M. J. Sadowksy and L. P. Wackett (1994) Metabolism of chlorofluorocarbons and polybrominated compounds by *Pseudomonas putida* G-786 (pHG-2) via an engineered metabolic pathway. *Appl. Environ. Microbiol.*

Janssen, D. B. and J. P. Schanstra (1994) Engineering proteins for environmental applications. *Curr. Op. Biotech.* 5:253–259.

Kukor, J. J., M. D. Mikesell and R. H. Olson (1994) Sequence analysis of catechol 2,3-dioxygenases functional under oxygen-limited conditions. *Abstr. Ann. Am. Soc. Microbiol. K-182*, p. 307, ASM Press, Washington, D.C.

LaRoche, S. D. and T. Leisinger (1991) Identification of dcmR, the regulatory gene governing expression of dichloromethane dehalogenase in *Methylobacterium* sp. strain DM4. *J. Bacteriol.* 173:6714–6721.

Lefever, M. R. and L. P. Wackett (1994) Oxidation of low molecular weight chloroalkanes by cytochrome P450$_{CAM}$. *Biochem. Biophys. Res. Comm.* 201:373–378.

Leisinger, T. (1983) Microorganisms and xenobiotic compounds. *Experientia* 39:1183–1191.

Li, S. and L. P. Wackett (1992) Trichloroethylene oxidation by toluene dioxygenase. *Biochem. Biophys. Res. Comm.* 185:443–451.

Li, S. and L. Wackett (1993) Reductive dehalogenation by cytochrome P450$_{CAM}$: Substrate binding and catalysis. *Biochemistry* 32:9355–9361.

Logan, M. S. P., L. M. Newman, C. A. Schanke and L. P. Wackett (1993) Co-substrate effects in reductive dehalogenation by *Pseudomonas putida* G786 expressing cytochrome P450$_{CAM}$. *Biodegradation* 4:39–50.

Maltoni, C. and G. Lefemine (1974) Carcinogenicity bioassays of vinylchloride: Research plan and early results. *Environ. Res.* 7:387–396.

Martin, A. (1994) Starvation promoters of *Escherichia coli*. Their function, regulation, and use in bioprocessing and bioremediation. *Ann. N.Y. Acad. Sci.* 721:277–291.

Molina, M. J. and F. S. Rowland (1974) Stratospheric sink for chlorofluoromethanes: Chlorine atom-catalyzed destruction of ozone. *Nature* 249: 810–812.

Oldenhuis, R., R. L Vink, J. M. Vink, D. B. Janssen and B. Witholt (1989) Degradation of chlorinated aliphatic hydrocarbons by *Methylosinus trichosporium* OB3b expressing soluble methane monooxygenase. *Appl. Environ. Microbiol.* 55:2819–2826.

Orser, C. S., C. C. Lange, L. Xun, T. C. Zahrt and B. J. Schneider (1993a) Cloning, sequence analysis, and expression of the *Flavobacterium* pentachlorophenol-4-monooxygenase gene in *Escherichia coli*. *J. Bacteriol.* 175:411–416.

Orser, C. S., J. Dutton, C. Lange, P. Jablonski, L. Xun and M. Hargis (1993b) Characterization of a *Flavobacterium* glutathione S-transferase gene involved in reductive dechlorination. *J. Bacteriol.* 175:2640–2644.

Pantoliano, M. W., R. C. Ladner, P. W. Bryan, M. L. Rollence, J. F. Wood and T. L. Poulos (1987) Protein engineering of subtilisin BPN': Enhanced stabilization through the introduction of two cysteines to form a disulfide bond. *Biochemistry* 26:2077–2082.

Paulsen, M. D. and R. L. Ornstein (1992) Predicting the product specificity and coupling of cytochrome P450$_{CAM}$. *J. Comput. Aided Mol. Des.* 6:449–460.

Paulsen, M. D. and R. L. Ornstein (1994) Active site mobility inhibits reductive dehalogenation of 1,1,1-trichloroethane by cytochrome P450$_{CAM}$. *J. Comput. Assis. Mol. Des.* 8:389–404.

Paulsen, M. D., D. Filipovic, S. G. Sligar and R. L. Ornstein (1993) Controlling the regiospecificity and coupling of cytochrome P450$_{CAM}$: T185F mutant increases coupling and abolishes 3-hydroxynorcamphor product. *Protein Sci.* 2:357–365.

Poulos, T. L. and R. Raag (1992) Cytochrome P450$_{CAM}$: Crystallography, oxygen activation, and electron transfer. *FASEB J.* 6:674–679.

Poulos, T. L., B. C. Finzel, I. C. Gunsalus, G. C. Wagner and J. Kraut (1985) The 2.6 Å crystal structure of *Pseudomonas putida* cytochrome P-450. *J. Biol. Chem.* 260:16122–16130.

Pries, F., A. J. vanden Wijngaard, R. Bos, M. Pentenga and D. B. Janssen (1994) The role of spontaneous cap domain mutations in haloalkane dehalogenase specificity and evolution. *J. Biol. Chem.* 269:17490–4.

Rasche, M. E., M. R. Hyman and D. J. Arp (1991) Factors limiting aliphatic chlorocarbon degradation by *Nitrosomonas europeae:* Cometabolic inactivation of ammonia monooxygenase and substrate specificity. *Appl. Environ. Microbiol.* 57:2986–2994.

Roberts, A. L., P. N. Sanborn and P. M. Gschwend (1992) Nucleophilic substitution reactions of dihalomethanes with the bisulfide ion HS$^-$. *Environ. Sci. Technol.* 26:2263–2274.

Rojo, E., D. H. Pieper, K.-H. Engesser, H.-J. Knackmuss and K. T. Timmis (1987) Assemblage of *ortho*-cleavage routes for simultaneous degradation of chloro- and methylaromatics. *Science* 238:1395–1398.

Shields, M. S., S. O. Montgomery, S. M. Cuskey, P. J. Chapman and P. H. Pritchard (1991) Mutants of *Pseudomonas cepacia* G4 defective in catabolism of aromatic compounds and trichloroethylene. *Appl. Environ. Microbiol.* 57:1935–1941.

Sligar, S. G., D. Fillipovic and P. S. Stayton (1991) Mutagenesis of cytochromes P450$_{CAM}$ and b$_5$. *Met. Enzymol.* 206:31–49.

Steiert, J. G., J. J. Pignatello and R. L. Crawford (1987) Degradation of chlorinated phenols by a pentachlorophenol-degrading bacterium. *Appl. Environ. Microbiol.* 53:907–910.

Timmis, K. N., R. J. Steffan and R. Unterman (1994) Designing microorganisms for the treatment of toxic wastes. *Ann. Rev. Microbiol.* 48:525–557.

Tsien, H-C., G. A. Brusseau, R. S. Hanson and L. P. Wackett (1989) Biodegradation of trichloroethylene by *Methylosinus trichosporium* OB3b. *Appl. Environ. Microbiol.* 55:3155–3161.

Unger, B. P., I. C. Gunsalus and S. C. Sligar (1986) Nucleotide sequence of the *Pseudomonas putida* cytochrome P450$_{CAM}$ gene and its expression in *Escherichia coli. J. Biol. Chem.* 261:1158–1163.

Wackett, L. P. (1991) Dehalogenation of organohalide pollutants by bacterial enzymes and coenzymes. In J. W. Kelly, T. O. Baldwin (eds.), *Applications of Enzyme Biotechnology,* New York, Plenum Press, pp. 191–200.

Wackett, L. P., G. A. Brusseau and R. S. Hanson (1989a) Survey of microbial oxygenases: Trichloroethylene degradation by propane-oxidizing bacteria. *Appl. Environ. Microbiol.* 55:2960–2964.

Wackett, L. P. and D. T. Gibson (1988) Degradation of trichloroethylene by toluene dioxygenase in whole cell studies with *Pseudomonas putida* F1. *Appl. Environ. Microbiol.* 54:1703–1708.

Wackett, L. P., W. H. Orme-Johnson and C. T. Walsh (1989b) Transition metal enzymes in bacterial metabolism, In Beveridge, T. J. and R. J. Doyle (eds.) *Metal Ions and Bacteria,* New York: John Wiley and Sons, pp. 165–206.

Wackett, L. P., M. J. Sadowsky, L. M. Newman, H-G. Hur and S. Li (1994) Metabolism of polyhalogenated compounds by a genetically engineered bacterium. *Nature* 368:627–629.

Winter, R. B., K-M. Yen and B. D. Ensley (1989) Efficient degradation of trichloroethylene by a recombinant *Escherichia coli. Biotechnology* 7:282–285.

Zylstra, G. J., W. R. McCombie, D. T. Gibson and B. A. Finette (1988) Toluene degradation by *Pseudomonas putida* F1: Genetic organization of the *tod* operon. *Appl. Environ. Microbiol.* 54:1498–1503.

Zylstra, G. J., L. P. Wackett and D. T. Gibson (1989) Trichloroethylene degradation by *Escherichia coli* containing the cloned *Pseudomonas putida* F1 toluene dioxygenase genes. *Appl. Environ. Microbiol.* 55:3162–3166.

Membrane Technology Fundamentals for Bioremediation

JOHN PELLEGRINO
National Institute of Standards and Technology
Thermophysics Division, 838.01
325 Broadway
Boulder, CO 80303, USA

SUBHAS K. SIKDAR
U.S. Environmental Protection Agency
National Risk Management Research Laboratory
26 W. Martin Luther King Drive
Cincinnati, OH 45268, USA

INTRODUCTION

Overview

Membrane processes are primarily used for separations. As a unit operation, they have several major attributes:

- They provide a well-defined, mass transfer surface area that is mostly independent of the operating conditions.
- Their surface area provides selective transport of specific components between two phases.
- Membrane process units are built of modules.
- As devices, membrane modules typically provide high surface area per unit volume (the highest being for hollow fiber forms) and are relatively easy to operate.
- The scaleup of membrane processes is usually linear with load.

The selective separating layer in the module is what is typically called "the membrane." Currently, membranes made from a broad spectrum of polymeric and inorganic compounds are commercially available. The fabrication and characterization of membrane materials are usually the domain of materials technologists. Within a membrane module, the membrane is often created from several layers that are built up. Each layer can be made from different materials and will often have very different structural or morphological characteristics from the other layers. The layering is usually necessary for both the manufacturing and the end-use requirements of the final module. There will be differences in mass transfer performance of membrane modules (both within a given manufacturer's products and between manufacturers) as a result of their choice of materials, manufacturing processes, and module design.

Once created, a membrane module is incorporated into a treatment train consisting of pumps, compressors, pretreatment processes, sensors, and controllers to perform a membrane process. The engineering performance of that process is usually defined by two notions: the productivity of a module and the degree of separation of key components. These performance criteria will depend on the actual operating conditions of the process, and there are design tools for estimating module performance from more fundamental data.

There are many membrane processes or unit operations. They are differentiated primarily on the basis of the driving force for mass transfer through the membrane, the predominant transport mechanism, and the phases that are present. Each membrane unit operation has its own specific nomenclature, engineering characteristics, and concerns, but it is also possible that a given membrane module can be used for different membrane unit operations. Table 1 provides a brief overview of the characteristics of major membrane unit operations.

There have been a number of developments in novel uses of membrane modules. One of these is what are called hybrid processes. In this case, membrane unit operations are used as part of a process design with other unit operations. This is an alternative to conventional design philosophy that would use either a membrane or another unit operation *alone* to accomplish the process objective. For example, the dehydration of organic solvents is done mostly by distillation. Membrane modules that can very selectively separate water from many organic liquids, including alcohols, are available. While it is tempting to consider completely replacing the distillation unit operation with one based on membranes, it is more advantageous to use both distillation and membrane unit operations in series. The hybrid can successfully increase the capacity of existing plants and improve product quality. A hybrid process is

Table 1. Summary overview of characteristics of membrane unit operations.

Unit Operation	Reverse Osmosis	Nanofiltration	Ultrafiltration
Phases of feed/permeate	Liquid/liquid	Liquid/liquid	Liquid/liquid
General purpose (illustrative application)	Produce pure water as permeate and/or concentrated solute in retentate (desalination of sea or brackish water)	Organic solute and/or salt fractionation (removal of phenol from groundwater)	Size-based fractionation of macromolecules; concentration of large molecules (recovery of proteins from cells and/or cell debris)
Permeate: types and sizes of species	H_2O, small polar solvents, salts, gases	H_2O, small polar solvents, salts, gases	H_2O, solvents, salts, gases
Retentate: types and sizes of species	Salts, nonpolar solvents, solutes >0.8 to 1.5 nm	Salts, nonpolar solvents, solutes >1.5 to 8 nm	Polymers, proteins, micelles, colloids particulates; components retained in range 1 to 100 nm (molecular masses 1 K to 500 K g/mol)
Primary transmembrane driving force (relative magnitude)	Hydrostatic pressure (1300 to 7000 kPa), chemical activity	Hydrostatic pressure (500 to 1400 kPa), chemical activity	Hydrostatic pressure (200 to 1400 kPa), chemical activity
Dominant mechanisms of transport and selectivity	Solubility or sorption on surface and in large free volume regions of membrane and capillary flow under a pressure gradient; diffusion also occurs; selectivity from differential sorption	Solubility or sorption on surface and in large free volume regions of membrane and capillary flow under a pressure gradient; diffusion also occurs; Donnan exclusion; selectivity from differential sorption	Hydrodynamic flow of solvent and hindered diffusion of solutes; selectivity due to size exclusion

(continued)

459

Table 1. (continued).

Unit Operation	Reverse Osmosis	Nanofiltration	Ultrafiltration
Characteristics of membranes and modules	Polyamide TFC, cellulose acetates and polysulfone polymers primarily in spiral wound elements; polyaramid, hollow, fine fibers; also tubular and plate and frame	Polyamide TFC, cellulose acetates, polysulfone polymers, polyvinyl alcohol and polyacrylonitrile; same elements and modules as RO	Large variety of materials and module designs, including inorganics, polyolefins, fluoropolymers
Typical engineering considerations and concerns	Pretreatment needed to remove particulates and colloids; free chlorine (<1 mg/l) and free oxygen are kept low; biological fouling; concentration polarization; membranes need to be kept wet	Same as RO; solvent resistance when fractionating or concentrating organics	Concentration polarization; fouling due to surface and pore sorption
Potential applications in bioremediation	Concentration of leachates prior to biological treatment	More selective control of compositions of organics available to biofilm; membrane bioreactor; oxygenator	More selective control of compositions of organics available to biofilm; membrane bioreactors; retention of organisms; oxygenator

Table 1. *(continued)*.

Unit Operation	Microfiltration	Pervaporation	Vapor Permeation	Gas Separation
Phases of feed/permeate	Liquid/liquid	Liquid/vapor	Vapor/vapor	Gas/gas
General purpose (illustrative application)	Size based fractionation or retention of particulates (concentration of cell biomass from fermentation)	Fractionation of liquid solutions (separating azeotropes); dewatering solvents; removal of organics from water (dehydrating ethanol)	Removal of trace condensible vapors from permanent gases; fractionation of vapor mixtures (separating azeotropes) (removal of VOCs from air)	Gas separation (production of enriched O_2)
Permeate: types and sizes of species	H_2O, solvents, salts, macromolecules, particulates, gases	Volatile species more soluble in membrane, gases	Volatile species more soluble in membrane, gases,	Gases ($\sim \leq 1$ nm) and polar vapors
Retentate: types and sizes of species	Macromolecules and particulates; components retained are larger than the range of 0.02 to 10 μm	Low volatility species; species less soluble in the membrane	Low volatility species; gas and vapor species less soluble in the membrane	Gases
Primary transmembrane driving force (relative magnitude)	Hydrostatic pressure (15–500 kPa)	Chemical activity mainly promoted by temperature and/or partial pressure gradient, usually from vacuum ($\Delta T \sim 10$ to 60 K, $\Delta P \sim 5$ to 90 kPa)	Chemical activity mainly promoted by partial pressure gradient ($\Delta P \sim 5$ to 200 kPa)	Chemical activity mainly promoted by partial pressure gradient ($\Delta P \sim 50$ to 800 kPa)

(continued)

Table 1. (continued).

Unit Operation	Microfiltration	Pervaporation	Vapor Permeation	Gas Separation
Dominant mechanisms of transport and selectivity	Hydrodynamic flow of solvent and hindered diffusion of solutes and suspended particulates; selectivity due to size exclusion and surface sieving	Selective solubility and diffusion in the membrane phase	Selective solubility and diffusion in the membrane phase	Selective solubility and differential diffusion rates in the membrane phase
Characteristics of membranes and modules	Large variety of materials and module designs, including inorganics, polyolefins, fluoropolymers; designs for dead-end and cross-flow processes	Large variety of cross-linked polymers; usually water and/or polar organic molecule selective; plate and frame (leaftype), spiral wound and hollow fiber (tube) modules	Primarily rubbery (elastomers) hydrophilic or hydrophobic polymers in a composite membrane form; spiral wound and hollow fiber modules	Primarily glassy polymers as asymmetric membranes; primarily hollow fine fibers and spiral wound modules
Typical engineering considerations and concerns	Cake buildup requires cleaning/regeneration; fouling due to surface and pore adsorption	Concentration polarization; cost of providing driving force; productivity and selectivity inversely related; high degree of swelling can cause difficulties in sealing	Concentration polarization; cost of providing driving force; productivity and selectivity inversely related; high degree of swelling can cause difficulties in sealing	Multiple stages needed to achieve desired purity; membrane lifetime; effects of feed impurities on the membrane
Potential applications in bioremediation	Membrane bioreactors; retention of organisms; oxygenators	Concentration of organic contaminants; membrane bioreactors; removal of inhibitory cometabolites; recycle oxygenators	Concentration of organic contaminants; membrane bioreactors; removal of inhibitory cometabolites; recycle oxygenators	Enriched oxygen from air for bioreactors

Unit Operation	Dialysis	Electrodialysis	Membrane Contactors	Immobilized Liquid Membranes
Phases of feed/permeate	Liquid/liquid	Liquid/liquid	Gas or liquid/gas or liquid	Gas or liquid/gas or liquid
General purpose (illustrative application)	Fractionation of solutes (blood purification after kidney failure)	Fractionation of small ions; removal of ions (separation of amino acids)	Deplete solute(s) from feed and concentrate them in the extractant stream; dispersion free mass transfer (extractive bioreactors)	Deplete solute(s) from feed and concentrate them in the extractant stream (concentration of heavy metals from aqueous streams)
Permeate: types and sizes of species	H_2O, solvents, salts, gases, microsolutes ≤ 5 nm	Small ions (≤ 0.8 nm) which are opposite charge as fixed charges in membrane; gases; H_2O; solvents	Gases, solutes, vapors soluble in the extractant	Gases, solutes, vapors soluble in the extractant
Retentate: types and sizes of species	Solutes ≥ 0.02 to $0.005\ \mu m$	Large ions (≥ 0.8 nm); small ions which are same charge as fixed charges in membrane; H_2O; solvents	Components of feed insoluble in extractant	Components of feed with low permeability in liquid membrane
Primary transmembrane driving force (relative magnitude)	Chemical activity by concentration gradient created by a sweep stream (dialysate) on the permeate side	Electrochemical potential provided by external electric field (voltage drop typically 1 to 2 volts per cell pair)	Chemical potential	Chemical potential
Dominant mechanisms of transport and selectivity	Hindered diffusion through membrane free volume primarily governed by molecular size	Electrokinetic mobility in the membrane; ion selectivity based on Donnan exclusion of co-ions	Diffusion across the interface and differential solubility in the extractant phase	Solution and diffusion in liquid membrane; differential solubility is primary determinant of selectivity; facilitated transport is often implemented

Table 1. (continued).

Unit Operation	Dialysis	Electrodialysis	Membrane Contactors	Immobilized Liquid Membranes
Characteristics of membranes and modules	Primarily cross-linked polymers; usually capable of being swollen by water and/or polar organic molecules; mostly hollow fiber (tube) modules	Primarily cross-linked copolymers of divinylbenzene and styrene or polytetrafluoroethylene and perfluorosulfonyl fluoride, fixed charges are either quaternary amine, carboxylate or sulfonate; modules are plate and frame	Microporous hollow fiber modules of polyolefins and other chemically robust polymers; may be coated with a thin, more selective barrier layer	Microporous flat sheet or hollow fiber membranes from robust polymers (i.e., PTFE and polyolefins); not generally commercialized but hollow fiber contained liquid membrane modules are available
Typical engineering considerations and concerns	Multiple stages needed to achieve desired purity; tradeoff between productivity and selectivity; boundary layer mass transfer resistances at interfaces	Concentration polarization in the feed channels; stability of anion exchange membranes; pressure losses in all flow channels	Stabilizing the interface between the two phases at one side of the membrane's pore; one liquid needs to be nonwetting of the membrane polymer; good mixing in the bulk phases	This unit operation is generally designed, engineered and constructed by the end user; it is an emerging technology; stability of the liquid membrane is the overwhelming concern
Potential applications in bioremediation	Control of nutrient environment for organisms	Selective concentration of charged species; removal of cometabolites	Multiphase membrane bio-reactors	Bioreactor organism immobilization and preconcentration of contaminants

successful if its overall cost and performance are an improvement over the individual alternatives.

Another novel development is the combination of membranes with reactions. In membrane reactors the basic attributes of the membrane—defined surface area, selective transport, high surface area/volume ratio—can be used in a more complex fashion than in a conventional membrane unit operation. In the most ideal definition of a membrane reactor, the reaction zone is created fairly close to the separating layer. In this context, there are two distinctions that should be made.

- The first and simplest distinction is facilitated transport membranes where a reversible chemical reaction/complexation is used to affect a separation. While this approach can be used to great advantage in separations, it is not a membrane reactor per se.
- The second, and more subtle, distinction is a reactive membrane where the membrane is used as an immobilization matrix for a catalyst. This can be considered a membrane reactor but does not always combine the functions of separation and reaction.

It is the development of hybrid processes, reactive membranes, and membrane (bio)reactors that is perhaps the most germane to novel bioremediation applications. Useful designs can result from efficient combinations of existing equipment, as well as novel constructions.

Membranes in Bioremediation

The use of membranes in bioremediation is clearly a nascent technology. There are already a number of straightforward uses of membrane unit operations as prefilters and preconcentrators and for composition control in suspended growth reactors on an external recycle loop (Chang et al., 1993). More novel uses include reactive membranes, membrane bioreactors (Conover, 1993), phase contactors, and internal composition control elements (Anderson et al., 1986; Chiemchaisri and Yamamoto, 1994).

The most significant emerging application of membranes in bioremediation involves the use of membranes to selectively separate hydrophobic organic pollutants from contaminated streams and to bring them in contact with microbes (aerobic and anaerobic). Commonly encountered pollutants of interest are acetone, simple aromatics (benzene, toluene, ethylbenzene, xylenes or BTEX), methyl ethyl ketone (MEK), methyl isobutyl ketone (MIBK), chlorinated aliphatics (such as trichloroethane

(TCE), methylene chloride, and chloroform), and semivolatiles such as phenolic compounds. The bioreaction destroys the target species and yields relatively benign species such as carbon dioxide, water, and hydrochloric acid. Tubular and hollow fiber configurations are logical choices for carrying out the treatment process. Depending on the circumstances, a microbial mass can be attached to the tube or shell side of the membrane to create a "biofilter," thus satisfying the ideal criteria for a membrane reactor (intimate contact between separating layer and reaction zone).

A somewhat simpler biofilter configuration that relies on the ability of a designed, in situ barrier to capture organic pollutants and destroy them, while letting water flow through, is being developed. For instance, these flat biomembranes can be inserted vertically in the ground as a passive system applied to a contaminated, groundwater plume.

In general, the desirable integrated process unit that separates and conducts bioremediation must be designed to exploit the destructive ability of the employed microbes. It must provide the needed oxygen for the aerobes and the chemistry to neutralize (i.e., the hydrochloric acid that results from the chlorinated organics) or selectively remove (i.e., CO_2) reaction products. The technical areas of membrane reactors, bioreactors, and reactive membranes resulted from the need to integrate reaction kinetics with the fundamental aspects of mass transfer through various types of membranes and the overall mass transfer performance of membrane modules. These areas are most applicable to bioremediation applications.

MEMBRANE UNIT OPERATIONS

A large number of useful references provide very detailed descriptions of the range of membrane unit operations and aspects of process design and synthesis. Two very useful pedagogical works (Wankat, 1990; Mulder, 1992) have been developed. The former is useful for a quick study, whereas the latter has more content, including coverage of material science topics. A number of references contain individual chapters for each of the full range of membrane unit operations, including separate sections devoted to specialty topics. These include Ho and Sirkar (1992), Belfort (1984), Hwang and Kammermeyer (1975), and Bungay et al. (1986). The last contains an especially useful chapter on process design and optimization (Rautenbach, 1986). For more coverage related to biotechnology and biomedical applications, Vieth (1988) provides a good starting point.

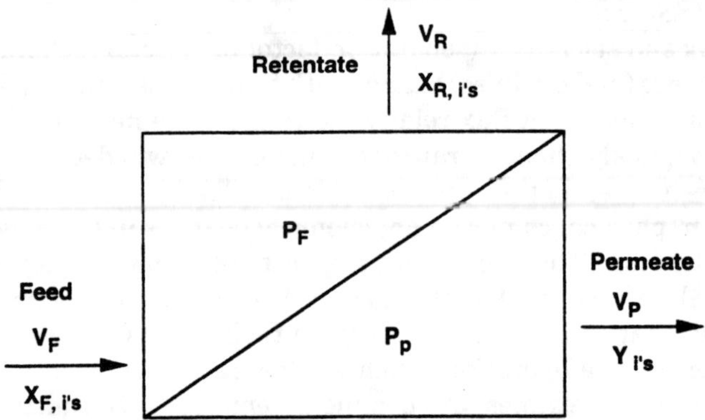

Figure 1. General schematic of a membrane unit operation.

General Schematic and Terminology

Figure 1 shows the general schematic of a membrane unit operation. The feed enters a module on the high-pressure side. The material that passes through the separating layer into the low-pressure side is the permeate. The part of the feed that stays on the high-pressure side is the retentate or concentrate. The general engineering parameters are

A_x membrane area, m^2

J flux, volumetric productivity per unit membrane area, V_P/A_x, $m^3\,m^{-2}\,s^{-1}$

P transmembrane pressure, usually taken as the difference between the average feed and permeate pressures, $P_F - P_P$, kPa

F permeance, the volumetric output per unit membrane area per unit transmembrane pressure, $m^3\,m^{-2}\,s^{-1}\,kPa^{-1}$

    stage cut, fraction of feed removed as permeate, V_P/V_F

V_P volumetric flow rate of the permeate, $m^3\,s^{-1}$

V_F volumetric flow rate of the feed, $m^3\,s^{-1}$

The flux of a membrane basically defines its productivity and, therefore, largely influences the capital cost (and to a lesser extent the operating cost). The selectivity, or specificity, of the membrane defines whether it will be useful or not. This quantity is called different things, depending on the specific membrane unit operation (i.e., separation factor in gas and vapor permeation, rejection in reverse osmosis, retention in ultrafiltration, etc.), and it influences how pure the product stream is and how much of it is recovered at the desired purity.

BROAD ISSUES

The flux and selectivity (separation factor or rejection) characteristics of membranes (and modules) change with time and use. In liquid systems, a significant change in flux relative to that for the new membrane will occur very quickly, then the rate of further change will decrease (depending on cleaning and other process options). Changes are due to fouling, adsorption, physical compression, chemical changes in the polymer, and physical trauma. The specific unit operations, types of polymers, and modules all influence what changes in flux and specificity are likely to occur and what can be done to mitigate their effects. Operating cycles, which can include operations such as flow reversals, back pulses, and periodic cleaning regimes, are commonly employed. Optimizing how to do these is an emerging discipline in its own right.

Membrane Module Forms

Membrane module designs have been in continual development as driven by four primary factors: (1) the need to obtain a high surface area in a small volume, (2) to minimize the mass transfer limitations from fluid flow at the membrane interfaces, (3) to maximize the ease and minimize the cost of module production, and (4) to minimize the module's contribution to the application's operating cost. Table 2 is a summary of the main characteristics of the different types of modules. Figures 2 to 5

Table 2. General comparison of various membrane module types.

Characteristic	Plate and Frame	Spiral Wound	Tubular	Capillary/ Hollow Fiber
Representative packing density (m²/m³)	500	800	70	6000
Capital cost	High	Med-high	High	Low
Fouling tendency	Low to moderate	Med-high	Low	High
Ease of cleaning	Good	Poor to good	Good to excellent	Poor
Membrane replacement (yes/no)	Yes	No	Sometimes	No
Operating cost	High	Moderate	High	Low
Feedstream prefiltration	Yes (10–25 μm)	Yes (10–25 μm)	No	Yes (5–10 μm)

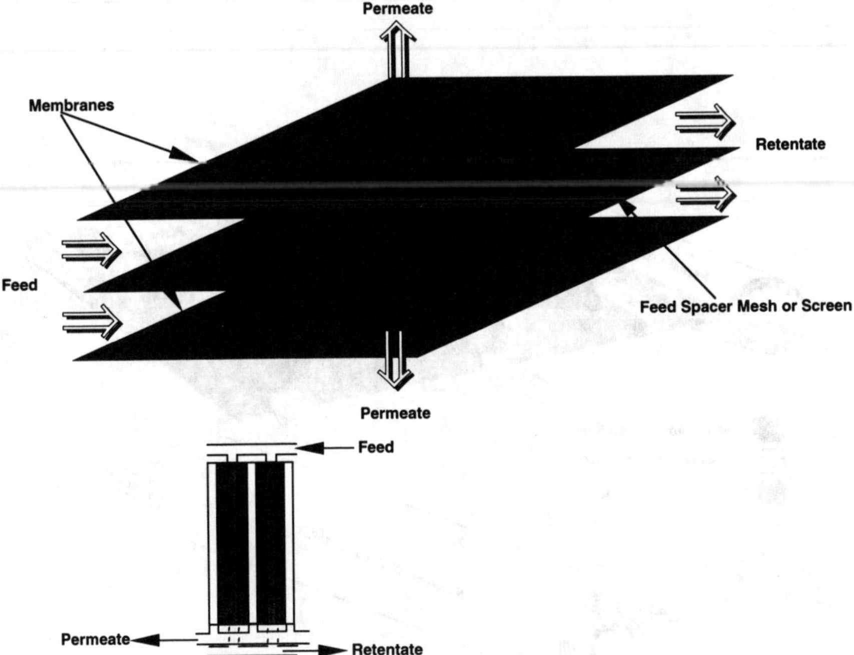

Figure 2. Schematic of plate and frame membrane module.

present schematics of several basic designs, and Figure 6 presents some recent innovations.

The plate and frame design (Figure 2) uses flat sheet membranes and is very versatile but expensive. It is commonly used in electrodialysis, pervaporation, and sometimes reverse osmosis applications. The modules can be constructed in a number of ways, including leafs extending perpendicular from a central shaft or disks stacked on a central core (Peters and Stanford, 1993), either of which would then be put in a pressure housing.

The spiral-wound module design (Figure 3) also uses flat sheets but is built up in a "jelly-roll" fashion to provide higher module surface area per unit volume and lower total cost. Several spiral-wound elements are mounted in a single pressure vessel, and several vessels are interconnected to provide the required overall surface area.

The range of membrane materials that can be used for tubular membranes is very broad. Ceramic membrane elements (Figure 4) are just one example. Tubular membranes have larger diameter and are mounted in housings in a variety of ways that are generically similar to the schematic

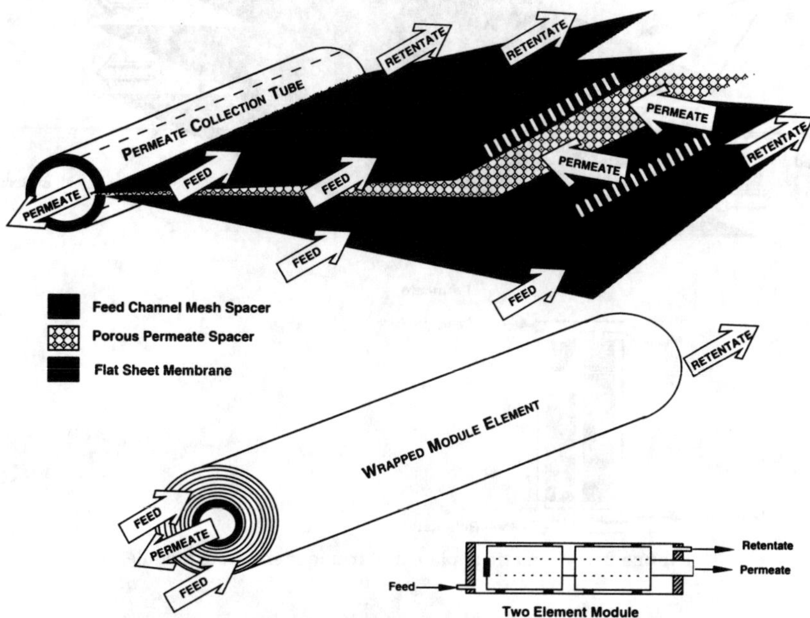

Figure 3. Schematic of spiral-wound membrane module.

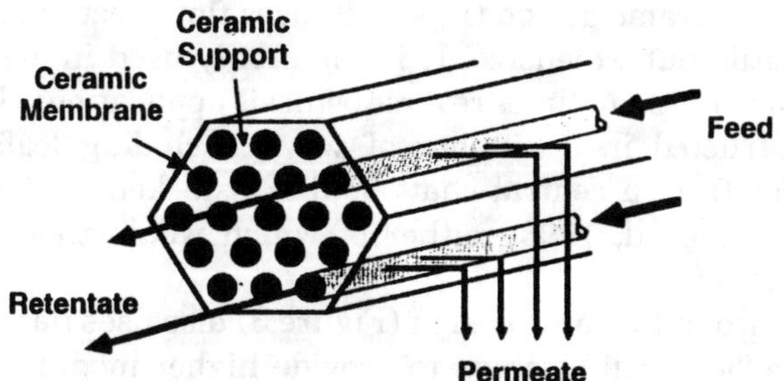

Figure 4. Schematic of tubular/ceramic membrane module.

470

in Figure 5 for capillary and hollow fiber membranes, that is, as parallel cylindrical channels with mass transfer through the walls of the cylinders. Due to their larger size, there are more options for securing them in a "header," including continuous 180° turnarounds. Tubular membrane modules are very popular for high solids or dirty feeds but do not provide as great a packing density for mass transfer.

Capillary/hollow fiber membranes provide extremely high mass transfer areas in a cost-effective module size. As indicated in Figure 5, there are a variety of options for feed and permeate flows. The small diameter of the fibers makes them more prone to plugging and fouling when the process feed is through the inside diameter (lumen); therefore, pretreatment/filtration requirements are usually more stringent in this case.

A very limited sampling of some of the innovations in module approaches over the past few years is presented in Figure 6 to reinforce the notion that a suitable design for specific bioremediation applications can be obtained or readily adapted.

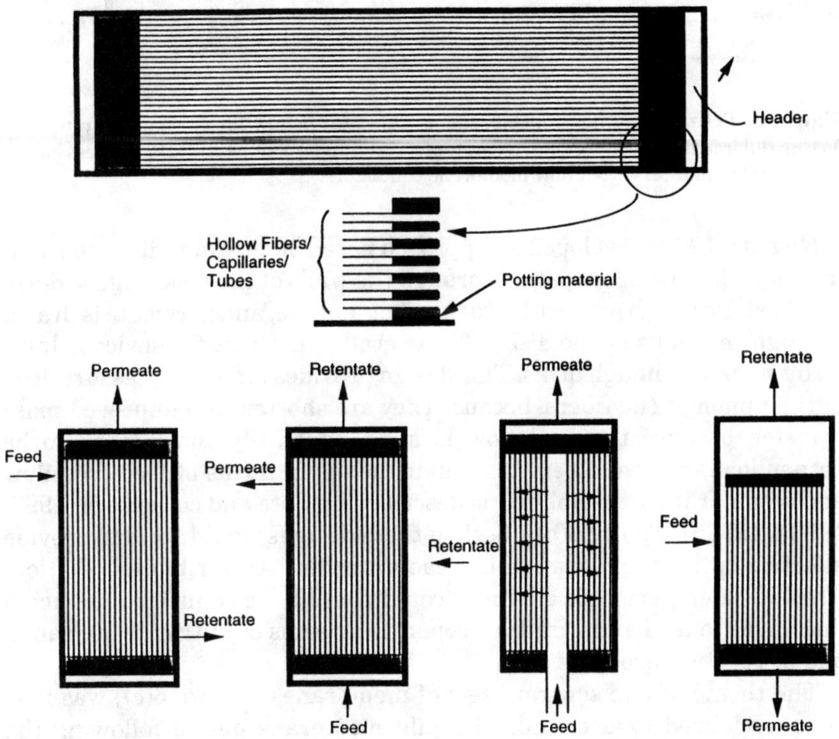

Figure 5. Schematic of hollow fiber/capillary membrane module.

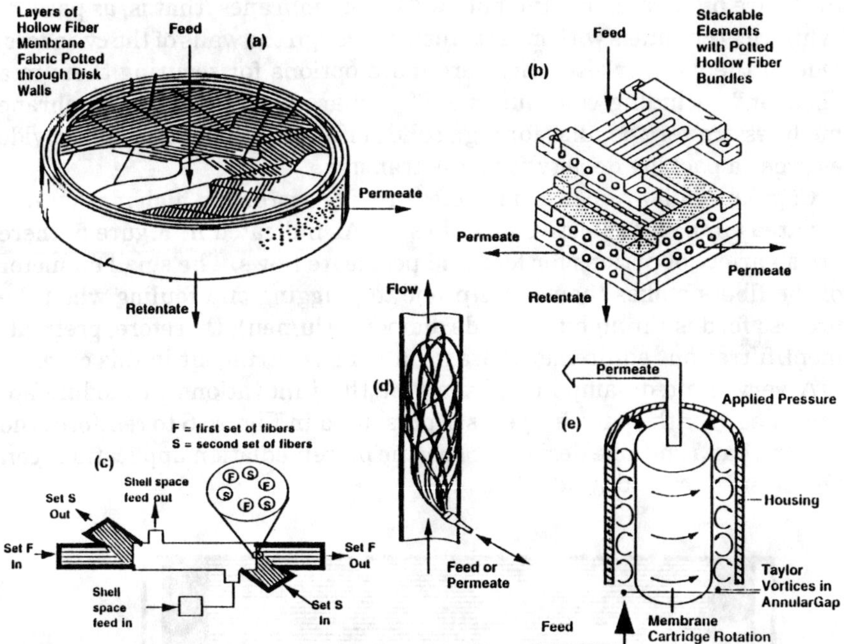

Figure 6. Schematic of novel membrane modules: (a) hollow fiber disks, (b) transverse flow modular hollow fiber bundles, (c) multiple fiber bundle contactors, (d) frameless hollow fiber bundles, and (e) vortex flow membrane contactors.

Nichols (1990) developed an approach based on modular disks containing short, hollow fibers potted through the walls of the disk [Figure 6(a)]. The feed flow is perpendicular to the fiber lengths, and permeate is drawn through the walls of the disks. The overall mass transfer device is built up by stacking enough disks. This design provides for lower pressure drop in the lumen of the fibers, because they are shorter, and improved mass transfer, because the feed flow is more effectively constrained to be perpendicular to the fibers. A similar theme, in terms of feed crossflow and modular fiber assemblies, is described by Cote and coworkers (1992, 1993a, 1993b) [Figure 6(b)]. Both of these designs would seem to provide effective options for creating bioreactors optimized for bioremediation. The indicated permeate channels could possibly be connected in such a fashion as to feed and withdraw separate streams of nutrients, oxidants, and/or reaction products.

The theme of two separate sets of membranes [Figure 6(c)] was first commercialized as a contained liquid membrane device following the work of Sengupta et al.(1988). It was initially conceived for liquid or gas separations where the shell side contains a solvent mixture with a selective complexing agent for one component that would permeate from fiber set F to set S. The membranes were microporous, hollow fibers that were unwetted by the solvent solution. This design is potentially useful for multiple feeds to bioremediation reactors.

Atkin and Gorsuch (1990, 1992a, 1992b, 1993) and Atkin et al. (1993) describe a plasma separation membrane for direct insertion in a vein. We illustrate it [Figure 6(d)] as an example of a versatile, general design that could be adapted to a variety of applications. In fact, a product with similar features is offered as a membrane aerator (Nomura, 1989a, 1989b). The net result of the loose fibers in the flowstream is that they wave around, providing mixing or turbulence to disrupt any mass transfer boundary layers. It can also inhibit the accumulation of foulants.

The design schematically represented in Figure 6(e) was commercialized in laboratory and small plant sizes (Rolchigo et al., 1988; Kroner and Nissinen, 1988). The inner housing rotates, generating flow instabilities (or Taylor vortices) in the annular space between the membrane and the housing. Therefore, the flow has a spiraling character that helps minimize mass transfer boundary layers and scours the surface. The original equipment designs were efficient at producing a high solids content retentate due to the screw-type movement through the device. More recently, module designs based on spiral housings that induce another type of flow instability, Dean vortices (Brewster et al., 1993; Chung et al., 1993), have been disclosed and also will serve to effectively reduce mass transfer boundary layers and deposition on membrane surfaces.

Pressure-Driven Membrane Processes (Liquids)

Under this topic, we will present a summary description of the membrane processes commonly encountered for liquid separations. There are two main sections. First, we will describe each of the membrane processes, define their main engineering variables, and then list a flux model. These are phenomenological flux models for any point along the membrane. They can also be used as an integrated representation, assuming some average (or weighted) values for the parameters. In these models, the osmotic pressure gradient is between the permeate and the solution at the membrane interface.

In the second section, we will illustrate a simple modeling approach for estimating the overall performance (mass flows, exit concentrations, and/or area requirements) of a pressure-driven, liquid filtration membrane module. Simple models are useful for exploring the effects of changes in membrane characteristics and module operating conditions.

A "simple approach" means that the solution can be obtained without complex numerical computation. Ideally, a spreadsheet program could be used to perform "what if" case studies. The reader is cautioned at the outset that, because of the assumptions required in a simple model, there will probably be significant differences between the actual performance and this estimate. More detailed models can be found in the appropriate chapters of the general references, and these should be used for detailed design projects.

GENERAL PHYSICAL MODEL

Figure 7 presents a simplified schematic of transport for pressure-driven liquid separations using membranes. The main driving force for mass transfer is the hydrostatic pressure gradient, which is applied across the membrane's thickness. This figure and the following models depict the case when a single component is retained by the membrane. The mathematics for the more realistic cases of multicomponent retention follow a similar reasoning. The process is described in at least two parts:

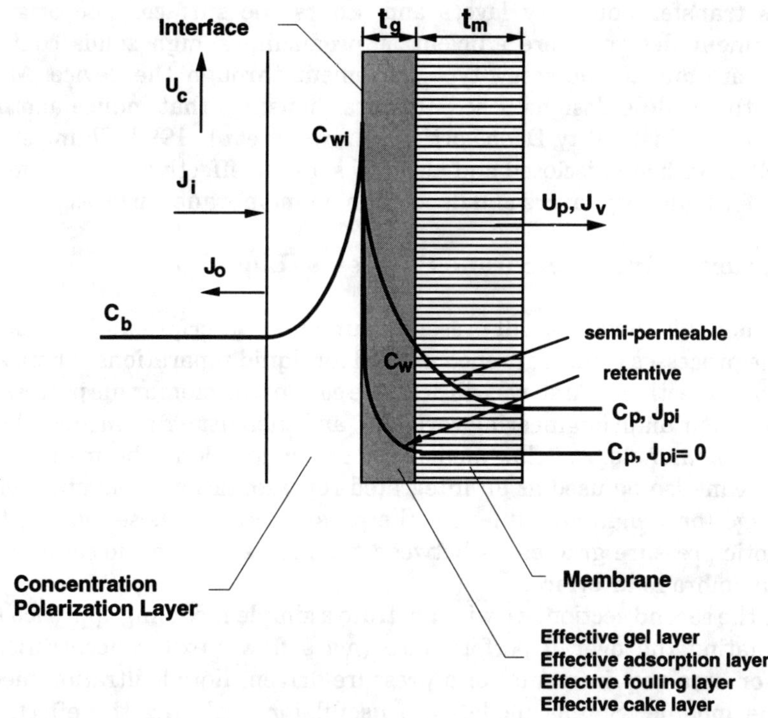

Figure 7. Schematic of pressure-driven membrane processes for liquids.

1. The flux of solvent (and other components) that can freely pass through the membrane
2. The flux of each component that is either partially or completely retained by the membrane.

Resistances in Series

Transport from the bulk feed to the bulk permeate phase is the result of permeation through a number of resistances in series. In many cases, the overall mass transfer may be controlled by a single, rate-limiting resistance.

1. The first resistance is in the hydrodynamic mass transfer boundary layer between the bulk solution and the first interface where partitioning (separation) can occur. That layer is usually referred to as the concentration boundary layer.
2. Next, there are several possible resistances that can be difficult to separate into individual contributions if they are all present at the same time:
   * a gel layer associated with aggregation of macromolecules and/or colloids
   * a cake layer from particles that are retained by the membrane
   * an adsorption layer from reversible equilibrium partitioning onto the surface of the membrane
   * a fouling layer that is an irreversible adsorption layer
   The mass transfer resistance from all these phenomena can often fruitfully be combined into a single parametric term.
3. Next, the mass transfer through the membrane itself is encountered. Flux models for each type of membrane are available to describe this mass transport process. The models run the gamut from totally empirical to quite fundamental. In a later section on simple process design, we will present empirical transport models.
4. Finally, another hydrodynamic mass transfer resistance is encountered, this time between the permeate interface and the bulk permeate fluid.

If a reaction is occurring, such as in a biofilter, that mass transfer step (or more exactly—mass transfer with reaction) would need to be included somewhere in the series outlined above.

PROCESSES

Hyperfiltration
Hyperfiltration is a pressure-driven process for producing a purified

liquid by retaining small molecules, typically salts from water. *Reverse osmosis* (RO) is often differentiated, based on its use for either seawater or brackish water purification. When the objective is primarily the removal of other small solutes (phenols, halomethanes or ethanes, sugars, amino acids, for example), the process is called *nanofiltration* (NF). It is good for multivalent ions but has poor rejection of nonpolar molecules unless they are rather large. Both of these are collectively referred to as hyperfiltration.

There are many membrane materials used for the separating layer. These include aramids, polyamides, cellulose acetates, polyvinyl alcohol, polysulfones, and polyacrylonitriles. The most common hyperfiltration membrane forms are either as a single layer (an asymmetric type) or as a "thin film composite" (TFC) that is made by an interfacial polymerization. The module elements are usually either hollow fiber or spiral-wound, but all polymers are not necessarily available in all configurations.

For hyperfiltration the typical engineering parameters are

J_v the water (solvent) flux, $m^3\ m^{-2}\ s^{-1}$

P_v specific water permeability, $m^3\ m^{-1}\ s^{-1}\ kPa^{-1}$

t_m membrane thickness, m

ΔP applied, transmembrane, mechanical pressure (at a specific point), kPa

$\Delta\pi$ actual osmotic pressure gradient based on C_w and C_p (at a specific point), kPa

A a lumped parameter $(= P_v/t_m)$ frequently tabulated as the water (solvent) transport coefficient, $m\ s^{-1}\ kPa^{-1}$

J_i the rate of solute i's (i.e., NaCl) permeation per unit time, $mol\ m^{-2}\ s^{-1}$

K_{si} the specific solute's permeability coefficient, $m\ s^{-1}$

R observed solute rejection, $1 - C_p/C_b$, dimensionless

R° intrinsic solute rejection, $1 - C_p/C_w$, dimensionless

C_{pi} concentration in permeate, $mol\ m^{-3}$

C_{bi} concentration in bulk feed, $mol\ m^{-3}$

C_{wi} concentration at wall-membrane interface on the feed side, $mol\ m^{-3}$

Flux Expressions for Hyperfiltration (with Intrinsic Rejection, R°, <1)

$$J_v = \frac{P_v}{t_m}(\Delta P - \Delta\pi) \tag{1}$$

$$J_i = K_{si}(C_{wi} - C_{pi}) \tag{2}$$

Ultrafiltration (UF)

Ultrafiltration is a pressure-driven filtration process for fractionating and/or separating macromolecules from water or other solvents. Size-based sieving is the commonly viewed separation principle, but thermodynamically based partitioning of macromolecules between the exterior solution and the membrane is also a mechanism. The characteristic dimensions of pores (or void volumes) are in the range of 1 to 100 nm. Every membrane will actually have a distribution of sizes around some mean value. The membranes are characterized by a molecular weight cutoff (MWCO) usually in the range of 300–500,000 g/mol. The MWCO value is defined as the smallest molecular mass species for which the membrane exhibits a 90% rejection. For a single component, the MWCO will largely depend on the largest pores in the membrane's pore size distribution. In actual multicomponent separations, the highly rejected species and any adsorbed foulants will form a "dynamic" rejection layer. This layer will cause rejection of molecules that would otherwise pass through a "fresh" membrane. Thus, the apparent MWCO may be lower than expected.

Module elements are available in all the classic forms. Most of the popular materials are available, including polysulfones (polyethersulfone, etc.), regenerated cellulose, cellulose acetates, and polyacrylonitriles. Inorganic membranes including alumina and zirconia/stainless steel are also commercially available.

For ultrafiltration, typical engineering parameters are

J_v the water (solvent) flux, $m^3 \ m^{-2} \ s^{-1}$

P_v' specific water conductance of a clean membrane, $m^3 \ m^{-1}$

t_m membrane thickness, m

P_g' specific water conductance of the gel layer, $m^3 \ m^{-1}$

t_g gel layer thickness, m

ΔP applied, transmembrane, mechanical pressure (at a specific point), kPa

$\Delta \pi$ actual osmotic pressure gradient based on C_w and C_p (at a specific point), kPa

η water (solvent) viscosity, kPa s

J_i the rate of solute i's (i.e., a protein) permeation per unit time, $mol \ m^{-2} \ s^{-1}$

P_{mi} the specific solute's permeability coefficient, $m \ s^{-1}$

σ_i the Staverman reflection coefficient ($\sigma_i = 1$, no solute permeation), dimensionless

R observed solute rejection, $1 - C_p/C_b$, dimensionless

$R°$ intrinsic solute rejection, $1 - C_p/C_w$, dimensionless

Flux Expressions for Ultrafiltration

$$J_v = \frac{\Delta P - \Delta \pi}{\left| \dfrac{t_m}{P'_v} + \dfrac{t_g}{P'_g} \right| \eta} \tag{3}$$

$$J_i = J_v(1 - \sigma_i)C_{wi} - \frac{J_v(1 - \sigma_i)(C_{pi} - C_{wi})}{\exp(J_v(1 - \sigma_i)/P_{mi}) - 1} \tag{4}$$

Microfiltration (MF)

Microfiltration is a filtration process for retaining and/or fractionating particulates, colloids, emulsions, cells, cell fragments, viruses, and most other suspended solids from liquids, gases, and vapors. Microfiltration is closely related to conventional coarse filtration. Size-based sieving at either the surface or in the depth of the membrane is the separation mechanism. The pore sizes of microfiltration membranes range from 10 to $0.05\,\mu$m. Some membrane processes that commonly use microfiltration membranes are dead-end and cross-flow microfiltration and phase contactors (i.e., oxygenators, liquid-liquid extractors, and contained liquid membranes). For bioremediation applications, cross-flow MF is the most likely process configuration.

MF membranes are available in a very large variety of materials (fluoropolymers, polyolefins, cellulosics, polycarbonate, polysulfones, polyimides, alumina, sintered metals, etc.) and module types. In fact, some of the most novel materials and module forms have been developed initially in microfiltration membranes. In most process applications, the key engineering consideration is flux decline due to buildup of the cake on the membrane surface. Adsorption on the membrane's surface and within its pores may also occur, depending on the process fluid's composition. A strategy of periodic flow reversal (pulsing the fluid) back from the permeate side into the feed side, to loosen up the cake layer, is often used to maximize overall onstream productivity. Moreover, many MF membranes (and modules) are robust enough to withstand vigorous cleaning procedures to regain productivity after extended periods of use.

In MF, it can be more straightforward to choose the pore size of the membrane to be completely retentive. A variety of models to predict particle capture efficiency depends on geometric and physical properties of the particles and filter medium, as well as the hydrodynamic properties of the bulk fluid flow through the filter. Recent theoretical treatments on particle retention in microfilters can be found in Tein (1989), Grant et al. (1989), and Grant and Zahka (1990). If there are soluble components in the feedstream that are only partially retained by the membrane, then an ultrafiltration solute flux model can be used for those components.

For microfiltration, typical engineering parameters are

J_v the water (solvent) flux, $\mathrm{m^3\,m^{-2}\,s^{-1}}$

P_v' specific water permeability of a clean membrane, $\mathrm{m^3\,m^{-1}}$

t_m membrane thickness, m

P_c' specific water permeability of the cake layer, $\mathrm{m^3\,m^{-1}}$

t_c cake layer thickness, m

ΔP applied, transmembrane, mechanical pressure (at a specific point), kPa

η water (solvent) viscosity, kPa s

Flux Expression for Microfiltration

$$J_v = \frac{\Delta P}{\left| \dfrac{t_m}{P_v'} + \dfrac{t_g}{P_g'} \right| \eta} \tag{5}$$

Often, further lumping of parameters is done in the form:

$$J_v = \frac{\Delta P}{[R_m + R_g]\eta} \tag{6}$$

where

R_m = clean membrane resistance

R_g = additional resistance from gels, cakes, and adsorption (fouling)

SIMPLE PROCESS DESIGN NOTES

For all of these filtration processes, $\Delta \pi$ and R_g will vary along the length of the module and also with time. For hyperfiltration and ultrafiltration, both $\Delta \pi$ and R_g can be significant, while for microfiltration, $\Delta \pi$ is likely to be negligible unless significant retention of smaller molecules occurs due to a "g" layer. This can be the case in biotechnology, as in fermentation broth separations. The osmotic pressure is often estimated by using the van't Hoff equation ($\pi = C_{wi}RT$, where R is the gas constant and T is absolute temperature), even though this is valid only for dilute, ideal solutions. If experimental data on solution vapor pressures or freezing points are available, then more accurate methods for estimating π are found in Reid (1966).

Estimation of Permeability Parameters

P'_v and P'_g can be estimated from some simple geometric viewpoints.

- for a group of pores containing laminar flow:

$$P' = \frac{\pi}{8} \sum_{i=0}^{i} n_i d_i^4 \tag{7}$$

where

n_i = number of pores with diameter i per unit area
d_i = diameter of pore i

- for packed particles (Carmen-Kozeny rigid particles):

$$P' = \frac{\varepsilon^3}{5(1-\varepsilon)^2} \left(\frac{v_p}{s_p} \right)^2 \tag{8}$$

ε = void fraction
v_p = volume of particles
s_p = surface area of particles

Estimation of Mass Transfer Boundary Layer Resistances

In the real case of a semi-permeable membrane (or any selective barrier), there is an accumulation of rejected solutes at the interface. Since the permeation properties of the membrane will depend on the solute(s) concentration at the wall (C_w), a way of estimating the ratio C_w/C_b is needed. At steady state, it is assumed that the rate of solute accumulation is zero; therefore, the rate of incoming solute is balanced by its back diffusion and permeation through the membrane. For the simple case of a single solute and Fickian diffusion, this is represented by the following mass balance:

$$J_v C_p - J_v C - D \frac{dC}{dy} = 0 \tag{9}$$

The boundary conditions are $C = C_w$ at $y = 0$ and $C = C_b$ at $y = \delta$ (δ is the thickness of the concentration boundary layer). With the further definition of the boundary layer mass transfer coefficient, $k = D/\delta$, the solution of this set of equations is

$$M = \frac{C_w}{C_b} = \frac{\exp(J_v/k)}{R^\circ + (1 - R^\circ)\exp(J_v/k)} \tag{10}$$

The boundary layer mass transfer coefficient k is often estimated by correlations that usually assume that $J_v \ll U_c$, where U_c is the average cross-flow velocity. Also, the definition $k = D/\delta$ assumes that the rejection is 100%. Therefore, for high flux membranes, defining $k = D/\delta$ is suspect. It is important to note that each specific application needs to be analyzed to determine the number of boundary layer resistances to consider and which correlations are appropriate. The following correlations are based on flow inside of a channel or tube; for many applications, the flow can be in the shell, and a variety of nonuniformities can exist. The module manufacturer can be a source for correlations based on their experiences.

General Correlations for Estimation of Mass Transfer Coefficient k
Definitions:

Reynolds No.

$$Re = \frac{dU_c\rho}{\mu} \tag{11}$$

Schmidt No.

$$Sc = \frac{\mu}{\rho D} \tag{12}$$

d = representative channel or tube dimension for flow (i.e., diameter), or vessel diameter, m
U_c = average cross-flow velocity, m s^{-1}
ρ = density, g m^{-3}
μ = shear viscosity, g m^{-1} s^{-1}
D = solute's binary diffusion coefficient in the solvent, m^2 s

For turbulent flow in tubes ($Re > 20,000$ in general, or >2000 for UF):

$$k = U_c\, 0.0396\, Re^{-1/4} Sc^{-2/3} \tag{13}$$

or

$$k = 0.023\,\frac{D}{d}\, Re^{0.83} Sc^{1/3} \tag{14}$$

For turbulent flow in batch, stirred vessels:

$$k = 0.0443 \frac{D}{d} \left\{ \frac{\rho}{\mu D} \right\}^{1/3} \left\{ \frac{\omega \mu d^2}{\rho} \right\}^{3/4} \tag{15}$$

where ω is the stirrer speed (rad/s).

For laminar flow in round tubes:

$$k = 1.295 \left(\frac{2U_c D^2}{dL} \right)^{1/3} \tag{16}$$

where L is the distance along the tube's length.

For laminar flow between parallel plates spaced at $2h$:

$$k = 1.177 \left(\frac{U_c D^2}{hL} \right)^{1/3} \tag{17}$$

Remember that, in laminar flow, k varies along the length of module. The correlations estimate only what exists at a given point in the fluid's path. Also, the appropriate viscosity to use is for the fluid at the membrane interface, not the bulk fluid. Failure to account for the increase of viscosity in the boundary layer as the solute concentration increases is a common source of error in estimating k.

Module Mass Balance

The model we present is for the case of perfectly mixed feed and permeate compartments, with concentration polarization on the feed side. It follows the method developed by Rao and Sirkar (1978). Perfectly mixed reflects the assumption that there are no concentration gradients in the flow (parallel to the membrane surface) direction, but that concentration polarization results because there are gradients in the direction perpendicular to the membrane surface. This model is the simplest to use and provides a conservative estimate of module performance. In the general case, an iterative (trial and error) method is required for solution of the design problem, but in special circumstances, an algebraic solution can be obtained.

This module model can be used for RO, NF, UF, and MF with appropriate modifications. The overall material balances and boundary layer aspects are essentially the same for all these applications. The definition of the flux, membrane resistances, and the concentrations can be modified to suit the specific type of filtration. Figure 8 helps define the nomenclature.

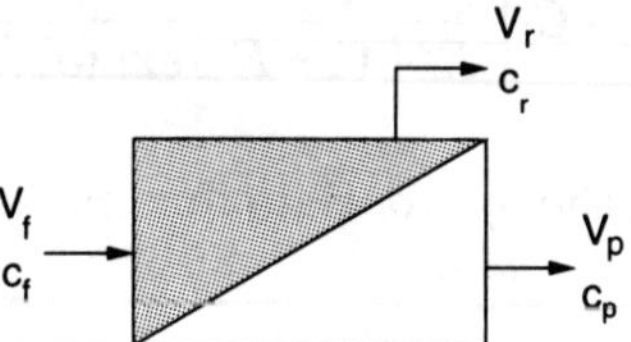

Figure 8. Schematic membrane module, V = volumetric flow, C = concentration of "key" component in the mixture. Subscripts: f = feed, r = retentate, and p = permeate.

External mass balance:

$$V_f = V_p + V_r$$

$$\Theta = V_p/V_f = \text{stage cut or recovery factor} \tag{18}$$

External key component balance:

$$V_f C_f = V_p C_p + V_r C_r \tag{19}$$

The overall volumetric flux through the membrane is J_v; therefore, $V_p = J_v^* A_x$, where A_x = the membrane area.

Let C_w = boundary layer concentration at the membrane interface due to concentration polarization. Futhermore, assume that there is no gel formation. Since the feed side is perfectly mixed, then $C_r = C_b$, where C_b = bulk concentration.

Then, the definition of the intrinsic rejection of a membrane is $R^\circ = 1 - C_p/C_w$.

From the simple boundary layer model for concentration polarization and assuming that R° is constant, the following relationship for C_p is obtained:

$$C_p = \frac{C_b(1 - R^\circ)\exp(J_v/k)}{R^\circ + (1 - R^\circ)\exp(J_v/k)} \tag{20}$$

And from the material balance assumption $C_r = C_b$ and is defined by

$$C_r = C_b = \frac{C_f}{(1 - \Theta) + \dfrac{\Theta(1 - R^\circ)\exp(J_v/k)}{R^\circ + (1 - R^\circ)\exp(J_v/k)}} \tag{21}$$

The wall concentration C_w is defined by

$$C_w = C_r \left(\frac{\exp(J_v/k)}{R^\circ + (1 - R^\circ)\exp(J_v/k)} \right) \qquad (22)$$

J_v is included by using the appropriate flux model; that is, for RO,

$$J_v = \frac{P_v}{t_m}(\Delta P - \Delta \pi) \qquad (23)$$

For engineering estimation purposes, a linear form for the concentration dependence of the osmotic pressure can be used: $\pi = aC$, where a is either a constant, fitted to experimental data in the expected range of concentrations, or RT based on the van't Hoff equation.

Using the definition of the intrinsic rejection,

$$R^\circ = (C_w - C_p)/C_w \qquad (24)$$

the result is

$$\pi = a(C_w - C_p) = aR^\circ C_w \qquad (25)$$

and

$$J_v = \frac{P_v}{t_m}(\Delta P - aR^\circ C_w) \qquad (26)$$

In the case of other filtration processes, the same general method can be followed, with the major exception being the definition of the solvent flux J_v. Lastly, an estimate for k is obtained from the appropriate boundary layer correlation and physical dimensions of the system.

This group of relationships is nonlinear because of the exponential term and can be solved iteratively (using successive substitutions), in a spreadsheet, or in a general programming language algorithm (i.e., FORTRAN, C, database, etc.). For an iterative solution using the RO flux model with concentration polarization, an objective function is

$$F(J_v) = 0$$

$$= \frac{P_v}{t_m}\left[\Delta P - aR^\circ \left\{ \frac{C_f \exp(J_v/k)}{(1 - \Theta)R^\circ + (1 - R^\circ)\exp(J_v/k)} \right\} \right] - J_v \qquad (27)$$

If $J_v/k \ll 1$, then the problem can be solved explicitly because the approximation that $\exp(J_v/k) \sim 1 + J_v/k$ leads to an algebraic solution in the following form:

$$a'J_v^2 + b'J_v + c' = 0 \tag{28}$$

where

$$a' = (1 - R^\circ)/k \tag{29}$$

$$b' = 1 - \Theta R^\circ - \frac{P_v(1 - R^\circ)\Delta P}{t_m k} + \frac{P_v C_f a R^\circ}{t_m k} \tag{30}$$

$$c' = \frac{P_v C_f a R^\circ}{t_m} - (1 - \Theta R^\circ)\frac{P_v \Delta P}{t_m} \tag{31}$$

Once J_v is known, then C_p and C_r can be calculated. The component material balance equation yields the flux of the key solute. The required membrane area can be determined based on the overall material balance and the previously cited relationship that $V_p = J_v^* A_x$.

The membrane processes covered in this section might be used to recover and concentrate solvent from soil washing, concentrate pollutants in a water stream before treatment with a biofilter, or concentrate and segregate different consortia of bacteria that are part of a treatment train, among many other applications. In all cases, this simple process model can be used to estimate the membrane area required to process a given amount of fluid or how much fluid can be processed with a given membrane area.

Activity-Driven Membrane Processes (Gases and Vapors)

GENERAL PHYSICAL MODEL

Figure 9 presents a schematic of gas and vapor permeation [Figure 9(a)] and pervaporation [Figure 9(b)] using membranes. In these processes, the main driving force for mass transfer is the activity gradient that exists across the membrane's thickness. This gradient is usually the result of increasing the pressure on the feed side and/or pulling a vacuum on the permeate side. The figure and following models depict the case of a binary mixture. As before, these are phenomenological flux models for any point along the membrane. They describe single-component permeation and

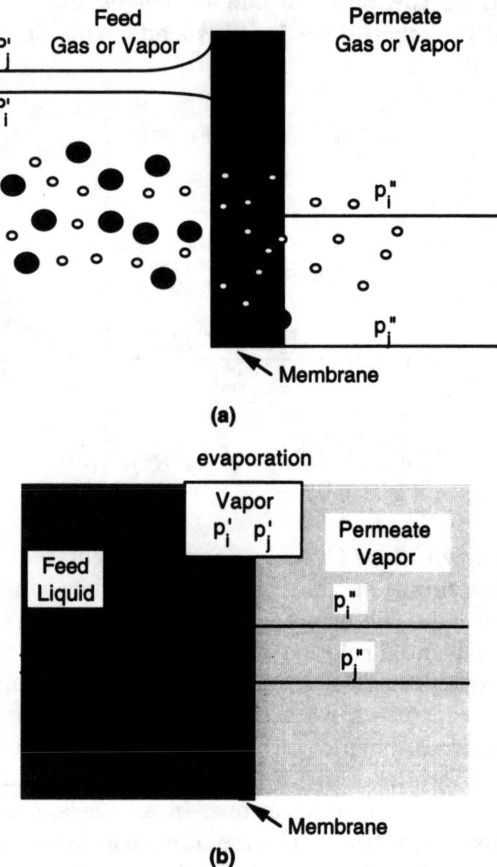

Figure 9. Schematic representation of (a) gas and vapor permeation and (b) pervaporation processes.

therefore do not represent interactions that occur in real multicomponent processes. These flux models can be used in an integrated representation assuming some average (or weighted) values for the parameters. More detailed models are in the appropriate chapters of the general references mentioned earlier.

Resistances in Series

As in the pressure-driven membrane processes for liquids, there can be a number of resistances in series with the membrane. Usually, the mass transfer boundary layer resistance on the gas or vapor side of the membrane is considered small with respect to the diffusional resistance in the membrane, and only the liquid side boundary layer is considered (if a liquid is present). For membranes with high selectivity and flux, ignoring the possibility of a gas (vapor) side boundary layer can be an erroneous assumption. In general, one should again reason out the approximate mass transfer resistances for all the potential interfaces in the application. In general,

1. The first resistance is the hydrodynamic mass transfer boundary layer between the bulk solution and the first interface where partitioning (separation) can occur. That layer is analogous to the concentration boundary layer mentioned previously.
2. With respect to fouling, in gas and vapor separations, we may need to consider the possibility of
    - an adsorption layer from reversible equilibrium partitioning onto the surface of the membrane that changes the membrane's properties
    - possibly a fouling layer from irreversible adsorption or deposition
3. Next, the mass transfer through the membrane itself is encountered, and the appropriate flux equation would be used.
4. Finally, the hydrodynamic mass transfer resistance between the permeate interface and the bulk permeate fluid is encountered. In bioremediation applications of gas and vapor permeation, the permeate side of the membrane is likely to be a biofilm or liquid phase. In the latter case, a liquid phase mass transfer resistance will most likely be needed. If a biofilm* is on the permeate side, the techniques discussed in the section on reactive membranes are germane.

Processes

GAS PERMEATION

In this process, there is a relatively large transmembrane total pressure gradient across the membrane. This creates a partial pressure (fugacity) driving force for diffusion of each gas-phase species. The rate of transport (permeation) of each solute is proportional to its solubility and diffusivity in the membrane. The product of these two is referred to as the gas's permeability. The ratio of two gases' permeabilities is their separation factor or the membrane's selectivity between those two species. Gas permeation membranes are a well established and growing industrial technology. For example, membrane systems that are effective at sepa-

---

*The aggregates of cells, extracellular polymer substance, and other particulate matter accumulated at surfaces is termed a biofilm. See the chapter on Biofilm Process Fundamentals in this volume, p. 511.

rating O_2 from N_2, CO_2 and H_2S from CH_4 and higher alkanes, and H_2 from N_2 and NH_3 are in place.

A very broad range of polymeric materials is available, including polyimides, polysulfones, and cellulose acetates. For the most part, glassy polymers are optimum for providing highly selective membranes. The intrinsic permeation rate through these materials is typically quite low, requiring very thin separating layers on a more open support structure. Relatively few defects in the skin layer of gas separation membranes can cause a complete loss of selectivity due to the orders of magnitude higher mass transfer from viscous flow through flaws versus diffusion through the solid film.

Module elements are usually hollow fiber or spiral-wound. Significant material science development continues in gas separation membranes, including the evolution of inorganic, ultramicroporous ceramic elements (Brinker et al., 1993); carbogenic molecular sieve materials (Rao and Sircar, 1993); metal membranes (Edlund et al., 1994); "robust" inorganic polymers (Peterson et al., 1995); and "tunable" polymers (Mattes et al., 1993). Though the development of the first three of these materials was spurred by the desire for high-temperature separations, some of the currently available ceramic membrane elements can provide useful characteristics for low temperature applications, primarily due to their ability to withstand harsh process and cleaning environments.

For gas permeation, typical engineering parameters are

J_i solute flux, m^3 m^{-2} s^{-1}, where gas volume is at STP**

P_i solute permeability, an intrinsic property of the membrane material, m^3 m m^{-2} s^{-1} Pa^{-1} (more typically solute permeability is listed in barrer units, 1 barrer = 10^{-10} cm^3 (STP) cm cm^{-2} s^{-1} $cmHg^{-1}$), where gas volume is at STP

P_i/l solute permeance (in the jargon this is called "pee over el"), flux divided by partial pressure driving force, m^3 m^{-2} s^{-1} Pa^{-1}, where gas volume is at STP

l thickness of the membrane separating layer, m

α_{ij} selectivity between components i and j, P_i/P_j, dimensionless

Flux Expression for Gas Separation

$$J_i = \frac{P_i}{l}(p_i^f - p_i^p) \tag{32}$$

**STP means volume measured at standard conditions of temperature (273.15 K) and pressure (101.32 kPa).

where p_i^f is the partial pressure of component i in the feed and p_i^p is its partial pressure in the permeate.

Pervaporation

A transmembrane partial pressure (fugacity) gradient is used to drive the permeation of volatile components from a liquid stream on the feed side into the vapor phase on the permeate side. Typically, a low-temperature condenser and vacuum source are used to maintain the low solute fugacity on the permeate side of the membrane. This gradient can also be supplied by sweeping with an inert gas (such as N_2) on the permeate side. In this case, the sweep gas would need to be regenerated and recirculated. Pervaporation is generally envisioned as three sequential steps: selective sorption into the membrane on the feed side, selective diffusion through the membrane, and desorption into the vapor phase on the permeate side.

One of the major attractions of pervaporation is the prospect of easily moving through an azeotropic composition that would bottleneck conventional vapor-liquid separations. The membrane is a mass-separating agent and acts as an additional component in the vapor-liquid equilibrium condition. Thus, it "breaks" the azeotrope on the feed side of the membrane. Currently, pervaporation is an emerging unit operation with the majority of applications in ethanol/water separation. Cross-linked, elastomeric or other highly swellable polymers [e.g., poly(vinyl alcohol), polydimethylsiloxane, poly(vinyl pyrollidone)] are typically used for pervaporation, but glassy and ion-exchange polymers can also be used. A very complete summary of experimental pervaporation data for a wide range of membranes and aqueous mixtures is provided by Leeper (1992).

Due to the phase change into the permeate side, net cooling is experienced. Module designs need to address the heating requirements to keep the feed at the appropriate temperature. Using an isothermal sweeping gas is one approach. An additional concern in creating a module is maintaining the integrity of membrane seals and gaskets in contact with the concentrated solvent components in the mixture. A variety of manufacturers have addressed these needs with the result that modules using plate and frame, spiral-wound, tubular, and hollow fibers are available in a variety of polymers.

For pervaporation, typical engineering parameters are

J_i　solute flux, $m^3 \ m^{-2} \ s^{-1}$, where gas volume is at STP

α_{perv}　pervaporation selectivity (ratio of the compositions of i and j in the permeate to their composition in the feed), dimensionless

β pervaporation enrichment factor (ratio of compression of i in the permeate to its composition in the feed), dimensionless

p_i' partial pressure of component i in equilibrium with its feed side, liquid phase concentration, kPa

p_i'' partial pressure of component i on the permeate side, kPa

Flux Expression for Pervaporation

$$J_i = \frac{D_{i0}}{A_i l}[\exp(A_i S_i p_i') - \exp(A_i S_i p_i'')] \tag{33}$$

$$D_i = D_{i0}\exp(A_i c_i) \tag{34}$$

$$c_i = S_i p_i \tag{35}$$

The adiabatic temperature change in a pervaporation process containing two primary components (A and B) is given by

$$\Delta T = \frac{\overline{J}(\Delta H_{v,B} + (\Delta H_{v,A} - \Delta H_{v,B})\overline{w}_{Ap})}{Q_f(C_B + (C_A - C_B)w_{Af})} \tag{36}$$

with

ΔT temperature drop on the feed side between inlet and outlet of module, K

$\Delta H_{v,i}$ latent heat of vaporization per unit mass, j kg^{-1}

w_{Af} inlet mass fraction of component A in the feed, dimensionless

Q_f inlet feed flow rate per unit of membrane area, kg m^{-2} s^{-1}

C_i heat capacity of component i, j kg^{-1} K^{-1}

$\overline{J}$ average transmembrane flux, kg m^{-2} s^{-1}

$\overline{w}_{Ap}$ average mass fraction of component A in the permeate, dimensionless

Vapor Permeation

This process combines aspects of both gas separation and pervaporation. Both the feed and permeate are vapors, so no phase change occurs, but lower transmembrane partial pressure gradients are sufficient to obtain adequate membrane fluxes. The membrane materials and modules are often similar to those for pervaporation. The mechanism of selectivity is also similar to that of pervaporation systems in that preferential sorption/swelling is the dominant phenomenon. High flux and selectivity in the membrane can cause a feed side mass transfer boundary layer.

Many materials, including rubbers, have been successfully applied to a variety of applications, including recovery of refrigerants and blowing agents from purge airstreams, air dehydration, and separation of higher alkanes from methane. Spiral-wound, hollow fiber, and tubular modules are all useful. Driving forces are typically provided by combinations of feed compression, vacuum pumps, sweep streams, and low-temperature condensers.

For vapor permeation, typical engineering parameters are

J_i solute flux, m³ m⁻² s⁻¹, where gas volume is at STP

P_i solute permeability, an intrinsic property of the membrane material, m³ m m⁻² s⁻¹ Pa⁻¹, where gas volume is at STP

P_i/l solute permeance, flux divided by partial pressure driving force, m³ m⁻² s⁻¹ Pa⁻¹, where gas volume is at STP

l thickness of the membrane separating layer, m

α_{ij} selectivity between components i and j, P_i/P_j, dimensionless

Flux Expression for Vapor Permeation

$$J_i = \frac{P_i}{l}(p_i^f - p_i^P) \tag{37}$$

SIMPLE PROCESS DESIGN NOTES

Module Mass Balance

The purpose of this section is to illustrate a simple modeling approach for estimating the overall performance of an activity-driven, gas, and vapor membrane module. The same mass transfer boundary layer models as previously presented for the pressure-driven liquid processes can be applied to gas and vapor membrane processes as well. We use the same module modeling concept as previously described—that both the feed and permeate sides are perfectly mixed. Figure 10(a) illustrates the nomenclature for gas and vapor permeation and Figure 10(b) for pervaporation.

External mass balance for gas and vapor permeation:

$$F_f = F_p + F_r \tag{38}$$

$$\Theta = F_p/F_f = \text{stage cut or recovery factor}$$

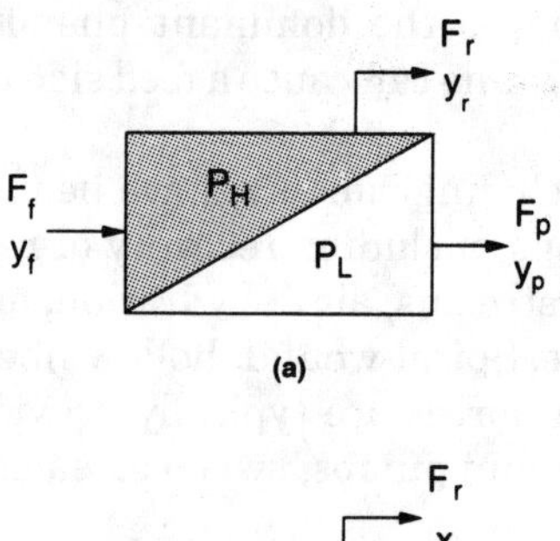

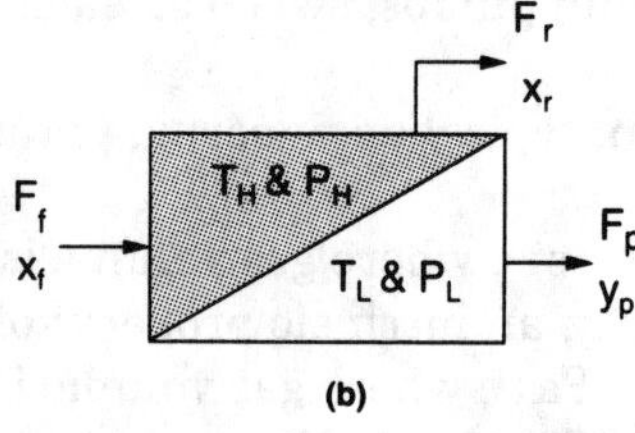

Figure 10. Schematic membrane modules for: (a) gas and vapor permeation and (b) pervaporation. F = total molar flow rate, $y(x)$ = gas (liquid) phase mole fraction of the more permeable component, i, in the mixture. Subscripts: f = feed, r = retentate, and p = permeate. P_H and P_L are the total pressures, and T_H and T_L are temperatures, on the feed (high pressure) and permeate (low pressure) sides, respectively.

External key component balance:

$$F_f y_f = F_p y_p + F_r y_r \tag{39}$$

$$y_r = (y_f - \Theta y_p)/(1 - \Theta) \tag{40}$$

The overall molar flux of i through the membrane is given by the following expression:

$$J_i \rho_i = P_i/I * (P_H y_r - P_L y_p) * \rho_i \tag{41}$$

and

$$F_p y_p = A_x J_i \rho_i \tag{42}$$

Similar expressions can be written for the less permeable component j; the two can be divided, and after clearing fractions we obtain the quadratic expression in y_p:

$$0 = a y_p^2 + b y_p + c \tag{43}$$

$$a = \left(\frac{\Theta}{1-\Theta} + \frac{P_L}{P_H}\right)\left(\alpha_{ij}\frac{\rho_i}{\rho_j} - 1\right) \tag{44}$$

$$b = (1-\alpha_{ij})\frac{\rho_i}{\rho_j}\left(\frac{P_L}{P_H} + \frac{\Theta}{1-\Theta} + \frac{y_f}{1-\Theta}\right) - \left(\frac{1}{1-\Theta}\right) \tag{45}$$

$$c = \alpha_{ij}\frac{\rho_i}{\rho_j}\left(\frac{y_f}{1-\Theta}\right) \tag{46}$$

and the positive root between 0 and 1 is the physically significant one.

For pervaporation without feed side concentration polarization effects, the definition of the separation factor yields

$$\alpha'_{ij} = \frac{y_{pi}/x_{ri}}{y_{pj}/x_{rj}} = \frac{y_{pi}(1-x_{ri})}{x_{ri}(1-y_{pi})} \tag{47}$$

The familiar material balance approach results in

$$x_{ri} = (x_{fi} - \Theta y_{pi})/(1-\Theta) \tag{48}$$

$$y_{pi} = \frac{\alpha'_{ij}x_{ri}}{1+(\alpha'_{ij} - 1)x_{ri}} \tag{49}$$

$$0 = ay_{pi}^2 + by_{pi} + c \tag{50}$$

$$a = (\alpha'_{ij} - 1)\Theta \tag{51}$$

$$b = (\Theta - 1) + (1-\alpha'_{ij})x_{fi} - \alpha'_{ij}\Theta$$

$$c = \alpha'_{ij}x_{fi} \tag{52}$$

For pervaporation, the calculation sequence should be trial and error because α'_{ij} is a strong function of x_{ri} (as well as potentially dependent on the energy balance). An approach like the following is standard:

- Guess x_{ri}.
- Determine α'_{ij}.
- Use known or defined values of Θ and x_{fi}.
- Calculate y_{pi}.
- Iterate based on agreement between guessed and recalculated x_{ri}.

Membrane Process Design Summary

The simple material balance calculations illustrated in the prior sections allow us to obtain a coarse estimate of the membrane area required for specific mass transfer requirements. The module model assumed perfect mixing on each side of the membrane. The same approach and sets of relationships can be applied to finite increments of either membrane area or composition changes, and, therefore, we can build up a problem solution in stepwise fashion with somewhat more accuracy. This latter approach can be implemented very easily with a spreadsheet program.

Much more sophisticated theoretical, design modeling approaches can be applied to every membrane unit operation. These models often include the full conservation of mass and momentum equations numerically integrated through the module. Such an approach can more accurately represent the real flow patterns and mass transfer driving forces that exist in the variety of modules.

MEMBRANE REACTORS, BIOREACTORS, AND REACTIVE MEMBRANES

The membrane reactor can increase the efficiency (selectivity and yield) of cellular metabolic reactions by influencing the concentration of primary substrate(s) in the immediate vicinity of the biomass, by controlling the concentration of favorable and unfavorable coreactant species, and by removing inhibitory metabolites. Additionally, the membrane can act as a controlled interfacial support that promotes more uniform dispersion of catalyst and reactants. Figure 11 provides a schematic of three possible types of membrane reactors. There are obviously many other possible permutations on the general theme.

Even in bioremediation applications, the combination of chemical reaction and separation in a single, relatively thin layer follows the general paradigm of the catalytic membrane reactors. Sun and Khang (1988, 1990) provide a detailed modeling perspective of catalytic membrane reactors with the focus on comparison to other types of reactors. A complete review of catalytic membrane reactors, including detailed models, configurations, data, perspectives, and a very complete bibliography, is provided by Tsotsis et al. (1993). A theoretical comparison of membrane reactors for catalytic series and series-parallel reactions is presented by Bernstein and Lund (1993). All of these references contain general modeling approaches that would need to be adapted for specific bioreaction rate expressions.

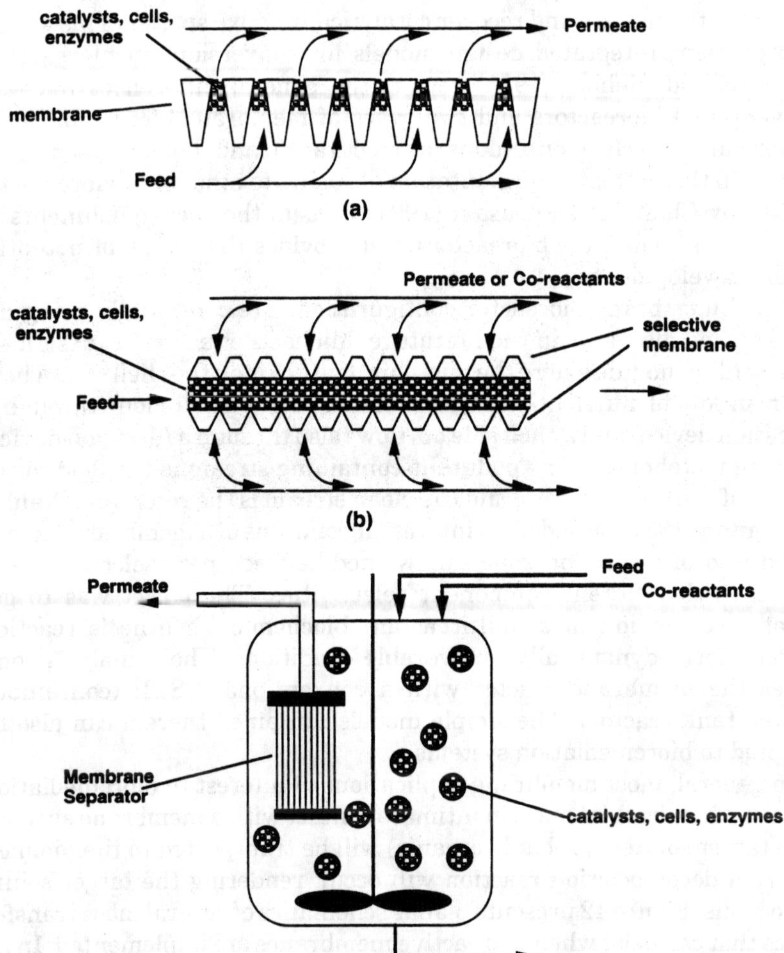

Figure 11. Schematic of three types of membrane reactors: (a) catalytic membrane reactor (CMR) or catalytic, nonselective membrane reactor (CNMR); (b) packed-bed membrane reactor (PMR); and (c) continuous stirred-tank membrane reactor (CSTMR).

Spurred by the growth of biotechnology and bioprocess development, a number of reviews of membrane bioreactors have also been published. Flaschel et al. (1983) presented a detailed discussion of the use of UF processes to separate and recover catalytically active species. This review also contains integrated design models for conversions in bioreactors. Cheryan and Mehaia (1986) review the general differences between conventional bioreactors and two types of membrane bioreactors—the membrane recycle (continuous stirred tank) and hollow fiber (plug flow)—in the context of fermentation of sugars to ethanol. A more recent review by Chang and Furusaki (1991) revisits the accomplishments in the area of membrane bioreactors and provides their view of potential future developments.

Novel membrane bioreactor configurations are continually being developed and discussed in the literature. Michaels et al. (1991) describe a hollow fiber module where the microbial culture is on the shell side. There is an inflow of nutrients and coreactants to the shell side (through a filtration device) and a shell side outflow (also through a filtration device) bearing metabolites. The pollutant-containing stream is the feed to the lumen of the hollow fibers, and the clean stream is the retentate. Stanley and Quinn (1987) provided an interesting analysis of a membrane reactor comprised of a reaction zone sandwiched between permselective membranes with different component selectivities. The object was to get greater conversion in a multireactant biochemical synthesis reaction under thermodynamically unfavorable conditions. Their analysis compares the membrane reactor with a conventional CSTR (continuous stirred tank reactor). The simple models contained therein can also be adapted to bioremediation systems.

In general, most membrane applications of interest to bioremediation consist of a microbial mass in intimate contact with a membrane surface. The target solutes (probably organic) will be transported to the biomass where a decomposition reaction will occur, rendering the target solute innocuous. Figure 12 presents planar schematics of several mass transfer cases that can exist when bioreactive membranes are implemented. In an actual application, all of them would coexist. The shapes of the profiles are stylized and will change significantly, depending on the actual system parameters.

In case I, solute A passes through a permselective membrane (with or without a feed side boundary layer) and then diffuses through a layer with thickness d in which the reaction occurs. The concentration C_{Ap} of A in the permeate is controlled by several design parameters, including the rate expression r_A, the membrane's properties, and A's diffusion coefficients in various regions.

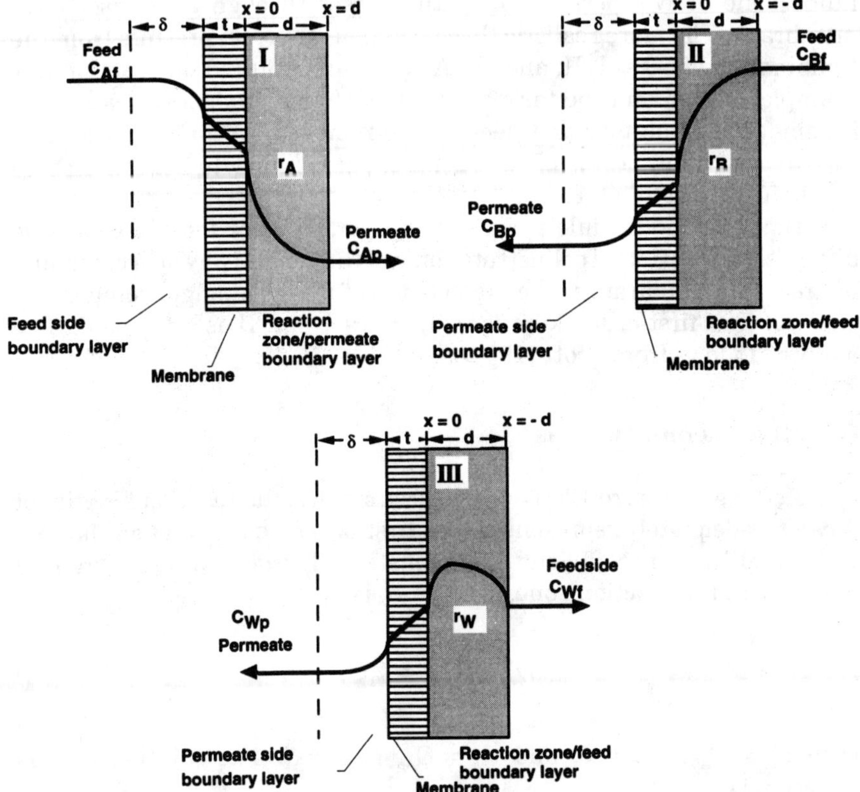

Figure 12. Membrane mass transfer with reaction. Case I refers to permselective membrane controlling solute passage from feed into the reaction zone. Case II refers to permselective membrane controlling solute passage out of the reaction zone into the permeate stream. Case III refers to mass transfer from the reaction zone into both permeate and feedstreams with membrane between reaction zone and permeate. Note that, depending on the particular membrane process being used, there could be discontinuities in the concentration profiles at the membrane interfaces due to partitioning.

In case II, solute B is now separated from the membrane by a reaction zone. It permeates through the reaction layer, the membrane, and possibly a boundary layer into its permeate stream. Its concentration C_{Bp} in the permeate stream is controlled by the same family of parameters as mentioned above.

Cases I and II represent idealized ways in which single reactants can be presented to a reactive layer that is in intimate contact with a semi-permeable membrane. We can also envision case III in which solute W is formed in the reaction layer and permeates into two different bulk fluids. One of W's permeation paths passes through a permselective membrane. The more realistic three-component process results from the combination of cases I, II, and III. A and B are required for reaction (for example, oxygen and the target pollutant). Some product of reaction W is somewhat inhibitory and needs to be removed. In the simplest case, the concentration profile of each component is linked to the other three through the reaction term.

Estimating the module's engineering performance for these cases is difficult, but possible. To illustrate some of the general physical reasoning for reactive membranes, the steady-state flux for a single component, experiencing first-order kinetics, for cases I and II is presented. The analysis follows from Scott (1994).

Reactive Membrane Case I

Referring to Figure 12(I), we will assume that the reaction kinetics of A can be adequately represented by a first-order expression and that A's effective diffusion coefficient D_{Ar} is constant. The governing differential equation in the reaction zone, $0 \leq x \leq d$, is

$$D_{Ar} \frac{d^2 C_A}{dx^2} = k_1 C_A \tag{53}$$

with $C_A = C_{A0}$ at $x = 0$, and $C_A = C_{Ap}$ at $x = d$. The solution of this equation is

$$C_A = \left(C_{A0} \sinh\left(\phi \left(1 - \frac{x}{d} \right) \right) + C_{Ap} \sinh(\phi x/d) \right) \Big/ \sinh \phi \tag{54}$$

with $\phi = \sqrt{(k_1 d^2 / D_{Ar})}$. The flux at the membrane/reaction zone interface, $x = 0$, is

$$N_{A,x=0} = \left(1 - \frac{C_{Ap}}{C_{A0} \cosh \phi} \right) \frac{D_{Ar}}{d} C_{A0} \frac{\phi}{\tanh \phi} \tag{55}$$

At steady state, this flux into the reaction/permeate boundary region will equal the flux from the bulk liquid feed through the feed side boundary layer and the membrane. By using the techniques of matching resistances in series, we can eliminate C_{A0} and express the flux from the feed side in terms of the feed and permeate sides' bulk concentrations.

The flux through the feed side boundary layer is

$$N_A = h_f(C_{Af} - C_{At}) \tag{56}$$

where h_f is the mass transfer coefficient $\approx D_A/\delta$ in the bulk feed and C_{At} is the concentration in the bulk liquid at the membrane interface on the feed side. The flux of solute through the membrane is given by the general mass transfer form:

$$J_A = \frac{D_{Am}m}{t}(C_{At} - C_{A0}) = P_A(C_{At} - C_{A0}) \tag{57}$$

where D_{Am} is the diffusion coefficient of A in the membrane; m = partition coefficient of A between bulk phase and membrane phase, $m = C_m/C$; and t = thickness of membrane separating layer. In principle, an effective permeance P_A, a property of the membrane as made, can be defined.

After appropriate algebra, the steady-state flux into the reaction zone is given by

$$N_{A,x=0} = \left(\frac{1}{P_A} + \frac{1}{h_f} + \frac{1}{K_0}\right)^{-1}\left(C_{Af} - \frac{1}{\cosh\phi}C_{Ap}\right) \tag{58}$$

where $K_0 = (D_{Ar} \cdot \phi)/(d \cdot \tanh\phi)$.

In order to complete a mass balance around the reaction zone, we need to calculate the flux leaving the reaction zone, at $x = d$. This is given by

$$N_{A,x=d} = K_2(P_A'C_{Af} + K_3C_{Ap}) \tag{59}$$

where

$K_2 = K_0/([K_0 + P_A']\cosh\phi)$
$P_A' = (1/P_A + 1/h_f)^{-1}$
$K_3 = [K_0(1 - \cosh^2\phi) - P_A'\cosh^2\phi]/\cosh\phi$

Reactive Membrane Case II

A similar approach is taken for case II in Figure 12. We will call the solute B in order to maintain consistency with the figure. In this case, the steady-state flux of B at $x = 0$ is leaving the reactive/boundary layer $(-d \leq x \leq 0)$ and permeating into the membrane $(0 \leq x \leq t)$. The governing equation is

$$D_{Br}\frac{d^2C_B}{dx^2} = kC_B \tag{60}$$

with $C_B = C_{Bf}$ at $x = -d$, $C_B = C_{B0}$ at $x = 0$, and $D_{Br} =$ the effective diffusion coefficient of B in the reaction/boundary layer. The flux at the membrane/reaction interface is

$$N_{B,x=0} = \frac{\phi}{d}\coth\phi\, D_{Br}\left(C_{Bf}\left\{\cosh\phi - \frac{\sinh\phi}{\coth\phi}\right\} - C_{B0}\right) \tag{61}$$

Again, using resistances in series, we arrive at the flux of B through the membrane into the permeate as

$$N_{B,x=0} = \left[\frac{1}{P_B} + \frac{1}{h_p} + \frac{1}{H_0}\right]^{-1}(K_1C_{Bf} - C_{Bp}) \tag{62}$$

with

$P_B =$ the membrane permeance for B
$h_p =$ permeate side mass transfer coefficient $\approx D_B/\delta$
$H_0 = (\phi/d)D_{Br}\coth\phi$
$K_1 = \cosh\phi - (\sinh\phi/\coth\phi)$
$\phi = \sqrt{(kd^2/D_{Br})}$

Again, in order to complete the mass balance around the reaction zone, the flux into the reactive/boundary layer (at $x = -d$) is given by

$$N_{B,x=-d} = H_1C_{Bf} - H_2C_{Bp} \tag{63}$$

where

$H_2 = (1/\cosh\phi)[(1/P'_B) + (1/H_0)]^{-1}$
$P'_B = [(1/P_B) + (1/h_p)]^{-1}$
$H_1 = H_0(1 - K_1H_2/P'_B)$

Reactive Membrane Case I: Stage Mass Balance

We will again use the simple module mass balance to illustrate the effects of the various physical parameters on the efficiency of removal of a solute from the feedstream for the case I configuration. We assume that the feed and permeate side are well mixed so that the composition in each compartment is the same as that leaving it. We then write the mass balance for each side and solve for the compositions. (A similar approach can be developed for case II.)

For the feed side,

$$Q_f C_{Af,in} - Q_f C_{Af} = a_m N_{A,x=0} \tag{64}$$

where Q_f is the volumetric flow rate on the feed side and a_m is the membrane area. For the permeate side,

$$Q_p C_{Ap} - Q_p C_{Ap,in} = a_m N_{A,x=d} \tag{65}$$

where Q_p is the volumetric flow rate of the permeate.

After substituting the equations for the fluxes and algebraic manipulation, we can derive explicit expressions for C_{Af} and C_{Ap} in terms of known parameters. We have used these equations in a simple spreadsheet model that increments the membrane area in small steps (0.1 m) and uses the exit streams from the current step as the inlet to the next. This technique is a conservative approximation of a cocurrent process but is sufficient to illustrate the effects of the key process variables, h_f, P_A, d, k_1, and D_{ar}. The last three combine to yield the Hatta modulus, ϕ (defined previously), a ratio of the time scales for mass transfer to reaction—the higher the Hatta modulus the greater the extent of reaction.

Figure 13 illustrates that, as the Hatta modulus increases, the transport through the reaction zone becomes slower, and the total movement of solute out of the feedstream per unit membrane area is less. But the increased residence time in the reaction zone means that more of the permeated solute is being reacted versus just passing through into the permeate.

The effect of changing the membrane's permeance is shown in Figure 14. Clearly, for the base conditions used in this model, the rate-limiting mass transfer step is in the membrane. As the overall flux out of the feed zone increases, more of the solute reaches the permeate stream unreacted in the reaction zone. It is important to remember that, in this idealized model, the reaction rate is assumed to be first-order. Therefore, it does not saturate (when the reaction rate does not change with further increase of reactant concentration) as it might in many systems of interest in bioremediation. The amount of unreacted solute could very likely increase to a greater extent than indicated in this figure.

When the membrane permeance is high enough ($P_A = 1 \times 10^{-2}$ cm/s), we can illustrate the effect of changing the mass transfer boundary layer resistance. This is shown in Figure 15. The major effect on solute removal

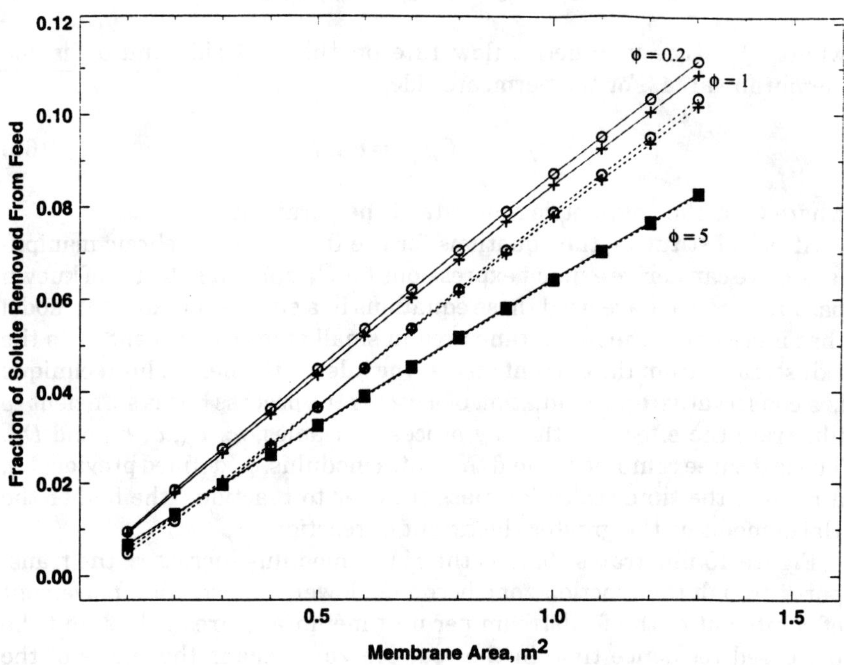

Figure 13. The effect of changing the Hatta modulus ϕ by varying the diffusion coefficient in the reaction boundary layer. The ratio of the feed (Q_f) to permeate (Q_p) side volumetric flow rates is 10, with $Q_f = 100$ cm³/s; the membrane permeance (cm/s), $P_A = 1 \times 100^{-3}$ cm/s; the feed side mass transfer coefficient, $h_f = 1$ cm/s; the pseudo first-order rate constant, $k^1 = 10$ s^{-1}; and the thickness of the reaction boundary layer, $d = 1 \times 10^{-3}$ cm. For any plotting symbol, the dashed line represents the fraction of the initial solute feed that has been reacted in the reaction zone, and the solid line is the total fraction that has left the feedstream.

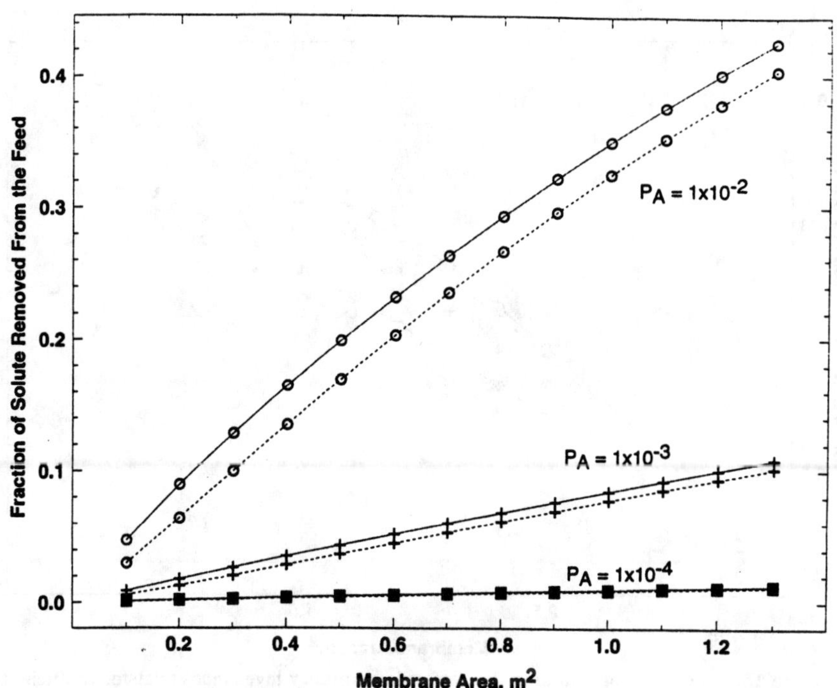

Figure 14. The effect of changing the membrane permeance P_A at a constant Hatta modulus, $\phi = 1$ (the diffusion coefficient in the reaction zone, $D_{Ar} = 1 \times 10^{-5}$). All other parameters and plot label conventions are as in Figure 13.

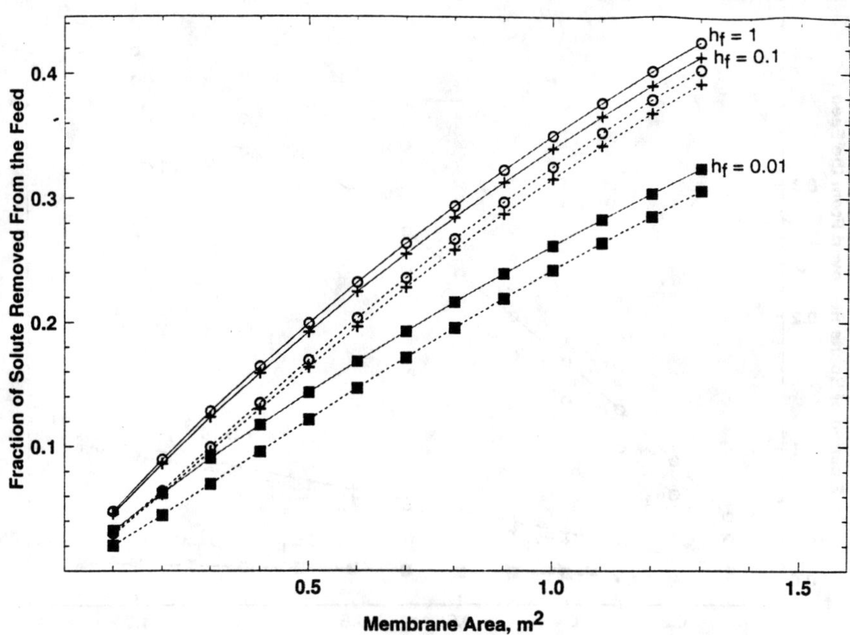

Figure 15. The effect of changing the feed side boundary layer mass transfer coefficient h_f at a constant membrane permeance, $P_A = 1 \times 10^{-2}$, and a constant Hatta modulus, $\phi = 1$ (the diffusion coefficient in the reaction zone, $D_{Ar} = 1 \times 10^{-5}$). All other parameters and plot label conventions are as in Figure 13.

occurs when h_f is on the same order as the membrane permeance ($h_f =$ 0.01). In all cases, the overall mass transfer, including through the reactive layer, is fast enough that a significant fraction of the solute remains unreacted.

Recently, Freitas dos Santos and Livingston (1995) described experimental and theoretical studies on a membrane bioremediation reactor for the oxidation of 1,2-dichloroethane (DCE). Their modeling combines the physics of cases I and II with a Monod reaction model for the biofilm growth kinetics resulting from the O_2 and DCE concentrations. The oxygen availability is represented as B in our case II schematic and the DCE by A in case I. Their model constitutes a nonlinear boundary value problem that must be solved numerically. The general conclusion from their numerical modeling (that was substantiated by experimental results) was that the DCE mass transfer with reaction through the membrane and biofilm into the airstream was very low (the fraction of solute permeated $\approx$ the fraction of solute reacted, i.e., higher ϕ). This results in less loss of DCE into the air-stripping stream—a desirable result—but also a slower rate of removal of DCE from the wastewater. They did not vary any of the mass transfer design parameters in their reported modeling results.

CONCLUSIONS

We have summarized the general mass transfer characteristics and design considerations for several membrane unit operations that can be useful as parts of an overall bioremediation process. These include reverse osmosis, vapor permeation, and pervaporation membrane devices that can be used to concentrate target contaminants, gas separation membranes that can be used to control oxygenation and to remove gaseous products, and microfiltration and ultrafiltration that can be used in a variety of ways including recovery and recycle of cell mass.

The role of membranes in creating a very efficient reaction/separation environment was illustrated in our discussion of membrane reactors and reactive membranes. This is arguably the most significant emerging role for membranes in bioremediation. There is tremendous growth in the use of permselective, thin, mass transfer films—that also provide a defined support structure—to control the inflow of substrates and nutrients and the outflow of reaction products in biological reactions. For example, a recent review by Hubbell and Langer (1995) provides numerous illustrations of the design requirements for membranes in implantable medical applications. There are many parallels between these and applications in industrial bioreactors and bioremediation processes. The material science advances in creating suitable membrane structures and module designs are certain to be cross-cutting.

REFERENCES

Anderson, G. K, Saw, C. B., and Fernandes, M. I. A. P. 1986. "Application of Porous Membranes for Biomass Retention in Biological Wastewater Treatment Processes," *Process Biochem.*, 21(6):174–182.

Atkin, J. and Gorsuch, R. G. F. 1990. "Apparatus and Method for in vivo Plasma Separation," US Patent 4,950,224.

Atkin, J. and Gorsuch, R. G. F. 1992a. "Apparatus and Method for Kidney Dialysis Using Plasma in Lieu of Blood," US Patent 5,151,082.

Atkin, J. and Gorsuch, R. G. F. 1992b. "Apparatus and Method for Selective Separation of Blood Cholesterol; in vivo Separation of Plasma from Blood," US Patent 5,152,743.

Atkin, J. and Gorsuch, R. G. F. 1993. "Transvivo Extraction Catheter Device," US Patent 5,224,926.

Atkin, J., Cooper, T. G., and Gorsuch, R. G. F. 1993. "Apparatus and Method for Direct Measurement of Blood Parameters," US Patent 5,242,382.

Belfort, G. ed. 1984. *Synthetic Membrane Processes, Fundamentals and Water Applications.* Orlando, FL: Academic Press, Inc.

Bernstein, L. A. and Lund, C. R. F. 1993. "Membrane Reactors for Catalytic Series and Series-Parallel Reactions," *J. of Membrane Sci.*, 77:155–164.

Brewster, M. E., Chung, K-Y., and Belfort, G. 1993. "Dean Vortices with Wall Flux in a Curved Channel Membrane System 1. A New Approach to Membrane Module Design," *J. Membrane Sci.*, 81:127–137.

Brinker, C. J., Ward, T. L., Sehgal, R., Raman, N. K., and Hietala, S. L. 1993. "Ultramicroporous Silica-based Supported Inorganic Membranes," *J. of Membrane Sci.*, 77:165–175.

Bungay, P. M., Lonsdale, H. K., and de Pinho, M. N. eds. 1986. *Proceedings of the NATO Advanced Institute on Synthetic Membranes: Science, Engineering and Applications. Series C, Vol. 181.* Dordrecht, NL: D. Reidel Publishing Co.

Chang, H. N. and Furusaki, S. 1991. "Membrane Reactors: Present and Prospects," in Fiechter, A., ed. *Advances in Biochemical Engineering/BioTechnology-44.* Berlin, GDR: Springer-Verlag, 27–64.

Chang, J., Manem, J., and Beaubien, A. 1993. "Membrane Bioprocesses for the Denitrification of Drinking Water Supplies," *J. Membrane Sci.*, 80:233–239.

Cheryan, M. and Mehaia, M. A. 1986. "Membrane Bioreactors," *Chemtech*, Nov.:676–681.

Chiemchaisri, C. and Yamamoto, K. 1994. "Performance of Membrane Separation Bioreactor at Various Temperatures for Domestic Wastewater Treatment," *J. Membrane Sci.*, 87:119–129.

Chung, K-Y., Bates, R., and Belfort, G. 1993. "Dean Vortices with Wall Flux in a Curved Channel Membrane System 4. Effects of Vortices on Permeation Fluxes of Suspensions in Microporous Membranes," *J. Membrane Sci.,* 81:139–150.

Conover, S. B. 1993. "Membrane Supported Biofilm Reactors—Phototrophs to Anaerobes," in *Proceedings of the Eleventh Annual Membrane Technology/Separations Planning Conference.* Norwalk, CT: Business Communications Co., Inc.

Cote, P. L, Deutschmann, A. A., Pedersen, S. K., Rodrigues, C. F., and Smith, B. M. 1993. "Frameless Array of Hollow Fiber Membranes and Method of Maintaining Clean Fiber Surfaces While Filtering a Substrate to Withdraw a Permeate; Headers and Fibers with Accumulation," US Patent 5,248,424.

Cote, P. L, Lipski, C. L., and Maurion, R. P. 1992a. "Frameless Array of Hollow Fiber Membranes and Module Containing a Stack of Arrays; Shell with Removable Fluid Couplings; Laminated Split-Clip Headers; Fluid Tight Conduit; Transverse Feedstream Flow; Dialysis; Ultrafiltration; Reverse Osmosis," US Patent 5,104,535.

Cote, P. L, Lipski, C. L., Maurion, R. P., and Pedersen, S. K. 1993. "Cartridge of Hybrid Frameless Arrays of Hollow Fiber Membranes and Module Containing an Assembly of Cartridges; Channels and Grooves Traversing Fibers, a Shell of Potting Resin, Feeding Flow over Fibers to Effect Separation and Removal of Concentrate," US Patent 5,182,019.

Edlund, D., Friesen, D., Johnson, B., and Pledger, W. 1994. "Hydrogen-Permeable Metal Membranes for High-Temperature Gas Separations," *Gas Sep. & Pur.,* 8:131–136.

Flaschel, E., Wandrey, C., and Kula, M-R. 1983. "Ultrafiltration for the Separation of Biocatalysts," in Fiechter, A., ed. *Advances in Biochemical Engineering/BioTechnology-26.* Berlin, GDR: Springer-Verlag, 73–142.

Freitas dos Santos, L. M. and Livingston, A. G. 1995. "Novel Membrane Bioreactor for Detoxification of VOC Wastewaters: Biodegradation of 1,2-dichloroethane," *Wat. Res.,* 29:179–194.

Grant, D. C., Liu, B. Y. H., Fischer, W. G., and Bowling, R. A. 1989. "Particle Capture Mechanisms in Gases and Liquids: An Analysis of Operative Mechanisms," *J. Environ. Sci.* 42:43–51.

Grant, D. C. and Zahka, J. G. 1990. "Sieving Capture of Particles by Membrane Filters from Clean Liquids," *Swiss Contamination Control,* 3:160–164.

Ho, W. S. W. and Sirkar, K. K., ed. 1992. *Membrane Handbook.* New York, NY: Van Nostrand Reinhold.

Hwang, S.-T. and Kammermeyer, K. 1975. *Membranes in Separations,* New York, NY: Wiley.

Hubbell, J. A. and Langer, R. 1995. "Tissue Engineering," *C&EN* (March 13, 1995) 73:42–53.

Kroner, K. H. and Nissinen, V. 1988. "Dynamic Filtration of Microbial Suspensions Using an Axially Rotating Filter," *J. Membrane Sci.*, 36:85–100.

Leeper, S. A. 1992. "Membrane Separations in the Recovery of Biofuels and Biochemicals: An Update Review." in Li, N. N. and Calo, J. M., ed. *Separation and Purification Technology*, New York, NY: Marcel Dekker, Inc., Chap 5.

Mattes, B. R., Anderson, M. R., Reiss, H., and Kaner, R. B. 1993. "The Separation of Gases Using Conducting Polymer Films," in Aldissi, M. ed. *Intrinsically Conducting Polymers: An Emerging Technology;* NATO ASI Series E: Applied Sciences—Vol. 246. Dordrecht NL: Kluwer Academic Publishers, Chapter 7, 61–74.

Michaels, A. S., Peretti, S. W., and Tompkins, C. J. 1991. "Bioreactor Process for Continuous Removal of Organic Toxicants and Other Oleophilic Solutes from an Aqueous Process Stream: Permeable, Hollow Fiber Membrane & Microorganisms for Metabolism of Poisons," US Patent 4,988,443.

Mulder, M. 1992. *Basic Principles of Membrane Technology.* Dordrecht, NL: Kluwer Academic Publishers, Inc.

Nichols, R. W. 1990. "Hollow Fiber Separation Module and Method for the Use Thereof," US Patent 4,959,152.

Nomura, H. 1989a. "Pore Size Control Using Plasma Polymerization Techniques; Using Low Molecular Monomer for Coating," US Patent 4,806,246.

Nomura, H. 1989b. "Gas Permselective Composite Membrane Prepared by Plasma Polymerization Coating Techniques; Diloxanes-Polysilanes Copolymer Blends," US Patent 4,824,444.

Peters, T. A. and Stanford, P. 1993. "Reverse Osmosis and DT-Module for Purification of Landfill Leachate," in *Proceedings Eleventh Annual Membrane Technology/Planning Conference,* Norwalk, CT, BCC Comm. Co., Inc.

Peterson, E. S., Stone, M. L., Orme, C. J., and Reavill III, D. A. 1995. "Chemical Separations Using Shell and Tube Composite Polyphosphazene Membranes," *Sep. Sci. Technol.*, 30:1573–1587.

Rao, G. and Sirkar K. K. 1978. "Explicit Flux Expressions in Tubular Reverse Osmosis Desalination," *Desalination,* 27:99–116.

Rao, M. B. and Sircar, S. 1993. "Nanoporous Carbon Membranes for Separation of Gas Mixtures by Selective Surface Flow," *J. of Membrane Sci.*, 85:253.

Rautenbach, R. 1986. "Process Design and Optimization," in Bungay, P. M., Lonsdale, H. K., and de Pinho, M. N. (eds.). *Proceedings of the NATO Advanced Institute on Synthetic Membranes: Science, Engineering and Applications. Series C, Vol. 181.* Dordrecht, NL: D. Reidel Publishing Co, 457–521.

Reid, C. E. 1966. "Principles of Reverse Osmosis," in Merten, U., ed. *Desalination by Reverse Osmosis.* Cambridge, MA: MIT Press, Chap 1.

Rolchigo, P. M., Raymond, W. A., and Hildebrandt, J. R. 1988. "The Improved Control of Ultrafiltration with the Use of Vorticular Hydrodynamics and

Ultra-Hydrophilic Membranes," product literature, Garfield, NJ, Membrex, Inc.

Scott, K. 1994. "Membrane Reactors for Electrochemical Synthesis Processes," *J. Membrane Sci.*, 90:161–172.

Sengupta, A., Basu, R., and Sirkar, K. K. 1988. "Separation of Solutes from Aqueous Solutions by Contained Liquid Membranes," *AIChE*, 34:1698–1708.

Stanley, T. J. and Quinn, J. A. 1987. "Efficient Mixing in Membrane Bioreactors: Minimizing Dilution Effects," *J. Membrane Sci.*, 30:243–256.

Sun, Y-M. and Khang, S-J. 1988. "Catalytic Membrane for Simultaneous Chemical Reaction and Separation Applied to a Dehydrogenation Reaction," *Ind. Eng. Chem. Res.*, 27:1136–1142.

Sun, Y-M. and Khang, S-J. 1990. "A Catalytic Membrane Reactor: Its Performance in Comparison with Other Types of Reactors," *Ind. Eng. Chem. Res.*, 29:232–238.

Tein, C. 1989. *Granular Filtration of Aerosols and Hydrosols,* Boston, MA: Butterworths.

Tsotsis, T. T., Minet, R. G., Champagnie, A. M., and Liu, P. K. T. 1993. "Catalytic Membrane Reactors," in Becker, E. R. and Pereira, C. J. eds. *Computer Aided Design of Catalysts, Chemical Industry Vol. 51.* New York, NY: Marcel Dekker, Chapter 12, 471–551.

Vieth, W. R. 1988. *Membrane Systems: Analysis and Design: Applications in Biotechnology, Biomedicine, and Polymer Science,* Munich: Carl Hanser Verlag.

Wankat, P. C. 1990. *Rate-Controlled Separations.* New York, NY: Elsevier Applied Science, Chaps. 12 and 13.

Biofilm Process Fundamentals

Z. LEWANDOWSKI AND A. B. CUNNINGHAM
Center for Biofilm Engineering
Montana State University
Bozeman, MT 59715, USA

INTRODUCTION

Biofilm processes are intrinsically related to biotransformation of contaminants in groundwater and soil, as well as in engineered reactors. In these systems, suspended microbial cells are transported to solid surfaces where they may adsorb. Some fraction of these cells desorb and return into suspension, perhaps through a diffusion-like process. If environmental conditions are favorable, the adsorbed cells grow and replicate and produce an extracellular polymer substance (EPS), which binds the cells together. The aggregate of cells, EPS, and other particulate matter accumulated at surfaces is termed a biofilm (Characklis and Marshall, 1990).

Biofilm morphology can be highly variable, ranging from patchy, discontinuous colonies to thick, continuous films. For example, subsurface biofilms capable of biotransformation of trace organic compounds generally consist of isolated microcolonies of native soil bacteria in the range of 10^5 to 10^7 cfm/g of soil. At the other extreme, packed bed bioreactors for treating volatile organic emissions may develop biofilms several millimeters thick. Once a biofilm is established, additional cells and particulates may attach to and detach from the biofilm surface. In porous media systems, biofilms may bridge across pore channels, thereby enhancing net accumulation by way of filtration.

Biofilms are metabolically active and can transform many contaminants into less environmentally harmful forms. Biofilm processes can be estimated based on knowledge of suspended cultures; however, fundamental differences between the two systems may compromise such estimates. These differences include mass transport mechanisms, reaction

511

kinetics, and the organization of physiological groups within the system. Characteristic features of biofilm versus suspended culture systems include the following.

Biotransformation Rate Limitation

While the biotransformation kinetics in suspended cultures are mostly reaction rate–limited, the biofilm systems are mainly diffusion-limited. To increase the substrate conversion rate in a suspended growth reactor, one should increase the substrate concentration. To achieve the same effect in a biofilm reactor, one should increase the mass transport rate.

Reaction Kinetics

The reaction kinetics in undefined suspended cultures is frequently described using "average" kinetic parameters for the entire population. Mixed population biofilms form consortia of highly specialized microorganisms. The substrate utilized by different fractions of the biofilm consortia may change with location inside the film along with substrate uptake rate. Consequently, the "average" kinetic parameters are difficult to define. This also breaks the link between the microscale microsensor measurements, and macroscale process modeling poses problems in scaling up the observations.

Rate-Limiting Nutrient

A suspended culture reactor can be operated as an oxygen-limited or carbon-limited reactor. The operation can be controlled based on relatively simple kinetic calculations and bulk substrate concentration. In biofilms, the distinction between oxygen-limited and carbon-limited system is not that clear. The abundance of substrate in the bulk does not necessarily imply its abundance within the biofilm. Availability of substrate is limited by its concentration in the bulk and by the mass transport rate. Stratification of substrate concentration causes stratification of physiological groups within the system. Biofilm systems are intrinsically heterogeneous.

This chapter summarizes biofilm fundamentals and is intended to provide relevant background for bioremediation practitioners. Since biofilm processes, mass transport, and biotransformation are highly interrelated, it is important to understand biofilm process fundamentals in order to improve decision making regarding the entire spectrum of bioremediation technologies.

BIOFILM FORMATION

Surface Characteristics

The nature and the state of surfaces subjected to microbial colonization can be studied from many branches of physical sciences, including thermodynamics, kinetics, electrochemistry, and solid-state physics. Two parameters are often evaluated with respect to their influence on the surface microbial colonization: surface hydrophobicity and surface charge. Both parameters can be evaluated from relatively simple measurements.

SURFACE ENERGY

The hydrophobicity (water dislike) causes the particles suspended in water to prefer to interact with particles of their own kind rather than with water. Because of this dislike, a nonpolar particle immersed in a polar solvent creates a "hole" in the bulk solution. Water is a highly structured polar solvent due to hydrogen bonds between the particles and the electrical dipoles associated with each particle. Forming a "hole" deforms the structure; breaking the hydrogen bonds between water particles to form such a hole requires energy. When two hydrophobic particles suspended in water adhere, then the surface of the new particle is smaller than the sum of surfaces of the component particles. Consequently, the energy associated with the deformation of water, which is proportional to the combined surfaces of the component particles, decreases. When two hydrophobic surfaces approach each other, the free energy of the system decreases, favoring adhesion. Conversely, hydrophilic surfaces form structures with water particles that may be stronger than those due to hydrogen bonds between water particles. Removing water particles associated with such surfaces requires energy, which increases the total free energy of the system and makes adhesion thermodynamically unfavorable (Rosenberg and Kjelleberg, 1986). In summary, the more hydrophobic the adhering surfaces are, the stronger the adhesion. Unless there is a particular affinity of the surface for an adhering particle (e.g., electrostatic attraction), the hydrophobic effects dominate the accumulation at surfaces (Stumm, 1992); therefore, much attention has been devoted to the relationship between the surface hydrophobicity and microbial adhesion.

Since the hydrophobicity of the surface influences adhesion, it is important to quantify the factors determining the energetic state of the surface. Such information can be extracted from the measurements of the contact angle formed between a liquid drop and a solid surface. When a liquid phase contacts another fluid phase and a solid surface, the liquid-fluid interface assumes an equilibrium orientation with respect to the solid surface. The angle between the liquid and the solid phases in such an arrangement is called the contact angle. For practical purposes, if the contact angle is more than 90°, the liquid does not wet the surface. In the case of water, such surfaces are called hydrophobic. On hydrophobic surfaces, water does not enter capillary pores. The other extreme is the contact angle 0°. In such a case, the water spreads over the surface without forming droplets.

Measurements of contact angle can be interpreted from the thermodynamics. Contact angle is correlated with the interfacial free energies of the liquid, vapor, and solid phase. Liquid-air (vapor) interfacial free energy is equivalent to the surface tension of the phase boundary (Loeb, 1985). The surface tension or the surface energy is a measure of work required to increase the surface area by a unit amount. In liquid-vapor systems, the liquid state has a higher density than the vapor, causing the surface molecules to be pulled inward and thereby causing the surface to contract. The relationship between the surface free energy and surface tension in solids is more complex. Solids cannot change the internal arrangements of particles as easily as fluids to acquire a new equilibrium position after applying some force to their surfaces (Myers, 1991). Young's equation presents the horizontal components of the various surface forces expressed as surface tensions:

$$\gamma_{SV} = \gamma_{SL} + \gamma_{LV} \cos\theta$$

where γ is surface tension or free energy on the phase boundary; subscripts S, V, and L are the solid and two fluid phases (V is used because, frequently, the vapors of the liquid, L, are used as the gas phase); and Θ is the contact angle.

Zisman (1964) developed a useful system to characterize the wettability of surfaces based on measurements of contact angle. It was found that, for solid surfaces with surface tension below 100 mN m^{-1} (low energy surfaces), the contact angle formed by a drop of liquid on the solid surface is primarily a function of the surface tension of the liquid (γ_{LV}). For a given solid and a homologous series of related liquids (e.g., alkanes), cos Θ can be approximated as a linear function of γ_{LV}. The intercept of such a plot at cos $\Theta = 1$ (at $\Theta = 0$) is termed the critical surface tension of wetting, γ_c, for this particular solid. It is defined as the surface tension of a liquid that would just spread on the surface of the solid to give complete wetting. Based on this concept, materials can be classified in terms of γ_c, which further induces the terminology "high- and low-surface tension (energy) surfaces." Low-surface energy is associated with high hydrophobicity.

SURFACE CHARGE

Adhesion of microorganisms to inert surfaces is governed by both electrostatic and hydrophobic interactions (Mozes and Rouxhet, 1992). Surfaces acquire electrical charges due to preferential dissolution of surface ions, preferential adsorption of ions, electrolytic dissociation of surface groups, and charges deriving from specific crystal structure. Electrical charge of colloidal particles is often measured using microelectrophoretic techniques. The results of the electrophoretic measurements are expressed in the form of ζ (zeta) potential from which, using some simplifying assumptions, surface charge can be evaluated. Microelectrophoretic measurements indicate that most bacteria, algae, and detritus in natural waters carry a net electronegative surface charge (Neihof and Loeb, 1972). Surface charge effects can be studied by applying known potentials to metal surfaces. An enhanced adsorption was observed when an anodic (positive) potential was applied to the titanium surface (Little, 1985). Thus, the electrostatic interactions between adhering surfaces influence the adsorption. This effect is somewhat mitigated by the presence of the adhesive substances produced by the microorganisms. These extracellular polymers (EPS) may bridge the microbial cell with the surface and allow negatively charged bacteria to adhere onto both negatively and positively charged surfaces.

CELL SURFACE HYDROPHOBICITY

Hydrophobicity influences adhesion when both the surface and the adhering particle exhibit a certain degree of hydrophobicity. Bacterial cell surfaces are mostly hydrophilic for practical reasons. Cells have to exchange materials with the surrounding water environment. However, the adhesion may be mediated by a small portion of the bacterial cell surface, which is hydrophobic or less hydrophilic. Despite the widely acknowledged importance of the hydrophobicity on adhesion, there is no uniform measure of bacterial hydrophobicity nor any definitive scale of values to compare hydrophobicities measured in different systems (Rosenberg and Doyle, 1990; Rosenberg and Kjelleberg, 1986). One way to evaluate bacterial hydrophobicity is by measuring the kinetics of microbial adhesion to surfaces of immiscible hydrophobic liquids (Bar-Ness et al., 1988), or solid surfaces (Fletcher and Loeb, 1979; Dexter et al., 1975).

Conditioning Film

The process of biofilm formation consists of a sequence of steps and begins with adsorption of macromolecules (proteins, polysaccharides, nucleic acids, and humic acids) and smaller molecules (fatty acids, lipids, and pollutants like polyaromatic hydrocarbons and polychlorinated biphenyls) to the surfaces. These adsorbed molecules form conditioning films that may have multiple effects: alter the physicochemical characteristics of the surface, act as a concentrated nutrient source for microorganisms, suppress or enhance the release of toxic metal ions, detoxify the bulk solution by adsorption of inhibitory substances, supply the required nutrients and trace elements, and trigger sloughing mechanisms (Chamberlain, 1992; Marshall et al., 1994).

The conditioning film changes surface hydrophobicity and the surface electrical charge. Both parameters have proven effects on the extent and kinetics of microbial attachment. Loeb and Neihof (1976) demonstrated by electrophoretic studies that platinum particles suspended in seawater depleted of organic matter were electropositive and become electronegative upon immersion in natural water. Surface charge is modified because of dissociation of chemical groups in the conditioning film and/or because of adsorption of ions from the solution. The amount of adsorbed organic material on surfaces is a function of ionic strength and can be enhanced by the anodic polarization of metal surfaces (Little, 1985).

Although there is no doubt that the conditioning film influences microbial colonization, the effect is difficult to evaluate. As accurately pointed out by Chamberlain (1992), the conditioning film is formed on surfaces immediately upon immersion, and consequently, there are no surfaces immersed in water that would be without such film to serve as a reference. However, this condition can be reasonably approximated under laboratory conditions.

Initial Colonization

Microbial cells transported with the stream of fluid above the surface interact with the conditioning film adhering tightly to the surface. It has been demonstrated that the conditioning film influences the surface charge, surface free energy, and the attachment of microorganisms (Baier, 1980). Initial colonization begins with the transport of microorganisms to the surface. This may be due to at least three different mechanisms: (1) diffusive transport due to Brownian motion, (2) convective transport due to the liquid flow, and (3) active movement of motile bacteria near the surface (van Loosdrecht et al., 1990). The most intensive mechanism is convective transport, which may exceed the other two by several orders of magnitude. Once the microbial cell is in contact with the surface, it may adhere to it or not. The probability of adsorption is termed "sticking efficiency" and depends on many factors, including surface properties, physiological state of organisms, and hydrodynamic shear stress near the surface (Escher and Characklis, 1990):

$$\text{sticking efficiency} = \frac{\text{number of cells binding to the surface}}{\text{number of cells transported to the surface}}$$

Biofilm Growth

Adherence of bacteria to a surface is usually followed by production of slimy adhesive substances, predominantly exopolysaccharides (EPS). Although the association of EPS with attached bacteria has been well documented (Fletcher and Floodgate, 1973), there is little evidence to suggest that EPS participates in the initial stages of adhesion. However, EPS definitely assists the formation of microcolonies and microbial films (Allison and Sutherland, 1987). According to Silverman et al. (1984), the synthesis of these adhesive substances is genetically controlled and is influenced by environmental factors. The EPS bridges the microbial cell with the substratum, and consequently, negatively charged bacteria can adsorb onto both negatively and positively charged surfaces.

Colonies of microorganisms initiating the biofilm formation do not necessarily consist of the same species but, instead, can be organized as communities (Gibbons and Nygard, 1970). It is possible that cells of one species associate with those of another species for mutual physiological benefit (Costerton et al., 1987). This phenomenon is presently a subject of intensive studies. The initial stage of biofilm formation is illustrated in Figure 1, where the cells adhered to a glass surface forming discrete microcolonies. As time progresses, the discrete microcolonies grow in size (Figure 2), forming heterogeneous layers of microorganisms embedded in the polymer matrix covering the entire surface. A series of events during the biofilm formation on glass surfaces is presented in Figure 3. Eventually, in the natural environment, mature biofilms form that are even more heterogeneous in nature and contain a large fraction of adsorbed and entrapped materials such as inorganic particles, e.g., clay, silt, etc., forming characteristic biofouling deposits.

Biofilm Accumulation Rate

Biofilm accumulation is the net result of microbial attachment, growth, decay, and detachment. More rigorous treatments introduce adsorption and desorption (Bryers and Characklis, 1992; Escher, 1986). However, because of complexity of microbial binding to surfaces, the terms *attachment* and *detachment* are frequently used without referring to any specific physical process as adsorption or desorption. Attachment is due to microbial transport and subsequent binding to surfaces. Growth is due to

Figure 1. Initial attachment. Bacterial cells (*Pseudomonas aeruginosa, Pseudomonas fluorescens,* and *Klebsiella pneumoniae*) are attaching to a glass surface (courtesy of Paul Stoodley, Center for Biofilm Engineering).

microbial replication and extracellular polymer production. Microbial growth rate is traditionally described by Monod kinetics:

$$\mu = \frac{\mu_{max} * S}{K_s + S}$$

where μ_{max} is maximum specific growth rate (t^{-1}), K_s is half saturation coefficient (ML^{-3}), and S is substrate concentration (ML^{-3}). Each species in the biofilm has its own growth parameters. Unlike in suspended growth reactors, in biofilms, the growth rate may depend on the spatial position of the microorganisms. Assumption that all microorganisms of the same species have the same growth parameters in biofilms appears to be overly simplistic. What should be expected, instead, is some spatial distribution of growth parameters. Despite the awareness of this situation, little is known how to handle this complicating factor and its consequences. Extracellular polymer production rate depends on the microbial species and environmental parameters. Detachment includes erosion and sloughing. Sloughing is the process due to which large pieces of biofilm are rapidly removed, frequently exposing the surface. The reasons for biofilm sloughing are not well understood. Biofilm erosion is defined as continuous removal of single cells or small groups of cells from the biofilm surface. Erosion is related to shear stress at the biofilm-fluid interface. Frequently, detachment is identified with erosion, especially in conduits. An increase in shear stress increases erosion rate and decreases biofilm accumulation rate. Empirical observations indicate that erosion rate is related to biofilm thickness and density. It is not quite clear how these parameters are related. The most frequently encountered empirical form of equation describing erosion rate E_r is

$$E_r = bX^m L^n$$

where b is the erosion coefficient, X is biofilm concentration (mass/volume), L is biofilm thickness, and m and n are empirical coefficients. Rittmann (1982) describes detachment rate as a first-order reaction with

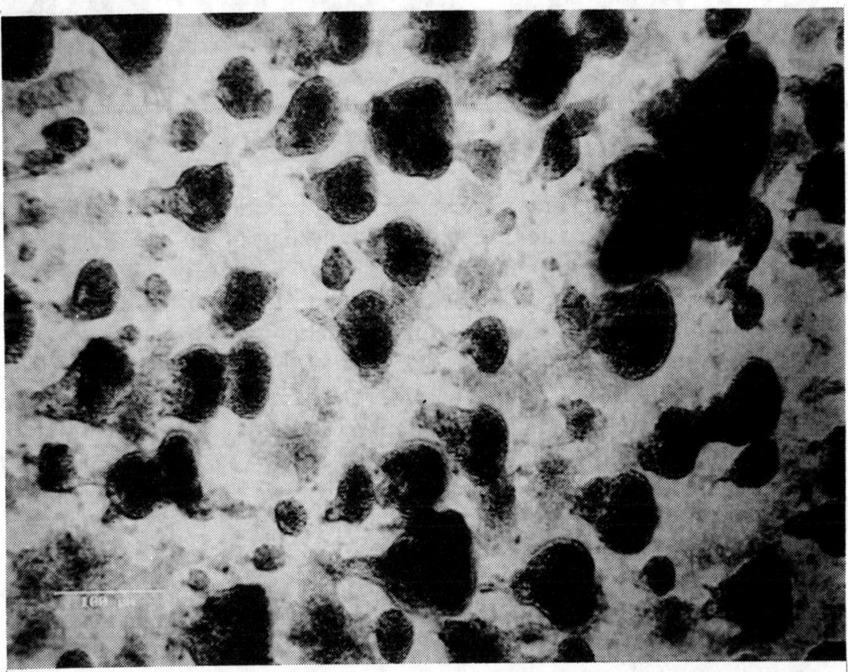

Figure 2. Bacterial cells multiply on the surface forming microcolonies. Microbial composition the same as in Figure 1 (courtesy of Paul Stoodley, Center for Biofilm Engineering).

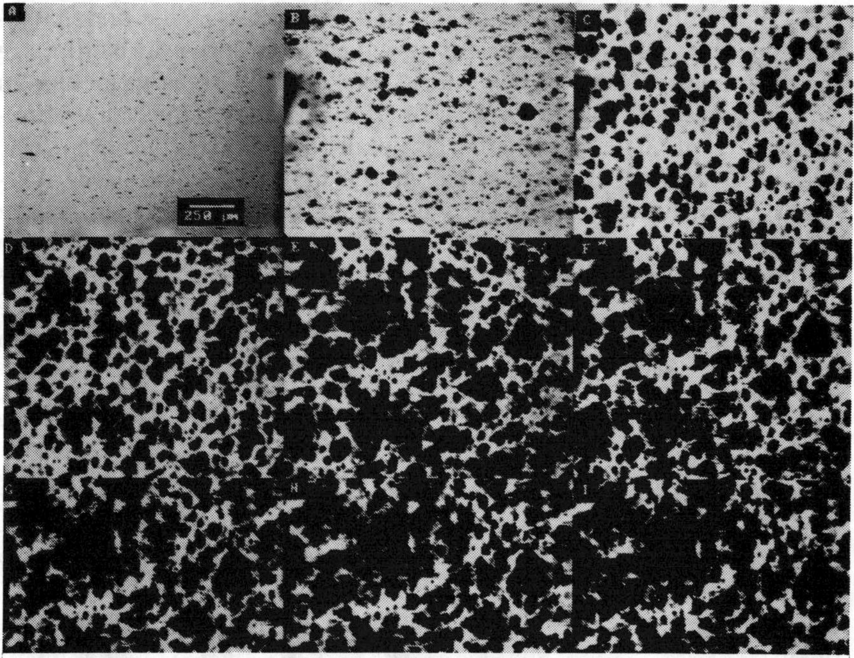

Figure 3. Microcolonies grow forming thicker deposits, the biofilm. Microbial composition the same as in Figure 1. Images taken every second day (courtesy of Paul Stoodley, Center for Biofilm Engineering.)

regard to biomass concentration ($m = 1, n = 1$). Bakke et al. (1984) report $m = 1$ and $n = 0$. Characklis et al. (1990) suggest $m = 2$ and $n = 0$. Wanner and Gujer (1986) assume $m = 1$ and $n = 2$. The nature of the coefficient (b) is not well recognized. Rittmann (1989) postulates that the coefficient (b) represents all detachment mechanisms, while Bakke et al. (1984), Wanner and Gujer (1986), and Characklis (1990) relate it only to erosion. Rittmann (1989) reported different expressions for the relation between shear stress and detachment coefficient, depending on the biofilm thickness. This further reinforces the view that the complexity of the process of biofilm erosion considerably exceeds the simplistic empirical expression.

Biofilm Structure

Mathematical models (Rittmann and Manem, 1992; Wanner and Gujer, 1986) assume that biofilms are planar structures with homogeneous cell distribution. Mass transfer through the mass boundary layer and within the biofilm is assumed to be diffusional and perpendicular to the surface to which it is attached (the substratum). However, microscopic observations indicate that biofilms are not flat and the distribution of microorganisms is not uniform. Instead, multi-species biofilms form highly complex structures containing "voids," "channels," "cavities," "pores," and "filaments," with cells arranged in "clusters" or "layers." Such complex structures were found in a wide variety of biofilms such as methanogenic films from fixed-bed reactors (Robinson et al., 1984), aerobic films from wastewater plants (Eighmy et al., 1983; Mack et al., 1975), nitrifying biofilms (Kugaprasatham et al., 1992), and pure culture biofilms of *Vibrio parahaemolyticus* (Lawrence et al., 1991) and *Pseudomonas aeruginosa* (Steward et al., 1993).

Studies of the structure of biofilms using Scanning Confocal Laser Microscopy (CSLM), followed by the studies of flow near microbially colonized surfaces using Nuclear Magnetic Resonance Imaging (NMRI) (Lewandowski et al., 1992, 1993 and 1994a) and CSLM (DeBeer et al., 1994a, 1994b; Stoodley et al., 1994), delivered detailed information on the structure of biofilms and the nature of the water flow in biofilm systems. The conceptual image of biofilms became much more complex than the uniform layer with imbedded microorganisms that dominated the early studies (Massol-Deya et al., 1995; Gjaltema et al., 1994; Zhang and Bishop, 1994a, 1994b). Microorganisms in biofilms are aggregated in cell clusters or microcolonies separated by interstitial voids. The new conceptual model assumes an inherent biofilm heterogeneity and constitutes the foundation for studying biofilm structure and the consequences of this structure to the physical and chemical microenvironments.

It has been hypothesized that the biofilm structure is not a chance occurrence, but represents an optimal arrangement for the influx of nutrients; however, no direct evidence has been presented (Lawrence et al., 1991; Robinson et al., 1984). The effective diffusion coefficient in aerobic biofilms was found to be dependent on flow conditions and biofilm structure, implicating convection through pores in the biofilm (Siegrist and Gujer, 1985). Voids might enhance substrate and product fluxes throughout the biofilm by decreasing diffusional resistance or by facilitating convection. It has been demonstrated that the substrate concentration in voids is higher than that in adjacent biomass (DeBeer et al., 1994a). Integrating confocal scanning laser microscopy (CSLM) and microelectrode techniques permitted assessment of the relationship between the internal structure of biofilms and oxygen concentration profiles. CSLM enhances visualization of biofilm structures by eliminating the interference from out-of-focus objects (Caldwell et al., 1992; Wilson,

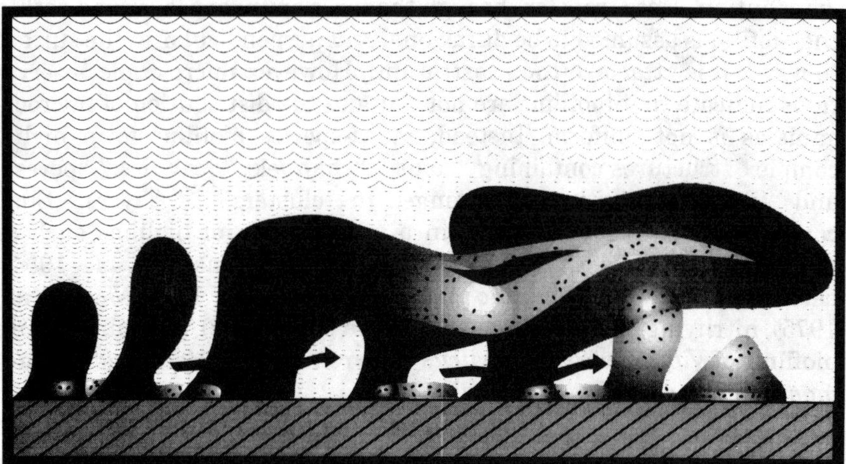

Figure 4. Structure of the hypothetical bacterial film drawn from CSLM microscopy examination of a large number of monospecies and mixed-species biofilms. Note the relatively open channels between discrete microcolonies in which bacterial cells are enclosed in a dense exopolysaccharide matrix. The arrows indicate convective flow within the water channels.

1990). Observations were done under growth conditions, while the biofilm remained physiologically active.

Based on those images, a new conceptual model of biofilm structure emerged (Figure 4). The biofilm matrix is densely concentrated around the microcolonies of cells that have produced the biopolymer and less densely distributed in the interstitial voids.

HYDRODYNAMICS IN BIOFILM SYSTEMS

Biofilm accumulation changes the surface and modifies the hydrodynamics of the system. This process seems to be particularly effective once the biofilm thickness becomes comparable to the thickness of the hydrodynamic boundary layer (Picologlou et al., 1980). Biofilm accumulation at surfaces is an autocatalytic process. Initial colonization of surface increases surface irregularity and promotes further formation of biofilm. Bouwer (1987) points out that increased surface irregularity due to biofilm formation can influence particle transport rate and biofilm attachment rate by (1) increasing convective mass transport near the surface, (2) providing shelter from shear forces, and (3) increasing surface area for attachment. A study by Siegrist and Gujer (1985) showed that biofilm irregularity can increase eddy diffusion and the external mass transfer rate into the biofilm. Lewandowski and Walser (1991), using a rotating disc reactor, demonstrated that biofilm thickness reaches maximum within the transition zone between the laminar and turbulent flow. It was postulated that, for laminar flow, the biofilm thickness was limited by transport of substrate and, for turbulent flow, by erosion due to increased shear stress.

Hydrodynamics influences biofilm accumulation in many ways. During the initial events of biofilm formation, it controls the transport of microorganisms to the surface (Duddridge et al., 1982). After the biofilm is formed, hydrodynamics controls the transport of substrate to the biofilm and the transport of metabolites from the biofilm. Shear stress is responsible for biofilm erosion. There are indications that hydrodynamics may be involved in biofilm sloughing. Biofilms developed under different hydrodynamic conditions have different physical structure and different density; those growing under high shear stress conditions tend to be more dense than those growing under low shear stress conditions.

Quantification of the relationships between hydrodynamics and biofilm accumulation is not trivial. Fluid mechanics typically describes conditions near rigid surfaces, but a biofilm consists mainly of viscoelastic polymers and its surface is soft and often filamentous. In fact, the nature of the surface can be a function of the hydrodynamics; for example, soft, filamentous biofilm surfaces can be more convoluted for low flow velocities than for higher velocities. With increasing fluid flow velocities, the filaments may yield to the flow, resulting in a smoother surface. There is no theoretical treatment that could be directly applied to describe the flow near such surfaces. Despite these difficulties, the needs of engineering practice requires that some approximation be accepted to design and to evaluate the performance of biofilm reactors. Since the only available description is the classical hydrodynamics, the suggested solutions are analogous to those described for rigid surfaces.

An influential work, presenting such a simplifying image of hydrodynamics near biofilms was published by Picologlou et al. (1980). The authors postulated that biofilm accumulation has the same effect as formation of rigid roughness elements. When the flow velocity increases, the thickness of the hydrodynamic boundary layer decreases, and these roughness elements protrude through the hydrodynamic boundary layer, causing an increase in friction resistance. Such a concept parallels the classical study by Nikuradse (1933) who studied the effects of rigid roughness elements in pipes. The analogy was familiar to the engineering community and was, therefore, readily accepted. Consequently, the hydrodynamics near surfaces covered with biofilms were described with the assumption that the biofilm formed rigid, rough surfaces.

Lewandowski and Stoodley (1995) presented a concept of an intricate interplay between hydrodynamics and viscoelastic biopolymer matrix, which explains some of the experimental observations. Initial stages of biofilm accumulation lead to the formation of discrete microcolonies firmly attached to the surface. Each of these microcolonies behaves as a bluff body attached to the reactor's wall. As water particles flow toward the leading edge of such a body, the pressure in the water increases from the free stream pressure to the stagnation pressure. The high pressure near the leading edge induces the development of boundary layers on both sides of the microcolony. At higher flow velocities, the viscosity forces are not sufficient to force the boundary layers around the back (downstream) side of the microcolony. Near the widest section of the microcolony, the boundary layers separate from each side and form two shear layers. These two shear layers in turn form a wake. Since the innermost part of the shear layers moves slower than the outermost portions, which are in contact with the free stream, the free shear layers tend to roll up into discrete vortices forming Karman's vortex trail. A regular pattern of vortices is formed in a wake that interacts with body motion and is the source of the effect called vortex-induced vibration (Griffin and Ramberg, 1974). In biofilms, it causes the development of pressure drag. The biopolymers holding the microcolony together are viscoelastic and, subjected to external force, assume elongated shapes, called streamers. Streamers are the filamentous parts of biofilm surface frequently observed by many authors. Vortices cause the streamer to move. This movement is transferred back to the underlying biofilm causing the cluster to oscillate. When a bluff cylinder is excited into resonant oscillations by an incident flow, the cylinder and its shed vortices have the same frequency near one of the characteristic frequencies of the body. This coincidence of resonance of the shedding and vibration frequencies, termed lock-on (Hall and Griffin, 1993), causes the cylinder to vibrate. Such vibration, particularly near the shedding frequency, may increase the vortex strength and result in the increase of the drag force (Bishop and Hassan, 1964). The oscillating forces cause vibrations in elastically mounted cylinders, and the system behaves as a self-excited oscillator. Such vibrations induced in elastic structures by vortex shedding can have destructive effects for the entire structure. It is interesting to speculate that a similar effect in biofilms may induce localized sloughing events. Experimentally documented flow-induced vibrations of the entire biofilm are responsible for a large portion of kinetic energy dissipation and pressure drop in conduits covered with biofilm.

Flow Velocity Profiles

In conduits, a set of average hydrodynamic parameters can be easily calculated from simple measurements as the average flow velocity and knowledge of the wall roughness. In order to evaluate the effects of hydrodynamics on biofilms, and vice versa, such a description has to be relevant for the near surface environment. There is no theoretical treatment that could be directly applied to describe the flow near biofilm covered surfaces. Therefore, the empirical description of such flow is important. One hopes that accumulation of empirical results will, in the future, help establish equations describing the flow near biofilms. Presently, such effort helps verification of existing models of biofilm accumulation. Two techniques seem to be useful for this purpose: the Nuclear Magnetic Resonance Imaging (NMRI) and Particle Image Velocimetry (PIV).

NUCLEAR MAGNETIC RESONANCE IMAGING

A versatile set of noninvasive techniques using Nuclear Magnetic Resonance (NMR) has been developed to study chemical and physical properties of small samples (see Abragam, 1961) and to allow spatial mapping, or imaging, of larger systems. Morris (1986) provides detailed descriptions of current NMR imaging methods. Most NMR imaging (and all of the imaging described in this chapter) deals with protons ^{1}H in the liquid state. While many physical parameters of liquids may be imaged, proton concentration and velocity are emphasized here. Concentration images show where fluid is within the imaging volume; given knowledge of system boundaries, the amount of fluid displaced by an intruding solid phase may also be determined by subtraction (see Majors et al., 1989 and Altobelli et al., 1991). Velocity images provide maps of selected velocity components. NMR experiments are performed in a strong, uniform, static magnetic field. Radiofrequency magnetic fields are then applied to the sample to elicit the NMR signal, and spatially linear magnetic field gradients are superposed on the static field to encode position and velocity information into the NMR signal. In addition to the usual concentration images, it is possible to use NMR to acquire information about hydrodynamic parameters of flowing systems, as reviewed by Caprihan and Fukushima (1990). In particular, it is now possible to obtain detailed velocity images of steady flows in 2-D, and there have even been exploratory studies of 3-D flows and nonsteady flows. Velocity images provide maps of selected velocity components.

Our most recent set of experiments (Lewandowski et al., 1993) employed a reactor of rectangular cross section (2.0 cm wide by .25 cm deep by 38 cm long). NMRI velocity measurements were performed at a series of flow rates in a reactor with and without biofilm. Four flow rates are presented in Figure 5, and for each flow rate, side-by-side surface-rendered plots are shown. These plots depict the velocity as a function of the spatial, cross-sectional coordinates. The cross-sectional coordinates are not shown to scale, since the height of the reactor channel is much less than the width. The profiles were obtained at 15 cm (15/.25 = 60 times the channel height) from the inlet. The results indicate that the increas-

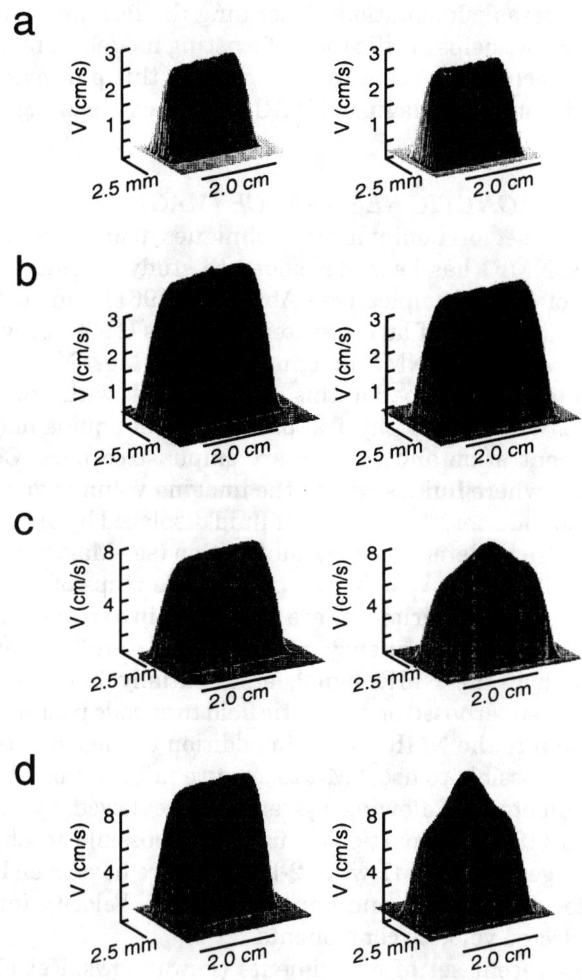

Figure 5. Flow velocity profiles in a flat plate biofilm reactor.

ing flow velocity did not affect the character of the velocity profiles in the reactor with biofilm. On the other hand, the same increase in velocity had a pronounced effect on the reactor without biofilm. For average flow velocities of 1.5 and 2.4 cm/s [Figure 5(a) and 5(b)], the reactor colonized with biofilm had the same flow velocity distribution as the reactor without the biofilm. Further increasing the average flow velocity to 4.0 cm/s resulted in formation of a jet in the reactor without the biofilm [Figure 5(c)]. The further increase in the average flow velocity to 4.6 cm/s made the jet even more pronounced in the reactor without the biofilm [Figure 5(d)].

The reason for such behavior can be explained in terms of the "entry length" required for development of the velocity profile. The entry length required for development of viscous flow was compared for reactors with and without the biofilm operated at the same flow velocity. The results in Figure 5 indicate that biofilm accumulation decreased the entry length required for fully developed flow, implying that biofilm formation was actually smoothing the walls. Although the results were obtained under laminar flow conditions and such flow is unusual in industrial settings, they showed that the biofilm accumulation may, under certain circumstances, actually stabilize the flow and that the relations between the surface roughness and biofilm formation can be quite complex. Consequently, the declaration that biofilm accumulation always increases surface roughness may be in need of revision. Based on a conceptual model of hydrodynamics presented by Lewandowski and Stoodley (1995), such behavior can be prescribed to the viscoelastic nature of the biofilms and forming complying surfaces. Water flowing over such surfaces causes the hydrodynamic boundary layer to interact with the surface. The most evident example of such interaction is that the flow remains laminar for higher Reynolds numbers, which would be possible in the case of rigid surfaces. The complying coating can also modify the turbulence in the boundary layer. An analogous example of a similar surface may be the one described by Klinzing et al. (1969). Water was pumped through thin-walled flexible tubes covered inside with polyurethane foam as a dampening medium. Using such a system, they were able to obtain hydraulically smooth surfaces up to the Reynolds number of 20,000. Such a phenomenon may explain why we have seen that the conduit covered with biofilm was more hydraulically smooth than a conduit without biofilm, at least for small flow velocities.

We also made cross sections through the velocity profiles obtained at higher flow rates, measured on Day 3 and Day 4 of biofilm growth (Figure 6). Small nonuniformities ("kinks") are apparent near the reactor walls, indicating that there is motion of the bulk solution at a level that is

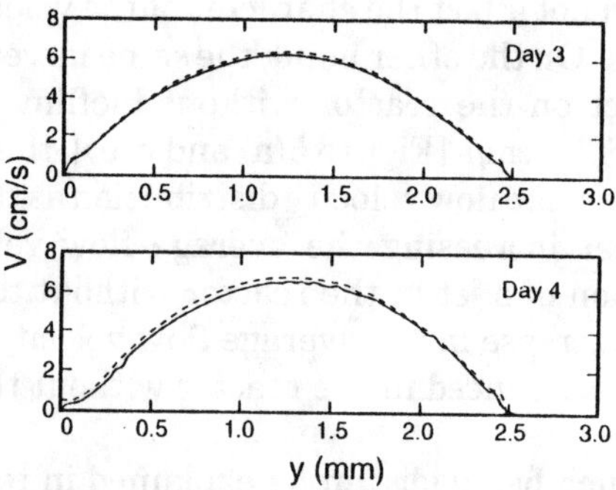

Figure 6. NMR velocity profiles across the 2.5 mm depth of the reactor.

partially occupied by the biofilm. Such a conclusion has important consequences for the mass transport mechanism in biofilm systems. It is customary to assume that the mass transport mechanism above the biofilm surface is dominated by convection and below biofilm surface by molecular diffusion. In such a case, the biofilm surface should behave as a rigid surface and, consequently, the flow velocity should reach zero at the biofilm surface. The results presented in Figure 6 indicate water movement within the space occupied by the biofilm and imply existence of convection below the biofilm surface.

PARTICLE IMAGE VELOCIMETRY

Particle image velocimetry is a technique that has been used successfully in a number of studies for flow visualization (Nielsen et al., 1993; Saleh et al., 1993; Wan et al., 1993). In our laboratory, a Bio-Rad MRC600 confocal scanning laser microscope (CSLM) was used in conjunction with an Olympus BH2 light microscope for particle tracking and simultaneous biofilm visualization. Neutral density fluorescent latex spheres, $0.282\,\mu$m diameter, were added to the biofilm reactor to achieve a final concentration of 1×10^7 particles/ml. Velocity profiles in the reactor were obtained by capturing images at various focal depths by raising or lowering the motorized stage. Particles traveling across the field of view appeared as dashed lines. The particle velocity was calculated from the distance between the dashes and the time taken to scan the number of frame lines crossed by the particle. Particle image velocimetry due to the high resolution can demonstrate and quantify the water flow in the biofilms. Figure 7 shows a series of CLM superimposed images of a single fluorescent bead, 0.3 micron in diameter, moving within the biofilm (Stoodley et al., 1994). The profiles of flow velocity can be studied both outside and inside the biofilm using this technique. Figure 8 provides a conceptual image of flow field near biofilm surfaces.

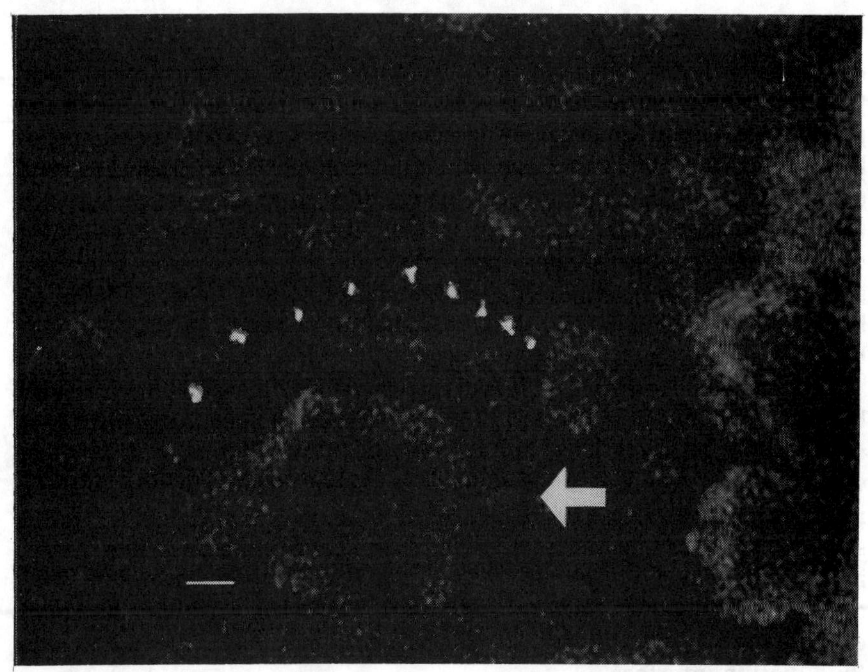

Figure 7. Fluorescent bead, 0.3 micron diameter, moving along the interstitial voids of a biofilm.

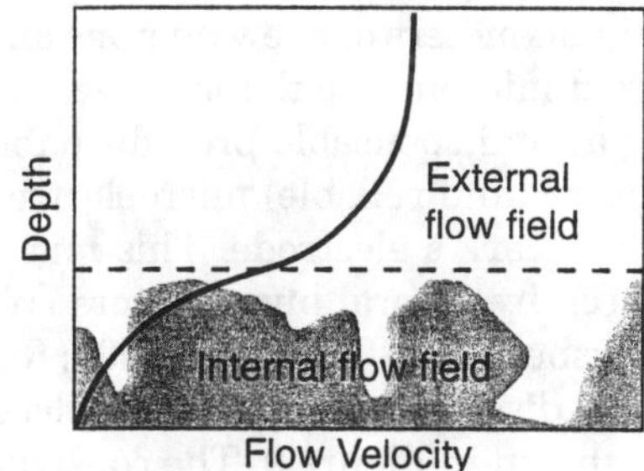

Figure 8. Flow velocity fields outside and inside the biofilm.

REACTION KINETICS IN BIOFILM SYSTEMS

Chemical Gradients in Biofilms

In biofilms more than a few microns thick, most of the chemical reactions are diffusion-limited. Consequently, each chemical compound that is consumed or produced in the biofilm system forms a concentration profile. The substrate supplied from the bulk fluid is consumed by the biofilm, thereby forming a concentration gradient that decreases with depth into the film. However, other compounds that are generated within the biofilm may diffuse outward into the bulk fluid. Each point on a concentration profile reflects local equilibrium between consumption and supply. The shape of the profile is influenced by many factors, including biofilm activity, hydrodynamics, temperature, presence or absence of biocides, reaction with substratum, etc.

The transport of dissolved substrates to the biofilm interface from the bulk solution is due to diffusion and convection. Convection is strongest in the bulk solution and decreases towards the biofilm along with the flow velocity. Consequently, the dissolved substrate concentration profile is nonlinear, and the concentration gradient is maximum at the biofilm surface. Substrate concentration within the biofilm decreases due to the internal diffusion resistance, as well as microbial consumption. The combined influence of these two processes causes the substrate profile to change slope in such a way as to produce a point of inflection at the fluid-biofilm interface.

Measuring chemical gradients in biofilms requires specialized instrumentation. Microelectrodes (microsensors) that are sensitive to specific ions, compounds, or dissolved gases are particularly useful. Microsensors are constructed with tip diameters of a few microns and are driven across the biofilm by motorized micromanipulators. Measurement of chemical gradients is becoming an indispensable procedure for biofilm research. Perhaps the most popular (and reliable) microelectrode is the dissolved oxygen sensor based on Clark's electrode. This microsensor was introduced to biofilm research by several investigators (Whalen et al., 1969; Bungay et al., 1969; Revsbech and Jorgensen, 1986; Riethues et al., 1986; Baumgartl, 1987). The dissolved oxygen microelectrode used in the authors' laboratory is shown in Figure 9. The construction is based on a design by Revsbech (1989). Electrodes with tip diameters of 10 microns or less are required because the tip should penetrate the biofilm without physical damage to its structure. Larger electrodes produce dents in the biofilm, allowing bulk water to penetrate and influence the measurement. Some measurements, which rely on mass transfer between the environ-

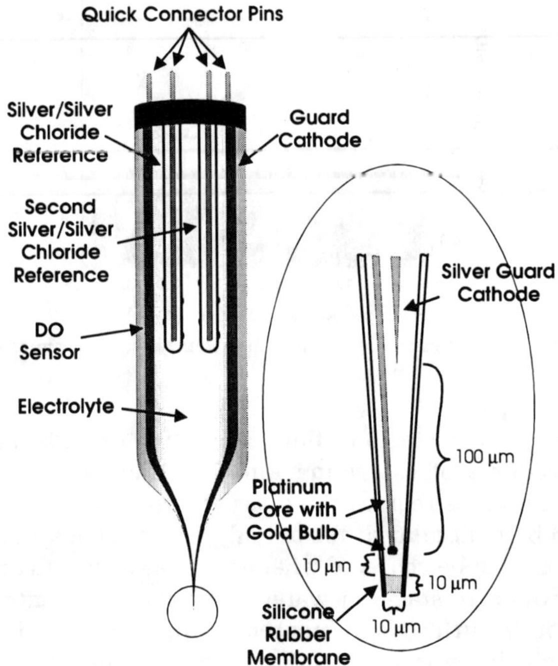

Figure 9. Dissolved oxygen microelectrode.

ment and the sensor (e.g., dissolved oxygen), actually consume the substrate during measurement with a rate that is proportional to sensor surface. A probe with significant surface area may substantially interfere with the system.

An example of a dissolved oxygen microprofile is presented in Figure 10. The oxygen concentration was measured with a microsensor across the water phase and across the biofilm. Successful measurement of the substrate profile is just the beginning of the process of extracting useful kinetic parameters entangled in complex relationships. The shape of the profile is determined by: (1) microbial substrate uptake rate, (2) substrate transport rate through the film, and (3) substrate transport rate to the biofilm.

Each point of a concentration profile reflects an equilibrium between substrate transport and consumption. The transport of substrate in moving liquid is governed by molecular diffusion and convection. Transport of substrate to the biofilm interface is due to the combination of these processes. In stagnant water, transport of substrate is almost entirely due to molecular diffusion, neglecting any contribution from natural convec-

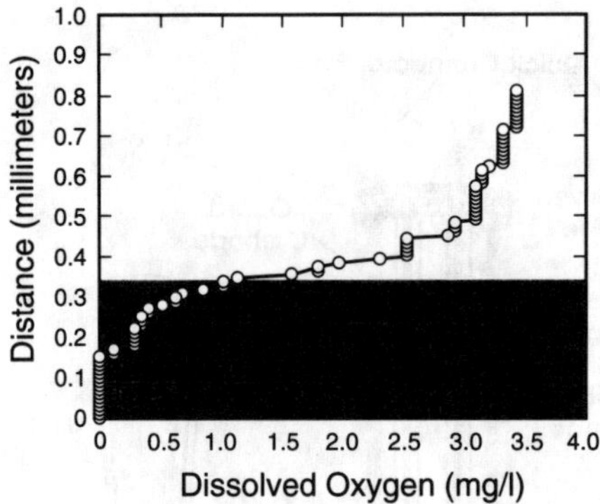

Figure 10. Dissolved oxygen profile in a biofilm system taken with a microelectrode. Note the inflection point at the biofilm surface.

tion. When the water starts to flow, convection exceeds diffusion. The existing procedures of extracting kinetic parameters from substrate concentration profiles require the mass transport within the biofilm to be controlled by molecular diffusion only. Reasonable approximation of such conditions can be achieved when the measurements are conducted in stagnant water. In stagnant water, a substrate concentration profile below the biofilm surface is influenced by the molecular diffusivity and by the microbial reaction rate. Assuming one-dimensional diffusion (Lewandowski, 1994), the gradient of substrate concentration at the biofilm surface is

$$\left(\frac{dC}{dx}\right)_f = \sqrt{2\frac{V_{\max}}{D_f}\left(C - C_0 - K_s * \ln\frac{K_s + C}{K_s + C_0}\right)}$$

where D_f is the diffusion coefficient for the substrate in the biofilm, C is the substrate concentration at a point x, C_0 is the substrate concentration at film/solid surface interface, and $V_{\max}$ and K_s have the usual meaning in the Michaelis-Menten equation. Reaction rate (R) can be directly calculated by recognizing that it is equal to the mass transfer rate across the biofilm surface:

$$R = FD_f\left(\frac{dC}{dx}\right)_f$$

where F is the biofilm surface area. Finally, the reaction rate at the biofilm surface is calculated as

$$R = F\sqrt{2V_{\max}D_f\left(C_s - C_0 - K_s * \ln\frac{K_s + C_s}{K_s + C_0}\right)}$$

where C_s is the substrate concentration at the biofilm-bulk water interface. This equation permits calculating the reaction rate when the concentrations of substrate at two interfaces—between bulk water and biofilm and between biofilm and substratum—are known from microelectrode measurements. It also permits extracting kinetic parameters for the substrate consumption, which may be further used to verify biofilm models (Cunningham et al., 1995).

Hydrodynamics and Kinetics

It was already discussed that the shape of the concentration profile is simultaneously influenced by hydrodynamics and kinetics. Figure 11 shows a conceptual model of a biofilm system where these relationships are depicted. The model is the result of previous experiments that lead us to the conclusion that biofilms are constructed of discrete microcolonies separated by interstitial voids and that flow below the virtual biofilm surface is possible. Such representation permits mathematical analysis of the relationship between hydrodynamics and kinetics in biofilm systems. Further on, it visualizes the mechanism of mass transport in such systems.

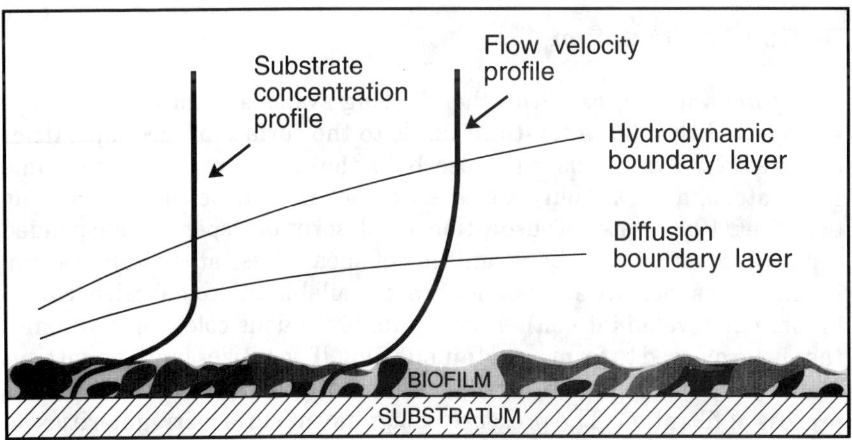

Figure 11. Hydrodynamics and kinetics in a biofilm system. The shape of the concentration profile is simultaneously influenced by hydrodynamics and kinetics.

Computational Tools

Biofilm systems can be analyzed computationally using published biofilm process models such as BIOSIM (Reichert et al., 1990). BIOSIM uses a conservation of mass approach to describe the behavior of all system constituents, including dissolved solutes in the bulk fluid phase, as well as attached and suspended biomass. The code calculates mass balances on all system constituents, including substrates, electron acceptors, multiple species, and attached and suspended biomass. Attached biomass is calculated as a uniform biofilm thickness on a flat surface of prescribed surface area and length and, therefore, is considered to be a one-dimensional model. BIOSIM is well documented and has been previously used to analyze biofilms in various system geometries, including river beds and surfaces of tanks and conduits. Studies that apply BIOSIM to porous media biofilm systems are presently under way by the authors.

BIOSIM and other mathematical models of biofilm systems assume that biofilms form continuous layers and can be conceptually represented by a limited number of compartments positioned parallel to the substratum. This assumption is in disagreement with the experimental evidence that biofilms are porous structures composed of microcolonies separated with a network of interstitial voids. It is not clear to what extent this difference in conceptual approach influences the predictions of BIOSIM and other mathematical models.

BIOFILMS IN POROUS MEDIA

Biofilm Accumulation

In porous media, as with other flowing systems, microbial cells are transported through interstitial fluids to the surface of media particles (Figure 12). Some cells will adsorb to the media particles, consume substrate, and grow. Many cells desorb back into suspension. According to Escher (1986), rates of adsorption and desorption depend on suspended cell concentration and the magnitude of shear stress at the substratum. If sufficient substrate and nutrients are available, microbial cell accumulation will develop into either patchy discontinuous colonies or colonies that have merged to form a continuous film. If pore sizes (with or without biofilm) are small enough, biofilm accumulation will be enhanced by the process of filtration. The presence of significant surface irregularities on the biofilm will most likely increase filtration potential. As the biofilm continues to develop, suspended cells will attach to and detach from the biofilm surface.

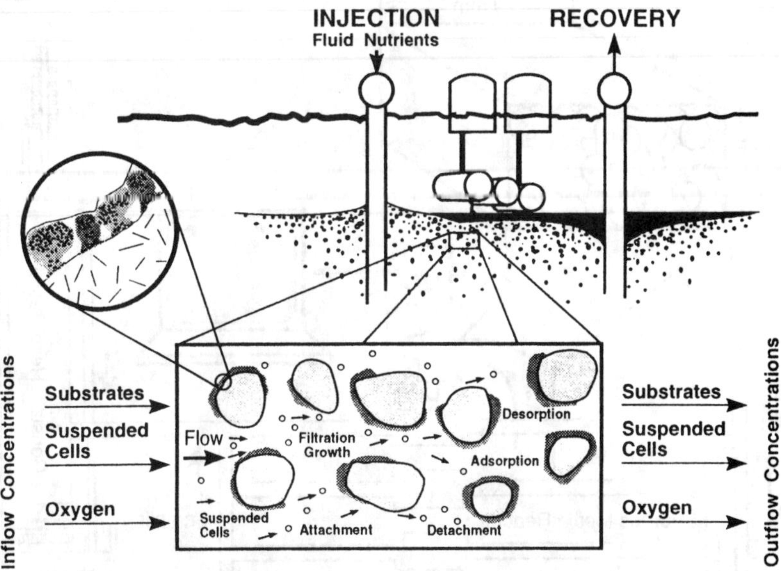

Figure 12. Injection/recovery scheme for enhancing in situ bioremediation of contaminated aquifer. Insert identifies individual microbial transport processes, which contribute to biofilm accumulation and activity in porous media.

The microscale phenomena that govern biofilm accumulation and activity (i.e., chemical and hydrodynamic gradients) are difficult to measure in a porous media environment. Consequently, most investigations to date have focused attention on quantifying the effects of biofilm development rather than on the behavior of individual biofilm processes. However, some insights have been developed by visual observation of biomass accumulation patterns occurring in specially designed packed-bed reactors. In experiments reported by Cunningham et al. (1991), average biofilm thickness was the variable used to estimate net accumulation. The experimental porous media reactor (Figure 13) allowed optical measurements of biofilm thickness throughout the experiment. The experimental system consisted of five parallel, horizontally mounted rectangular porous media reactors. Nutrients were supplied to the reactors under gravity flow from a constant head mixing chamber. A constant piezometric head drop was maintained across the system throughout each experiment by locating the downstream end of each reactor effluent tube at an elevation ranging between 2.7 cm and 16 cm below the level in the constant head mixing chamber. Flow rate through each reactor varied in proportion to reactor permeability, which decreased substantially during the experiments due to biofilm accumulation.

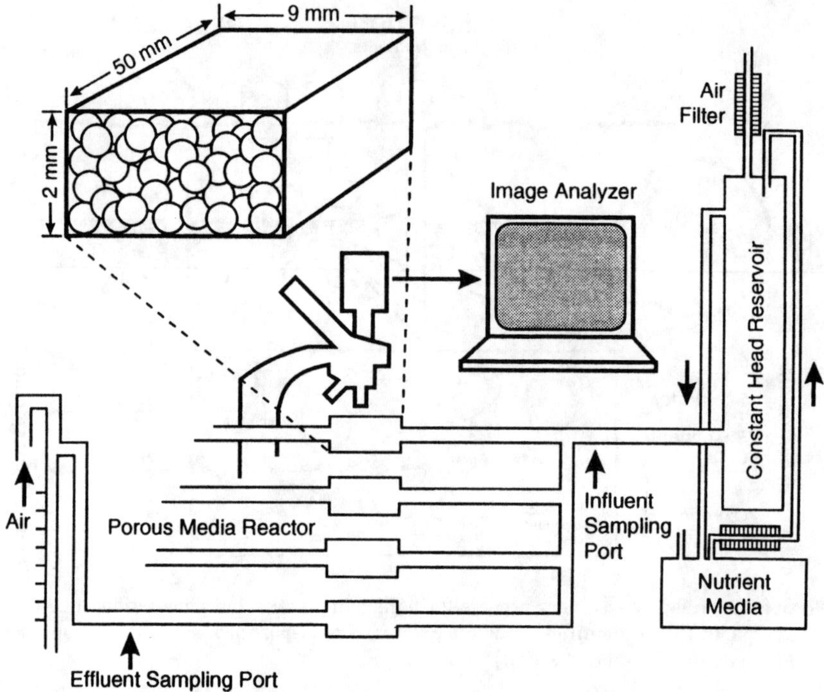

Figure 13. Experimental apparatus for measurement of biofilm thickness and corresponding permeability reduction in porous media.

The progression of biofilm thickness for porous media of various size and composition was generated by simultaneously running five parallel reactors under a constant piezometric head gradient of 0.5 cm/cm (Figure 14). Comparison of these accumulation curves indicates that the ultimate biofilm thickness varied directly with media pore space size and was not influenced significantly by media composition. A similar observation was made using "empty" rectangular reactors (i.e., containing no porous media) of dimension 0.2 × 4 mm on 0.4 × 4 mm; the reactor with the larger cross section (0.4 × 4 mm) and, hence, the larger mass flux exhibited a greater maximum biofilm thickness.

Permeability Reduction

As biofilm thickness increases, the diffusional path length within the biofilm increases, thereby decreasing nutrient concentrations in the base film. Providing that the piezometric head gradient remains constant, increased thickness will also result in decreased pore velocity. Decreased pore velocities will reduce both advective and dispersive transport, thereby lowering nutrient concentrations at the film-water interface, which will subsequently reduce growth rate. Decreased pore velocities will also reduce shear stress, thereby reducing the rate of detachment. Accumulation of biofilm will continue until specific growth rate is balanced by detachment rate. These interactions give rise to the sigmoidal shape exhibited by the accumulation progressions, as measured by biofilm thickness, shown in Figure 14. After biofilm thickness reaches quasi-steady state, the volume of effective pore space (permeability) appears to stabilize and remain constant until operating conditions are disturbed.

Investigations documenting permeability reduction due to biofilm accumulation have been widely reported (Macleod et al., 1988; Shaw et al., 1985; Taylor and Jaffe, 1990). There appear to be two distinct mechanisms by which porous media biofilms reduce permeability. In case one, continuous films are formed, resulting in pore space reduction, causing both porosity and permeability to decrease, as demonstrated by Cunningham et al. (1991). An example of the permeability decrease, corresponding

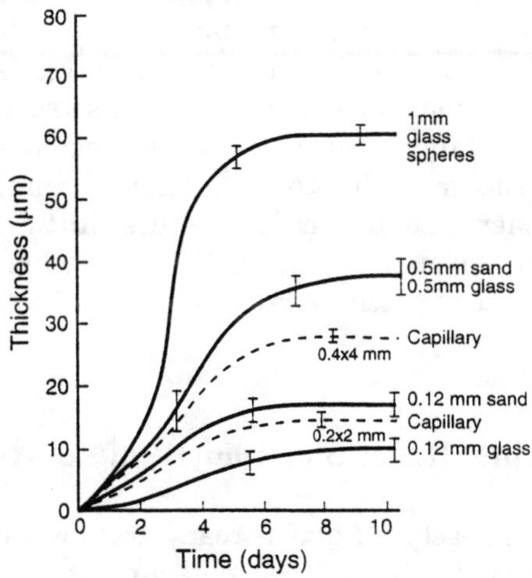

Figure 14. Progression of biofilm thickness (*P. aeruginosa*) for media of different diameter and composition. All reactors were run in parallel under a piezometric gradient of 0.5. Data indicate a quasi-steady-state thickness is reached after about five days of reactor operation. Maximum biofilm thickness values of 63 μm, 40 μm, and 9–14 μm, were observed for media particle diameters of 1 mm, 0.54 mm, and 0.12 mm. Clean surface permeability values of 2.1×10^{-5}, 2.17×10^{-6}, and 9.7×10^{-7} cm^2 correspond to these particle diameters, suggesting a direct relationship between media permeability and maximum biofilm thickness.

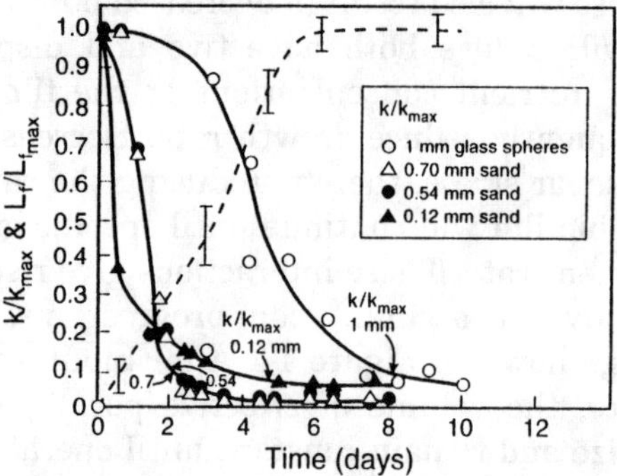

Figure 15. Porous media permeability decrease corresponding to increased biofilm thickness. L_f/L_{fmax} is a single composite dimensionless curve representing all thickness curves from Figure 14. K_{max} values are 2.1×10^{-5} cm^2 (1 mm glass spheres), 3.19×10^{-6} cm^2 (0.70 mm sand), 2.17×10^{-6} cm^2 (0.54 mm sand) and 9.7×10^{-7} cm^2 (0.12 mm sand). After five days of reactor operation, the media/biofilm permeability stabilized and remained essentially constant in the range of 3 to 7×10^{-8} cm^2.

to increasing biofilm thickness for a laboratory porous media reactor system, is shown in Figure 15. Here it is seen that biofilm accumulation results in a 95–99% reduction in the original clean-surface permeability.

In the second case, permeability is reduced by patchy biofilm aggregates accumulating mainly in pore throats, as seen in Figure 16. Under this condition, relatively small amounts of biomass (relative to free pore volume) can cause large reductions in permeability. According to Rittmann (1993), substrate flux to the biofilm is mainly responsible for determining whether case 1 or case 2 occurs. Rittmann developed an index variable, the normalized surface loading, which can be used to predict the occurrence of patchy versus continuous biofilm accumulation in porous media.

Mass Transport and Reaction under High Substrate Loading

Biofilm thickness reaches a quasi-steady state when the net specific growth rate is equal to the net cell detachment rate. An analysis of experimental results shown in Figure 16 reveals that mass transport, biofilm accumulation, and uptake of growth-limiting substrate [Figure 17(b)] are highly interrelated. These laboratory observations were obtained from packed-bed biofilm reactors inoculated with pure culture *Pseudomonas aeruginosa*, and run under constant flow rate conditions. The cylindrical reactor was 5 cm in length, had a 3.1-cm diameter, and was packed with 1-mm glass beads. Substrate uptake here is taken as the difference between the influent glucose concentration (approximately 8 gC M^{-3}) and the daily effluent concentrations shown in Figure 17(a).

It is clear that substrate uptake increased significantly with the onset of biofilm accumulation (i.e., through day 6 when biofilm thickness reached about 20 microns) and yet did not increase further as the biofilm grew to over 100 microns. This observation indicates that, initially, the thin biofilm structure provides a favorable transport environment for encouraging substrate biotransformation. The diffusional path length through the film is relatively low, thereby permitting significant specific growth rates at depth inside the film. Under these conditions, substrate biotransformation is limited by the intrinsic growth kinetic of the micro-organisms. As thickness increased, however, the diffusional path length increased through the film, thereby reducing growth and biotransformation rates in the lower film layers. Under these conditions, the biotransformation process was considered to be transport (i.e., diffusion) limited. The net result was that, for biofilm thicknesses exceeding 20 microns,

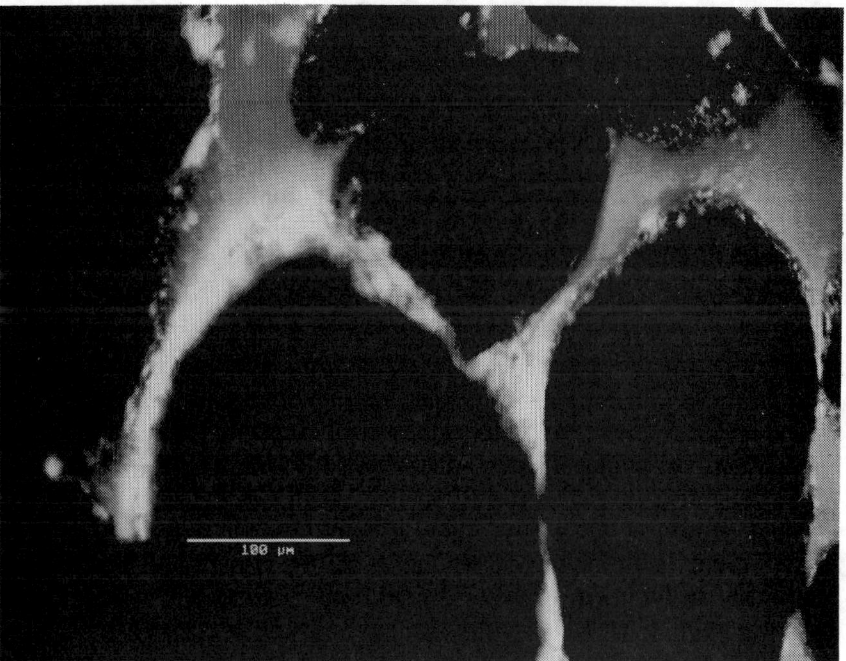

Figure 16. Patchy biofilm forming a partial plug across a sandstone pore throat (courtesy of Paul Stoodley, Center for Biofilm Engineering).

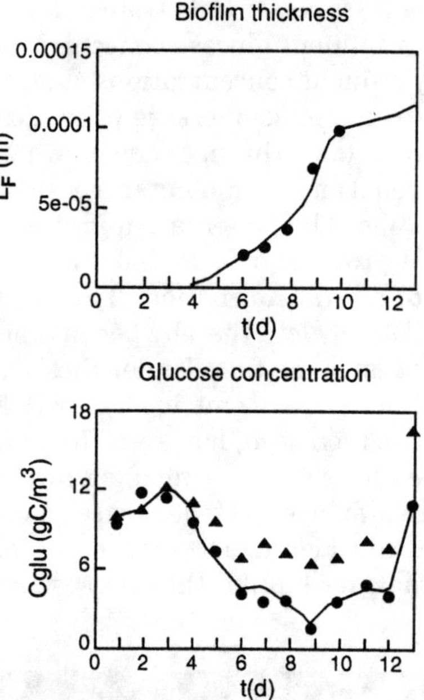

Figure 17. Measured and calculated time series for biofilm thickness and effluent glucose concentrations.

the biotransformation potential for the reactor (as measured by substrate uptake) became independent of the amount of biomass inside the reactor.

Patterns of Biofilm Accumulation

Individual processes that are important for different levels of biofilm accumulation include

1. Sparse Accumulation: Patchy, discontinuous biofilms are characteristic in soils and during early stages of accumulation in engineered reactors. For this condition, the important processes that influence biotransformation are mass transport in the bulk fluid and biofilm kinetic parameters.
2. Thin Continuous Films: Isolated colonies may grow together to form continuous biofilms. In porous media, thin biofilms may develop that completely cover the substratum yet do not appreciably alter the permeability and porosity of the media. In these systems, mass transport across the fluid-biofilm interface and within the biofilm become important, in addition to bulk mass transport and biofilm kinetics.

3. Thick Biofilms: If biofilms develop to more than a few microns in thickness, mass transport (particularly diffusion gradients) within the biofilm become important to biotransformation of dissolved substrates. The presence of flow channels within the film will have a significant effect on spatial gradients in substrate and will result in a heterogeneous distribution in microbial activity. If thick biofilms reduce the free pore space media permeability, mass transport in the bulk phase will also be reduced.

REFERENCES

Abragam A. 1961. *The Principles of Nuclear Magnetism*. Clarendon, Oxford.

Allison D. and Sutherland I. W. 1987. "The role of exopolysaccharides in adhesion of freshwater bacteria," *Journal of General Microbiology,* 133:1319–1327.

Altobelli S. A., Givler R. C., and Fukushima E. 1991. "Velocity and concentration measurements of suspensions by nuclear magnetic resonance imaging," *J. Rheol.,* 35(5):721–734.

Baier R. E. 1980. Substrate influence on adhesion of microorganisms and their resultant new surface properties. In: *Adsorption of Microorganisms to Surfaces,* G. Bitton and K. C. Marshall (Eds.), Wiley-Interscience, New York, Chap. 3.

Bakke R., Trulear, M. G., Robinson, J. A. and Characklis, W. G. 1984. "Activity of *Pseudomonas aeruginosa* in biofilms," *Biotech. Bioeng.,* 26:1418–1424.

Bar-Ness R., Avrahamy N., Matsuyama T., Rosenberg M. 1988. "Increased cell surface hydrophobicity of a *Serratia marcescens* NS 38 mutant lacking wetting activity," *Journal of Bacteriology,* 170:4361–4364.

Baumgartl, H. 1987. Systematic Investigations of Needle Electrode Properties in Polarographic Measurements of Local Tissue PO2. In: *Clinical Oxygen Pressure Measurement.* A. M. Ehrly, J. Hauss, R. Huch (Eds.). Springer, Berlin. pp. 17–42.

Bouwer, E. J. 1987. "Theoretical investigation of particle deposition in biofilm systems." *Water Research,* 21:1489–1498.

Bryers J. D. and Characklis W. 1992. Biofilm Accumulation & Activity: A Process Analysis Approach. In: *Biofilms—Science and Technology.* L. F. Melo, T. R. Bott, M. Fletcher, B. Capdeville (Eds.). Kluwer Academic Publishers. Dordrecht, Boston, London. pp. 221–237.

Bungay H. R. III, Whalen W. J. and Sanders W. M. 1969. "Microprobe techniques for determining diffusivities and respiration rates in microbial slime systems." *Biotechnology and Bioengineering,* 11:765–772.

Bishop R. E. D. and Hassan A. Y. 1964. "The lift and drag forces on a circular cylinder in a flowing field." *Proc. Roy. Soc. (London) Ser. A,* 277:51–75.

Caldwell D. E., Korber D. R. and Lawrence J. R. 1992. Confocal laser microscopy and computer image analysis in microbial ecology. In: *Advances in Microbial Ecology. Vol. 12.* K. C. Marshall ed. Plemum Press, New York, p. 1.

Caprihan A. and Fukushima E. 1990. "Flow measurement by NMR," *Phys. Rep.* 198:195–235.

Chamberlain A. H. L. 1992. The role of adsorbed layers in bacterial adhesion. In: *Biofilms—Science and Technology.* L. F. Melo, T. R. Bott, M. Fletcher, B. Capdeville (Eds.). Kluwer Academic Publishers. Dordrecht, Boston, London. pp. 59–67.

Characklis W. G. and Marshall K. C. (Eds.) 1990. *Biofilms.* John Wiley & Sons, Inc. New York.

Costerton J. W., Cheng K. J., Geesey G. G., Ladd T. I., Nickel J. C., Dasgupta M. and Marrie T. J. 1987. "Bacterial biofilms in nature and disease," *Annu. Rev. Microbiol.,* 41:435–464.

Cunningham, A. B. 1989. Hydrodynamics and solute transport at the fluid-biofilm interface. In: *Structure and Function of Biofilms.* John Wiley. W. G. Characklis and P. A. Wilderer (Eds.). pp. 19–31.

Cunningham A. B., Characklis W. G., Abedeen F. and Crawford D. 1991. "Influence of Biofilm Accumulation on Porous Media Hydrodynamics," *Environmental Science & Technology,* 25(7):1305–1310.

Cunningham A. B., Visser E., Lewandowski Z. and Abrahamson M. 1995. Evaluation of a coupled mass transport–biofilm process model using dissolved oxygen microsensor. *Biofilm Structure, Growth, and Dynamics.* 30 Aug–1 Sep. Meeting of the International Association on Water Quality. Leeuwenhorst, The Netherlands.

DeBeer D., Stoodley P., Roe F. and Lewandowski Z. 1994a. "Effects of biofilm structures on oxygen distribution and mass transport," *Biotechnology and Bioengineering* 43:1131–1138.

DeBeer D., Stoodley P. and Lewandowski Z. 1994b. "Liquid flow in heterogeneous biofilms," *Biotechnology and Bioengineering,* 44:636–641.

Dexter S. C, Sullivan J. D., Jr., Williams J. III and Watson S. W. 1975. "Influence of substrate wettability on the attachment of marine bacteria to various surfaces," *Applied Microbiology.* 30:298–308.

Duddridge J. E., Kent C. A. and Laws J. F. 1982. "Effect of surface shear stress on the attachment of *Pseudomonas fluorescens* to stainless steel under defined flow conditions," *Biotechnology and Bioengineering,* 26:153–164.

Eighmy T. T., Maratea D. and Bishop P. L. 1983. "Electron microscopic examination of wastewater biofilm formation and structural components," *Appl. Environm. Microbiol.* 45:1921–1931.

Escher A. R. 1986. "Colonization of a smooth surface by *Pseudomonas aeruginosa:* Image analysis methods," Ph.D. Dissertation, Montana State University, Bozeman, MT.

Escher A. and Characklis W. G. 1990. Modeling the initial events in biofilm accumulation. In: *Biofilms,* W. G. Characklis and K. C. Marshall (Eds.). John Wiley & Sons, Inc. New York 1990. pp. 445–486.

Fletcher M. and Floodgate G. D. 1973. "An electron-microscopic demonstration of an acidic polysaccharide involved in the adhesion of a marine bacterium to solid surfaces," *Journal of General Microbiology,* 74:325–334.

Fletcher M. and Loeb G. I. 1979. "Influence of substratum characteristics on the attachment of marine pseudomonad to solid surfaces," *Applied Environmental Microbiology.* 37:67–72.

Fukushima, E. and Roeder, S. B. W., 1981. *Experimental Pulse NMR.* Addison-Wesley.

Gibbons R. J. and Nygard M. 1970. "Interbacterial aggregation of plaque bacteria," *Archives Oral Biology,* 15:1397–1400.

Gjaltema P. A., Arts P. A. M., Loosdrecht M. C. M., Kuenen J. G. and Heijnen J. J. 1994. *Biotechnology and Bioengineering* 44:194–204.

Griffin O. M. and Ramberg S. E. 1974. "The vortex-street wakes of vibrating cylinders," *Jour. Fluid. Mech.* 66:553–576.

Hall S. M. and Griffin O. M. 1993. "Vortex shedding and lock-on in a perturbed flow," *Journal of Fluids Engineering. Transactions of the ASME* 115:283–291.

Klinzing G. E., Kubovcik R. J. and Marmo J. F. 1969. "Frictional losses in foam-damped flexible tubes," *I&EC Process Design and Development.* 8:112–114.

Kugaprasatham S., Nagaoka H. and Ohgaki S. 1992. "Effect of turbulence on nitrifying biofilms at non-limiting substrate conditions," *Wat. Res.* 26(12): 1629–1638.

Lawrence J. R., Korber D. R., Hoyle B. D., Costerton J. W. and Caldwell D. E. 1991. "Optical sectioning of microbial biofilms," *J. Bacteriol.* 173:6558.

Lewandowski Z. 1994. Dissolved oxygen gradients near microbially colonized surfaces. In: *Biofouling and Biocorrosion in Industrial Water Systems.* G. Geesey, Z. Lewandowski, H-C. Flemming (Ed.). CRC Press Inc. Lewis Publishers. Boca Raton, pp. 175–188.

Lewandowski Z. and Walser G. 1991. "Influence of hydrodynamics on biofilm accumulation," *Environmental Engineering Proceedings.* EE Div/ASCE. Reno, Nevada. July 8–10, 1991. pp. 619–624.

Lewandowski Z., Altobelli S. A., Majors P. D. and Fukushima E. 1992. "NMR imaging of hydrodynamics near microbially colonized surfaces." *Wat. Sci. Tech.* 26:577–584.

Lewandowski Z., Altobelli S. A. and Fukushima E. 1993. "NMR and microelectrode studies of hydrodynamics and kinetics in biofilms," *Biotechnology Progress,* 9:40–45.

Lewandowski Z., Stoodley P., Altobelli S. and Fukushima E. 1994. "Hydrodynamics and kinetics in biofilm systems—Recent advances and new problems," *Water Science and Technology.* 29:223–229.

Lewandowski Z. and Stoodley P. 1995. "Flow induced vibrations, drag force, and pressure drop in conduits covered with biofilm," *Biofilm Structure, Growth, and Dynamics*. 30 Aug–1 Sep. Meeting of the International Association on Water Quality. Leeuwenhorst, The Netherlands.

Little B. 1985. "Factors influencing the adsorption of dissolved organic material from natural waters," *Journal of Colloid and Interface Science*, 108:331–339.

Loeb G. 1985. The properties of nonbiological surfaces and their characterization. In: *Bacterial adhesion, mechanisms and physiological significance*. Ed. Savage D. and Fletcher M. Plenum Press. New York, 1985. pp. 111–129.

Loeb G. I. and Neihof R. A. 1976. "Adsorption of an organic film at the platinum-seawater interface." *Journal of Marine Research*. 35(2):283–291.

Mack W. N., Mack J. P. and Ackerson A. O. 1975. "Microbial film development in trickling filters," *Microb. Ecol.*, 2:215–316.

MacLeod F. A., Lappin-Scott H. M. and Costerton J. W. 1988. "Plugging of a model rock system by using starved bacteria," *Appl. Environ. Microbiol.* 54:1365–1372.

Majors P. D., Givler R. C. and Fukushima E. 1989. "Velocity and concentration measurements in multiphase flows by NMR," *J. Magn. Reson.* 85:235–243.

Marshall K. C., Power K. N., Angles M. L., Schneider R. P. and Goodman A. E. 1994. Analysis of bacterial behavior during biofouling of surfaces. In: *Biofouling and Biocorrosion in Industrial Water Systems*. G. Geesey, Z. Lewandowski, H-C. Flemming (Ed.). CRC Press Inc. Lewis Publishers. Boca Raton, pp. 15–26.

Massol-Deya A. A., Whallon J., Hickey R. F. and Tiedje J. M. 1995. "Channel structures in aerobic biofilms of fixed-film reactors treating contaminated groundwater," *Applied and Environmental Microbiology*. 61:769–777.

Morris P. G. 1986. *Nuclear Magnetic Resonance Imaging in Medicine and Biology*. Clarendon, Oxford.

Mozes N. and Rouxhet P. G. 1992. Modification of surfaces for promoting cell immobilization. In: *Biofilms—Science and Technology*. L. F. Melo, T. R. Bott, M. Fletcher, B. Capdeville. (Eds.). Kluwer Academic Publishers. Dordrecht, Boston, London. pp. 69–85.

Myers D. 1991. *Surfaces, interfaces and coloids. Principles and applications*. VCH Publishers. p. 116.

Neihof R. A. and Loeb G. I. 1972. "The surface charge of particular matter in sea water," *Limnology and Oceanography*, 17(1):7–16.

Nikuradse, J. 1933. "Stromungsgeretze in rauchen Rohren," *VD1—Forschungsh.*, No 361.

Nielsen N. F., Larsen P. S., Riisgard H. U. and Jorgensen C. B. 1993. "Fluid motion and particle retention in the gill of *Mytilus edulis*: Video recordings and numerical modelling," *Marine Biology*. 116:61–71.

Picologlou B. F., Zelver N. and Characklis W. G. 1980. "Biofilm growth and hydraulic performance," *Journal of the Hydraulic Division, ASCE,* 106:No HY5 733–746.

Reichert P., Ruchti J. and Wanner O. 1990. "BIOSIM—An interactive program for the simulation of mixed culture biofilm systems on a personal computer," Swiss Federal Institute for Enviromental Science and Technology, CH-8600 Duebendorf, Switzerland.

Revsbech N. P., 1989. "An oxygen microelectrode with a guard cathode," *Limnol. Oceanogr.* 34:474–478.

Revsbech N. P. and Jorgensen B. B. 1986. Microelectrodes: Their use in microbial ecology. In: *Advances in Microbial Ecology,* vol. 9, K. C. Marshall (Ed.), pp. 293–352.

Riethues M., Buchholtz R., Onken O., Baumgartl H. and Lubbers D.W. 1986. "Determination of oxygen transfer from single air bubbles to liquids by oxygen microelectrodes," *Chem. Eng. Process,* 20:331–337.

Rittmann B. E. 1993. "Modeling biofilms in porous media," *Wat. Resour. Resear.* 9(3):331–337.

Rittmann B. E. and Manem J. A. 1992. "Development and experimental evaluation of a steady-state, multispecies biofilm model," *Biotechnol. Bioeng.,* 39: 914–922.

Rittmann B. E. 1989. Detachment from biofilms. In: Characklis W. G. and Wilderer P. A. (Eds.). *Structure and function of biofilms.* Berlin, Germany: Dahlem Konferenzen. pp. 49–58.

Rittmann B. E. 1982. "The effect of shear stress on biofilm loss rate," *Biotechnol. Bioeng.,* 24:501–506.

Robinson R. W., Akin D. E., Nordstedt R. A., Thomas M. and Aldrich H. C. 1984. "Light and electron microscopic examinations of methane-producing biofilms from anaerobic fixed-bed reactors." *Appl. Environm. Microbiol.,* 48:127–136.

Rosenberg M. and Kjelleberg S. 1986. Hydrophobic interactions: Role in bacterial adhesion. In: *Adv. Microbial Ecology.* K. C. Marshall (Ed). Plenum Press, NY pp. 353–393.

Rosenberg M. and Doyle R. J. 1990. Microbial cell surface hydrophobicity. History, measurement, and significance. In: *Microbial cell surface hydrophobicity.* R.J. Doyle and M. Rosenberg. (Eds.). American Society for Microbiology. Washington, D.C. pp. 1–39.

Saleh S., Throvert J. F. and Adler P. M. 1993. "Flow along porous media by particle image velocimetry," *AIChe Journal.* 39(11):1765–1776.

Shaw J. C., Bramhill B., Wardlaw N. C. and Costerton J. W. 1985. "Bacterial fouling in a model core system," *Applied and Envir. Microbiol.,* 49:693.

Siegrist H. and Gujer W. 1985. "Mass transfer mechanisms in a heterotrophic biofilm," *Wat. Res.,* 19(11):1369–1378.

Silverman M., Belas R., Simon M. (Ed) and K. C. Marshall 1984. Genetic control of bacterial adhesion. In: *Microbial Adhesion and Aggregation*. Dahlem Konferenzen. Springer-Verlag. Berlin, Heidelberg, New York, Tokyo, pp. 95–107.

Steward P. S., Peyton B. M., Drury W. J. and Murga R. 1993. "Quantitative observations of heterogeneities in *Pseudomonas aeruginosa* biofilms," *Appl. Environm. Microbiol.*, 59(1):327–329.

Stoodley P., DeBeer D. and Lewandowski Z. 1994. "Liquid flow in biofilm systems," *Applied and Environmental Microbiology*, 60:2711–2716.

Stumm W. 1992. *Chemistry of the solid-water interface. Processes at the mineral-water and particle-water interface in natural systems*. John Wiley & Sons, Inc. p. 115.

Taylor S. W. and Jaffe P. R. 1990. "Biofilm growth and related changes in the physical properties of a porous medium," *Wat. Res. Resear.*, 26(9):2161–2169.

van Loosdercht M. C. M., Lyklema J., Norde W. and Zehnder A. J. B. 1990. "Influence of interfaces on microbial activity," *Microbiological Reviews*. Mar.:75–87.

Wan, J. and J. L. Wilson. 1993. Visualization of colloid transport during single and two-fluid flow in porous media. In J. F. McCarthy and F. J. Wobber (ed.), *Manipulation of groundwater colloids for environmental restoration*. Lewis, Boca Raton. pp. 335–339.

Wanner O. and Gujer W. 1986. "A multi-species biofilm model." *Biotech. Bioeng.*, 28:314–328.

Whalen W. J., Bungay H. R. III and Sanders, W. M. III 1969. "Microelectrode determination of oxygen profiles in microbial slime systems." *Environmental Science and Technology*, 3:1297–1298.

Wilson, T. 1990. *Confocal microscopy*. Academic Press, London, UK.

Zhang T. C. and Bishop P. 1994a. "Density, porosity, and pore structure of biofilms," *Water Research*. 28:2267–2277.

Zhang T. C. and Bishop P. L. 1994b. "Evaluation of tortuosity factors and effective diffusivities in biofilms," *Water Research*. 28:2279–2287.

Zisman W. A. 1964. Relation of the equilibrium contact angle to liquid and solid constitution. In: *Contact angle, wettability and adhesion*. R. F. Gould (Ed). American Chemical Society. Washington, D.C. pp. 1–51.

Aeration through Gas-Permeable Membranes in Sequencing Batch Biofilm Reactors

Robert Chozick and Robert L. Irvine
Department of Civil Engineering and Geological Sciences
University of Notre Dame
Notre Dame, IN 46556, USA

INTRODUCTION

The **Sequencing Batch** Biofilm Reactor (SBBR) is a system for wastewater treatment that combines the operating and performance advantages of the Sequencing Batch Reactor (SBR), aeration through gas-permeable membranes, and biofilm operation. This hybrid system provides the operational flexibility necessary for biological treatment of wastewaters containing volatile organic compounds (VOCs) with minimal fugitive emissions, low-strength wastewaters that are difficult to treat in suspended growth reactors, and wastewaters containing surfactants. The SBBR also maintains the operational flexibility of the SBR for nutrient removal. The primary focus of this chapter is the development and evaluation of a model of the membrane aeration system. A brief overview of the major components of the SBBR is presented in the following sections.

Sequencing Batch Reactors

The Sequencing Batch Reactor (SBR) is a suspended growth, periodic wastewater treatment system that has been shown to be effective for the treatment of both municipal (Irvine et al., 1983) and industrial (Irvine et al., 1984) wastewaters since its inception in the 1980s. As a time-oriented

Please direct communication to Robert Chozick at Foster Wheeler Environmental Corporation, 1290 Wall Street West, Lyndhurst, NJ 07071-0661.

process, simple operating modifications can be implemented in an SBR to alter the extent of nitrification and denitrification (Alleman and Irvine, 1980a, 1980b; Palis and Irvine, 1985), achieve biological phosphorus removal (Manning and Irvine, 1985; Ketchum et al., 1987), and control bulking sludge (Chiesa and Irvine, 1985). Steady-state, suspended growth wastewater treatment systems [e.g., those that utilize continuous flow stirred tank reactors (CFSTRs)] offer less operational flexibility and are therefore less amenable to biological nutrient removal than SBRs. In addition, steady-state wastewater treatment systems are more susceptible to bulking sludge and poor performance since prevailing, low-substrate concentrations are among the factors that contribute to the growth of filamentous organisms (Chudoba, 1973).

In conventional SBRs (and CFSTRs), oxygen is typically provided to the suspended biomass by diffused and/or mechanical aeration (Metcalf & Eddy, Inc., 1991). These methods of aeration have been used extensively and are shown to be very effective for oxygen supply to the biomass for biodegradation. Unfortunately, they are equally effective at stripping VOCs (e.g., benzene, toluene, trichloroethylene, etc.) from wastewaters that contain such components. Although such components are found at very low concentrations in many applications (e.g., less than 1 mg/l in domestic wastewater), the large volume of wastewater that passes through a treatment system can result in the release of significant quantities of these compounds to the atmosphere.

Another operational difficulty often associated with mechanical and diffused aeration is reactor foaming, which occurs when wastewaters containing surfactants are introduced into wastewater treatment systems using these modes of aeration. Reactor foaming can result in reduced oxygen transfer, undesirable anoxic or anaerobic reactor conditions, production of nonbiodegradable intermediates, undesirable population shifts, and reactor failure (Wilderer et al., 1985).

The operational difficulties discussed above can be overcome by replacing mechanical and/or diffused aeration systems with alternatives that minimize stripping of volatile compounds and provide oxygen under less vigorous mixing conditions to prevent reactor foaming.

Membrane Aeration Systems

One alternative to diffused and mechanical aeration that has received some attention is aeration by diffusion through gas-permeable membranes. In 1960, Schaffer et al. first investigated the possibility of using permeable plastic films (e.g., polyethylene, ethyl cellulose, and polystyrene) to supply oxygen to biomass for wastewater treatment. Oxygen transfer through the membranes was reported; however, the delicate nature and low oxygen permeabilities of the plastic films used for this study led to the conclusion that this method of aeration would have only limited applications.

In 1968, Robb investigated the permeation properties of silicone membranes, which have much better physical properties than the membranes mentioned above, and reported that silicone membranes had permeabilities 30–600 times those of the membranes investigated by Schaffer et al. (1960) and exhibited selectivity (i.e., different permeabilities for different compounds). Applications in oxygen enrichment, air regeneration in an isolated chamber, and underwater air regeneration were reported (Robb, 1968).

In 1972, Yasuda and Lamaze compared the oxygen transfer properties of porous (e.g., polysulfone) and nonporous (e.g., silicone) membranes. In gas-membrane-gas transfer experiments (i.e., experiments in which both sides of a membrane are exposed to gases and mass transfer is measured), which reveal the "true" membrane permeability, porous membranes have permeabilities approximately three to four orders of magnitude greater than nonporous membranes. In gas-membrane-liquid experiments (i.e., one side of the membrane is exposed to a liquid phase), however, the permeabilities of the two types of membranes were found to be comparable due to liquid film resistance. The ability to predict the oxygen transfer through nonporous membranes in gas-membrane-liquid applications using data collected in gas-membrane-gas experiments and liquid parameters (e.g., Reynolds number) suggests that, from a design perspective, nonporous membranes are a better choice for wastewater treatment applications.

In the mid-1980s, renewed interest arose in applying gas-permeable membranes in wastewater treatment applications. Wilderer et al. (1985) used a submerged silicone membrane for auxiliary oxygenation of an SBR that performed poorly using diffused aeration alone due to reactor foaming. Muollo et al. (unpublished) used silicone membrane aeration to control VOC emissions from wastewater treatment plants, and Smith and Wilderer (1987) aerated SBRs with submerged silicone membranes to treat landfill leachate.

Biofilm Reactors

The first biofilm wastewater treatment system, a trickling filter, was put into operation approximately 100 years ago. Since that time, trickling filters and other fixed film processes, including roughing filters, rotating biological contactors, and nitrification units, have been put into operation for wastewater treatment applications (Metcalf & Eddy, Inc., 1991).

Although these biofilm systems are not as widely used as suspended growth wastewater treatment systems (e.g., CFSTRs), biofilm systems offer some advantages over suspended growth reactors in certain applications. Biofilm reactors have been shown to be highly efficient for the treatment of low-strength wastewaters (i.e., wastewaters containing small concentrations of organic compounds), due to the ability of these systems to maintain high concentrations of slow-growing (i.e., low yield and/or low specific growth rate) and/or poor-settling microorganisms that would tend to wash out of suspended growth systems. In addition, as a result of diffusional resistances through the biofilm, microorganisms growing in a biofilm are less susceptible than suspended organisms to the toxic effects of shock loadings, transients, and the introduction of toxic compounds (Bryers and Characklis, 1990).

Mathematical Description of Oxygen Transfer through Gas-Permeable Membranes

Applications of gas-permeable membranes in medicine (e.g., heart-lung machines), biology (e.g., growth of shear-sensitive cell cultures), chemical processes (e.g., adsorption and extraction), etc., have led to much research on the properties of membranes and, specifically, the design and optimization of membrane contactors and aerators (e.g., Yang and Cussler, 1986; Wickramasinghe et al., 1991; Su et al., 1992). In much of this work, mathematical models were developed for the aeration system, assuming steady-state conditions prevailed on both sides of the membrane. In wastewater treatment, such conditions might occur in a membrane-aerated CFSTR. However, in an SBBR-like system, conditions on the liquid side of the membrane and, consequently, on the gas side are continuously changing. This requires that unsteady-state models be developed for membrane aeration of such systems. A comprehensive unsteady-state waste treatment model was developed for the SBBR (Chatzopoulos, 1994). The model included both biological treatment and adsorption with granular activated carbon.

Mass transfer through a nonporous, gas-permeable membrane occurs by a sorption-diffusion-desorption mechanism (Yasuda and Lamaze, 1972). Using a resistance in series model, the following equation was developed by Côté et al. (1988, 1989) to describe oxygen flux through a gas-permeable membrane:

$$J = \frac{1}{\frac{1}{K_G} + \frac{1}{K_M S_{MG}} + \frac{1}{K_L(S_{MG}/S_{ML})}} \left(C^* - \frac{C_L}{S_{MG}/S_{ML}} \right) \tag{1}$$

where

J = flux, mg/cm^2-min
K_G = gas phase mass transfer coefficient, cm/min
K_M = membrane mass transfer coefficient, cm/min
S_{MG} = gas-membrane partition coefficient, dimensionless
K_L = liquid phase mass transfer coefficient, cm/min
S_{ML} = liquid-membrane partition coefficient, dimensionless
C^* = liquid phase saturation concentration, mg/cm^3
C_L = liquid phase concentration, mg/cm^3

Neglecting the gas film mass transfer resistance ($1/K_G$) and using Henry's Law ($p_G = HC_L$) and the ideal gas law ($p_G = C^*RT/M_G$) to replace gas concentrations with partial pressures, the following simplified equation is obtained:

$$J = \frac{1}{\dfrac{1}{K_M S'_{MG}} + \dfrac{H}{K_L}}(p_G - HC_L) \tag{2}$$

where

S'_{MG} = gas-membrane partition coefficient, mg/cm^3-atm
H = Henry's Law coefficient, cm^3-atm/mg
p_G = partial pressure, atm
R = gas law constant, cm^3-atm/mol-K
T = temperature, K
M_G = molecular weight, mg/mol

The overall mass transfer coefficient is given by

$$\frac{H}{K} = \frac{1}{K_M S'_{MG}} + \frac{H}{K_L} \tag{3}$$

where K = overall mass transfer coefficient, cm/min.

Thus, the overall mass transfer resistance is primarily a function of the membrane mass transfer resistance and the liquid film mass transfer resistance. The membrane mass transfer resistance is a function of the membrane material and configuration (e.g., thickness, surface area, etc.), while the liquid phase mass transfer is a function of the hydrodynamic properties on the liquid side of the membrane (e.g., Reynolds number). Equation (3) can also be written as

$$\frac{1}{K} = \frac{\tau_e}{P_M H} + \frac{1}{K_L} \tag{4}$$

where

τ_e = equivalent membrane thickness, cm
P_M = membrane permeability, mg-cm/cm^2-atm-min

in which K_M has been replaced by D_M/τ, the membrane diffusivity divided by the membrane thickness; the product $S'_{MG}D_M$ has been replaced by P_M; and τ, the membrane thickness, has been replaced by the equivalent thickness τ_e, given by the following expression:

$$\tau_e = r_{\text{out}} \ln\left(\frac{r_{\text{out}}}{r_{\text{in}}}\right) \tag{5}$$

where

r_{out} = membrane outer radius, cm
r_{in} = membrane inner radius, cm

The replacement of τ by τ_e results from the derivation of the flux equations in cylindrical coordinates for hollow fiber membranes. In general, if the membrane thickness, $r_{\text{out}} - r_{\text{in}}$, is small compared to the outer radius of the membrane, r_{out}, τ_e is approximately equal to τ (Côté et al., 1989). Multiplying Equation (2) by the specific surface area and rearranging gives

$$\frac{dC_L}{dt} = Ka\left(\frac{p_G}{H} - C_L\right) \tag{6}$$

where a = outer specific surface area, cm^2/cm^3.

For constant p_G, Equation (6) integrates to

$$\left(\frac{p_G/H - C_L(t)}{p_G/H - C_L(0)}\right) = e^{-Kat} \tag{7}$$

Côté et al. (1989) calculated overall mass transfer coefficients through silicone membranes using Equation (7) and reported the effects of oxygen partial pressure, gas composition, gas flow regime (i.e., the flow rate of gas through the lumen of the submerged tubing, with a flow rate of zero

Table 1. Effects of operating parameters on K (from Côté et al., 1989).

Operating Parameter	Observed Effect on K
Oxygen partial pressure	No effect at low pressures. Slight reduction when bubbles form on tubing surface.
Composition of gas	Mass transfer reduced when air was used instead of oxygen.
Gas flow regime	Marked reduction in dead-end mode.
Surfactant addition	Slight increase in oxygen transfer.
Liquid velocity	Increase with increasing Re.

indicating dead-end operation), addition of surfactant, and liquid velocity on the overall mass transfer coefficient. A summary of these results is presented in Table 1.

MATERIALS AND METHODS

Model Development

The development of the model for the aeration system begins with the assumption that the submerged membrane acts like a tubular reactor with axial diffusion. This assumption was tested and verified by Chozick (1992). With reference to Figure 1, a mass balance on the shaded region, which has a volume $A\Delta z$, is given by

$$\frac{\partial}{\partial t}(A\Delta z C_i) = Q(C_z - C_{z+\Delta z}) + AD_{ab}\left(\frac{\partial C}{\partial z}\bigg|_{z+\Delta z} - \frac{\partial C}{\partial z}\bigg|_z\right)$$

$$- \frac{V_L Ka(C_i - C_L)\Delta z}{L} \tag{8}$$

where

A = area, cm²
Δz = length of element, cm
C_i = gas phase concentration in element i, mg/cm³
Q = flow, cm³/min
C_z = gas phase concentration at position z, mg/cm³
D_{ab} = diffusion coefficient, cm²/min

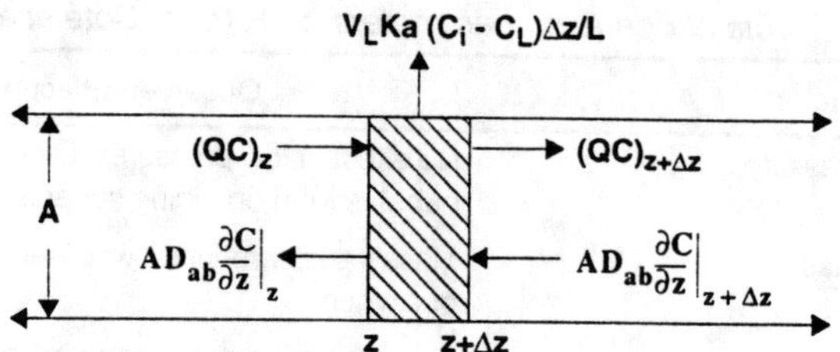

Figure 1. Tubular reactor with axial diffusion.

V_L = liquid volume, cm³
L = length, cm

The last term of the equation is multiplied by $\Delta z/L$ to correct for the fact that K will be determined experimentally for the entire length, L, of silicone tubing in the reactor. Dividing both sides of this equation by $A\Delta z$, which is assumed constant, and taking the limit as Δz goes to zero, the following differential equation is obtained:

$$\frac{\partial C_i}{\partial t} = -\frac{\partial}{\partial z}(vC_i) + D_{ab}\frac{\partial^2 C_i}{\partial z^2} - \frac{V_L Ka(C_i - C_L)}{AL} \tag{9}$$

v = velocity, cm/min.

Although Equation (9) is generally valid for a tubular reactor with diffusion, a more useful form of this equation is obtained by replacing the gas phase concentrations, C_i, with partial pressures, p_G, by use of the ideal gas law (i.e., $C_i = p_G(M_G)/RT$) and Henry's Law (i.e., $C_i = p_G/H$). Upon substitution and rearrangement, the following equation, valid for any gas in the tube, is obtained:

$$\frac{\partial p_G}{\partial t} = -\frac{\partial}{\partial z}(vp_G) + D_{ab}\frac{\partial^2 p_G}{\partial z^2} - \frac{RTV_L Ka(p_G/H - C_L)}{M_G AL} \tag{10}$$

In order to calculate the change in dissolved gas concentration in the reactor, the following equation is developed, again referring to Figure 1, and with the tube divided into n equal-sized elements of length Δz:

$$\frac{d}{dt}(V_L C_L) = \sum_{i=1}^{n}\left(\frac{V_L Ka(C_i - C_L)\Delta z}{L}\right) \tag{11}$$

Dividing both sides of this equation by the constant liquid volume, V_L, taking the limit as Δz goes to zero, and making the substitution described above for C_i, the following equation, valid for the dissolved concentration of any gas, is obtained:

$$\frac{dC_L}{dt} = \int_0^L \left(\frac{Ka(p_G/H - C_L)dz}{L} \right) \tag{12}$$

In this equation, the partial pressure of the gas, p_G, is a function of both time and space and is given by Equation (10). However, Equation (10) cannot be solved until the velocity of the gas in the tube is known. This value is estimated by performing an overall mass balance on the shaded region in Figure 1 and assuming that the total pressure in the tube is constant over time and space. This assumption is validated by previously described experiments (Chozick, 1992) where it was found that, by controlling the pressure at the inlet of the tubing and by restricting the flow at the tubing outlet with a flowmeter, the pressure change over the length of the tube is negligible (e.g., at 10 psig inlet pressure and with an outlet flow of 15 ml/min, the lowest outlet pressure observed was 9.97 psig). Using this assumption, the following equations are developed:

$$P = \Sigma p_G \tag{13}$$

where P = total pressure, atm.

$$\frac{\partial P}{\partial t} = \Sigma \frac{\partial p_G}{\partial t} = \Sigma \left(-\frac{\partial}{\partial z}(vp_G) + D_{ab}\frac{\partial^2 p_G}{\partial z^2} - \frac{RTV_L Ka(p_G/H - C_L)}{M_G AL} \right) \tag{14}$$

Expanding the first term on the right-hand side of the equation gives

$$\Sigma \left(-\frac{\partial}{\partial z}(vp_G) \right) = \Sigma \left(-v\frac{\partial p_G}{\partial z} - p_G\frac{\partial v}{\partial z} \right) \tag{15}$$

The assumption that the total pressure, P, is not a function of time or space provides the following equalities:

$$\frac{\partial P}{\partial t} = \frac{\partial P}{\partial z} = \Sigma\frac{\partial p_G}{\partial t} = \Sigma\frac{\partial p_G}{\partial z} = \Sigma\frac{\partial^2 p_G}{\partial z^2} = 0 \tag{16}$$

and thus, upon rearrangement, Equation (14) simplifies to

$$\frac{\partial v}{\partial z} = \frac{1}{P}\Sigma\left(-\frac{RTV_L Ka(p_G/H - C_L)}{M_G AL}\right) \tag{17}$$

A complete mathematical description of oxygen transfer to the reactor and the gas composition in the tube is thus given by Equations (10), (12), and (17), where Equations (10) and (12) must be computed for each gas in the system, while Equation (17) represents the sum of all gases present. In addition to these equations, initial and boundary conditions for Equation (10) for each of the gases in the tube, initial conditions for Equation (12) for each of the gases in the liquid phase, and boundary conditions for Equation (17) are necessary to solve the equations. For the current model, which only allows for oxygen and nitrogen in the tube, the initial and boundary conditions are given by the following equations:

For oxygen:

$$p_G(0,z) = 0.21 \times P \tag{18}$$

$$p_G(t,0) = f_O \times P \tag{19}$$

where f_O = fraction of oxygen in inlet gas stream.

$$C_L(0) = 0.21/H = 9.2 \tag{20}$$

For nitrogen:

$$p_G(0,z) = 0.79 \times P \tag{21}$$

$$p_G(t,0) = f_N \times P \tag{22}$$

where f_N = fraction of nitrogen in inlet gas stream.

$$C_L(0) = 0.79/H = 14.3 \tag{23}$$

For the velocity in Equation (17):

$$v(L) = Q_G/A \tag{24}$$

where Q_G = outlet gas flow rate, ml/min.

Numerical Solution of Mathematical Model

The mathematical model developed above does not have an analytical solution. In this section, a simple numerical solution using finite difference techniques will be developed for the model. Using the following finite difference approximations:

$$\frac{\partial p}{\partial z} = \frac{p_{z+\Delta z,t} - p_{z,t}}{\Delta z} \tag{25}$$

$$\frac{\partial p}{\partial t} = \frac{p_{z,t+\Delta t} - p_{z,t}}{\Delta t} \tag{26}$$

$$\frac{\partial^2 p}{\partial z^2} = \frac{p_{z+\Delta z,t} - 2p_{z,t} + p_{z-\Delta z,t}}{(\Delta z)^2} \tag{27}$$

the following explicit solutions are found for Equations (10), (12), and (17), respectively:

$$p_{z,t+\Delta t} = p_{z,t} - \left(\frac{(vp)_{z+\Delta z,t} - (vp)_{z,t}}{\Delta z} \right) \Delta t$$

$$+ \left(D_{ab} \frac{p_{z+\Delta z,t} - 2p_{z,t} + p_{z-\Delta z,t}}{(\Delta z)^2} \right)(\Delta t) - \left(\frac{RTV_L Ka(p_{z,t}/H - C_L)}{M_G AL} \right) \Delta t \tag{28}$$

$$C_{L,t+\Delta t} = C_{L,t} + \sum_z \left(\frac{Ka(p_{z,t}/H - C_L)\Delta z}{L} \right) \Delta t \tag{29}$$

$$v_{z,t} = v_{z+\Delta z,t} + \frac{1}{P} \sum \left(\frac{RTV_L Ka(p_{z+\Delta z,t}/H - C_L)}{M_G AL} \right) \Delta z \tag{30}$$

Note that Equation (30) was solved for $v_{z,t}$ in terms of $v_{z+\Delta z,t}$, which is the reverse of the other solutions. This was done because the boundary condition for this equation is the velocity at the outlet of the tubing.

An original Fortran program using the following solution algorithm was used to solve Equations (28) through (30):

1. After initializing all variables, the user is prompted to enter trial-specific parameters, including output file names and print intervals, flow rates, trial times, total pressure, inlet gas composition, and the status of the trial (i.e., flow, dead-end, purge). If a purge trial, additional parameters for the purge frequency and duration are requested. The program also allows for a constant oxygen demand in the liquid phase and prompts for such a value.

2. Internal to the program are system-specific parameters, including the mass transfer coefficients for any gases present in the system, tubing dimensions, reactor volume, equations to calculate gas velocities based on user-specified flow rates, and physical constants. The model uses an average mass transfer coefficient for oxygen developed from 58 experimental trials conducted using a flow strategy (Chozick, 1992). Based on the nitrogen permeability of silicone tubing, a value equal to one-half the value obtained for oxygen was used for nitrogen (Robb, 1968).

3. Initial and boundary conditions are calculated from user-specified flow rates and inlet gas compositions and assuming that the aeration system is initially filled with air (i.e., 21% oxygen, 79% nitrogen) at the user-specified total pressure. The initial liquid concentrations are assumed to be at equilibrium with the atmosphere. At a specified time, the dissolved oxygen concentration is reduced to zero.

4. The algorithm uses time steps of 0.001 min and the aeration system is divided into 100 equal-sized elements. At each time step, the velocity profile through the entire system is calculated using Equation (30) and the user-specified outlet flow rate. For purge periods, the outlet flow rate is adjusted to the purge flow rate, and a new velocity profile is calculated.

5. Dissolved gas concentrations and partial pressure profiles for the next time step are calculated using Equations (28) and (29).

6. The program returns to calculate a new velocity profile based on the partial pressure distribution in the tube and the dissolved gas concentrations and continues until the desired end time for the trial is reached.

It should be noted that, as written, the computer algorithm can only predict oxygen transfer when the gas in the aeration system is a binary mixture of oxygen and nitrogen. This limitation is imposed for several reasons, including the fact that the effect of carbon dioxide transfer into the liquid phase has not been thoroughly investigated and that diffusion coefficients for mixtures other than binary gas mixtures are very difficult to estimate.

Experimental Confirmation of Model

In Figure 2, the experimental apparatus used for experimental oxygen transfer studies is shown. The glass reactor, which had a total liquid volume of 7.725 L with zero headspace, was sealed with a plexiglass cover containing ports for the DO probe and the shaft of the overhead stirrer. The shaft contained three propellers equally spaced over the depth of the reactor for mixing. There was also a vent in the cover to allow gas displacement during fill and draw operation and a port on the side of the reactor for liquid sampling.

The aeration system consisted of 18 lengths of 1/8" ID × 3/16" OD silicone rubber tubing, each approximately 40 cm long, which provides for a total surface area of 1070 cm². The silicone tubes were connected in series using 4.5 cm lengths of 1/8" ID × 1/4" OD tygon tubing and arranged in a circle in the reactor. The inlet and outlet gas lines were also tygon tubing.

External equipment on the reactor included oxygen and nitrogen cylinders connected to a gas blending flowmeter to supply the desired gas composition to the reactor, an in-line pressure regulator downstream of the flowmeter to control the inlet pressure to the silicone tubing, sampling ports and pressure transducers on the inlet and outlet gas lines to allow measurement of the gas composition and total pressure of the gas entering and exiting the membrane aeration system, a voltmeter to which the inlet and outlet transducers were connected, a flowmeter on the outlet gas line to control the gas flow through the aeration system, a DO meter and strip chart recorder to continuously monitor the dissolved oxygen concentration in the reactor during oxygen transfer experiments, and a GOW-MAC gas chromatograph with a thermal conductivity detector to analyze gas samples collected from the aeration system. A schematic diagram of the experimental setup is shown in Figure 3.

Experimental Verification of Model for Flow Operation

Clean-water oxygen transfer experiments were conducted as follows:

1. The outlet gas flowmeter was set to the desired flow rate for the trial.
2. The gas blending system was set to approximately the desired ratio of oxygen and nitrogen.
3. The pressure regulator was set to the desired pressure for the trial using the inlet pressure transducer as an indicator.
4. The gas blending system was readjusted, if necessary, to maintain the desired gas composition.

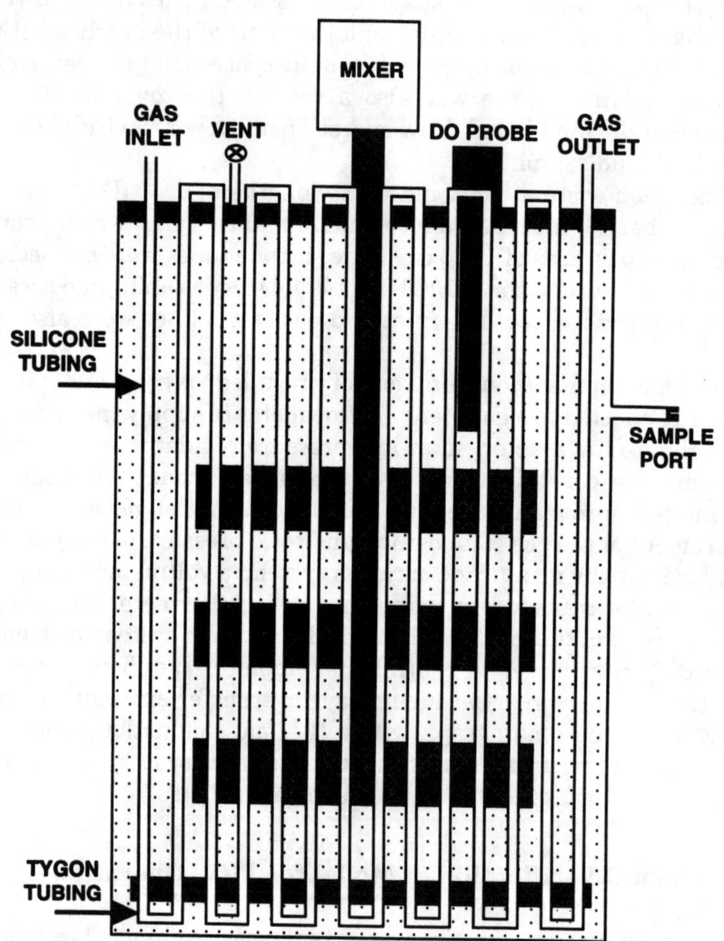

Figure 2. Reactor design for oxygen transfer experiments.

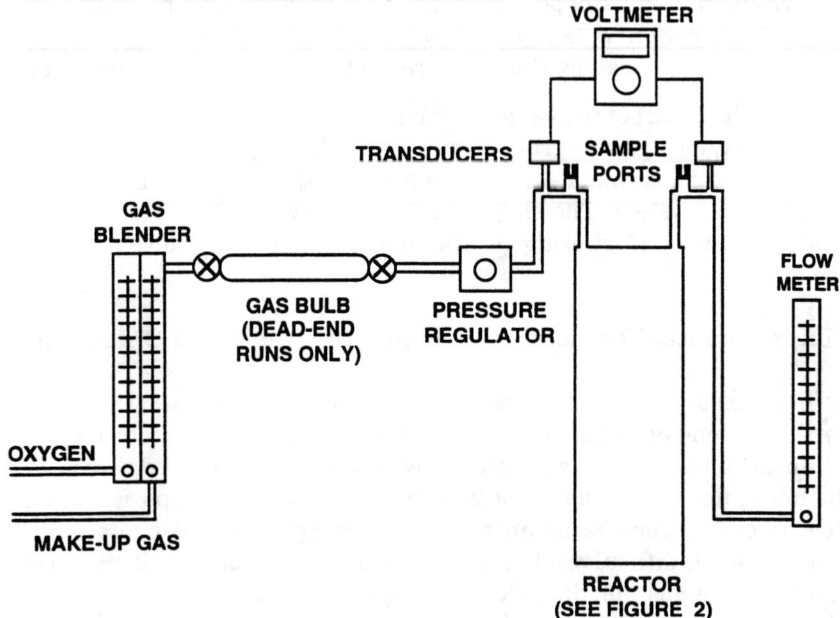

Figure 3. Experimental apparatus for oxygen transfer experiments.

5. Gas was allowed to flow through the silicone membrane for approximately 20 minutes to allow the gas composition in the tubing to reach a "steady-state" profile.
6. The reactor was filled with 7.725 L of double distilled water and the mixer was turned on to a speed of 250 rpm.
7. Approximately 0.75 g Na_2SO_3 and 0.02 g $CoCl_2$ were added to the reactor to reduce the dissolved oxygen concentration to zero. The $CoCl_2$ acts as a catalyst for the removal of oxygen by oxidation of Na_2SO_3.
8. The DO concentration in the reactor was monitored with the DO meter until it dropped to zero, at which time the strip chart recorder was turned on.
9. Inlet and outlet gas samples were collected through the gas sampling ports with gastight syringes and analyzed using the gas chromatograph. Samples were collected periodically throughout each trial so that between three and five inlet and outlet samples were collected for each trial. Due to the short length of some trials and the gas analysis time, only one or two gas samples was collected for a small number of trials.

10. Each trial was stopped when the DO concentration reached 80% of air saturation. A few runs in which the partial pressure of oxygen was too low to allow the DO to reach this level were stopped sooner.

In Table 2, a list of the values chosen for each of the variables of interest in the oxygen transfer trials is presented. The values chosen for total pressure and gas composition are representative of the useful operating range of the system. The outlet gas flow rates were chosen arbitrarily to test the validity of ignoring gas film mass transfer resistance in the data analysis.

Experimental Verification of Model for Dead-End Operation

In order to test the oxygen transfer capability of the aeration system with dead-end operation (i.e., outlet gas flow rate = 0), a modification to the experimental apparatus was necessary. It was discovered that the gas blending flowmeter could not maintain a stable gas composition at the low inlet gas flow rates encountered during dead-end operation. To overcome this difficulty, a 1-L stainless steel gas sampling bulb was placed between the gas blending flowmeter and the pressure regulator. Each dead-end trial was then conducted as follows:

1. The outlet gas flowmeter was set to approximately 20 ml/min.
2.–4. See above for flow trials.
5. Gas was allowed to flow through the silicone membrane for approximately 60 minutes to allow the gas composition in the tubing to reach a "steady-state" profile and to allow the gas composition in the sampling bulb to reach the desired composition.
6.–7. See above for flow trials.
8. The DO concentration in the reactor was monitored with the DO meter until it dropped to zero, at which time the strip chart recorder was turned on, the outlet flowmeter was set to zero, and the inlet valve on the gas sampling bulb was closed.
9.–10. See above for flow trials.

Table 2. Range of variables examined in clean-water oxygen transfer experiments.

Variable	Selected Values
Total pressure (psig)	1, 5, 10
Gas composition (percent oxygen)	Variable (10–100)
Outlet gas flow rate (ml/min)	5, 10, 15
Balance gas	Nitrogen

Table 3. Conditions for model runs.

Parameter	Flow	Purge	Dead-End
Pressure (psig)	1	1	1
Oxygen in gas stream (%)	100	100	100
Initial outlet flow (ml/min)	10	10	10
Time for initial flow (min)	20	20	20
Trial flow (ml/min)	10	0	0
Purge frequency (h^{-1})	0	1	0
Purge flow (ml/min)	NA	10	NA
Purge duration (min)	NA	1	NA

NA—not applicable.

The data collected for each trial included gas chromatograms, inlet and outlet gas pressures, and a continuous record of DO concentration versus time. The gas chromatograph was calibrated using pure oxygen and nitrogen, and the gas chromatograms were analyzed using peak height. The percentages obtained were then normalized to 100% to eliminate error resulting from the injection of different sample sizes. It should be pointed out that this also eliminates the effects of other components interfering with oxygen transfer (e.g., water vapor). The oxygen concentration (in percent) of each sample was then converted to a partial pressure using the gas pressure measured with the transducers, at the time each sample was collected.

Oxygen Transfer Predictions Using Model

Having confirmed that the model accurately predicts oxygen transfer into the reactor under most conditions, several model runs were conducted to examine the changing partial pressure profiles in the tube and to determine what the model predicts will happen when alternative operating strategies are used for the aeration system (i.e., flow, purge, and dead-end). The conditions for each of these model runs is given in Table 3.

RESULTS

Experimental Verification of Model for Flow Operation

In total, 58 oxygen transfer trials were conducted using a flow strategy for the aeration system. The average mass transfer coefficient obtained for all the trials was 0.0087 min⁻¹, with a standard deviation of 0.00051. In Table 4, the trials are divided by total pressure, gas composition, and

Table 4. Oxygen transfer results by variable.

	Number of Trials	Average *Ka* (min^{-1})	Standard Deviation
Pressure (psig)			
1	18	.0085	.00027
5	20	.0089	.00073
10	20	.0087	.00038
Gas Composition			
(% oxygen)			
0–25	12	.0085	.00077
25–50	17	.0087	.00037
50–75	13	.0088	.00049
75–100	16	.0088	.00049
Flow Rate (ml/min)			
5	22	.0088	.00060
10	18	.0087	.00062
15	18	.0086	.00022

flow rate. Gas composition, which varied over the range 0 to 100% oxygen, was arbitrarily divided into four groups to estimate differences over these ranges, since specific values of this variable were not used.

In Figures 4 through 7, the (time, DO) data collected from several of the flow trials, representative of a variety of experimental conditions, are shown. Model-predicted profiles and theoretical profiles are also shown in each of these figures. For the model predictions, the specified inlet gas composition was allowed to flow through the tube for 20 minutes before the DO was reduced to zero and the DO profile recorded. This represents the same conditions used in the experimental trials. As is evident in these figures, the model predicts the DO profile in the reactor very accurately. Theoretical profiles were obtained by solving Equation (7) for $C_L(t)$, with $C_L(0) = 0$, which gives

$$C_L(t) = \frac{p_G}{H}(1 - e^{-Kat}) \tag{31}$$

and using the average mass transfer coefficient from all experimental trials. The theoretical profiles also match the experimental data very closely. This is not surprising since, for a flow trial, the partial pressure profiles established in the tube after 20 minutes are assumed to remain approximately constant for the remainder of the trial, and changes in the mass transfer coefficient are not expected.

The gas analysis for each trial confirmed this assumption. The selected inlet gas oxygen concentration for each trial was essentially constant

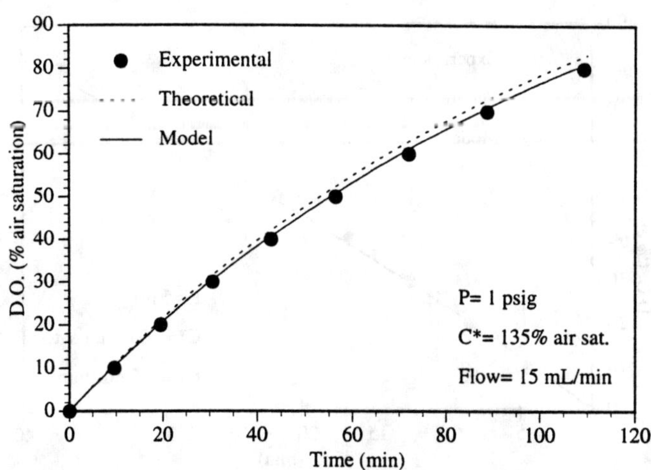

Figure 4. Experimental results and model predictions for dissolved oxygen profile during flow trial 2.

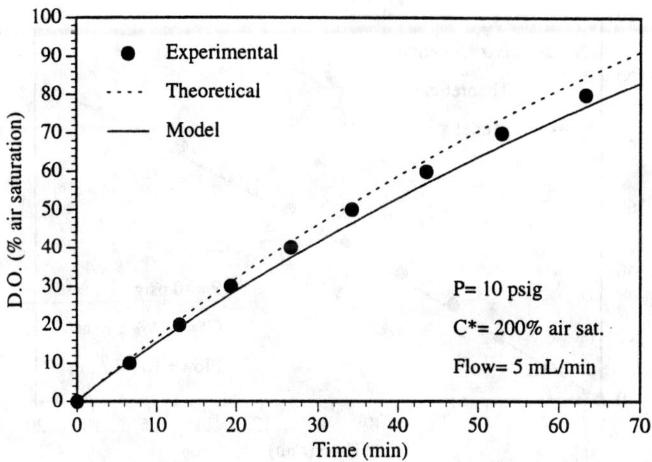

Figure 5. Experimental results and model predictions for dissolved oxygen profile during flow trial 52.

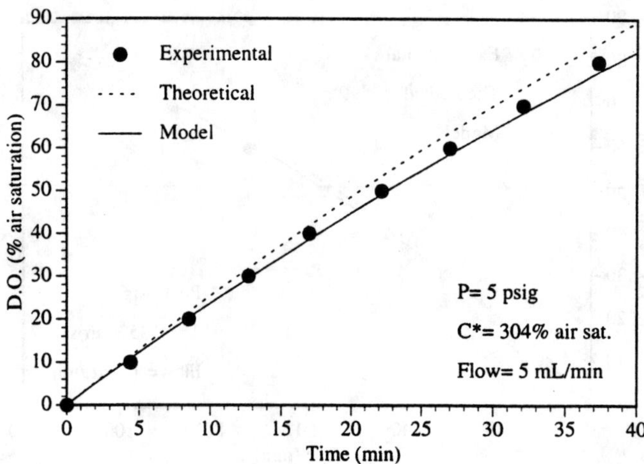

Figure 6. Experimental results and model predictions for dissolved oxygen profile during flow trial 46.

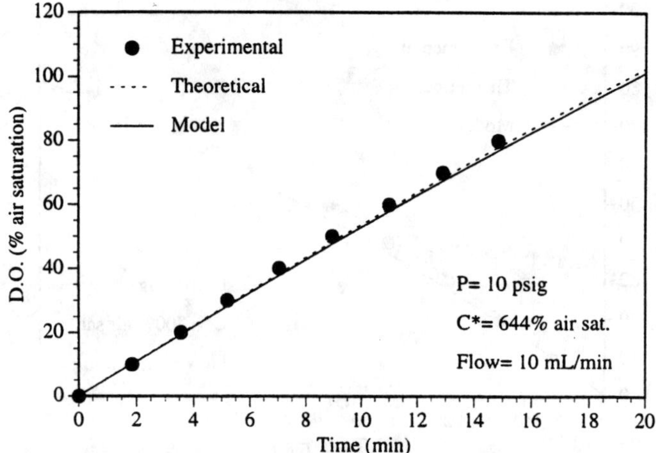

Figure 7. Experimental results and model predictions for dissolved oxygen profile during flow trial 35.

566

throughout the trial; the outlet gas oxygen concentration, which was slightly lower than the inlet gas oxygen concentration at the start of each run, was either constant or increased slightly over the course of the trial as less oxygen was transferred due to increasing DO and a decreased driving force. The outlet gas oxygen concentration was always within a few percent of the inlet gas concentration.

Experimental Verification of Model for Dead-End Operation

In total, 10 dead-end trials were conducted. In Figures 8 through 12, the results of several of the dead-end trials, as well as the model and theoretical predictions, are presented. As shown in these figures, the model predictions for dead-end trials were considerably more accurate than those given by the analytical solution [Equation (31)].

Since the gas mixture was allowed to flow through the tube until immediately before the start of each trial, the mass transfer coefficient at the beginning of the trials should be unaffected by the dead-end operation during each trial. The data (not presented) and the figures demonstrate that the initial mass transfer coefficients were essentially the same for all 10 dead-end trials. The average of these values was 0.0084 min^{-1}, and the standard deviation was 0.0005. A statistical analysis comparing these results to those of the 58 flow trials indicates that, at the 95% confidence level, the initial mass transfer coefficients during dead-end trials were not significantly different than the average mass transfer coefficients during flow trials.

As discussed above, a partial pressure gradient was established through the tube, with the outlet gas initially containing slightly lower oxygen concentrations than the inlet gas. The inlet gas oxygen concentration was essentially constant over the course of each trial, as expected. However, due to transfer of oxygen and the dead-end operation, it was expected that the oxygen concentration on the downstream side of the reactor would be lower than the inlet concentration and would decline over the course of each run. This was not observed in the current experiment because outlet oxygen concentrations were also essentially constant throughout each of the dead-end trials.

Oxygen Transfer Predictions Using Model

In Figure 13, the results of three model runs using pure oxygen as the inlet gas, each of the aeration system operating strategies, and exerting a constant oxygen demand of 10 mg/l-h are presented. This oxygen demand was chosen to demonstrate the large effect a small oxygen

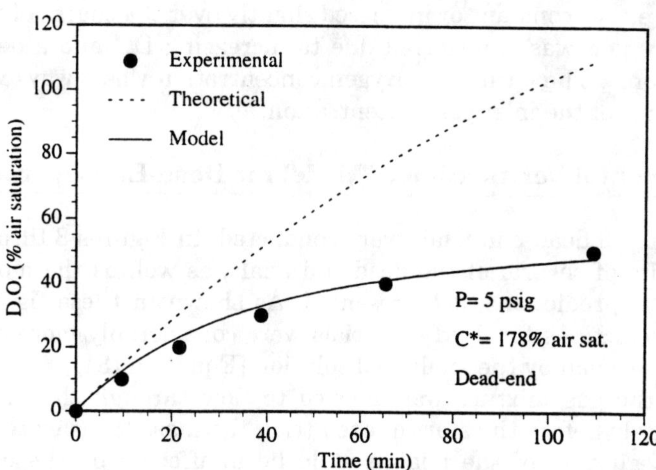

Figure 8. Experimental results and model predictions for dissolved oxygen profile during dead-end trial 2.

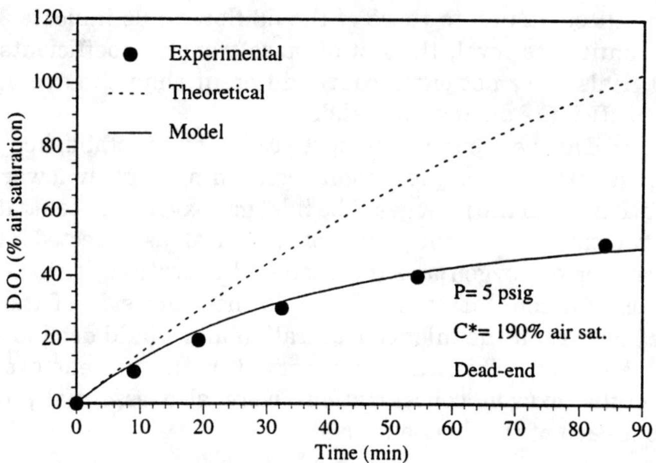

Figure 9. Experimental results and model predictions for dissolved oxygen profile during dead-end trial 3.

568

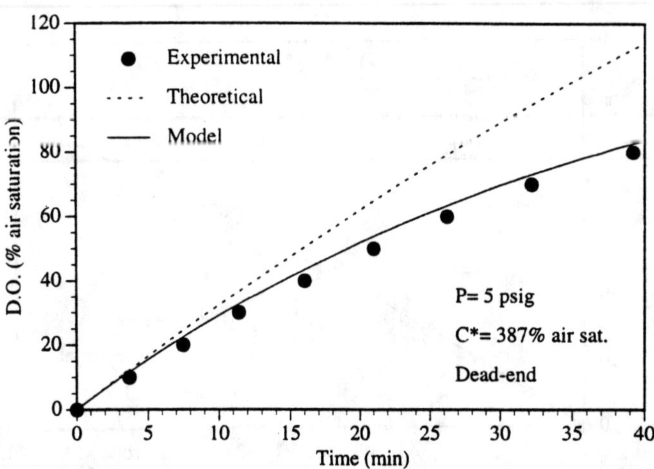

Figure 10. Experimental results and model predictions for dissolved oxygen profile during dead-end trial 5.

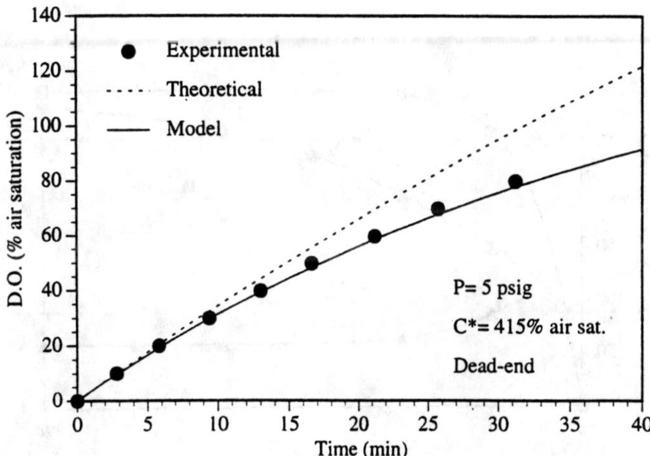

Figure 11. Experimental results and model predictions for dissolved oxygen profile during dead-end trial 6.

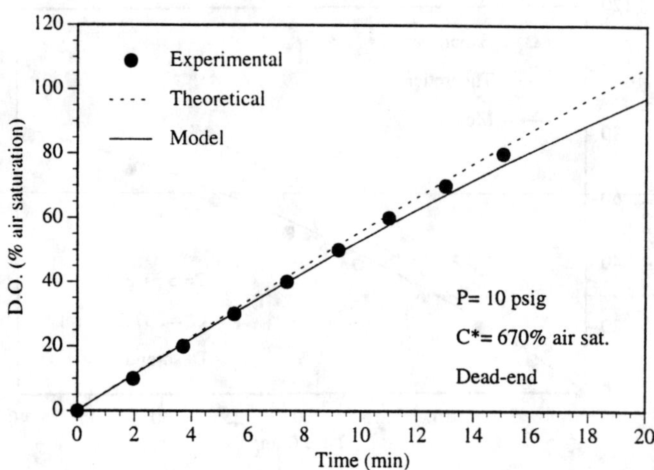

Figure 12. Experimental results and model predictions for dissolved oxygen profile during dead-end trial 9.

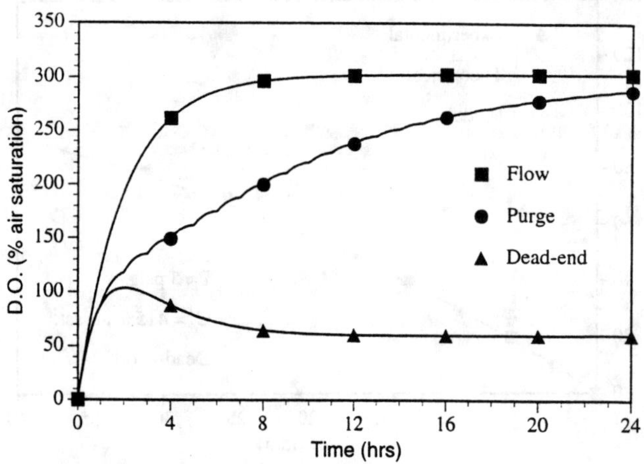

Figure 13. Predicted dissolved oxygen profiles with pure oxygen and an oxygen demand of 10 mg/l-h.

570

demand (e.g., an endogenous respiration rate) can have during purge and dead-end operation.

DISCUSSION

The results of the mass transfer experiments presented in Table 4 indicate little difference across the variables analyzed (i.e., total pressure, gas composition, and flow rate). The finding that the composition of the gas in the membrane does not affect the oxygen mass transfer coefficient (in flow trials) does not agree with the results presented in Table 1, where it was reported that the use of air instead of oxygen led to reduced mass transfer coefficients. Côté et al. (1989) attributed this observation to the formation of bubbles of nitrogen that formed on the outside of the tubing and stripped oxygen from the liquid phase. The high turbulence in the reactor caused by mixing at 250 rpm in the current studies apparently prevented the formation of gas bubbles on the surface of the tube, since none were observed during these trials and, therefore, prevented oxygen stripping by nitrogen.

The representative trials shown in Figures 4 through 7 indicate that the oxygen transfer through the membrane during flow trials was well described by both the model developed herein and by the theoretical solution given by Equation (31), indicating that the assumption that Ka and p_G were constant during flow trials was valid. Notice that, for the model predictions, the inlet oxygen partial pressure is provided rather than the average partial pressure over the length of the tubing, since the model calculates a theoretical partial pressure profile based on flow, diffusion, and oxygen transfer. The theoretical partial pressure profiles (not presented) were fairly stable during the flow trials, with little variation along the length of the tube (i.e., the outlet gas composition was nearly identical to the inlet), even at very low flows.

Markedly different results were obtained for dead-end trials. Figures 8 through 12 indicate that the theoretical solution given by Equation (31) did not accurately predict oxygen transfer during dead-end trials, except at very high oxygen partial pressures, when decreases in p_G would have a less noticeable effect. This results from a violation of the assumption that Ka and p_G are constant. The experimental apparatus used for these experiments was not specifically designed to measure the gas profiles in the tube during dead-end operation. Gas samples were collected downstream of the silicone tubing where the gas composition could only change by diffusion and, thus, did not detect changes in p_G associated with oxygen transfer. Nonetheless, it is evident that p_G would change during a dead-end oxygen transfer experiment, since there is essentially no advective transport (i.e., flow) in the tube, and diffusion would not be sufficient to replace the oxygen transferred through the tube.

In addition to p_G changing over the course of a dead-end trial, Ka might also change over the course of a trial due to the accumulation of water vapor inside the tube and a resulting increase in the "gas"-phase mass transfer coefficient. Such an accumulation of water in the tubing would also not be detected in the current apparatus.

The initial mass transfer coefficients during dead-end trials were calculated using the initial gas analyses, which were collected immediately after switching to dead-end operation, and should have provided an accurate estimate of the partial pressure profile in the tube, since little oxygen transfer had taken place. These values indicate that, at least initially, the mass transfer coefficient during dead-end trials was the same as that in flow trials.

In the model described in this chapter, Ka is assumed constant (and equal to the value determined during flow trials), and p_G is allowed to change over the course of each trial. With the exception of one dead-end trial, which had a very low inlet oxygen partial pressure that may not have been accurately maintained over the course of the trial, this approach predicts the DO profiles in the reactor fairly well, suggesting that the changing partial pressure profile in the tube may be responsible for most, if not all, of the observed reduction in oxygen transfer for dead-end trials.

Using the model, oxygen transfer was predicted over extended operating periods using flow, dead-end, and purge operating strategies to determine the effect of the operating strategies on oxygen transfer. In addition, the effect of a constant oxygen demand on the performance of a membrane aeration system using each of these operating strategies was examined. These results, presented in Figure 13, indicate that oxygen transfer over extended periods of time is significantly affected by operating strategy.

With pure oxygen and a flow strategy, steady-state oxygen concentrations are rapidly achieved in the reactor. Purge-operating strategies with pure oxygen cause reduced oxygen transfer rates; however, the DO slowly approaches values close to those obtained using flow operation. The length and duration of purge cycles will have a dramatic affect on these profiles, with longer and/or more frequent purges causing these profiles to approach those obtained with flow operation.

Dead-end operation causes a marked reduction in oxygen transfer. Diffusion of oxygen into the tube from the supply and very slow advective transport necessary to maintain the total pressure in the tube result in an oxygen transfer rate that is unable to maintain elevated DO levels when a small oxygen demand (e.g., 10 mg/l-h) is exerted.

The jagged appearance of the DO profiles with purge operation in Figure 13 represents a phenomenon that was generally observed during purge operation of the SBBR. When the tubing is purged, even for a relatively short period of time (e.g., 10 seconds), the oxygen partial pressure in the tube increases enough to cause a noticeable difference in the oxygen transfer rate.

CONCLUSIONS

Oxygen transfer through gas permeable silicone membranes was investigated, and it was found that the mass transfer coefficient was independent of the gas flow rate (as long as the flow rate was not zero) and the oxygen partial pressure in the tube.

Dead-end operation (i.e., zero flow at the tubing outlet) of the membrane aeration system led to significant reductions in oxygen transfer, and more importantly, the oxygen transfer was not adequately described by a simple oxygen transfer expression that accurately predicted the dissolved oxygen profiles for all trials using flow operation. The failure of this equation to predict dead-end oxygen transfer in the SBBR and the cause of reduced oxygen transfer during dead-end operation can be attributed to decreasing oxygen partial pressure in the tubing as dead-end operation progresses.

A mathematical model for the membrane aeration system in an SBBR was developed to overcome some of the difficulties previously encountered when simple mass transfer equations were used to estimate oxygen transfer through silicone membranes and as a general tool for design of membrane aeration systems.

With the exception of a single experimental trial at very low oxygen partial pressure, the accuracy of the model was confirmed by comparison to experimental results. The model was then used to predict the behavior of the aeration system using flow, purge, and dead-end operating strategies with air and oxygen and with an oxygen demand. The model predictions indicate that flow strategies provide the most stable operation, purge strategies may be used if chosen appropriately (i.e., duration and frequency of purges) to meet an oxygen demand, and dead-end strategies cannot be utilized in systems with even very small oxygen demands. Purge strategies have been used effectively for oxygen supply in SBBRs (Chozick and Irvine, 1991; Chozick, 1992).

Future development of the model should include a predictive aspect for the mass transfer coefficient, including the effect of biofilm growth, and allowing for the inclusion of additional gases that may be present in an operational system (e.g., carbon dioxide, volatile organic compounds, etc.).

REFERENCES

Alleman, J. E. and R. L. Irvine, "Nitrification in the sequencing batch reactor," *J. Water Pollut. Control Fed.*, 52, 2747 (1980a).

Alleman, J. E. and R. L. Irvine, "Storage-induced denitrification using sequencing batch reactor operation," *Wat. Res.*, 14, 1483 (1980b).

Bryers, J. D. and W. G. Characklis, "Biofilms in water and wastewater treatment," in *Biofilms* (ed. by Characklis, W. G. and K. C. Marshall), John Wiley & Sons, Inc., New York, pp. 671–696 (1990).

Chatzopoulos, D., "Design and Operation of Hybrid Granular Activated Carbon-Sequencing Batch Biofilm Reactors for the Treatment of Wastewaters Containing Volatile Aromatic Compounds," Ph.D. Dissertation, University of Notre Dame, Notre Dame, IN (1994).

Chiesa, S. C. and R. L. Irvine, "Growth and control of filamentous microbes in activated sludge: An integrated hypothesis," *Wat. Res.*, 19, 471 (1985).

Chozick, R. "Operation and Performance of a Membrane-Aerated Sequencing Batch Biofilm Reactor for the Treatment of Wastewater Containing Volatile Organic Compounds," Ph.D. Dissertation, University of Notre Dame, Notre Dame, Indiana (1992).

Chozick, R. and R. L. Irvine, "Preliminary Studies on the Granular Activated Carbon–Sequencing Batch Biofilm Reactor," *Environmental Progress*, 10, 282 (1991).

Chudoba, J., V. Ottavá, and V. Madera, "Control of activated sludge filamentous bulking—I. Effect of the hydraulic regime or degree of mixing in an aeration tank," *Wat. Res.*, 7, 1163 (1973).

Côté, P., J.-L. Bersillon, and A. Huyard, "Bubble-free aeration using membranes: Mass transfer analysis," *J. Membrane Sci.*, 47, 91 (1989).

Côte, P., J.-L. Bersillon, A. Huyard, and G. Faup, "Bubble-free aeration using membranes: Process analysis," *J. Water Pollut. Control Fed.*, 60, 1986 (1988).

Irvine, R. L., L. H. Ketchum, Jr., R. E., Breyfogle, and E. F. Barth, "Municipal application of sequencing batch treatment at Culver, Indiana," *J. Water Pollut. Control Fed.*, 55, 484 (1983).

Irvine, R. L., S. A. Sojka, and J. F. Colaruotolo, "Enhanced Biological Treatment of Leachates for Industrial Landfills," *Haz. Waste*, 1, 123 (1984).

Ketchum, L. H., Jr., R. L. Irvine, R. E. Breyfogle, and J. F. Manning, Jr., "A comparison of biological and chemical phosphorus removals in continuous and sequencing batch reactor," *J. Water Pollut. Control Fed.*, 59, 13 (1987).

Manning, J. F., Jr. and R. L. Irvine, "The biological removal of phosphorus in a sequencing batch reactor," *J. Water Pollut. Control Fed.*, 57, 87 (1985).

Metcalf & Eddy, Inc., *Wastewater Engineering: Treatment, Disposal, and Reuse*, 3rd Ed., McGraw-Hill, New York (1991).

Muollo, S. J., R. G. Smith, and E. D. Schroeder, "Control of VOC emissions from wastewater treatment plants using membrane aeration" (unpublished).

Palis, J. C. and R. L. Irvine, "Nitrogen removal in a low-loaded single tank sequencing batch reactor," *J. Water Pollut. Control Fed.*, 57, 82 (1985).

Robb, W. L., "Thin silicone membranes—Their permeation properties and some applications," *Ann. N.Y. Acad. Sci.*, 146, 119 (1968).

Schaffer, R. B., F. J. Ludzack, and M. B. Ettinger, "Sewage treatment by oxygenation through permeable plastic films," *J. Water Pollut. Control Fed.*, 32, 939 (1960).

Smith, R. G. and P. A. Wilderer, "Treatment of hazardous landfill leachate using sequencing batch reactors," *Proceedings of the 41st Industrial Waste Conference*, 41, 272 (1987).

Su, W. W., H. S. Caram, and A. E. Humphrey, "Optimal design of the tubular microporous membrane aerator for shear-sensitive cell cultures," *Biotechnol. Prog.*, 8, 19 (1992).

Wickramasinghe, S. R., M. J. Semmens, and E. L. Cussler, "Better hollow fiber contactors," *J. Membrane Sci.*, 62, 371 (1991).

Wilderer, P. A., J. Brautigam, and I. Sekoulov, "Application of gas permeable membranes for auxiliary oxygenation of sequencing batch reactors," *Conservation & Recycling*, 8, 181 (1985).

Yang, M.-C. and E. L. Cussler, "Designing hollow-fiber contactors," *AIChE J.*, 32, 1910 (1986).

Yasuda, H. and C. E. Lamaze, "Transfer of gas to dissolved oxygen in water via porous and nonporous polymer membranes," *J. Appl. Polym. Sci.*, 16, 595 (1972).

Fundamentals of Bioremediation Treatability Studies

STEVEN I. SAFFERMAN
*Department of Civil and Environmental Engineering
and Engineering Mechanics
The University of Dayton
Dayton, OH 45469-0343, USA*

INTRODUCTION

Treatability studies enhance the chance of choosing an optimal remediation technology and provide pertinent data required for implementation of the selected technology. The design and operation of a treatability study can be complex, especially for bioremediation. A clear understanding of the technology and the objectives and limitations of the treatability study is the primary requirement for success. If an understanding is not established during the planning stage, the collected data may be of little or no value in achieving the study's goals. This chapter emphasizes setting objectives and general approaches to achieve them.

The early sections in this chapter discuss general issues associated with bioremediation treatability studies for polluted soil. The latter sections pertain to specific tiers of treatability studies (screening, selection, and remedial design/action), all of which build upon each other in terms of operational complexity, cost, and the amount of information gained. Often, approaches in higher level tiers (remedy selection and remedy design/action) require the use of a lower tier (remedy screening) to aid in evaluating results.

The intention of this chapter is not to provide comprehensive guidance or protocols on how to conduct treatability studies but, rather, to furnish the foundation, describe the decision processes, and present approaches for conducting treatability studies.

CATEGORIES OF TREATABILITY STUDIES

General Objectives

Treatability studies are conducted to provide data that cannot be obtained from process theory and/or past experience. They are used to answer questions concerning the applicability of a technology to a specific site under specific conditions. The U.S. Environmental Protection Agency (1992) has established three tiers of treatability testing: remedy screening, remedy selection, and remedial design/action. Screening-level studies are designed to determine if bioremediation has any chance of successfully remediating a site. Selection studies aid in determining the probability of reaching cleanup goals within the desired time frame and, therefore, are typically used to assess a remediation technology. Remedial design/action studies provide design and cost data needed for construction specifications, schedules, and budgets and uncover potential full-scale operational problems. The successive tiers provide an increasing, comprehensive understanding of the application of a technology to site-specific conditions. Each tier, however, requires additional cost and time.

Schedule and Cost Considerations

The establishment of specific guidelines on cost and time requirements for each of the treatability study tiers is difficult. The specific bioremediation technology to be simulated and study objectives must be considered, especially for selection and remedial design/action studies. This is illustrated by comparing selection-level studies for a slurry reactor and composting technology. A slurry reactor is conceptually simple and has well-defined operational parameters that are easily simulated, even at a small scale. Conditions for bioremediation are optimized; therefore, the time for bioremediation is minimized. Consequently, the study is relatively inexpensive and short. Composting, although also conceptually simple, is greatly influenced by the transport of oxygen and other amendments. Larger reactors or full-scale soil piles are required to accurately simulate transport mechanisms. In addition, longer treatment times are required for this technology because of nonoptimal mixing. The result is a relatively longer and more expensive treatability study.

Several references are available that provide general costs and schedule issues involved in planning and conducting treatability studies (Rogers et al., 1993; U.S. Environmental Protection Agency, 1991, 1993c). In addition, the U.S. Environmental Protection Agency Superfund Innovative Technology Evaluation Program Technology Profiles document provides abstracts on emerging technologies that are being, or have been, recently demonstrated (U.S. Environmental Protection Agency, 1994c). The document is generally updated annually, and specifics concerning each completed demonstration, including cost and schedule data, are obtained from documentation associated with each specific demonstration.

USE OF TREATABILITY STUDIES

The use of all three tiers of treatability studies should not be automatic. Instead, a strategy for selecting and designing a remediation technology study should be developed based on site-specific conditions, the database of experience, and a theoretical understanding of the technology.

Treatability testing is typically excluded or minimized if the application of the technology at a specific type of site is very well established. Alternatively, a treatability study would not be conducted if its projected cost is greater than the projected consequence of implementing a less than optimal technology. As an example, if a particular bioremediation technology appears to have the best potential based on a wealth of experience, only a screening level or a hybrid of a screening and selection study would be conducted. The objectives would be to determine if the site matrix has an adverse effect on what has been a successful utilization of the technology in the past.

The background information required to develop a strategy can be largely obtained from numerous sources of literature on specific bioremediation technologies and treatability studies and specialists in bioremediation. Of particular use are the following databases and documents that are periodically updated: *The Bioremediation in the Field Search System* (U.S. Environmental Protection Agency, 1994a), the *Vender Information System for Innovative Treatment Technologies* (U.S. Environmental Protection Agency, 1995c), the *RREL Treatability Data Base* (U.S. Environmental Protection Agency, 1993d), the *North American Wetlands for Water Quality Treatment Database* (U.S. Environmental Protection Agency, 1994b), *Bioremediation in the Field* newsletter (U.S. Environmental Protection Agency, 1995a), and the *Bioremediation Resource Guide* (U.S. Environmental Protection Agency, 1993a). Several substantial review articles and documents are also available (Alexander, 1981; Bourquin, 1989; Kobayashi and Rittmann, 1982; Norris et al., 1994; U.S. Environmental Protection Agency, 1986, 1988, 1989b).

DESIGN OF TREATABILITY STUDIES

The importance of a strategy for the overall approach of selecting a remediation technology was discussed above. Of equal importance is the experimental design for a specific treatability study. Considerations general to all bioremediation treatability studies are presented below.

Detailed, all-encompassing, cookbook procedures for conducting bioremediation treatability studies are not available. Each study must be customized to the study's objectives and the site under investigation. Several general design issues, however, are important to all bioremediation studies, including the analysis for the pollutant, the fate of the pollutant, and the interpretation of the data.

Analysis for Pollutants

High variability of pollutant concentrations can mask results, making conclusions meaningless. This is of particular concern for soil because of its heterogeneity. Soil preparation steps can reduce the variability. Typical procedures include screening, crushing, sieving, removing debris, and removing the larger soil fractions (Leitzinger et al., 1995). Sample replications, spike recoveries, blanks, and evaluation using multiple analytical procedures are all tools that can aid in evaluating the variability of data (U.S. Environmental Protection Agency, 1979). A statistical design can be used to determine the appropriate level of replication; however, costs often inhibit the collection and analysis of the appropriate number of samples.

Contaminant Fate Analysis

The evaluation of a treatability study based solely on the disappearance of the pollutant might not indicate an overall reduction in exposure to toxicity (the goal of any remediation project). The fate of the pollutant must also be determined. Manipulations and amendments used in bioremediation often encourage abiotic pollutant removal (volatilization, extraction, and adsorption). Volatilization can occur from the intense energy input used by some bioremediation technologies to achieve contact between the pollutants and microorganisms and/or for aeration. Extraction of pollutants from soil can result when certain nutrients (especially oleophilic ones) and surfactants are added. Adsorption of pollutants to amendments or biomass can be enhanced due to mixing.

In addition, bioremediation does not always result in the mineralization of complex pollutants. Transformation products that are generally less toxic are typically formed. Many exceptions, however, can occur (Kincannon et al., 1983; Park et al., 1990). A common example is the anaerobic transformation of trichloroethylene (TCE) to vinyl chloride (Bouwer et al., 1981). Vinyl chloride has a time-weighted average threshold limit value that is an order of magnitude lower than TCE (ACGIH, 1993–1994). Even if the pollutant is biotransformed to a less toxic compound, it could be more mobile and bioavailable, resulting in an overall increase in toxicity. High-molecular-weight polynuclear aromatic hydrocarbons are typically transformed to lower ones, with less toxicity. The lower-molecular-weight compounds, however, are more polar than the parent ones and can be transported in water before further degradation occurs (Baud-Grasset et al., 1993; Metosh, 1991).

Fate analysis to determine abiotic losses and biodegradation products can be a major undertaking. A detailed analysis is often not feasible because of cost, time, and the lack of available protocols. Several simplified approaches are available and are discussed below.

TOXICITY TESTING

Several ecological assays are available to measure toxicity changes in soil as a result of remediation. These assays measure either chronic or geno toxicity and include one that measures a pollutant's effect on bacteria (Ames Test, Microtox™), plants (root elongation, seed germination, Tradescantia), or small organisms (minnows, earthworms, shrimp, water fleas, amphipod, larvae). Caution in using these assays must be exercised, however, because the toxicity in one organism cannot be easily correlated to toxicity in another organism (Metosh, 1991; U.S. Environmental Protection Agency, 1995b).

QUALITATIVE ANALYSIS

A qualitative fate analysis for a compound can be conducted by using the properties of the compounds that can predict transport and transformation. These properties include the compound's water solubility, vapor pressure (and Henry's coefficient), photolytic half-life, octanol water partitioning coefficient, adsorption partitioning coefficient, and biodegradation kinetic coefficient (Ney, 1995; U.S. Environmental Protection Agency, 1993a). In addition, the characteristics of the matrix, including hydraulic conductivity, soil moisture, water potential, salinity, pH, nutrients, and ion exchange capacity, must also be considered (U.S. Environmental Protection Agency, 1991, 1993b, 1993c).

One further consideration, however, is that the expected properties of a pollutant can be modified for the purpose of mobilizing or demobilizing the compound. DDT was frequently combined with an emulsifier so that it could be carried in water and applied by spraying. Consequently, the commercial use of the compound must be known to conduct a complete qualitative fate analyses. Safferman et al. (1995) describe a soil pan study in which this concern was accounted for in a treatability study.

MASS BALANCE APPROACH

A mass balance accounts for all possible routes of input, exit, and transformation when characterizing the behavior of a compound in a process. A mass balance approach for determining the fate of a pollutant uses this principle; however, complete accountability (closure) is not possible. Specifically, bioremediation cannot be directly determined. The removal caused by volatilization, extraction and runoff from the process, and adsorption (abiotic mechanisms) can be measured and converted to a mass value. Bioremediation can then be calculated by taking the difference between the overall removal and that by the abiotic processes.

This approach can be difficult to institute because of the difficulty of quantitating all of the abiotic removals. Of specific concern is the accounting of all volatiles from a large-scale system and the extraction of compounds adsorbed to bulking agents. Nevertheless, the approach is useful, especially if a particular abiotic loss is more likely to be significant based on a qualitative fate analysis (Safferman et al., 1995).

KILLED CONTROL

Killed controls inhibit microorganisms so that biodegradation cannot occur. Abiotic mechanisms are solely responsible for pollutant removal. Microbial inhibition can be accomplished by gassing the soil with a toxic gas (ethylene bromide) or mixing the soil with a toxic salt such as mercuric chloride, cadmium, or formaldehyde (Parkinson et al., 1971). Alternatively, lowering the pH to below 2 will also be effective in deactivating microorganisms. The use of any deactivation agent, especially pH, will change the physical and chemical structure of the soil and could bias the results. In addition, complete deactivation of soil microorganisms is very difficult. The particles of soil can shelter microorganisms, and complete distribution of the inhibiting agent is difficult. Therefore, monitoring for microorganisms must accompany killed controls.

Controls must be operated identically to the test reactors if valid comparisons are to be made. Microorganisms are very sensitive to their environment, and slight variances can have a major effect on the microbial population and invalidate comparisons between test and control reactors.

Interpretation of Data

Treatability study results must ultimately be related to the study's objectives. Values obtained from analytical analysis must be corrected for dilution, and quality assurance indicators must be reviewed. Calculations should be scrutinized because many errors can occur in relating concentrations of extracts to concentrations in the original soil. Conclusions on the effectiveness of the process should provide a discussion of fate analysis.

Related to data evaluation is the interpretation of conclusions. Because bioremediation is an umbrella of diverse technologies, specific applications have few applied similarities. A small slurry reactor could be effectively used as a screening tool to determine if bioremediation has potential for a specific pollutant in a specific matrix, but it will not be predictive of the required time and obtainable cleanup levels for an in situ application. A positive, simple screening study would, however, lead to the conclusion that further selection or remedial design/action studies are desired or that proceeding with a full-scale installation is warranted.

BIOREMEDIATION SCREENING-LEVEL STUDIES

Bioremediation screening-level studies are designed to determine if a technology has any chance of successfully bioremediating (biotransformation or mineralization) a contaminated site. They often consist of simple analyses and have little resemblance to the full-scale installation that will ultimately remediate the site.

To design a screening level treatability study, the basic principles of microbial biodegradation must be understood. An organic pollutant will not biodegrade, nor an inorganic compound be biotransformed, when one of the required components for a living microorganism is not present. These components include the pollutant's inherent biodegradability, an energy or carbon source (often the pollutant but not exclusively), macro- and micronutrients, moisture, contact between the microorganism and pollutant, an electron acceptor (oxygen for aerobic biodegradation), the appropriate temperature range, the appropriate pH range, and freedom from toxicity (caused either by high pollutant concentrations or by other materials in the polluted matrix).

Bioremediation provides for or supplements a limiting condition. The inherent biodegradability of common pollutants can often be determined by consulting the scientific literature (Alexander, 1981; Bourquin, 1989; Ghosal et al., 1985; Kobayashi and Rittmann, 1982; U.S. Environmental Protection Agency, 1988, 1989b). If references are not available, screening-level treatability studies can be used to determine the biodegradation potential of a compound. A screening-level study is, however, more typically used to determine if matrix interference or high pollutant concentrations will cause inhibition.

Three general approaches are often used to conduct screening-level treatability studies, direct measurement of the microorganisms, measurement of microbial activity, and disappearance of the pollutant. Several tests for each of these general categories are discussed.

Microorganism Measurements

If bioremediation is to be considered, the polluted matrix must support microbial growth. All soils, except those containing no organics, should contain indigenous microorganisms. While the microorganisms might not be flourishing because of limiting conditions, a baseline population should be present. The best measurement of the baseline microbial level is unpolluted in the vicinity of the site. A relatively low level of microorganisms in a polluted matrix could indicate a potential toxicity problem. Because of the vast number of variables that can adversely affect bioremediation, characterization of the soil for components that would indicate microbial inhibition would be very difficult and expensive.

PLATE COUNTING

The quantity of microorganisms in soil can be determined by plate counting. In this procedure, a petri dish containing a soil extract type agar is combined with a dilute slurried soil. The agar contains nutrients and substrate required for microbial growth. After the plates are incubated for the appropriate time and temperature, microorganism colonies become visible and can be counted (Johnson et al., 1959).

Alternatively, noble agar can be used in place of soil extract agar. Noble agar contains nutrients but no sources of energy for microbial growth. The pollutant provides the energy. If the pollutant is a semivolatile, it is added directly to the agar, but if it is volatile, it is maintained in the incubation atmosphere by placing an open container of the pollutant in the incubator. The plates are incubated and evaluated as above. Although the use of noble agar determines if microorganisms can be supported exclusively on the pollutant, some pollutants do not serve as a sole energy source, yet they still biodegrade because of cometabolism. Therefore, the lack of microbial growth on noble agar does not indicate that bioremediation will not occur.

Plate counting is rapid, inexpensive, and relatively simple to conduct. A major disadvantage, however, is the artificial nature of the test. The petri dish does not represent the soil and even has been found to be nonoptimal. The number of microorganisms can be underestimated by as much as two to four orders of magnitude (Karl, 1986). Consequently, very low bacteria numbers might be interpreted as a toxicity problem while, in reality, the medium just does not support the population.

PHOSPHOLIPID ANALYSIS

Phospholipids are fatty acids contained in cell walls. Once the cell is no longer viable, phospholipids generally have a half-life of two days (White et al., 1979). Phospholipids can be directly extracted from the sample (Bligh and Dyer, 1959) and then measured by simple colorimetric techniques (Findlay et al., 1989). The artificial condition of measuring the microbial population outside its natural environment, as in plate counting, is avoided. Correlations between phospholipids and the quantity of bacteria, as measured by plate counting, have been made but are dependent on the type of microorganism and site-specific conditions (White et al., 1979).

Phospholipids can be speculated by gas-liquid chromatography. Specific phospholipid species correspond to a specific category of organisms. The category can be helpful in understanding the organism's biodegradation capacity (Vestal and White, 1989).

Phospholipids are good indicators of the presence or absence of microorganisms and their relative quantity. The analyses, however, can be variable, and several replicates are required in order to find a mean value that has a reasonable confidence limit (Herrmann and Safferman, 1995).

Other substances associated with microbial growth can be used to indicate microbial population size. The most common is the measurement of adenosine triphosphate (ATP), the high-molecular-weight energy compound. Phospholipids, however, are the more advantageous because they are short-lived once the cell is no longer viable.

DIRECT VISUAL EVALUATIONS

The scanning electron microscope (SEM) and the optical microscope can be used to directly observe the microbial population. The SEM provides a three-dimensional view of a soil particle. A semiquantitative counting and characteristic evaluation approach has been presented in the literature (Safferman et al., 1993). Specialized equipment and skilled operators with specific experience in examining samples for microbial characteristics are required. Sample preparation can be extensive, taking up to two hours (process time, not labor), and artifacts can result (Chang and Rittmann, 1986; Little et al., 1991). In addition, only very small samples can be examined at one time.

For the examination of a sample using an optical microscope, the samples must be first properly prepared and treated with a fluorescent dye. The result is points of fluorescence that can be visually observed and counted. The dye and preparation steps can be selected for specific categories of microorganisms (Karl, 1986). This procedure is much simpler than the SEM method and requires less specialized equipment. Attachment and surface characteristics cannot be observed, however, because of the lower magnifications associated with an optical microscope and the lack of a three-dimensional perspective. Like the SEM analysis, only small samples can be examined at one time.

MOLECULAR-LEVEL IDENTIFICATION TECHNIQUES

Several molecular features of microorganisms can be related to the quantity of microorganisms. Included are molecular probes and antibody production (immunoassay technologies). Molecular probes select for a genotype (typically a DNA sequence) that corresponds to specific metabolic activity (Haugland et al., 1990). DNA can be directly extracted from the soil (Tsai and Olson, 1991). Radiolabeled tracers have been typically used to determine hybridization although nonlabeled tracers have been developed (Haugland et al., 1990).

Unique antibodies are produced in response to toxic exposures. They can be measured by radiolabeling or colorimetry. Immunoassays have been typically used as field screening tools for the identification of pollutants (VanEmon and Gerlack, 1994). The principles can also be applied to finding evidence of a microbial population that has responded to a pollutant.

Both of the molecular-level techniques are similar to phospholipid analysis in that samples are examined directly and artificial incubation is not required. Both, however, are very specific to the microorganism and compound of interest. These features are highly desirable for some applications but not for screening-level treatability studies.

Because these technologies are relatively recent, comprehensive, commercial genotype (probes) libraries do not exist. Highly skilled scientists are needed to effectively develop probes, prepare samples, and use the probes. Due to the rapid advancement in these fields, these tools will become further refined and more common in the near future.

Microbial Activity

An alternative to searching for the presence, quantity, and/or category of microorganisms is to examine their activity. An active microbial population and the rate of activity will indicate whether a polluted matrix can support bioremediation. Activity is typically determined based on the disappearance of a substrate (the target pollutant) and/or the formation of a by-product or respiration (oxygen utilization or carbon dioxide evolution for aerobic microorganisms). Microbial activity studies vary from fast, inexpensive assays to small-scale reactor runs. They generally take longer to conduct than microbial quantity tests but also provide more insight into the biodegradation process. The decision to use an activity approach, instead of quantity, depends on how it will support the next stage of the remediation technology selection process.

ESTERFIED DYES

Incubation of a slurried sample with an esterfied dye enables microbial activity to be correlated to the production of color. A biodegradable substrate/dye complex (fluorescein diacetate, also known as 3′,6′-diacetylfluorescein, as an example) is added to a slurried sample followed by incubation under optimal environmental conditions. The substrate is a readily degradable organic (such as acetate or glucose) (Schnurer and Rosswall, 1982). As the substrate degrades, the dye is released, resulting in a color change of the incubation solution. Activity is expressed as the concentration of dye produced over a period of time. To gain an understanding of the results, comparisons must be made to a condition that is known not to inhibit microbial growth. If available, noncontaminated soil with characteristics similar to the polluted soil is ideal.

This analysis is rapid, precise, and does not require extensive skills and equipment. This type of analysis, however, does not mimic site-specific conditions.

OPTIMIZED SLURRY REACTOR

A slurry reactor is commonly used when microbial activity is evaluated by the disappearance of the pollutant. The slurry reactor provides optimal contact between microorganisms and pollutants and a convenient mechanism to add amendments. The water content of the slurry can be between 40 and 80% (U.S. Environmental Protection Agency, 1993e). Contents closer to 80% are more common.

These studies have usefulness beyond simply examining the potential of bioremediation because they can provide insight into cleanup levels and kinetics. Consequently, this approach might be used as a hybrid between a screening and selection-level study. Along with the increased level of achievable knowledge is also higher cost and time requirements.

The potential for pollutant removal by abiotic mechanisms is of particular concern for a slurry reactor. The potential for volatilization due to the turbulence created by mixing is of great importance. Killed control studies can be used effectively to determine abiotic losses. The mixing in the reactor provides good distribution of the sterilizing agent and increases the potential for successful deactivation of the microorganisms.

RESPIRATION

The utilization of oxygen, the production of carbon dioxide (or methane and carbon dioxide for anaerobic conditions), or a combination of both is measured in respirometry. The choice is dependent on whether activity is the only focus of the study or kinetic rates are desired. This technique is very effective in measuring microbial activity for both slurries and solids and for both aerobic and anaerobic conditions. To measure slurries, electrolytic respirometry is used (Jasvinder et al., 1989; Volskay and Grady, 1990). These reactors often resemble slurry reactors. Biometers are typically used for soils that are not slurried (CFR, 1987; Muller et al., 1992). They have the benefit of not diluting the soil, which potentially masks toxicity associated with high pollutant concentrations. The potential for abiotic losses by volatilization and adsorption is also reduced. The lack of optimal incubation conditions, however, greatly adds to the required incubation time.

Activity, as measured by respirometry, does not automatically indicate that the pollutant is being biodegraded. Other organic material within the soil might be responsible for the activity. Regardless, evidence of respiration still indicates that a matrix can support bioremediation.

Respiration, combined with pollutant compound disappearance and a gross measurement of organic material, can be related to theoretical oxygen demand or carbon dioxide evaluation. The removal of the pollutant can then be related to biodegradation.

BIOREMEDIATION SELECTION-LEVEL STUDIES

Selection-level studies mimic full-scale bioremediation technologies. They are designed to determine if the pollutant can be removed to target levels within the desired time period and to provide design data for scaleup. The pollutant's fate is also often determined as part of a selection-level study to verify that detoxification occurs.

This section discusses simulation theory, experimental apparatus, and operation and analysis for selection-level treatability studies for each general category of bioremediation.

Slurry Reactors

Slurry reactors provide optimized conditions for bioremediation to occur. Excellent contact between the microorganisms and the slurry is achieved. Amendments can be easily added and distributed. These can include pH adjusters, nutrients, electron acceptors, an easily degradable energy source if cometabolism is required, surfactants, and heat. If the pollutant won't biodegrade in a slurry reactor, it most likely won't be degraded by any other bioremediation technology.

SIMULATION THEORY

Slurry reactors can be easily simulated because the soil is slurried and completely mixed. Wall effects are of minimal concern and, therefore, the reactor size need only be adequate to permit intermediate sampling without a significant reduction in the overall quantity of the polluted matrix. Mixing intensity can be directly scaled down from a full-scale unit by keeping the velocity gradient constant. The velocity gradient is calculated from the power input for mixing, fluid viscosity, and the volume of the reactor (Metcalf & Eddy, Inc., 1991).

APPARATUS

The shape of the slurry reactor should limit dead zones by avoiding sharp inside corners. Mixing is typically achieved with an impeller attached to a motor. Alternatively, the reactor can contain a stir bar and can be placed on a magnetic mixer. This however, is more common for a screening-level study. The materials used to fabricate the reactor are typically glass or stainless steel. This minimizes interference with the pollutants and analysis. Alternatively, construction materials used in full-scale installation could be used so they can be examined for process interference. Lauch et al. (1992) and the U.S. Environmental Protection Agency (1993f) provide a detailed diagram and description of a pilot-scale slurry reactor.

Provisions for collecting off-gases to determine if parent compounds are removed by volatilization can be easily built into the reactor by channeling the gaseous exit stream and analyzing for the volatile pollutant. To achieve this, however, the bearings holding the impeller's driveshaft must be airtight. In installation of a shaft that is not true or if excessive vibrations occur, the bearings will not be gastight, and calculations to determine the amount of volatilization will be inaccurate.

If aerobic conditions are to be maintained, provisions must be made for aeration. Any air entering the reactor should be filtered to remove microorganisms, oil, and grease.

OPERATIONS AND ANALYSIS

The soil used for a treatability study must have the same size distribution and slurry moisture content as that used at field scale. This is particularly important because of the limitation of this technology to treat larger diameter soil particles and gritty materials that cause abrasion.

Obtaining homogeneous samples from the reactor is relatively easy because of the completely mixed nature of the reactor. Segregation, however, can occur if a wide distribution of soil particle diameters exists. Therefore, confirmation that a homogeneous sample was obtained might be required. If the reactor is opened to obtain the slurry sample, interference with the analysis for volatiles occurs. To minimize the effect, samples can be obtained from ports on the side of the reactor so that interference with the head space is minimized. To compensate for the vacuum created when the sample is withdrawn, inlet air is required.

Both the solid and liquid phase of a slurry sample should be analyzed individually so the partitioning of the compound between the soil and water can be determined. This is especially important for polar compounds with high water solubility. If the microbial population is also being examined, both phases should be analyzed to determine in which phase biodegradation is occurring. This results in an increase in analytical load that must be considered during the planning stages of the experiment.

Solid Phase

The contaminated soil in solid-phase treatment (also known as land farming) is treated as a crop in which microorganisms are cultivated to biodegrade the pollutants. Conventional farming techniques are used to aerate the soil and add amendments. This soil depth is based on what can be easily cultivated by conventional farm equipment.

SIMULATION THEORY

The depth of soil in a solid-phase treatability study must be the same as the full-scale installation, 20 to 30 cm, so that oxygen transfer and contact between microorganisms and pollutants are accurately represented. The surface area is relatively unimportant as long as wall effects are minimized and intermediate samples can be collected without a substantial decrease in the overall volume of the polluted matrix.

APPARATUS

In the laboratory, stainless steel soil pans are typically used for the treatability studies. Perforated, small-diameter pipes placed within pea gravel permit excess water drainage. A shallow layer of sand is often placed above the pea gravel to prevent the soil from exiting through the drain pipes. Leachate can be easily collected from the drain and analyzed to determine if pollutants are extracted from the soil. Measurement of pollutant volatilization in soil pans can be challenging because of their overall large size and the need for access to routinely cultivate the soil. The pans can be housed in a glovebox so that gases can be channeled out and analyzed to determine the amount of pollutant volatilization. Placement of the pans in a glovebox also has the added benefit of providing a barrier between operators and the contaminated soil. Personnel protective equipment needs are, therefore, minimized for everyday maintenance.

If the treatability study is to be conducted in the field, soil plots or large soil pans can be used (McGinnis et al., 1991). The design is similar to field-scale installations. Embankment structures containing gravel berms or straw are used to contain the soil. The entire plot is lined with high-density polyethylene. A gravel base with perforated drainage pipes is used with an overlaying sand layer. Measurement of leachate for extracted pollutants can be easily achieved by collecting samples from the drains. Measurements of pollutant volatilization, however, are difficult and expensive although large containment-type structures could be installed above the plots. This also provides better temperature and moisture control.

OPERATIONS AND ANALYSIS

Although solid-phase treatment is tolerant of fairly large soil particle sizes, they impact the representativeness of the samples collected to evaluate the performance in a treatability study. Therefore, oversized material should be removed for the treatability study as long as it occupies only a small percentage of the total quantity of the soil.

Operation of the soil pans or plots mimics full-scale installations. Moisture is sprayed onto the soil, amendments are spread on the top, and mixing and aeration are achieved by tilling. Typically untreated tap water is not used because of its chlorine content. The most effective technique to till the soil in a soil pan is to use common hand-held garden tools to overturn the soil. For soil plots, a garden rototiller is convenient. The schedule of tillage and amendment addition should be based on what would be done in a full-scale operation.

Even with frequent tilling, samples will still be heterogeneous. Therefore, multiple samples must be taken to insure precision. A randomized block design provides a nonbiased basis for determining the location in which to collect multiple samples.

Composting

For this chapter, composting is defined as any bioremediation technology that treats soil in large piles; adds bulking agents and, potentially, other amendments; provides aeration; and enhances microbial/pollutant contact by periodic mixing. These technologies use frontloaders, windrowing equipment, or mechanical devices within a reactor for mixing. The operational modes for each type are similar, and they are therefore discussed together.

SIMULATION THEORY

Composting often uses bulking agents that have fairly large dimensions. The use of square wood chips (typically 2 to 5 sq cm), crop waste, straw, and manure is common. This size dictates a fairly large treatability study reactor. If a composter is thought of as a static bed reactor, the minimal diameter should be at least ten times the largest diameter amendment. In addition, compost soil piles are typically not confined within a reactor (except for a mechanical one). Consequently, large reactors must be used to minimize the effect of side and top boundary conditions and to accurately simulate transport.

Limited references are available on the simulation of composting. Potter et al. (1995) used 55-gallon stainless steel rotating drums that were periodically aerated. Others have used soil pans or small piles. The most accurate technique, however, is to set up piles similar in size to full-scale installation. Because of the large space requirements, the study would need to be done in the field or a high bay facility.

APPARATUS

If reactors are used to simulate composting, issues similar to those discussed above apply. Mixing is achieved by tumbling. Provisions that allow easy aeration and venting lines are needed.

Requirements for soil piles are limited to a staging area and, if required, an underdrain system. Runoff can be virtually avoided by the careful addition of moisture. An underdrain system, similar in design to that described for solid phase, however, can be easily added. Measurements for volatilization of the pollutant can be difficult because of the size of the piles. Containment-type structures could, however, be used. These structures would need to be very large to accommodate the mixing process.

OPERATIONS AND ANALYSIS

As with any ex situ technology, the soil must have the characteristics of that which will be used in the full-scale installation. As in solid phase, dilute, oversized soil particles should be removed.

The primary operational control is mixing. If reactors are used, mixing is achieved by either rotating the entire reactor or with hand-held garden tools. Mixing of piles can be achieved with standard garden tools or frontloaders, depending on the size of the pile. The moisture content should be maintained at the level expected in a full-scale installation.

Techniques previously described can be used to determine the pollutant's fate if a reactor is used. The determination of volatilization, however, is more difficult for soil piles, because of their large surface area. The use of bulking agents could also increase the potential for pollutant removal by adsorption. Consequently, extractions of the bulking agent would be advisable.

The use of killed controls is not a practical alternative to measure abiotic removal for soil piles. The sterilization of the required quantity of soil is close to impossible. In addition, disposal is difficult because of the presence of toxic sterilizing agents.

A large number of replicates is needed to evaluate the treatment process because of soil pile heterogeneity. The added cost must be considered in the budget.

Static Piles

Static pile (soil heaping) technologies are similar to composting, except mixing is only conducted initially to distribute amendments. Aeration is provided by either forcing or vacuuming air through the pile.

SIMULATION THEORY

The complexity of oxygen, nutrient, and microorganism transport within a soil pile inhibit small-scale simulations. Full-scale piles must be used.

APPARATUS

Issues discussed previously concerning composting apply to static piles. An air manifold or vacuum distribution system, however, is required for aeration in place of mixing and would be typically constructed from perforated PVC embedded in pea gravel and sand. Equalization of air or vacuum is difficult because of the heterogeneity of a soil pile.

OPERATION AND ANALYSIS

Operational requirements for a static pile are minimum. If air is distributed through the pile, filtration for microorganisms, oil, and grease is required. Pollutant removal caused by volatilization is difficult to measure for a static soil pile with forced air. The concerns are similar to that of a compost pile. The measurement of volatilization for a soil pile that is being aerated by vacuuming is relatively simple because all of the air is channeled through a vacuum pump and can be easily collected for analysis.

In situ

In situ technologies include enrichment by injection or surface distribution of amendments (referred to as in situ bioremediation) and in situ bioventing. The principles and operation of both are different; however, selection-level treatability approaches are similar. Therefore, these technologies are discussed together.

SIMULATION THEORY

Simulating in situ technologies in the laboratory is close to impossible. Soil that is undisturbed cannot be collected for use in the treatability study. Its structure will change due to the collection process and the removal of overburden pressure. Operations would also be difficult to simulate because of transport phenomenon and boundary conditions.

Simple field tests can be used to aid in predicting the potential of bioremediation and provide design data on in situ technologies. These tend to resemble screening tools because of their simplicity, even though their objectives are to aid in the design of a remediation technology.

APPARATUS

Although laboratory simulations are not representative of in situ conditions, many have been reported in the literature. Intact soil cores have been collected and incorporated directly into a laboratory apparatus or packed into columns or microcosms (Boyer, 1986; IT Corporation, 1995; Lund and Gudehus, 1990; U.S. Environmental Protection Agency, 1989a, 1990).

Several field tests are commonly used to aid in determining the potential and provide design data for in situ technology. Two common ones are the percolation tests, designed to determine the rate at which amendments can be injected into the soil, and in situ respirometry, which is used to measure air permeability, important in the design of a bioventing system (Ong, 1991).

OPERATION AND ANALYSIS

Operation of an in situ treatability study is very specific to the technology and the study. For simple field tests, the analysis is often a simple measurement of a flow rate or oxygen level.

Issues discussed previously for reactors are also true for in situ studies. Of particular concern, however, is obtaining representative initial samples from a long core. Composite samples can be produced from each end of the core or from a neighboring core not used in the experiment. Both techniques, however, have the potential to bias results because of potential variations in pollutant concentrations.

BIOREMEDIATION REMEDIAL DESIGN/ACTION-LEVEL STUDIES

Remedial design/action treatability studies are designed to provide biodegradation kinetic rates, costs, and design data and to reveal potential operational difficulties. These are typically full-scale studies using equipment manufactured by the vender who will ultimately remediate the site. They are not designed for the selection of a remediation technology but, rather, for installation and initiation of the full-scale process. Often, the treatability study is conducted with a full-scale mobile unit.

The decision and type of treatability studies to be conducted at this level are, therefore, very specific to the site, the technology, and manufacturer and are beyond the scope of this chapter.

CONCLUSIONS

Because bioremediation is emerging and the enzymatic processes are complex and poorly understood, treatability studies are crucial for almost all large remediations. In addition, the fate of the pollutant must be determined because of the potential transfer from soil to another matrix and/or biotransformation to an undesirable end product.

No one treatability study will be universally representative of bioremediation. Bioremediation encompasses many different technologies and treatability studies must be customized for the technology and the site. Even screening-level studies, generally designed to determine if bioremediation is at all feasible, can take many forms, depending on site-specific issues. Although dividing treatability studies into tiers based on their general objective, screening, selection, or remedial design/action is convenient for discussion, a tidy fit does not always occur. A slurry test might be a screening-level study for solid phase or in situ but can accurately represent a full-scale installation in a selection study. Therefore, the tier approach should be used with caution with the realization that some treatability study approaches are hybrids of more than one tier.

REFERENCES

ACGIH (American Conference of Government Industrial Hygienists). 1993–1994. *Threshold Limit Values.* Cincinnati, Ohio.

Alexander, M. January 9, 1981. "Biodegradation of Chemicals of Environmental Concern." *Science,* 211:132–211.

Baud-Grasset, S., Baud-Grasset, F., Bifulco, J. M., Meier, J. R., and Ma, T. 1993. "Reduction of Genotoxicity of a Creosote-Contaminated Soil after Fungal Treatment Determined by the Trade Scantia-Micronucleus Test." *Mutation Research,* 303:77–82.

Bligh, E. G. and Dyer, W. J. August 1959. "A Rapid Method of Total Lipid Extraction and Purification." *Canadian Journal of Biochemistry and Physiology,* 37(8):911–917.

Bourquin, A. W. September/October 1989. "Bioremediaiton of Hazardous Waste." *Hazardous Materials Controls:* 16–58.

Bouwer, E. J., Rittmann, B. E., and McCarty, P. L. 1981. "Anaerobic Degradation of Halogenated 1- and 2-Carbon Organic Compounds." *Environmental Science Technology,* 15:596–599.

Boyer, J. D., Kosson, D. S., and Ahlert, R. C. 1986. "Microbial Remediation of Soil Contaminated with 1,1,1-Trichloroethane." *Proceedings of the 1986 Hazardous Material Spills Conference, Preparedness, Prevention, Control and Cleanup of Releases,* May 5–8, 1986, St. Louis, MO, pp. 98–107.

CFR (*Code of Federal Register*). July 1, 1987. 40 CFR Ch. 1, 796.3400, pp. 545–550.

Chang, H. T. and Rittmann, B. E. 1986. "Biofilm Loss during Sample Preparation for Scanning Electron Microscopy." 1986. *Water Res.,* 20(11):1451–1456.

Findlay, R. H., King, G. M., and Watling, L. November 1989. "Efficacy of Phospholipid Analysis in Determining Microbial Biomass in Sediments." *Applied and Environmental Microbiology:* 2888–2893.

Ghosal, D, You, I.-S., Chatterjee, D. K., and Chakrabarty, A. M. April 12, 1985. "Microbial Degradation of Halogenated Compounds." *Science,* 228(4696): 135–142.

Haugland, R. A., Rope, A. F., Strange, C., Loper, J. C., Tabak, H. C., and Sferra, P. R. August 1990. "Use of DNA Hybridization Probes in the Monitoring of Biotreatment Systems." *Proceedings of Remedial Treatment Action and Disposal of Hazardous Waste 16th Annual RREL Hazardous Waste Research Symposium,* Cincinnati, OH. pp. 303–312.

Herrmann, A. M. and Safferman, S. I. March 1995. "Phospholipid and FDA Activity Measurements Adapted to Biological GAC." *Journal of Environmental Science and Health,* A30(2):263–280.

IT Corporation. 1995. Laboratory Evaluation of in-situ Biodegradation of Hazardous Pollutants in Landfills. Internal Report prepared for the U.S. Environmental Protection Agency, Contract No. 68-C2-0108, Work Assignment No. 3-46, Work Assignment Manager, Carl Potter, Cincinnati, Ohio.

Jasvinder, S., Dang, D., Harvey, M., Jobbaby, A., and Grady, C. P. 1989. "Evaluation of Biodegradation Kinetics with Respirometric Data." *Research Journal WPCF,* 61(11/12):1711–1721.

Johnson, L., Curl, E. A., Bond, J. H., and Fribourgh, H. A. 1959. *Methods for Studying Soil Microflora–Plant Disease Relationships.* Minneapolis, MN: Burgess Publishing Company, pp. 1–22.

Karl, D. M. 1986. "Determination of in situ Microbial Biomass, Viability, Metabolism, and Growth." *Bacteria in Nature,* 2.

Kincannon, D. F., Stover, Enos L., Nichols, V., and Medley, D. 1983. "Removal Mechanisms for Toxic Priority Pollutants." *Journal of WPCF,* 55(2):157–163.

Kobayashi, H. and Rittmann, B. E. 1982. "Microbial Removal of Hazardous Organic Compounds." *Environmental Science and Technology,* 16(3):170A–183A.

Lauch, R. P., Herrmann, J. G., Mahaffery, W. R., Jones, A. B., Dosani, M., and Hessling, J. November 1992. "Removal of Creosote from Soil by Bioslurry Reactors." *Environmental Progress,* 11(4):265–271.

Leitzinger, A., Evans, J., Hayes, S., Partymiller, K., and Swanson, G. April 1995. "Representative Sampling and Analysis of Heterogeneous Soils." *Proceedings of the 21st Annual RREL Research Symposium,* Cincinnati, OH.

Little, B., Wagoner, P., Ray, R., Pope, R., and Scheetz, R. 1991. "Biofilms: An ESEM Evaluation of Artifacts Introduced during SEM Preparation." *Journal of Industrial Microbiology,* 8:213–222.

Lund, Dipl.-Ing. N.-Ch. and Gudehus, Ing. G. 1990. "Large Scale Sample Tests for a Biological In-Situ Remediation of Soils Polluted by Hydrocarbons." In *Contaminated Soil,* Kluwer Academic Press, Netherlands, Eds. F. Arendt M. Hinsenveld and W. J. van den Brink, pp. 463–471.

McGinnis, G. D., Matthews, J., and Pope, D. 1991. June 16–21, 1991. "Bioremediation Studies at a Northern California Superfund Site." *Proceedings of the Air & Waste Management Association 84th Annual Meeting & Exhibition,* Vancouver, B.C.

Metcalf & Eddy, Inc. 1991. *Wastewater Engineering, Treatment Disposal Reuse.* New York: McGraw-Hill, Inc.

Metosh, C. A. 1991. Biomonitoring of Bioremediated, Oil-Based Sludge Contaminated Soil. M.S. Thesis. Louisiana State University. Agricultural and Mechanical College.

Muller, J. G., Resnick, S. M., Shelton, M. E., and Pritchard, P. H. 1992. "Effect of Inoculation on the Biodegradation of Weathered Prudhoe Bay Crude Oil." *Journal of Industrial Microbiology,* 10:92–102.

Ney, R. E. 1995. *Fate and Transport of Organic Chemicals in the Environment, A Practical Guide,* Second Edition. Rockville, MD: Government Institutes, Inc.

Norris, R. D., Hinchee, R. E., Brown, R., McCarty, P. L., Semprini, L., Wilson, J. T., Kempel, D. H., Reinhardt, M., Bower, E. J., Borden, R. C., Vogel, T. M., Thomas, J. M. and Ward, C. H. 1994. *Handbook of Bioremediation.* Boca Raton: Lewis Publishers.

Ong, S. K., Hinchee, R. E., Hoeppel, R. and Schultz, R. 1991. "In situ Respirometry for Determining Aerobic Degradation Rates." In *In situ Bioremediation.* Butterworth-Heinemann, Boston, Eds. R. E. Hinchee and R. F. Olfenbuttel, pp. 541–545.

Park, Kap S., Sims, Ronald C., Dupont, R. Ryan, Doucette, William, J., and Matthews, John E. 1990. "Fate of PAH Compounds in Two Soil Types: Influence of Volatilization, Abiotic Loss and Biological Activity." *Environmental Toxicology and Chemistry,* 9:187–195.

Parkinson, D., Gray, T. R. G., Williams, S. T. 1971. *Methods for Studying the Ecology of Soil Micro-Organisms.* Oxford and Edinburgh: Blackwell Scientific Publications, pp. 46–47.

Potter, C. L., Glaser, J. A., Dosani, M. A., Krishnan, S., Deets, T. A., and Krishnan. E. R. April 1995. "Design and Testing of an Experimental In-Vessel Composting System." *Proceedings of the 21st Annual RREL Research Symposium,* Cincinnati, OH.

Rogers, J. A., Tedaldi, D. J., Kavanaugh, M. C. May 1993. "A Screening Protocol for Bioremediation of Contaminated Soil." *Environmental Progress,* 12(2):146–156.

Safferman, S. I., Baud-Grasset, S., Bracjett, K. A., Clark, P. J., and Bishop, P. L. December 1993. "A Systematic Scanning Electron Microscopy Technique for Evaluating Combined Biological/Granular Activated Carbon Treatment Processes." *Journal of Environmental Science and Health,* A28(10): 2239–2262.

Safferman, S. I., Lamar, R. T., Vonderhaar, S., Neorgy, R., Haught, R. C., and Krishnan, E. R. 1995. "Treatability Study Using *Phanerochaete sordida* for the Bioremediation of DDT Contaminated Soil." *Toxicological and Environmental Chemistry,* 50:237–251.

Schnurer, J. and Rosswall, T. June 1982. "Fluorescein Diacetate Hydrolysis as a Measure of Total Microbial Activity in Soil and Litter." *Applied and Environmental Microbiology.* 1256–1261.

Tsai, Y. L. and Olson, B. H. April 1991. "Rapid Method for Direct Extraction of DNA from Soil and Sediments." *Applied and Environmental Microbiology,* 57(4):1070–1074.

U.S. Environmental Protection Agency. March 1979. *Handbook for Analytical Quality Control in Water and Wastewater Laboratories.* EPA Document No. EPA-600/4-74-019.

U.S. Environmental Protection Agency. September 1986. *Microbial Decomposition of Chlorinated Aromatic Compounds.* EPA Document No. EPA/600/2-86/090.

U.S. Environmental Protection Agency. February 1988. *Treatment Potential for*

56 EPA Listed Hazardous Chemicals in Soil. EPA Document No. EPA/600/6-88/001.

U.S. Environmental Protection Agency. September 1989a. In situ Bioremediation of Spills from Underground Storage Tanks: New Approaches for Site Characterization Project Design, and Evaluation of Performance. EPA Document No. EPA/600/S2-89/042.

U.S. Environmental Protection Agency. September 1989b. *Treatability Potential for EPA Listed Hazardous Wastes in Soil.* EPA Document No. EPA/600/S2-89/011.

U.S. Environmental Protection Agency. April 1990. *Enhanced Bioremediation Utilizing Hydrogen Peroxide as a Supplemental Source of Oxygen: A Laboratory and Field Study.* EPA Document No. EPA/600/S2-90/006.

U.S. Environmental Protection Agency. 1991. *Guide for Conducting Treatability Studies under CERCLA: Aerobic Biodegradation Remedy Screening, Interim Guidance.* EPA Document No. EPA/540/2-91/013A.

U.S. Environmental Protection Agency. 1992. *Guide for Conducting Treatability Studies under CERCLA.* EPA Document No. EPA/540/R-92/017a.

U.S. Environmental Protection Agency. September 1993a. *Bioremediation Resource Guide.* EPA Document No. EPA/542-B-93-004.

U.S. Environmental Protection Agency. August 1993b. *Bioremediation Using Land Treatment Concept.* EPA Document No. EPA/600/R-93/164.

U.S. Environmental Protection Agency. August 1993c. *Guide for Conducting Treatability Studies under CERCLA, Biodegradation Remedy Selection, Interim Guidance.* EPA Document No. EPA/540/R-93/519a, pp. 50–51.

U.S. Environmental Protection Agency. 1993d. *RREL Treatability Data Base, Version 5.0.* EPA Document No. EPA-600/C-93/003a.

U.S. Environmental Protection Agency. June 1993e. *Selecting Remediation Techniques for Contaminated Sediment.* EPA Document No. EPA-823-B93-001.

U.S. Environmental Protection Agency. September 1993f. *Technology Demonstration Summary Pilot-Scale Demonstration of a Slurry-Phase Biological Reactor for Creosote-Contaminated Soil.* EPA Document No. EPA/540/S5-91/009.

U.S. Environmental Protection Agency. June 1994a. *The Bioremediation in the Field Search System.* EPA Document No. EPA/540/R-95/508A and B.

U.S. Environmental Protection Agency. July 1994b. *North American Wetlands for Water Quality Treatment Database, Version 1.0.* Contact Don Brown, U.S. EPA, 26 W. Martin Luther King Drive, Cincinnati, OH, 45268.

U.S. Environmental Protection Agency. November 1994c. *Superfund Innovative Technology Evaluation Program Technology Profiles, Seventh Edition.* EPA Document No. EPA/540/R-94/526.

U.S. Environmental Protection Agency. 1995a. *Bioremediation in the Field.* EPA Document No. EPA/540/N-94/501.

U.S. Environmental Protection Agency. July 1995b. *Engineering Bulletin Biological Toxicity Testing.* EPA Document No. EPA/540/S-95/501.

U.S. Environmental Protection Agency. 1995c. *Vender Information System for Innovative Treatment Technologies.* EPA Document No. EPA-542-C-95-001.

VanEmon, J. M. and Gerlack, C. L. July/August 1994. "Immunoassay Technology." *Environmental Testing and Analysis.* 18–21.

Vestal, J. R. and White, D. C. September 1989. "Lipid Analysis in Microbial Ecology, Quantitative Approaches to the Study of Microbial Communities." *BioScience,* 39(8):535–541.

Volskay, V. T. and Grady, C. P. L. 1990. "Respiration Inhibition Kinetic Analysis." *Wat. Res.* 24(2):863–874.

White, D. C., Davis, W. M., Nickels, J. S., King, J. D., and Bobbie, R. J. 1979. "Determination of the Sedimentary Microbial Biomass by Extractable Lipid Phosphate." *Oceologia,* 40:51–62.

Biodegradation of Vapor-Phase Contaminants

KERRY A. KINNEY, WILLIAM WRIGHT, DANIEL P. Y. CHANG,
AND EDWARD D. SCHROEDER
Department of Civil and Environmental Engineering
University of California, Davis
Davis, CA 95616, USA

INTRODUCTION

Biodegradable contaminants are often found in airstreams
emitted from industrial processes, commercial operations, and waste
management and treatment systems. Industrial sources include gases
from coating and printing operations, production of adhesives, fiberglass,
pharmaceuticals, plastics, polymers, solvents, and wood products. Commercial operations such as chemical and petroleum storage, tank cleaning, baking, cocoa and coffee roasting, and meat smoking produce contaminated gas streams. Waste management and treatment sources
include off-gases from wastewater treatment plants, gases from landfills
and other solid waste management operations, and gases produced during
soil and groundwater remediation by methods such as soil vapor extraction. The most common groups of volatile materials found in airstreams
are solvents (including chlorinated compounds), gasoline and other petroleum products, fermentation products, polymer components such as
styrene, and sulfides. Contaminant concentrations may range from a few
parts per billion by volume (ppb_v) to several thousand parts per million
by volume (ppm_v) depending on the source.

Biological treatment was first applied to airstreams for the purpose of
odor control (Pomeroy, 1957, 1982; Carlson and Leiser, 1966). In the early
odor treatment systems, contaminated gases were passed through soil
beds and the units were termed *soil filters*. Similar systems in which bed
packing was composed of compost, peat, or synthetic materials are
generally called *biofilters,* and any process using a solid phase for supporting microbial cultures treating contaminated gases is currently re-

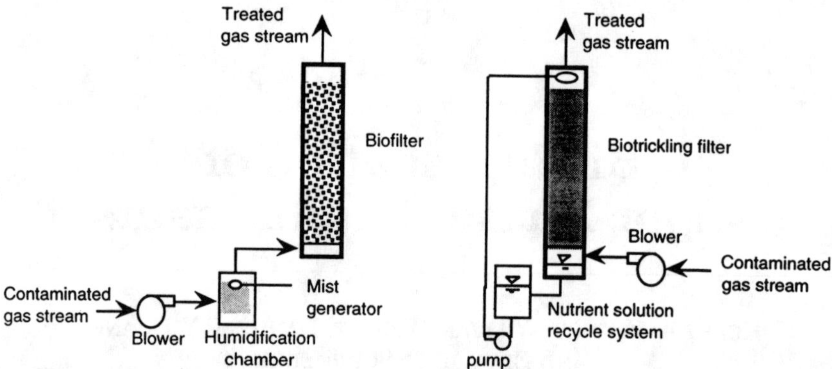

Figure 1. Schematic diagrams of typical biofilter and biotrickling filter systems for biological treatment of airstreams contaminated with biodegradable compounds. Note that flow direction is optional and that many biofilters are designed as downflow units, and many biotrickling filters are designed to operate cocurrently with the nutrient solution.

ferred to as biofiltration. In the 1980s Ottengraf and his coworkers (1983, 1984, 1986) conducted extensive studies on biological treatment of exhaust gases using organic based solid-phase packing mediums, such as compost and peat, and developed kinetics-based process models. Commercial application of biofiltration has been extensive in Europe since the mid-1980s. Since 1990, research and application of biological gas treatment have greatly increased in the United States. The two general process configurations in use, biofilters in which nutrients are stored in the packing material and biotrickling filters in which nutrients are supplied by a recycled liquid stream, are shown schematically in Figure 1.

CONSTRUCTION AND OPERATION

Biofiltration systems have three basic components: the packed bed reactor, a system for maintaining moisture levels in the reactor, and a blower to push or pull contaminated air through the porous medium. Granular activated carbon (GAC) may be required downstream of the reactor to ensure that emission requirements are consistently met. Pressure losses across the packing media have varied greatly. Wright et al. (1995) operated a pilot-scale compost biofilter treating gasoline vapor from a soil vapor extraction operation for 200 days, during which head losses did not exceed 25 mm of water. Inlet concentrations ranged from 100 to 700 ppm$_v$ as TPH and empty bed contact times of one and two minutes. Morgenroth (1994) operated a compost biofilter at loadings of 0. 4 to 0.6 kg C/m$^3 \cdot$d and observed no clogging over 60 days of operation.

Kinney (1996) operated a synthetic medium (Celite™ G-635) biofilter that did not observe head losses greater than 50 mm of H_2O in 200 days of operation with toluene at a loading of 0.53 kg C/m^3·d. Sorial et al. (1995) and Smith et al. (1995), working with biotrickling filters, reported interstitial clogging and headlosses of 100 to 500 mm of H_2O developing over periods of a few days in the loading rate range of 1.2 to 2 kg C/m^3·d. Decreases in removal from 99.9% to about 96% were attributed to channelization associated with clogging. A backwashing program was instituted to remove accumulated biomass. Holubar et al. (1995) limited potassium in the feed to a biotrickling filter to control clogging. Apparently clogging is related to the amount of moisture in the reactor, as well as organic loading rates, and biotrickling filters are susceptible to clogging unless control measures are used. Reports of clogging of compost biofilters are few, even when treating loads comparable to biotrickling filters. In biofilters, system pressure losses are dominated by frictional losses in piping, losses at junctions, entrances and exits, and, if required, in GAC polishing beds.

Reactors

Reactors can be constructed of fiberglass, polyethylene, mild steel, or concrete, although corrosive conditions must be considered, and protection of component materials will be necessary in some cases. Flow direction is optional (upflow, downflow, horizontal) and provision for flow reversal may be advantageous. The inlet section should consist of a plenum approximately 150 mm in depth to allow good flow distribution, as indicated in Figure 2. Usually, the outlet section is of similar depth. In vertical configurations, the medium is most often supported by a metal grid, although slotted ceramic tile is used in some proprietary units. In natural media systems, a 150-mm layer of coarse material (e.g., 25 to 50 mm redwood bark) is placed on top of the supports, as shown in Figure 2(a). A small pressure drop across the coarse material layer is desirable to prevent channeling.

Corrosion is a significant problem in cases where the contaminated air contains sulfides or halogenated compounds. Sulfides are of particular significance in municipal wastewater treatment plants where inlet structure off-gas sulfide concentrations sometimes exceed 100 ppm$_v$. Microbial oxidation of sulfides usually occurs in the first 150 mm of packing, and pH values of drainage may drop below 2. Upflow operation is necessary when sulfides are present to prevent acidification of the complete bed. Biodegradation of chlorinated VOCs (e.g., dichloromethane) results in stoichiometric production of HCl, and buffering of the entire system is

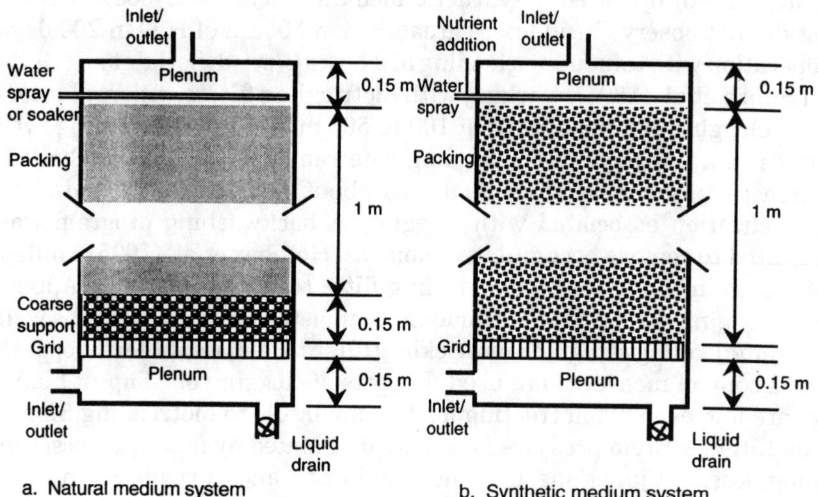

Figure 2. Typical layout of natural and synthetic medium biofiltration systems. Biotrickling filters are typically constructed using synthetic media and with the water spray and liquid drain connected through a recycle tank and pump as shown in Figure 1. In natural media systems, water is added as needed to keep moisture content in the range of 1 to 1.5 g/g dry medium.

required. In natural medium systems, buffering can be provided by including $CaCO_3$ in the packing. Nutrient solutions used in synthetic medium systems can be buffered as required. Protection of the reactor from corrosion is a difficult problem and areas particularly at risk are the media supports, lower plenum walls, and drains.

Construction and operation of biofiltration systems are based on the requirements necessary to maintain microbial growth in the particular environment. Biofiltration systems can be classified as packed bed or attached growth biological processes. The microbial communities involved are composed principally of bacteria, although fungi play a role in most natural medium systems. Protozoans and invertebrates may, through predation, be important in controlling plugging of the pores between media particles. Factors needed to support microbial growth that constrain or impact the construction and operation of biofiltration systems include

* packing characteristics
* moisture content
* inorganic nutrients supply

- liquid phase pH
- temperatures

Packing

Packing used in biofiltration systems, as noted above, is classified as "natural" or "synthetic." Natural media in use include soil, peat, and compost, with compost being by far the most common because of its high porosity, high surface-to-volume ratio, the presence of a diverse and active microbial population, and presence of trace nutrients.

NATURAL PACKING

Soil porosities are usually so low that air flux/head loss relationships are uneconomical. The density of soil results in heavy loads on media support relative to peat and compost. Finally, soils rarely contain significant quantities of nutrients. Satisfactory treatment of inorganic volatiles, such as H_2S may be possible because of the low yields and associated low nutrient requirements of chemoautotrophic organisms. However, treatment of air containing high concentrations of organic compounds (>1 ppm$_v$) will result in rapid exhaustion of the available inorganic nutrients. Moisture content of soil filters should be maintained at approximately 75% of field capacity, the amount held by capillary action and sorption to particle surfaces. Wetting of soil is difficult, and thus, addition of nutrients to soil as a solution following the initial setup is even more difficult.

Peat has physical characteristics similar to compost but is lacking in nutrients. The porous structure of peat tends to collapse, when wet and bulking agents such as perlite or vermiculite are required to maintain porosities in the bed. Nutrients can be added as part of the initial wetting solution. Later addition of nutrients requires system shutdown and soaking the bed with a nutrient solution.

Compost varies considerably in character by origin of the materials used. Compost sources that have been used in biofilters include yard wastes, redwood and other woods, sewage sludge, and manure (Allen and Yang, 1991; Joyce et al., 1992; Eitner, 1989). A bulking agent must be used with compost to maintain the porous structure of the bed, and as with peat biofilters, perlite, rice hulls, and vermiculite have been most commonly used. The packing is normally constructed using equal volumes of compost and bulking agent. Nutrients and buffer, often in the form of crushed oyster shells (Ergas et al., 1992b), are added in dry form, and water is added to bring the moisture content to the desired level.

Microbes, isolated from the packing material of active columns, can be grown up in liquid culture with the contaminant compounds as the sole carbon and energy source to speed up column acclimation.

SYNTHETIC PACKING

Extruded diatomaceous earth pellets, polypropylene pall rings, expanded polystyrene spheres, sintered glass disks, and granular activated carbon have been used in experimental or full-scale biofilters and biotrickling filters, with the diatomaceous earth pellets being most commonly used. The porous nature of diatomaceous earth pellets, sintered glass, and activated carbon are desirable because liquid-containing nutrients can be stored and bacterial colonies can grow in the granule pores. Thus, the effective surface-to-volume ratio is relatively large compared to nonporous materials used in pall rings. Bacteria in biofilters range from approximately 0.2 μm to 2 μm in dimension. Thus, internal pores smaller than 10 μm will be subject to rapid plugging with microbial growth, as shown in Figure 3. Diffusivities in air are approximately three orders of magnitude greater than diffusivities in water or microbial slimes; therefore, filled pores effectively act as smooth surfaces in terms of contaminant removal. The differences can be seen quite clearly in the scanning electron micrographs of Celite™ G-635, diatomaceous earth pellets shown in Figure 3.

MOISTURE CONTENT

Bacteria obtain all of their nutrition from the liquid phase, and a liquid film must be provided into which energy sources, carbon sources, and inorganic nutrients can dissolve. Maintaining bed moisture content is particularly important in natural medium systems because excessive drying may result in crack formation and channeling that remains after rewetting of the bed. In natural medium systems, water is usually added to the dry packing material during construction. Moisture content (θ) is usually reported on a wet weight basis and set at 50% to 60%.

$$\theta = \frac{\text{mass water}}{\text{mass water} + \text{mass dry packing}} \tag{1}$$

The corresponding mass ratios of water to dry packing are between 1.0 and 1.5 g/g.

NATURAL MEDIUM SYSTEMS

The moisture content is maintained during operation by saturating the contaminated air feed stream with water vapor. Maintaining a slightly

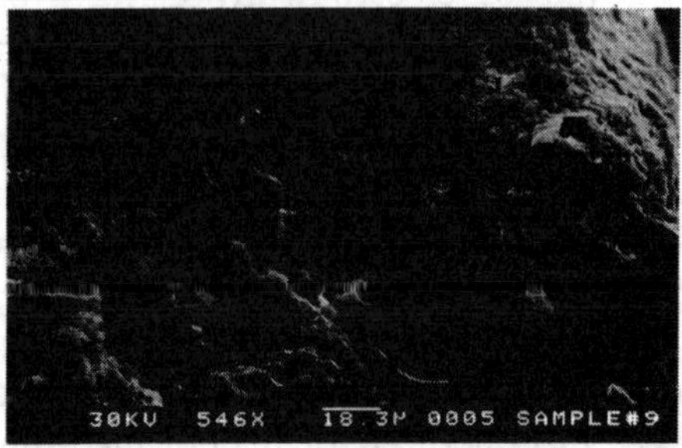

a. Cross section of unused Celite™ G-635 pellet.

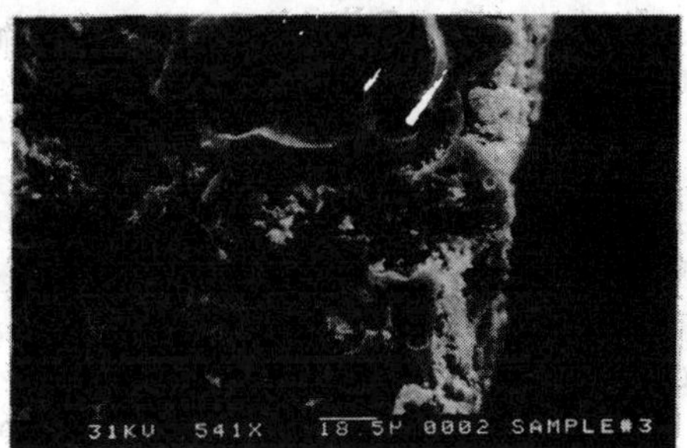

b. Cross section of Celite™ G-635 pellet after extensive period
of operation.

Figure 3. Scanning electron micrographs of biofilter medium (a) prior to use and (b) following an extended period of operation with a 100 ppmv inlet toluene concentration and a 1 m/min superficial velocity.

cooler temperature in the bed or at the outlet, relative to the feed air temperature, will result in condensation and maintain the moisture content of the bed. Conversely, drying will result if the bed temperature is higher than the air feed stream. Good control of the bed moisture content can normally be maintained through heating the humidification water and/or cooling the air leaving the bed. Many air feed streams are quite warm (e.g., soil vapor extraction gas is typically greater than 40°C), and auxiliary heating is unnecessary. Cooling occurring in the humidification chamber and solar heating of the reactor may result in conditions that cause drying. A supplemental water source is always advisable. In most cases, a "soaker" similar to that used for lawns or flowerbeds can be arranged at the top of the packing and controlled automatically or manually. Bed moisture content can be monitored by manual sampling and gravimetric analysis or electronically using a conductivity-based probe.*

SYNTHETIC MEDIUM SYSTEMS

Synthetic medium processes can be operated as biofilters with packing medium particles' pores being initially soaked with a nutrient solution. The drained packing is then fed a saturated airstream, as in the case of natural medium systems. Periodic addition of nutrients by shutting the unit down and soaking the medium with nutrient solution is required. Kinney (1996) successfully operated a synthetic medium biofilter treating toluene without periodic shutdowns by application of a nutrient-containing aerosol.

Operation as a biotrickling filter is accomplished by continuous spraying of water and nutrient solution over the top of the packing. To minimize nutrient requirements and discharge, drainage is recycled, as shown in Figure 1. The result is a thin liquid film continuously flowing over the packing, which is recycled. Air feed humidification is unnecessary when processes are operated as biotrickling filters with continuous liquid addition. Water application rates in use are approximately $1 \text{ m}^3/\text{m}^2 \cdot \text{d}$ (Sorial et al., 1995). Application can be intermittent or continuous, but application rates must be low enough to prevent blockage of the packing pores and increased head losses. An alternative method of maintaining a moist environment is to add a microbe-free aerosol having particle diameters in the 10 to 25 μm range to the contaminated air feed. Contaminants are sorbed into the aerosol particles, as well as onto the attached liquid film. The particles are removed by impaction and other filtration mechanisms.

Inorganic Nutrients

Nitrogen, phosphorus, and trace nutrients such as sulfur, iron, and magnesium are required for microbial growth. In natural medium filters, trace nutrients are usually available in excess. However, nitrogen and phosphorus must usually be added at regular intervals. Nutrients are recycled through growth, death, and release. Simple stoichiometric estimates of nutrient requirements for operating systems will be very conservative. In systems operated as biofilters, the medium is soaked in a concentrated nutrient solution. After a period of soaking that allows all of the medium-packing particle pores to fill, the nutrient solution is

---

*A unit used experimentally at UC Davis with some success is Watermark™, Irrometer Co., Inc., Riverside, CA.

drained and the amount of nutrients available is calculated by the difference. Conservative estimates of nutrient requirements can be made using yield factors of 0.5 g cells produced per gram carbon oxidized, 0.06 g N required per g carbon oxidized, and 0.01 g P per g carbon oxidized. Sample calculations are provided in Example 1.

EXAMPLE 1: DETERMINATION OF NUTRIENT SOLUTION NITROGEN CONCENTRATION

An air stream contaminated with 50 ppm_v of toluene is to be treated in a compost biofilter at a temperature of 25°C. The biofilter packing is to be 1 m deep, the design moisture content is to be 1.2 g water/g dry medium, and the air flux is to be 2 $m^3/m^2 \cdot min$. Dry bulk density of the packing is 500 kg/m^3. Select a nitrogen concentration for the nutrient solution that will provide for six months of operation.

Solution

1. Determine the mass concentration of toluene in the air:

$$C_{Tg} = \frac{MW_T p_T}{RT}$$

where

C_{Tg} = mass concentration of toluene in gas phase, g/m^3
MW_T = molecular weight of toluene, 92.1 g/mol
p_T = partial pressure of toluene, atm (numerically equal to the volumetric concentration, 5×10^{-5} atm)
R = gas constant, 8.2057×10^{-5} $m^3 \cdot atm/mol \cdot K$
T = temperature, K

$$C_{Tg} = \frac{(92 \text{ g/mol})(5 \times 10^{-5} \text{ atm})}{(8.2057 \times 10^{-5} \text{ m}^3 \cdot \text{atm/mol} \cdot \text{K})(298 \text{ K})} = 0.188 \text{ g/m}^3$$

2. Determine the mass of toluene, m_T, fed to the biofilter in six months for a unit area:

$$m_T = Q_g C_{Tg} t = (2 \text{ m}^3/\text{min})(0.188 \text{ g/m}^3)(180 \text{ d})(1440 \text{ min/d})$$

$$= 97,460 \text{ g}$$

3. Estimate nitrogen mass, N_T, requirement:

$$m_N = (0.06 \text{ g N/g carbon})(97{,}460 \text{ g toluene})(0.75 \text{ g carbon/g toluene})$$

$$= 4{,}386 \text{ g N}$$

4. Estimate the mass of water added to the bed:

$$m_W = (500 \text{ kg})(1.2 \text{ g water/g dry packing}) = 600 \text{ kg}$$

5. Determine nitrogen concentration in nutrient solution:

$$C_N = \frac{4386 \text{ g}}{600 \text{ l}} = 7.31 \text{ g/l}$$

Note that the value calculated is as N and that nitrogen should be added in the nitrate form as $NaNO_3$ or KNO_3. The corresponding concentrations of $NaNO_3$ and KNO_3 would be 44.4 g/l and 52.8 g/l. Thus, the liquid phase will be quite saline.

Liquid-Phase pH

The microbial population in biofilters and biotrickling filters is attached to the solid packing surface but is in direct contact with the liquid film. Thus, the liquid film pH impacts microbial growth and metabolism. Most bacteria that utilize organic materials grow best in the pH range of 6 to 9, with individual species having somewhat narrower ranges. Measuring pH in a biofilter requires extraction of liquid from the medium, usually by dilution. The diluted mixture pH is determined, and the pH of the pore water is then estimated by calculation. In biotrickling filters, the pH of the recycle stream is probably a reasonable estimate of the pH in the pores.

Temperature

Optimal growth temperature for most bacterial species in soil, and hence in biofilters, is in the range of 15°C to 40°C. Thermophilic operation (50°C to 60°C) may be possible for some compounds, but interphase transport limitations and instability resulting from limited population diversity may result. For particular systems, the optimal temperature can be determined experimentally if temperature control is feasible. In biofilters, as opposed to biotrickling filters, temperature may be of more importance in relation to maintaining moisture levels than reaction rates. When water-saturated air is fed to a biofilter, the unit will not dry out if the temperature in the packing is slightly cooler than the inlet temperature. It is often stated in the literature that the heat released from the organic matter on oxidation will result in heating and drying of the bed. However, most organic substrates contain hydrogen, as well as carbon, so that water also forms during the oxidation process. At the temperature of typical biofilter operation, that water will condense. A cooling coil, placed at the packing outlet, through which cool tapwater is continuously circulated will be satisfactory for maintaining moisture content control in many circumstances.

REMOVAL CONCEPTS

Both biofilters and biotrickling filters are packed-bed, heterogeneous reaction systems in which gas-phase contaminants are transported across a liquid film and metabolized by bacteria and fungi growing on the packing. The liquid film is very shallow (probably of the order of 20 μm or less in depth in biofilters, although possibly deeper in biotrickling filters). Based on published experimental data, transport into/through the liquid film is rarely a rate-limiting factor. Contaminants diffuse through the liquid film and into the biofilm. Packing media used in biofilters and biotrickling filters are usually irregularly shaped and porous, although some work has been done with Rashig rings, pall rings, and gridded packing used in wastewater treatment. A typical medium has a specific surface area of 6 to 10 m^2/g, and as suggested in Figure 4, the attached microbial population is probably discontinuous. Conceptual models developed for trickling filters used in wastewater treatment can be applied with relatively minor modifications, and to date, nearly all modeling has been based on a two-phase structure (gas-biofilm) rather than a three-phase structure (gas-liquid film-biofilm) (Ottengraf and Van Den Oever, 1983; Ergas et al., 1993; Morgenroth et al., 1995).

In packed-bed biological gas treatment systems, the fluid of interest is air, and the physical properties of air (low viscosity, heat capacity, and density) are extremely important (see Table 1). Biofiltration systems operate as saturated porous mediums. Even in biotrickling filters where a liquid film is moving downward over the media, the dominant fluid is air. Typical gas flux rates in use are between 0.5 and 2 $m^3/m^2 \cdot min$ and flow in the pores is laminar. Resistance to flow is unlikely to be spatially homogeneous and channeling is a potential problem that may limit process performance.

Both upflow and downflow systems are in use, and differences in performance related to flow direction have not been reported. Biotrickling filters are typically operated as downflow units so that gas and liquid

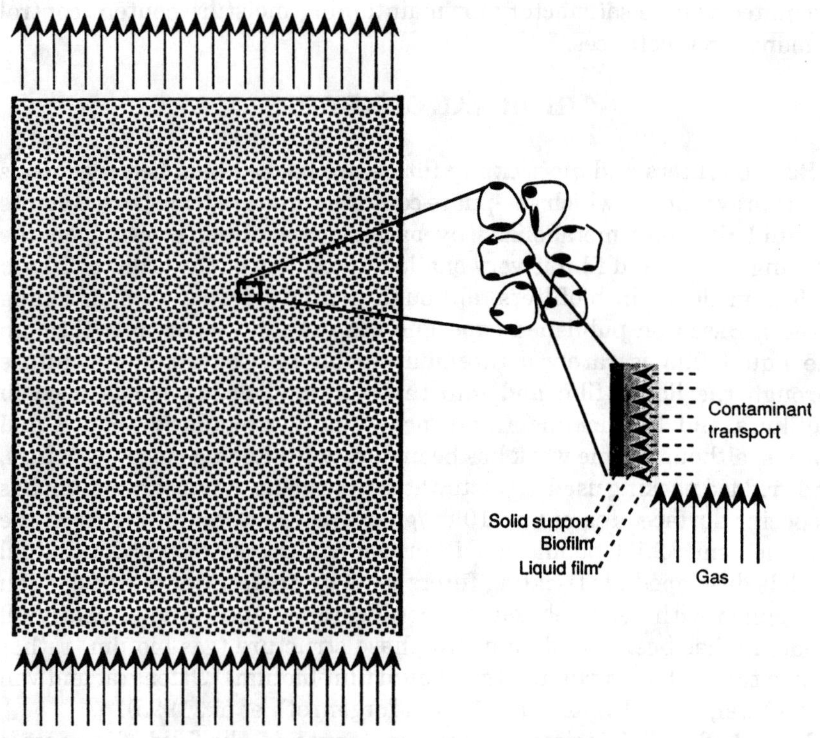

Figure 4. Conceptualization of packed-bed system for treatment of contaminated gases.

Table 1. Physical properties of air.

T (°C)	Density (g/l)	Viscosity (N·s/m^2)
−20	1.395	1.61×10^{-5}
0	1.293	1.71×10^{-5}
10	1.248	1.76×10^{-5}
20	1.205	1.81×10^{-5}
30	1.165	1.86×10^{-5}
40	1.128	1.90×10^{-5}
60	1.060	2.00×10^{-5}
80	1.000	2.09×10^{-5}
100	0.946	2.18×10^{-5}

flows are concurrent. Gas sources in biofiltration systems are usually controlled venting systems, and hence, flow rates are steady. However, contaminant concentrations may vary considerably due to source emission variation, and large transients in contaminant mass loading rates are common (Ergas et al., 1992a, 1995a, 1995b; Togna and Frisch, 1993).

Mathematical Description of Biofiltration

Mathematical description of the biofiltration process requires consideration of three phases: transport in the gas phase, transport and reaction in the liquid phase, and transport and reaction in the biofilm. Models published to date have been based on the assumptions of longitudinal plug flow through the pores, no lateral (radial) resistance to transport in the gas phase, and steady state, as suggested in Equations (2) and (3). Conventional two-film models have been applied to gas-liquid transport with equilibrium assumed at the phase interface and steady-state conditions, as described in Equations (4) and (5) and shown in Figure 5.

$$U_{gz} \frac{\partial C_g}{\partial z} = a N_{gy} \tag{2}$$

$$N_{gy} = N_{Ly} = -D_L \frac{\partial C_L}{\partial y} \tag{3a}$$

$$N_{Ly} = D_L \frac{(C_s - C_{Lbf})}{\delta_L} \tag{3b}$$

where

U_{gz} = gas velocity in pores, m/s
C_g = gas-phase contaminant concentration, kg/m³
a = specific surface area, m²/m³
N_{gy} = lateral diffusive contaminant flux in gas phase, kg/m²·s
N_{Ly} = lateral diffusive contaminant flux in liquid phase, kg/m²·s
C_S = liquid contaminant concentration at equilibrium with the gas phase, kg/m³
C_{Lbf} = liquid phase contaminant concentration at biofilm interface, kg/m³
δ_L = thickness of liquid film, m

$$U_{gz}\frac{\partial C_g}{\partial z} = \frac{D_L}{\delta_L}(C_S - C_{Lbf})a = k_L(C_S - C_{Lbf})a \tag{4}$$

where k_L = gas-liquid mass transfer coefficient, m/s.

$$HC_S = C_g \tag{5}$$

where H = dimensionless Henry's law coefficient.

The liquid and biofilm phases are often combined because of the difficulty of measuring depths and sensing the transition between the two zones. In biofilters, there should be no reaction in the liquid phase. However, in biotrickling filters, microbial cells may accumulate in the recirculated nutrient fluid, and the liquid phase reaction rate may be significant . The simpler case of no reaction in the liquid phase, shown in Figure 5, is described by Equation (6):

$$k_L(C_S - C_{Lbf})a = -\bar{r}_{BF} \tag{6}$$

where $\bar{r}_{BF}$ = average reaction rate in biofilm, kg/m³·s.

Use of an average microbial transformation rate of the contaminants in the biofilm, as is done in Equation (6), is probably satisfactory. Biofilms in most biofiltration operations are very shallow and may be only a few cells deep. However, under conditions of high organic loading and continuous nutrient recycle, biofilm thickness can become great enough to result in clogging of the interstices (Sorial et al., 1993, 1995), and a spatially variable reaction model [Equation (7)] may be appropriate.

$$D_{BF}\frac{\partial^2 C_{BF}}{\partial y^2} = r_{BF} \tag{7}$$

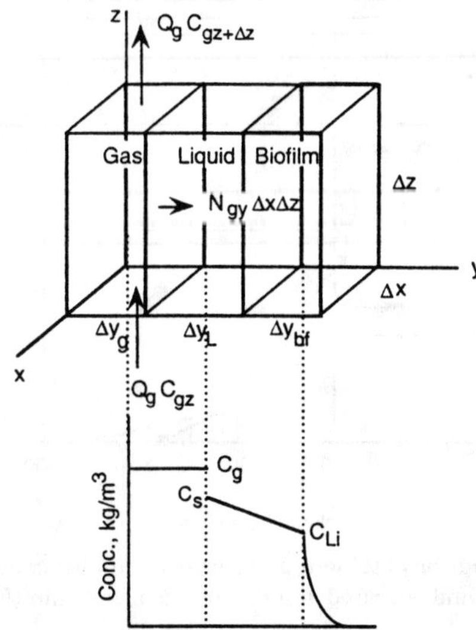

Figure 5. Conceptual mass transfer and reaction model for biofiltration systems without liquid-phase reaction.

where

D_{BF} = diffusivity in biofilm, m²/s
C_{BF} = contaminant concentration in biofilm, kg/m³
r_{BF} = reaction rate in biofilm, kg/m³·s

Reaction Rate Models

Rate models used in describing biofilter performance have included zero-order, first-order, and Monod-type expressions (Ottengraf and Van Den Oever, 1983; Ergas, 1993; Morgenroth et al., 1995). Because measurement of the biofilm density or cell mass concentration is extremely difficult, most workers have assumed a constant value through the unit. Such an assumption is generally unreasonable because contaminant concentrations vary a great deal longitudinally, and microbial populations should reflect this variation. Support for this assumption is provided by the data of Ergas et al. (1994a), shown in Figure 6, for a biofilter oxidizing toluene. Morgenroth et al. (1995) demonstrated the importance of incorporating a spatially varying biofilm density in dynamic modeling of biofiltration processes.

In most cases, the liquid-phase contaminant concentrations will be

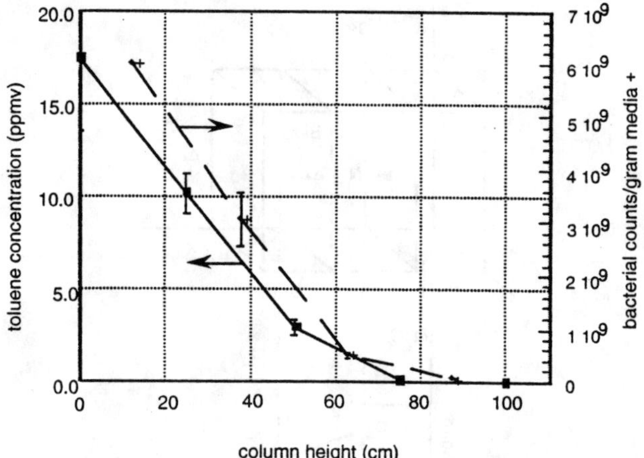

Figure 6. Spatial distribution of toluene and bacterial number concentration in a compost biofilter treating toluene and operated at a flux of 1.5 m³/m²-min (Ergas et al., 1994a).

quite small. For example, air at 20°C, having a 200 ppm$_v$ (0.77 mg/l) toluene concentration, will result in a liquid phase concentration of 3.3 mg/l at equilibrium. The toluene concentration in the biofilm would be substantially less than 3.3 mg/l and quite likely well below the Monod saturation constant. In such a system, measured reaction rates would be apparently first-order with respect to contaminant concentration, as was reported by Ergas (1993).

Augmentation of Microbial Populations

Population augmentation has two general meanings in biological treatment: provision of microorganisms during the initiation of process operation and provision of organisms capable of enhancing removals (usually of a specific compound). Biofilters are open systems, and stable population mixtures, providing optimal removal of contaminants present in the inlet air, can be expected to develop over time. Augmentation of microbial populations in biofiltration is usually focused on decreasing the time to optimal performance. In most cases, a mixture of organisms developed by growing up mixed cultures on the contaminants known to be in the air, activated sludge from local wastewater treatment plants, and pure or enriched cultures thought to be applicable are used. Because of the open nature of biofilters and biotrickling filters, the final population mixture is virtually uncontrollable, and the focus in augmentation is providing large, active populations with a wide range of degradation capabilities.

PERFORMANCE OF BIOFILTRATION SYSTEMS

Biofiltration systems are subjected to five fundamental perturbations: startup, transient loading, nutrient limitation, temperature variation, and competition resulting from mixtures of VOCs present. Startup conditions include the initial operational phase of a newly constructed unit and the initial operation following a shutdown of several hours or more. Transient loadings are nearly always due to variation in inlet concentration because volumetric flow rates are positively controlled. Concentration variation occurs often in industrial operations, and in extreme cases, mass fluxes may vary by an order of magnitude in short periods of time.

Nutrient limitations result from volatilization of ammonia or from permanent storage of nitrogen in humic materials formed in the bed. Denitrification is possible if pores become anaerobic or if the feed is oxygen-deficient (Apel et al., 1995) and could be a cause of nutrient limitation in cases where pore clogging with biomass occurs. Microorganisms respond very quickly to variations in ambient temperature, and biofilters subjected to extreme short-term temperature variation would be expected to have significant variations in removal rates. Effects of long-term temperature variations may be mitigated by adaption of species selection in attached microbial mass. The response of biofilters to VOC mixtures is interesting because, in some cases, completely different groups of microorganisms appear to be involved. In other cases, VOCs appear to be removed in relation to their solubility, Henry's law coefficients, and structure.

Performance during Startup

Performance during the start-up period for toluene, an easily degraded VOC, and for dichloromethane (DCM, or methylene chloride), a somewhat recalcitrant VOC, are shown in Figures 7 and 8, respectively. Data presented in Figures 7 and 8 were taken from a single laboratory column having an inside diameter of 150 mm, a packing depth of 1 m, and an air flux of 1 m/min. Contaminant concentrations were each 3 ppm$_v$ (0.012 g/m^3 of toluene and 0.013 g/m^3 of DCM). Activated sludge mixed liquor inoculum was used as a component of the water added to bring the equal volumes of compost and perlite to approximately 55% by wet weight moisture content [Equation (1)].

Toluene and other members of the BTEX group (benzene, toluene, ethylbenzene and *o*-, *m*-, and *p*-xylene) are characteristically easy compounds to remove from air by biofiltration. The pattern shown in Figure 7 is typical of those often reported in the literature. Toluene removal was

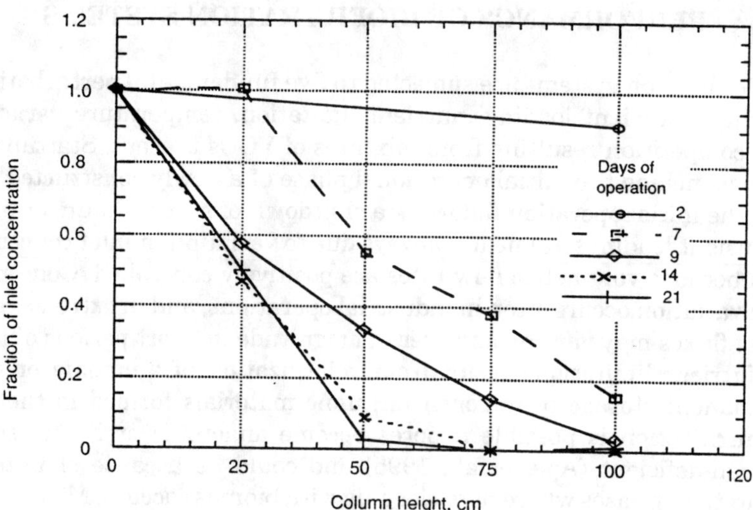

Figure 7. Removal of toluene during startup of compost biofilter operated at an air flux of 1 m/min and an inlet toluene concentration of 3 ppm$_v$. The feedstream also contained 3 ppm$_v$ of dichloromethane (Source: Ergas et al., 1993).

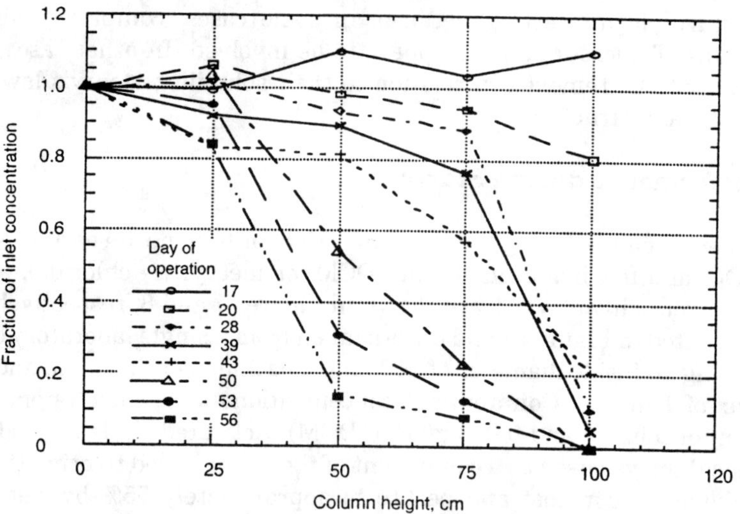

Figure 8. Removal of dichloromethane during startup of compost biofilter operated at an air flux of 1 m/min and an inlet dichloromethane concentration of 3 ppm$_v$. The feedstream also contained 3 ppm$_v$ of toluene.

618

observed after two days of operation, and outlet concentrations of toluene were at the detection limit (approximately 50 ppb$_v$) after 14 days. The point of nearly complete removal progressed toward the inlet and eventually reached a condition where more than 90% toluene removal occurred upstream of the first sampling point at 250 mm.

Microbial degradation of DCM has been well documented in the literature (Stucki et al., 1981; Rittman and McCarty, 1980; Hartmans and Tramper, 1991; Dicks and Ottengraf, 1991). However, dichloromethane is difficult to degrade for two reasons: a limited number of species of microorganisms is capable of metabolizing single carbon compounds and the two chlorine atoms inhibit biological transformation. Growth rates are usually slow when single-carbon compounds are the sole carbon sources, and both startup and recovery from process upsets may be time-consuming. Competition for nutrients or medium surface sites may be a factor where mixtures of VOCs are present, as was the case in the system from which the data shown in Figures 7 and 8 was observed. The four-week period required for development of the DCM-degrading population may have been partly the result of competition with faster growing toluene-degrading organisms. Removal of DCM in the first 25 cm of the column never exceeded 20% of the inlet concentration, whereas toluene removals were greater than 90% in the same section at steady state. Experiments in which DCM was the sole VOC in the inlet air have been conducted using virtually identical experimental systems, and 90% removal of DCM occurred in the first 25 cm (Veir, 1995).

MICROBIAL AUGMENTATION

The time from startup to steady-state operation can be shortened through microbial augmentation. Inclusion of actively growing microbial cultures capable of degrading VOCs present in the contaminated airstream has been shown to be effective (Ergas, 1993; Ergas et al., 1995b). The principal problems in microbial augmentation to improve startup performance are providing enough cells and avoiding significant time delays following addition of the culture to the packing medium. Addition of approximately 10^6 cells per gram of dry medium appears to be satisfactory based on a limited number of experiments. The data shown in Figure 9 were obtained from a laboratory compost biofilter constructed of four 25-cm sections. Each section was separated by a 50-mm plenum. After three weeks of operation (the before inoculation curve), a culture of *Pseudomonas putida* strain TOL1A was added at the top of each section. Samples taken three days after inoculation were little different from those prior to inoculation. However, at seven days after inoculation, toluene removal in the first 25 cm had increased from 65% to 90%.

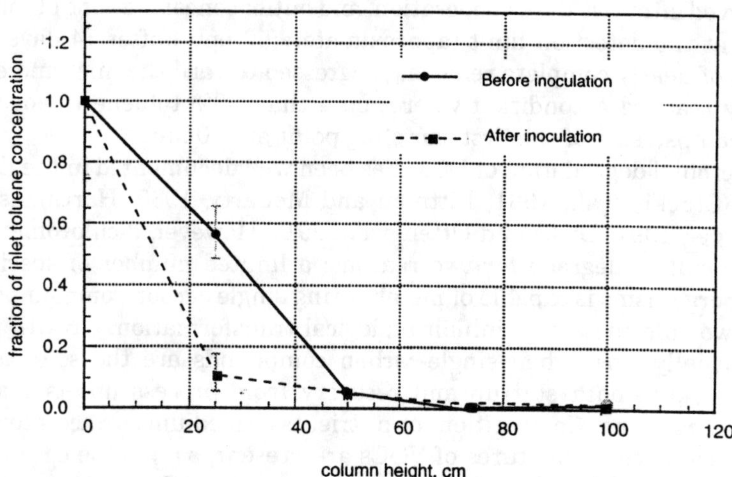

Figure 9. Results of inoculation of biofilter with *Pseudomonas putida* strain TOL1A (Source: Ergas, 1993; Ergas et al., 1995b).

Continued operation would be expected to result in a depleted microbial population in the final 75 cm of the column due to lack of toluene and an inability to maintain high performance during transient loadings.

ACCUMULATION OF BIOMASS

Increased loadings on attached growth processes, such as biofilters and biotrickling filters, results in greater accumulation of microbial mass. In practice, increasing the inlet concentration should result in a larger microbial population density in the inlet zone and little change in fractional removal pattern until the mass transfer rate becomes limiting. When media particle pores are filled (see Figure 3), the transport rate of VOCs to cells located within the pellets decreases. The removal rate decreases, and higher inlet concentrations will result in plots of fractional concentration versus column height having lower slopes. The data shown in Figure 10 support this concept in that the fractional concentration profiles for inlet concentrations of 3 and 50 ppm$_v$ are virtually identical. When medium particle pores, as opposed to packing interstices, become plugged, cells in the inner portions of the particles can receive VOCs only by diffusion through the microbial mass. Diffusivities in the microbial slime are approximately three orders of magnitude lower than those in air, and the concentration gradients will be lower because of the decreased availability of VOCs within the pellets. A considerable amount of the reactive mass may be lost quite suddenly as the result of overgrowth, and unless the particles are cleaned in some manner, the system may not recover completely when loadings are reduced.

Increases in air flux at constant inlet concentration result in a change in loading rate without a corresponding change in the concentration gradient, as can be seen from combining Equations (2) and (3).

$$-\frac{dC_g}{dz} = \frac{k_L}{U_{gz}}\left(\frac{C_g}{H} - C_{Lbf}\right)a \qquad (8)$$

Using the most desirable condition, a fast reaction rate that results in $C_{Lbf} \ll C_g$, the effect of increasing the air flux, U_g, will be a decrease in the slope of the removal curve. In actual systems, C_{Lbf} will not be negligible and the result will be an even lower slope. Results of experiments with a 150 mm diameter, 1 m packing length compost biofilter treating air containing 3 ppm$_v$ of toluene support the concept, as shown in Figure 11.

Experiments with hexane conducted using a compost biofilter identical to that used for the studies of Figure 10 resulted in a less clear picture relative to effect of loading rate on biofilter performance. Hexane is quite biodegradable but considerably less soluble than toluene (14 mg/l versus 550 mg/l, respectively). The expectation in using hexane was that gas/liquid mass transport might become limiting and that breakthrough was likely. Inlet concentrations were used in the hexane experiments and were two orders of magnitude greater than those used in the comparable toluene experiments and the mass flux of hexane was kept constant by

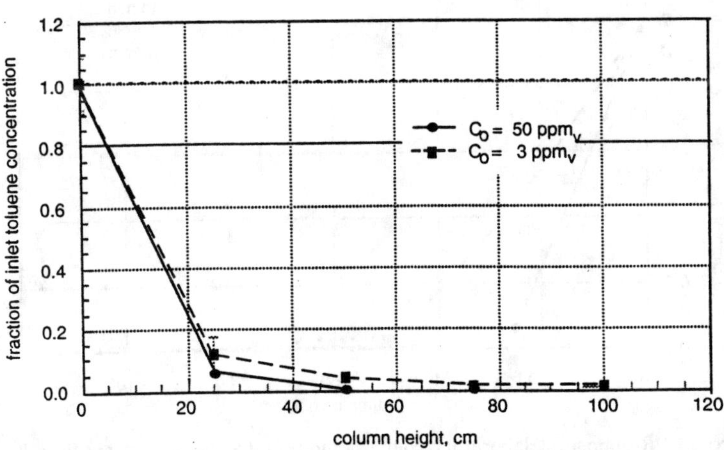

Figure 10. Steady-state operation of compost biofilter at an air flux of 1 m/min and inlet toluene concentrations of 3 and 50 ppm$_v$ (Source: Ergas, 1993; Ergas et al., 1995b).

decreasing the inlet concentration as the air flux was increased. The result of decreasing the concentration would be a decrease in the saturation concentration and a corresponding decrease in the mass transfer rate. The experiments were hampered by unexpected nutrient limitations in the compost, and considerable time was expended in operation of the systems prior to establishing that nitrogen needed to be added. Once nitrogen was added, hexane removals increased rapidly over a two-week period as shown in Figure 12. At the end of the two-week period, the air flux was doubled from 1 m/min to 2 m/min, and the inlet hexane concentration was decreased from 200 ppm$_v$ to 100 ppm$_v$. Removal rates did decrease, as would be predicted from Equation (7), but the change was not particularly dramatic. One reason for the small change in performance may have been that the level of microbial activity in the middle section of the column was high at the time of the change in loading. A longer period of operation at 1 m/min may have resulted in a lower microbial population and a larger change in performance.

Nutrient Limitation

Nutrient limitation results in cessation of growth and metabolism. In compost biofilters, inorganic nutrients are partially derived from the packing medium, and the effects of nutrient limitation are damped. The

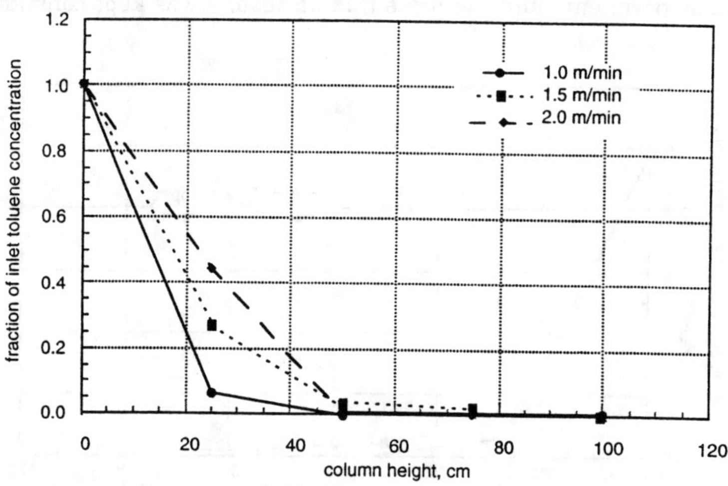

Figure 11. Response of laboratory compost biofilter to changes in air flux. The inlet concentration of toluene was 3 ppm$_v$ and curves represent steady-state operation (Source: Ergas, 1993; Ergas et al., 1995).

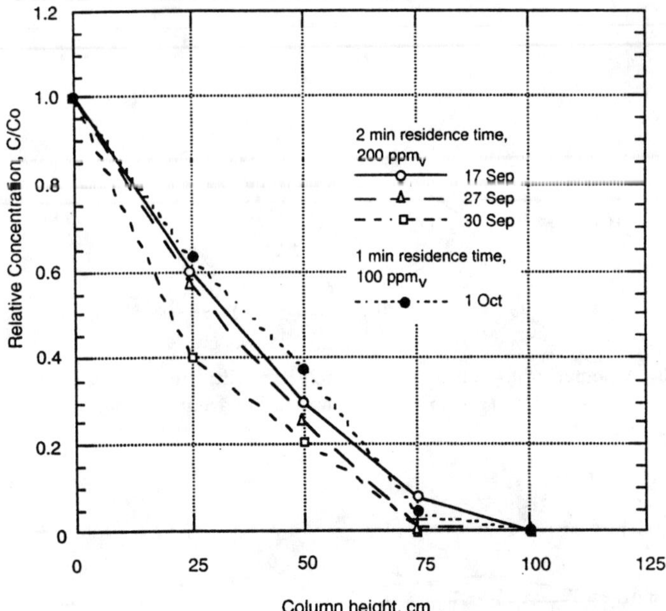

Figure 12. Hexane removal in compost biofilter. September 17, 27, and 30 data nominal residence time was 2 minutes and the inlet hexane concentration was 203 ppm$_v$. October 1 represents first day at nominal residence time of 1 minute and inlet hexane concentration of 102 ppm$_v$ (Source: Morgenroth, 1994).

composting process results in ammonia volatilization and well-composted sewage sludge, yard waste, and wood pulp will, in general, be nitrogen-depleted. Addition of nitrogen in the NO_3^- form is preferable to the NH_4 form because ammonia is quickly stripped from the system. Nitrogen is lost to humic material formation, and in some cases, local anaerobic zones may generate losses in nitrogen through denitrification. Cell death and decay results in release of ammonia, which has three fates: assimilation into new cell growth, volatilization, and nitrification. Eventually, nitrogen added will be depleted from a biofilter, but time estimates made using the method described in Example 1 will be conservative because of partial cycling, as depicted in Figure 13.

In synthetic medium systems, nutrient limitations result in very rapid decreases in performance, as can be seen in Figure 14. The data shown in Figure 14 were obtained in a laboratory biofilter packed with Isolite CG-6,** an extruded diatomaceous earth pellet similar to the Celite™

**Innova Corp. Denver, CO.

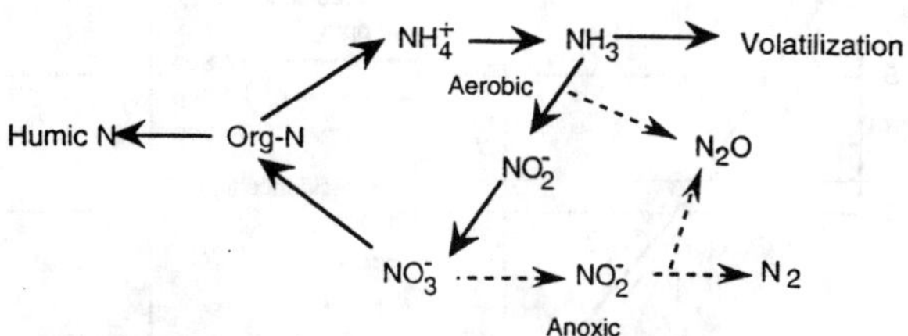

Figure 13. Schematic diagram of nitrogen cycle in biofiltration. Losses to humic material, which are essentially nonretrievable, and volatilization eventually deplete nitrogen stores added at process startup.

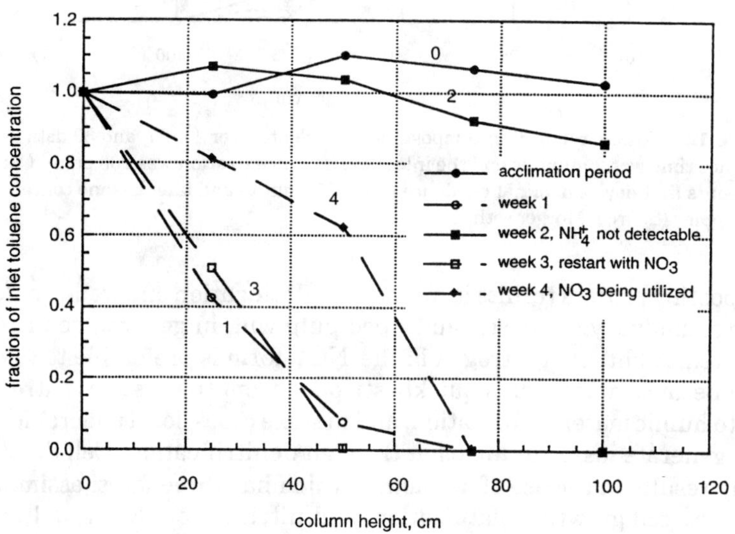

Figure 14. Toluene profiles for the first four weeks of Isolite column operation showing acclimation period and toluene utilization under nitrogen-limited and nitrogen-rich conditions. The toluene concentration was 50 ppm$_v$ and the air flux rate was 1 m/min.

624

shown in Figure 3. Prior to startup, the packing was soaked in a solution containing a culture of *Pseudomonas putida* TOL1A, NH_4Cl, phosphate buffer, and UC Davis tapwater, which is highly mineralized and contains a wide range of trace nutrients. Following a one-week acclimation period (curve 0), excellent toluene removals were observed (curve 1). However, by the second week of operation, removals were negligible, and ammonia nitrogen was absent from the column (curve 2). An assumption was made that most of the ammonia nitrogen added at startup had been lost by volatilization. The column was then soaked overnight in a solution containing KNO_3, and performance returned to that of the first week of operation (curve 3). The amount of nitrate N added was small and as expected the column began to fail after an additional week of operation (curve 4).

Transient Loading

The response of biofilters and biotrickling filters to transients is apparently controlled by the biomass density and activity. In biofilters operated at steady state for extended periods, the microbial population density decreases along the direction of flow (see Figure 6). Because biodegradation always results in net growth, bacteria accumulate, or "overgrow," in the active removal zone, and the local potential reaction capacity exceeds the steady-state loading rate. The "overgrowth" explains the very sharp decrease in VOC concentrations nearly always observed in biofilters. When transient loadings occur, the excess reaction capacity is utilized, and the result is that biofilters have a good deal of resilience in responding to transients. However, when the reaction capacity is exceeded, breakthrough is likely to occur because of the depleted microbial concentrations beyond the active removal zone. Step increases in loading will result in rapid growth of microorganisms, and the system performance will improve over a period of approximately a few hours to a few days in most cases. The response to a step function transient is illustrated in Figure 15 for a system treating toluene. The new steady state corresponding to the 35 ppm_v inlet concentration would be expected to be very similar to the values for 10 ppm_v, as suggested by the results shown in Figure 10.

In cases where transients occur on a regular basis, the response of biofilters is quite good. When short-term transients of one to 24 hours are associated with normal operation, a microbial population distribution should develop that corresponds to the requirements for good contaminant removal. Togna and Folsom (1992) found that, shortly after startup of a biofilter treating contaminated air from a paint spray booth, break-

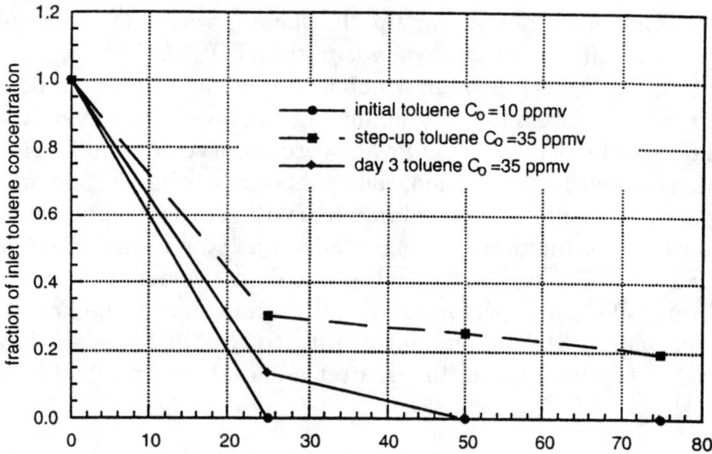

Figure 15. Response of compost biofilter operated with air flux of 1 m/min and an initial steady-state inlet concentration of 10 ppm$_v$. Breakthrough occurred following step increase in inlet toluene concentration to 35 ppm$_v$. Performance improved considerably over a three-day period with response probably resulting from increase in microorganism concentrations along the column.

through occurred following lunchbreaks and overnight shutdowns. Over a period of weeks, performance improved to the point that variations in outlet concentration were minimal.

Treatment of VOC Mixtures

Each component of a VOC mixture is removed in an individual chain of rate processes. The rate of transfer from the gas to liquid is a function of the gas- and liquid-phase diffusivities, the Henry's law coefficients, and the gas- and liquid-phase concentrations of each species. Biodegradation involves transport of a compound into the microbial cells and transformation in a series of oxidation-reduction reactions. Transport across the cell boundary usually involves an enzyme-catalyzed reaction, and each of the oxidation-reduction reactions is catalyzed by a specific enzyme as well. Rates of every step of the transport and transformation processes are affected by structural characteristics of the compounds. Moreover, the capabilities of microorganisms to metabolize organic and inorganic compounds is limited in terms of range. Some microbial species, such as the methanotrophic bacteria, can metabolize a very few species of organic compounds and occupy a small niche in the total microbial ecosystem. Others, such as members of the genus *Pseudomonas*, metabolize an amazing number of aliphatic, alicyclic, and aromatic compounds, and a number of species in this group are capable of utilizing nitrate, as well as oxygen, as a terminal electron acceptor. Uniform removal rates for all compounds in a VOC mixture would not be expected and based on observations reported to date, do not occur.

When VOC mixtures are fed to biofilters and biotrickling filters, spatial separation of the compounds occurs. Compounds that are rapidly transported and easily degraded by the resident microorganisms are quickly removed, and compounds that are less rapidly transported and less easily degraded move farther into the column. An example is provided by the toluene-DCM mixture of Figures 7 and 8. The toluene and DCM removal profiles for this system on the 56th day of operation are shown in Figure 16. Note that the removal of DCM in the second 25-cm column segment is quite large and approximately 3.5 times the DCM removal in the first 25 cm. Biodegradation of DCM is restricted to a limited number of bacterial species with the most commonly observed organisms being of the genuses *Hyphomicrobium* and *Methylobacterium* (Hartman and Tramper, 1991) and *Flavobacterium* (Ergas et al., 1994b). A tentative

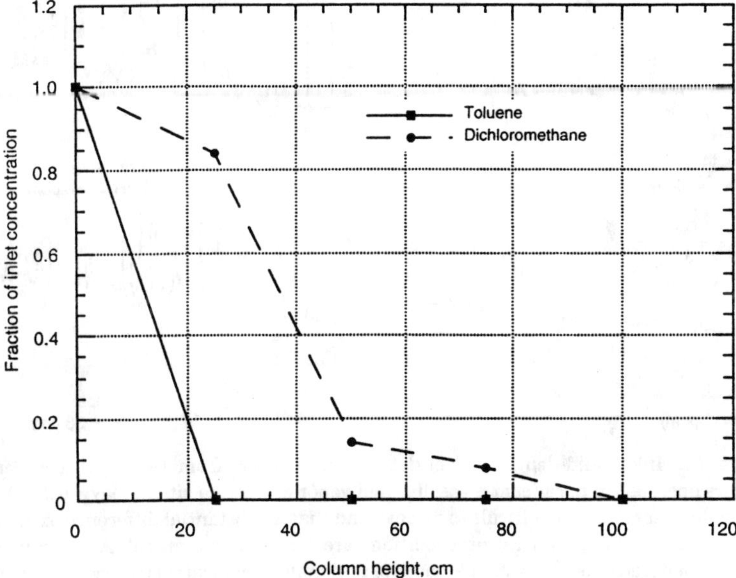

Figure 16. Removal profiles of toluene and dichloromethane 56 days after startup of laboratory compost biofilter. Toluene concentrations were below the detection limit of 50 ppbv at all sample points beyond the inlet while a distinct concentration profile was observed for dichloromethane. Inlet concentrations of both compounds was 3 ppm$_v$ and the air flux was 1 m/min.

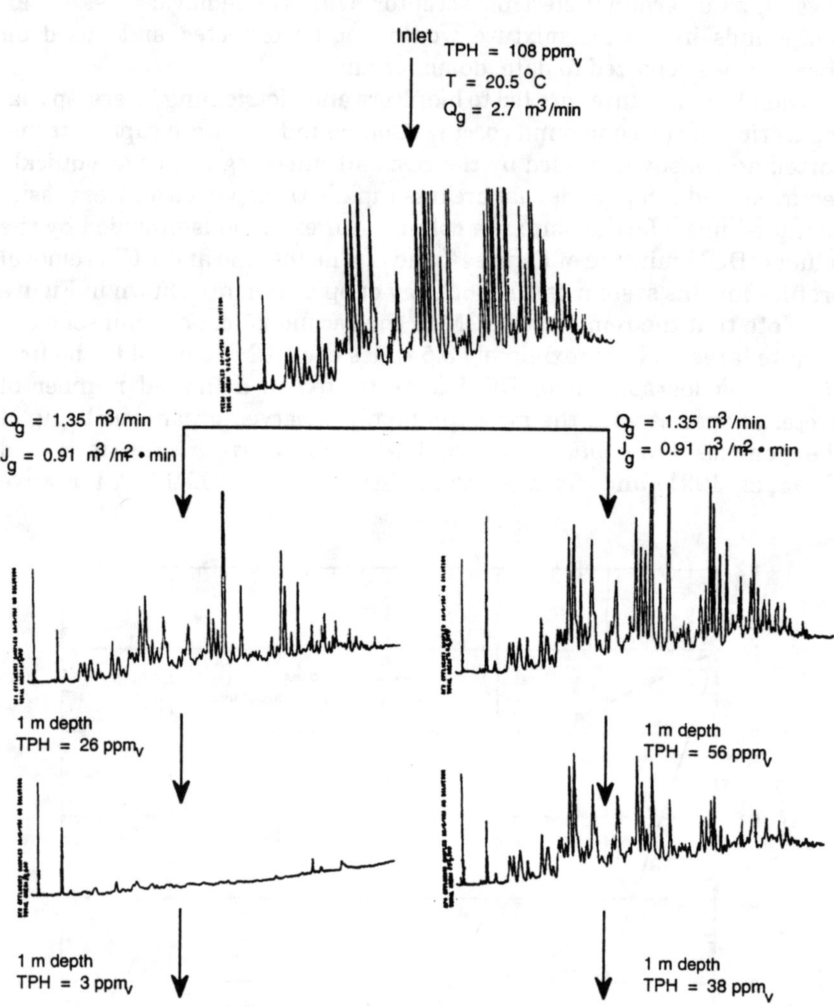

Figure 17. Inlet, mid-depth, and outlet chromatograms from two 2-m deep compost biofilters operated in parallel at a gasoline soil vapor extraction site in Hayward, CA. Note that the light alkanes are difficult to remove and that a substantial difference exists in the parallel units, although operating conditions are essentially identical. A lower moisture content is believed to be the principal cause of the smaller removals of the poorer performing unit.

conclusion drawn from the concurrent biodegradation studies is that the DCM degraders are less competitive than the toluene degraders for space (i.e., surface area), nutrients, or perhaps oxygen.

Selective removal of VOC species observed in treating gasoline vapor appears to be related more to structural characteristics of the compounds than microbial species involved. Gasoline is typically composed of approximately 60 hydrocarbons ranging in size from C_5 to C_{12}. Branched chain aliphatic and substituted alicyclic compounds are the most difficult to degrade. The light paraffins, such as n-butane and n-pentane may be missing from the "weathered gasoline" extracted from soils contaminated by leaking underground storage tanks. This was the case at a gasoline soil vapor extraction site in Hayward, California, where biofiltration was used. Inlet, mid-depth, and outlet chromatograms from two 2-m deep biofilters operated in parallel are shown in Figure 17. Note that light alkanes are difficult to remove and that a substantial difference in removal exists in the parallel units, although operating conditions are essentially identical. A lower moisture content of the compost packing media is believed to be the principal reason for smaller removals in one system. The compounds appearing between 7 and 10 minutes on the chromatograms have been tentatively identified as branched chain aliphatics and substituted alicyclics in the C_7 and C_8 range (e.g., dimethylpentane and propylcyclopropane).

SUMMARY

Biodegradation of vapor-phase contaminants using packed-bed systems such as biofilters and biotrickling filters is an extremely promising technology. A significant number of applications have been made for both odor control and VOC degradation. The processes have been most effectively applied for odor control, but the number of successful VOC removal operations is growing rapidly. Advantages of biofiltration include low capital and operating cost relative to alternatives such as combustion and adsorption, biodegradation of contaminants to CO_2, and simplicity. Disadvantages include a relatively large footprint compared to catalytic oxidation, potential pH control problems if sulfides or chlorinated organics are being oxidized, difficulty of controlling moisture content of natural medium systems, and a relatively poor record of process stability. All of the problems associated with biofilter and biotrickling filter operation appear to be solvable. However, until satisfactory performance records have been established, there will be a need for extensive monitoring of full-scale applications.

REFERENCES

Allen, Eric R., Y. Yang (1992) "Operational Parameters for the Control of Hydrogen Sulfide Emissions Using Biofiltration," *Proceedings of the 85th Annual Meeting of the Air and Waste Management Association,* Kansas City, MO.

Apel, W. A., J. M. Barnes, K. B. Barrett (1995) "Biofiltration of nitrogen oxides from fuel combustion gas streams," *Proceedings of the 88th Annual Meeting & Exhibition, Air & Waste Management Association,* San Antonio, TX, June 18–23.

Carlson, D. A., C. P. Leiser (1966) "Soil Beds for the Control of Sewage Odors," *Journal of the Water Pollution Control Federation,* vol 38: 5, pp. 429–440.

Dicks R. M. M., S. P. P. Ottengraf (1991) Verification studies of a simplified model for the removal of dichloromethane from waste gases using a biological trickling filter (Part 1 and Part 2). *Bioprocess Eng.,* vol. 6, pp. 93–99, 131–140.

Eitner, D. (1989) "Biofilter in Flue Gas Cleaning: Biomasses, Design, Costs, and Applications," *Brennst.-Waerme-Kraft,* vol. 41: 3.

Ergas S. J. (1993) Control of Low Concentration Volatile Organic Compound Emissions Using Biofiltration. Ph.D. Thesis. University of California, Davis.

Ergas S. J., E. D. Schroeder, D. P. Y. Chang (1991) VOC Emission Control from Wastewater Treatment Facilities Using Biofiltration. *Proceedings 84th Annual Meeting Air & Waste Man. Assoc.,* Vancouver, British Columbia.

Ergas, S. J., E. D. Schroeder, D. P. Y. Chang, R. Morton (1992a) "Control of Volatile Organic Compound Emissions from a POTW Using a Compost Biofilter," *Proceedings 85th Annual Meeting of the Air and Waste Management Association,* 92-116.02, Kansas City, MO.

Ergas, S. J., E. D. Schroeder, D. P. Y. Chang (1992b) Biodegradation Technology for VOC Removal from Airstreams: Phase I: Performance Verification, Department of Civil & Environmental Engineering, University of California, Davis, California Air Resources Board Contract AO -32-137, May.

Ergas, S. J., E. D. Schroeder, D. P. Y. Chang (1993) "Control of Air Emissions of Dichloromethane, Trichloroethene, and Toluene by Biofiltration," *Proceedings 86th Annual Meeting of the Air and Waste Management Association,* Denver, CO.

Ergas S. J., E. D. Schroeder, D. P. Y. Chang, K. M. Scow (1994a) Spatial Distribution of Microbial Population in Biofilters. *87th Annual Meeting, Air & Waste Man. Assoc.,* Denver, CO.

Ergas, S. J., K. Kinney, M. E. Fuller, K. M. Scow (1994b) "Characterization of a compost biofiltration system degrading dichloromethane," *Biotechnology and Bioengineering,* vol. 44, pp. 1048–1054.

Ergas, S. J., K. Kinney, M. E. Fuller, K. M. Scow (1994c) "Control of dichloromethane emissions using biofiltration," Presented at the *I&EC Special Symposium,* American Chemical Society, Atlanta, GA, September 19–21.

Ergas, S. J., E. D. Schroeder, D. P. Y. Chang, R. Morton (1995a) "Control of VOC Emissions from a POTW Using a Compost Biofilter," *Water Environment Research,* vol. 67, pp. 816–821.

Ergas, S. J., K. A. Kinney, E. Morgenroth, Y. Davidova, K. M. Scow, D. P. Y. Chang, E. D. Schroeder (1995b) Biodegradation Technology for VOC Removal from Airstreams Phase II: Determination of Process Design Parameters and Constraints, Center for Environmental and Water Resources Engineering, University of California, Davis.

Harremoes, P. (1978) "Biofilm Kinetics," in *Water Pollution Microbiology, vol. 2,* Wiley, New York, 71–109.

Hartman, S., J. Tramper (1991) "Dichloromethane Removal from Waste Gases with a Trickle Bed Reactor," *Bioprocess Engineering,* vol. 6, pp. 83–92.

Holubar, P., C. Andorfer, R. Braun (1995) "Prevention of Clogging in Trickling Filters for Purification of Hydrocarbon Contaminated Wastewater Air," *Proceedings, USC-TRG Conference on Biofiltration,* University of Southern California, October 5–6.

Joyce, M., N. Wadsworth, B. Howard, and J. Bewley (1992) "Control of Sulfide and VOC Emissions with a Soil Bed Scrubber," Presented at the *California Water Pollution Control Association Conference,* Sacramento, CA.

Kinney, K. A. (1996) Control of Clogging and Transient Loadings in a Biofilter Using Alternating Directional Flow, Ph.D. Dissertation, Department of Civil and Environmental Engineering, University of California, Davis.

Morgenroth, E. (1994) Degradation Kinetics of Hexane in a Biofilter, M.S. Thesis, Department of Civil and Environmental Engineering, University of California, Davis.

Morgenroth, E., E. D. Schroeder, D. P. Y. Chang, K. M. Scow (1995) "Modeling of VOC removal in biofilters incorporating cell density," *Innovative Technologies for Site Remediation and Hazardous Waste Management,* R. D. Vidic and F. G. Pohland, ed., ASCE, pp. 473–480, American Society of Civil Engineers, NY.

Ottengraf, S. P. P., A. H. C. Van den Oever, F. J. C. M. Kempenaars (1984) "Waste Gas Purification in a Biological Filter Bed," *Prog. Ind. Microbiol.,* vol. 20, pp. 157–167.

Ottengraf, S. P. P., J. J. P. Meesters, A. H. C. Van den Oever and H. R. Rozema (1986) "Biological Elimination of Volatile Xenobiotic Compounds in Biofilters," *Bioprocess Engineering,* vol. 1, pp. 61–69.

Ottengraf, S. P. P., A. H. C. Van Den Oever (1983) "Kinetics of Organic Compound Removal from Waste Gases with a Biological Filter," *Biotechnology and Bioengineering,* vol. 25, pp. 3089–3103.

Pomeroy, R. D. (1957) "Deodorizing Gas Streams by the Use of Microbiological Growths," U.S. Patent # 2,793,096.

Pomeroy, R. D. (1982) "Biological Treatment of Odorous Air," *Journal of the Water Pollution Control Association,* 54, pp. 1541–1545.

Rittman, B. E., P. L. McCarty (1980) "Utilization of Dichloromethane by Suspended and Fixed-Film Bacteria," *Applied and Environmental Microbiology*, vol. 39, pp. 1225–1226.

Smith, F. L., G. A. Sorial, M. T. Suidan, P. Biswas, R. C. Brenner (1995) "Management of High-Rate Trickle Bed Biofilters," *Innovative Technologies for Site Remediation and Hazardous Waste Management*, R. D. Vidic and F. G. Pohland, ed., ASCE, pp. 489–496, American Society of Civil Engineers, NY.

Sorial, G. A., F. L. Smith, A. Pandit, M. T. Suidan, P. Biswas, R. C. Brenner (1995) "Performance of Trickle Bed Biofilters under High Toluene Loading," *Proceedings, 88th Annual Meeting & Exhibition, Air & Waste Management Association*, San Antonio, TX, June 18–23.

Sorial, G. A., F. L. Smith, M. T. Suidan, P. Biswas, R. C. Brenner (1993) "Evaluation of Biofilter Media for Treatment of Air Streams Containing VOCs," *Proceedings Water Environment Federation 66th Annual Conference and Exposition, vol. 10, Facility Operations*, pp. 429–439.

Stucki, G., R. Galli, H. R. Ebersold, T. Leisinger (1981) "Dehalogenation of Dichloromethane by Cell Extracts of Hyphomicrobium DM2," *ARCH Microbiol*, vol. 130, pp. 366–371.

Togna, A. P., B. R. Folsom (1992) "Removal of Styrene from Air Using Bench-Scale Biofilter and Biotrickling Filter Reactors," *Proceedings of the 85th Annual Meeting of the Air and Waste Management Association*, Kansas City, MO.

Togna, A. P., S. Frisch (1993) "Field-Pilot Study of Styrene Biodegradation Using Biofiltration," *Proceedings of the 86th Annual Meeting of the Air and Waste Management Association*, Denver, CO.

van Lith, C. (1989) "Design Criteria for Biofilters," *Proceedings of the 82nd Annual Meeting of the Air and Waste Management Association*, Anaheim, CA.

van Lith, C., S. P. P. Ottengraf, B. Osinga, R. M. M. Diks (1993) "The Biological Purification of Multi-Component Waste-Gas Discharge in Artificial Glass Production," *Proceedings of the 86th Annual Meeting of the Air and Waste Management Association*, Denver, CO.

Veir, J. (1995), personal communication.

Wright, W. F., Y. Davidova, E. D. Schroeder, D. P. Chang (1995) "Performance of a Compost Biofilter Treating Gasoline from a Soil Vapor Extraction Operation," *Proceedings, USC-TRG Conference on Biofiltration, University of Southern California*, October 5–6.

In situ Bioremediation Process Engineering Concepts

A. B. CUNNINGHAM AND P. J. STURMAN
Center for Biofilm Engineering
Montana State University
Bozeman, MT 59717, USA

INTRODUCTION

In situ bioremediation attempts to stimulate the growth of indigenous or introduced microorganisms in regions of subsurface contamination, thereby encouraging biotransformation of sorbed and dissolved contaminant. In some cases, existing conditions are sufficient at a site to permit natural biotransformation. In most cases, bioremediation requires installation of engineered systems that inject or infiltrate fluid, nutrients, and one or more electron acceptors to improve the rate of contaminant degradation.

Microorganisms growing as biofilms in the subsurface have an advantage over suspended species in that they can remain near the source of fresh substrate and nutrients that flow by them (Characklis and Marshall, 1990). The rate of biofilm growth, and thus the rate of biotransformation, is therefore strongly influenced by transport processes, including pore velocity distribution, dispersivity, molecular diffusivity, sorption phenomena, and other processes that affect the delivery rate of substrate and nutrients to growing cells.

Bioremediation of these compounds involves complex interactions of biological, chemical, and physical processes and requires integration of phenomena operating at scales ranging from that of the microbial cell (10^{-6} m) to that of the geological site (10–1000 m). Laboratory investigations of biodegradation are usually performed at a relatively small scale that is governed by convenience, cost, and expediency. However, the extension of results from a laboratory-scale experimental system to the design and operation of field-scale systems introduces (1) additional mass transport mechanisms and limitations; (2) the presence of multiple

633

phases, contaminants, and competing microorganisms; (3) spatial hetero-geneities; and (4) subsurface environmental factors that may inhibit bacterial growth such as temperature, pH, nutrient, or redox conditions. Field bioremediation rates may be limited by the availability of one of the necessary constituents for biotransformation: substrate (contaminant), electron acceptor, nutrients, or microorganisms capable of degrading the target compound. The limiting factor for bioremediation may not be the same in the laboratory as it is in the field, thereby leading to development of unsuccessful remediation strategies (Goldstein et al., 1985; Lee et al., 1988).

The volume of current bioremediation literature is dominated by inves-tigation of individual phenomena, usually at the bench scale. Relatively few studies address interactions among different bioremediation phe-nomena or consider how phenomenological effects can be integrated to make predictions of field-scale process behavior. The relative absence of practitioner-oriented tools for decision making suggests that process engineering methods are needed to improve the state of the art of bioremediation practice.

Process engineering, in the context of in situ bioremediation, involves the integration of data from different scales of observation in order to make predictions and design decisions. Ultimately, a priori predictions need to be compared with field observations of remediation system performance. Evaluation studies of this kind need to be published in current literature to better define the overall uncertainty surrounding bioremediation decision making. This chapter uses the scales of observa-tion approach, along with other process engineering concepts, to provide a framework for understanding factors affecting biotransformation rate and extent with the intent of improving the design of subsurface biore-mediation systems. Several sections of this chapter have been condensed from a review of current bioremediation literature by Sturman et al. (1995).

SCALE DEFINITIONS

The engineering of bioremediation systems is aided by defining three scales of observation: microscale, mesoscale and macroscale. The scale definitions are arbitrary but serve as a useful conceptual structure for approaching the engineering scale-up problem.

The microscale is taken as the scale at which chemical and microbio-logical species and reactions can be characterized independently of any transport phenomena. Examples of microscale features would be the composition of the microbial population and kinetics and stoichiometry of transformation reactions. The physical scale of these phenomena is the dimension of the microbial cell, on the order of 10^{-6} to 10^{-5} m. The mesoscale is defined as the scale at which transport phenomena and system geometry are first apparent, with the exclusion of advective or mixing processes. Mesoscale phenomena include diffusion, nonequilibrium sorption, and interphase mass transfer. Possible physical scales for mesoscale phenomena include the size of pore channels or soil particles, the characteristic diffusion length, or the dimension of microbial aggregates. The length scale for such features ranges from approximately 10^{-5} to 10^{-2} m. The ability to discern advective or mixing phenomena defines the macroscale. Advection, dispersion, and geologic spatial heterogeneity are examples of macroscale phenomena. The corresponding physical scale for these phenomena ranges from approximately 10^{-2} to 10^2 m or even larger. Table 1 summarizes by scale many of the phenomena influencing

Table 1. Phenomena influencing bioremediation.

Scale	Representative Characterization Methods
Microscale	
Microorganisms	Plate counts, gene probes
Degradation pathways	Batch reaction studies
Reaction stoichiometry	Batch reaction studies
Reaction kinetics	Batch reaction studies
Electron acceptors	Chemical analysis for N, P
Nutrients	Chemical analysis
Inhibitors, toxicity	Batch reaction studies
Water activity, pH	Electrochemical probes
Reactions with soil or aquifer matrix	Abiotic reaction studies
Chemical equilibria	
Sorption (equilibrium)	Abiotic batch sorption studies
Mesoscale	
Sorption (nonequilibrium)	Abiotic batch and column sorption studies
Attachment/detachment (micro-organisms)	Biofilm studies, attached microbe enumeration
Diffusion	
Plugging/filtration	Column studies, pressure drop and flow rate
Interphase transport	Multiphase column studies
Macroscale	
Advection	Well elevations, pump tests, tracer studies
Dispersion	Conservative tracer studies
Spatial heterogeneity	Well logs, core permeabilities
Hydrologic properties and boundary conditions	As for advection and dispersion

bioremediation. Phenomena are classified according to the smallest scale at which they can be observed.

SCALEUP PROCEDURES

Results from bench-scale treatability studies, pilot- or field-scale demonstrations (Wilson et al., 1986; Rogers et al., 1993), and simulation modeling studies provide the tools for performing bioremediation scaleup. The scaleup process involves investigating each scale of observation to identify and quantify the biotransformation rate-limiting processes and to integrate information from each scale to obtain overall predictions. At the microscale, this information includes the need to (1) quantify the activity of contaminant-degrading microorganisms at the site, (2) determine which chemical species limits the extent of biotransformation, and (3) estimate the intrinsic kinetics of biotransformation. At the mesoscale, the influence of sorption and interphase transport on biotransformation must be established. Macroscale issues include whether advective-dispersive transport, spatial heterogeneity, and subsurface environmental conditions limit biodegradation rates.

As Table 1 indicates, the in situ bioremediation of contaminated groundwater and soil is a complex undertaking that requires an understanding of many physical, chemical, and biological phenomena. The three scales of observation defined above are intended to help organize these phenomena in a way that facilitates evaluation of bioremediation potential at a field site. Some important points here are: (1) observations made at the micro- or mesoscale may not necessarily apply at the field (macro) scale. Observed contaminant loss rates, for example, depend on scale. Table 2 illustrates this point with a compilation of several reports from the scientific literature in which lab-scale and field-scale rates are compared. Field-measured half-lives tend to be four to ten times longer than laboratory-determined values. The lower rates measured in the field presumably reflect the effects of mesoscale and macroscale phenomena. (2) For a given set of environmental conditions, there is one phenomenon that will limit the rate at which bioremediation can proceed. The basic approach to bioremediation engineering must therefore begin with a thorough analysis of all relevant phenomena in Table 1 to determine which will limit contaminant biotransformation rate for subsurface environmental conditions representative of the site in question. (3) Since the subsurface environment may vary considerably across a site due to heterogeneities, this analysis may conclude that several different phenomena are limiting, depending on the exact location. (4) Once a particular remediation strategy is proposed (e.g., pump-and-treat or nutrient

Table 2. Scale dependence of contaminant half-lives.

Laboratory Half-Life (days)	Field Half-Life (days)	Ratio of Field/Lab Half-Life	Contaminant	Reference
42	397	9.5	Aviation gas as TPH	Wilson, 1991, Huling and Bledsoe, 1990
3.6	23	6.4	Gasoline, No. 2 fuel oil as TPH	Block et al.
28	111	4.0	Benzene	Barker et al., 1987
6.1	42	6.8	Toluene	Hutchins et al., 1991
5.6	55	9.8	*m,p*-xylenes	Hutchins et al., 1991
10	73	7.3	BTEX	Chiang et al., 1989

injection/recovery), the situation must be reassessed to determine if rate-limiting phenomena are affected. Once the limiting rate information has been assembled, the integrated effect can be used to address engineering feasibility questions such as: is there potential for successful application of bioremediation? Can the rate of biodegradation be enhanced? How can the bioremediation processes be engineered? How can bioremediation be monitored and verified? None of these questions can be properly answered by observations made at a single scale alone.

MICROSCALE PHENOMENA

Microbial Activity

Successful biodegradation depends on the presence and activity of a microbial population capable of degrading the target contaminant. Hydrocarbon degrader numbers have been correlated with prior exposure of an ecosystem to contamination. Atlas (1981) reports that hydrocarbon degraders may constitute less than 0.1% of total organisms in unpolluted systems but up to 100% of total counts in the vicinity of a hydrocarbon release. However, the presence of hydrocarbons in groundwater and soil does not a priori indicate the presence of a population capable of hydrocarbon degradation, particularly if more easily metabolized substrates are present (Swindoll et al., 1988; Leahy et al., 1990). Methods of bacterial enumeration from soil/water systems include heterotrophic plate counts, epifluorescence direct counts, hydrocarbon degrader plate counts, most probable number (MPN) techniques, DNA probes, lipid assays, metabolic indicators, and radioisotopic methods. Direct counts include both metabolically active and inactive organisms, while heterotrophic plate counts theoretically measure only the fraction capable of growth. MPN techniques and DNA probes indirectly measure microbial density through chemical changes brought about due to metabolite production (MPN) or extraction and quantification of cell genetic material (DNA probes). Lipid assays quantify the fatty acid content of cell suspensions and relate this to the total microbial population. Radioisotopic methods measure the kinetics of cell uptake of tritiated or ^{14}C-labeled compounds (tritiated-thymidine or -acetate) and translate this uptake into cell activity estimates (O'Carroll, 1988).

Though no clear consensus exists, several researchers have compared the relative usefulness of one technique over another in qualitative assessment of biodegradation capacity. In isolating hydrocarbon degraders from an environmental sample, Wyndham and Costerton (1981) found that plate count enumerations more accurately reflected in situ microbial activity than direct counts and MPN techniques. Conversely, Atlas (1981) reports better enumerations using MPN procedures over plate counts. The ratio of total organisms to hydrocarbon degraders has also been used as a measure of a consortium's fitness to degrade hydrocarbon contamination. Because of their selectiveness for single species, DNA probes are useful mainly in the laboratory or in a field situation where the fate of an individual organism is tracked. These probes are sensitive to 4.3×10^4 cells g^{-1} soil (Holben et al., 1988) and can be constructed to test for the genes encoding hydrocarbon-catabolic pathways (Trevors, 1985).

An important consideration for laboratory and particularly field work is the enumeration of sessile versus planktonic bacteria. Attached bacteria are normally more abundant than free-floating cells (Jensen et al., 1986); however, many in situ sampling techniques neglect attached organisms because of the difficulty of retrieving subsurface soil samples from existing monitoring wells.

Microbial enumeration techniques are similar for all scales of observation, but the accuracy of results may not be. Commonly used microbial enumeration techniques are listed in Table 3. Methods that depend on bacterial growth on a laboratory medium such as agar or MPN buffer may not accurately measure the in situ community due to selective growth on the media (Slater and Lovatt, 1984). The laboratory medium may be free from selective pressures such as contaminant interactions, oxygen depletion, and consortia symbiosis, which exert a strong influence on adaptation and survival in the field. Caution should therefore be exercised in applying laboratory-scale results to the field-scale. The research consensus is that microscale tests can accurately determine the presence, but not the activity, of hydrocarbon-degrading organisms at the field scale.

Table 3. Microbial enumeration methods.

Method	Description	Advantages/Disadvantages
Heterotrophic plate counts	Aqueous suspension is inoculated onto sterile agar either while still liquid (pour plate) or after agar has dried (spread plate).	*Adv:* Easy to perform, bacterial identification possible. *Disadv:* Counts only organisms viable on media (agar) used. Incubation takes 48–72 hours. Faster growers overshadow slower growers.
Hydrocarbon de-grader plate counts	Similar to heterotrophic plate counts, except specific hydrocarbon (or group of hydrocarbons) is only carbon source for growth.	*Adv:* Determines presence and quantity of bacteria capable of degrading contaminant mix of concern. *Disadv:* Introduction of more easily degraded carbon source in trace amounts; may allow growth of nontarget organisms.
Epifluorescence direct counts	Sample is stained with one of several fluorescent stains. Counts are then made directly under the microscope.	*Adv:* Fast—no waiting period. Viability-related phenomena are excluded. *Disadv:* Difficult to distinguish cells from debris (especially in soil samples). Differentiation not possible on basis of taxonomy or activity.
Most probable number techniques	Samples are incubated with reaction specific indicators. Color change, gas, or precipitate formation indicates the presence/activity level of target culture. Bacterial density is predicted by the observed distribution of positive and negative results.	*Adv:* Offers quantitative evidence for bacterial activity. Useful for enumeration of electron acceptor specific cultures such as sulfate-reducing bacteria. *Disadv:* Large number of sample portions required for significance of results. Generally low precision. Labor-intensive.

(continued)

Table 3. (continued).

Method	Description	Advantages/Disadvantages
Oligonucleotide probes	Labeled (fluorochrome) DNA or RNA complementary sequences for a target sequence in the organism are added to an environmental sample. Observe cells by microscopy for label (fluorochrome or others).	*Adv:* Nonculturable organisms can be detected. Specific species can be enumerated. *Disadv:* Probe is specific for a certain sequence. Penetration of probe into cell may not be complete. Observation by microscopy may be difficult due to debris (see direct counts).
Lipid assays	Lipids are extracted from pure culture and analyzed in a gas chromatograph. Chromatogram is compared to standard library for species match.	*Adv:* Relatively fast. Useful in tracking known organisms in field or laboratory systems. *Disadv:* Not useful in quantifying unknown organisms. Lipid concentration may change with growth conditions. Organisms must be isolated and purified prior to use.
Metabolic indicators	Tests utilize the capacity of a specific species to make a measurable color or size change due to the degradation of a specific substrate.	*Adv:* Visual evidence of microbial activity. *Disadv:* Not useful in field applications. Only useful for single species. An activity indicator, not always applicable to enumeration.
Radioisotopic methods	Uptake of radioactive isotopes (such as tritiated thymidine or a ^{14}C-labeled carbon source) by microorganisms is measured as indicative of microbial activity.	*Adv:* Measures the growth on a specific substrate. Relatively fast. Allows quantitation of microbial kinetics. *Disadv:* Analysis limited to labeled substrate, cannot be released into the environment.

Reaction Kinetics

The rate of contaminant removal is of prime significance in field situations because it determines the practical feasibility and ultimate cost of a treatment scheme. When biodegradation is the primary removal mechanism, the kinetics of the microbially catalyzed reactions determine how long a microbial community will take to degrade a contaminant to acceptable levels. According to Larson (1984), kinetic data are necessary for application of bioremediation computer models, extrapolation of degradation results from one scale to another, and objective evaluation of the persistence of a chemical in a particular environment.

Previous research has focused on the determination of single species or consortia kinetics for growth on a single contaminant compound in aqueous or soil slurry microcosms. Results show wide variation in removal rates for like compounds using different microbial inocula (Figure 1). Most kinetic data are reported either in the form of mass of contaminant removed per biotic mass per day (e.g., mg toluene/mg cells/day) or mass of contaminant removed per volume per day (mg toluene/liter/day). Where biotic mass is not reported, comparison of rates is compromised by not knowing the amount of active biomass responsible for the degradation rate. In the case of alkylbenzenes, aqueous microcosms from a

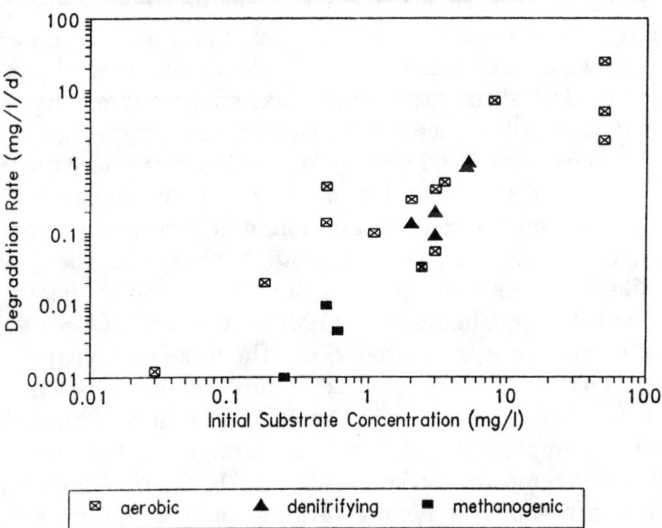

Figure 1. Observed degradation rate dependence of aromatic hydrocarbons (BTEX) on initial substrate concentration under aerobic, denitrifying, and methanogenic conditions. All points were measured in the laboratory.

landfill leachate plume degraded toluene at a rate of 1.2 μg l^{-1} d^{-1} (Armstrong et al., 1991) whereas soil slurry microcosms have shown toluene degradation varying from 30 μg l^{-1} d^{-1} (Barker et al., 1987) to 50 μg l^{-1} d^{-1} (Alvarez and Vogel, 1991; Hutchins, 1991). The variation between these observations is probably due in part to inocula concentration differences, but such data underscore the need for site-specific experimentation with the aquifer material in question. These data also indicate that microcosms that only consider the biota suspended in the groundwater may seriously underestimate the biodegradation potential of an aquifer. In evaluating the importance of attached (biofilm) versus suspended (planktonic) microbial growth, Jensen et al. (1986) reported that naphthalene was degraded by suspended cells about four times faster than attached cells (biofilm thickness is not reported). While most soil systems are thought to contain many more attached than suspended cells, this indicates the suspended cells may be disproportionately responsible for contaminant removal.

Reaction order is also subject to some variability, though most researchers show biodegradation following either zero-, first-, or some fractional order dependence (Jensen et al., 1986; Barker et al., 1987; Major et al., 1988; Hutchins, 1991; Hutchins et al., 1991). Jensen et al. (1986) reported a very low half-saturation coefficient for their consortia ($K_s = 0.1$–$1.0\,\mu$g l^{-1}). If this is representative of in situ consortia degrading aromatics, then zero-order kinetics should be expected for these compounds at the concentrations commonly found at contaminated sites. Figure 2 shows a compilation of BTEX degradation rates plotted as a function of the initial substrate concentration. Separate plots are shown for lab- and field-scale remediation systems. The roughly linear progression of lab-scale rates is predicted by Monod kinetics where the substrate concentration S is on the order of the half-saturation coefficient K_s, however, there is an apparent stagnation of field-scale degradation rates as concentration increases. This suggests that reaction order in situ may be controlled by some other process, such as oxygen availability in the aquifer. For example, Chiang et al. (1989a) observed field-scale degradation kinetics to be oxygen-limited. Since the process of oxygen diffusion to the reaction site is not represented in most microcosm studies, it must be somehow incorporated into the scaleup procedure. Plume geometry may also be an important factor in determining both the rate and order of contaminant removal (Barker et al., 1987). Where plumes have little surface area exposed to oxygenated groundwater, a relatively small fraction of the plume may contain active microorganisms. Alternate electron acceptors that exhibit higher water solubilities, such as nitrate, will be more available than oxygen under transport-limiting conditions if they

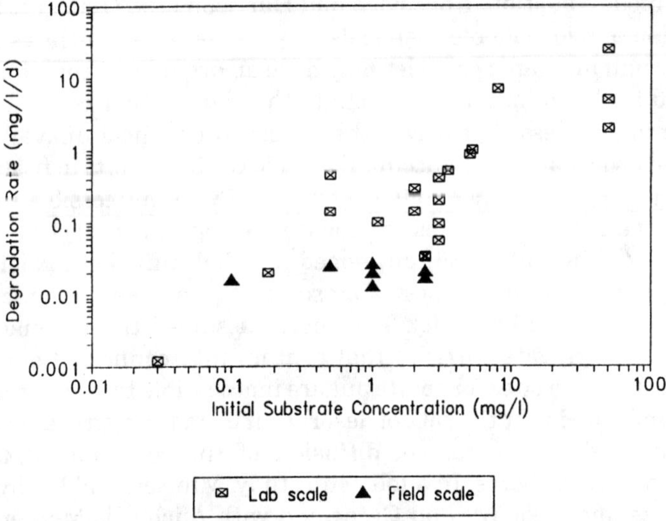

Figure 2. Observed degradation rate dependence of aromatic hydrocarbons (BTEX) on initial substrate concentration under laboratory and field measurement conditions. All points were measured under aerobic or denitrifying conditions.

are supplied in adequate quantities. The extrapolation of kinetic rates between observational scales is necessary for scaleup. The potential for major discrepancies in these rates is high, particularly where contamination exists in a concentrated plume or a nonaqueous-phase liquid. Successes in micro- to macroscale rate prediction have been achieved (Chiang et al., 1989a) through consideration of as many of the macroscale factors as possible.

MESOSCALE PHENOMENA

Influence of Sorption

Many organic contaminants are hydrophobic and tend to sorb onto soil such that only a small fraction of the compound may actually be in the bulk water phase. Most evidence indicates that biotransformation occurs via the liquid phase, and consequently, processes such as sorption and volatilization, which reduce the aqueous-phase concentration, tend to reduce the biotransformation rate (Bouwer et al., 1994).

The issue of sorbed solute bioavailability has been widely investigated. Steen et al. (1979) investigated the desorption/degradation kinetics of di-*n*-butyl phthalate and concluded that sorbed contaminant was unavailable for biodegradation. Knezovich and Harrison (1987) suggest that the adsorption of hydrophobic chemicals on pure inorganic surfaces is negligible, though inorganic particles may have an organic coating that allows significant adsorption. They conclude that adsorption is principally a partitioning process that leaves the adsorbed chemical unavailable for microbial degradation. The accumulation of contaminants in fissures and cavities within subsurface soils may render them inaccessible to microorganisms and their enzymes as proposed by Ogram et al. (1985).

Wu and Gschwend (1986) developed a radial diffusion model, which quantified desorption kinetics using solute $K_{o/w}$ and soil organic matter content, density, and porosity. This model assumed the soil matrix was made up of aggregate particles that contain interconnected micropores that offer sorption sites for solute but are inaccessible to microorganisms. Results indicated that the rate of desorption from the aggregate particles was controlled by the rate of diffusion of the contaminant through intraparticle micropores. In a followup study, Mihelcic and Luthy (1991) coupled the model of Wu and Gschwend with Michaelis-Menten biodegradation kinetics to predict the desorption and biodegradation of naphthalene in a soil system with aggregate micropores. Results indicated that the naphthalene adsorption/desorption process was rapid compared to biodegradation and therefore did not limit substrate utilization. Using a similar model, Rijnaarts et al. (1990) modeled biodegradation in a mixed soil system that was desorption limited. In this case, contaminant diffusion through the soil micropores was found to limit the biodegradation rate. Such contradictory results are not unexpected considering the potential for differences in the systems studied. Many contaminant, soil, and microbial consortium variables affect the desorption/biodegradation process. Solute hydrophobicity, solute diffusivity, equilibrium partitioning coefficient, soil organic matter content, soil particle geometry, soil porosity, soil/water ratio, microbial kinetics, and the presence or absence of a biofilm may all influence either the desorption or biodegradation rates.

Another possible reason for contradictory conclusions in desorption limitation experiments may be the description of adsorption/desorption as a two-stage process. In studying toluene degradation in an organic soil, Robinson et al. (1990) reported that the first stage reflects the rapid adsorption/desorption of approximately 90% of the toluene, while the remaining 10% adsorbs/desorbs more slowly. The significance of the two apparent rates is that the microbial degradation reaction was limited only by the slow desorption rate.

In a comprehensive review of sorptive phenomena, Pignatello (1989) further suggests that the interactions of (1) soil organic matter amount and composition, (2) contaminant hydrophobicity, and (3) the length of time the contaminant has been in place, make generalizations about sorptive behavior less certain. Organic polymers within the soil attract organic contaminants via their nonpolar gel structure. Once within the polymer matrix, contaminants may bond (covalent or hydrogen bonding) to specific sites or may diffuse through micropores in the gel. The longer the exposure time to the contaminant, the greater the probability that contaminant bonding and deep-matrix diffusion will occur. Pignatello suggests that, if such tightly bound contaminants desorb at all, it is at a rate that certainly limits biodegradation. In studies with naphthalene, Guerin and Boyd (1992) found desorption rate limitations to be organism-specific, which suggests that the bioavailability of contaminants may also vary, depending on the consortia present.

Progress in the area of sorptive behavior has been considerable, however, the above citations indicate the need for further description of the soil-, organism-, and contaminant-specific interactions of sorbed solute bioavailability and desorption kinetics. In a recent review, Fogarty and Tuovinen (1991) concluded that, in general, sorption isotherms in relation to contaminant bioavailability to microorganisms are poorly understood and have yet to be experimentally addressed.

Interphase Transport

Organic contaminants in a soil system can exist in four phases: (1) dissolved in a nonaqueous phase liquid (NAPL) either free-floating on the water table (LNAPL), below the water table (DNAPL), or bound to soil particles; (2) dissolved in soil water (aqueous phase); (3) vapor phase existing in soil pores in the unsaturated zone, and (4) adsorbed to soil particles or soil organic matter. Most petroleum products exhibit low solubilities in water, facilitating the formation of free-floating LNAPL (provided contaminant specific gravity < 1) which moves slowly down gradient, but acts essentially as a stationary source of contaminant recharge to the groundwater. If soil conditions permit, the NAPL will also act as a source of contamination for the vadose zone, as the more volatile hydrocarbon fractions move into the vapor phase. Biodegradation of subsurface hydrocarbons occurs in the aqueous phase, either below the capillary fringe or in hydrated microsites within the vadose zone (Morgan and Watkinson, 1989). Hydrocarbon movement from the NAPL to the aqueous phase is therefore a prerequisite for biodegradation to occur. If this dissolution process occurs at a slower rate than site microorganisms are capable of utilizing the contaminant, interphase transport will limit biodegradation. Several researchers have noted such limitations for low solubility compounds such as long chain alkanes (C_{18}) and naphthalene at the microscale (Thomas et al., 1986; Miller and Bartha, 1989). Lower molecular weight hydrocarbons and alkylbenzenes typically exhibit much higher solubilities, which facilitates dissolution and may preclude this type of limitation.

Mass transport limitations evident at the microscale may be magnified in the field, particularly in soils with fine texture and low porosity. Advective transport may be minimal, further slowing the dissolution process. Macroscale features such as soil type and porosity will also affect the NAPL surface area/volume ratio. As dissolution is a linear function of interphase surface area, such features may determine whether dissolution limits biodegradation (Spain and Van Veld, 1983; Wilson et al., 1989).

Accurate quantification of the nonaqueous phase is necessary both for electron acceptor demand calculations and for estimating the contaminant recharge to the groundwater and vapor phases (Wilson et al., 1989). Where macroscale research has been performed, results are predictably site dependent. Octadecane mineralization has been observed to be dissolution controlled (Thomas et al., 1986), while naphthalene dissolution has been found to be faster than biodegradation (Foght and Westlake, 1982).

Efforts to increase the rate of contaminant dissolution from a NAPL to the aqueous phase have centered on the addition of surfactants to soil systems. Surfactants have been used extensively in marine oil spill treatment for many years and have recently been used successfully for terrestrial applications as well (Hoeppel et al., 1991). Site microbial communities may also be responsible for the production of surfactants to speed contaminant dissolution (Bossert and Bartha, 1984). Ammonium nitrate and peptone have been used to stimulate microbial production of surfactants to release a recoverable oil emulsion from the columns (Vanloocke et al., 1979).

Whether NAPL dissolution limits biodegradation can be measured in the laboratory, but the result obtained may not be representative of the field-scale unless the site biotic, soil, and NAPL characteristics were fully considered in the microscale experimentation.

MACROSCALE CONCERNS

Advective-Dispersive Transport Limitations

Mass transport by advection and dispersion may have a significant effect on contaminant distribution and substrate/electron acceptor availability to microorganisms. Although controlled experimentation in the field is difficult, there are certain measurements that may reveal how mass transport can limit the rate of in situ biodegradation. For example, contaminant degradation rates at the edge of a subsurface plume have been observed to be much higher than those in the plume center due to the maintenance of aerobic conditions at the edge (Chiang et al., 1989a; Morgan and Watkinson, 1989). Such transport limitations apply to nutrients and other electron acceptors as well but are most often observed with oxygen due to its common stoichiometric limitation. This is usually not observed at the microscale, where mass transport limitations are often minimized or eliminated by design.

The overall effect of increasing advective-dispersive transport within an aquifer is to enhance the process of mixing contaminants, electron acceptor, and cells capable of contaminant biodegradation. Only where all three of these constituents are present concurrently can biodegradation occur. Lee et al. (1988) observed that this mixing is frequently confounded in aquifers in which contaminants adhere to solids within areas of low advective flow, while dissolved oxygen and nutrients flow within adjacent higher permeability zones. The influence of advective and dispersive transport are now routinely included in biodegradation models. MacQuarrie and Sudicky (1990) have illustrated that variation in aquifer physical characteristics such as dispersivity can significantly affect advective flow and rates of biotransformation over comparatively small spatial separations.

The Impact of Spatial Heterogeneity

Spatial heterogeneity at contaminated field sites can significantly influence contaminant movement and rate of degradation. Subsurface properties that are subject to significant spatial variation include porosity, permeability, degree of microbial colonization, and chemical properties such as nutrient and electron acceptor conditions. If significant heterogeneities exist, it is possible that different transport phenomena may limit the rate and extent of biodegradation across the site. For example, an often encountered soil textural heterogeneity is the presence of clay lenses in otherwise permeable sandy soils, which can reduce the local permeability by up to five orders of magnitude (Todd, 1980). As mentioned above, drastically reduced groundwater flow velocities generally make diffusive transport predominant within clay lenses. This may cause such lenses to act as reservoirs for contaminants, replenishing groundwater concentrations for long periods after more permeable zones have been cleansed. In such a case, biotransformation within the clay may be limited by molecular diffusion while the higher permeability zones are advection limited. Evidence indicating contaminant persistence in areas of lower advective flow velocity have been noted in sandy soils with underlying clay lenses (Sutton and Barker, 1985).

Contaminants also may exhibit significant spatial heterogeneity with regard to flow channels and contaminant phase. An example is the weathering process of a spilled gasoline or crude oil, where the lighter hydrocarbon fraction may volatilize quickly after introduction, leaving progressively heavier fractions down the path of migration. In addition, dispersive activity and the progression of biodegradation from the edges into the center of a plume usually causes the plume to contain significantly higher (three or more orders of magnitude) contaminant concentrations in its center than at the edges, causing differences in the rate degradation proceeds. Where biodegrading microorganisms are primarily fixed to aquifer solids, the residence time of the contaminated groundwater in the system must be sufficient to allow the reaction to proceed to completion. Aquifer physical heterogeneities that affect groundwater flow rates therefore may impact degradation rates.

The effects of nutrient conditions in soil and aquifer system petroleum degradation has been studied and reviewed extensively; however, consideration of the impact of spatial heterogeneities on nutrient availability has not. This issue is important mainly in nutrient deficient aquifers where the addition of nutrients is conducted via injection or surface application. Added nutrients must flow to the site of active microorganisms and therefore are subject to transport limitations imposed by aquifer heterogeneities.

Also relevant to the discussion of spatial heterogeneities is the phenomena of localized plugging of a subsurface formation, creating zones of very low permeability and significantly reduced transport of oxygen and nutrients. Such plugging has been observed from a variety of unrelated sources, such as the formation of gas bubbles, the accumulation of microbial biomass, or chemical precipitation within soil pores. Microbial plugging of aquifers has been convincingly demonstrated in laboratory systems (Taylor and Jaffe 1990; Cunningham et al., 1991) wherein observed porosity and permeability decreases of up to 96% and 98%, respectively, occurred in 1-mm glass spheres when microbial growth conditions were optimized. Several researchers have found that bacteria injected into model core systems deposited close to the point of injection. Copious exopolysaccharide production of attached organisms subsequently plugged the core at the inlet point and caused permeability reductions ranging from 70–99% (Shaw et al., 1985; Raiders et al., 1986; Torbati et al., 1986). Using starved bacteria, MacLeod et al. (1988) found injected cells to be uniformly distributed in artificial rock cores and subsequently found significant permeability reductions when growth conditions were enhanced (Lappin-Scott et al., 1988).

Hydrogen peroxide is sometimes added to augment the dissolved oxygen supply in O_2 deficient groundwater. H_2O_2 is more water-soluble than molecular oxygen, allowing much greater oxygen transport to the subsurface; however, H_2O_2 rapidly reacts to form elemental oxygen and water. As discussed earlier, liberated oxygen may form bubbles in the aquifer, decreasing permeability and preventing transport (Thomas et al., 1987; Wilson and Ward, 1987; Pardieck et al., 1992). Aquifer plugging due to precipitation of inorganic insoluble salts is well documented, particularly where biodegradation in the aquifer leads to oxygen-poor conditions. Subsequent lower pH and higher alkalinity may result in increased solubility of redox sensitive elements, such as iron and manganese. The addition of oxygen- and phosphate-enriched nutrient solutions may alter the soil-groundwater chemical equilibrium to the point of precipitation of Fe, Mn, and Ca phosphates (Aggarwal et al., 1991). Biomass buildup and inorganic precipitation may occur concurrently, exacerbating plugging problems. Microbial production of extracellular polysaccharides may entrap precipitates, further decreasing aquifer permeability.

Estimating Bioremediation Rates

Contaminant disappearance from a field site may occur for a variety of reasons, including volatilization, off-site migration, sorption onto soil particles, and abiotic transformation, thereby complicating the measurement of in situ bioremediation rates. The highly controlled conditions of the microcosm allow substantiation of biodegradation, but the addition of unmeasurable contaminant sinks in field-scale systems have led practitioners to use a variety of other methods to corroborate contaminant disappearance data. Madsen (1991) suggests an approach to proving biodegradation that integrates chemical factors such as selective disappearance of microbiologically labeled isotopes or nonreactive tracers with biotic factors such as increased activity in degrader microorganisms or predators that feed upon them.

Many researchers have relied on contaminant disappearance and concurrent hydrocarbon degrader population increases to prove biodegradation. The most common laboratory technique is the use of ^{14}C-labeled contaminants with a known microbial consortia. Microcosms using native microbial populations have shown good correlation with the disappearance of crude oil (Mulkins-Phillips and Stewart, 1974), jet fuel (Song and Bartha, 1990; Hutchins et al., 1991), aromatics (Arvin et al., 1989;

Karlson and Frankenberger, 1989), and phenolics (Spain et al., 1980) under aerobic and anaerobic conditions. Several researchers have suggested the measurement of protozoan predators as a useful indicator of bacterial vigor (Rogerson and Berger, 1981; Madsen et al., 1991).

The transition from the laboratory to the field necessarily involves the use of several concurrent confirmatory tools to prove bioremediation is responsible for contaminant removal. Several field-scale studies have shown both increased bacterial numbers and contaminant removal in crude oil; however, more macroscale work has focused on linking contaminant removal with electron acceptor disappearance (Barker et al., 1987; Chiang et al., 1989b). The choice of methods to monitor and verify bioremediation is dependent on the level of control that may be imposed upon the experimental system. Field sites will always involve uncertainty; the challenge, therefore, is to prove bioremediation through a battery of methods that, taken together, offer compelling evidence that microorganisms are responsible for the observed contaminant disappearance. Several methods demonstrating bioremediation are listed in Table 4.

The Influence of Temperature

Although temperature is usually beyond the control of the practitioner, it remains valuable to understand how diurnal and seasonal changes in temperature impact bioremediation performance. In addition, in certain cases, such as covered land farm operations, the practitioner may be able to exert some influence on the temperature.

Temperature can be expected to influence process rates at every scale of observation; however, such factors as microbial origin and seasonal variation have been shown to be important considerations when evaluating the effects of temperature in laboratory studies. It is well documented that mixed culture degradation of petroleum hydrocarbons is most efficient in a temperature range from 20–30°C (Dibble and Bartha, 1979; Song et al., 1990; Atlas, 1981). Bacterial consortia freshly removed from the environment, however, utilize organics fastest when incubated at the ambient temperature from which they were removed (Morgan and Watkinson, 1989; Atlas, 1981). Furthermore, biotic samples from the same sites taken at different times of year show differences in species predominance. This indicates that psychrophillic organisms may predominate in surface soils in winter months, while not being measurable in the summer. Several researchers have reported the maintenance of contaminant removal at very low temperatures in situ (Morgan and Watkinson, 1989), though microbial activity does appear to cease in frozen soils.

Table 4. Methods of demonstrating biodegradation.

Demonstration Method	Reliability/Problems	References
Contaminant attenuation	*Lab*: Reliable, depending on controls. Volatilization often a problem. *Field*: Not usually reliable, often many uncontrolled abiotic loss mechanisms.	Dibble, 1979; Barles, 1979; Kilbane, 1983; many others
Electron acceptor loss	*Lab & Field*: May indicate noncontaminant organic matter is degrading (particularly in the presence of enhanced N, P). Adequate controls often not possible in the field.	Zeyer, 1986; Hutchins, 1991; Chiang, 1989
$^{14}CO_2$ capture	*Lab*: Usually the most reliable indicator of mineralization if good controls are used. *Field*: Not practical for use in field situation—capture of CO_2 very difficult.	Pfaender, 1982; Spain, 1983; Kuhn, 1988; many others
Growth of contaminant degrading microbes	*Lab & Field*: Offers reliable supporting evidence of biodegradation, though not conclusive proof alone, esp. at field scale.	Spain, 1983; Wyndham, 1981
Nutrient removal (N, P)	*Lab*: Reliable indicator of biotic activity where no completing substrates and adequate controls are used. *Field*: Supporting evidence only. Not reliable where alternate substrates are available.	Odu, 1987; Jamison, 1975; Swindoll, 1988
Intermediate metabolite production	*Lab & Field*: Reliable indicator of biodegradation. Controls to insure metabolite was not indigenous to the site are necessary.	Dasappa, 1991
Degradable/nondegradable contaminant ratios	*Lab & Field*: Most common is the C_{17}/pristane ratio. Reliability hinges on pristane persistence in medium.	Pritchard, 1991; Wilson, 1989
Growth of noncontaminant degrading microbes	*Lab & Field*: Growth of protozoan grazers has been used as an indirect measure of bacterial growth. Reliability compromised by protozoan feeding on noncontaminant degrading bacteria.	Madsen, 1991

The net effect of temperature on the overall rate of bioremediation depends on which processes are limiting. To illustrate the possible effects of temperature on bioremediation, Figure 3 compares the predicted relative temperature dependency of microbial growth, molecular diffusion, and advective solute transport. Increasing the temperature generally increases process rates of growth, diffusion, and advection. It is therefore important to consider not only the response of the site consortium to temperature changes but also the effects temperature changes may have on other phenomena important to the bioremediation reaction.

Modeling in situ Bioremediation

Numerous mathematical models have been proposed with application to in situ bioremediation. Such models integrate processes occurring at different scales of observation to predict spatial and temporal patterns of contaminant distribution. Table 5 provides a summary of transport/reaction models that are potentially useful for bioremediation applications.

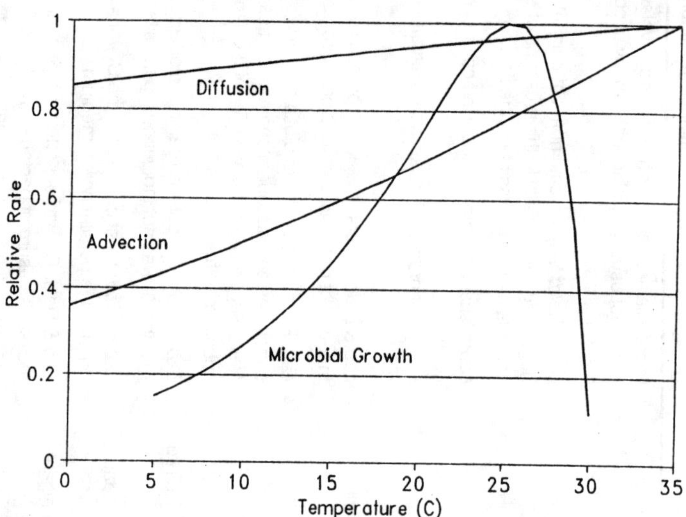

Figure 3. Temperature dependency of constituent processes of bioremediation. Process rates, normalized to the maximum rate in the temperature interval shown, for advection, diffusion, and microbial growth are plotted as a function of temperature. These predictions are based on the following assumptions: advective flux of oxygen $\propto$ (advective velocity) $\times$ (oxygen solubility). Velocity $\propto$ (density)/(viscosity). Diffusion coefficient $\propto$ (viscosity)/(temperature). Net microbial growth rate was modeled as the sum of growth and decay terms.

Table 5. *Phenomenological models for in situ bioremediation.*

Authors	Year	Transport[a]	Kinetics[b]	Species[c]	Sorption[d]	Phases[e]	Data	Other
Van Genuchten	1981	1D	1	S	E	D	—	
Enfield et al.	1982	1D	1	S	E	D	Field	
Sykes et al.	1982	2D	M	SM	—	D	Field	
Borden et al.	1984	2D	DM	SEM	E	D	—	
Borden and Bedient	1986	2D	DM, F	SEM	E	DG	—	f,h
Borden et al.	1986	2D	F	SE	E	D	Field	f
Wheeler et al.	1986	2D	DM	SEM	E	D	—	h
Rifai et al.	1988	2D	F	SE	E	D	Field	f
Chiang et al.	1989	2D	F	SE	E	D	Field	
Corapcioglu and Haridas	1984	3D	M	SM	E	D	—	g,h
Corapcioglu and Haridas	1985	2D	M	SM	E	D	—	
Angelakis and Rolston	1985	1D	1	S	E	DIG	Lab	
Huyakorn et al.	1985	2D	1	S	E	D	Field	
Molz et al.	1986	1D	DM	SEM	E	D	—	i
Molz and Widdowson	1988	2D	DM	SEM	E	D	—	
Widdowson et al.	1988	1D	TM	SEM	E	D	—	i
Wagenet and Hutson	1986	1D	1	S	E	D	Field	n
Dawson, Wheeler and Chiang et al.	1986–1992	2D	DM	SEM	E	D	—	o
Wheeler et al.	1992	3D	DM	SEM	E	D	—	o
Corapcioglu and Baehr	1987	3D	F	SE	E	DIG	—	g,j
Baehr and Corapcioglu	1987	1D	F	SE	E	DIG	—	
Domenico	1987	3D	1	S	—	D	—	
Kosson et al.	1987	1D	M	SM	E	D	Lab	
Srinivasan and Mercer	1988	1D	DM	S	E	D	Lab	

(continued)

653

Table 5. (continued).

Authors	Year	Transport[a]	Kinetics[b]	Species[c]	Sorption[d]	Phases[e]	Data	Other
Klečka et al.	1990	1D	DM	S	E	D	—	
Baveye and Valocchi	1989	1D	M	SM	—	D	—	g,i
Celia et al.	1989	1D	gen	SEM	E	D	—	g,j,l
Kindred and Celia	1989	1D	M, C, I	SEM	E	D	—	j,m
Tim and Mostaghimi	1989	1D	1	S	E	D	Lab	j
Illangasekare and Döll	1989	2D	1	S	NE	DI	—	k
Van Genuchten and Wagenet	1989	1D	1	S	NE	D	—	k
MacQuarrie et al.	1990	2D	DM	SEM	E	D	Lab	moment
MacQuarrie and Sudicky	1990	2D	DM	SEM	E	D	Field	o
Semprini and McCarty	1991	1D	DM, C	SEM	NE	D	Field	
Brusseau et al.	1992	1D	1	S	NE	D	Lab	

[a]1D, 2D, 3D—spatial dimensionality.
[b]1—first-order; M—Monod; DM—double Monod; TM—triple Monod; F—instantaneously fast reaction between substrate and electron acceptor; I—inhibition; C—cometabolism.
[c]S—substrate (contaminant); E—electron acceptor or nutrient; M—microorganisms.
[d]E—equilibrium sorption; NE—nonequilibrium sorption; — = not considered.
[e]D—dissolved; I—insoluble; G—gaseous.
[f]Oxygen transfer from the unsaturated to saturated zone.
[g]Theoretical development.
[h]Microbial transport.
[i]External mass transfer between bulk liquid and biological phases.
[j]Multiple substrates or metabolites.
[k]Two-zone model.
[l]Multiple microorganisms.
[m]Internal mass transfer resistance within a biofilm.
[n]Water loss by plant transpiration.
[o]Spatial heterogeneity in hydraulic conductivity.

The mass transport component of these models is based on advection and dispersion of a dissolved contaminant. Equilibrium partitioning of the contaminant between the aqueous and solid phases is usually assumed. Reaction is handled by a wide variety of rate expressions, the most common being first-order, Monod, and double Monod kinetics. About half of the models combine material balances on the electron acceptor and biomass in addition to the substrate. Those models that only include a substrate mass balance lack the ability to describe rate limitation by an electron acceptor or to account for the change in intrinsic reaction rate that may accompany biomass growth or decay. Interphase transport and/or nonequilibrium sorption are contained in only about 20% of the models shown in Table 5. Additional transport phenomena that have been incorporated in models include diffusive transport between the bulk liquid and a surface-associated biological phase (Molz et al., 1986; Widdowson et al., 1988; Baveye and Valocchi, 1989), oxygen transfer from the unsaturated to the saturated zone (Borden and Bedient, 1986; Borden et al., 1986; Rifai et al., 1988), hydrologic spatial heterogeneity (Dawson et al., 1986; Wheeler and Dawson, 1987; Chiang et al., 1989b; MacQuarrie and Sudicky, 1990; Dawson and Wheeler, 1992), and microbial transport (Corapcioglu and Haridas, 1984; Borden and Bedient, 1986; Wheeler et al., 1987).

Predictive uncertainty of bioremediation models has not been well established. Only a few of the models in Table 5 involve comparison of simulations with field measured gradients in space (Sykes et al., 1982; Enfield et al., 1982; Huyakorn et al., 1985; Borden et al., 1986; Rifai et al., 1988) or time (Wagenet and Hutson, 1986; MacQuarrie and Sudicky, 1990; Semprini and McCarty, 1991a, 1991b). Only the BIOPLUME model has been applied to more than one site (Borden et al., 1986; Rifai et al., 1988; Chiang et al., 1989b). There is clearly a need for additional model evaluation studies in which a priori model predictions are compared with subsequent field observations of bioremediation phenomena. These studies are needed to define predictive uncertainty and to confirm that model structure incorporates all essential processes.

SUMMARY

Review of current literature clearly points out the complexities faced by those who must make decisions concerning the feasibility of in situ bioremediation of groundwater and soil. It is important for decision makers to be aware of all of the important phenomena that influence bioremediation, as well as the way in which these phenomena interact.

This chapter has identified important rate-controlling phenomena and classified them according to the scale at which they occur. At the microscale, microbial activity (i.e., kinetic reaction rate) is the primary concern, while at the mesoscale, the partitioning processes of sorption and interphase transport are of primary interest. At the macroscale, flow-related processes of advection and dispersion, along with the effects of field heterogeneities, become important as phenomena that can influence, if not control, the overall rate and extent of in situ bioremediation at a given field site.

All of these phenomena must be considered in the decision-making process. It must be determined which phenomena limit the actual rate of biotransformation on a case by case basis. For example, if the concentration of active biomass is too low, the (microscale) kinetic reaction rate will limit the overall rate of contaminant bioremediation. If soil organic carbon content is high, the rate of (mesoscale) desorption of contaminant from within the soil matrix may limit bioremediation. Alternatively, if low pore velocities prevail (macroscale), advection/dispersion may be the rate-limiting processes. It is also clear that heterogeneities in the field can substantially complicate the issue, in that different phenomena may limit the biotransformation, depending on location at a given site. The assessment of feasibility of an in situ bioremediation project is dominated by the need to identify and estimate the appropriate rate-controlling phenomena.

The approach to scaleup should therefore include the following steps: (1) a thorough analysis of all relevant phenomena to determine which will limit contaminant biotransformation rate for prevailing subsurface environmental conditions (because the subsurface environment may vary considerably across site due to heterogeneities, this analysis may conclude that several different phenomena are limiting depending on location); (2) once a particular remediation strategy is proposed (i.e., pump-and-treat or nutrient injection/recovery), the situation must be reassessed to determine a new set of probable biotransformation rates and the phenomena that limit; and (3) the outcome of steps 1 and 2 can be used to assess project feasibility.

While the experimental protocols for bench-, pilot-, and field-scale treatability are rapidly emerging, there exist very few case studies in which predicted rates are actually compared with observed rates in the field. Evaluation of the uncertainty with which a priori predictions of bioremediation processes compare with subsequent field observations needs to be carried out in order to develop improved bioremediation process engineering methods.

ACKNOWLEDGEMENTS

This work was supported by Cooperative Agreement EEC-8907039 between the National Science Foundation and Montana State University. Funding was also provided by Conoco Inc. and by agreement R-815709 through the Hazardous Substance Research Center for U.S. EPA regions 7 and 8 at Kansas State University.

REFERENCES

Aggarwal, P. K., Means, J. L. and Hinchee, R. E., 1991. Formulation of nutrient solutions for in situ bioremediation. In: R. E. Hinchee and R. F. Olfenbuttel (eds.), *In Situ Bioreclamation*. Butterworth-Heinemann, Stoneham, MA, pp. 51–66.

Angelakis, A. N. and Rolston, D. E. 1985. Transient movement and transformation of carbon species in soil during wastewater application. *Water Resour. Res.*, 21:1141.

Alvarez, P. J. J. and Vogel, T. M., 1991. Substrate interactions of benzene, toluene, and para-xylene during microbial degradation by pure cultures and mixed culture aquifer slurries. *Appl. Environ. Micro.*, 57:2981–2985.

Armstrong, A. Q., Hodson, R. E., Hwang, H. M. and Lewis, D. L., 1991. Environmental factors affecting toluene degradation in ground water at a hazardous waste site. *Environ. Toxic. Chem.*, 10:147–158.

Arvin, E., Jensen, B. and Gundersen, A. T., 1989. Substrate interactions during aerobic biodegradation of benzene. *Appl. Environ. Micro.*, 55:3221–3225.

Atlas, R. M., 1981. Microbial degradation of petroleum hydrocarbons: an environmental perspective. *Microbiol. Rev.*, 45:180–209.

Baehr, A. L. and Corapcioglu, M. Y., 1987. A compositional multiphase model for groundwater contamination by petroleum products 2. Numerical solution. *Water Resour. Res.*, 23:201–213.

Barles, R. W., Daughton, C. G. and Hsieh, D. P. H., 1979. Accelerated parthion degradation in soil inoculated with acclimated bacteria under field conditions. *Arch. Environm. Contam. Toxicol.*, 8:647–660.

Barker, J. F., Patrick G. C. and Major, D. W., 1987. Natural attenuation of aromatic hydrocarbons in a shallow sand aquifer. *GWMR* Winter:64–71.

Baveye, P. and Valocchi, A., 1989. An evaluation of mathematical models of the transport of biologically reacting solutes in saturated soils and aquifers. *Wat. Resour. Res.* 25:1413–1421.

Block, R. N., Bishop, M. and Clark, T. P., 1989. Biological remediation of petroleum hydrocarbons. *Proc: Sixth National Conference on Hazardous Wastes and Hazardous Materials*, Hazardous Materials Control Institute, New Orleans, LA.

Borden, R. C. and Bedient, P. B., 1986. Transport of dissolved hydrocarbons influenced by reaeration and oxygen limited biodegradation. I. Theoretical development. *Water Resour. Res.*, 22:1973.

Borden, R. C., Bedient, P. B., Lee, M. D., Ward, C. H. and Wilson, J. T., 1986. Transport of dissolved hydrocarbons influenced by reaeration and oxygen limited biodegradation. II. Field application. *Water Resour. Res.*, 22:1983.

Borden, R. C., Lee, M. D., Wilson, J. T., Ward, C. H. and Bedient, P. B., 1984. Modeling the migration and biodegradation of hydrocarbons derived from a wood-creosoting process waste. In: *Proceedings, Petroleum Hydrocarbons and Organic Chemicals in Ground Water: Prevention, Detection, and Restoration*, Nov. 5–7, 1984, Houston, TX, pp. 130–143, National Water Well Association, Worthington, OH.

Bossert, I. and Bartha, R., 1984. The fate of petroleum in ecosystems. In: R. M. Atlas (ed.), *Petroleum Microbiology.* Macmillan Publishing Co., New York.

Bouwer, E., Durant, N., Wilson, L., Zhang, W. and Cunningham, A., 1994. Degradation of xenobiotic compounds in situ: Capabilities and limits. *FEMS Microbiology Reviews,* 15:307–317.

Brusseau, M. L., Jessup, R. E. and Rao, P. S. C., 1992. Modeling solute transport influenced by multiprocess nonequilibrium and transformation reactions. *Water Resour. Res.*, 28:175–182.

Celia, M. A., Kindred, J. S. and Herrera, I., 1989. Contaminant transport and biodegradation 1. A numerical model for reactive transport in porous media. *Wat. Resour. Res.*, 25:1141–1148.

Characklis, W. G. and K. C. Marshall, 1990. Biofilms. John Wiley and Sons.

Chiang, C. Y., Salanitro, J. P., Chai, E. Y., Colthart, J. D. and Klein, C. L., 1989a. Aerobic biodegradation of benzene, toluene, and xylene in a sandy aquifer—data analysis and computer modeling. *Ground Water,* 27:823–834.

Chiang, C. Y., Wheeler, M. F. and Bedient, P. B., 1989b. A modified method of characteristics technique and mixed finite elements method for simulation of groundwater solute transport. *Wat. Resour. Res.*, 25:1541–1549.

Corapcioglu, M. Y. and Haridas, A., 1984. Transport and fate of microorganisms in porous media: a theoretical investigation. *J. Hydrol.*, 72:149–169.

Corapcioglu, M. Y. and Haridas, A., 1985. Microbial transport in soils and groundwater: A numerical model. *Adv. Water. Resour.*, 8:188–200.

Corapcioglu, M. Y. and Baehr, A. L., 1987. A compositional multiphase model for groundwater contamination by petroleum products 1. Theoretical considerations. *Water Resour. Res.*, 23:191–200.

Cunningham, A. B., Characklis, W. G., Abedeen, F. and Crawford, D., 1991. Influence of biofilm accumulation on porous media hydrodynamics. *Environ. Sci. Technol.*, 25:1305–1311.

Dasappa, S. M. and Loehr, R. C., 1991. Toxicity reduction in contaminated soil bioremediation processes. *Wat. Res.*, 25:1121–1130.

Dawson, C. N., Wheeler, M. F. and Borden, R. C., 1986. Numerical simulation of microbial biodegradation of hydrocarbons in groundwater. In: *Proc. Conf. on Finite Elements in Flow Problems VI*, Antibes, France.

Dawson, C. N. and Wheeler, M. F., 1992. Time splitting methods for advection-diffusion-reaction equations arising in contaminant transport. *Proc. ICIAM '92*, Society for Industrial and Applied Mathematics, Philadelphia, PA.

Dibble, J. T. and Bartha, R., 1979. Effect of environmental parameters on the biodegradation of oil sludge. *Appl. Environ. Micro.*, 37:729–739.

Domenico, P. A., 1987. An analytical model for multidimensional transport of a decaying contaminant species. *J. Hydrol.* 91:49–58.

Ehrhardt, H. M. and Rehm, H. J., 1985. Phenol degradation by microorganisms adsorbed on activated carbon. *Appl. Microbiol. Biotechnol.*, 21:32–36.

Enfield, C. G., Carsel, R. F., Cohen, S. Z., Phan, T. and Walters, D. M., 1982. Approximating pollutant transport to ground water. *Ground Water*, 20:711–722.

Fogarty, A. M. and Tuovinen, O. H., 1991. Microbial degradation of pesticides in yard waste composting. *Appl. Environ. Microbiol.*, 55:225–233.

Foght, J. M. and Westlake, D. W. S., 1982. Effect of dispersant Corexit 9527 on the microbial degradation of Prudoe Bay oil. *Can. J. Microbiol.*, 28:117–122.

Goldstein, R. M., Mallory, L. M. and Alexander, M., 1985. Reasons for possible failure of inoculation to enhance biodegradation. *Appl. Environ. Micro.*, 50:977–983.

Guerin, W. F. and Boyd, S. A., 1992. Differential bioavailability of soil-sorbed naphthalene to two bacterial species. *Appl. Environ. Micro.*, 58:1142–1152.

Hoeppel, R. E., Hinchee, R. E. and Arthur, M. F., 1991. Bioventing soils contaminated with petroleum hydrocarbons. *J. Indust. Microbiol.*, 8:141–146.

Holben, W. E., Jansson, J. K., Chelm, B. K. and Tiedje, J. M., 1988. DNA probe method for the detection of specific microorganisms in the soil bacterial community. *Appl. Environ. Micro.*, 54:703–711.

Huling, S. G. and Bledsoe, B. E., 1990. *Enhanced bioremediation utilizing hydrogen peroxide as a supplemental source of oxygen: a laboratory and field study.* EPA R. S. Kerr Environmental Research Lab. EPA/600/2-90/006:1–47.

Hutchins, S. R., 1991. Biodegradation of monoaromatic hydrocarbons by aquifer microorganisms using oxygen, nitrate, or nitrous oxide as the terminal electron acceptor. *Appl. Environ. Micro.*, 57:2403–2407.

Huyakorn, P. S., Mercer, J. W. and Ward, D. S., 1985. Finite element matrix and mass balance computational schemes for transport in variably saturated porous media. *Water Resources Res.*, 21:346–358.

Illangasekare, T. H. and Döll, P., 1989. A discrete kernel method of characteristics model of solute transport in water table aquifers. *Water Resour. Res.*, 25:857–867.

Jamison, V. W., Raymond, R. L. and Hudson, J. O., 1975. Biodegradation of high-octane gasoline in groundwater. *Dev. Ind. Microbiol.*, 16:305–312.

Jensen, B., Arvin, E. and Gundersen, A. T., 1986. The degradation of aromatic hydrocarbons with bacteria from oil contaminated aquifers. In: *Proceedings, NWWA/API Conference on Petroleum Hydrocarbons and Organic Chemicals in Ground Water—Prevention, Detection, and Restoration*. National Water Well Association, Worthington, OH, pp. 421.

Karlson, U. and Frankenberger, W. T., 1989. Microbial degradation of benzene and toluene in groundwater. *Environ. Contam. Toxicol.*, 43:505–510.

Keely, J. F., Piwoni, M. D. and Wilson, J. T., 1986. Evolving concepts of subsurface contaminant transport. *J. Wat. Pol. Cont. Fed.*, 349–357.

Kilbane, J. J., Chatterjee, D. K. and Chakrabarty, A. M., 1983. Detoxification of 2,4,5-trichlorophenoxyacetic acid from contaminated soil by *Pseudomonas cepacia. Appl. Environ. Micro.*, 45:1697–1700.

Kindred, J. S. and Celia, M. A., 1989. Contaminant transport and biodegradation 2. Conceptual model and test simulations. *Wat. Resour. Res.*, 25:1141–1148.

Klečka, G. M., Davis, J. W., Gray, D. R. and Madsen, S. S., 1990. *Ground Water,* 28:534–543.

Knezovich, J. P. and Harrison, F. L., 1987. The bioavailability of sediment-sorbed organic chemicals: a review. *Water, Air, and Soil Pollution,* 32:233–245.

Kosson, D. S., Agnihotri, G. C. and Ahlert, R. C., 1987. Modeling and simulation of a soil-based microbial treatment process. *J. Haz. Matl.*, 14:191–211.

Kuhlmeier, P. D. and Sunderland, G. L., 1985. Biotransformation of petroleum hydrocarbons in deep unsaturated sediments. In *NWWA/API Conference on Petroleum Hydrocarbons and Organic Chemicals in Ground Water—Prevention, Detection, and Restoration:* November 13–15, 1985, American Petroleum Institute—National Water Works Association, Dublin, OH, pp. 445–462.

Kuhn, E. P., Zeyer, J., Eicher, P. and Schwarzenbach, R. P., 1988. Anaerobic degradation of alkylated benzenes in denitrifying laboratory aquifer columns. *Appl. Environ. Micro.*, 54:490–496.

Lappin-Scott, H. M., Cusack, F. and Costerton, J. W., 1988. Nutrient resuscitation and growth of starved cells in sandstone cores: a novel approach to enhanced oil recovery. *Appl. Environ. Micro.*, 54:1373–1382.

Larson, R. J., 1984. Kinetic and ecological approaches for predicting biodegradation rates of xenobiotic organic chemicals in natural ecosystems. In: M. J. Kluge and C. A. Reddy (Eds.) *Current perspectives in microbial ecology. Proceedings Third National Symposium on Microbial Ecology.* ASM. pp. 677–686.

Leahy, J. G. and Colwell, R. R., 1990. Microbial degradation of hydrocarbons in the environment. *Microbiol. Rev.*, 54:305–315.

Lee, M. D., Thomas, J. M., Borden, R. C., Bedient, P. B., Wilson, J. T. and Ward, C. H., 1988. Biorestoration of aquifers contaminated with organic compounds. *CRC Critical Reviews in Environmental Control,* 18:29–89.

MacLeod, F. A., Lappin-Scott, H. M. and Costerton, J. W., 1988. Plugging of a model rock system by using starved bacteria. *Appl. Environ. Micro.*, 54:1365–1372.

MacQuarrie, K. T. B., Sudiky, E. A. and Frind, E. O., 1990. Simulation of biodegradable organic contaminants in groundwater 1. Numerical formulation in principal directions. *Water Resour. Res.*, 26:207–222.

MacQuarrie, K. T. B. and Sudicky, E. A., 1990. Simulation of biodegradable organic contaminants in groundwater 2. Plume behavior in uniform and random flow fields. *Water Resources Research*, 26:223–239.

Madsen, E. L., 1991. Determining in situ biodegradation; facts and challenges. *Environ. Sci. Tech.*, 25:1663–1673.

Madsen, E. L., Sinclair, J. L. and Ghiorse, W. C., 1991. In situ biodegradation: Microbial patterns in a contaminated aquifer. *Science*, 252:830–833.

Major, D. W., Mayfield, C. I. and Barker, J. F., 1988. Biotransformation of benzene by denitrification in aquifer sand. *Ground Water*, 26:8–14.

Mihelcic, J. R. and Luthy, R. G., 1991. Sorption and microbial degradation of naphthalene in soil-water suspensions under denitrification conditions. *Environ. Sci. Tech.*, 25:169–177.

Miller, R. M. and Bartha, R., 1989. Evidence from liposome encapsulation for transport-limited microbial metabolism of solid alkanes. *Appl. Environ. Micro.*, 55:269–274.

Molz, F. J., Widdowson, M. A. and Benefield, L. D., 1986. Simulation of microbial growth dynamics coupled to nutrient and oxygen transport in porous media. *Water Resour. Res.*, 22:1207.

Molz, F. J. and Widdowson, M. A., 1988. Internal inconsistencies in dispersion-dominated models that incorporate chemical and microbial kinetics. *Water Resour. Res.*, 24:615–619.

Morgan, P. and Watkinson, R. J., 1989. Hydrocarbon degradation in soils and methods for soil biotreatment. *CRC Crit. Rev. Biotech.*, 8:305–333.

Mulkins-Phillips, G. C. and Stewart, J. E., 1974. Effect of four dispersants on biodegradation and growth of bacteria on crude oil. *Appl. Microbiol.*, 28:547–552.

O'Carroll, K., 1988. Assessment of bacterial activity. In: B. Austin (Ed.) *Methods in Aquatic Bacteriology*. John Wiley and Sons, New York, NY, pp. 353–358.

Odu, C. T. I., 1978. The effect of nutrient applications and aeration on oil degradation in soil. *Environ. Pollut.*, 15:235–240.

Ogram, A. V., Jessup, R. E., Ou, L. T. and Rao, P. S. C., 1985. Effects of sorption on biological degradation rates of (2,4-dichlorophenoxy)acetic acid in soils. *Appl. Environ. Microbiol.*, 49:582–587.

Pardieck, D. L., Bouwer, E. J. and Stone, A. T., 1992. Hydrogen peroxide use to increase oxidant capacity for in situ bioremediation of contaminated soils and aquifers: A review. *J. Contam. Hydrol.*, 9:221–242.

Pfaender, F. K. and Bartholomew, G. W., 1982. Measurement of aquatic biodegradation rates by determining heterotrophic uptake of radiolabeled pollutants. *Appl. Environ. Micro.*, 44:159–164.

Pignatello, J. J., 1989. Sorption dynamics of organic compounds in soils and sediments. In: B. L. Sawhney and K. Brown (Eds.), *Reactions and Movement of Organic Chemicals in Soils.* Soil Science Society of America, Madison, WI., pp. 45–80.

Pritchard, P. H. and Costa, C. F., 1991. EPA's Alaska oil spill bioremediation project. *Environ. Sci. Tech.*, 25:372–379.

Raiders, R. A., McInerney, M. J., Revus, D. E., Torbati, H. M., Knapp, R. M. and Jenneman, G. E., 1986. Selectivity and depth of microbial plugging in Berea sandstone cores. *J. Ind. Microbiol.*, 1:195–203.

Rifai, H. S., Bedient, P. B., Wilson, J. T., Miller, K. M. and Armstrong, J. M., 1988. Biodegradation modeling at aviation fuel spill site. *J. Env. Eng. ASCE* 114:1007–1029.

Rijnaarts, H. H. M., Bachmann, A., Jumelet, J. C. and Zehnder, A. J. B., 1990. Effect of desorption and intraparticle mass transfer on the aerobic biomineralization of α-hexachlorocyclohexane in a contaminated calcareous soil. *Environ. Sci. Technol.*, 24:1349–1354.

Robinson, K. G., Farmer, W. S. and Novak, J. T., 1990. Availability of sorbed toluene in soils for biodegradation by acclimated bacteria. *Wat. Res.*, 24:345–350.

Rogers, J. A., Tedaldi, D. J. and Kavanaugh, M. C., 1993. A screening protocol for bioremediation of contaminated soil. *Environ. Prog.*, 12:146–156.

Rogerson, A. and Berger, J., 1981. Effect of crude oil and petroleum-degrading microorganisms on the growth of freshwater and soil protozoa. *J. Gen. Microbiol.*, 124:53–59.

Semprini, L. and McCarty, P. L., 1991a. Comparison between model simulations and field results for in-situ biorestoration of chlorinated aliphatics: Part 1. Biostimulation of methanotrophic bacteria. *Ground Water,* 29:365–374.

Semprini, L. and McCarty, P. L., 1991b. Comparison between model simulations and field results for in-situ biorestoration of chlorinated aliphatics: Part 2. Cometabolic transformations. *Ground Water,* 30:37–44.

Shaw, J. C., Bramhill, B., Wardlaw, N. C. and Costerton, J. W., 1985. Bacterial fouling on a model core system. *Appl. Environ. Micro.*, 49:693–701.

Slater, H. J., and Lovatt, D., 1984. Biodegradation and the significance of microbial communities. In D. T. Gibson (Ed.), *Microbial degradation of organic compounds.* Marcel Dekker, Inc. New York, pp. 439–455.

Song, H. G. and Bartha, R., 1990. Effects of jet fuel spills on the microbial community of soil. *Appl. Environ. Micro.*, 56:646–651.

Song, H. G., Wang, X. and Bartha, R., 1990. Bioremediation potential of terrestrial fuel spills. *Appl. Environ. Micro.*, 56:652–656.

Spain, J. C., Pritchard, P. H. and Bourquin, A. W., 1980. Effects of adaptation on biodegradation rates in sediment/water cores from estuarine and freshwater environments. *Appl. Environ. Micro.,* 40:726–734.

Spain, J. C. and Van Veld, P. A., 1983. Adaptation of natural microbial communities to degradation of xenobiotic components: effects of concentration, exposure time, and chemical structure. *Appl. Environ. Micro.,* 45:428–435.

Srinivasan, P. and Mercer, J. W., 1988. Simulation of biodegradation and sorption processes in ground water. *Ground Water,* 26:475–487.

Steen, W. C., Paris, D. F. and Baughman, G. L., 1979. Effects of sediment sorption on microbial degradation of toxic substances. In: *Proceedings National American Chemical Society Meeting,* April 1979.

Sturman, P. J., Cunningham, A., Stewart, P., and Bouwer, E. 1995. Engineering scale-up of in situ bioremediation processes: a review. *Journal of Contaminant Hydrology,* 19:171–203.

Sutton, P. A. and Barker, J. F., 1985. Migration and attenuation of selected organics in a sandy aquifer—A natural gradient experiment. *Ground Water,* 23:10–16.

Swindoll, C. M., Aelion, C. M. and Pfaender, F. K., 1988. Influence of inorganic and organic nutrients on aerobic biodegradation and on the adaptation response of subsurface microbial communities. *Appl. Environ. Micro.,* 54: 212–217.

Sykes, J. F., Soyupak, S. and Farquhar, G. J., 1982. Modeling of leachate organic migration and attenuation in ground waters below sanitary landfills. *Water Resour. Res.,* 18:135.

Taylor, S. W. and Jaffe, P. R., 1990. Biofilm growth and the related changes in the physical properties of a porous medium, 1, experimental investigation. *Water Resources Research,* 26:2153–2160.

Thomas, J. M., Lee, M. D., Bedient, P. B., Borden, R. C., Carter, L. W. and Ward, C. H., 1987. *Leaking underground storage tanks: Remediation with emphasis on in situ reclamation.* EPA R. S. Kerr Environmental Research Lab. EPA/600/S2-87/008.

Thomas, J. M., Yordy, J. R., Amador, J. A. and Alexander, M., 1986. Rates of dissolution and biodegradation of water-insoluble organic compounds. *Appl. Environ. Micro.,* 52:290–296.

Tim, U. S. and Mostaghimi, S., 1989. Modeling transport of a degradable chemical and its metabolites in the unsaturated zone. *Ground Water,* 27:672–681.

Todd, D. K., 1980. *Groundwater Hydrology.* John Wiley and Sons. New York, pp. 68–81.

Torbati, H. M., Raiders, R. A., Donaldson, E. C., McInerney, M. J., Jenneman, G. E. and Knapp, R. M., 1986. Effect of microbial growth on pore entrance size distribution in sandstone cores. *J. Ind. Microbiol.,* 1:227–234.

Trevors, J. T., 1985. DNA probes for the detection of specific genes in bacteria isolated from the environment. *Trends Biotechnol.,* 3:291–293.

Van Genuchten, M. Th., 1981. Analytical solutions for chemical transport with simultaneous adsorption, zero-order production and first-order decay. *J. Hydrol.* 49:213–233.

Van Genuchten, M. Th. and Wagenet, R. J., 1989. Two-site/two-region models for pesticide transport and degradation: theoretical development and analytical solutions. *Soil Sci. Soc. Am. J.,* 53:1303–1310.

Vanloocke, R., Verlinde, A. and Verstraete, W., 1979. Microbial release of oil from soil columns. *Environ. Sci. Tech.,* 13:346–348.

Wagenet, R. J. and Hutson, J. L., 1986. Predicting the fate of nonvolatile pesticides in the unsaturated zone. *J. Environ. Qual.,* 15:315–322.

Wheeler, M. F., Dawson, C. N., Bedient, P. B., Chiang, C. Y., Borden, R. C. and Rifai, H. S., 1987. Numerical simulation of microbial biodegradation of hydrocarbons in ground water. *Proc. NWWA/IGWMC Conference "Solving Ground Water Problems with Models,"* February 10–12, 1987. Denver, CO.

Wheeler, M. F. and Dawson, C. N., 1986. An operator-splitting method for advection-diffusion-reaction problems. In: *Proc. MAFELAP IV,* Whitman, J. A. (Ed.), Academic Press, pp. 463–483.

Widdowson, M. A., Molz, F. J. and Benefield, L. D., 1988. A numerical transport model for oxygen- and nitrate-based respiration linked to substrate and nutrient availability in porous media. *Water Resour. Res.,* 24:1553–1565.

Wilson, B. H., Smith, G. B. and Rees, J. F., 1986. Biotransformations of selected alkylbenzenes and halogenated aliphatic hydrocarbons in methanogenic aquifer material: a microcosm study. *Environ. Sci. Tech.,* 20:997–1002.

Wilson, J. T., Leach, L. E., Michalowski, J., Vendegrift, S. and Callaway, R., 1989. *In situ bioremediation of spills from underground storage tanks: New approaches for site characterization, project design and evaluation of performance.* EPA R. S. Kerr Environmental Research Lab, EPA/600/S2-89/042.

Wilson, J. T. and Ward, C. H., 1987. Opportunities for bioreclamation of aquifers contaminated with petroleum hydrocarbons. *Dev. Ind. Microbiol.,* 27:109–116.

Wu, S. and Gschwend, P. M., 1986. Sorption kinetics of hydrophobic organic compounds to natural sediments and soils. *Environ. Sci. Technol.,* 20:717–723.

Wyndham, R. C. and Costerton, J. W., 1981. Heterotrophic potentials and hydrocarbon potentials of sediment microorganisms within the Athabasca oil sands deposit. *Appl. Environ. Micro.,* 41:783–790.

Zeyer, J., Kuhn, E. P. and Schwarzenbach, R. P., 1986. Rapid microbial mineralization of toluene and 1,3-dimethylbenzene in the absence of molecular oxygen. *Appl. Environ. Micro.,* 52:944–947.

Test Systems for Balancing and Optimizing the Biodegradation of Contaminated Soils: A German Perspective

K. Hupe, J.-C. Lüth, J. Heerenklage and R. Stegmann
Department of Waste Management
Technical University of Hamburg-Harburg
Harburger Schloßstrasse 37
21079 Hamburg, Germany

INTRODUCTION

During recent years, bioreclamation of contaminated soils has been established as a treatment technology for contaminated sites. In the field of ex situ bioreclamation, biopile or windrow technique, as well as bioreactors, have been applied and developed.

Nevertheless, there are still a lot of unsolved problems. The processes inside biopiles, windrows, and bioreactors are largely unknown. For this reason, it is difficult to control the degradation processes specifically. First of all, it is necessary to understand the degradation processes in general in order to be able to optimize these processes. The second step is to increase the scale and then apply them in practice.

Measuring of contaminant concentrations alone is not sufficient to describe the processes of degradation; for this reason, balancing of the contaminants' turnover is of central importance in the treatment of contaminated soils. In order to give a complete balance of the contaminant turnover, it is necessary to measure all gaseous and liquid emissions, as well as the residues remaining in the soil.

The mechanisms and processes taking place during biological degradation are investigated and described; the results obtained are the basis to optimize the methods of treatment. In addition, the complex questions of

665

biology and analytics were addressed. The following questions are of particular interest:

- optimization of biological soil treatment in bioreactors (water content, temperature, nutrients, oxygen supply, surfactants)
- detailed and distinct analyses to describe the processes
- balance of contaminant degradation (mineralization, degassing, fixation)
- enhancement of biological processes by using mixing reactors, additives (e.g., biocompost), etc.

METHODS APPLIED

Soils from an A_h-horizon, B_t-horizon, and clay-peat, which had been contaminated with diesel fuel and lubricating oil in a quantitative way, were used. These soil samples, as well as real-contaminated soil (abbreviation: NMS), were characterized and described under soil scientific aspects. Table 1 shows some characteristics of the soils that were sieved to $\leq 2000 \, \mu$m.

When artificially contaminated soils are investigated, the oil concentration is normally adjusted to 1% by weight of soil dry matter. The treatment of the soil takes place at high solid contents ("dry process"). The optimum water content is about 60% of the maximum water capacity (determined by Alef, 1991).

The compost added was biocompost with a degree of maturity at V. It

Table 1. Characteristics of the solid materials.

Solid Material	pH	WC_{max} (g H_2O/100 g dwt)	TOC (% of dwt)	Comments: Texture Class [clay, silt, sand (%)]
A_h	4.5	28.8	1.1	Slightly loamy sand (6.4, 15.4, 78.2)
B_t	4.5	31.8	0.14	Loamy sand (12.3, 15.3, 72.4)
Clay-peat	3.3	187.1	11.0	Silty loam (33.1, 52.8, 14.1)
NMS[a]	8.2	22.6	0.86	Slightly silty sand (0.5, 17.6, 81.9)
Biocompost	7.4	150	15.1	Degree of maturity: V

[a]NMS: real-contaminated with high-boiling oil (1% of dwt soil) from Neumünster.

was derived from kitchen and yard waste separately collected in households and was produced at a small windrow plant in Hamburg, Germany. For investigations only, the fraction $\leq 4000\,\mu$m was used.

TEST UNITS

In order to be able to make mass balances, only closed systems were used to evaluate and optimize biological soil treatment processes. Different test systems and bench-scale reactors were available (Hinchee and Olfenbuttel, 1991; Dott, 1992; Klein, 1992; Tabak et al., 1995). In this chapter, only test units and experiments, where the degradation of unlabeled contaminants was observed, are described. The test units used in experiments with ^{14}C-labeled hydrocarbons are described in a series of publications (e.g., Kästner et al., 1995b; Eschenbach et al., 1995).

For the investigations to optimize the milieu conditions (temperature, water content, additives), special glass vessels and respirometers were used.

Glass Vessels (Volume 1.5 L)

Gas-tight sealed glass vessels (batch setup) represent a particularly simple system to investigate degradation processes (Figure 1).

The discontinuous CO_2 release was determined via absorption to NaOH by means of titration with HCl. Contaminants emitted in the gas phase were analyzed using gas chromatography (GC). Also soda lime tubes were useful for the semiquantitative measurement of CO_2 production (Schaefer et al., 1992). In addition, the soil samples were analyzed for contaminant relevant parameters. The test procedure being relatively simple, it was possible to investigate a large number of setups simultaneously.

Respirometers (Volume 250 ml)

The biochemical O_2 consumption is continuously measured in the respirometer. This setup and its operation principle are presented in Figure 2.

Due to biochemical degradation processes, the generated CO_2 is adsorbed by the soda lime, causing a vacuum in the closed glass vessels; via a pressure gauge, this vacuum affects the oxygen, which is electrolytically produced from $CuSO_4$. By connecting the measuring unit to a personal computer, the oxygen consumption can be continuously recorded. To

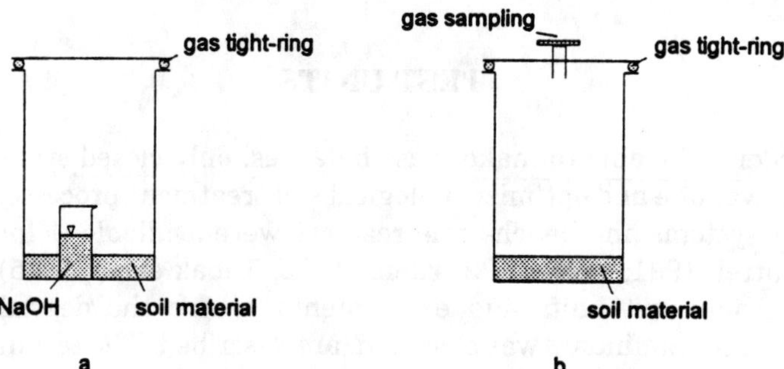

Figure 1. 1.5 L batch setup; (a) determination of CO_2 release by absorption to NaOH and titration with HCl; (b) determination of substances emitted in gaseous form by means of GC measuring.

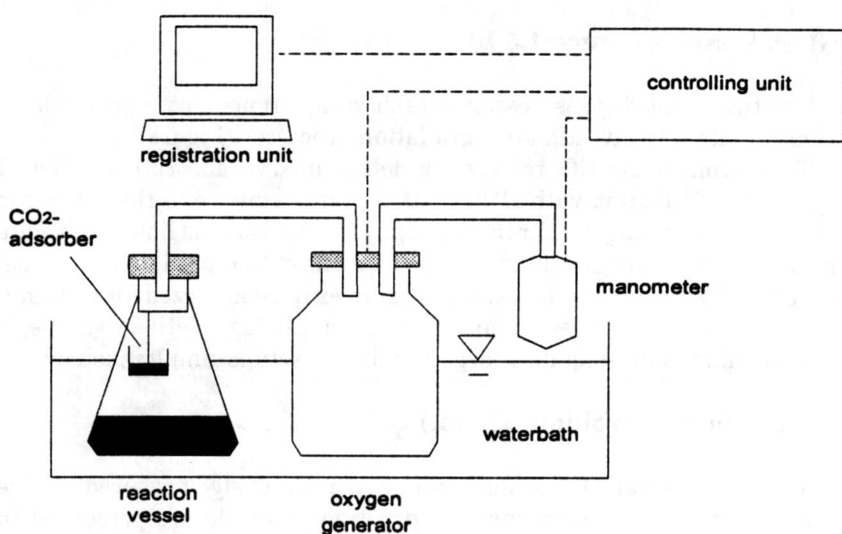

Figure 2. Operation principle of the respirometers (Voith—Sapromat, Germany).

monitor the degradation in the respirometer, it is possible to remove vessels during the test period and to analyze the soil material. Usually, this system has room for 12 testing vessels.

In addition to the described test system, other respirometers with direct measurement of the CO_2 production are also available (Berg et al., 1992).

Reactor Systems

Reactor systems are applied at different scales (volume 3 L, 6 L or 90 L) to simulate the conditions in aerated windrow or large-scale reactors. Aeration is conducted by compressed air, where the degasification of the volatiles can easily be measured. Figure 3 shows the scheme of a static and a dynamic bioreactor, including continuous recording and control of the measured values.

STATIC BIOREACTOR SYSTEMS (VOLUME 3 L, 6 L, 90 L)

The bioreactors developed by the authors were continuously improved on the basis of the experiences gained during the test series. In the lower part of the reactor, there was a sieve on which soil material was placed. At different heights, sampling pipes were installed at the reactor. Controlled aeration was conducted from the bottom to the top, where the feed air was led through a wash bottle filled with water in order to avoid drying out the soil. In the exhaust air, the CO_2 concentrations were continuously

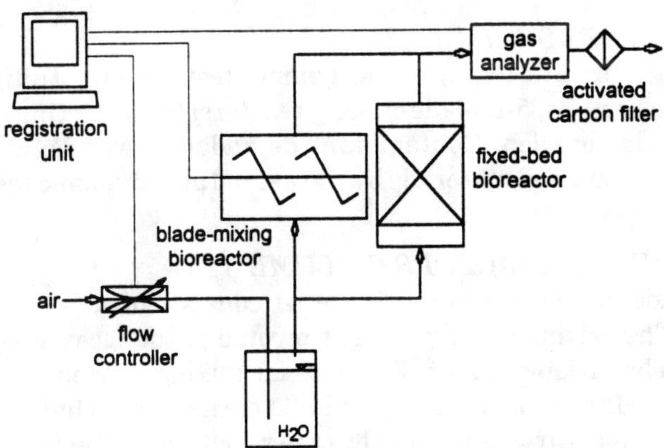

Figure 3. Principle of the bioreactor test system (scheme of a static bioreactor and a blade-mixing reactor).

Figure 4. Blade-mixing reactor (volume: 5 L).

measured by IR (infrared photospectrometer) and the TOC concentrations by FID (flame ionization detector). Further gas samples could be taken with a syringe through a septum from the upper part of the reactor and could then be injected into a GC (gas chromatograph).

DYNAMIC TEST SYSTEMS

In addition to static test systems, dynamic test systems of different scale (1.5-L batch setups, 5-L blade-mixing reactors, 90-L rotating drum reactor) were also applied. Contaminant degradation was improved, especially for cohesive soil materials by the use of these dynamic test systems.

BLADE-MIXING REACTORS (VOLUME 5 L)

The blade-mixing reactor (Figure 4) consists of a horizontal glass cylinder. The driving shaft, gas-tight pivoted at the faces, was driven by an infinitely variable motor. For optimal mixing, various mixing tools could be fixed to the shaft. In the middle of the glass cylinder, there was a sampling pipe. Air was supplied to the reactor over the face areas. CO_2 and TOC in the exhaust airstream of the reactor were measured continuously.

ANALYTICS

Contamination

For measuring contaminants in the soil samples, the extraction of the contaminants should be as complete as possible. In order to quantitatively determine the oil content in the soil, gravimetric, chromatographic, and infrared photospectrometric methods were used. Which of the different analysis methods is applied depends on the individual formulation of the question.

The gravimetric determination of the hydrocarbon content in soil samples is carried out after extraction in a Soxhlet apparatus using various solvents (toluene, petroleum ether, etc.). The advantage of the chromatographic analytical method (gas chromatography, high-pressure liquid chromatography) is its suitability to determine individual components. In cooperation with the Institute of Food Chemistry at the University of Hamburg, a gas chromatographic method was developed, by which both individual substances and substance groups can be quantified. The individual n-alkanes C_{10} to C_{24}, six relevant iso-alkanes, and the sum of the individual components were determined.

Following DIN 38 409 Teil 18 (abbreviated: H18; Anonymous, 1981), the total amount of hydrocarbons was quantified after ultrasonic extraction. Compared with the Soxhlet extraction, this method is more environmentally friendly since fewer solvents are used. Moreover, it is far less time-consuming, and higher extraction rates are obtained. The quantitative analysis is done by means of IR spectroscopy.

Gas Phase

In the gas phase, the parameters CO_2 and TOC (total organic carbon), as well as volatile oil components, were measured. To determine the degree of degradation, it is of utmost importance to record exactly the concentration and the total quantity of CO_2 in the exhaust air. The quantity of CO_2 released from the reactor was measured by means of GC (discontinuously) or IR (continuously). When these values are measured discontinuously in the GC, the samples taken from the exhaust airstream are directly injected into the GC; the amount of CO_2 is determined by a thermal conductivity detector (TCD). The quantity of CO_2 released from the bioreactors is now continuously measured by means of IR.

For the determination of TOC, the sample taken from the exhaust airstream was also directly injected into the GC. The calibration and assessment was made in accordance with a VDI-Richtlinie (VDI-guide-line) on exhaust air control (Anonymous, 1980). During a control using octadecane, an extraction rate of 100% was found. The continuous TOC monitoring was achieved by means of an FID.

Soil

Dependent upon the objectives of the individual test series, the soil analyses comprise the parameters TOC, COD (chemical oxygen demand), TKN (total Kjeldahl nitrogen), ammonia and nitrate, total phosphorus, heavy metals (Zn, Pb, etc.), as well as water content, maximum water capacity (WC_{max}), pH value, and conductivity of the soil.

Biomass

The biomass was determined in the respirometer by means of the substrate-induced respiration method (SIR) (Anderson and Domsch, 1978), where the amount of biomass is calculated via the increased CO_2 production after glucose addition. The SIR method is only partly acceptable for biomass determination. Its successful determination in contaminated soils is still in the process of development (Kaiser et al., 1992; Joergensen et al., 1994).

INVESTIGATIONS AND RESULTS

Comparison of the Test Systems

The three static test systems—respirometer, batch setup, and bioreactor—can be applied for preinvestigations and the optimization of the biological degradation of organic contaminants in soil under "dry" conditions. The results should be comparable. Each system has its specific fields of application and advantages:

- The respirometer is applied when it is important to continuously and precisely record the oxygen consumption (preinvestigations, optimization, determination of influencing factors).
- Batch setups are a sensible system for long-term investigations. In gas-tight sealed vessels the CO_2 production can be determined (NaOH absorption or GC measuring). In addition, soil samples can be analyzed. Here, the sufficient O_2 supply has to be secured.
- The static bioreactors simulate soil treatment with closer relation to practical application. Especially the volatilization of contaminants

can be recorded by appropriate measurements of the exhaust air. This system enables intensive analytical monitoring by measuring the exhaust air quality and by frequent soil sampling through the lateral pipes. Therefore, this test system is particularly suitable for long-term investigations.

The comparability of the three static systems was investigated in test series at 30°C. The artificially contaminated soil material was mixed with compost in the ratio 2:1. The water content was adjusted to 60% of the maximum water capacity. The following parameters were compared: oxygen demand in the respirometer, CO_2 production in the batch setups and the bioreactors, and the hydrocarbon concentration in all three systems, as well as the biomass in the respirometer and the bioreactors. After 21 days, the hydrocarbon concentrations were between 400 and 500 mg/kg of the soil dry matter. Taken into account the problems of soil sampling and of the inhomogeneities of the samples the values taken on the same days correspond with each other quite well.

Determination of Substrate Respiration

The test series to optimize and balance the biological degradation processes are often evaluated by means of parameter oxygen consumption per g hydrocarbons (mg O_2/g HC). In the following, assumptions for the calculation of "substrate respiration" are explained.

The specific oxygen consumption (substrate respiration) is the difference between the oxygen consumption in the contaminated and the noncontaminated test setup. This value is related to the amount of hydrocarbons (oil) degraded (Kästner et al., 1995a). A typical example using soil/compost-mixture contaminated with diesel fuel is represented in Figure 5. The cumulative curve of the O_2 consumption of the uncontaminated setup is linear, whereas the curve typical of substrate decomposition represents the mineralization of the oil.

Carbon Balance of the Oil Degradation

Based on the results of extensive test series carried out in static bioreactors, an approach was developed to determine the carbon balance of oil degradation (Lotter et al., 1992). During the investigations, the parameters listed in Table 2 were measured.

Using the data of the contaminated samples and the uncontaminated control for each parameter (Table 2), it is possible to calculate an oil carbon balance. This approach only partly describes the changing condi-

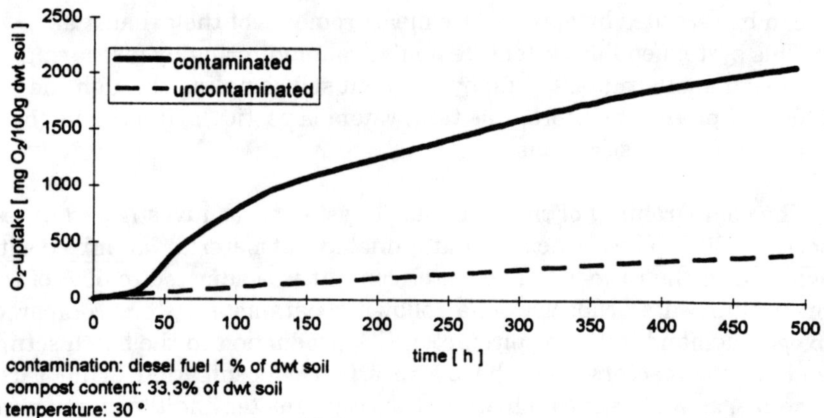

Figure 5. Cumulative curves of the O_2 consumption of a soil-compost mixture (contaminated and uncontaminated).

tions in the soil after oil addition; on the other hand, it provides a good first approximation on the decomposition of the contaminants. The C content of the measured hydrocarbon concentrations is calculated on the basis of the carbon content of the initial oil. The analysis of the oil showed a carbon content of 86.1% for diesel fuel and 83.8% for lubricating oil.

Results from various test series in Figure 6 show the C balance of the carbon content of the oil during treatment in the static bioreactor.

It was observed that there was always a gap in the balance sum. The analytically measured reduction of hydrocarbon cannot be quantitatively explained by the decomposition into CO_2, by volatilization, or the production of biomass. In this regard, the weak and/or strong interactions of the

Table 2. Investigated parameters and analytical methods for the balancing approach (Lotter et al., 1992).

Parameter	Analyzing Method
Hydrocarbon content in the soil	Analogous H18 after ultrasonic extraction
Biomass	SIR-method in respirometer
CO_2 in exhaust	Discontinuously: GC-TCD Quasicontinuously: IR
VOC in exhaust	Discontinuously: GC-FID Quasicontinuously: FID

H18: DIN 38 409 H18 (Anonymous, 1981); SIR: substrate-induced respiration (Anderson and Domsch, 1978); GC: gas chromatograph; TCD: thermal conductive detector; IR: infrared photospectrometer; VOC: volatile organic carbon; FID: flame ionization detector.

contaminants or the metabolites with the humus matrix of the compost are of particular importance (Lotter et al., 1992).

The possible interactions were investigated and described by various research groups (Gerth et al., 1990; Kästner et al., 1995a, 1995b; Eschenbach et al., 1995). For biological remediation, it is important to know whether the adsorption to the soil particles and/or to the humus matrix is reversible or irreversible.

In case of a permanent gap in the C balance, the formation of bound residues may be assumed. The analysis to determine the incorporation of the contaminant into the humus matrix is still being developed (Richnow et al., 1994). If the balance works out, even after a certain period of time, it can be concluded that the adsorbed/incorporated substances have possibly been set free during humus degradation and are subsequently biologically degraded.

If a C balance is made for oil-contaminated soil without additional compost, the balance works out within the scale of measuring accuracy ($100 \pm 10\%$) (Stegmann et al., 1991). Since the humus content in the soil is clearly lower than in the compost, this is a further indication of the importance of the interactions between compost and contaminants.

To record the CO_2 emitted from the bioreactors, a quasi-continuous measuring device was installed. By this means, it was possible to close the balance gap partly. In addition, due to the temporal variation of the CO_2 production, the metabolism or the kinetics could be better understood.

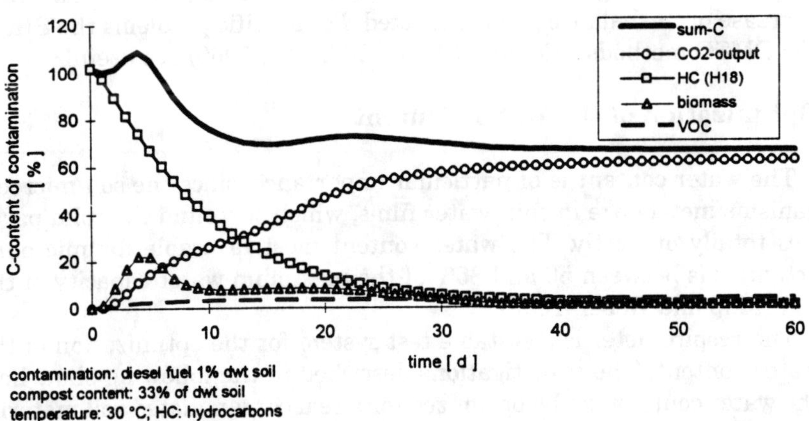

Figure 6. Model of a carbon balance of the C content of the oil during treatment of an artificially oil-contaminated soil material with compost addition in a static bioreactor at 30°C (Lotter et al., 1992).

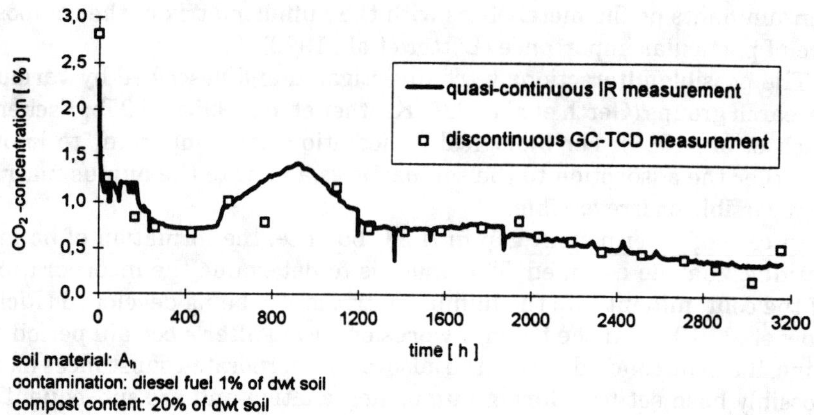

Figure 7. Comparison of different CO_2 measuring methods in bioreactors: quasi-continuously by means of IR; discontinuously by means of TCD.

Figure 7 shows a typical result of a test series to measure CO_2 in the exhaust air in a discontinuous and also quasi-continuous way. From this example, it becomes obvious that the interpolation of the values in the case of discontinuous CO_2 recording results in a different pattern. Calculating the cumulative CO_2 production from the results presented in Figure 7 in the case of discontinuous CO_2 recording, the total sum was 7% less than in the case of quasi-continuous recording.

In further investigations, a method for measuring and calculating the contamination degradated biomass had to be developed. In this way a decrease of the balance gap is expected. For specific problems the SIR or the DMSO methods (Alef and Klein, 1989; Alef, 1990) are useful.

Optimization of the Water Content

The water content is of particular importance, since the soil microorganisms metabolize in thin water films, which surround the solid particles totally or partly. The water content most favorable for microbial activities is between 50 and 80% of the maximum water capacity of the soil (Filip and Weber, 1990).

The respirometer is a suitable test system for the optimization of the water content. The investigations described in the following show how the water content can be optimized for a real contaminated soil material and a synthetically contaminated soil material.

The test material NMS (slightly silty sand) from a former gas station was contaminated with high-boiling oil (Table 1). The other test material was the A_h-soil material (Table 1). Both soils were mixed with 20% (of dwt soil) compost. The maximum water capacity (determined by Alef, 1991) of the soil/compost mixture with the material NMS was 49.3 g H_2O/100 g dwt and 66.1 g H_2O/100 g dwt with the model soil A_h.

The results show that the optimum water content (at 22°C) of the used soil/compost mixtures was different (Figure 8). In the case of the NMS sample, the optimum water content was at about 70% of the maximum water capacity and, in the case of the A_h sample, at about 53% of the maximum water capacity. This example shows that the optimization of the water content of the soil material during the preinvestigation phase is of utmost importance.

Temperature

For the microbial degradation of hydrocarbons, the influence of the temperature on the biochemical oxidation rate is very distinct. The optimum temperature for mesophilic soil microorganisms is between 25 and 40°C (Filip and Weber, 1990). In the literature, the change of degradation rates with temperature is often expressed in terms of Q_{10} values, denoting the factor by which the rates increase for each 10° rise in temperature. But studies show that, in soils, this correlation is not simply describable by the concept of Q_{10} value and that it is dependent on the kind of soil and also on the temperature range (Bossert et al., 1984; Dalyan et al., 1990).

Our investigations concerning the influence of temperature on microbial degradation of hydrocarbons in different soil materials verified the results of Dalyan et al. (1990). In order to investigate the influence of the temperature on the aerobic degradation activity of hydrocarbons, test jars were placed at different temperatures (10°C, 20°C, and 30°C). Soil material of the A_h horizon and the B_t horizon was artificially contaminated with diesel fuel (1% of dwt soil) and mixed with mature compost (20% of dwt soil). The measured parameters were the total hydrocarbon content (method DIN H18) and the CO_2 production.

Figure 9 presents the cumulative values of CO_2 production during these test series. These results show that temperature increase results in a higher microbial activity and, consequently, in a decrease of the hydrocarbon concentration in the soil materials. After 49 days, the decrease of hydrocarbons in the test series with the A_h material amounted to 62% of the original content at 10°C, 86% at 20°C, and 97% at 30°C; the actual degradation into CO_2 amounted to the following percentages depending on the temperature: at 10°C, 19%; at 20°C, 41%; and at 30°C, 62%.

The influence of temperature on the biological activity of the soil from

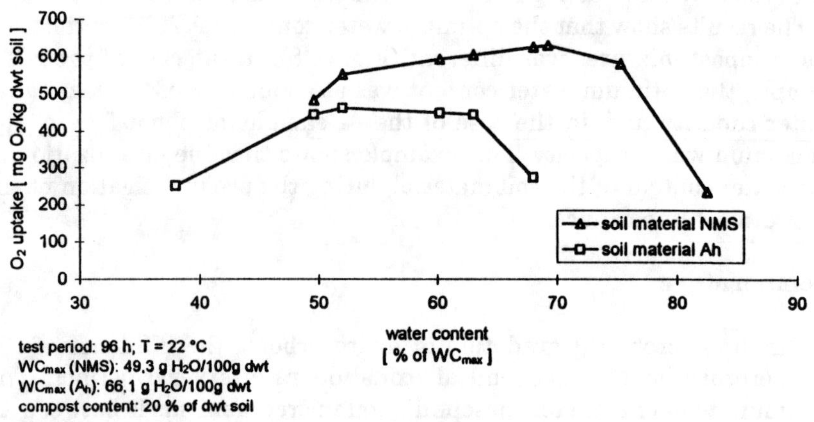

Figure 8. Optimization of the water content in the respirometer at 22°C for two different soils: NMS and A_h.

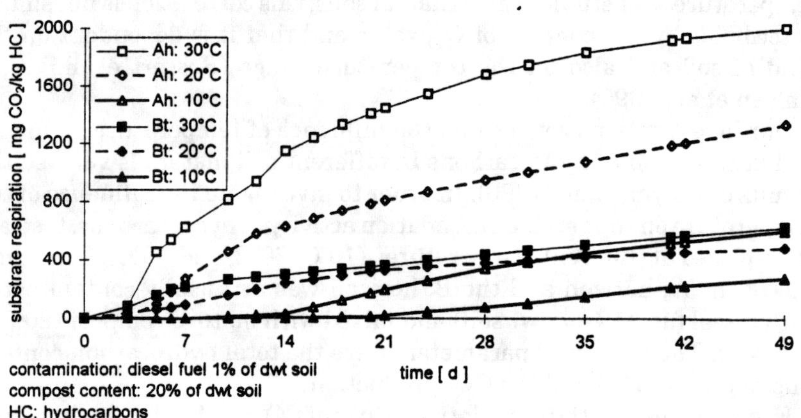

Figure 9. Investigation of the influence of temperature on the microbial degradation of diesel fuel in the soil material of an A_h and B_t horizon in jars.

678

the A_h horizon was totally different from that observed in the test series using material from the B_t horizon. In this case, the relationship between the degradation rate and the increase of the temperature per 10°C was not constant.

It could be observed that, at higher temperatures (>20°C), the initial lag phase could be reduced by about 2 to 3 weeks. Due to the latter observation, a reduction of the biological degradation period by a defined thermal pretreatment of the soil material may be considered.

Oxygen Content

For the aerobic biological treatment of contaminated soil, an optimization of oxygen supply is very important regarding energy input minimization and reduction of exhaust air. In order to control the aeration in the windrow technique or in the bioreactors, it was important to know the optimal range of the oxygen concentration in the soil material, including the minimum oxygen concentration allowing microbial degradation of the contaminants in the soil.

There are only a few published investigations about the consequences of partial pressures of oxygen in unsaturated soil materials. In the case of water-saturated systems (e.g., sludge suspension), studies have been published on the influence of different oxygen concentrations on the microbial degradation of hydrocarbons (e.g., Berthe-Corti et al., 1991; Nitschke, 1994).

Tests were carried out to investigate the influence of partial pressure of oxygen on the biological degradation of diesel fuel in artificially contaminated soil. For these investigations, the soil bioreactors were aerated with synthetic air (different O_2/N_2 mixtures; Table 3).

In the test series, BVS artificially contaminated soil material from the A_h horizon (1% diesel fuel of dwt soil) was mixed with 20% resp. 10% (of dwt soil) matured compost. The results of the test series BVS 14 are summarized in Table 4. They confirm the relevance of the oxygen supply for the biodegradation of hydrocarbons; as expected under anaerobic conditions, no oil degradation was observed. Overall, the results did not show a significant stimulating or limiting effect of the oxygen concentration range (1 to 80%) in the supply air on the biodegradation of the hydrocarbons. First results from test series BVS 16 and BVS 17 (see Table 3) show that an oxygen concentration below 1% in the supply air limit the degradation of the contaminants (Bollow, 1995). In order to come to a final conclusion, additional investigations are necessary.

Table 3. Investigations on the influence of oxygen content on the biological degradation of diesel fuel in soil bioreactors.

Test	O_2 Concentration (%)	Compost Content (% of dwt soil)	Source
BVS 14	0; 5; 10; 21; 40; 80	20	Woyczechowski, 1993
BVS 16	0; 0.1; 0.2; 1; 5; 21	10	Bollow, 1995
BVS 17	1; 2; 3; 4; 5; 21	10	Bollow, 1995

O_2 concentration: percentage by volume; dwt: dry weight.

Table 4. Influence of partial oxygen pressure in the supply air on the oil carbon balance after a test period of 48 days (BVS 14; Woyczechowski, 1993).

Test (% O_2)	Σ VOC (% C)	Σ CO_2 (% C)	Σ HC (% C)	Balance Gap after 48 Days (% C)
0	6	2	79	15
5	2	67	12	17
10	2	56	15	26
21	2	54	13	29
40	2	56	12	30
80	2	52	15	30

C: carbon of the contamination; % O_2: volumetric percent of oxygen in the feed; HC: hydrocarbons (method H18), measured discontinuously; CO_2: GC analysis, measured discontinuously.

Effect of Compost Addition

Several investigations have shown that the degradation of organic contaminants may be enhanced by means of biocompost addition (Lotter et al., 1993; Pennerstorfer et al., 1993; Pettersen et al., 1993; Mahro et al., 1994; Civilini, 1994; Civilini et al., 1995; Castaldi et al., 1995; Diaz et al., 1995; Lilja and Votila, 1995; Cole and Liu, 1995). This effect was further investigated in several respirometer test series using compost at different degrees of maturity. The compost added was biocompost derived from kitchen and garden waste separately collected in households, and it showed very low contamination. The following parameters of influence derived from the biocompost may be of importance:

- compost as a bulking agent to improve aeration particularly in cohesive soil materials (Civilini, 1994)
- compost as a supplier of a great variety of microorganisms (Kästner et al., 1995b; Castaldi et al., 1995)
- improvement of the pH buffer capacity (Dalyan et al., 1990) and water storage capacity due to compost addition
- compost as a structural material to reduce pellet formation when cohesive soil materials are treated in mixing reactors ("dynamic treatment")
- compost as a source of nutrients and trace components (particularly nitrogen and phosphorus); compost as a depot fertilizer (Ebertseder et al., 1995)
- compost as a cosubstrate
- interaction of the organic matrix of compost with the contaminant (incorporation of the contaminants in the soil/humus matrix)

Moreover, by means of compost addition, a decrease in ecotoxicology in oil contaminated soils could be determined. This was measured by bacterial activity, as well as by algae and plant toxicity tests (Ahlf et al., 1993).

In Germany, the degree of maturity has been ranked from I to V (measured as self-heating or respiration; Kehres and Maile, 1994). The composting process itself can be divided into three stages: high-rate decomposition, stabilization, and curing. During high-rate decomposition, fresh organic matter is transformed into so-called "fresh compost," which should be hygienically acceptable, and the most significant odor caused by organics are degraded (about 2–4 weeks, degree of maturity: II). In the stabilization stage, the material is further decomposed and stabilized (2–4 weeks, degree of maturity: III–IV). The final curing formation takes place over a period of several weeks, as long as 18 to 20 weeks (degree of maturity: V) (Krogmann, 1992).

In order to determine the most effective biocompost as an additive, the optimization of the compost age, the influence of different compost quantities added, and the addition of nitrogen were investigated.

AMOUNT OF COMPOST ADDED

It could be shown that the addition of compost in the ratio of 2:1 soil/compost enhanced the oil degradation significantly. For remedial actions in practice, however, this compost ratio is too high. In Germany, when windrow techniques are applied, a volume rate of about 9:1 (soil/additive) is frequently used. For this reason, soil/compost mixtures of 2:1, 4:1, and 8:1 (referred to as dry matter) were investigated. The diesel fuel addition was, in all cases, 1% of dwt soil. The temperature was adjusted at 22°C.

Figure 10 presents the cumulative oxygen consumption for the test series described above. For all setups, the lag phase was about 2 days. It was found that, with decreasing compost content, the cumulative O_2 consumption caused by the oil degradation decreased. Moreover, the maximum oxygen consumption related to the oil degradation per hour could be reached when the compost addition was increased at an earlier stage.

ABIOTIC INFLUENCE OF COMPOST

It has been mentioned earlier that, during balancing investigations with artificially contaminated soil material, a balance gap was detected when compost was added to the soil. Without compost addition to the soil,

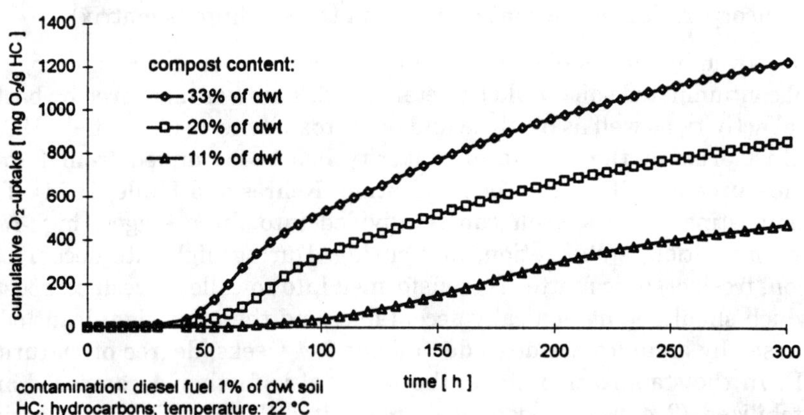

Figure 10. Cumulative O_2 consumption from the degradation of the oil contaminant in the soil materials using various soil/compost mixtures (Stegmann et al., 1991).

no gap in the carbon balance was observed (accuracy of measurement 100 ± 10%). The interaction between compost and contaminants in the soil was of utmost importance in regard to the gap in the carbon balance (Lotter et al., 1992). In order to find out more about the phenomenon, matured compost artificially contaminated with diesel fuel was used in tests for balancing the degradation of the contamination in bioreactors (Lotter, 1995). For test series A, B, and C, the compost was contaminated with diesel fuel (1% of dwt compost). The compost in test series D was uncontaminated for measuring the endogenous respiration ("control"). In test series A, activated biomass was used to investigate the degradation of oil in the compost; in test series B and C, the inhibition of the microbial activity was achieved to a high degree by the addition of $CaCl_2$ (10% of dwt compost) in order to determine the abiotic interactions between compost and contaminants. The influence of oxidation processes by aeration was investigated indirectly by using nitrogen instead of air in test series C.

After 21 days in test series A (basic setup), only 8% of the carbon content of the original contamination was extractable, 59% was mineralized, 5% was stripped, and about 4% of the original contamination was transformed into biomass. Thus, a gap of 24% in the carbon balance was determined. In comparison to this result, in the "inhibited" series, only 14% of the contamination was mineralized, stripped, or transformed into biomass, a balance gap of 16% was observed, and about 70% of the original contamination was extractable after a test period of 21 days.

This experiment demonstrated that the incorporation of the contaminants in the compost also could be referred to abiotic interactions between these compounds like sorption and/or diffusion effects. Apparently, the incorporation of contaminants in the compost matrix was independent of an oxidative milieu since the incorporating effect in the oxygen atmosphere (test A: "basic setup") was approximately identical with that taking place in an atmosphere lacking in oxygen (test C: setup with $CaCl_2$ and N_2 aeration).

INFLUENCE OF VARYING AGE OF COMPOST ON THE CARBON BALANCE

In order to examine the influence of various ages of compost on the biological degradation of hydrocarbons, soil material of an A_h horizon was contaminated with 1% (of dwt soil) diesel fuel. As additives, 20% (of dwt soil) compost of different age (2 weeks, 2.5 months, 6 months, 13 months) were added in the different setups (Lotter, 1995). In Table 5, the compost materials used are described. The results (Figure 11) show that the reduction of the hydrocarbon content of the diesel fuel after 60 days was

Table 5. Description of the characteristics of the used compost materials.

	Age of Compost			
Parameter	0.5 Months	2.5 Months	6 Months	13 Months
Biomass (g C/kg dwt)	17.6	14.1	1.4	0.4
TC (% dwt)	26.8	19.4	12.7	10.6
Nitrogen (TKN) (% dwt)	1.2	2.4	1.4	1.0
Ignition loss (% dwt)	44.8	41	24.4	20.8
Degree of maturity	I	III	IV	V
Reduction of hydrocarbon content after 60 days (%)	95	95	94	93

C: carbon; dwt: dry weight; TC: total carbon; degree of maturity: I–V, determined by Anonymous, 1981.

more or less independent of the compost age and amounted to approximately 94% of the original contamination. The determined carbon balances indicate different balance gaps. The size of the balance gap decreased with increasing compost age.

In the case of fresh compost (degree of maturity < IV), it was very difficult to determine the influence of the contamination on the CO_2 and biomass production. This was due to the high endogenous respiration of the fresh compost material. So the differences of these parameters between contaminated and uncontaminated setups were small. The size of the balance gap could be reduced for the setups with fresh compost when the concentration of the diesel fuel added was increased (to 2.5% of dwt soil). The higher amount of oil made it easier to observe the degradation of oil into CO_2 and biomass. The results show that the

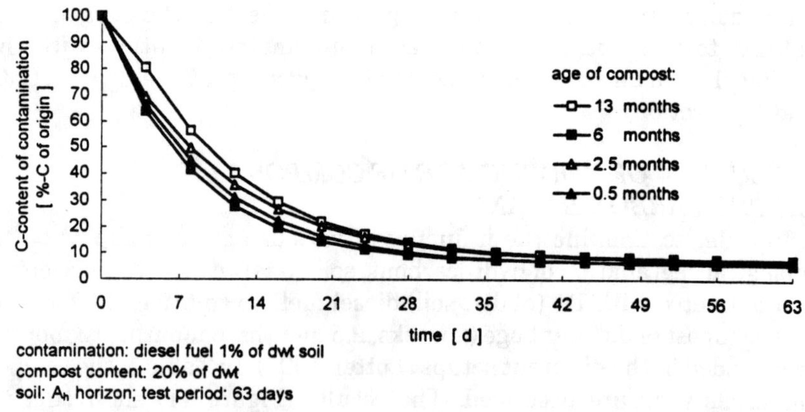

contamination: diesel fuel 1% of dwt soil
compost content: 20% of dwt
soil: A_h horizon; test period: 63 days

Figure 11. Influence of varying compost age on the degradation of diesel fuel.

balance system is not useful when fresh compost/organic matter is used as an additive.

Treatment in Dynamic Reactors

Dynamic reactors gently mix the contaminated soil. During recent years, bioreactors have increasingly been applied or developed for the biological treatment of contaminated soils. Compared to the windrow technique, the advantage of bioreactors is that all emissions can be controlled and completely recorded (Kleijntjens, 1991; Irvine, 1994; Nitschke, 1994). Most of the bioreactors developed are slurry reactors (Klein, 1994). Their advantage is that they can be operated in a simple technical way, and pellet formation during "dry treatment" (water content < maximum water capacity of the soil material) is a priori avoided (Parthen, 1992). The main disadvantage of slurry treatment is the necessary posttreatment: drying of the treated soil material and wastewater treatment. In order to avoid the posttreatment step, the dynamic treatment of dry soil material (water content < maximum water capacity of the soil) was investigated.

The first investigations on the application of dynamic reactor systems for the treatment of soils contaminated with oil were carried out in batch test setups at different temperatures (10°C, 20°C, 30°C). Soil material artificially contaminated with diesel oil (soil of an A_h horizon and of a B_t horizon) was mixed with 10% by weight of compost material (degree of maturity: V). The water content was adjusted at 60% of the maximum water capacity. To secure sufficient oxygen supply, the setups were artificially aerated daily for a period of about 90 s. In the "dynamic" setups, the material was stirred for the same time with a laboratory spatula. The test period was 49 days. During this time, the CO_2 production and the hydrocarbon content in the soil were determined according to DIN H18 after ultrasonic extraction.

The temperature influence on the decrease of hydrocarbon content indicates that dynamic treatment of contaminated soil material has a positive effect, particularly at low temperatures (see Figure 12). Further investigations in this area are necessary to find out the reason for the effect described above.

OPTIMIZATION OF THE APPLICATION
OF BLADE-MIXING REACTORS

Preinvestigations showed that the dynamic treatment had a positive effect on the contaminant's turnover, especially at low temperatures (<20°C). But these tests also showed that, during turning and mixing, the soil in the reactor has a tendency to build up pellets. It was found out that pellet formation could be reduced when the water content was relatively low [<55% WC_{max} (maximum water capacity)], the portion of organic structural material was high, the structural material contained only a low portion of fines (d_p < 630 μm; <10%) and the rotational speed of the mixer was low (8 rpm).

INFLUENCE OF DYNAMIC TREATMENT DEPENDENT UPON THE KIND OF SOIL AND THE CONTAMINATION

Due to the preinvestigation results, the tests were run at a constant temperature (20°C). Results from various test series using blade-mixing reactors show that the effect of the dynamic treatment of contaminated soils varies, depending on the kind of soil and the contamination. An investigation was carried out where both reactors were aerated with the same volume per time.

The effect of static treatment (no stirring or mixing) was compared with that of dynamic treatment. Two different soil types were tested, one of an A_h horizon contaminated with diesel fuel and a clay-peat sample contaminated with lubricating oil. The soil samples were contaminated with oil (1% of weight); then compost (20% of weight, degree of maturity: V) was added, and the water content was adjusted at 50% WC_{max}. The samples were placed into the reactors (static: fixed-bed reactor; dynamic: blade-mixing reactor). The CO_2 production was continuously measured during the period of treatment.

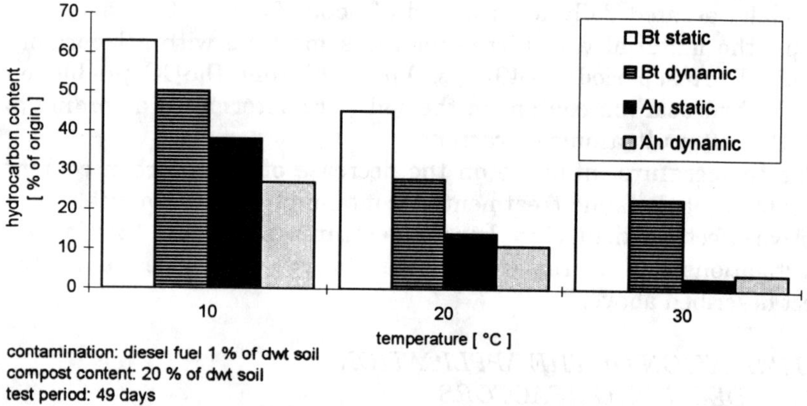

Figure 12. Influence of temperature on the decrease of the hydrocarbon content during static and dynamic treatment in batch setups.

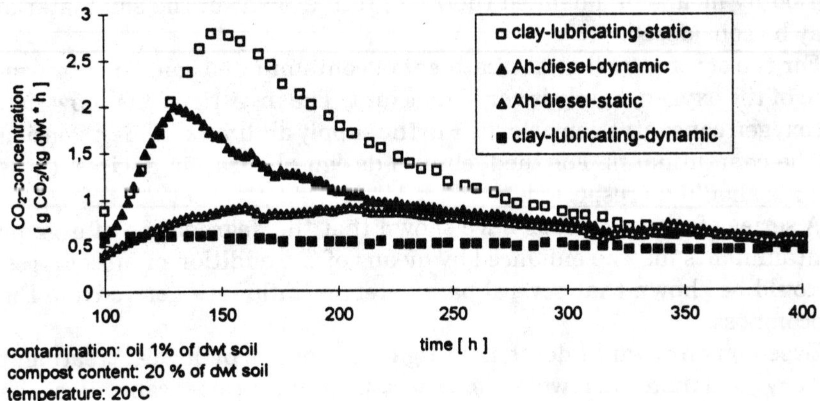

Figure 13. Influence of dynamic treatment dependent upon the kind of soil and the contamination.

The curves in Figure 13 represent the CO_2 concentrations in the exhaust gas from the bioreactors during two test series. Whereas the turnover of diesel fuel in a soil (A_h horizon and clay-peat)/compost mixture was enhanced by the dynamic treatment, the opposite effect could be observed when lubricating oil was treated.

SUMMARY AND OUTLOOK

The investigations carried out so far using batch setups as well as respirometers proved to be very suitable, particularly for preinvestigations on the optimization of milieu conditions (water content, temperature, additives, compost quantity, compost age, etc.). To determine the degradation rates of contaminants, closed aerated glass bioreactor systems with continuous analytical measuring of the exhaust airstream (CO_2, TOC) are recommended. Using this method the fate of the contaminants can be quantified (degradation, volatilization, biomass formation, adsorption, and/or incorporation into the humic matrix).

It could be shown that the optimization of the water content of the soil material during the preinvestigation phase is of utmost importance. For biological "dry treatment" processes, the water content of the soil should be below 70% of the maximum water capacity of the soil. At higher temperatures (>20°C), the initial lag phase could be reduced by about 2 to 3 weeks. As a consequence, a reduction of the biological degradation period by means of a defined thermal pretreatment of the soil material may be considered.

For the aerobic biological treatment of contaminated soil, an optimization of the oxygen supply is very important. The investigations show that an oxygen concentration below 1% in the supply air limits the degradation of the contaminants. For the technical design of aeration devices, these results should be respected.

A series of investigations have shown that the degradation of organic contaminants may be enhanced by means of the addition of biocompost. It could be shown that several parameters of influence derive from the biocompost.

Based on our own in-depth investigations for the biological treatment of "dry" soil materials (water content < maximum water capacity of the soil material), mixing reactors (dynamic reactors) cannot be recommended in general. The reasons are

- It is difficult to avoid pellet formation, which leads to the reduction of the microbial activity and, as a consequence, to the reduction of the contaminant turnover.
- The higher treatment costs resulting from higher investment and operating costs cannot be justified due to the relatively short reduction of the treatment period (only a couple of days can be saved by dynamic treatment of dry soil during a treatment period over several months).

Although the dynamic bioreactors did not show the expected advantages in the "dry treatment," it may be useful to apply the mixers to mechanically pretreat soils prior to biological treatment. Ploughshare mixers with installed cutterheads are suitable for the pretreatment of soils or in the treatment of soils during remedial actions for desagglomeration, homogenization, disintegration (aeration), and admixing of additives.

One must be careful that the water content is adjusted at about 50% of the maximum water capacity. In many cases, the water content can be reduced, and the disintegration of the soil materials can be achieved by adding compost or sawdust. However, it may not be useful to apply the mixers, when mycel-forming organisms are of significant influence on the degradation process. But also in this regard, investigations on process optimization are necessary.

The formation of nonbioavailable (not extractable) residual contaminants shall be investigated (interaction of the contaminants with humus/compost), where the processes of carbon turnover in the soil are of specific interest.

REFERENCES

Ahlf, W., Gunkel, J., Rönnpagel, K., Stegmann, R. eds. 1993. *Bodenreinigung,* Bonn: Economica Verlag, pp. 275–286.

Alef, K. 1990. "Bestimmung mikrobieller Aktivität und Biomasse in Boden und Kompost mit der Dimethylsulfoxid-Reduktion," *UWSF-Z Umweltchem. Ökotox.,* 2(2):76–78.

Alef, K. 1991. *Methodenhandbuch Bodenmikrobiologie,* Landsberg/Lech: Ecomed Verlagsgesellschaft GmbH.

Alef, K., Kleiner, D. 1989. "Rapid and sensitive determination of microbial activity in soils and in soil aggregates by dimethylsulfoxide reduction," *Biol. Fertil. Soils,* 8:349–355.

Anderson, J. P. E., Domsch, K. H. 1978. "A physiological method for the quantitative measurement of microbial biomass in soils," *Soil Biol. Biochem.,* 10:215–221.

Anonymous 1980. "Gaseous emission measurement. Determination of volatile organic compounds, especially solvents. Flame ionization detector (FID)," VDI 3481, Blatt 3, VDI-Kommission Reinhaltung der Luft, Düsseldorf, Germany.

Anonymous 1981. "German standard methods for the examination of water, waste water and sludge; general measures of effects and substances (group H); determination of hydrocarbons (H18)," DIN 38 409, Beuth Verlag GmbH, Berlin, Germany.

Berg, J. D., Bennett, T. E., Nesgard, B. S., Mueller, J. G. 1992. *Proceedings of the international symposium on environmental contamination in central and eastern Europe,* October 12–16, 1992, Budapest, Hungary.

Berthe-Corti, L., Höpner, T., Michaelsen, M. 1991. "Einflüsse auf Kinetik und Bilanz des biologischen Kohlenwasserstoff-Abbaus im Wattsediment," Forschungsbericht, Bundesministerium für Forschung und Technologie, Germany.

Bhandari, A., Dove, D. C., Novak, J. T. 1994. "Soil washing and biotreatment of petroleum-contaminated soils," *Journal of Environmental Engineering,* 120(5):1151–1169.

Bollow, S. 1995. "Weiterführende Untersuchungen zum Einfluß der Sauerstoffkonzentration auf die biologische Umsetzung von Kohlenwasserstoffen im Boden," Diplomarbeit, Fachhoch-schule Hamburg, Germany.

Bossert, I., Bartha, R./Atlas, R.M., eds. 1984. *Petroleum Microbiology.* New York: Macmillan Publishing Company, pp. 435–473.

Castaldi, F. J., Bombaugh, K. J., McFarland, B. 1995. *Proceedings of the third international symposium on in situ and on-site bioreclamation,* April 24–27, 1995, San Diego, USA.

Civilini, M. 1994. "Fate of creosote compounds during composting," *Microbiology Europe,* 6(2):16–24.

Civilini, M., Domenis, C., de Bertoldi, M. 1995. *Proceedings of the international symposium on the science of composting*, 30 May–2 June, 1995, Bologna, Italy.

Cole, M. A., Liu, X. 1995. *Proceedings of the international symposium on the science of composting*, 30 May–2 June, 1995, Bologna, Italy.

Dalyan, U., Harder, H., Höpner, T. 1990. "Hydrocarbon biodegradation in sediments and soils, a systematic examination of physical and chemical conditions—Part III, Temperature," *Erdöl und Kohle-Erdgas-Petrochemie vereinigt mit Brennstoff-Chemie*, 43(11):437.

Dalyan, U., Harder, H., Höpner, T. 1991. "Hydrocarbon biodegradation in sediments and soils, a systematic examination of physical and chemical conditions—Part II. pH values," *Erdöl und Kohle-Erdgas-Petrochemie vereinigt mit Brennstoff-Chemie*, 43(9):337–342.

Diaz, L. F., Savage, G. M., Golueke, C. G. 1995. *Proceedings of the international syposium on the science of composting*, 30 May–2 June, 1995, Bologna, Italy.

Dott, W. 1992. *Proceeding of the international symposium on soil decontamination using biological processes*, December 6–9, 1992, Karlsruhe, Germany, pp. 133–135.

Ebertseder, T., Gutser, R./Faulstich, M., Kolb, F. R., Netter, R. E., eds. 1995. *Praxis der biologischen Abfallbehandlung*, München: Wassergüte- und Abfallwirtschaft der TU München, Germany, pp. 139–149.

Eschenbach, A., Kästner, M., Wienberg, R., Mahro, B. 1995. *Proceedings of the DECHEMA-Jahrestagung*, 30 May–1 June, 1995, Wiesbaden, Germany. Band II—Umwelttechnik, pp. 33–34.

Filip, Z./Weber, H. H., eds. 1990. Altlasten: Erkennen, Bewerten, Sanieren, Berlin: Springer-Verlag, pp. 300–328.

Gerth, J., Förstner, U./Stegmann, R., Franzius, V., eds. 1990. Reinigung kontaminierter Böden, Bonn: Economica Verlag, pp. 13–26.

Hinchee, R. E., Olfenbuttel, R. F., eds. 1991. *Proceedings of the first international symposium on in situ and on-site bioreclamation*, March 19–21, 1991, San Diego, USA.

Irvine, R. L. 1994. "Soil Bioreactors—Perspectives in the USA." Oral presentation at 418th DECHEMA-Kolloquium on "Biologische Bodenreinigung in Reaktoren," March 4, 1994, Frankfurt a.M., Germany.

Joergensen, R. G., Schmädeke, F., Windhorst, K., Meyer, B. 1994. *Proceedings of Ecoinforma '94*, September 5–9, 1994, Wien, Austria. Volume 6, pp. 225–235.

Kaiser, E.-A., Mueller, T., Joergensen, R. G., Insam, H., Heinemeyer, O. 1992. "Evaluation of methods to estimate the soil microbial biomass and the relationship with soil texture and organic matter," *Soil Biol. Biochem.*, 24(7):675–683.

Kästner, M., Eschenbach, A., Schaefer, G., Hupe, K., Wienberg, R., Stegmann, R., Mahro, B. 1995a. *Proceedings of the DECHEMA-Jahrestagung '95*, 30 May–1 June, 1995, Wiesbaden, Germany. Band II - Umwelttechnik, pp. 87–88.

Kästner, M., Lotter, S., Heerenklage, J., Breuer-Jammeli, M., Stegmann, R.,

Mahro, B. 1995b. "Fate of ^{14}C-labeled anthracene and hexadecane in compost manured soil," *Applied Microbiology and Biotechnology,* in press.

Kehres, B., A. Maile, eds. 1994. *Methodenbuch zur Analyse von Kompost,* Bundes-gütegemeinschaft Kompost e.V., Verlag Abfall Now e.V., Stuttgart, Germany.

Klaus, C. 1994. "Vergleichende Untersuchungen zur Biomassebestimmung in kontaminierten Böden," Diplomarbeit, Fachhochschule Hamburg, Germany.

Kleijntjens, R. H. 1991. "Biotechnological slurry process for the decontamination of excavated polluted soils." Ph.D. Thesis, Technical University of Delft, The Netherlands.

Klein, J. ed. 1992. Laboratory methods for the evaluation of biological soil cleanup processes, DECHEMA, Frankfurt a.M., Germany.

Klein, J. 1994. "Entwicklungsperspektiven von Reaktorverfahren zur biologischen Boden-reinigung. Protokoll der achten Sitzung des Interdisziplinären Arbeitskreises 'Umweltbio-technologie-Boden,' " March 4, 1994, Frankfurt a.M., Germany.

Krogmann, U. 1992. *Proceedings of the second international forum on resource recovery from waste,* September 21–24, 1992, Imola, Italy.

Lilja, R., Votila, J. 1995. *Proceedings of the international symposium on the science of composting,* 30 May–2 June, 1995, Bologna, Italy.

Lotter, S. 1995. "Kohlenstoffbilanzierung und Kohlenstoffumsetzung bei der biologischen Bodensanierung." Ph.D. Thesis, Technical University of Hamburg-Harburg, Germany.

Lotter, S., Brumm, A., Bundt, J., Heerenklage, J., Paschke, A., Steinhart, H., Stegmann, R. 1993. *Proceedings of fourth international KfK/TNO conference on contaminated soil,* May 3–7, 1993, Berlin, Germany. pp. 1235–1246.

Lotter, S., Stegmann, R., Heerenklage, J. 1992. *Proceedings of international symposium on soil decontamination using biological processes,* December 6–9, 1992, Karlsruhe, Germany. pp 219–227.

Mahro, B., Eschenbach, A., Kästner, M., Schaefer, G. 1994. *Proceedings of the symposium on "Biologischer Abbau von polycyclischen aromatischen Kohlen-wasserstoffen,"* November 18–19, 1993, Berlin, Germany. pp. 55–68.

Nitschke, V. 1994. Entwicklung eines Verfahrens zur mikrobiologischen Reinigung feinkörniger, mit polyzyklischen aromatischen Kohlenwasserstoffen belasteter Böden. Ph.D. Thesis, University of Paderborn, Germany.

Parthen, J. 1992. "Untersuchungen zur mikrobiologischen Sanierung von mit PAK verunreinigten feinkörnigen Böden im Drehtrommelreaktor." Ph.D. Thesis, University of Hannover, Germany.

Pennerstorfer, C., Bauer, E., Kandeler, E., Braun, R. 1993. *Proceedings of the first European conference on integrated research for soil and sediment protection and remediation (EUROSOL),* September 6–12, 1993, Maastricht, The Netherlands. pp. 682–683.

Pettersen, B. W., Andersen, B. M., Baggesgard, H., Jensen, L. H., Lyngsø, B. 1993.

Proceedings of the first European conference on integrated research for soil and sediment protection and remediation (EUROSOL), September 6–12, 1993, Maastricht, The Netherlands. pp. 684–685.

Richnow, H. H., Seifert, R., Hefter, J., Mahro, B., Kästner, M., Michaelis, W. 1994. "Metabolites of xenobiotica and mineral oil constituents linked to macromolecular organic matter in polluted environments," *Organic Geochemistry.*

Schaefer, G., Kästner, M., Kasche, V., Mahro, B. 1992. *Proceedings of the 10th Dechema Fachgesprächs Umweltschutz,* June 24–26, 1992, Leipzig, Germany. pp. 385–388.

Stegmann, R., Lotter, S., Heerenklage, J. 1991. *Proceedings of the first international symposium on in situ and on site bioreclamation,* March 19–21, 1991, San Diego, USA. On-Site Bioreclamation, pp. 188–208.

Tabak, H. H., Govind, R., Fu, C., Yan, X., Pfanstiel, S., Gao, C. 1995. *Proceedings of the 21st annual risk reduction engineering laboratory (RREL) research symposium,* April 4–6, 1995, Cincinnati, USA. pp. 49–55.

Woyczechowski, H. 1993. "Einfluß der Sauerstoffkonzentration auf den biologischen Abbau von Kohlenwasserstoffen im Boden," Diplomarbeit, Fachhochschule Hamburg, Germany.

Surfactant-Enhanced Aquifer Remediation: Fundamental Processes and Practical Applications

KURT D. PENNELL

School of Civil & Environmental Engineering
Georgia Institute of Technology
Atlanta, GA 30332-0512, USA

LINDA M. ABRIOLA

Department of Civil & Environmental Engineering
University of Michigan
Ann Arbor, MI 48109, USA

INTRODUCTION

The widespread detection of organic contaminants in groundwater has prompted the initiation of remediation efforts at numerous sites throughout the United States. The most common technology employed for aquifer remediation is pump-and-treat, in which contaminated groundwater is extracted from the subsurface using wells or drainage systems and then treated aboveground. Frequently, this approach results in a rapid decrease in aqueous-phase concentrations of the contaminant and a reduction in the size of the contaminated groundwater plume. This initial success, however, is often followed by a leveling-off and gradual decline in contaminant concentrations, which may persist for years or decades (Mackay and Cherry, 1989; NRC, 1994). Thus, the remediation of an aquifer to a specific health-based standard (e.g., aqueous- or solid-phase concentration) is often unachievable within an acceptable time frame. In addition, the costs associated with maintaining a pump-and-treat system, monitoring contaminant levels in groundwater, and litigation proceedings over prolonged periods of time can be prohibitive.

693

The most recalcitrant remediation scenarios addressed by regulatory agencies often involve aquifers containing organic compounds existing as a separate liquid phase or nonaqueous phase liquid (NAPL). Classic examples of NAPLs include chlorinated solvents, such as trichloroethylene (TCE), and petroleum-based products, such as gasoline. Upon entering the unsaturated zone, NAPLs will migrate downward as a result of gravitational and capillary forces. If the volume of NAPL released to the subsurface is sufficient to reach the water table, an organic liquid that is more dense than water (DNAPL) will tend to migrate vertically through the saturated zone until a confining or low-permeability layer is reached (Figure 1). In contrast, an NAPL that is less dense than water (LNAPL) will tend to spread laterally within the capillary fringe, forming a "lens" of free product. Subsequent fluctuations in the water table, however, may result in vertical displacement of the LNAPL lens and consequent redistribution within the saturated zone. The overall distribution of an NAPL within an aquifer formation may be extremely complex due to the presence of small- and large-scale heterogeneities and temporal flow variations.

As the NAPL is transported through the subsurface, a portion of the organic liquid will be retained within soil pores as ganglia or globules due to interfacial forces (Figure 1). Under normal flow regimes, the entrapped NAPL will not be displaced from the porous media. This immobile fraction of NAPL is commonly referred to as "residual" and may occupy between 5 and 40% of the pore volume (Schwille, 1984; Hunt et al., 1988). Due to the low solubility of most organic liquids, the residual NAPL frequently represents a long-term source of groundwater contamination. Concentrations of NAPLs in groundwater are rarely observed to exceed 10% of their aqueous solubility. This phenomenon has been attributed to irregular NAPL distributions, nonuniform flow patterns, and dilution effects (Gillham and Rao, 1990), as well as rate-limited mass transfer between the organic and aqueous phases (Powers et al., 1992; Gellar and Hunt, 1993). Thus, conventional pump-and-treat systems, which are based on NAPL dissolution into groundwater, have proven to be an ineffective and costly method of aquifer restoration. In the presence of NAPLs, the initial success of pump-and-treat systems can be attributed to removal of the contaminated groundwater plume, while the persistent low levels of contaminant in extraction wells are due to NAPL dissolution into "clean" groundwater passing through the zone of residual contamination.

To overcome such limitations, surfactants have been proposed as a means for enhancing the performance of conventional pump-and-treat systems. This approach is based on two recovery processes: (a) micellar

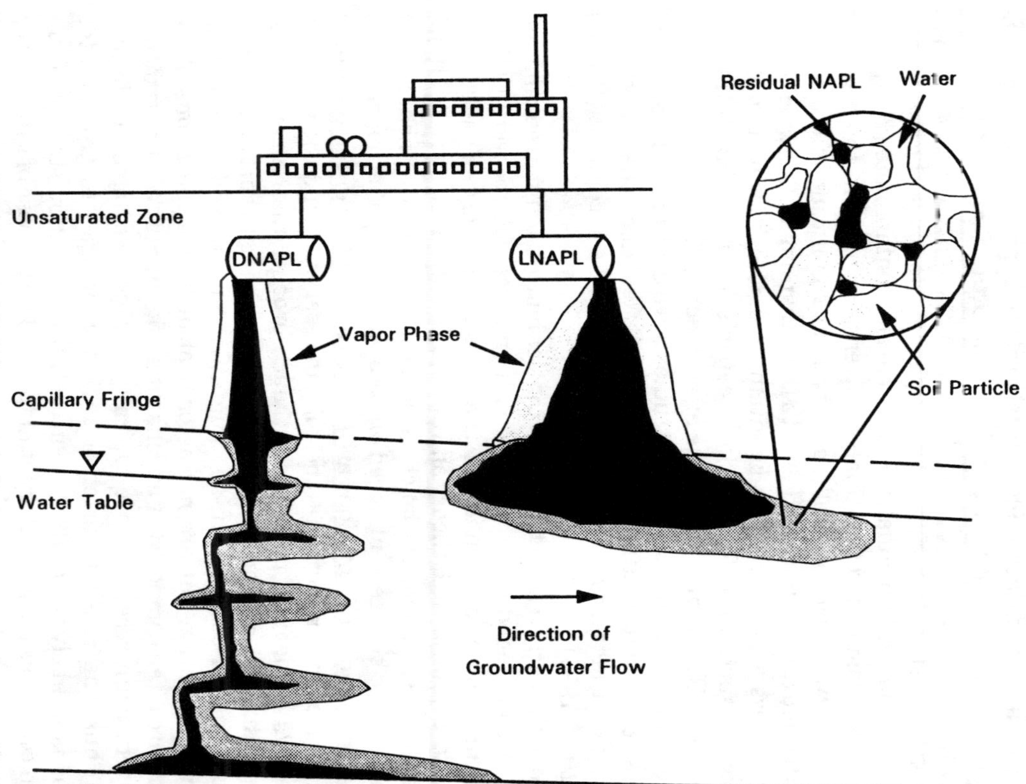

Figure 1. Schematic diagram of a DNAPL and LNAPL spill showing the entrapped organic liquid.

solubilization of the residual NAPL and (b) mobilization of entrapped NAPL as free product through interfacial tension reductions. Initial studies designed to evaluate the use of surfactants for enhanced NAPL recovery have yielded mixed results. Early success was realized by the Texas Research Institute (1985) in laboratory experiments, where approximately 80% recovery of gasoline from Ottawa sand was attained using a mixture of 2% Richonate YLA, an alkyl benzene sulfonate, and 2% Hyonic PE-90, an ethoxylated nonylphenol. However, the application of similar surfactant formulations at a field site contaminated by JP-4 jet fuel, fire retardant, and chlorinated solvents resulted in the complete plugging of two injection wells and had no significant effect on the concentration of contaminants beneath the remaining wells (Nash, 1987). Better success was reported by Fountain et al. (1991), who identified a number of surfactant formulations capable of removing tetrachloroethylene (PCE) from soil columns after injecting 7 to 14 pore volumes of surfactant solution. More recent studies have confirmed the ability of aqueous surfactant solutions to dramatically enhance the recovery of residual dodecane and PCE from one-dimensional soil columns (Pennell et al., 1993, 1994).

Much of the early research in surfactant-enhanced aquifer remediation (SEAR) was founded on experience with enhanced oil recovery (EOR), where surfactants are applied in the tertiary recovery of petroleum. Excellent reviews of EOR theory and application are given by Lake (1989) and Taber (1981). There are, however, several important distinctions that should be made between enhanced oil recovery and aquifer remediation. For example, aquifer formations are subject to much lower temperatures and pressures in comparison to petroleum reservoirs, and the variation among NAPL chemical properties is much broader than among petroleum deposits. In addition, regulatory issues relating to surfactant toxicity, recovery, and potential off-site migration must be addressed during aquifer remediation but are not as critical in EOR applications. Also, aquifers are often composed of unconsolidated, unconfined porous media while petroleum reservoirs are typically confined, sedimentary formations. Thus, the direct transfer of EOR technologies to groundwater systems is unlikely to result in a successful remediation strategy, from both an economic and regulatory standpoint. The purpose of this chapter is to provide a concise description of the fundamental principles, current research efforts, and future directions of surfactant-enhanced aquifer remediation. Specific topics to be addressed include micellar solubilization, NAPL mobilization, surfactant sorption, preliminary laboratory- and field-scale studies, and numerical modeling of SEAR. Although the primary focus of this chapter will be on the remediation of NAPL-contaminated aquifers, many of the processes discussed herein are applicable to sorbed-phase contaminants.

SURFACTANT CLASSIFICATION

Surfactants are amphiphilic compounds, meaning that they possess both polar (hydrophilic) and nonpolar (hydrophobic) groups or moieties. The hydrophobic group is typically a long-chain or halogenated hydrocarbon, while the hydrophilic group is ionic or highly polar. This unique chemical structure results in the tendency for surfactants to adsorb at the interface between two phases and markedly alter interfacial properties. For example, surfactants will tend to accumulate at an oil-water interface, with the hydrophobic portion of surfactant associated with the oil phase and the hydrophilic portion associated with the aqueous phase. This characteristic behavior gives rise to the term *surfactant*, which is a contraction of surface-active agent. There are literally thousands of different types of surfactants, with applications that range from food additives to soaps and detergents. Commercial surfactants are typically grouped into the following classes based on the nature of their hydrophilic group: anionic, cationic, nonionic, and amphoteric (Table 1). Biosurfactants are often treated as a separate class of surfactant because they are naturally occurring and possess unique properties. A brief description of each surfactant type is given below. More detailed reviews of the surfactant classes can be found in the *Surfactant Science Series* published by Marcel Dekker (e.g., see Vol. 12, 23, 48, 53, and 56).

Anionic

Anionic surfactants are characterized by a negatively charged head group (typically sulfate or sulfonate) associated with a positively charged metal cation (such as sodium). In aqueous solution, the surfactant molecule ionizes to yield a free metal cation and the anionic surfactant monomer. Anionic surfactants are widely used as detergents and have been the surfactant of choice in much of the EOR research. Their selection for EOR applications can be attributed to their relatively low tendency to adsorb, stability at high temperature and pressure, and low cost (Lake, 1989). Anionic surfactants are, however, extremely sensitive to background electrolyte concentrations. This property may be used to great advantage, for example, when changes in phase behavior brought about by increased electrolyte concentration result in enhanced oil solubilization capacity. Conversely, the presence of a background electrolyte above a critical level may lead to anionic surfactant precipitation.

Table 1. Representative examples of different surfactant types.

Anionic

Sodium dodecyl sulfate (SDS)

$$C_{12}H_{25}OSO_3^-Na^+$$

Sodium dihexylsulfosuccinate (Aerosol MA-100)

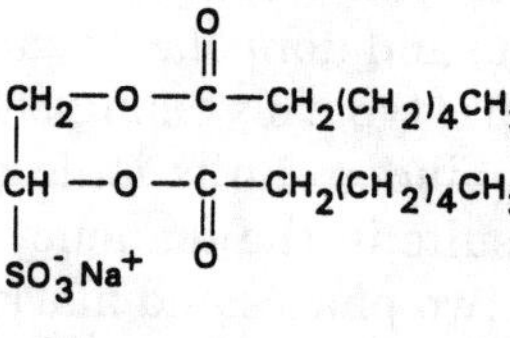

$$CH_2-O-\overset{\overset{\displaystyle O}{\|}}{C}-CH_2(CH_2)_4CH_3$$
$$CH-O-\underset{\underset{\displaystyle O}{\|}}{C}-CH_2(CH_2)_4CH_3$$
$$SO_3^-Na^+$$

Cationic

Cetyltrimethylammonium bromide (CTAB)

$$C_{16}H_{33}N(CH_3)_3^+Br^-$$

Trimethylphenylammonium chloride (TMPA)

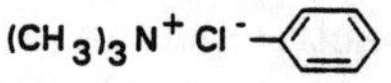

$$(CH_3)_3N^+Cl^- -\bigcirc$$

Nonionic

POE (20) Sorbitan Monooleate (Witconol 2722, Tween 80)

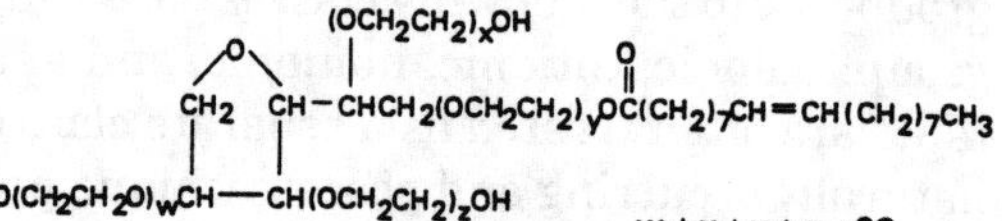

$$CH_2 \quad CH-CHCH_2(OCH_2CH_2)_yOC(CH_2)_7CH=CH(CH_2)_7CH_3$$
$$(OCH_2CH_2)_xOH$$
$$HO(CH_2CH_2O)_wCH-CH(OCH_2CH_2)_zOH$$
$$w+x+y+z=20$$

Dodecyl Alcohol Ethoxylate (n=9) (Witconol SN-120)

$$C_{12}H_{25}O(CH_2CH_2O)_9H$$

Amphoteric

N-dodecyl dimethyl betaine

$$C_{12}H_{25}N^+(CH_3)_2CH_2SO_3^-Na^+$$

N-dodecyl aminopropionic acid

$$C_{12}H_{25}N^+H_2CH_2CH_2COO^-$$

Biosurfactant

Rhamnolipid 1

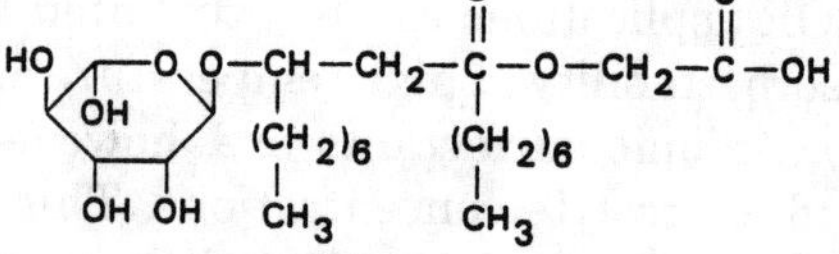

Cationic

Cationic surfactants possess a positive charge that is typically associated with an amine or quaternary ammonium group (Table 1). Due to their positive charge, cationic surfactants are strongly sorbed to soil minerals, which generally possess a net negative charge. This property makes cationic surfactants unsuitable for most remediation systems involving subsurface flushing through contaminated zones. However, cationic surfactants hold promise as a means for removing contaminants from aqueous waste streams. For example, cationic surfactants have been utilized to increase the sorptive capacity of clay minerals (Lee et al., 1989; Srinivasan and Fogler, 1990) and zeolites (Haggerty and Bowman, 1994) and could potentially alter soil capillary pressure-saturation relationships in subsurface systems (Desai et al., 1992). Many cationic surfactants, however, are toxic to bacteria and fungi. Although this property makes them ideally suited for disinfection and other medicinal purposes, injection of cationic surfactants into the subsurface may result in the elimination of native soil microorganisms and significant changes in microbial ecology.

Nonionic Surfactants

Nonionic surfactants are widely used in food products, pharmaceuticals, and detergents. In the past few years, they have received considerable attention as potential candidates for use in subsurface remediation applications. In contrast to cationic and anionic surfactants, the water-soluble or hydrophilic moiety of a nonionic surfactant consists of either hydroxyl groups or an ethylene oxide (EO) chain (Table 1). A unique characteristic of nonionic surfactants is that the number of EO groups can be varied during synthesis, thereby imparting a greater or lesser tendency to associate with water. The relative proportions of hydrophobic and hydrophilic functional groups within a nonionic surfactant are often represented by the hydrophile-lipophile balance (HLB). The HLB of a nonionic surfactant can be calculated using the following general equation (Rosen, 1989):

$$\text{HLB} = 20 \times \frac{M_H}{M_H + M_L} \tag{1}$$

where M_H and M_L are the molecular weights of the hydrophilic and lipophilic portions of the surfactant, respectively. For surfactants containing only ethylene oxide as the hydrophilic group, Equation (1) may be written as

$$HLB = \frac{wt\% \ EO}{5} \qquad (2)$$

Although the behavior of nonionic surfactants in solution is not strongly influenced by changes in pH or electrolyte concentration, compared to ionic surfactants, nonionic surfactants are sensitive to temperature changes. As temperature increases, aqueous solutions of nonionic surfactants typically separate into a surfactant-rich and surfactant-depleted phase. The temperature at which this separation occurs is referred to as the cloud point. The presence of organic compounds may either increase or decrease the cloud point and should be considered in the design of subsurface remediation systems.

Amphoteric

Amphoteric surfactants are unique in that they may function as either cationic, anionic, or nonionic species, depending upon the pH of the solution and the specific structure of the surfactant. The alkylaminopropionic acid shown in Table 1 has an isoelectric point at pH ~4 and will exhibit properties of a cationic surfactant below pH 4, an anionic surfactant above pH 4, and a zwitterionic (both anionic and cationic) surfactant at pH ~4 (Rosen, 1989). In contrast, the dimethylbetaine shown in Table 1 will behave as zwitterionic regardless of the pH. Amphoteric surfactants are known for their mildness to skin and eyes and, thus, are widely used in eyedrops and cosmetics (Bluestein and Hilton, 1982). However, because these surfactants are relatively expensive, tend to adsorb strongly, and are produced in small quantities, they have not been considered as candidates for subsurface remediation applications.

Biosurfactants

Biosurfactants can be broadly categorized as naturally occurring surface active agents derived from microbial, plant, or animal sources. From a remediation perspective, much of the recent work has focused on surfactants produced by various bacterial strains (e.g., *Pseudomonas, Rhodococcus, Bacillus,* and *Acinetobacter* spp.) growing on glucose and/or water-insoluble substrates such as hexadecane (Van Dyke et al. 1993; Finnerty, 1994). The resulting microbial surfactants represent a diverse range of biomolecules, including fatty acids, mono- and diglycerides, glycolipids, and phospholipids. The structure of a representative biosurfactant, rhamnolipid 1, is shown in Table 1 (Sreekala and Shreve, 1994). In general, the isoelectric point of hydrophilic functional groups associated with biosurfactants is on the order of pH 4 to 5, and thus, most biosurfactants are considered to be anionic in neutral pH environments. However, the structure and functionality of biosurfactants may be more analogous to synthetic nonionic surfactants, such as ethoxylated sorbitol esters and propylene glycol esters. The use of biosurfactants in remediation efforts has gained attention because they can be produced naturally, have been shown to enhance the biodegradation of hydrocarbon contaminants (Zhang and Miller, 1992), and could potentially be produced in situ. Due to their limited availability, however, biosurfactants are unlikely to be economically or practically feasible for use in flushing technologies. Nevertheless, biosurfactants could play an important role in the desorption and bioavailability of contaminants in subsurface systems.

Although surfactants are classified based upon their hydrophilic head group, differences in surfactant behavior are also related to the nature of the hydrophobic moiety or tail. Traditionally, the hydrophobic tail consisted of a mixture of linear alkyl groups (C_{12}–C_{20}) derived from natural fats and oils. Other common hydrophobic groups found in commercial surfactants include branched alkyls (C_8–C_{20}), linear alkylbenzenes (C_8–C_{15}), polypropylene oxide derivatives, and alkylphenols. In general, as the length of the hydrophobic tail increases, the solubility of the surfactant in water will decrease, interfacial adsorption will be more prevalent, and for ionic surfactants, there will be a greater tendency to precipitate in the presence of salts (Rosen, 1989). With respect to surfactant biodegradation, the presence of an aromatic group or branching within the hydrophobic moiety typically inhibits surfactant transformation. For example, octylphenol ethoxylates (nonionic), such as the Triton X series, have been shown to be extremely recalcitrant to biodegradation in wastewater treatment systems and natural bodies of water.

Although the behavior of a particular surfactant can be inferred from its chemical structure, it should be recognized that surfactants are rarely available in a pure form. In fact, the chemical structure given for most nonionic surfactants (e.g., see Table 1) represents an "average structure" that varies in both the degree of ethoxylation and the nature of the hydrophobic group. This type of surfactant is typically referred to as polydisperse. Thus, the physical and chemical properties given for a "pure" or homogeneous surfactant may be quite different from the commercially available form of the surfactant.

MICELLAR SOLUBILIZATION

Micelle Formation and Structure

In addition to interfacial adsorption, surfactants possess the unique property of aggregating in solution to form micelles. In very dilute solutions, surfactants only exist as individual molecules or monomers. As the surfactant concentration increases, however, the hydrophobic moieties of the surfactant monomers associate with one another to form aggregates or micelles. The surfactant concentration corresponding to the initiation of micelle formation is referred to as the critical micelle concentration (CMC). In general, any surfactant property that reduces water solubility will increase the tendency for the surfactant monomers to associate with one another, thereby lowering the concentration required to initiate micelle formation. For example, nonionic surfactants exhibit much lower CMCs than ionic surfactants containing similar hydrophobic groups. Likewise, the CMC of surfactants within the same class or containing similar hydrophilic groups decreases as the number of carbon atoms in the hydrophobic group increases. Above the CMC, the concentration of surfactant monomers in solution remains constant, while the number of micelles increases with increasing surfactant concentration. Thus, the aggregation number (N_A), or number of surfactant monomers per micelle, remains essentially unchanged until the solubility limit of the surfactant is approached. It should be noted, however, that surfactant micelles are dynamic structures, with surfactant monomers continuously joining or leaving the micelle (Clint, 1992). Thus, reported aggregation numbers actually represent an average value over some period of time. The aggregation number, CMC, and related properties of several representative nonionic, anionic and cationic surfactants are given in Table 2.

The size and shape of surfactant micelles can have a significant impact on the phase behavior and capacity of a surfactant to solubilize organic compounds. Traditionally, surfactant micelles were considered to be spherical in shape, although both ellipsoid and vesicle structures have been reported in the literature. In aqueous micellar systems, the hydrophilic (polar) head group of the surfactant is oriented toward the aqueous phase, while the hydrophobic tail is oriented inward, away from the aqueous phase. Thus, an idealized micelle can be envisioned as a hydrophilic mantle or shell surrounding a hydrophobic core. The hydrophilic shell acts to exclude water from the interior of the micelle, resulting in a hydrophobic core that, in many respects, functions as a liquid hydrocarbon phase (Attwood and Florence, 1983). In nonpolar solvents the sur-

Table 2. *Critical micelle concentration (CMC) and related properties of representative nonionic, anionic, and cationic surfactants.*

Surfactant Trade Name	Ave. Mol. Formula	Ave. MW (g/mole)	HLB[a]	CMC[b] (mg/L)	N_A[c]
Nonionic					
Witconol SN-120	$C_{10-12}H_{21-25}O(CH_2CH_2O)_9H$	569	13.6	54	105
Tergitol NP-15	$C_9H_{19}C_6H_4O(CH_2CH_2O)_{15}H$	882	14.3	97	100
Tween 80	$C_{18}H_{34}O_2C_6H_{10}O_4(CH_2CH_2O)_{20}$	1310	15.0	13	110
Anionic					
SDS	$C_{12}H_{25}OSO_3^-Na^+$	288	na	2100	64
Cationic					
CTAB	$C_{16}H_{33}N^+(CH_3)_3Br^-$	364	na	361	75

[a]Hydrophile-lipophile balance (HLB = wt% EO/5) (Rosen, 1989).
[b]Critical micelle concentration (CMC) (Becher, 1967).
[c]Aggregation number (N_A) (Schick et al., 1962; Arnarson and Elworthy, 1981).

factant orientation is reversed, with the outer portion of the micelle consisting of the hydrophobic moiety of the surfactant.

To estimate the structure of surfactant micelles, Israelachvili (1985) developed a critical packing parameter based on the ratio of the volume of the hydrophobic group (V_H) to the product of the hydrophobic chain length (l_c) and the cross-sectional area occupied by the hydrophilic group (a_o) (Table 3). From Tanford (1980), the volume of the hydrophobic group (Å^3) can be calculated using the following equation:

$$V_H = [27.4 + 26.9(N_C - 1)] \qquad (3)$$

where N_C is the number of carbon atoms in the hydrophobic chain. The data shown in Table 3 suggest that the spherical model may only be appropriate for anionic surfactants in which the effective area occupied by the hydrophilic head group (a_o) is quite large due to electrostatic repulsion. Nonionic surfactants and ionic surfactants in high-salt systems yield critical packing numbers in the range of one-third to one-half, which correspond to nonspherical or ellipsoid micelles (Table 3). The formation of vesicles, which consist of a circular bilayer with aqueous phase existing both internal and external to the structure, is typically limited to surfactants containing double-chain lipids and relatively large head groups.

Robson and Dennis (1977) conducted a more detailed geometrical analysis to evaluate three possible micelle models for Triton X-100, an octylphenol ethoxylate ($n = 9.5$) (Figure 2). The authors showed that a spherical micelle could only be possible if several of the polyoxyethylene head groups were located either partially or wholly within the hydrophobic core. For this reason, it was hypothesized that Triton X-100 micelles exist as oblate or prolate ellipsoids. Additional viscosity and X-ray diffraction measurements were consistent with an oblate (disc-like) rather than

Table 3. *Estimated structure of surfactant micelles based on volume of the hydrophobic core (V_H), length of the hydrophobic chain (l_c), and the cross-sectional area occupied by the hydrophilic group (a_o) (Israelachvili, 1985).*

$V_H/l_c a_o$	Micelle Structure
0 to 1/3	Spherical in aqueous solutions
1/3 to 1/2	Oblate or prolate ellipsoids in aqueous solutions
1/2 to 1	Vesicles or bilayers in aqueous solutions
>1	Inverse (reverse) micelles in nonpolar media

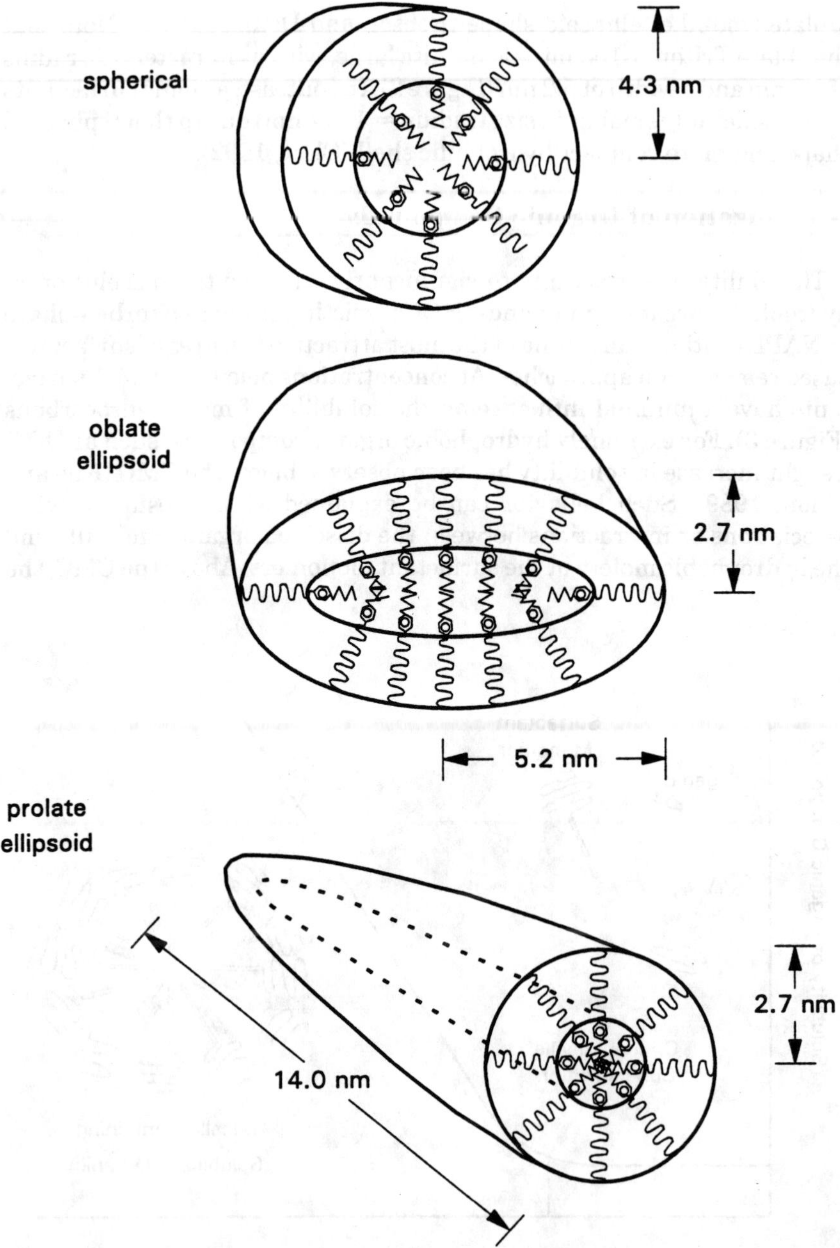

Figure 2. Cross-sectional diagrams of (a) spherical, (b) oblate ellipsoid, and (c) prolate ellipsoid micelles proposed for Triton X-100 (adapted from Attwood and Florence, 1983).

prolate (rod-like) ellipsoid shape (Robson and Dennis, 1977). Note that the oblate Triton X-100 micelle is quite large, with a characteristic radius of 2.7 nm and a width of 5.2 nm (Figure 2). In contrast, anionic surfactants are considerably smaller in size (radius $\approx$ 1 to 2 nm) due to their spherical shape and more compact hydrophilic shell (Clint, 1992).

Solubilization of Organic Compounds

The ability of surfactants to significantly enhance the dissolution of hydrophobic organic compounds (HOCs) existing either as a sorbed-phase or NAPL contaminant is one of the most attractive features of surfactant-based remediation approaches. At concentrations below the CMC, surfactants have a minimal influence on the solubility of most hydrocarbons (Figure 3). For extremely hydrophobic organic compounds, such as DDT, a slight increase in solubility has been observed below the CMC (Kile and Chiou, 1989). Such behavior can be explained as the result of weak associations or interactions between the dissolved organic molecule and the hydrophobic moiety of the surfactant monomers. Above the CMC, the

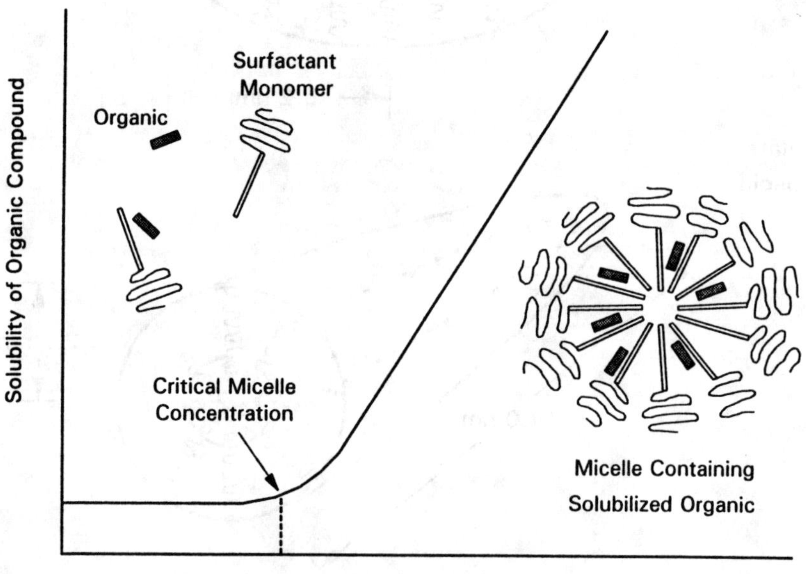

Figure 3. The effect of surfactants on the solubility of hydrophobic organic compounds above and below the critical micelle concentration (CMC).

solubility of HOCs in aqueous surfactant solutions has been consistently shown to increase in a linear fashion (Figure 3). The observed linear enhancement in solubility has been attributed to the incorporation or partitioning of HOCs within surfactant micelles (e.g., Klevens, 1950; Kile and Chiou, 1989; Edwards et al., 1991; Valsaraj and Thibodeaux, 1989). The resulting solution is thermodynamically stable, consisting of NAPL solubilized or dispersed within surfactant micelles that are on the order 10 to 100 nm in diameter. This type of system is often referred to as an NAPL-in-water (NAPL/W) microemulsion or Winsor Type I system. In a conceptual sense, micellar solubilization is similar to the partitioning of HOCs into soil organic matter, although it should be recognized that soil organic matter is a far less uniform and structured environment than a surfactant micelle.

The solubilization capacity of a particular surfactant can be represented as the ratio of the number of moles of hydrocarbon solubilized to the number of moles of surfactant in the micellar form. A molar solubilization ratio (MSR) may be defined as (Edwards et al., 1991)

$$\text{MSR} = \frac{C_o - C_{o,cmc}}{C_s - C_{s,cmc}} \tag{4}$$

where C_o is the molar concentration of organic in a surfactant solution, $C_{o,cmc}$ is the molar concentration of organic at the CMC, C_s is the molar concentration of surfactant, and $C_{s,cmc}$ is the molar concentration of surfactant at the CMC. Given the potential for slight solubility enhancements below the CMC and the fact that the presence of an organic solute may reduce the CMC (Rosen, 1989), the values of $C_{o,cmc}$ and $C_{s,cmc}$ are often not explicitly known. The MSR, however, can be calculated from the slope of the solubility curve above the CMC, when concentrations are expressed on a molar basis.

Although the MSR has been widely used to characterize the solubility of HOCs in micellar solutions, relatively few studies have addressed the effects of organic properties and surfactant structure on the solubilization of environmentally relevant NAPLs. Based on MSR data, it has been shown that micellar solutions have a greater capacity to solubilize polarizable hydrocarbons, such as tetrachloroethylene (40,000 mg/l), than extremely hydrophobic compounds, such as dodecane (3500 mg/l) (Pennell et al., 1996b). These trends can be attributed to the fact that strongly hydrophobic organic compounds are solubilized almost entirely within the hydrophobic core of surfactant micelles. In contrast, polarizable hydrocarbons may be distributed throughout the hydrophilic mantle and the hydrophobic core (Mukerjee and Cardinel, 1978; Nagarajan et al.,

1984). This difference in locus of solubilization results in the greater overall capacity of surfactants to solubilize polarizable hydrocarbons, although the relative magnitude of the solubility enhancement is likely to be greater for extremely hydrophobic compounds. To further investigate the locus of solubilization, Diallo et al. (1994) obtained MSR data for eleven NAPLs using a homologous series of dodecyl alcohol ethoxylates with HLBs ranging from 12 to 18. The authors were able to correlate a decrease in the solubilization of saturated hydrocarbons, such as dodecane, to the reduction in hydrophobic core volume. These findings indicate that it is possible to relate the solubilization behavior of a particular surfactant to changes in the size and structure of surfactant micelles.

A second approach frequently used to characterize the solubilization of hydrophobic organic compounds is based on the partitioning of organic compounds between surfactant micelles and the aqueous phase. The micelle-aqueous phase partition coefficient (K_{mw}) can be defined as

$$K_{mw} = \frac{X_m}{X_a} \tag{5}$$

where X_m is the mole fraction of organic in the micellar phase and X_a is the mole fraction of organic in the aqueous phase. The mole fraction of organic in the micellar phase can be calculated from the MSR (Edwards et al., 1991):

$$X_m = \frac{\text{MSR}}{(1 + \text{MSR})} \tag{6}$$

For dilute solutions, the mole fraction of organic in the aqueous phase can be estimated as follows:

$$X_a = \frac{C_{o,cmc}}{V_w} \tag{7}$$

where V_w is the molar volume of water (e.g., 55.402 mol/l at 25°C).

The K_{mw} values derived for surfactant-organic systems can be correlated to the partitioning behavior of organic compounds between octanol and water, represented by the octanol-water partition coefficient, K_{ow}. Edwards et al. (1991) obtained a linear relationship between $\log K_{mw}$ and $\log K_{ow}$ for the solubilization of several polycyclic aromatic hydrocarbons (PAHs) by Triton X-100 (Figure 4). A log-linear relationship between K_{mw} and K_{ow} was also reported by Valsaraj and Thibodeaux (1989) for the

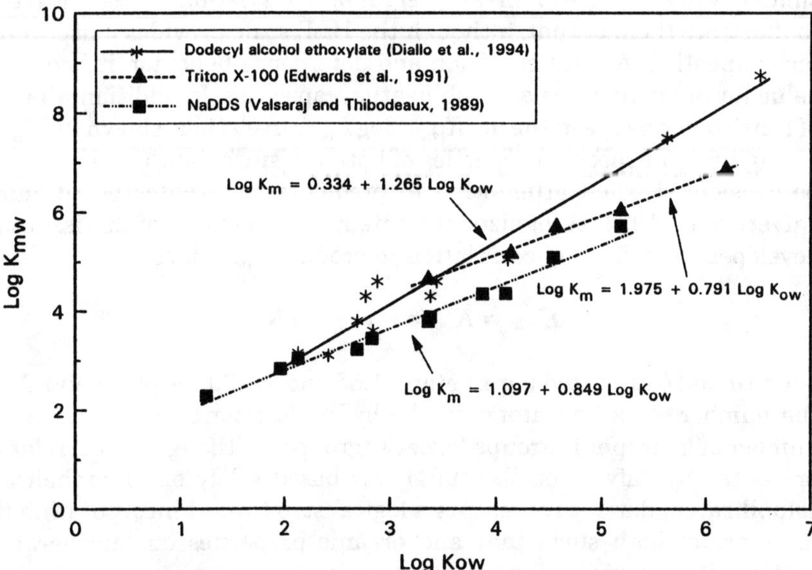

Figure 4. Relationship between micelle water (K_{mw}) and octanol-water (K_{ow}) partition coefficients for a range of surfactants and organic compounds.

solubilization of eleven HOCs in micellar solutions of sodium dodecyl sulfate (SDS). Although the slopes of these two relationships were similar, the intercept obtained by Edwards et al. (1991) was slightly larger than that reported by Valsaraj and Thibodeaux (1989), indicating that Triton X-100 has a greater solubilization capacity than SDS. It should be recognized that these correlations provide only a first-order approximation of solubilization capacity, since seemingly minor differences in log K_{mw} values can translate into sizable differences in solubility values when expressed on a mass per volume basis.

A third correlation was developed from solubilization data reported by Diallo et al. (1994) for a linear dodecyl ethoxylate (HLB = 14.2). Notice that the log-linear correlation has a significantly smaller intercept, but a larger slope than the other two correlations. Thus, the linear dodecyl ethoxylate exhibited a substantially greater capacity to solubilize extremely hydrophobic compounds, such as dodecane (log K_{ow} = 6.51), compared with Triton X-100 and SDS. Although differences in the solubilization behavior of an anionic (SDS) and a nonionic (linear dodecyl ethoxylate) surfactant would be expected, the large discrepancy between reported solubilization values for Triton X-100 (HLB = 15.0) and the linear dodecyl ethoxylate (HLB = 14.2) was not anticipated due to the similarity in surfactant HLB and degree of ethoxylation. These data serve to illustrate the fact that, although the HLB scale provides a useful tool for delineating general emulsion and detergency behavior, it is of little value for predicting actual solubilization capacities. In addition, the lack of consistency between the $\log K_{mw} - \log K_{ow}$ correlations shown in Figure 4 suggests that specific properties of both the surfactant and HOC must be considered when attempting to predict micelle-water partitioning. Jafvert et al. (1994) recognized the potential importance of this issue and developed the following correlation to predict K_{mw} values:

$$K_{mw} = K_{ow}(a * N_C - b * N_H)$$ (8)

where a and b are fitted parameters (1.65 and -0.30, respectively), N_C is the number of carbon atoms in the hydrophobic group, and N_H is the number of hydrophilic groups (e.g., EO groups). Although the correlation presented by Jafvert et al. (1994) was based solely on chlorobenzene solubilization data, it represents a logical step toward incorporating the influence of both surfactant and organic properties on micelle-water partitioning. Obviously, more complex approaches may be required to account for specific interactions between the POE shell of ethoxylated surfactants and polarizable hydrocarbons, such as PCE.

NAPL MOBILIZATION

Interfacial Tension (IFT) Reductions

Due to their amphiphilic nature, surfactants exhibit a marked tendency to accumulate at phase boundaries and consequently influence interfacial properties. The adsorption of a solute, in this case surfactant, at an interface can be described using the Gibbs adsorption equation:

$$\Gamma_i = -\frac{d\gamma}{d\mu_i}$$ (9)

where Γ_i is the surface excess of component i, γ is the interfacial tension, and μ_i is the chemical potential of component i. At equilibrium,

$$d\mu_i = RTd \ln a_i$$ (10)

where R is the gas constant, T is temperature in degrees Kelvin, and a_i is the activity of component i in the bulk phase. For a dilute solution

($<10^{-2}$ M) containing a single nonionic surfactant and no background electrolyte, the activities of the solute (surfactant) and solvent (water) can be considered to be constant. Under these conditions, the surface excess may be expressed as

$$\Gamma_s = -RT\,\frac{d\gamma}{d\ln C_s} \tag{11}$$

where Γ_s is the surface excess of surfactant and C_s is the molar concentration of surfactant. In gas-water systems, the surface tension of water decreases sharply as the surfactant concentration is increased up to the CMC. Above the CMC, further adsorption of surfactant monomers at the interface is minimal, and thus, the surface tension of aqueous surfactant solutions remains essentially constant beyond this level. For this reason, the CMC of surfactants is often determined based on the inflection point in log-log plots of surfactant concentration versus surface tension. In general, the surface tension of water containing surfactant micelles ranges from 30 to 50 dyne/cm at 25°C, and will depend upon surfactant properties (Rosen, 1989).

For systems consisting of two immiscible liquids, such as dodecane and water, the hydrophobic portion of a surfactant will be oriented toward the organic liquid, while the hydrophilic head group will be oriented toward the aqueous phase. The presence of an NAPL, as opposed to air, at the interface results in a greater tendency for the surfactant to adsorb at the interface. Thus, the reduction in interfacial tension observed for NAPL/water/surfactant systems is typically greater than surface tension reduction observed in air/water/surfactant systems. For example, the interfacial tension between nonionic surfactant solutions and PCE falls within the range of 1 to 5 dyne/cm, compared to a surface tension between the same surfactant solution and air of approximately 30 dyne/cm (Fountain et al., 1991; Pennell et al., 1994).

In enhanced oil recovery applications, it is often considered necessary to reach ultra-low interfacial tensions ($\leq 10^{-3}$ dyne/cm) in order to recover significant amounts of residual oil from petroleum reservoirs (Bourrel and Schecter, 1988). To further reduce interfacial tensions beyond those values typically observed in Winsor Type I systems described above, the concentration of electrolyte (for ionic surfactants) or temperature (for nonionics) is often increased. The influence of salt concentration and temperature on the phase behavior and interfacial tension of a representative LNAPL/surfactant system is shown in Figure 5. The left-hand side of the diagram represents a traditional Winsor Type I system consisting of a NAPL/W microemulsion. As salt is added to anionic surfactant solutions, interactions between hydrophilic head groups and water are reduced, and thus, the surfactant becomes more hydrophobic (Rosen, 1989). The increased hydrophobicity of the surfactant results in greater interactions between the surfactant and NAPL, which results in a lowering of the NAPL/W interfacial tension. As more salt is added to the system, the anionic surfactant becomes increasingly hydrophobic. The corresponding enhancement in interactions between the surfactant and NAPL leads to the formation of a three-phase system, consisting of an LNAPL phase, a middle-phase (MP) microemulsion, and a water phase (Figure 5). The middle-phase microemulsion appears as a clear to slightly opaque phase, containing surfactant, NAPL, and water. There is considerable debate over the exact structure of middle-phase microemulsions,

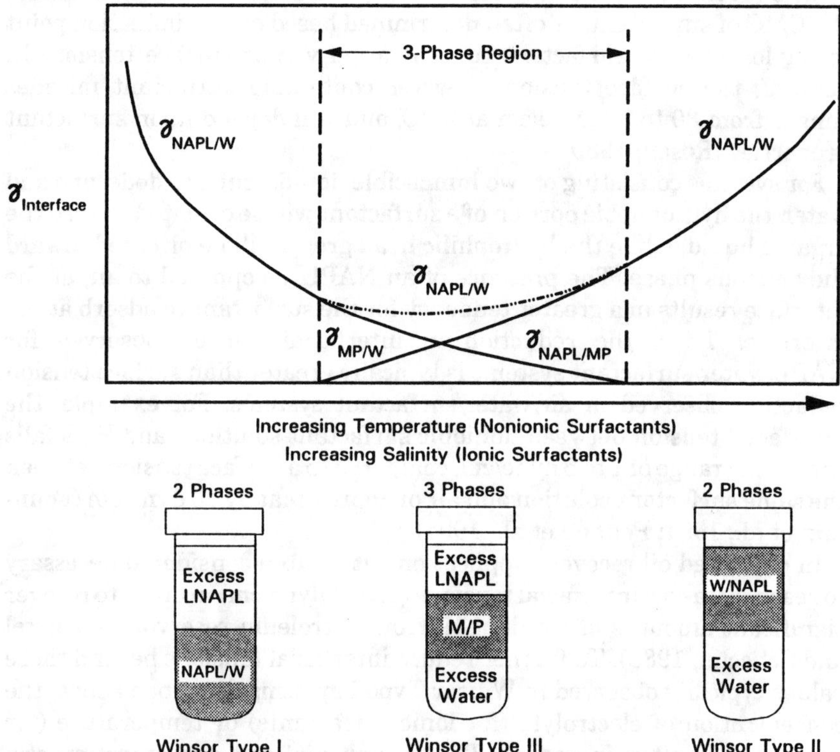

Figure 5. Effect of changes in temperature (nonionics) and salinity (ionics) on the interfacial tension and phase behavior of LNAPL/surfactant systems (adapted from Rosen, 1989).

which are thought to consist of bicontinuous lamellar structures or a mixture of normal and reverse micelles (Bourrel and Schecter, 1988). The NAPL/W interfacial tension typically reaches a minimum value in this three-phase region, which is approximately equal to the sum of the NAPL/ MP interfacial tension and MP/W interfacial tension (dashed line in Figure 5). As the salt concentration is increased further, the surfactant becomes extremely hydrophobic and tends to dissolve into the NAPL phase. This leads to the formation of reverse micelles within the organic liquid, which possess a hydrophilic core capable of enhancing the solubility of water in the NAPL. A similar type of behavior is observed for nonionic surfactants when temperature is increased. Elevated temperatures act to dehydrate the polyoxyethylene head group, which increases the hydrophobicity of nonionic surfactants. From a practical perspective, it is far easier to control the salt concentration of a surfactant solution, and thus, most enhanced oil recovery research has focused on anionic surfactants. Frequently, cosolvents, such as n-pentanol, are also added to anionic surfactant formulations to promote the formation of middle-phase microemulsions.

NAPL Displacement from Porous Media

In order for NAPL mobilization to occur during surfactant flushing, the reduction in interfacial tension must be sufficient to overcome the capillary forces acting to retain NAPL ganglia within a porous medium. For the case of a water-wetting porous medium, the conditions governing the onset of NAPL mobilization can be evaluated by considering the forces acting on a single NAPL "globule." Figure 6 depicts a single NAPL globule, entrapped within a pore based on the pore snap-off model (Melrose and Brander, 1974). The pore is oriented in a direction l, which makes an arbitrary angle, α, with the horizontal axis. Within this pore, pressure and gravity forces, which act to mobilize the globule, are balanced by capillary forces acting to retain the NAPL globule. Shear forces relevant to this system would be a function of the viscosity contrast between the aqueous and NAPL phases. It is generally accepted that shear forces do not play a substantial role in this system, and thus, they will be neglected herein (Ng et al., 1978).

To evaluate the potential for NAPL mobilization, let us consider the forces acting upon the NAPL globule shown in Figure 6. In order for the NAPL globule to be mobilized, the pressure and gravitational forces acting to displace the globule from the pore must exceed the capillary

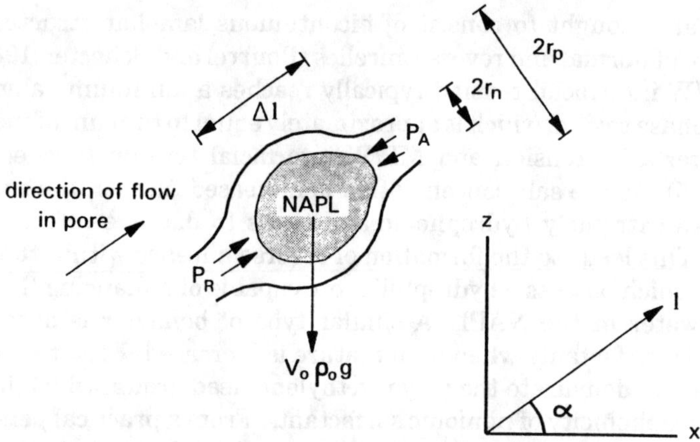

Figure 6. Schematic diagram of the forces acting upon an NAPL globule entrapped within a single pore.

forces acting to retain the globule within the pore. The summation of pressure and gravity forces acting on the globule in the l direction can be expressed as

$$P_R(\pi r_b^2) - P_A(\pi r_b^2) - \rho_o g(\pi r_b^2)\Delta l \sin \alpha \tag{12}$$

where P_R is the pressure force on the receding side (foot) of the globule, P_A is the pressure force on the advancing foot, Δl is the average length of the globule, ρ_o is the density of the organic liquid, g is the gravity acceleration constant, and πr_b^2 is the globule area normal to the vector l. The NAPL globule volume, V_o, has been approximated as $\pi r_b^2 \Delta l$. At residual saturation, the pressure and gravity forces [Equation (12)] will be balanced by the net capillary pressure force within the pore. This capillary pressure force can be approximated using Laplace's equation (Lake, 1989):

$$2\sigma_{ow}\left(\frac{\cos\theta_A}{r_n} - \frac{\cos\theta_R}{r_p}\right)\pi r_b^2 \tag{13}$$

where σ_{ow} is the interfacial tension between the organic liquid and water; θ_A and θ_R are the advancing and receding contact angles, respectively; r_p is the radius of the pore body; and r_n is the radius of the pore neck. Assuming that the contact angles of the advancing and receding ends of the NAPL globule are similar, Equation (13) may be rewritten as

$$\frac{2\beta\sigma_{ow}\cos\theta}{r_n}(\pi r_b^2) \qquad \text{where } \beta = 1 - \frac{r_n}{r_b} \tag{14}$$

Thus, balance of forces required to retain the NAPL globule within the pore can be found by equating Expressions (12) and (14) and dividing through by πr_h^2:

$$\frac{\partial P}{\partial l}\cdot\Delta l - \rho_o g\Delta l\sin\alpha = \frac{2\beta\sigma_{ow}\cos\theta}{r_n} \tag{15}$$

where the difference in pressures has been expressed in terms of the pressure gradient.

The left-hand side of Equation (15) can be reformulated directly in terms of viscous and buoyancy forces using an alternative representation of the pressure gradient obtained from Darcy's law:

$$q_{w_l} = -\frac{kk_{rw}}{\mu_w}\left(\frac{\partial P}{\partial l} + \rho_w g\sin\alpha\right) \tag{16}$$

where q_{wl} is the Darcy velocity of the aqueous phase in the l direction, k is the intrinsic permeability of the porous medium, k_{rw} is the relative permeability to the aqueous phase, and μ_w is the dynamic viscosity of the aqueous phase. Using Equation (16) to express P, and substituting it into Equation (15) yields

$$\frac{\mu_w\Delta l}{kk_{rw}}q_{w_l} + \Delta\rho g\sin\alpha\Delta l = \frac{2\beta\sigma_{ow}\cos\theta}{r_n} \tag{17}$$

Here $\Delta\rho = \rho_w - \rho_o$. Equation (17) may be written in terms of the capillary (N_{Ca}) and bond (N_B) number as

$$N_{Ca} + N_B\sin\alpha = \frac{2\beta kk_{rw}}{\Delta l r_n} \tag{18}$$

where

$$N_{Ca} = \frac{q_{w_l}\mu_w}{\sigma_{ow}\cos\theta} \qquad N_B = \frac{\Delta\rho g kk_{rw}}{\sigma_{ow}\cos\theta} \tag{19}$$

The capillary number relates viscous to capillary forces, while the bond number represents the ratio of buoyancy (gravity) to capillary forces.

Equation (18) presents a force balance at equilibrium. If the left-hand side of the equation becomes larger than the right-hand side, mobilization of the NAPL globule will occur. It is customary to assume that the contact angle, θ, is zero, and thus, this term falls out of Equation (19) (Lake, 1989). For a vertically oriented pore ($\alpha = 90°$), with NAPL displacement in the direction of the buoyancy force, the sum of the bond and capillary numbers controls displacement. This sum has been referred to as the total trapping number (N_T) by Pennell et al. (1996a):

$$N_T = |N_{Ca} + N_B| \tag{20}$$

Use of the total trapping number concept to describe the mobilization of residual NAPLs from soil columns will be presented in a later section of this chapter. The development presented above may be generalized to include a collection of pores. A complete derivation of the total trapping number for general flow and porous media conditions can be found in Pennell et al. (1996a).

SURFACTANT SORPTION

Sorption Processes

Over the past several years, surfactant sorption has received considerable attention within the environmental science and engineering literature. Much of this research effort has been driven by the concern over the potential negative impacts of sorption processes on surfactant delivery to the contaminated region of the subsurface (i.e., surfactant losses). For example, economic analyses of surfactant flushing have estimated losses due to sorption on the order of 20% of the first two pore volumes applied (Krebs-Yuill et al., 1995). Other studies have suggested that the selection of surfactants for subsurface remediation applications should be based, in large part, on minimizing surfactant losses due to sorption (Rouse et al., 1993). In their analyses, however, many investigators have failed to completely account for the fact that most surfactants, regardless of their head group type, exhibit a limiting or maximum sorption capacity at or near the CMC (Clunie and Ingram, 1983).

Due to their amphiphilic nature, surfactant interactions with the solid phase may be associated with both the hydrophilic and hydrophobic moieties of the surfactant. The lipophilic portion of a surfactant may interact with hydrophobic mineral surfaces or natural organic matter due to van der Waals attractions and partitioning phenomena, respectively. In general, as the length of the hydrocarbon chain becomes longer, this type of interaction will become more prominent. Because most soil minerals possess a net negative charge, interactions between the hydrophilic head group and the solid phase generally decrease in the following order: cationic > nonionic > anionic. For ethoxylated nonionic surfactants, the polyoxyethylene groups may interact with hydrophilic surfaces via hydrogen bonding with SiOH groups of the solid phase, whereas ionic surfactants are likely to be adsorbed as a result of ion exchange reactions (Rosen, 1989). At low surface coverages, surfactant monomers are thought to lie parallel to the sorbent surface (Clunie and Ingram, 1983). As the surfactant concentration increases, however, the hydrophobic portion of the surfactant may be displaced from mineral surfaces, allowing for lateral interactions between adjacent hydrophobic groups of sorbed monomers (Partyka et al., 1984). Above the CMC, the sorbed surfactant phase may exist as either a monolayer (Aston et al., 1982) or a bidimensional aggregate (bilayer) (Adeel and Luthy, 1995). Such behavior typically results in simple (L2) or stepped (L4) Langmuir isotherms, which reach a limiting value at or near the CMC.

A representative isotherm for the sorption of Witconol 2722 (Tween 80) on 20–30 mesh Ottawa sand is shown in Figure 7. Sorption data are expressed as the amount of surfactant sorbed per unit weight of sorbent (Q) versus the concentration of surfactant in the solution phase at equilibrium (C_e). These data conform to the Langmuir equation, which can be written in the form:

$$\frac{Q}{Q_m} = \frac{bC_e}{1 + bC_e} \tag{21}$$

where Q_m represents the maximum sorption capacity and b is a constant equal to the rate of adsorption divided by the rate of desorption. The experimental sorption data were fit to Equation (21) using a least-squares, nonlinear regression procedure (SYSTAT) to obtain the values of Q_m and b shown in Figure 7. Based on a calculated cross-sectional area of 300 Å^2 per surfactant molecule and specific surface area of 0.1 m^2/g, monolayer coverage was estimated to occur at a solution-phase concentration slightly above the reported CMC (13 mg/l). At the maximum sorption capacity of 0.14 mg/g, the mass of surfactant sorbed onto the sand was equivalent to approximately two monolayers. Thus, it was hypothesized that the adsorbed surfactant molecules formed a vertically oriented bilayer, with each molecule occupying approximately 150 Å^2 of surface area (Pennell et al., 1996c). A similar conceptual model was proposed by Adeel and Luthy (1995) to describe the equilibrium sorption of Triton X-100 onto Lincoln fine sand.

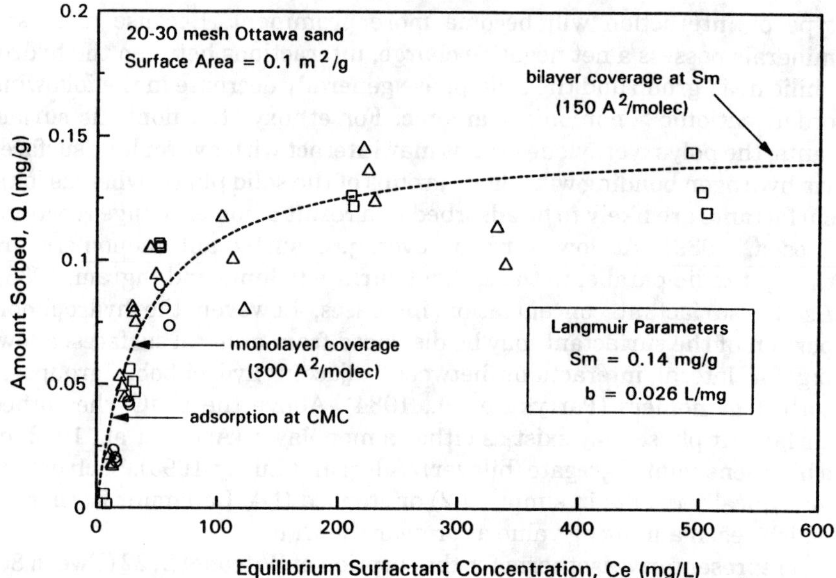

Figure 7. Equilibrium sorption of Witconol 2722 (Tween 80) by 20–30 mesh Ottawa sand. The points corresponding to monolayer and bilayer coverage are indicated based on a calculated surfactant cross-sectional area of 300 A²/molecule (adapted from Pennell et al., 1996c).

Surfactant Transport

In order to investigate the implications of sorption processes on surfactant delivery to a contaminated zone, it is necessary to couple surfactant sorption with transport. For illustrative purposes, we will consider surfactant transport in a simplified one-dimensional system. The approach used to describe the transport of a sorbing solute in a porous medium is based on the advection-dispersion reactive (ADR) transport equation. Assuming conditions of macroscale homogeneity, a one-dimensional form of the ADR equation may be written as

$$\frac{\partial C}{\partial t} + \frac{\rho_b}{\theta}\frac{\partial Q}{\partial t} = D_H\frac{\partial^2 C}{\partial x^2} - v\frac{\partial C}{\partial x} \tag{22}$$

where C is the solution-phase concentration, t is time, ρ_b is the soil bulk density, θ is the volumetric water content, Q is the amount sorbed, D_H is the hydrodynamic dispersion coefficient, x is distance, and v is the average pore-water velocity. To incorporate the Langmuir model into the ADR equation, assuming that local equilibrium exists between Q and C, Equation (21) can be differentiated with respect to time:

$$\frac{\partial Q}{\partial t} = \frac{\partial Q}{\partial C}\frac{\partial C}{\partial t} = \frac{d}{dC}\left(\frac{Q_m bC}{1+bC}\right)\frac{\partial C}{\partial t} = \frac{Q_m b}{(1+bC)^2}\frac{\partial C}{\partial t} \qquad (23)$$

Equation (23) can then be incorporated into the ADR equation using the traditional retardation factor concept (R_F):

$$R_F \frac{\partial C}{\partial t} = D_H \frac{\partial^2 C}{\partial x^2} - v\frac{\partial C}{\partial x} \qquad (24)$$

where

$$R_F = \left(1 + \frac{\rho_b Q_m b}{\theta(1+bC)^2}\right) \qquad (25)$$

A series surfactant transport simulations were performed using the Langmuir sorption parameters ($Q_m = 0.14$ mg/g; $b = 0.026$ L/mg) given in Figure 7. In these simulations, finite pulses (~2.5 pore volumes) of aqueous surfactant solution were introduced into a 1-D domain (soil column) containing 20–30 mesh Ottawa sand. The predicted time evolution of effluent concentration [breakthrough curves (BTCs)] are shown in Figure 8 for influent concentrations (C_o) of 100, 500, and 5000 mg/l. Here the pore volume represents the volume of fluid contained in the soil column. For the lowest influent concentration depicted (100 mg/l), the BTC exhibits a sharp leading edge and significant tailing in the distal portion of the pulse. Such behavior is characteristic of Langmuir-type sorption and can be attributed to a strong affinity of the solute for the solid phase at low concentrations (prior to attainment of monolayer coverage). Stronger sorption at low concentrations also influences the shape of the leading edge of the solute pulse, resulting in "self-sharpening" behavior in comparison with the symmetrical shape observed for solutes exhibiting linear sorption. The prolonged tailing of the BTC is due to release of strongly sorbed surfactant at levels below monolayer coverage. These simulations reveal that the BTCs of surfactants exhibiting Langmuir sorption isotherms are extremely sensitive to the applied (influent) concentration. As the influent concentration is increased, the solute pulse shifts to the left, and the number of pore volumes displaced at the time of initial solute appearance at the end of the domain approaches 1.0, confirming that sorption has minimal impact on solute

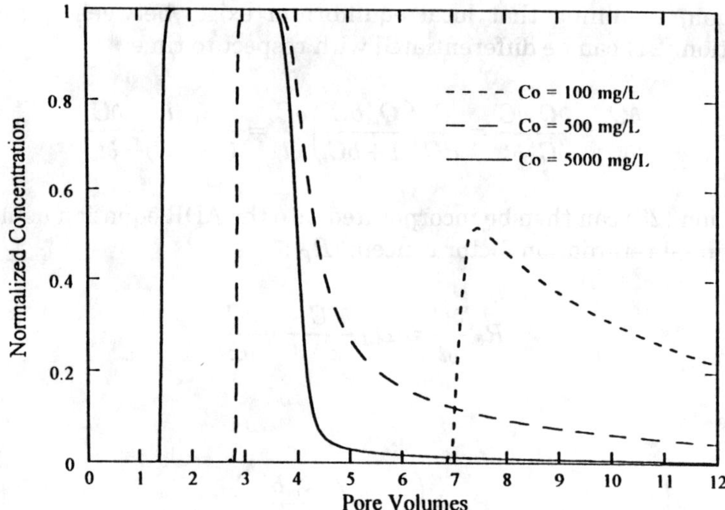

Figure 8. Effect of influent concentration on predictions of Witconol 2722 transport in Ottawa 20–30 mesh sand using the Langmuir sorption equation. Simulation parameters: column length = 12 cm; pulse width = 3 pore volumes; Peclet number = 200; pore-water velocity = 9.76 cm/h; porosity = 0.34.

transport at very high solution-phase concentrations. The Langmuir model is based upon a limiting sorption capacity, and thus, once this value is reached, sorption no longer significantly influences solute transport. Therefore, at the concentrations typically proposed for remediation applications (e.g., 4% or 40,000 mg/l), the impact of sorption on surfactant delivery may be minimal. The quantity of surfactant sorbed by the solid phase, however, will not be insignificant. Thus, sorption processes will undoubtedly influence surfactant recovery from the subsurface and the potential for enhanced sorption of hydrophobic organic compounds (HOCs) by the surfactant-coated solid phase, particularly if surfactant desorption is rate-limited (Adeel and Luthy, 1995).

LABORATORY INVESTIGATIONS OF SURFACTANT TREATMENTS

Micellar Solubilization of Hydrophobic Organic Compounds (HOCs)

A number of laboratory studies have been conducted to investigate the ability of surfactants to enhance the recovery of sorbed or deposited HOCs from soils and aquifer materials (Ellis et al., 1985; Gannon et al., 1989; Vigon and Rubin, 1989). Many early surfactant trials of this type examined the capacity of micellar solutions to solubilize strongly sorbed contaminants, such as polychlorinated biphenyls (PCBs) and polycyclic aromatic hydrocarbons (PAHs), in either batch (soil washing) or column studies (surfactant flushing). Abdul and Gibson (1991) and Abdul et al. (1992) have published articles in which they demonstrated that nonionic surfactants, primarily linear alcohol ethoxylates, could recover significant amounts of PCBs from soils in both batch and column systems. Similar studies confirmed that surfactant treatments (concentrations ranging from 1 to 4% wt) result in the removal of 90 to 99% of the PCBs from soils (Ellis et al., 1985; Vigon and Rubin, 1989; Clarke et al., 1991).

Several issues must be considered, however, when attempting to utilize surfactants to extract strongly sorbed compounds from soils. First, surfactant addition will substantially increase the aqueous solubilities of the HOC, often by several orders of magnitude (e.g., Kile and Chiou, 1989; Pennell et al., 1993). If one attempts to employ such an approach in situ, the resulting aqueous-phase concentration of contaminant is likely to be well above regulatory limits. Thus, contaminants that were relatively immobile within the soil profile, such as highly chlorinated PCBs, may be subject to advective transport after surfactant treatment. In order to minimize spreading and potential off-site migration of the contaminant plume, barrier walls and aggressive hydraulic control may be required. A second concern is that surfactant treatments could substantially reduce the hydraulic conductivity of the porous medium. Several authors have reported decreased infiltration rates and hydraulic conductivities following the application of aqueous surfactant solutions (e.g., Ziegenfus, 1987; Allred and Brown, 1994). The observed reductions in permeability have been attributed to dispersion of fine particles by micellar surfactant solutions and subsequent pore blockage or clogging of pore throats. This problem is of particular concern because extremely hydrophobic compounds tend to be strongly sorbed and, thus, are often found in surface soils containing a significant amount of fine materials and natural organic matter. For these reasons, the in situ application of surfactant solutions to contaminated surface soils must be approached with caution, and surfactant-based treatment of these materials may be most appropriately addressed by the use of ex situ soil washing procedures.

Influence of Surfactants on Bioavailability

Surfactants have also been proposed as a means to enhance the bioavailability of sorbed-phase contaminants, both for in situ (soil flushing) and ex situ (bioreactor) applications. A first step toward understanding this complex system is to develop a conceptual model of the possible mechanisms controlling coupled solubilization and biodegradation of HOCs in soil-water systems. A schematic diagram of such a system, in which both surfactant and HOC are distributed between the solid, aqueous, and micellar phases is given in Figure 9 (steps 1, 2, 3, and 4). Incorporation and biotransformation of the surfactant and HOC are represented by steps 5 and 6, respectively. When evaluating HOC sorption and desorption (step 4 at equilibrium), it is important to consider the role of both surfactant sorption (step 1) and micellar solubilization (step 3). Below the CMC, sorption of surfactant onto the solid phase has been shown to enhance the sorption of HOCs (Sun et al., 1995). In effect, the sorbed-phase surfactant acts to increase the organic carbon content of the solid phase and, in some cases, may have a greater capacity to sorb HOCs than natural organic matter (Edwards et al., 1994). At surfactant concentrations exceeding the CMC, micellar solubilization results in a linear increase in the aqueous solubility of HOCs. This leads to increased desorption of HOCs from the solid phase and a reduction in the apparent HOC distribution coefficient, K_D (Sun et al., 1995).

The rate of HOC desorption (step 4) may be influenced by the presence of surfactant. In general, rate-limited desorption of HOCs from contaminated soils and sediments has been attributed to mass transfer limitations within the solid phase, which could result from diffusion within natural organic matter or microporous mineral particles (Karickhoff, 1984; Brusseau and Rao, 1989). It has been proposed that sorption of surfactant by the solid phase could increase rates of contaminant desorption by swelling the organic matrix (Deitsch and Smith, 1995). A similar explanation was given by Nkedi-Kizza et al. (1989) to account for reductions in nonequilibrium sorption in the presence of a water-miscible cosolvent. Conversely, sorbed-phase surfactants may form bilayers on the sorbent surface (step 1), which could potentially reduce rates of contaminant mass transfer into the aqueous phase. It has also been postulated that surfactants could affect rates of HOC biotransformation by modifying cell wall properties (step 6). At low concentrations (sub-CMC), surfactant monomers may be incorporated into cell membranes and interact with membrane lipids (Van Hoof and Rogers, 1992; Rouse et al., 1994). This mechanism could potentially increase the rate of HOC mass transfer into the microbial cell due to membrane fluidization, thereby enhancing the rate of biotransformation. The influence of such a membrane modification may be dependent upon the surfactant concentration and stearic compatibility with the cell membrane. Surfactant concentrations that are too low may not yield any observable effects, whereas high-surfactant

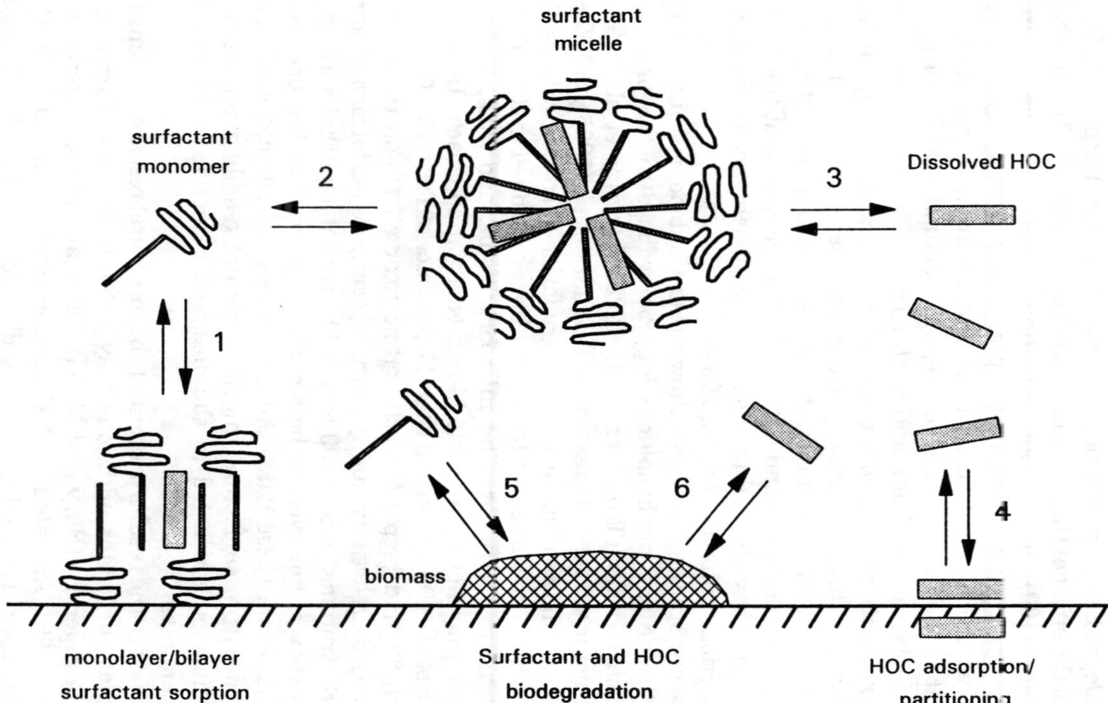

Figure 9. Conceptual model of coupled HOC sorption, micellar solubilization, and biodegradation.

concentrations may disrupt the cell membrane or lead to cell lysis, thereby decreasing the rate of HOC degradation.

One of the first detailed studies that coupled surfactant treatments with biodegradation was conducted by Laha and Luthy (1991), who evaluated the effect of three nonionic surfactants (Brij 30, Tergitol NP-10, and Triton X-100) on the transformation of phenanthrene. Results of these experiments indicated that the presence of surfactant at concentrations below the CMC had no effect on phenanthrene degradation, while surfactant concentrations above the CMC actually inhibited phenanthrene mineralization (Laha and Luthy, 1991). The authors considered several explanations to account for these results, including surfactant toxicity, phenanthrene toxicity, preferential use of surfactant as a substrate, and reduced bioavailability of solubilized phenanthrene. Subsequent experimentation and data analysis suggested that the observed inhibition may have been due to surfactant interference with cell membrane function (Laha and Luthy, 1992).

Two recent studies have also shown that the bioavailability of PAHs (phenanthrene and naphthalene) may be reduced in the presence of nonionic surfactants (Tsomides et al., 1995; Volkering et al., 1995). Of seven nonionic surfactants tested by Tsomides et al. (1995), only Triton X-100 did not inhibit phenanthrene degradation at concentrations above the CMC. Based on these results, the authors concluded that surfactants can be toxic to PAH-degrading microorganisms, and thus, surfactant treatments may not be a viable bioremediation approach. In contrast, Volkering et al. (1995) reported that the presence of Triton X-100, Tergitol NPX, Brij 35, and Igepal CA-720 had no toxic effects on the microbial activity based on activity and growth experiments performed at surfactant concentrations up to 10,000 mg/l. The rate of naphthalene degradation was shown, however, to be significantly lower in the presence of surfactant. From these data, the authors suggested that naphthalene solubilized within surfactant micelles was not readily available to the microorganisms, which led to the observed reduction in naphthalene degradation rates.

In contrast to the results presented above, several research groups have shown that surfactant additions may enhance the transformation of decane, tetradecane (Bury and Miller, 1993), and phenanthrene (Aronstein et al., 1991; Aronstein and Alexander, 1992; Guha and Jaffe, 1996). In the studies performed by Bury and Miller (1993), solubilization of decane and tetradecane by linear alcohol ethoxylates (Neodol 25-9, 25-7, and 25-3) was shown to greatly increase hydrocarbon degradation by pure strains of *Pseudomonas aeruginosa* and *Ochrabactrum anthropi*, both gram-negative bacteria. Experiments conducted by Aronstein and coworkers (1991, 1992) indicated enhanced degradation of phenanthrene and biphenyl at surfactant concentrations below the CMC. Because the aqueous solubility of the hydrocarbons was not increased under these conditions, the data suggest that the presence of surfactant may enhance cell permeability toward the organic species or limiting nutrients (Mihelcic et al., 1993). The recent studies demonstrating enhanced degradation of phenanthrene at relatively high surfactant concentrations (Guha and Jaffe, 1996) are of particular interest because they appear to directly contradict previous data showing inhibitory effects for similar surfactant-organic systems (e.g., Laha and Luthy, 1991).

The conflicting results reported to date suggest that the effect of surfactants on HOC bioavailability is highly system-specific and that careful characterization of all experimental parameters is necessary for proper interpretation of coupled solubilization and biodegradation data. A number of research groups are currently working in this area, and it is anticipated that results from these studies will help to provide a more definitive understanding of coupled solubilization and bioavailability. Additional references and information related to this topic can be found in recent review articles by Mihelcic et al. (1993) and Rouse et al. (1994).

Surfactant Flushing of NAPL-Contaminated Materials

Considerable research effort has also been directed toward evaluating the use of surfactants to recover petroleum hydrocarbons and chlorinated solvents existing as NAPLs in the subsurface (e.g., Fountain et al., 1991; Peters et al., 1992; Pennell et al., 1993). These compounds frequently represent a long-term source of aquifer contamination as the NAPL slowly dissolves into the groundwater. The intent of this remediation strategy is not to treat the entire aquifer but, rather, to remove the NAPL-contaminated zone that serves as the primary source of groundwater contamination. Early studies in this area often did not clearly distinguish between NAPL existing as a mobile phase and that entrapped as a residual (immobile) phase. For example, the Texas Research Institute (1985) was able to recover approximately 80% of the gasoline spilled in a 3-D tank after flushing with a 4% surfactant solution. Similar recovery rates, however, were obtained by water flushing alone, and thus, no significant advantage was realized from the use of surfactants. These data serve to illustrate the fact that mobile NAPLs (i.e., the fraction above residual saturation) can be recovered from the subsurface by standard extraction procedures without the addition of surfactant. Thus, surfactants are more appropriately used to recover immobile NAPLs, which typically occupy between 10 and 25% of the pore volume.

Pennell et al. (1993) conducted a series of column experiments to investigate the ability of Witconol 2722 to recover residual dodecane from Ottawa sand. The results of a representative column experiment containing an initial dodecane saturation of 19.7% are shown in Figure 10. The concentration of dodecane exiting the column is plotted versus the volume of surfactant solution (4% Witconol 2722) injected into the soil column. Following the introduction of surfactant solution, the effluent concentration of dodecane increased to a steady-state value of approximately 500 mg/l. Although this represents a five orders-of-magnitude enhancement in dodecane solubility ($3.7 \mu g/l$), the effluent concentration was still seven times less than the equilibrium value (3500 mg/l) measured in batch solubilization experiments. These findings suggest that micellar solubilization of residual dodecane was rate-limited. To further investigate this phenomenon, Pennell et al. (1993) varied the aqueous-phase (pore-water) velocity from 6 to 22 cm/h and interrupted flow for periods ranging from 3.5 to 100 hours (Figure 10).

The flow interruption technique involves the initial displacement of solution at a constant flow rate, interruption of flow for a specified period

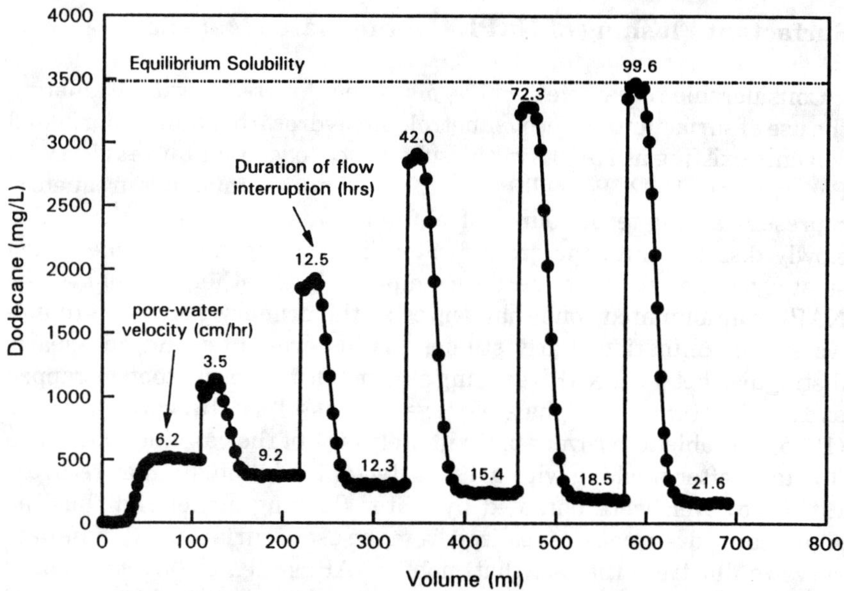

Figure 10. The effect of pore-water velocity and duration of flow interruption on effluent concentrations of dodecane after flushing with a 4% (43.2 g/l) solution of Witconol 2722. The soil column was packed with 20–30 mesh Ottawa sand and contained an initial dodecane saturation of 19.7% (adapted from Pennell et al., 1993).

of time, followed by the recommencement of flow. The eluted surfactant solution containing elevated concentrations of dodecane represents the pore volume residing within the column during the flow interruption period. The sizable increase in the effluent concentration of dodecane following flow interruption is indicative of rate-limited, rather than instantaneous, solubilization of residual dodecane (Figure 10). As the duration of flow interruption or residence time increased, the extent of dodecane solubilization within the soil column approached the equilibrium value measured in batch experiments. An interruption time of approximately 100 hours was required to achieve equilibrium solubilization of the residual dodecane. Following elution of the interrupted solution, the effluent concentration returned to a steady-state value corresponding to the applied pore-water velocity (Figure 10). In effect, the stoppage of flow allowed for greater contact time between the micellar solution and residual dodecane, which resulted in a corresponding increase in the aqueous-phase concentration of dodecane. In a similar manner, the effluent concentration of dodecane increased as the pore-water velocity of the surfactant solution was decreased. These results further demonstrate that the solubilization of residual dodecane increased as the residence time of the micellar solution within the soil column increased.

Although rate-limited or nonequilibrium solubilization occurred within the soil columns, flushing with a 4% solution of Witconol 2722 greatly enhanced the recovery of residual dodecane. Based on mass balance calculations, 481 mg of dodecane were solubilized after flushing the soil column with 0.7 L of surfactant solution. The recovery of equivalent amounts of dodecane by water flushing alone would have required 1.3 × 10^5 L of water, assuming that the equilibrium solubility of dodecane (3.7 μg/l) was reached. These data clearly demonstrate that substantial improvement in residual NAPL recovery can be realized by the use of aqueous surfactant solutions in a conventional pump-and-treat remediation framework. Based on the results of the flow interruption and steady-state experiments, the flow rate and schedule could be optimized to further improve the efficiency of dodecane solubilization, thereby minimizing the amount of surfactant required for dodecane recovery. Thus, the identification and assessment of rate-limited NAPL solubilization is an important consideration during the development of efficient and cost-effective surfactant-based remediation technologies on a larger scale.

Even though the majority of solubilization studies conducted to date have assumed equilibrium, rather than rate-limited conditions, two mechanistic models have been proposed to describe micellar solubiliza-

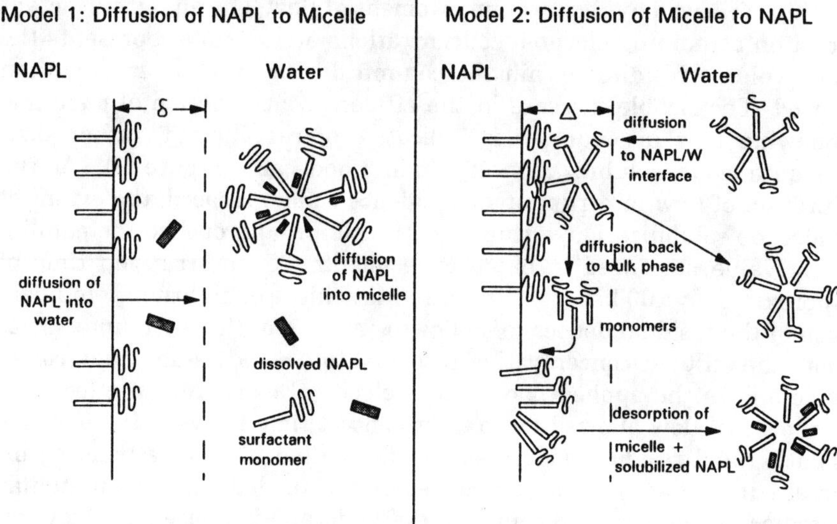

Figure 11. Schematic diagram of two conceptual models used to describe rate-limited micellar solubilization (adapted from Carroll, 1981).

tion of organic liquids (Figure 11). The first model involves the dissolution of an organic liquid into the aqueous phase, adsorption of the organic at the micelle/water interface, and subsequent incorporation within the micelle (Volkov, 1979; Vaumgardt and Kahlweit, 1980; Arytyunyan and Beileryan, 1983). In the second model, the micelles are proposed to diffuse to the organic/water interface, dissociate into monomers that are adsorbed at the interface, and then reform into micelles containing the associated organic liquid (Chan et al., 1976; Carroll, 1981; Bolsman et al., 1988). The rate of solubilization has been shown to increase with the surfactant concentration above the CMC and the polarity of the organic solute (Carroll et al., 1982). In addition, the rate of n-hexadecane solubilization in mixtures of linear alkylarenesulfates and an ethoxylated alcohol was found to be strongly correlated to the molecular structure of the surfactant formulation (Bolsman et al., 1988). These results have been used in support of the second model. The flow interruption studies presented by Pennell et al. (1993) demonstrate that the rate of dodecane solubilization decreased as the solubilization limit was approached. Similar kinetic relationships have been reported for the solubilization of naphthalene, anthracene, pyrene, and dibenzanthracene (Vaumgardt and Kahlweit, 1980). These results could be attributed to a reduction in the rate of organic incorporation within the micelle (Model 1) or a reduction in the rate of micelle diffusion to or from the organic/water interface (Model 2) as the micelles become saturated with dodecane. However, the inherent complexity of micellar solubilization precludes a definitive mechanistic interpretation of the flow interruption results presented to date.

NAPL Mobilization Experiments

Over the past several years, a number of research groups have demonstrated the capacity of surfactants to enhance the recovery of residual NAPLs from unconsolidated porous media through both micellar solubilization and NAPL mobilization (Abdul and Gibson, 1992; Fountain et al., 1991; Pennell et al., 1993). Although mobilization has been shown to be a far more efficient method than micellar solubilization for removing entrapped NAPLs from soil columns (Pennell et al., 1994), utilization of this approach in the field could lead to uncontrolled migration of the mobilized NAPL phase. This is of particular concern in the case of DNAPLs, which would tend to migrate downward through an aquifer formation due to gravitational forces. Although nonwetting phase entrapment and viscous displacement of NAPLs have received considerable attention in the enhanced oil recovery literature (e.g., Melrose and Brander, 1974; Stegemeier, 1977; Larson et al., 1981), few studies have considered the potential impact of buoyancy forces on NAPL mobilization (Ng et al., 1978; Morrow and Songkran, 1981). In most treatments of this subject, buoyancy forces were either neglected, or more commonly, the nonwetting and wetting phases were selected so that their respective densities were identical (e.g., Ng et al., 1978).

Pennell et al. (1996a) recently conducted a matrix of 1- and 2-D column experiments to assess the effect of both viscous and buoyancy forces on the mobilization of residual tetrachloroethylene (PCE) during surfactant flushing. To induce PCE mobilization, the interfacial tension (IFT) between the entrapped PCE and aqueous phase was reduced from 47.8 to 0.09 dyne/cm by injecting four different surfactant solutions (Table 4). The results of PCE mobilization experiments, conducted using four size fractions of Ottawa sand, are shown in Figure 12. In this graph, the residual saturation of PCE in the soil column is plotted against the total trapping number, where $N_T = N_{Ca} + N_B$ [Equation (20)]. As the interfacial tension between the displacing fluid and PCE was reduced from 47.8 dyne/cm (water) to 5.0 dyne/cm (Witconol 2722), there was little change in the saturation of PCE. However, when the interfacial tension was further reduced to 0.58 dyne/cm (Aerosol MA/OT) and 0.09 dyne/cm (Aerosol AY/OT), almost complete mobilization of PCE as free product

*Table 4. Relevant properties of aqueous solutions used
in the column experiments.*

Displacing Fluid	IFT (dyne/cm)	PCE Solubility (mg/l)	ρ (g/cm^3)	μ (cpi)
MilliQ Water	47.8	200	0.988	0.955
4% Witconol 2722	5.0	34,100	1.031	1.15
4% 4:1 Aerosol MA/OT	0.58	16,300	1.021	1.20
4% 1:1 Aerosol AY/OT	0.09	71,720	1.023	1.20

was observed. Thus, when the total trapping number was less than approximately 2×10^{-5}, the residual saturation of PCE was essentially unchanged. As the value of the total trapping number increased, however, the PCE saturation decreased sharply, with complete mobilization of residual PCE achieved as the trapping number approached a value of 1×10^{-3}. The desaturation curves shown in Figure 12 indicate that the critical value of N_T required for PCE mobilization in Ottawa sand falls within the range of approximately 2×10^{-5} to 5×10^{-5}, which corresponds to an interfacial tension range of 4.3 to 1.7 dyne/cm. These data are consistent with the observations of Fountain et al. (1991) and Pennell et al. (1994), who reported that surfactants yielding interfacial tensions below 1.0

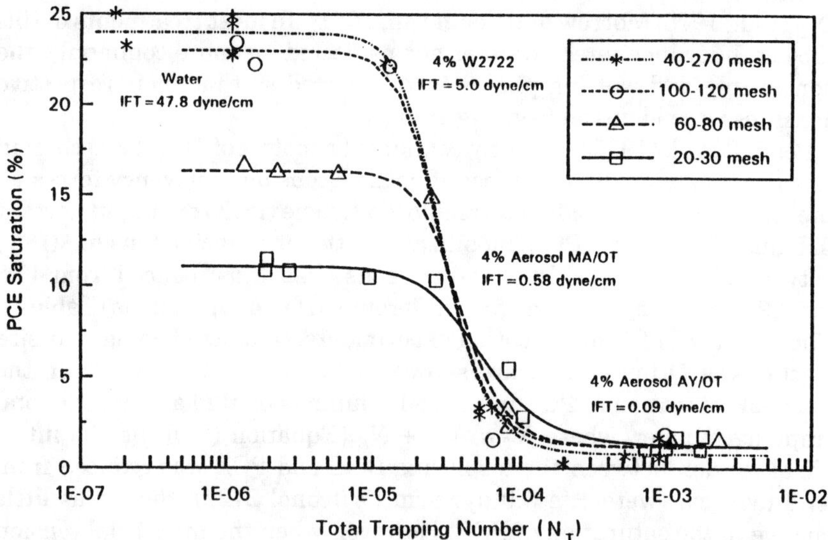

Figure 12. PCE desaturation curves for several size fractions of Ottawa expressed in terms of the total trapping number (adapted from Pennell et al., 1996a).

dyne/cm caused rapid downward migration of PCE in unconsolidated sands. It is important to recognize, however, that, in addition to the interfacial tension, the magnitude of the Bond number will be directly related to the density of the NAPL and the effective permeability $(k \cdot k_{rw})$ of the porous media. In contrast, the capillary number will also be governed by the viscosity and velocity of the displacing fluid.

The magnitude of the total trapping number corresponding to the onset of PCE mobilization (2×10^{-5} to 5×10^{-5}) is similar to critical capillary numbers reported for water-wet sandstones and unconsolidated sands (Melrose and Brander, 1974; Delshad et al., 1986). In these systems, however, the density difference between liquid phases was negligible, and thus, the bond number approached zero. The similarity in these data sets indicates that the total trapping number can be utilized for systems with or without significant buoyancy effects. It should also be noted that the total trapping number corresponding to complete PCE mobilization (1×10^{-3}) is consistent with capillary number values reported in the enhanced oil recovery literature (Delshad et al., 1986).

For some remediation scenarios, such as LNAPL contamination or in well-confined formations, it may be advantageous to promote NAPL mobilization. This recovery strategy could be far more efficient and economical than micellar solubilization alone, provided that the free product can be readily extracted. It will be particularly critical, however, to employ the total trapping number analysis when selecting surfactant formulations to minimize NAPL mobilization. The experimental results presented by Pennell et al. (1996a) indicate that the value of N_T should be less than 1×10^{-5} to minimize the potential for NAPL mobilization in both graded and well-sorted sands. Further refinement and validation of the total trapping number critical range will be required at both the laboratory and field scales prior to general application of this approach.

MATHEMATICAL MODELING

Model Development

Until very recently, relatively little attention had been directed toward the mathematical modeling of surfactant-enhanced aquifer remediation. The chemical and physical complexities of mixed organic/surfactant/aqueous/soil systems create severe challenges for the modeler. Those same complexities, however, ultimately make modeling a necessary component in the evaluation of laboratory column data and the design and assessment of the potential performance of various surfactant-based remediation technologies at the field scale. It is only through mathematical modeling that we can couple and assess the relative importance of individual physical, chemical, and biological processes that will be operative under natural remediation conditions.

Of the few SEAR models that have appeared in the literature, most have restricted their applicability to single-phase (aqueous) flow, assuming the recovery of sorbed or entrapped organic liquid residuals is achieved solely through micellar solubilization. A one-dimensional solubilization model was first presented by Wilson (1989) and subsequently extended to a homogeneous, isotropic, two-dimensional ideal flow regime in Wilson and Clarke (1991). Harwell et al. (1993) presented a similar one-dimensional modeling approach, analyzed in terms of shock wave propagation. In these models, the solubilization process was assumed to be described by local equilibrium relations. Consistent with laboratory column solubilization observations, some recent works have presented models that incorporate some form of rate-limited solubilization behavior (Abriola et al., 1993; Megehee et al. 1993; Dekker and Abriola, 1994).

As discussed in a previous section of this chapter, solubilization is not the only process that may act to enhance the recovery of entrapped NAPLs during surfactant flushing. Reductions in interfacial tension may result in the mobilization of the organic as a separate phase liquid. To adequately describe the process of organic mobilization, a full multiphase, multicomponent flow and transport modeling approach is necessary. Here, researchers have taken advantage of the experience of the petroleum industry in the modeling of enhanced oil recovery strategies. A number of models have been presented in the petroleum literature for the simulation of micellar and chemical reservoir flooding (e.g., Thomas et al., 1984; Datta Gupta et al., 1986; Scott et al., 1987). Brown et al. (1994) adapted such a simulator (UTCHEM) for the examination of SEAR performance. A complete description of this three-dimensional multiphase compositional simulator may be found in Delshad et al. (1995).

All SEAR modeling approaches are developed from fundamental mass balance principles. Within a particular phase:

$$\text{accumulation of component } i \quad = \quad \text{net advective and dispersive/diffusive flux} \quad + \quad \text{sources/sinks}$$

$$(26)$$

In mathematical terms, a basic balance equation for a chemical component i in phase α may be written as

$$\frac{\partial}{\partial t}(\varepsilon_\alpha \rho^\alpha \omega_i^\alpha) + \nabla \cdot (\varepsilon_\alpha \rho^\alpha \omega_i^\alpha v^\alpha) - \nabla \cdot (\varepsilon_\alpha \rho^\alpha D_{hi} \cdot \nabla \omega_i^\alpha) = \sum_{\beta \neq \alpha} E_i^{\alpha\beta} \qquad (27)$$

where ω_i^α is the mass fraction of component i in phase α, ε_α is the volume fraction of the α phase, ρ^α is the mass density of the α phase, v^α is the velocity of the α phase, D_{hi}^α is the hydrodynamic dispersion tensor for component i in phase α, and $E_i^{\alpha\beta}$ is the exchange of mass of component i between the α and β phases. If we restrict our attention to the saturated groundwater zone, a typical SEAR scenario may involve two fluid (organic and aqueous) and one solid (soil matrix) phase. A minimum of three components must be modeled: surfactant, organic, and water. A multi-component contaminant will require equations for each of its chemical constituents. Other components could also appear with the addition of electrolytes or cosolvents. Sources and sinks to be modeled include surfactant and organic sorption, organic-aqueous–phase mass transfer (e.g., dissolution and micellar solubilization), and chemical or biological degradation. Degradation losses have been neglected in all modeling efforts to date.

The phase velocity appearing in the previous equation is typically expressed in terms of the fluid-phase potential via a modified Darcy law expression:

$$v^\alpha = -\frac{k_{r\alpha} k}{\varepsilon_\alpha \mu_\alpha} \cdot (\nabla P^\alpha - \rho^\alpha g) \tag{28}$$

Here $k_{r\alpha}$ is the relative permeability of phase α to the other fluid phase, k is the intrinsic permeability tensor, μ_α is the α-phase viscosity, P is the pressure of the α phase, and g is the acceleration due to gravity. Intrinsic permeability is a function of soil type and texture, while relative permeability is a function of the phase volume fraction and the fluid saturation history.

Equations (27) and (28) represent a system of partial differential equations in two groups of primary unknowns: component mass fractions and fluid-phase pressures. Although conceptually straightforward, these equations have many inherent complexities that make solution for phase composition and distribution a difficult task. For example, phase composition will have a direct effect on interfacial tension, which, in turn, will affect fluid equilibrium saturation distributions and relative permeabilities. In a like manner, surfactant concentration will influence organic component sorption and solubilization behavior. These examples reveal that the individual component balance equations are nonlinear (their coefficients depend upon the primary unknowns) and that they are strongly coupled (through compositional effects on physical properties and interphase mass exchange).

In order to solve the system of equations presented above, parametric relations are required to quantify system properties (equation coefficients) in terms of the primary variables. It is customary to express relative permeabilities as power law functions of phase saturations, in accordance with semi-empirical models presented in the petroleum (Brooks and Corey, 1966) or soil science hydrology (van Genuchten, 1980; Lenhard and Parker, 1987) literature. Similarly, phase saturations are linked to fluid pressures through empirical capillary pressure relations. Interfacial tensions may be modeled as a function of phase composition through a correlation developed in the oil reservoir literature for the interfacial tension between oil and microemulsions (Huh, 1979). These interfacial tensions may then be used to adjust (scale) saturations–capillary pressure relations according to an approach developed by Leverett (1941). For further details on parametric relations, the interested reader is referred to Brown et al. (1994).

As noted in a previous section of this chapter, reductions in interfacial tension may also reduce the residual (entrapped) organic saturation. In oil reservoir simulators, it was previously common practice to adjust organic residual saturations as a function of the capillary number, N_{ca}, according to estimated or measured desaturation curves. The experimental studies described above, however, have revealed that this approach is inadequate to predict the onset of mobilization, particularly for DNAPLs. To simulate DNAPL recovery, then, traditional oil industry models must be modified to incorporate buoyancy forces through the use of the total trapping number concept.

The simplest way to treat interphase mass transfer is to assume that contiguous phases are in thermodynamic equilibrium. This is the approach commonly adopted in compositional oil reservoir modeling. If the phases are in equilibrium, then the mass fraction of a particular component in a particular phase will dictate its mass fraction in all other phases at the same location. Equilibrium expressions may then be employed to describe the mass fraction relation of a constituent among the system phases, thus reducing the total number of unknowns. The sorbed mass fraction of a component is customarily related to its aqueous concentration by an equilibrium isotherm. For many surfactants, batch data have been shown to conform to a Langmuir isotherm. A Freundlich expression is often used to represent organic component sorptive behavior. Equilibrium micellar solubility of the organic liquid in the aqueous phase is typically represented as a linear function of surfactant concentration above the CMC (Figure 3) (see also Abriola et al., 1993).

Given the equilibrium relations described above, equations of the form of Equation (27) may be summed over all phases to yield component transport equations. In oil reservoir simulators, it is customary to express such balance equations in terms of volume concentrations rather than mass fractions. Here, the implicit assumption is that volume is conserved in a mixture, i.e., that the volume of a mixture is equal to the sum of the volumes of its components. Governing equations may then be solved for the pressure of one of the phases and the total (volume) concentration of each component. Equilibrium and capillary relations may then be used to evaluate the corresponding phase saturations and component concentrations within each phase. An example of this simulation approach may be found in Brown et al. (1994).

If local thermodynamic equilibrium is too restrictive an assumption, an alternative solution approach must be sought. Under such conditions, it is necessary to develop constitutive expressions for the $E_i^{\alpha\beta}$ terms appearing in Equation (27), which describe the transfer of mass between phases. The simplest approach to representing this mass transfer is in the form of a linear driving force expression:

$$E_i^{\alpha\beta} = \hat{k}_{\alpha\beta}\rho^\alpha(\omega_{ei}^{\alpha\beta} - \omega_i^\alpha) \qquad (29)$$

Here $\hat{k}_{\alpha\beta}$ is a "lumped" mass transfer coefficient, which is the product of a mass transfer coefficient and the specific interfacial (contact) area between the α and β phases. The mass fraction, $\omega_{ei}^{\alpha\beta}$, is the α-phase mass fraction of component i, which is in equilibrium with the β phase. In Equation (29), the driving force for mass transfer is the deviation between the existing concentration and its equilibrium value. This expression is a linearization of Fick's law. Equations of the type of Equation (29) have been used successfully to model entrapped NAPL dissolution in sandy media (e.g., Powers et al., 1991, 1992) and to model rate-limited organic sorption behavior in aggregated porous media (e.g., DeSmedt and Weirenga, 1979). For the case of micellar solubilization, the equilibrium mass fraction of a component i in the aqueous phase becomes a function of surfactant concentration. Abriola et al. (1993) employed an organic-aqueous-phase mass transfer expression of the form of Equation (29) to model dodecane effluent concentration data obtained from a series of column surfactant flushing experiments. The linear driving force model was able to capture the important features of the experimental data. Although more sophisticated than the petroleum industry approach in terms of describing rate-limited mass transfer behavior, their model greatly simplified the multiphase flow aspects of the problem. The organic phase was assumed to be immobile, and the potential influence of interfacial tension reductions was ignored. To date, no model has been presented in the literature that couples full multiphase migration with rate-limited mass transfer.

Model Evaluation and Practical Implications

Mathematical modeling can be used to highlight and investigate phenomena that may be important for SEAR performance under field conditions. Few SEAR model applications have appeared in the literature to date, and constitutive relations are not yet well developed for these models. It is still useful, however, to present here some preliminary model investigations. These simulations serve to illustrate model capabilities and to suggest potential issues that will be important in SEAR performance and design.

Based on the numerical modeling approaches described in the previous section, Abriola et al. (1993) evaluated the effect of different pumping flow regimes on the recovery of dodecane from soil columns by micellar solubilization. The initial model evaluation and sensitivity analysis were based on column experiments reported by Pennell et al. (1993) (Figure 10). Due to the mass transfer limitations resulting from rate-limited solubilization, model predictions indicated that greater quantities of surfactant solution would be required to remove the entrapped dodecane as the flow velocity was increased. Utilization of low pumping rates, for which equilibrium solubilization would be approached, resulted in substantially increased cleanup times. Thus, there is a distinct trade-off between treatment time and the volume of surfactant solution required to reach a specified remediation goal. This relationship is illustrated by the solid line shown in Figure 13(a), where each point represents a specified constant pumping rate. The absolute minimum volume required to clean up the column, assuming equilibrium solubilization, is shown as a vertical line on the left-hand side of Figure 13(a). Obviously, if local equilibrium conditions were applicable, the flushing volume will be independent of flow rate, and thus, all mass transfer-limited predictions approach this line asymptotically. Further inspection of the solid line shown in Figure 13(a) indicates that, beyond a certain flow rate, in this case approximately 0.03 ml/min, an increase in flow rate only yields minimal reductions in remediation time. More importantly, though, the increased pumping rates result in sizable increases in the amount of surfactant solution required for cleanup. Thus, in order to minimize the cost of surfactant and still reach remediation goals within a reasonable amount of time, it is advisable to select a pumping rate that corresponds

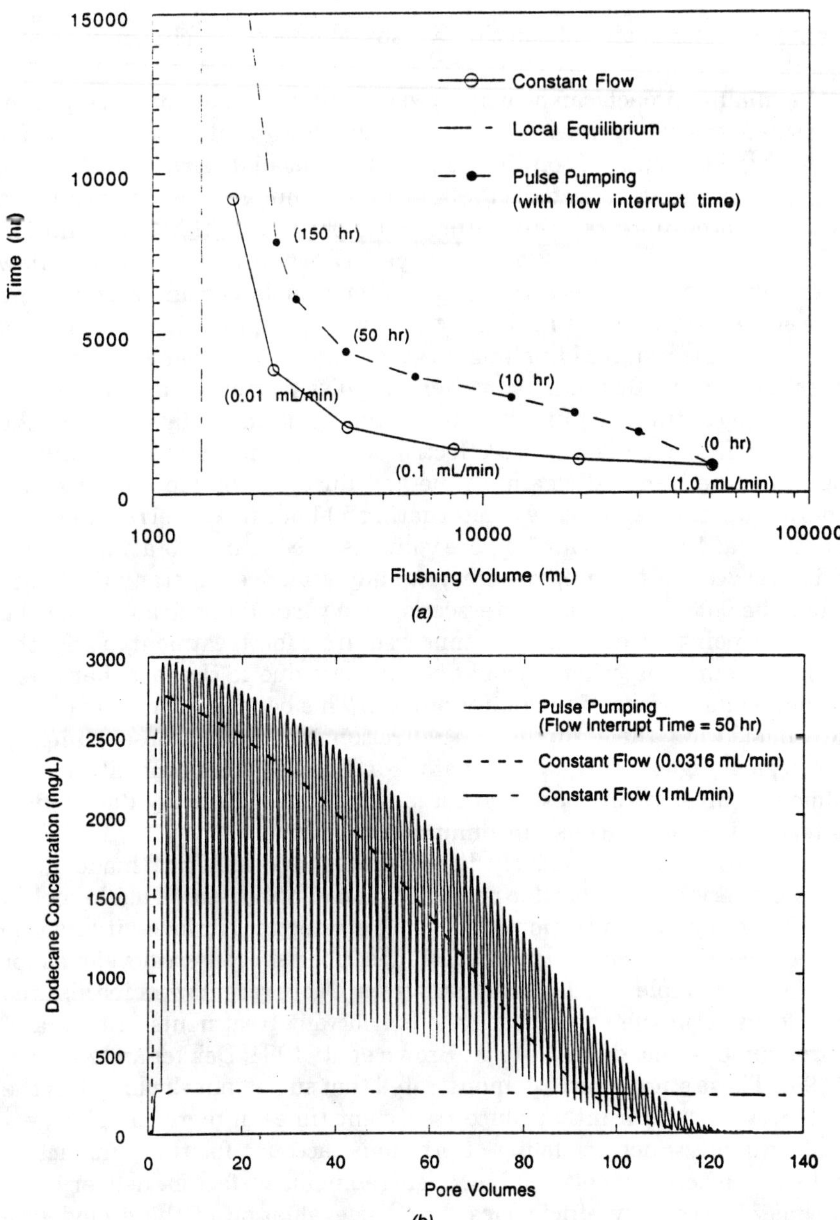

Figure 13. Comparison of surfactant flushing using pulse-pumping and constant flow regimes. Analysis of (a) recovery time versus volume of surfactant solution and (b) effluent dodecane concentrations (adapted from Abriola et al., 1993).

737

to the inflection point of the time versus volume curve shown in Figure 13(a).

A similar approach can be employed to evaluate the use of pulse-pumping strategies, which have received considerable attention as a means for reducing the volume of solution required to remediate a site. To simulate this technique, Abriola et al. (1993) utilized a pumping rate of 1.0 ml/min for 1.5 pore volumes, alternating with periods of flow interruption [dashed line in Figure 13(a)]. As expected, when the duration of flow interruption approaches zero, the plot intersects the continuous pumping line at the 1.0 ml/min point. When the flow interruption time is increased from zero, the required flushing volume decreases compared to continuous pumping at 1.0 ml/min. However, the pulse-pumping curve never falls below the continuous pumping curve, indicating that adjustment of the constant flow rate is a more effective strategy than pulse-pumping to optimize the trade-off cleanup time and flushing volume. To illustrate this point, consider the case of alternating 50-hour flow interruption with pumping at 1.0 ml/min for 1.5 pore volumes [see Figures 13(a) and 13(b)]. With respect to flushing volume, equally effective constant flow rate would be 0.03 ml/min (i.e., intersection of a vertical line drawn from the 50-hour point). However, the time required for treatment using the constant pumping scheme would be 50% less due to the fact that there would be no periods of flow interruption. This behavior can be explored by considering the effluent concentration curves for these pumping strategies [Figure 13(b)]. Note that the average effluent concentration is almost equivalent in these two cases, yet there is no delay due to flow interruption for the constant pumping scheme.

The above discussion clearly demonstrates that numerical modeling of surfactant-enhanced aquifer remediation can provide a valuable tool in the design and optimization of flushing strategies. However, it must be recognized that the model predictions described above were developed for a relatively simple one-dimensional system. More recent work in this area has focused on the consideration of surfactant treatments in two- and three-dimensional domains (e.g., Brown et al., 1994; Dekker and Abriola, 1994). These studies have demonstrated that surfactant flushing has the potential to significantly reduce treatment times in more complex geologic systems. Such modeling efforts must account for the influence of subsurface heterogeneity on NAPL entrapment, surfactant delivery, and ultimately, recovery efficiencies. Further evaluation of these modeling approaches, using both pilot- and field-scale data, will be required before these simulators can be applied in a general manner. Nevertheless, the importance of developing robust numerical models in conjunction with field demonstrations cannot be overemphasized.

FIELD-SCALE REMEDIATION TESTING

Field Trials

To date, only a limited number of studies have been conducted to evaluate the performance of surfactant-based remediation technologies in the field. Rather than provide an exhaustive description of each field trial, the purpose of this section is to highlight the important contributions of these studies and discuss some of the problems encountered when attempting to utilize surfactants for in situ subsurface remediation. One of the first pilot-scale tests was conducted at Volk Air Force Base in Douglas, Wisconsin (Nash, 1987). The site consisted of a fire-training area contaminated with JP-4 jet fuel, fire retardant, and chlorinated solvents. Laboratory tests indicated that 75 to 94% of the contaminants could be removed by flushing with 12 pore volumes of a 1:1 mixture of Richonate YLA and Hyonic PE-90. This and several other surfactant formulations were introduced into bore holes, at depths of 2–4 and 12–14 inches below the soil surface. The surfactant treatments, however, resulted in the complete plugging of two bore holes and had no significant effect on contaminant levels below the injection points. The failure of this study may have been due to reductions in soil permeability resulting from the surfactant-induced dispersion of fine materials or by the formation of highly viscous emulsions. The potential for pore blockage and clogging in surface soils suggests that laboratory infiltration and permeability tests should be conducted prior to surfactant applications in the unsaturated zone.

Abdul and coworkers (1991, 1992, 1994) used a combination of laboratory and field experiments to evaluate the use of surfactants to recover PCBs from a five-acre site containing contaminated fill material to a depth of 13–15 feet. Detailed laboratory studies, conducted before field application of surfactants, indicated that approximately 80% of the PCBs and associated oils could be removed from contaminated soil columns by flushing with 105 pore volumes of a 2.5% surfactant solution (Witconol SN-70) (Ang and Abdul, 1991). Based on these results, a two-phase field test was initiated in a 10-ft diameter by 6-ft deep cell located at the site. In Phase I of the project, 5.7 pore volumes (5375 gallons) of a 0.75% surfactant solution was applied to the surface of the test cell at a rate of 77 gal/day. This resulted in the removal of approximately 10% of the contaminants (1.6 kg PCBs and 1.69 kg oil) from the cell (Abdul et al., 1992). In Phase II, an additional 2.3 pore volumes of surfactant solution was applied to the cell, which resulted in the cumulative removal of 24% of the PCBs and oil contaminants after flushing with a total of 8 pore volumes of surfactant solution (Abdul et al., 1994). Although these results demonstrate that surfactants can be successfully employed for in situ recovery of contaminants from unsaturated soils, the solubility enhancement realized for extremely hydrophobic compounds, such as PCBs, may still not be sufficient to render this technology economically feasible. That is, large volumes of surfactant solution may be required over extremely long periods of time to achieve regulatory cleanup goals. Recognizing this problem, Ang and Abdul (1994) evaluated the use of ultrafiltration as a means to recover and reuse the surfactant. Using hollow-fiber membrane cartridges, they were able to retain approximately 90% of the contaminants, while approximately 70% of the surfactant passed through the membrane and could be recovered. Although the technical feasibility of surfactant flushing was demonstrated in the studies, detailed economic analyses and comparisons with conventional technologies will be required to justify the use of surfactants for the removal of low-solubility compounds (e.g., PCBs) from the unsaturated zone.

One of the first field trials focusing on the recovery of NAPLs from the saturated zone was conducted at a former wood-treating facility located in Laramie, Wyoming (Sale et al., 1989). The test cell, which was isolated by sheet-piling walls, consisted of an alluvial aquifer approximately 12-ft thick, containing a waste wood-treating oil. Initially, a water flood (28 pore volumes) was carried out to remove mobile oil (~1600 gallons recovered). After flushing with 8 pore volumes of surfactant solution, followed by 30 pore volumes of water, an additional 260 gallons of oil was recovered. Based on soil core analyses, approximately 67% of the residual oil was removed by surfactant flushing. The entire treatment process resulted in the recovery of 94% of the wood-treating oil from the test cell. The results of this field trail indicate that surfactants hold promise as a means for enhancing the recovery of NAPLs from aquifer formations.

Two field trials have also been conducted by Fountain and coworkers in conjunction with the University of Waterloo Center for Groundwater Research (Fountain, 1995). The first test was performed at the Borden Site at the Canadian Forces Base, Ontario, Canada. In this study, 231 L of PCE was released into a 3 m by 3 m cell enclosed by sheet piling. Initial pumping of the cell and excavation of the unsaturated zone recovered 99 L of PCE. This was followed by water flushing, which removed 12 L of PCE. The remaining "residual" PCE located in the saturated zone was then flushed with a 5% surfactant solution. This treatment led to the additional recovery of approximately 79 L of PCE from the cell. Thus, a total of 190 L of PCE were removed from the test cell, with surfactant flushing directed toward remediation of the residual PCE fraction entrapped within the saturated zone. The second surfactant flushing field trial conducted by Fountain and coworkers involved a carbon tetrachloride spill at a chlorocarbon manufacturing plant in Corpus Cristi, Texas (Fountain, 1995). This pilot-scale test was conducted in a 20 ft by 30 ft by 12 ft thick test cell, which contained six injection wells and a centrally located extraction well. In the first phase of the project, a 1% (~10,000 mg/l) solution of Witconol 2722 was injected into the contaminated zone, followed by groundwater extraction and air stripping to remove carbon tetrachloride. The treated groundwater was then supplemented with additional surfactant to maintain a 1% concentration and reinjected into the test cell. Unfortunately, the recovery of Witconol 2722 and solubilized carbon tetrachloride from the extraction well was quite low, which was attributed to surfactant losses due to sorption and/or biodegradation. This prompted the use of a different surfactant, Tergitol 15-S-12 (an alcohol ethoxylate), in the second phase of the project. Flushing with a 1% Tergitol solution resulted in the removal of approximately 73 gallons of carbon tetrachloride from the cell, with groundwater concentrations decreasing by about one order of magnitude. Within the high permeability zones of the test cell, which were flushed with approximately 6 pore volumes of surfactant solution, almost complete removal of the carbon tetrachloride was achieved. However, a significant amount of carbon tetrachloride remained within the lower permeabilities regions of the cell. Although this field test was reasonably successful, these results illustrate the importance of considering the effects of subsurface heterogeneity on surfactant delivery and NAPL recovery from the subsurface.

CONCLUSIONS

The results of laboratory and modeling studies discussed in this chapter clearly demonstrate the ability of surfactants to enhance the recovery of hydrophobic organic contaminants from the subsurface by (a) micellar solubilization of entrapped NAPLs or sorbed-phase contaminants and (b) NAPL mobilization due to interfacial tension reductions. Although mobilization has been shown to be a more efficient means of NAPL removal in laboratory tests, utilization of this approach in the field could lead to the uncontrolled migration of mobilized NAPL. This is of particular concern in the case of DNAPLs, which may be transported further downward within an aquifer formation due to gravitational forces. Therefore, it may be advisable to select surfactants that function primarily as solubilization agents (e.g., IFTs greater than approximately 1 dyne/cm) when attempting to remediate sites contaminated by DNAPLs. In contrast, it may be advantageous to utilize surfactants that promote mobilization when treating LNAPL-contaminated sites. In this scenario, the mobilized LNAPL would rise to the surface of the water table where it could be extracted by conventional procedures.

To date, only a limited number of field trials have been conducted to evaluate surfactant-based remediation technologies under natural conditions. The majority of these studies have demonstrated that surfactant flushing is an effective method for enhancing the recovery of residual NAPLs from aquifer formations. Currently, the U.S. EPA National Risk Management Research Laboratory (Ada, OK) is overseeing several surfactant field experiments at the Hill Air Force Base in Utah. Another surfactant field study is under way at the Thouin Sand Quarry in Quebec, Canada (Martel et al., 1995). It is anticipated that these ongoing field studies will further our appreciation for, and understanding of, the complexities associated with implementing surfactant-based remediation technologies in the field. Clearly, many issues will need to be resolved prior to widespread application of this technology, including the influence of subsurface heterogeneity on surfactant delivery to the contaminated zone, surfactant/organic separation and recycling, potential for hydraulic conductivity reductions, and the development of optimal pumping strategies for in situ surfactant flushing.

ACKNOWLEDGEMENTS

The authors wish to thank Gary Pope and Tim Dekker for their discussions of surfactant remediation technologies and their contributions to several of the concepts presented herein. Preparation of this chapter was supported by the United States Environmental Protection Agency through Grant No. R-815750 to the Great Lakes Mid-Atlantic Hazardous Substance Research Center. This work has not been subject to Agency review and therefore does not necessarily reflect the views of the Agency, and no official endorsement should be inferred.

REFERENCES

Abdul, A. S. and Gibson, T. L. 1991. Laboratory studies of surfactant-enhanced washing of polychlorinated biphenyl from sandy material. *Environ. Sci. Technol.*, 25:665–671.

Abdul, A. S., Gibson, T. L., Ang, C. C., Smith, J. C., and Sobczynski, R. E. 1992. In situ surfactant washing of polychlorinated biphenyls and oils from contaminated soils. *Ground Water* 30:219–231.

Abriola, L. M., Dekker, T. J. and Pennell, K. D. 1993. Surfactant enhanced solubilization of residual dodecane in soil columns 2. Mathematical Modeling. *Environ. Sci Technol.* 27:2341–2351.

Abriola, L. M., Pennell, K. D., Pope, G. A., Dekker, T. J. and Luning-Prak, D. J. 1995. Impact of surfactant flushing on the solubilization and mobilization of dense nonaqueous phase liquids. In: Sabatini, D. A., Knox, R. C. and Harwell, J. H. (editors) *Surfactant-Enhanced Subsurface Remediation: Emerging Technologies.* ACS Symposium Series No. 594. American Chemical Society, Washington, D.C.

Adeel, Z. and Luthy, R. G. 1995. Sorption and transport kinetics of a nonionic surfactant through an aquifer sediment. *Environ. Sci. Technol.* 29:1032–1042.

Allred, B. and Brown, G. O. 1994. Surfactant-induced reductions in soil hydraulic conductivity. *Ground Water Monitor. Rev.* 174–184.

Ang, C. C. and Abdul, A. S., 1991. Aqueous surfactant flushing of residual oil contamination from sandy soil. *Ground Water Monitor. Rev.,* 121–127.

Arnarson, T. and Elworthy, P. H. 1981. Effect of structural variations of nonionic surfactants on micellar properties and solubilization: Surfactants containing very large hydrocarbon chains. *J. Pharm. Pharmacol.* 33:141–144.

Aronstein, B. N. and Alexander, M. 1992. Surfactants at low concentrations stimulate the biodegradation of sorbed hydrocarbons in samples of aquifer sands and soil slurries. *Environ. Tox. Chem.* 11:1227–1233.

Aronstein, B. N., Calvillo, Y. M. and Alexander, M. 1991. Effect of surfactants at low concentrations on the desorption and biodegradation of sorbed aromatic compounds in soil. *Environ. Sci. Technol.* 25:1728–1731.

Arytunyan, R. S. and Beileryan, N. M. 1983. Kinetics of solubilization. *Colloid J. USSR* 45:355–360.

Aston, J. R., Furlong, D. N., Greiser, F., Scales, P. J., and Warr, G. G. 1982. In *Adsorption at the Gas-Liquid and Liquid-Solid Interface.* J. Rouquerol and K. S. W. Sing (Editors), Elsevier, Amsterdam.

Attwood D. and Florence, A. T. 1983. *Surfactant Systems: Their Chemistry, Pharmacy, and Biology.* Chapman and Hall, New York, NY.

Baumgardt, K. and Kahlweit, M. 1980. On the kinetics of dissolution of naphthalene particles in aqueous micellar solutions. *Tenside Deterg.* 17:135–139.

Becher, P. 1967. In Nonionic Surfactants. M. J. Schick (Editor) *Surfactant Science Series Vol. 1,* Marcel Dekker, New York, NY.

Bluestein, B. R. and Hilton, C. L. 1982. *Amphoteric Surfactants. Surfactant Science Series Vol. 12,* Marcel Dekker, New York, NY.

Bolsman, T. A. B. M., Veltmaat, F. T. G., and van Os, N. M. 1988. *J. Am. Oil Chem. Soc.* 65:280–283.

Bourrel M. and Schecter, R. S. 1988. *Microemulsions and Related Systems. Surfactant Science Series Vol. 30,* Marcel Dekker, New York, NY.

Brown, C. L., Pope, G. A., Abriola, L. M. and Sepehrnoori, K. 1994. Simulation of surfactant-enhanced aquifer remediation. *Water Resour. Res.* 30:2959–2977.

Brooks, R. H. and Corey, A. T. 1966. Properties of porous media affecting fluid flow. *J. Irrig. Drain. Div. Am. Soc. Civ. Eng.* 6:61–88.

Brusseau, M. L. and Rao, P. S. C. 1989. Sorption nonideality during organic contaminant transport in porous media. *CRC Critical Rev. Environ. Control* 19:33–99.

Bury, S. J. and Miller, C. A. 1993. Effect of micellar solubilization on biodegradation rates of hydrocarbons. *Environ. Sci. Technol.* 27:104–110.

Carroll, B. J. 1981. The kinetics of solubilization of nonpolar oils by nonionic surfactant solutions. *J. Colloid Interface Sci.* 79:126–135.

Carroll, B. J., O'Rourke, B. G. C. and Ward, A. J. I. 1982. The kinetics of solubilization of single component non-polar oils by a nonionic surfactant. *J. Pharm. Pharmacol.* 34:287–292.

Chan, A. F., Evans, D. F., and Cussler, E. L., 1976. Explaining solubilization kinetics. *AIChE J.* 22:1006–1012.

Clarke, A. N., Plumb, P. D., Subramanyan, T. K., and Wilson, D. J. 1991. Soil clean-up by surfactant washing 1. Laboratory results and mathematical modeling. *Sep. Sci. Technol.* 26:301–343.

Clint, J. H. 1992. *Surfactant Aggregation.* Chapman and Hall, New York, NY.

Clunie, J. S. and Ingram, B. T. 1983. *Adsorption of nonionic surfactants. In Adsorption from Solution at the Solid/Liquid Interface.* G. D. Parkitt and C. H. Rochester (Editors), Academic Press, New York, NY.

Datta Gupta, A., Pope, G. A., Sepehrnoori, K. and Thrasher R. L. 1986. A symmetric positive definite formulation of a three-dimensional micellar/polymer simulator. *Soc. Pet. Eng. Resev. Eng.* 1:622–632.

Delshad, M., Pope, G. A. and Sepehrnoori, K. 1996. A compositional simulator for modeling surfactant enhanced aquifer remediation. *J. Contam. Hydrol.* (In review).

Deitsch, J. J. and Smith, J. A. 1995. Effect of Triton X-100 on the rate of trichloroethene desorption from soil to water. *Environ. Sci. Technol.* 29:1069–1080.

Dekker, T. J. and Abriola, L. M. 1994. Dissolution and enhanced solubilization of entrapped NAPLs: Implications of laboratory scale observations on field scale remediation. In *Toxic Substances and the Hydrologic Sciences.* A. R. Dutten (Editor), American Institute of Hydrology, pp. 290–301.

Desai, F. N., Demond A. H. and Hayes, K. F., 1992. Influence of surfactant sorption on capillary pressure-saturation relationships. In *Transport and Remediation of Subsurface Contaminants,* D. A. Sabatini and R. C. Knox (Editors), ACS Symposium Series 491, American Chemical Society, Washington, D.C., pp. 133–148.

DeSmedt, F. and Wierenga, P. J. 1979. Mass transfer in porous media with immobile water. *J. Hydrol.* 41:59–67.

Diallo, M. S., Abriola, L. M. and Weber, W. J. Jr. 1994. Solubilization of nonaqueous phase liquid hydrocarbons in micellar solutions of dodecyl alcohol ethoxylates. *Environ. Sci. Technol.* 28:1829–1837.

Edwards, D. A., Luthy, R. G., and Liu, Z., 1991. Solubilization of polycyclic aromatic hydrocarbons in micellar nonionic surfactant solutions. *Environ. Sci. Technol.*, 25:127–133.

Edwards, D. A., Adeel, Z. and Luthy, R. G. 1994. Distribution of nonionic surfactant and phenanthrene in a sediment/aqueous system. *Environ. Sci. Technol.* 28:1550–1560.

Ellis, W. D., Payne, J. R. and McNabb, G. D. 1985. *Treatment of Contaminated Soils with Aqueous Surfactants.* U.S. Environmental Protection Agency, Cincinnati, OH. EPA/600/S2-85/129.

Finnerty, W. R. 1994. Biosurfactants in environmental technology. *Current Opin. Biotechnol.* 5:291–295.

Fountain, J. C. 1995. Field tests of surfactant flooding. In: Sabatini, D. A., Knox, R. C. and Harwell, J. H. (editors) *Surfactant-Enhanced Subsurface Remediation: Emerging Technologies.* ACS Symposium Series No. 594. American Chemical Society, Washington, D.C.

Fountain, J. C., Klimek, A., Beirkirch, M. G., and Middleton, T. M., 1991. The use of surfactants for in situ extraction of organic pollutants from a contaminated aquifer. *J. Hazardous Mater.,* 28:295–311.

Gannon, O. K., Bibring, P., Raney, K., Ward, A. Wilson, J., Underwood, J. L., and Debelak, K. A., 1989. Soil clean up by in-situ surfactant flushing III. Laboratory results. *Sep. Sci. Technol.,* 24:1073–1094.

Gellar, J. T. and Hunt, J. R. 1993. Mass transfer from nonaqueous phase liquids in water-saturated porous media. *Water Resour. Res.,* 29:833–845.

Gillham, R. W. and Rao, P. S. C., 1990. Transport, distribution, and fate of volatile organic compounds in groundwater. In: N. M. Ram, F. C. Russel, and K. P. Contor (Editors), Lewis Publishers, Chelsae, MI, pp. 141–181.

Guha, S. and Jaffe, P. S. 1996. Biodegradation kinetics of phenanthrene partitioned into the micellar phase of nonionic surfactants. *Environ. Sci. Technol.* 30:605–611.

Haggerty, G. M. and Bowman, R. S. 1994. Sorption of chromate and other inorganic anions by organo-zeolite. *Environ. Sci. Technol.* 28:452–458.

Harwell, J. H., Sabatini, D. A. and Soerens, T. S. 1993. In *Migration and Fate of Pollutants in Soils and Subsoils,* NATO ASI Series G, P. Petruzzelli and F. G. Helfferich (Editors), *Ecological Sciences, Vol. 32,* Springer-Verlag, Berlin.

Hunt, J. R., Sitar, N. and Udell, K. S. 1988. Nonaqueous phase liquid transport and cleanup 1. Analysis of mechanisms. *Wat. Resour. Res.* 24:1247–1258.

Huh, C. 1979. Interfacial tensions and solubilizing ability of a microemulsion phase that coexists with oil and brine. *J. Coll. Interface Sci.* 71:408–426.

Israelachvili, J. N. 1985. *Intermolecular and Surface Forces.* Academic Press. London, England.

Jafvert, C. T., van Hoof, P. L. and Heath, J. C. 1994. Solubilization of nonpolar compounds by nonionic surfactant micelles. *Water Res.* 28:1009–1017.

Karickhoff, S. W. 1984. Organic pollutant sorption in aquatic systems. *J. Hydraul. Eng.* 110:707–735.

Kile, D. E. and Chiou, C. T., 1989. Water solubility enhancements of DDT and trichlorobenzene by some surfactants below and above the critical micelle concentration. *Environ. Sci. Technol.*, 23:832–838.

Klevens, H. B. 1950. Solubilization of polycyclic hydrocarbons. *J. Phys. Colloid Chem.* 54:283–298.

Krebs-Yuill, B., Harwell, J. H., Sabatini, D. A. and Knox, R. C. 1995. Economic considerations in surfactant-enhanced pump-and-treat remediation. In: Sabatini, D. A., Knox, R. C. and Harwell, J. H. (Editors) *Surfactant-Enhanced Subsurface Remediation: Emerging Technologies.* ACS Symposium Series No. 594. American Chemical Society, Washington, D.C.

Laha, S. and Luthy, R. G. 1991. Inhibition of phenanthrene mineralization by nonionic surfactants in soil-water systems. *Environ. Sci. Technol.* 25:1920–1930.

Laha, S. and Luthy, R. G. 1992. Effects of nonionic surfactants on the solubilization and mineralization of phenanthrene in soil-water systems. *Biotechnol. Bioengin.* 40:1367–380.

Lake, L. 1989. *Enhanced Oil Recovery.* Prentice-Hall, Englewood Cliffs, NJ.

Lee, J.-F., Crum, J. R. and Boyd, S. A. 1989. Enhanced retention of organic contaminants by soils exchanged with organic cations. *Environ. Sci. Technol.* 23:1365–1372.

Lenhard, R. J. and Parker J. C. 1987. Measurement and prediction of saturation-pressure relationships in three-phase porous media systems. *J. Contam. Hydrol.* 1:407–424.

Leverett, M. C. 1941. Flow of oil-water mixtures through unconsolidated sands. *Trans. Am. Inst. Min. Metall. Pet. Eng.* 132:149–171.

Mackay, D. M. and Cherry, J. A. 1989. Groundwater contamination: Pump-and-treat remediation. *Environ. Sci. Technol.* 23:630–636.

Martel, R., Gelinas, P. J., and Saumure, L. 1996. In situ recovery of DNAPL in sand aquifers: Clean-up test using surfactants at Thouin Sand Quarry. *Ground Water* (in review).

Megehee, M. M., Clarke, A. N., Oma, K. H. and Wilson, D. J. 1993. Soil clean-up by surfactant washing IV. Modification and testing of mathematical models. *Sep. Sci. Technol.* 28:2507–2527.

Melrose, J. C. and Brander, C. F. Role of capillary forces in determining microscopic displacement efficiency for oil recovery by waterflooding. *J. Can. Petr. Technol.* 13:54–62.

Mihelcic, J. R., Lueking, D. R., Mitzell, R. J. and Stapleton, J. M. 1993. Bioavailability of sorbed- and separate-phase chemicals. *Biodegradation* 4:141–153.

Morrow, N. R. and Songkran, B., 1981. Effect of viscous and buoyancy forces on nonwetting phase trapping in porous media. In *Surface Phenomenon in Enhanced Oil Recovery*, D. O. Shah (Editor), Plenum Publ., New York, pp. 387–411.

Mukerjee, P. and Cardinel, J. R. 1978. Benzene derivatives and naphthalene solubilized in micelles. Polarity of microenvironment, location and distribution in micelles, and correlation with surface-activity in hydrocarbon-water systems. *J. Phys. Chem.* 82:1620–1627.

Nagarajan, R., Chaiko, M. A. and Ruckenstein, E. 1984. Locus of solubilization of benzene in surfactant micelles. *J. Phys. Chem.* 88:2916–2922.

Nash, J. H., 1987. *Field Studies of in-situ Washing*. EPA/600/2-87/110, U.S. Environmental Protection Agency, Cincinnati, OH.

National Research Council (NRC), 1994. *Alternatives for Ground Water Cleanup*. National Academy Press, Washington, D.C.

Ng, K. M., Davis, H. T., and Scriven, L. E., 1978. Visualization of blob mechanics in flow through porous media. *Chem. Eng. Sci.*, 33:1009–1017.

Nkedi-Kizza, P., Brusseau, M. L., Rao, P. S. C. and Hornsby, A. G. 1989. Nonequilibrium sorption during displacement of hydrophobic organic chemicals and ^{45}Ca through soil columns with aqueous and mixed solvents. *Environ. Sci. Technol.* 23:814–820.

Partyka, S., Zaini, S., Lindheimer, M. and Brun, B. 1984. The adsorption of nonionic surfactants on a silica gel. *Colloids Surf.* 12:255–270.

Pennell, K. D., Abriola, L. M., and Weber, W. J., Jr., 1993. Surfactant enhanced solubilization of residual dodecane in soil columns 1. Experimental investigation. *Environ. Sci. Technol.*, 27:2332–2340.

Pennell, K. D., Jin, M., Abriola, L. M., and Pope, G. A. 1994. Surfactant enhanced remediation of soil columns contaminated by residual tetrachloroethylene. *J. Contam. Hydrol.*, 16:35–53.

Pennell, K. D., Pope, G. A. and Abriola, L. M. 1996a. Influence of viscous and buoyancy forces on the mobilization of residual tetrachloroethylene during surfactant flushing. *Environ. Sci. Technol.* (In press).

Pennell, K. D., Adinolfi, A. M., Abriola, L. M. and Diallo, M. S. 1996b. Solubilization of dodecane, tetrachloroethylene and 1,2-dichlorobenzene in micellar solutions of ethoxylated nonionic surfactants. *Water Res.* (In review).

Pennell, K. D., Wigder, D. E., Dekker, T. J. and Abriola, L. M. 1996c. Influence of rate-limited sorption on the transport of an ethoxylated nonionic surfactant. *Ground Water.* (In review).

Peters, R. W., Montemagno, C. D., and Shem, L., 1993. Surfactant screening of diesel-contaminated soil. *Hazardous Waste & Hazardous Mater.*, 9:113–138.

Powers, S. E., Loureiro, C. O., Abriola, L. M. and Weber, W. J., Jr., 1991. Theoretical study of the significance of nonequilibrium dissolution on nonaqueous phase liquids in subsurface systems. *Water Resour. Res.* 27:463–477.

Powers, S. E., Abriola, L. M. and Weber, W. J., Jr., 1992. An experimental investigation of nonaqueous phase liquid dissolution in saturated subsurface systems: Steady state mass transfer rates. *Water Resour. Res.*, 28:2691–2705.

Powers, S. E., Abriola, L. M. and Weber, W. J., Jr., 1994. An experimental investigation of NAPL dissolution in saturated subsurface systems: Transient mass transfer rates. *Water Resour. Res.* 30:321–322.

Robson, R. J. and Dennis, E. 1977. *J. Phys. Chem.* 81:1075.

Rosen, M. J. 1989. *Surfactants and Interfacial Phenomena*, 2nd. Ed. J. Wiley & Sons, New York, NY.

Rouse, J. D., Sabatini, D. A. and Harwell, J. H. 1993. Minimizing surfactant losses using twin-head anionic surfactants in subsurface remediation. *Environ. Sci. Technol.* 27:2702–2078.

Rouse, J. D., Sabatini, D. A., Suflita, J. M. and Harwell, J. H. 1994. Influence of surfactants on microbial degradation of organic compounds. *Crit. Rev. Environ. Sci. Technol.* 24:325–370.

Saito, H. and Shinoda, K., 1967. The solubilization of hydrocarbons in aqueous solutions of nonionic surfactants. *J. Colloid Interface Sci.*, 24:10–15.

Sale, T., Piontek, K. and Malcolm, P. 1989. Chemically enhanced in-situ soil washing. In *Petroleum Hydrocarbons and Organic Chemicals in Ground Water: Prevention, Detection and Restoration.* NWWA/API, Houston, TX, pp 486–502.

Schick, M. J., Atlas, S. M., and Eirech, F. R. 1962. Miceller structure of nonionic detergents. *J. Phys. Chem.* 66:1326–1333.

Schwille, F. 1984. In *Pollutants in Porous Media, Ecology Studies,* Springer-Verlag, New York, 47:27–48.

Scott, T. Sharpe, S. R., Sorbie, K. S., Clifford, P. J., Roberts, L. J., Foulser, R. W. S. and Oakes, J. A. 1987. A general purpose chemical flood simulator. *Proc. 9th SPE Symposium on Reservoir Simulation.* San Antonio, TX, Feb. 1–4, 1987.

Sreekala, T. and Shreve, G. S. 1994. Effect of anionic biosurfactant on hexadecane partitioning in multiphase systems. *Environ. Sci. Technol.* 28:1993–2000.

Srinivasan, K. R. and Fogler, H. S. 1990. Use of inorgano-organo-clays in the removal of priority pollutants from industrial wastewaters: Adsorption of benzo(a)pyrene and chlorophenols from aqueous solutions. *Clays Clay Miner.* 38:287–293.

Stegemeier, G. L. 1976. Mechanisms of entrapment and mobilization of oil in porous media. In *Improved Oil Recovery by Surfactant and Polymer Flooding.* D. O. Shah and R. S. Schecter (Editors), Academic Press, New York, NY.

Sun, S., Inskeep, W. P. and Boyd, S. A. 1995. Sorption of nonionic organic compounds in soil-water systems containing a micelle-forming surfactant. *Environ. Sci. Technol.* 29:903–913.

Taber, J. J., 1981. Research on enhanced oil recovery: Past, present and future.

In: D. O. Shah (Editor), *Surface Phenomenon in Enhanced Oil Recovery*, Plenum Publ., New York, pp. 13–52.

Tanford, C. 1980. *The Hydrophobic Effect: Formation of Micelles and Biological Membranes*, 2nd Ed. J. Wiley & Sons, New York, NY.

Texas Research Institute, 1985. *Test Results of Surfactant Enhanced Gasoline Recovery in a Large-Scale Model Aquifer.* API Publication No. 4390, American Petroleum Institute, Washington, D.C.

Thomas, C. P., Fleming, P. D. and Winter, W. K. 1984. A ternary, two-phase mathematical model of oil recovery with surfactant systems. *Soc. Pet. Eng. J.* 24:606–616.

Tsomides, H. J., Hughes, J. B., Thomas, J. M., Ward, C. H. 1995. Effect of surfactant addition on phenanthrene biodegradation in sediments. *Environ. Tox. Chem.* 14:953–959.

Valsaraj, K. T. and Thibodeaux, L. J. 1989. Relationships between micelle-water and octanol-water partition constants for hydrophobic organics of environmental interest. *Water Res.* 23:183–189.

Van Dyke, M. I., Gulley, S. L., Lee, H., and Trevors, J. T. 1993. Evaluation of microbial surfactants for recovery of hydrophobic pollutants from soil. *J. Industrial Microbiol.* 11:163–170.

van Genuchten, M. T. 1980. A closed-form equation for predicting the hydraulic conductivity of unsaturated soils. *Soil Sci. Soc. Am. J.* 44:892–898.

van Hoof, P. L. and Rogers, J. E. 1992. Influence of low levels of nonionic surfactant on the anaerobic dechlorination of hexachlorobenzene. In *Bioremediation of Hazardous Wastes: Biosystems Technology Development Program.* EPA/600/R-92/126. U.S. Environmental Protection Agency, Washington, D.C.

Vaumgardt, K. and Kahlweit, M. 1980. *Tenside Deterg.* 17:135.

Vigon, B. W. and Rubin, A. J., 1989. Practical considerations in the surfactant-aided mobilization of contaminants in aquifers. *J. Water Poll. Control Fed.,* 61:1233–1240.

Volkering, F., Bruere, A. M., van Andel, J. G. and Rulkens, W. H. 1995. Influence of nonionic surfactants on bioavailability and biodegradation of polycyclic aromatic hydrocarbons. *Applied Environ. Microbiol.* 61:1699–1705.

Volkov, V. A., 1979. Kinetics of solubilization. *Colloid J. USSR* 36:200–205.

West, C. C., 1992. Surfactant-enhanced solubilization of tetrachloroethylene and degradation products in pump and treat remediation. In: D. A. Sabatini and R. C. Knox (Editors), *Transport and Remediation of Subsurface Contaminants*, ACS Symposium Series 491, American Chemical Society, Washington, D.C., pp. 149–158.

Wilson, D. J. 1989. Soil clean-up by in-situ surfactant flushing 1. Mathematical modeling. *Sep. Sci. Technol.* 24:863–892.

Wilson, D. J., Clarke, A. N. 1991. Soil clean up by in-situ surfactant flushing IV. A two-component mathematical model. *Sep. Sci. Technol.* 26:1177–1194.

Wilson, J. L., Conrad, S. H., Mason, W. R., Peplinski, W., and Hagan, E., 1990. *Laboratory Investigations of Residual Liquid Organics from Spills, Leaks and Disposal of Hazardous Wastes in Groundwater.* EPA/600/6-90/004, U.S. Environmental Protection Agency, Ada, OK.

Zhang, Y. and Miller, R. N. 1992. Enhanced octadecane dispersion and biodegradation by a *Pseudomonas* rhamnolipid surfactant (biosurfactant). *Applied Environ. Microbiol.* 58:3276–3282.

Ziegenfus, P. S. 1987. The Potential Use of Surfactant and Cosolvent Soil Washing as Adjuvant for In-Situ Aquifer Restoration. M.S. Thesis, Rice University, Houston, TX.

Index

751